Understanding
MECHANICS

A.J. Sadler   *D.W.S. Thorning*

Oxford University Press

OXFORD
UNIVERSITY PRESS
Great Clarendon Street, Oxford OX2 6DP

Oxford University Press is a department of the University of Oxford. It furthers the University's objective of excellence in research, scholarship, and education by publishing worldwide in

Oxford New York

Athens Auckland Bangkok Bogotá Buenos Aires Calcutta Cape Town Chennai Dar es Salaam Delhi Florence Hong Kong Istanbul Karachi Kuala Lumpur Madrid Melbourne Mexico City Mumbai Nairobi Paris São Paulo Singapore Taipei Tokyo Toronto Warsaw

with associated companies in *Berlin Ibadan*

Oxford is a registered trade mark of Oxford University Press in the UK and in certain other countries

First edition published 1983

Second edition published 1996

Reprinted (with minor corrections) 1997, 1998, 1999 (twice)

ISBN 019 914675 6 School and college edition

ISBN 019 914676 4 Bookshop edition

A CIP record for this book is available from the British Library.

Typeset and illustrated by Tech Set Ltd., Gateshead, Tyne and Wear
Printed and bound in Great Britain by Butler & Tanner Ltd., Frome and London

Preface to Second Edition

Understanding Mechanics has proved to be a popular text with students of A-level Mathematics. Syllabuses are varied at this level, but the aim has been to cover all topics that could appear on Single Mathematics papers, at the same time as providing an extension into several Further Mathematics topics.

Recent changes to syllabuses have brought about the need for this new edition, and there are two particular emphases in modern syllabuses that have been dealt with. One of these is on the use of vectors, and for this reason we have introduced vectors right at the beginning of the book. The other emphasis is on the use of modelling. Students are required to understand the relationship between real-life situations and mathematical models. They need to understand the limitations of many common assumptions so that they can evaluate their own assumptions when setting up mathematical models to solve problems. We have provided some commentary to help with this.

In each chapter the theory sections are followed by a number of worked examples which are typical of, and lead to the questions in the exercises. By reading the theory sections and following the worked examples, the reader should be able to make considerable progress with the exercise that follows. After the introductory vector work, each chapter closes with a comprehensive selection of recent examination questions, allowing the student to evaluate and apply the skills learnt in the preparatory sections. These exercises are carefully graded, progressing to some quite demanding questions at the end.

We are grateful to the following examination boards for permission to use their questions (specimen questions are denoted by 'Spec'). The answers provided for these questions are the sole responsibility of the authors.

University of London Examinations and Assessment Council (ULEAC)
University of Cambridge Local Examinations Syndicate (UCLES)
University of Oxford Delegacy of Local Examinations (UODLE)
Oxford and Cambridge Schools Examination Board (OCSEB)
Midland Examining Group (MEG)
Associated Examining Board (AEB)
Welsh Joint Education Committee (WJEC)
Northern Ireland Council for the Curriculum Examinations and Assessment (NICCEA)
Southern Universities' Joint Board (SUJB)

Finally, we wish to acknowledge the help of four experienced teachers who helped with the writing of the new text: Julian Berry (Bloxham School), Dr Dominic Jordan (University of Keele), Robert Smedley (Liverpool Hope University College), and Garry Wiseman (Radley College). Special thanks are due to Garry Wiseman, who also played a central editorial role in melding the new and existing material together.

We trust that new generations of students will indeed find that this text promotes their understanding.

AJ Sadler DWS Thorning 1996

Contents

1 Vectors

From simple arithmetic it is known that $3 + 2 = 5$.
It is possible however, in another context, to obtain a different answer when 3 and 2 are added.

Suppose a man walks 3 km due north and then 2 km due south. In order to find the total distance walked, the separate distances have to be added:

i.e. 3 km + 2 km = 5km ...[1]

This statement does not give the final position of the man at the end of his walk. In fact he is clearly 1 km due north of his starting point and this is his *displacement* from his original position:

i.e. 3 km due N + 2 km due S = 1 km due N ...[2]

These two statements give different information. Statement [1] adds *scalar* quantities, i.e. quantities which only have magnitude (or size). The answer gives the total distance travelled which is also a scalar.
Statement [2] deals with the addition of two *vector* quantities, i.e. quantities which have both magnitude and direction, and gives the vector displacement of the man at the end of his two walks.
The addition of vectors which have the same, or opposite, directions can be done quite simply.
The addition of more general vectors requires a more sophisticated approach and can be done either by scale drawing or by calculation.

Example 1

A man walks 6 km south-west and then 4 km due west. How far, and in what direction, is he then from his starting point?

There are two ways of solving this problem:
(a) By scale drawing
Draw a sketch showing the two stages, OA and AB of his journey. From this sketch make a scale drawing with OA = 6 cm in a direction south-west of his starting-point O, and AB = 4 cm due west.

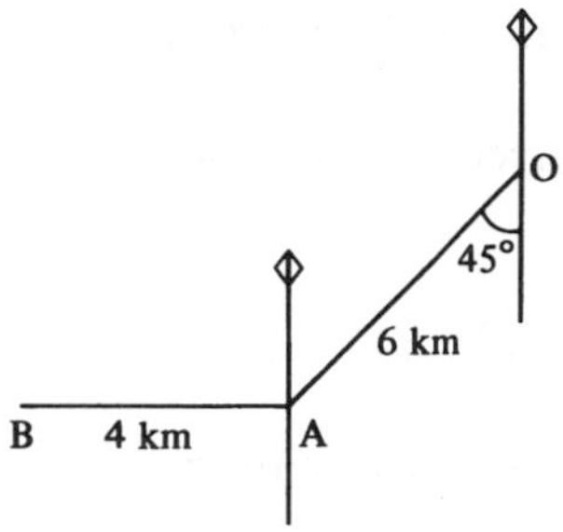

From the scale drawing, by measurement:

$$OB = 9{\cdot}3\,\text{cm} \qquad A\hat{O}B = 18^\circ$$

$$\therefore \qquad B\hat{O}S = 45^\circ + 18^\circ = 63^\circ$$

The man is 9·3 km from his starting point in a direction S 63° W.

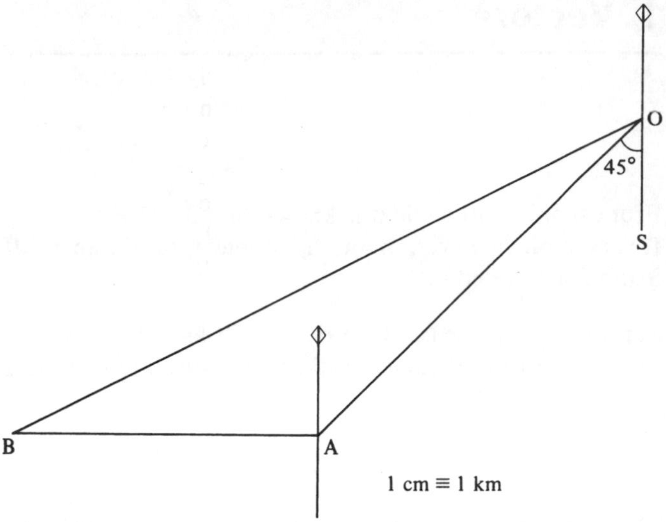

(b) By calculation
Draw a sketch showing the two stages, OA and AB, of his journey.

By the cosine rule:

$$OB^2 = OA^2 + AB^2 - 2 \times OA \times AB \cos B\hat{A}O$$
$$= 6^2 + 4^2 - 2(6)(4)\cos 135^\circ$$

Hence $\qquad OB = 9{\cdot}27$

By the sine rule:

$$\frac{OB}{\sin O\hat{A}B} = \frac{AB}{\sin B\hat{O}A}$$

Thus $\qquad \dfrac{9{\cdot}27}{\sin 135^\circ} = \dfrac{4}{\sin B\hat{O}A}$

Hence $\qquad B\hat{O}A = 17{\cdot}77^\circ$

and so $\qquad B\hat{O}S = 62{\cdot}77^\circ$

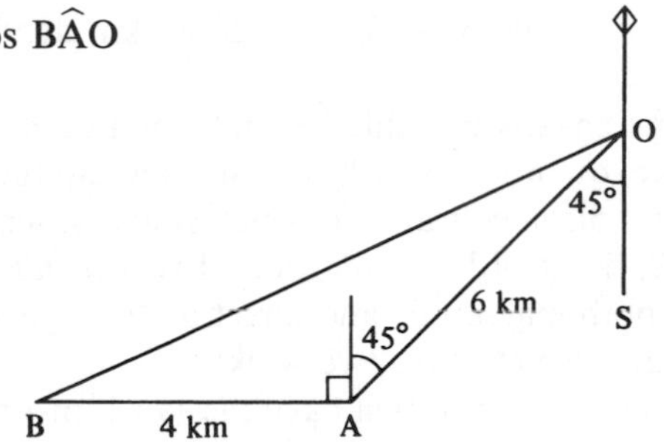

The man is 9·27 km from his starting point, in a direction S 62·77° W.

Exercise 1A

Solve each question by calculation or by scale drawing.

1. A woman cycles 5 km due east followed by 7 km due west.
   How far, and in what direction, is she then from her starting point?

2. A bird flies 40 km due south and then 30 km due east.
   Find the bird's distance and bearing from its original position.

3. A boat travels 6 km due east followed by 2·5 km due north.
   Find the distance the boat is then from its original position and the course it should set if it is to return by the shortest route.

4. A yacht sails 5 km in a direction N 30° E followed by 4 km due east.
   How far, and in what direction, is the yacht then from its original position?
   Would the yacht have reached the same position had it sailed 4 km due east followed by 5 km in a direction N 30° E?

5. I walk 800 m on a bearing 320° and then 500 m on a bearing 200°. Find how far I am then from my original position and the course I must set in order to return to my starting point by the shortest route.
6. An aeroplane flies from airport A to airport B 90 km away and on a bearing 070°. From B the aeroplane flies to airport C, 100 km from B on a bearing 210°. How far and on what course must the acroplanc now fly in order to return to A direct?
7. A ship travels 6 km north-east and then changes course and travels a further 3 km. If the ship is then 5 km from its original position, find the two possible directions for the course set by the ship on the second stage of its journey.
8. A man walks 4 km due east, 3 km due north and then 3 km on a bearing S 60° E. By making an accurate scale drawing, find the distance and bearing of the man's final position from his original position.

Resultant of vectors

In Example 1, the combined effect of two vectors was found by scale drawing and by calculation. This combined effect is said to be the *resultant* of the two vectors.
The resultant of two vectors is that single vector which could completely take the place of the two vectors, i.e. in Example 1, the man would have arrived at the same position had he walked 9·3 km in a direction S 63° W.

Example 2

Find the resultant of a vector of magnitude 5 units, direction 320°, and a vector of magnitude 8 units, direction 055°.

Draw a sketch combining the two vectors:

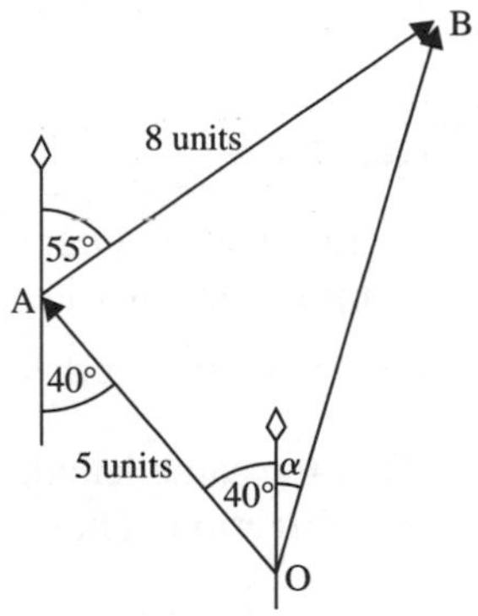

By the cosine rule:

$$OB^2 = OA^2 + AB^2 - 2 \times OA \times AB \cos B\hat{A}O$$
$$= 5^2 + 8^2 - 2(5)(8)\cos 85°$$

Hence $OB = 9{\cdot}06$

By the sine rule:

$$\frac{OB}{\sin O\hat{A}B} = \frac{AB}{\sin B\hat{O}A}$$
$$\frac{9{\cdot}06}{\sin 85°} = \frac{8}{\sin B\hat{O}A}$$

Hence $B\hat{O}A = 61{\cdot}6°$

Therefore $\alpha = 61{\cdot}6° - 40°$
$= 22°$ to the nearest degree.

The resultant vector has magnitude 9·06 units in a direction 022°.

Alternatively, the resultant could be determined by scale drawing.

Vector representation

Note that in Examples 1 and 2, when using a scale drawing or a sketch, we represented a vector quantity by a line segment of an appropriate length in a particular direction.
In order to distinguish between the distance OA and the vector OA, an arrow is placed over the letters of the vector. Thus $\overrightarrow{OA}$ represents the vector with magnitude and direction given by the line segment joining O to A.
Thus if a man walks from O to A and then from A to B, this could be written as a vector equation:

$$\overrightarrow{OA} + \overrightarrow{AB} = \overrightarrow{OB},$$

since the vector represented by the line segment OB is clearly the resultant of $\overrightarrow{OA}$ and $\overrightarrow{AB}$.

Note carefully that the direction of the arrows on the diagram correspond to the order of the letters of the vectors which they represent.

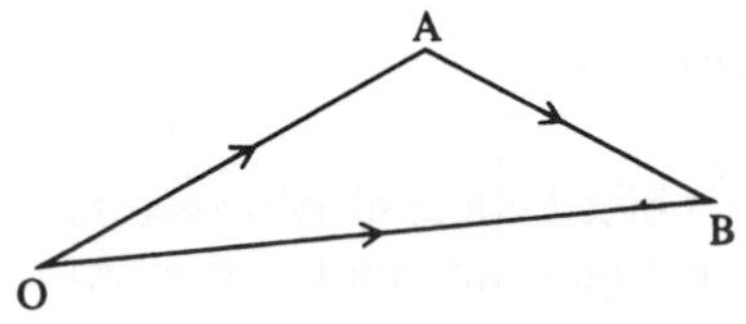

Vectors may also be written using single letters, and in this case heavy type is used. Thus:

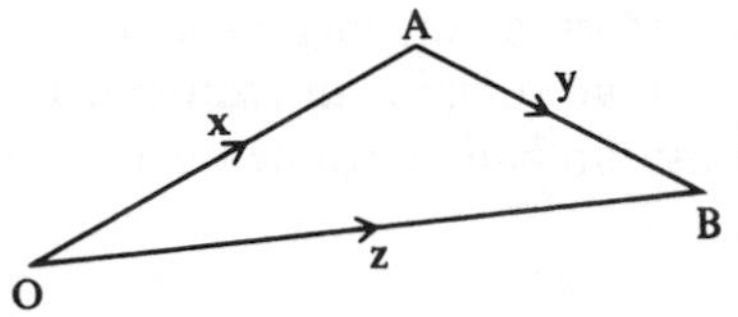

$$\mathbf{x} + \mathbf{y} = \mathbf{z}$$

Since $\overrightarrow{OA} = \mathbf{x}$, it follows that $\overrightarrow{AO} = -\mathbf{x}$ because $\overrightarrow{AO}$ has the same length as $\overrightarrow{OA}$, but it is in the opposite direction.

Example 3

The diagram shows a parallelogram ABCD with $\overrightarrow{AB} = \mathbf{a}$ and $\overrightarrow{BC} = \mathbf{b}$.
E is the mid-point of CD. Express the following vectors in terms of **a** and **b**.
(a) $\overrightarrow{AD}$ (b) $\overrightarrow{DC}$ (c) $\overrightarrow{CD}$ (d) $\overrightarrow{DE}$ (e) $\overrightarrow{AE}$

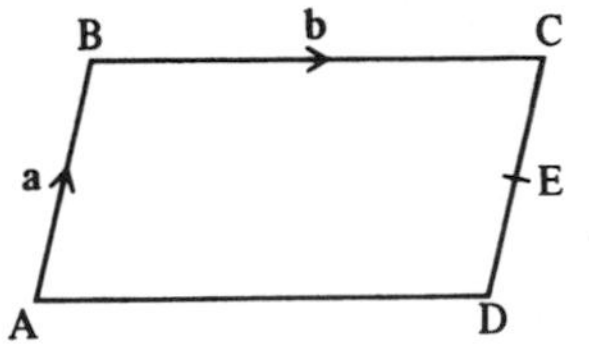

(a) AD is the same length as BC and in the same direction. Thus:

$$\overrightarrow{AD} = \overrightarrow{BC}$$

$$\therefore \quad \overrightarrow{AD} = \mathbf{b}$$

(b) In a similar way,

$$\overrightarrow{DC} = \mathbf{a}$$

(c) CD is the same length as AB but in the opposite direction.

$$\therefore \quad \overrightarrow{CD} = -\mathbf{a}$$

(d) $\overrightarrow{DE} = \frac{1}{2}\overrightarrow{DC}$

$\therefore \quad \overrightarrow{DE} = \frac{1}{2}\mathbf{a}$

(e) $\overrightarrow{AE} = \overrightarrow{AD} + \overrightarrow{DE}$

$\therefore \quad \overrightarrow{AE} = \mathbf{b} + \frac{1}{2}\mathbf{a}$

Exercise 1B

In questions **1** to **6**, the directions of the vectors are given as bearings, i.e. the angle the vector makes with the direction of north, measured clockwise from north.

1. Find the single vector that is the resultant of a vector of magnitude 7 units, direction 050°, and a vector of magnitude 4 units, direction 160°.

2. Find the single vector that is the resultant of two vectors of magnitude 6 units and 5 units and directions 240° and 260° respectively.

3. Find the resultant of a vector of magnitude 4 units, direction 040°, and a vector of magnitude 7 units, direction 130°.

4. Find the resultant of the vectors **a** and **b** if **a** has magnitude 6 units and direction 160° and **b** has magnitude 11 units and direction 320°.

5. By making a scale drawing, find the resultant of the vectors **a**, **b** and **c** given that **a** has magnitude 6 units and direction 060°, **b** has magnitude 7 units and direction 140° and **c** has magnitude 4 units and direction 020°.

6 Find the resultant of three vectors of magnitude 6 units, 9 units and 10 units and direction 330°, 200° and 080° respectively.

7. The diagram shows a parallelogram OABC with $\overrightarrow{OA} = \mathbf{a}$ and $\overrightarrow{OC} = \mathbf{b}$. Express the following vectors in terms of **a** and **b**:

(a) $\overrightarrow{AB}$ (b) $\overrightarrow{BA}$ (c) $\overrightarrow{CB}$ (d) $\overrightarrow{BC}$
(e) $\overrightarrow{OB}$ (f) $\overrightarrow{BO}$ (g) $\overrightarrow{AC}$ (h) $\overrightarrow{CA}$

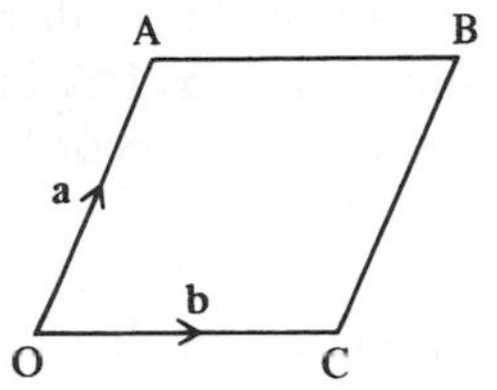

8. The diagram shows a triangle OAB with $\overrightarrow{OA} = \mathbf{a}$ and $\overrightarrow{OB} = \mathbf{b}$. C is the mid-point of AB. Express the following vectors in terms of **a** and **b**:

(a) $\overrightarrow{AB}$ (b) $\overrightarrow{BA}$ (c) $\overrightarrow{AC}$ (d) $\overrightarrow{OC}$

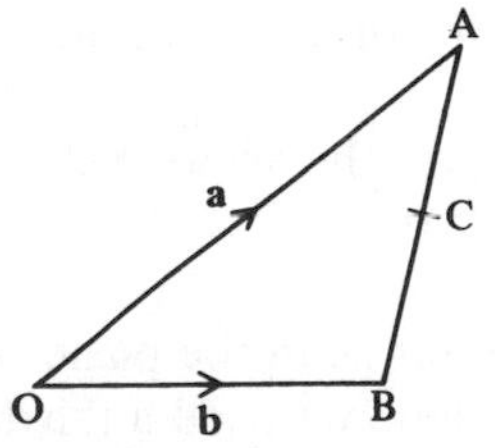

9. The diagram shows a trapezium OABC with $\overrightarrow{OA} = \mathbf{a}$ and $\overrightarrow{OC} = 2\mathbf{b}$. AB is parallel to and half as long as OC. Express the following vectors in terms of **a** and **b**:

(a) $\overrightarrow{AB}$ (b) $\overrightarrow{OB}$ (c) $\overrightarrow{BC}$

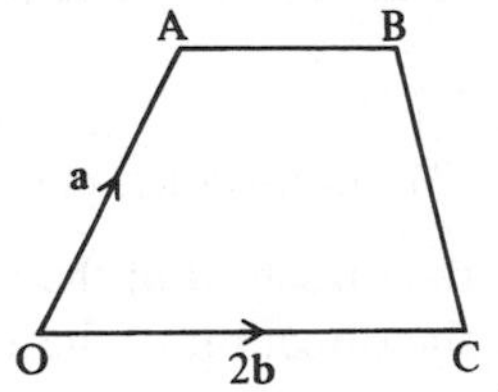

10. The diagram shows a triangle OAB with $\overrightarrow{OA} = \mathbf{a}$ and $\overrightarrow{OB} = \mathbf{b}$. C is a point on AB such that BC : CA = 1 : 2. Express the following vectors in terms of **a** and **b**:

(a) $\overrightarrow{AB}$ (b) $\overrightarrow{AC}$ (c) $\overrightarrow{BC}$ (d) $\overrightarrow{OC}$

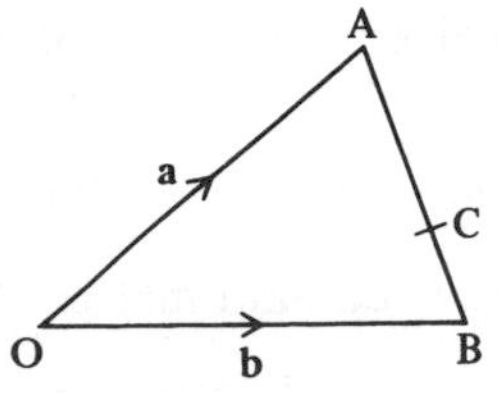

11. The diagram shows a parallelogram OABC with $\overrightarrow{OA} = \mathbf{a}$ and $\overrightarrow{OC} = \mathbf{c}$. E is a point on CB such that CE : EB = 1 : 3. D is a point on AB such that AD : DB = 1 : 2. Express the following vectors in terms of **a** and **c**:

(a) $\overrightarrow{AD}$ (b) $\overrightarrow{CE}$ (c) $\overrightarrow{OD}$
(d) $\overrightarrow{OE}$ (e) $\overrightarrow{AE}$ (f) $\overrightarrow{DE}$

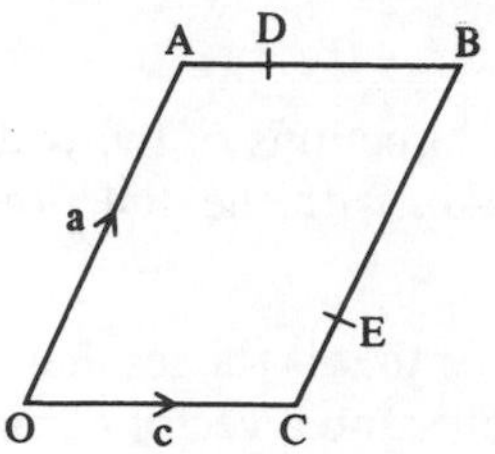

12. In a triangle OAB, the point C lies at the mid-point of OA and the point D lies on AB such that AD : DB = 3 : 1. If $\overrightarrow{OA} = \mathbf{a}$ and $\overrightarrow{OB} = \mathbf{b}$, express the following vectors in terms of **a** and **b**:

(a) $\overrightarrow{OC}$ (b) $\overrightarrow{AB}$ (c) $\overrightarrow{AD}$ (d) $\overrightarrow{OD}$ (e) $\overrightarrow{CB}$ (f) $\overrightarrow{CD}$

Unit vectors

A unit vector is one with a magnitude of 1 unit. Unit vectors may be in any direction, but it is usual to denote a unit vector in the direction of the positive **x**-coordinate axis by **i**. A unit vector in the direction of the positive *y*-coordinate axis is denoted by **j**.

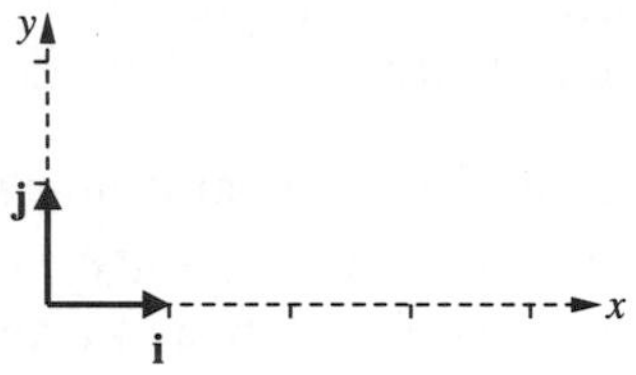

A vector $(4\mathbf{i} + 2\mathbf{j})$ units consists of:

4 units in the direction of the unit vector **i**
and 2 units in the direction of the unit vector **j**.

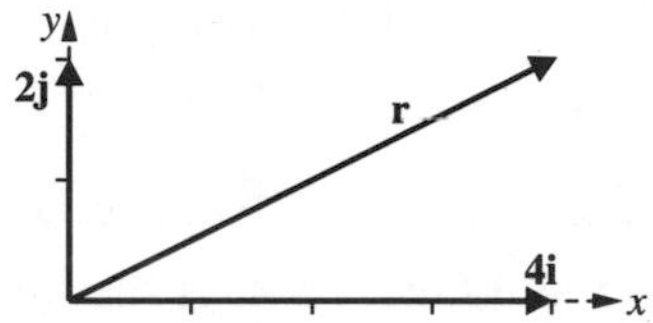

These combine to give the vector **r** shown in the diagram.

Some vectors may not lie in the plane of the *x* and *y* axes. For these we need a third axis at right angles to the other two. This is referred to as the *z*-axis and a unit vector in this direction is denoted by **k**.

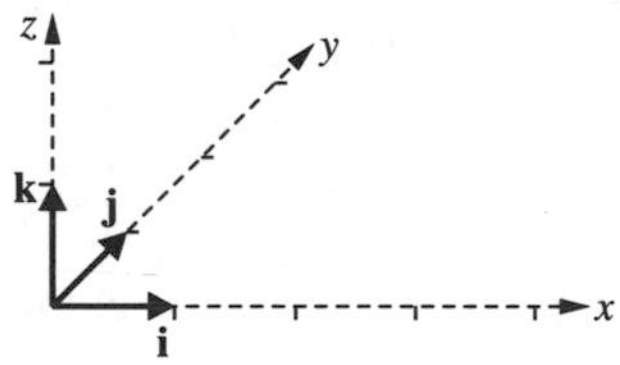

A vector $(4\mathbf{i} + 2\mathbf{j} + 3\mathbf{k})$ units consists of:

4 units in the direction of the unit vector **i**
2 units in the direction of the unit vector **j**
and 3 units in the direction of the unit vector **k**.

These combine to give the vector **p** shown in the diagram.

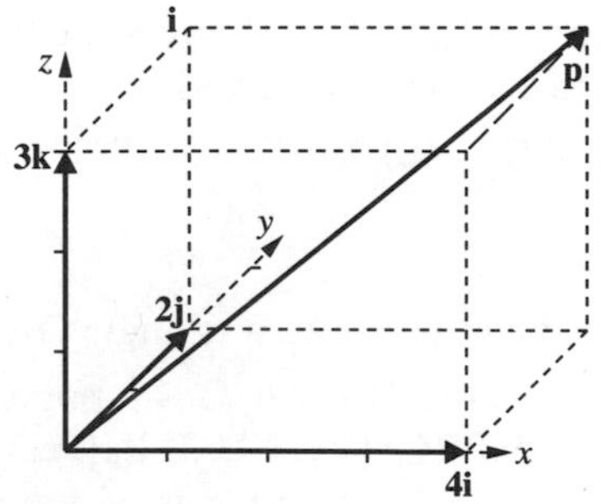

Note. An alternative way of writing the vector $a\mathbf{i} + b\mathbf{j}$ is to use the "column matrix" form $\begin{pmatrix} a \\ b \end{pmatrix}$. Example 5 uses this notation.

Addition of vectors

When vectors are given in terms of unit vectors, their addition is straightforward.

Example 4

Given $\mathbf{a} = 3\mathbf{i} + 2\mathbf{j}$ and $\mathbf{b} = 5\mathbf{i} - 6\mathbf{j}$, find the resultant of **a** and **b**.

$$\begin{aligned}\mathbf{a} + \mathbf{b} &= (3\mathbf{i} + 2\mathbf{j}) + (5\mathbf{i} - 6\mathbf{j}) \\ &= 8\mathbf{i} - 4\mathbf{j}\end{aligned}$$

The resultant is $8\mathbf{i} - 4\mathbf{j}$.

Example 5

Find the resultant of $\mathbf{a} = \begin{pmatrix} 7 \\ -4 \end{pmatrix}$ and $\mathbf{b} = \begin{pmatrix} 5 \\ 6 \end{pmatrix}$.

$$\begin{aligned}\mathbf{a} + \mathbf{b} &= \begin{pmatrix} 7 \\ -4 \end{pmatrix} + \begin{pmatrix} 5 \\ 6 \end{pmatrix} \\ &= \begin{pmatrix} 7 + 5 \\ -4 + 6 \end{pmatrix} \\ &= \begin{pmatrix} 12 \\ 2 \end{pmatrix}\end{aligned}$$

The resultant is $\begin{pmatrix} 12 \\ 2 \end{pmatrix}$.

Note that the answer is stated using the same notation as used in the question, i.e. in column form.

Example 6

Given $\mathbf{a} = 2\mathbf{i} + 3\mathbf{j} - 2\mathbf{k}$ and $\mathbf{b} = \mathbf{i} - 4\mathbf{j} + \mathbf{k}$, find the resultant of **a** and **b**.

$$\begin{aligned}\mathbf{a} + \mathbf{b} &= (2\mathbf{i} + 3\mathbf{j} - 2\mathbf{k}) + (\mathbf{i} - 4\mathbf{j} + \mathbf{k}) \\ &= 3\mathbf{i} - \mathbf{j} - \mathbf{k}\end{aligned}$$

The resultant is $3\mathbf{i} - \mathbf{j} - \mathbf{k}$.

The magnitude of a vector

The magnitude of the vector $\mathbf{v} = a\mathbf{i} + b\mathbf{j}$ can be determined using Pythagoras' theorem. From the diagram on the right we see that the magnitude of the vector **v**, written $|\mathbf{v}|$, is represented by the length OV.

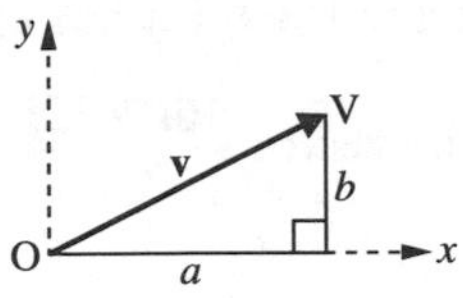

By Pythagoras, $\quad \mathrm{OV}^2 = a^2 + b^2$

i.e. $\quad \mathrm{OV} = \sqrt{a^2 + b^2}$

Thus if $\mathbf{v} = a\mathbf{i} + b\mathbf{j}$ then $\quad |\mathbf{v}| = \sqrt{a^2 + b^2}$.

Extending this to three dimensions, with $\mathbf{v} = a\mathbf{i} + b\mathbf{j} + c\mathbf{k}$, gives:

$$|\mathbf{v}| = \text{OV} \quad \text{(see diagram)}$$

But
$$\begin{aligned}\text{OV}^2 &= \text{OQ}^2 + \text{QV}^2 \\ &= \text{OP}^2 + \text{PQ}^2 + \text{QV}^2 \\ &= a^2 + b^2 + c^2\end{aligned}$$

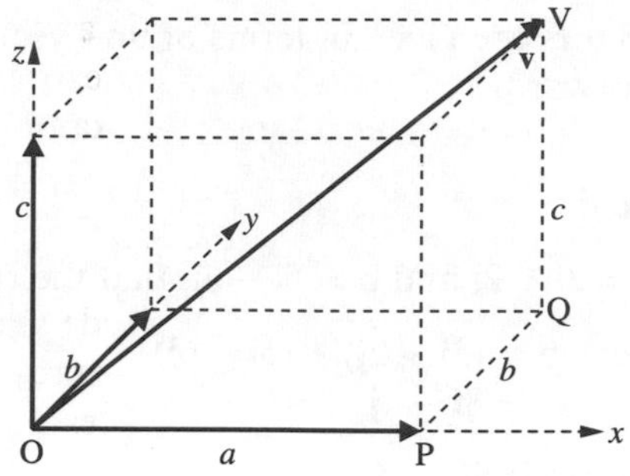

Thus if $\mathbf{v} = a\mathbf{i} + b\mathbf{j} + c\mathbf{k}$, then:

$$|\mathbf{v}| = \sqrt{a^2 + b^2 + c^2}.$$

Example 7

Find the magnitude and direction of the vector $\mathbf{v} = (3\mathbf{i} + 4\mathbf{j})$ units.

First sketch the vector:

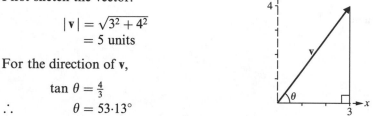

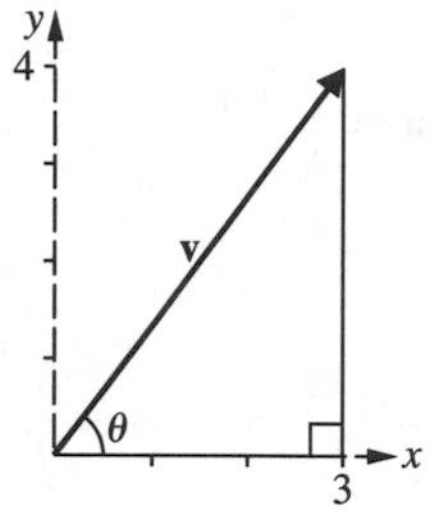

$$\begin{aligned}|\mathbf{v}| &= \sqrt{3^2 + 4^2} \\ &= 5 \text{ units}\end{aligned}$$

For the direction of $\mathbf{v}$,

$$\tan\theta = \tfrac{4}{3}$$

$\therefore \quad \theta = 53{\cdot}13°$

The magnitude of $\mathbf{v}$ is 5 units and it is at an angle of 53·13° to the x-axis.

Example 8

Given $\mathbf{a} = 5\mathbf{i} + 2\mathbf{j} - \mathbf{k}$ and $\mathbf{b} = \mathbf{i} - 6\mathbf{j} + 2\mathbf{k}$, find the magnitude of the resultant of $\mathbf{a}$ and $\mathbf{b}$.

The resultant is given by
$$\begin{aligned}\mathbf{r} &= \mathbf{a} + \mathbf{b} \\ &= (5\mathbf{i} + 2\mathbf{j} - \mathbf{k}) + (\mathbf{i} - 6\mathbf{j} + 2\mathbf{k}) \\ &= 6\mathbf{i} - 4\mathbf{j} + \mathbf{k}\end{aligned}$$

Thus
$$\begin{aligned}|\mathbf{r}| &= \sqrt{6^2 + (-4)^2 + (1)^2} \\ &= 7{\cdot}28\end{aligned}$$

The magnitude of the resultant of $\mathbf{a}$ and $\mathbf{b}$ is 7·28 units.

Example 9

Find the vector which has magnitude 15 units and is parallel to $16\mathbf{i} + 12\mathbf{j}$.

The vector $16\mathbf{i} + 12\mathbf{j}$ has magnitude $\sqrt{16^2 + 12^2} = 20$ units.

Thus the vector $\dfrac{16\mathbf{i} + 12\mathbf{j}}{20}$ will be parallel to $16\mathbf{i} + 12\mathbf{j}$ but will be of unit length.

Thus the required vector will be $15 \times \dfrac{16\mathbf{i} + 12\mathbf{j}}{20}$.

The required vector is $12\mathbf{i} + 9\mathbf{j}$.

Resolving a vector

In Example 7 we found the magnitude and direction of the vector $3\mathbf{i} + 4\mathbf{j}$. We can also do the reverse of this, i.e. express a vector in **i**–**j** form given its magnitude and direction, as the next example shows.

Example 10

The diagrams below show the magnitude and directions of two vectors, **p** and **q**, each lying in the x–y plane. Express each in the form $a\mathbf{i} + b\mathbf{j}$. (Give a and b correct to 2 decimal places if rounding is necessary.)

(a)

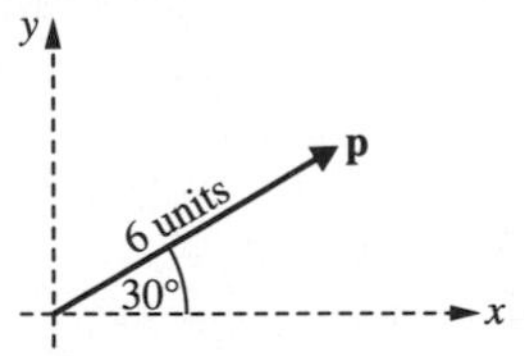

(b)

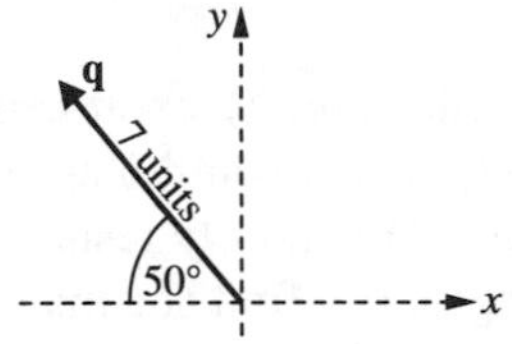

(a)

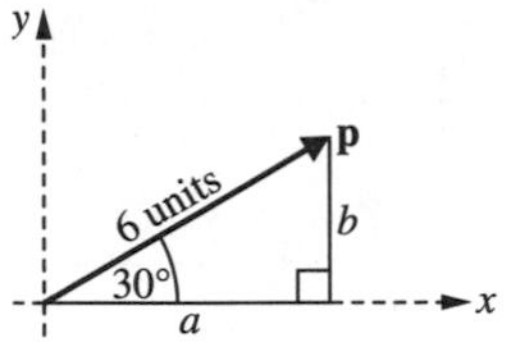

Using trigonometry gives:

$$\cos 30° = \frac{a}{6} \quad \text{and} \quad \sin 30° = \frac{b}{6}$$

$\therefore \quad a = 3\sqrt{3}$ and $b = 3$

Thus $\mathbf{p} = 3\sqrt{3}\,\mathbf{i} + 3\mathbf{j}$

(b)

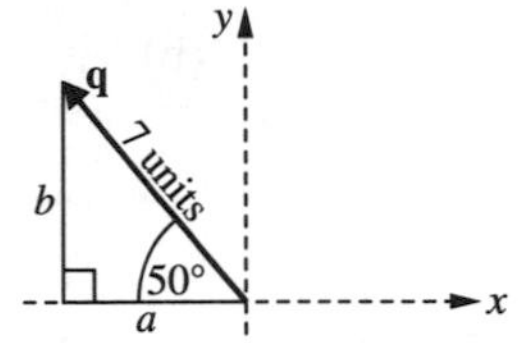

Using trigonometry gives:

$$\cos 50° = \frac{a}{7} \quad \text{and} \quad \sin 50° = \frac{b}{7}$$

$\therefore \quad a = 4{\cdot}50$ and $b = 5{\cdot}36$

Thus $\mathbf{q} = -4{\cdot}50\mathbf{i} + 5{\cdot}36\mathbf{j}$

To generalise:
If vector **v** lies in the x–y plane, is of magnitude v and makes an angle θ with the x-axis (see diagram), then:

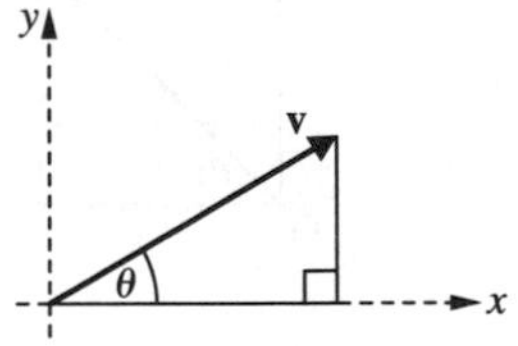

$$\mathbf{v} = v \cos \theta\,\mathbf{i} + v \sin \theta\,\mathbf{j}$$

($v \cos \theta$) and ($v \sin \theta$) are the *resolved parts* or *components* of **v** in the directions of the x-axis and y-axis respectively.

Example 11

The diagram on the right shows vector **p** of magnitude 8 units and making angles of 75°, 64° and 32° with the positive x, y and z axes respectively. Express **p** in the form $a\mathbf{i} + b\mathbf{j} + c\mathbf{k}$ with a, b and c given correct to one decimal place.

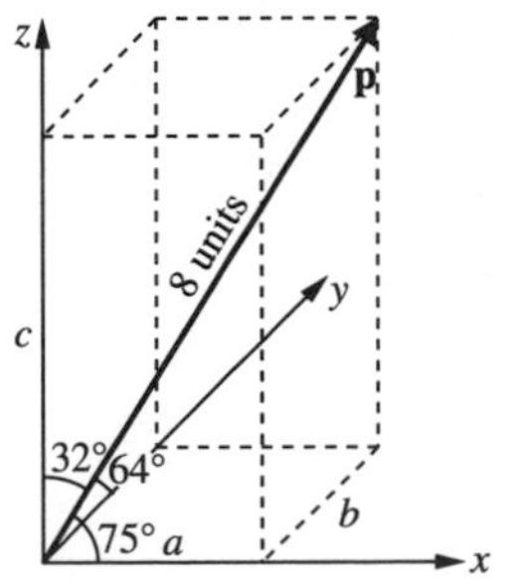

Using trigonometry gives:

$$\cos 75° = \frac{a}{8}, \quad \cos 64° = \frac{b}{8}, \quad \cos 32° = \frac{c}{8}.$$

$$\therefore \qquad a = 2{\cdot}1 \qquad b = 3{\cdot}5 \qquad c = 6{\cdot}8$$

Thus $\quad \mathbf{p} = 2{\cdot}1\mathbf{i} + 3{\cdot}5\mathbf{j} + 6{\cdot}8\mathbf{k}$ (all components correct to one decimal place)

Exercise 1C

1. For each part of this question, (i) express the vector (shown as a heavy line) in the form $a\mathbf{i} + b\mathbf{j}$ where a and b are numbers, $\mathbf{i}$ and $\mathbf{j}$ are unit vectors in the directions Ox and Oy respectively, and the squares in each grid are of unit of length, (ii) find the magnitude of the vector, (iii) find the angle θ.

(a)
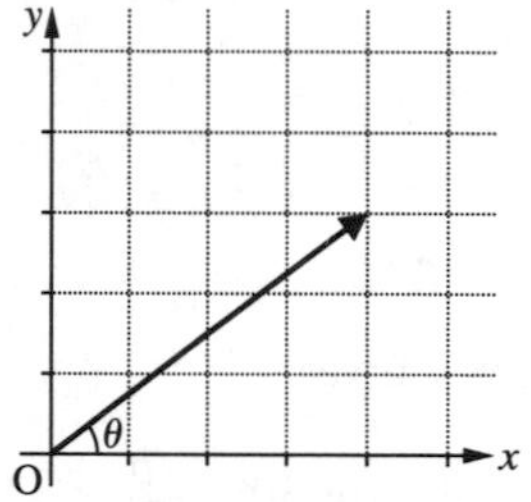

(b)
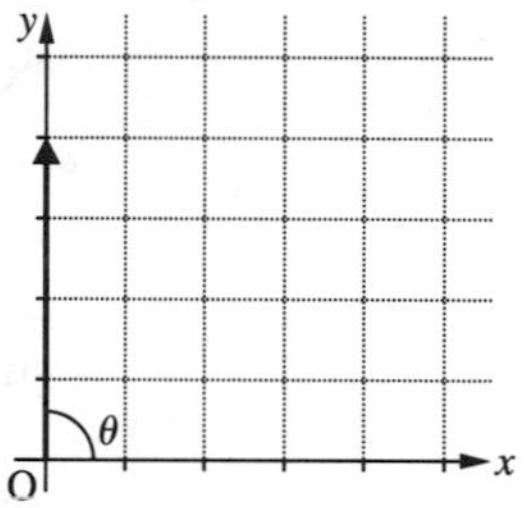

(c)
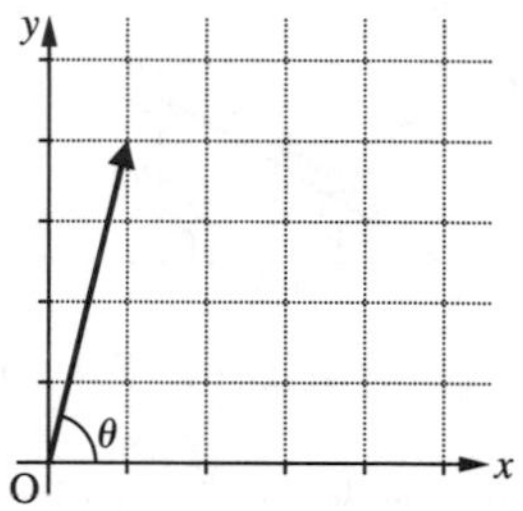

(d)
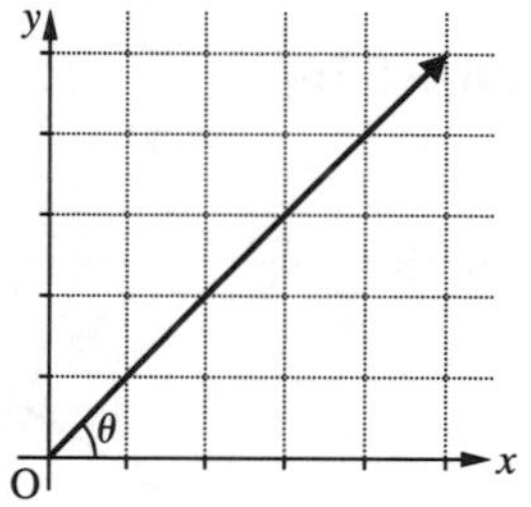

(e)
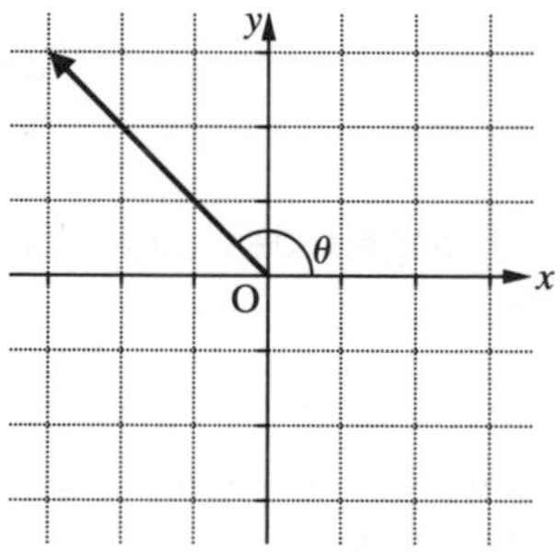

(f)
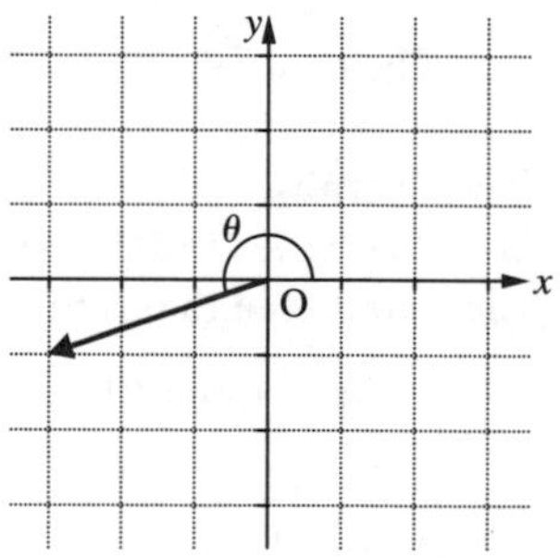

(g)
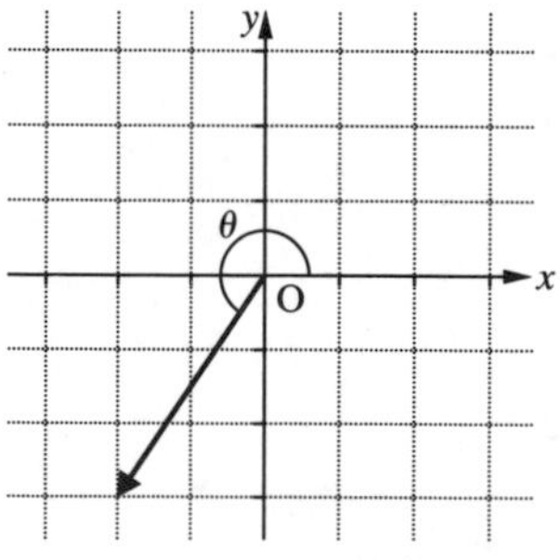

(h)
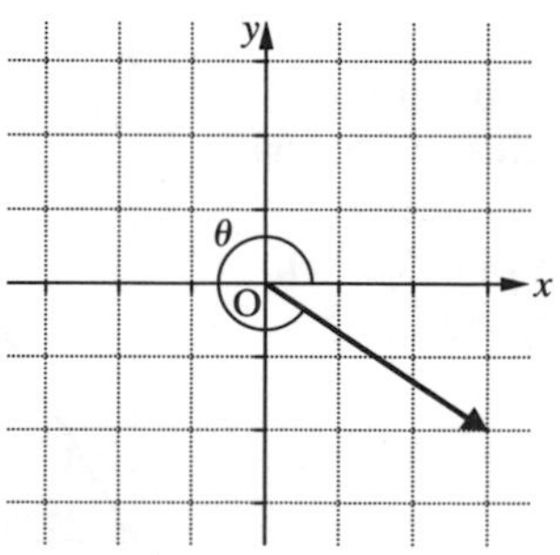

(i)
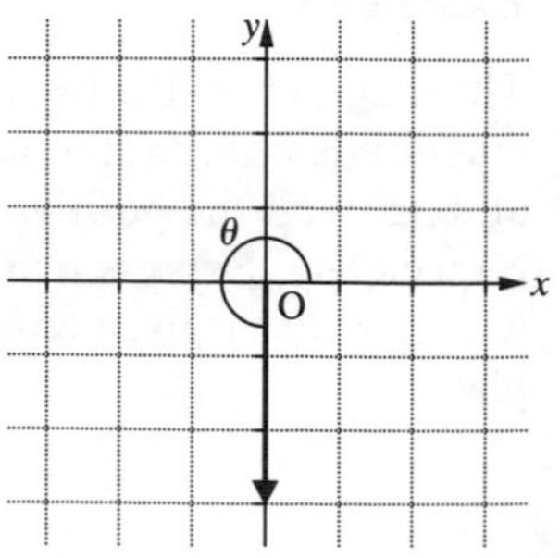

2. The diagrams below show the magnitude and directions of vectors **a** to **h**, each lying in the x–y plane. Express each in the form $a\mathbf{i} + b\mathbf{j}$, where **i** and **j** are unit vectors in the directions Ox and Oy respectively.

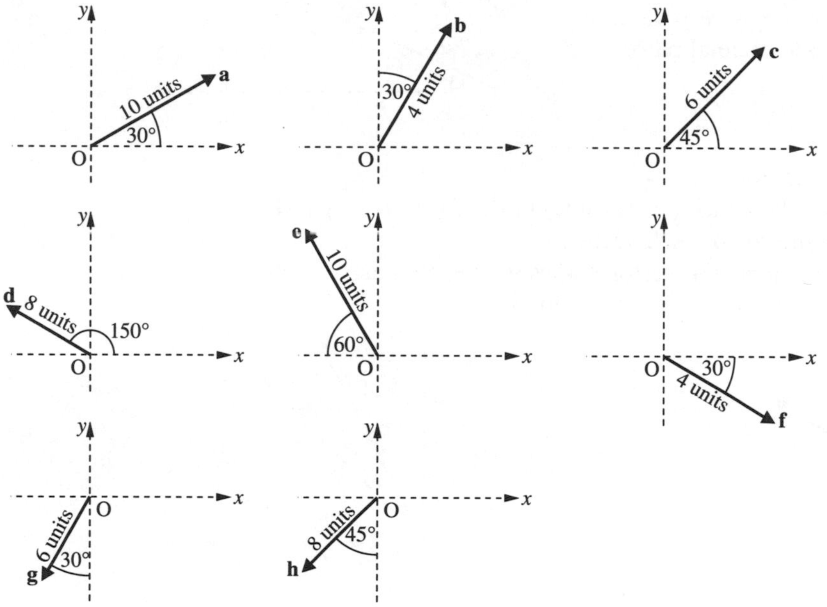

3. The following table gives the magnitude and direction of six vectors. Express each vector in the **i**–**j** form where **i** is a unit vector due east and **j** a unit vector due north.

vector	magnitude	direction (given as a bearing measured clockwise from north)
a	4 units	090°
b	7 units	180°
c	$5\sqrt{2}$ units	045°
d	10 units	060°
e	6 units	240°
f	10 units	335°

4. The diagram shows vector **p** of magnitude 6 units and making angles of 66°, 53° and 37° with the positive x, y and z axes respectively. Express **p** in the form $a\mathbf{i} + b\mathbf{j} + c\mathbf{k}$ with a, b and c given correct to one decimal place.

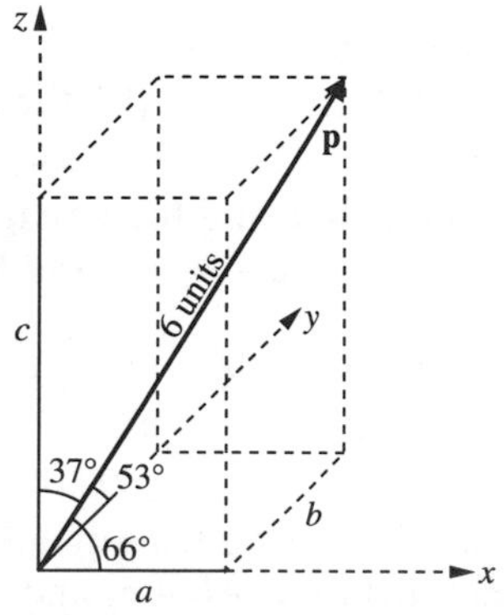

5. The diagram on the right shows vector **q** of magnitude 6 units and making angles of 35°, 60° and 70° with the positive x, y and z axes respectively.
Express **p** in the form $a\mathbf{i} + b\mathbf{j} + c\mathbf{k}$ with a, b and c given correct to one decimal place.

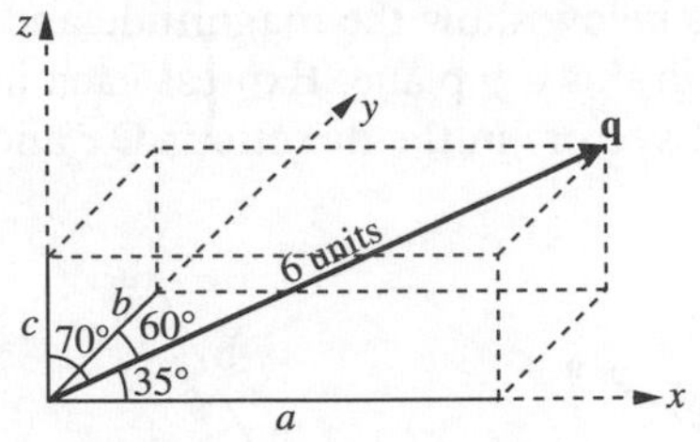

6. For each part of this question
(i) Express the vector (shown as a heavy line) in the form $a\mathbf{i} + b\mathbf{j} + c\mathbf{k}$.
(ii) Determine the magnitude of each vector.
(iii) Determine the *acute* angle the vector makes with each axis, x, y and z, giving your answers correct to one decimal place.

(a)

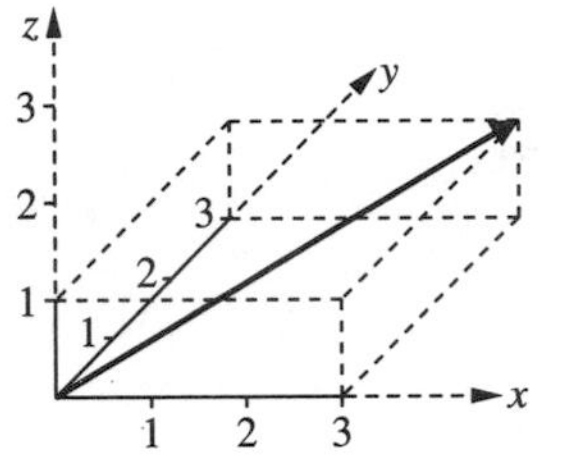

(b)

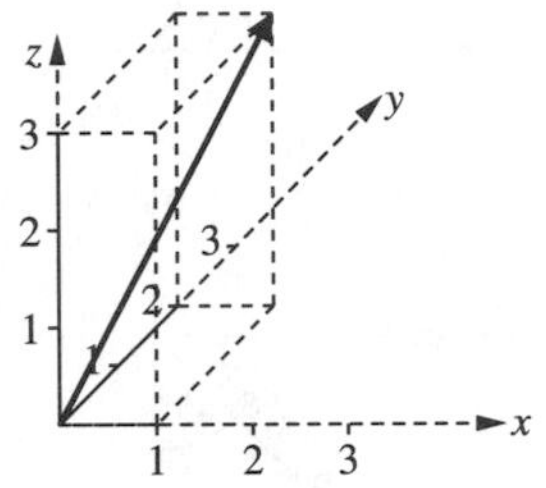

(c)

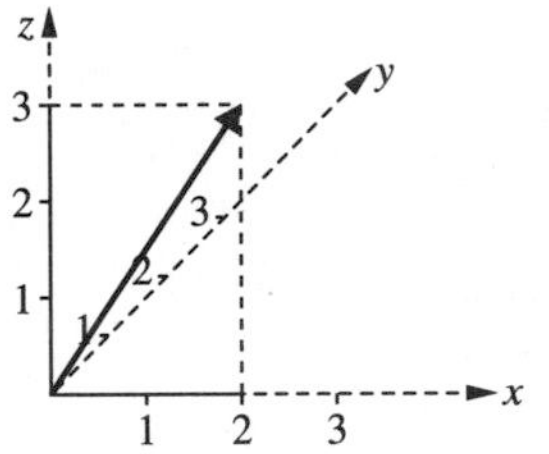

(d)

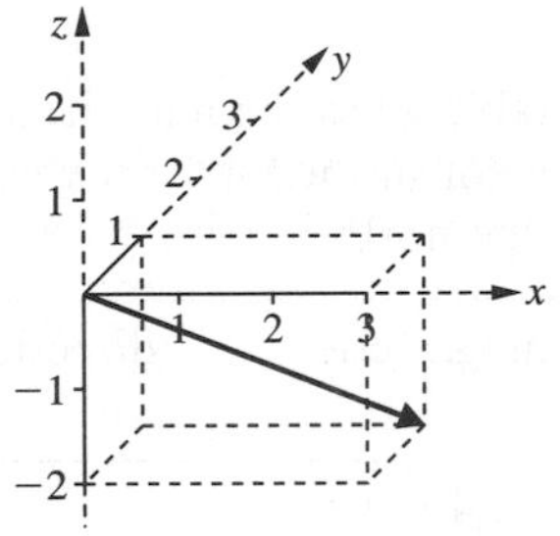

7. If $\mathbf{a} = 3\mathbf{i} + 4\mathbf{j}$, $\mathbf{b} = 4\mathbf{i} + 20\mathbf{j}$ and $\mathbf{c} = 5\mathbf{i} - 19\mathbf{j}$, find:
(a) the resultant of **a** and **b**, (b) the resultant of **a** and **c**,
(c) $|\mathbf{a}|$, (d) $|\mathbf{b}|$, (e) $|\mathbf{a} + \mathbf{b}|$,
(f) a vector that is parallel to **a** and has a magnitude of 15 units,
(g) a vector that is parallel to $(\mathbf{a} + \mathbf{b})$ and has a magnitude of 100 units.

8. If $\mathbf{a} = 2\mathbf{i} + 5\mathbf{j}$, $\mathbf{b} = -7\mathbf{i} + 7\mathbf{j}$ and $\mathbf{c} = 14\mathbf{i}$, find:
(a) the resultant of **a** and **b**, (b) the resultant of **a**, **b** and **c**,
(c) $|\mathbf{a}|$, (d) $|\mathbf{b}|$, (e) $|\mathbf{c}|$, (f) $|\mathbf{a} + \mathbf{b} + \mathbf{c}|$,
(g) a vector that is parallel to **a** and has a magnitude of $5\sqrt{29}$ units,
(h) a vector that is parallel to $(\mathbf{a} + \mathbf{b} + \mathbf{c})$ and has a magnitude of 90 units.

9. If $\mathbf{a} = \mathbf{i} - 3\mathbf{j} + 2\mathbf{k}$, $\mathbf{b} = 5\mathbf{i} + 4\mathbf{j}$ and $\mathbf{c} = 3\mathbf{i} + \mathbf{j} + 4\mathbf{k}$, find:
(a) the resultant of **a** and **b**, (b) the resultant of **a**, **b** and **c**,
(c) $|\mathbf{a}|$, (d) $|\mathbf{b}|$, (e) $|\mathbf{c}|$, (f) $|\mathbf{a} + \mathbf{b} + \mathbf{c}|$,
(g) a vector that is parallel to **a** and has a magnitude of 28 units,
(h) a vector that is parallel to $(\mathbf{a} + \mathbf{b} + \mathbf{c})$ and has a magnitude of 5 units.

10. If $\mathbf{a} = \begin{pmatrix} 2 \\ 7 \\ 7 \end{pmatrix}$, $\mathbf{b} = \begin{pmatrix} 6 \\ -3 \\ 2 \end{pmatrix}$, and $\mathbf{c} = \begin{pmatrix} 0 \\ -4 \\ -3 \end{pmatrix}$, find:

(a) the resultant of **a** and **b** (b) the resultant of **a** and **c**
(c) $|\mathbf{a}|$ (d) $|\mathbf{b}|$ (e) $|\mathbf{c}|$ (f) $|\mathbf{a}+\mathbf{b}+\mathbf{c}|$
(g) a vector that is parallel to $(\mathbf{a}+\mathbf{b}+\mathbf{c})$ and has a magnitude of 50 units

Scalar product

The idea of multiplying two vectors together may seem a little peculiar at first. How are we going to multiply together vectors such as 6 km in direction 040° and 3 km in direction 100°?
Could the answer possibly be 18 km in direction 4000°? Well of course we could define the technique of vector multiplication in this way, but if the answer such a process gives is of no use, there would be little point performing multiplication in this way. There are in fact two ways to define vector multiplication that do prove to be useful.
One method produces an answer that is a scalar. We call this the *scalar product* and we will consider this in just a moment.
The other method produces an answer that is a vector. We call this the *vector product* and it is beyond the scope of this book.
We define the scalar product of two vectors, **a** and **b**, to be the product of the magnitude of **a**, the magnitude of **b**, and the cosine of the angle between **a** and **b**.
We write this as: $\mathbf{a}\,.\,\mathbf{b}$ pronounced "**a** dot **b**".
For this reason the scalar product is also referred to as the *dot product*.

Thus $\boxed{\mathbf{a}\,.\,\mathbf{b} = |\mathbf{a}|\ |\mathbf{b}| \cos\theta \qquad \text{where } \theta \text{ is the angle between } \mathbf{a} \text{ and } \mathbf{b}.}$

Notice that when θ is acute $\cos\theta > 0$ and so $\mathbf{a}\,.\,\mathbf{b} > 0$,
and when θ is obtuse $\cos\theta < 0$ and so $\mathbf{a}\,.\,\mathbf{b} < 0$.

Also if **a** and **b** are perpendicular, $\theta = 90°$, $\cos\theta = 0$, and so $\mathbf{a}\,.\,\mathbf{b} = 0$.

Thus: *if two vectors are perpendicular, their scalar product is zero.*

And, provided neither **a** nor **b** have zero magnitude, it also follows that:

If the scalar product of two vectors is zero, the vectors are perpendicular.

Note 1. The phrase "the angle between two vectors" always refers to the angle between the directions of the vectors when these directions are either both towards their point of intersection or both away from their point of intersection. Thus, in each of the following diagrams, θ is the angle between the two vectors.

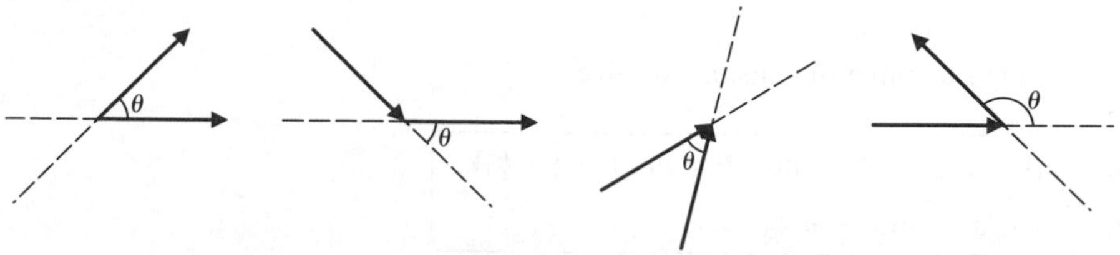

Note 2. The following properties, not proved here, follow from the scalar product definition:

$$\mathbf{a}\,.\,\mathbf{b} = \mathbf{b}\,.\,\mathbf{a}$$
$$\mathbf{a}\,.\,\mathbf{a} = a^2 \quad (\text{where} \quad a = |\mathbf{a}|)$$
$$\mathbf{a}\,.\,(\lambda\mathbf{b}) = \lambda(\mathbf{a}\,.\,\mathbf{b}) = (\lambda\mathbf{a})\,.\,\mathbf{b}$$
$$\mathbf{a}\,.\,(\mathbf{b}+\mathbf{c}) = \mathbf{a}\,.\,\mathbf{b}+\mathbf{a}\,.\,\mathbf{c}$$
$$(\mathbf{a}+\mathbf{b})\,.\,\mathbf{c} = \mathbf{a}\,.\,\mathbf{c}+\mathbf{b}\,.\,\mathbf{c}$$
$$(\mathbf{a}+\mathbf{b})\,.\,(\mathbf{c}+\mathbf{d}) = \mathbf{a}\,.\,\mathbf{c}+\mathbf{a}\,.\,\mathbf{d}+\mathbf{b}\,.\,\mathbf{c}+\mathbf{b}\,.\,\mathbf{d}$$

Example 12

Find $\mathbf{a}\,.\,\mathbf{b}$ for each of the following:

(a)

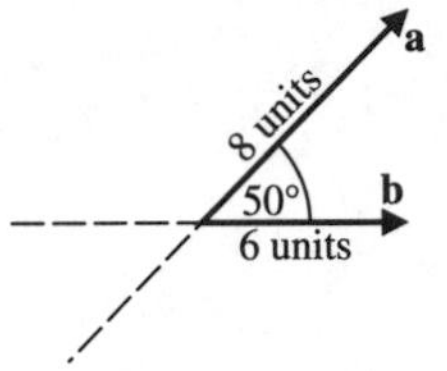

(b)

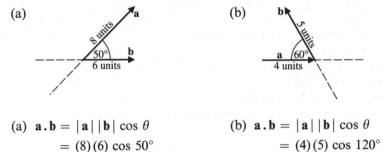

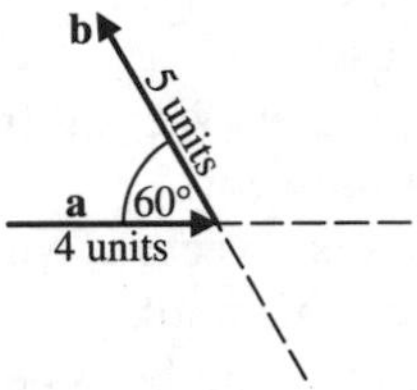

(a) $\mathbf{a}\,.\,\mathbf{b} = |\mathbf{a}|\,|\mathbf{b}|\cos\theta$
$= (8)(6)\cos 50°$
$= 30{\cdot}9$ (correct to 1 d.p.)

(b) $\mathbf{a}\,.\,\mathbf{b} = |\mathbf{a}|\,|\mathbf{b}|\cos\theta$
$= (4)(5)\cos 120°$
$= -10$

Scalar product from vectors in component form

Consider the vectors $\mathbf{a} = a_1\mathbf{i} + a_2\mathbf{j}$ and $\mathbf{b} = b_1\mathbf{i} + b_2\mathbf{j}$.

It follows that
$$\begin{aligned}\mathbf{a}\,.\,\mathbf{b} &= (a_1\mathbf{i}+a_2\mathbf{j})\,.\,(b_1\mathbf{i}+b_2\mathbf{j})\\ &= a_1b_1\mathbf{i}\,.\,\mathbf{i}+a_1b_2\mathbf{i}\,.\,\mathbf{j}+a_2b_1\mathbf{j}\,.\,\mathbf{i}+a_2b_2\mathbf{j}\,.\,\mathbf{j}\\ &= a_1b_1(1)+a_1b_2(0)+a_2b_1(0)+a_2b_2(1)\\ &= a_1b_1+a_2b_2\end{aligned}$$

Thus if $\mathbf{a} = a_1\mathbf{i} + a_2\mathbf{j}$ and $\mathbf{b} = b_1\mathbf{i} + b_2\mathbf{j}$, then $\mathbf{a}\,.\,\mathbf{b} = a_1b_1 + a_2b_2$.

The result can be extended to three dimensions to give:

If $\mathbf{a} = a_1\mathbf{i} + a_2\mathbf{j} + a_3\mathbf{k}$ and $\mathbf{b} = b_1\mathbf{i} + b_2\mathbf{j} + b_3\mathbf{k}$,
then $\mathbf{a}\,.\,\mathbf{b} = a_1b_1 + a_2b_2 + a_3b_3$.

Example 13

Find $\mathbf{p}.\mathbf{q}$ for each of the following:

(a) $\mathbf{p} = 2\mathbf{i} + 3\mathbf{j}$
$\mathbf{q} = \mathbf{i} - 4\mathbf{j}$

(b) $\mathbf{p} = \begin{pmatrix} 3 \\ 2 \\ -1 \end{pmatrix}, \mathbf{q} = \begin{pmatrix} 5 \\ -1 \\ 2 \end{pmatrix}$

(c) $\mathbf{p} = 4\mathbf{i} - 6\mathbf{j} + \mathbf{k}$
$\mathbf{q} = \mathbf{i} - 2\mathbf{j} - 3\mathbf{k}$

(a) $\mathbf{p}.\mathbf{q} = (2)(1) + (3)(-4)$
$= -10$

(b) $\mathbf{p}.\mathbf{q} = \begin{pmatrix} 3 \\ 2 \\ -1 \end{pmatrix} . \begin{pmatrix} 5 \\ -1 \\ 2 \end{pmatrix}$
$= 15 - 2 - 2$
$= 11$

(c) $\mathbf{p}.\mathbf{q} = (4)(1) + (-6)(-2) + (1)(-3)$
$= 13$

Example 14

Find the angle between $\mathbf{a} = 2\mathbf{i} - \mathbf{j} + 3\mathbf{k}$ and $\mathbf{b} = \mathbf{i} + 4\mathbf{j} + 3\mathbf{k}$.

$$\begin{aligned} \mathbf{a}.\mathbf{b} &= (2\mathbf{i} - \mathbf{j} + 3\mathbf{k}).(\mathbf{i} + 4\mathbf{j} + 3\mathbf{k}) \\ &= (2)(1) + (-1)(4) + (3)(3) \\ &= 7 \end{aligned}$$

But
$$\begin{aligned} \mathbf{a}.\mathbf{b} &= |\mathbf{a}|\,|\mathbf{b}| \cos\theta \\ &= \sqrt{2^2 + (-1)^2 + 3^2}\,\sqrt{1^2 + 4^2 + 3^2} \cos\theta \\ &= \sqrt{14}\,\sqrt{26} \cos\theta \end{aligned}$$

$\therefore \quad \sqrt{14}\,\sqrt{26} \cos\theta = 7$

Thus $\quad \theta = 68°$ to the nearest degree.

Exercise 1D

1. Find $\mathbf{a}.\mathbf{b}$ for each of the following:

(a)
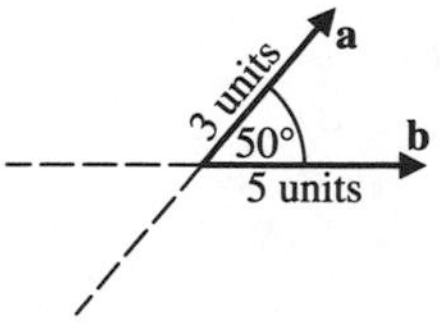

(b)
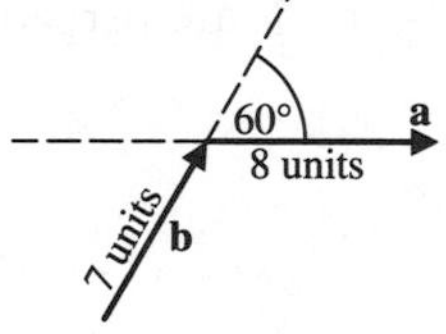

(c)
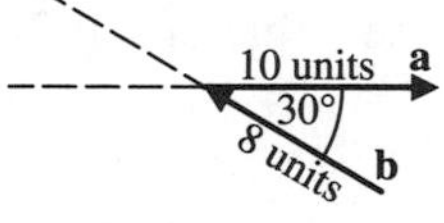

(d)
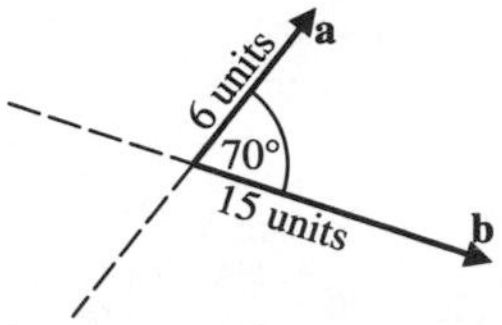

(e)
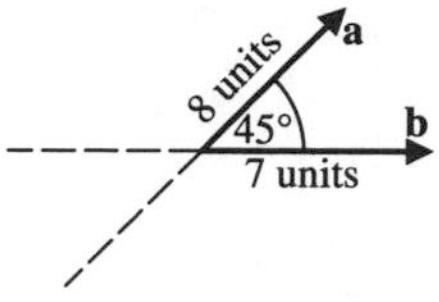

(f)
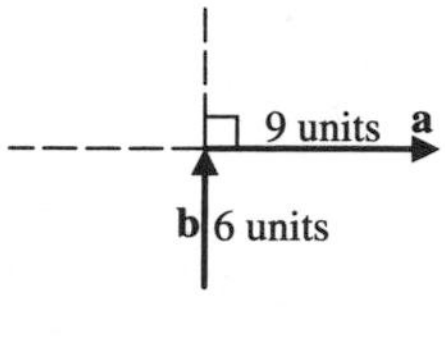

2. State whether each of the following are scalars or vectors:

(a) $\mathbf{a}.\mathbf{b}$ (b) $2\mathbf{a}$ (c) $\mathbf{a}+\mathbf{b}$ (d) $\mathbf{a}.(\mathbf{b}+\mathbf{c})$
(e) $6\mathbf{a}.\mathbf{b}$ (f) $(3\mathbf{a}).(2\mathbf{b})$ (g) $\mathbf{a}-\mathbf{b}$ (h) $(\mathbf{a}+\mathbf{b}).(\mathbf{c}+\mathbf{d})$
(i) $3\mathbf{i}-4\mathbf{j}+\mathbf{k}$ (j) $|\mathbf{a}+\mathbf{b}|$ (k) $2\mathbf{a}+3\mathbf{b}$ (l) $(\mathbf{a}+\mathbf{b}).(\mathbf{a}-\mathbf{b})$

3. Given that **i**, **j** and **k** are unit vectors in the direction of the positive x, y and z axes respectively find:

(a) $\mathbf{i}.\mathbf{j}$ (b) $\mathbf{i}.\mathbf{k}$ (c) $\mathbf{k}.\mathbf{j}$ (d) $\mathbf{i}.\mathbf{i}$
(e) $\mathbf{j}.\mathbf{j}$ (f) $\mathbf{k}.\mathbf{k}$ (g) $(3\mathbf{i}).(2\mathbf{i})$ (h) $(3\mathbf{i}).(2\mathbf{k})$

Find the scalar product for each of the following pairs of vectors:

4. $2\mathbf{i}+\mathbf{j}$ and $\mathbf{i}-3\mathbf{j}$
5. $4\mathbf{i}+2\mathbf{j}$ and $\mathbf{i}+6\mathbf{j}$
6. $3\mathbf{i}$ and $-2\mathbf{i}+\mathbf{j}$
7. $3\mathbf{i}-\mathbf{j}+\mathbf{k}$ and $2\mathbf{i}+\mathbf{j}+2\mathbf{k}$
8. $5\mathbf{i}+\mathbf{j}-2\mathbf{k}$ and $4\mathbf{i}+3\mathbf{j}-8\mathbf{k}$
9. $2\mathbf{i}+4\mathbf{j}-15\mathbf{k}$ and $-8\mathbf{i}+2\mathbf{j}-\mathbf{k}$
10. $-\mathbf{i}-2\mathbf{j}$ and $5\mathbf{i}+4\mathbf{j}+10\mathbf{k}$
11. $2\mathbf{i}+3\mathbf{j}+\mathbf{k}$ and $\mathbf{i}+2\mathbf{j}+5\mathbf{k}$
12. $\begin{pmatrix}2\\1\end{pmatrix}$ and $\begin{pmatrix}3\\-2\end{pmatrix}$
13. $\begin{pmatrix}5\\-1\end{pmatrix}$ and $\begin{pmatrix}2\\4\end{pmatrix}$
14. $\begin{pmatrix}0\\5\\-2\end{pmatrix}$ and $\begin{pmatrix}-3\\2\\1\end{pmatrix}$
15. $\begin{pmatrix}5\\2\\7\end{pmatrix}$ and $\begin{pmatrix}1\\2\\-1\end{pmatrix}$

Find the angle between each of the following pairs of vectors (giving your answers correct to the nearest degree):

16. $3\mathbf{i}+4\mathbf{j}$ and $5\mathbf{i}-12\mathbf{j}$
17. $2\mathbf{i}-3\mathbf{j}$ and $6\mathbf{i}+4\mathbf{j}$
18. $3\mathbf{i}$ and $-2\mathbf{j}$
19. $2\mathbf{i}+3\mathbf{j}+6\mathbf{k}$ and $2\mathbf{i}+\mathbf{j}+2\mathbf{k}$
20. $2\mathbf{i}+3\mathbf{j}-6\mathbf{k}$ and $2\mathbf{i}+\mathbf{j}+2\mathbf{k}$
21. $\mathbf{i}+2\mathbf{j}-\mathbf{k}$ and $-\mathbf{i}+2\mathbf{j}-\mathbf{k}$
22. $-4\mathbf{i}+2\mathbf{j}-4\mathbf{k}$ and $2\mathbf{i}-\mathbf{j}+2\mathbf{k}$
23. $2\mathbf{i}+3\mathbf{j}+2\mathbf{k}$ and $\mathbf{i}+4\mathbf{k}$
24. $\begin{pmatrix}3\\1\end{pmatrix}$ and $\begin{pmatrix}1\\-2\end{pmatrix}$
25. $\begin{pmatrix}6\\-8\end{pmatrix}$ and $\begin{pmatrix}5\\4\end{pmatrix}$
26. $\begin{pmatrix}0\\1\\-1\end{pmatrix}$ and $\begin{pmatrix}-1\\0\\1\end{pmatrix}$
27. $\begin{pmatrix}2\\2\\3\end{pmatrix}$ and $\begin{pmatrix}2\\1\\-1\end{pmatrix}$

28. If $\mathbf{a}=2\mathbf{i}-3\mathbf{j}$, determine which of the following vectors are perpendicular to **a**.

$\mathbf{b}=-6\mathbf{i}+4\mathbf{j}$ $\mathbf{c}=-6\mathbf{i}-4\mathbf{j}$ $\mathbf{d}=12\mathbf{i}+8\mathbf{j}$ $\mathbf{e}=-2\mathbf{i}+3\mathbf{j}$

29. Find the value of λ if $\lambda\mathbf{i}+2\mathbf{j}-\mathbf{k}$ and $5\mathbf{i}-\lambda\mathbf{j}+\mathbf{k}$ are perpendicular vectors.

30. Find the value of λ if $2\mathbf{i}+\lambda\mathbf{j}+\lambda\mathbf{k}$ and $-\lambda\mathbf{i}-\mathbf{k}$ are perpendicular vectors.

2 Distance, velocity and acceleration

Constant speed and constant velocity

The statement that the *speed* of a car is 40 kilometres per hour (written 40 km/h or 40 km h^{-1}) means that, if the speed remains unchanged, the car will travel 40 km in each hour. The speed of the car is then said to be *uniform* or *constant*. At the same speed the car would travel 80 km in 2 hours, 120 km in 3 hours, etc. Thus:

$$\text{distance travelled} = \text{speed} \times \text{time}$$

or

$$s = v \times t$$

The *velocity* of a car is a measure of the speed at which it is travelling in a particular direction.
If a car has constant, or uniform velocity, then both the speed and the direction of motion of the car remain unchanged.
Thus the *velocity* of a car may be stated as 50 km h^{-1} due north and the *speed* of this car is then 50 km h^{-1}.

So it is seen that speed is a scalar quantity, whereas velocity is a vector quantity and

$$\text{distance travelled in a particular direction} = \text{velocity} \times \text{time taken}$$

or

$$\mathbf{s} = \mathbf{v} \times t$$

The distance travelled in a particular direction may be referred to as the *displacement* of the body from some fixed point.
The letter v is used to denote both speed and velocity. This need cause no confusion provided that the difference between them is remembered, and it is clearly understood which is being used in a particular example.
In most cases, only linear motion will be considered, i.e. motion along a straight line.
Therefore, the velocity can only be in one of two directions. The direction of the velocity can then be distinguished by the use of positive and negative.

For example: 5 m s^{-1} $\longrightarrow$ denoted by velocity of 5 m s^{-1}

5 m s^{-1} $\longleftarrow$ denoted by velocity of -5 m s^{-1}

Change of units

The car which is travelling at 40 km h^{-1} is, of course, travelling a certain number of metres each second.

Example 1

Express a speed of $40\,\text{km}\,\text{h}^{-1}$ in $\text{m}\,\text{s}^{-1}$.

$$40\,\text{km}\,\text{h}^{-1} = 40 \times 1000\,\text{m}\,\text{h}^{-1} = \frac{40 \times 1000}{60 \times 60}\,\text{m}\,\text{s}^{-1} = 11\tfrac{1}{9}\,\text{m}\,\text{s}^{-1}$$

A speed of $40\,\text{km}\,\text{h}^{-1}$ is equivalent to a speed of $11\frac{1}{9}\,\text{m}\,\text{s}^{-1}$.

Use of $s = vt$

When the relationship $s = vt$ is used, the units of the quantities involved must be consistent. If the speed is in $\text{km}\,\text{h}^{-1}$, the time must be in hours and the distance will then be in km.

Example 2

Find the distance travelled in 3 minutes by a body moving with a constant speed of $15\,\text{km}\,\text{h}^{-1}$.
Find also the time taken by this body to travel 200 m at the same speed.

$v = 15\,\text{km}\,\text{h}^{-1}$

$t = 3 \text{ minutes} = \frac{1}{20}\text{ h}$

Using $s = vt$ gives:

$s = 15 \times \frac{1}{20}$

$\therefore \quad s = \frac{3}{4}\text{ km or } 750\text{ m}$

The distance travelled is 750 m.

To find the time taken to travel 200 m

$s = 200\text{ m} \qquad v = 15\,\text{km}\,\text{h}^{-1}$

$$= \frac{15 \times 1000}{60 \times 60}\,\text{m}\,\text{s}^{-1} = \frac{25}{6}\,\text{m}\,\text{s}^{-1}$$

Using $s = vt$ gives:

$200 = \frac{25}{6} \times t$ where t is measured in seconds

$\therefore \quad t = 48\text{ s}$

The time taken to travel 200 m at $15\,\text{km}\,\text{h}^{-1}$ is 48 s.

Average speed

In practice the speed and velocity of a body are seldom constant. When a car travels 40 km in one hour, it is unlikely that its speed is constant. It is probable that for part of the time the car is travelling at more than $40\,\text{km}\,\text{h}^{-1}$, and for some of the time the car's speed is less than $40\,\text{km}\,\text{h}^{-1}$. Thus we often refer to the *average* speed, or the *average* velocity of a body:

$$\text{average speed} = \frac{\text{total distance travelled}}{\text{total time taken}}$$

and $$\text{average velocity} = \frac{\text{total distance travelled in a particular direction}}{\text{total time taken}}$$

The distance travelled in a particular direction can more conveniently be referred to as the displacement from some fixed initial position.

Example 3

A, B and C are three points, in that order, on a straight road with AB = 40 km and BC = 90 km.
A woman travels from A to B at $10\,\text{km}\,\text{h}^{-1}$ and then from B to C at $15\,\text{km}\,\text{h}^{-1}$.
Calculate:
(a) the time taken to travel from A to B
(b) the time taken to travel from B to C
(c) the average speed of the woman for the journey from A to C.

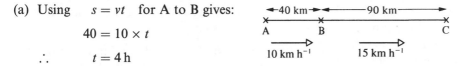

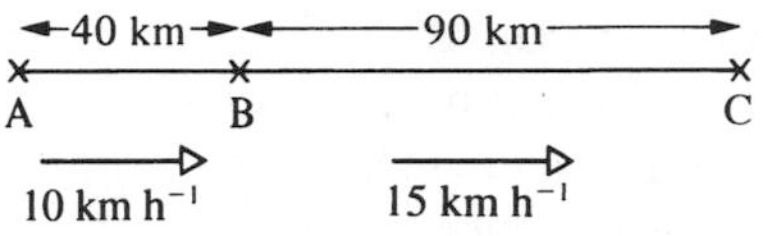

(a) Using $s = vt$ for A to B gives:

$$40 = 10 \times t$$

$\therefore$ $t = 4\,\text{h}$

The time taken to travel from A to B is 4 h.

(b) Using $s = vt$ for B to C gives:

$$90 = 15 \times t$$

$\therefore$ $t = 6\,\text{h}$

The time taken to travel from B to C is 6 h.

(c) Using $\text{average speed} = \dfrac{\text{total distance travelled}}{\text{total time taken}}$ for A to C gives:

$$v = \frac{40 + 90}{4 + 6}$$

$\therefore$ $v = 13\,\text{km}\,\text{h}^{-1}$

The average speed for the whole journey is $13\,\text{km}\,\text{h}^{-1}$.

Example 4

A man walks 400 m due east in a time of 190 s, and then 100 m due west in a time of 50 s.
Calculate:
(a) his average speed
(b) his average velocity, for the whole journey.

(a) Using $\text{average speed} = \dfrac{\text{total distance}}{\text{total time}}$ gives:

$$\text{average speed} = \frac{400 + 100}{190 + 50}$$

$$= \tfrac{500}{240} = 2\tfrac{1}{12}\,\text{m s}^{-1}$$

The average speed is $2\frac{1}{12}\,\text{m s}^{-1}$.

(b) Using $\text{average velocity} = \dfrac{\text{displacement}}{\text{total time}}$ gives:

$$\text{average velocity} = \frac{400\text{ m E} + 100\text{ m W}}{240} = \frac{300\text{ m E}}{240}$$

$$= 1\tfrac{1}{4}\,\text{m s}^{-1}\text{ E}$$

The average velocity is $1\frac{1}{4}\,\text{m s}^{-1}$ east.

Exercise 2A

1. Express a speed of $36\,\text{km h}^{-1}$ in m s^{-1}.
2. Express a speed of $81\,\text{km h}^{-1}$ in m s^{-1}.
3. Express a speed of $35\,\text{m s}^{-1}$ in km h^{-1}.
4. Express a speed of $22\,\text{m s}^{-1}$ in km h^{-1}.
5. Express a speed of $6\,\text{km min}^{-1}$ in m s^{-1}.
6. A body travelling at a constant speed covers a distance of 200 m in 8 seconds. Find the speed of the body.
7. A body travelling at a constant speed covers a distance of 3 km in 2 minutes. Find the speed of the body.
8. Find the distance travelled in 5 seconds by a body moving with a constant speed of $3{\cdot}2\,\text{m s}^{-1}$.
9. Find the distance travelled in 2 minutes by a body moving with a constant speed of $6\,\text{km h}^{-1}$.
10. Find the time taken by a body, moving with a constant speed $3{\cdot}5\,\text{m s}^{-1}$, to travel a distance of 21 m.
11. At time $t = 0$ a body passes through a point A and is moving with a constant velocity of $4\,\text{m s}^{-1}$.
    (a) Find how far the body is from A when $t = 3$ s.
    (b) What is the value of t when the body is 22 m from A?
12. The spacecraft Voyager II travels at a constant velocity of $80\,000\,\text{km h}^{-1}$. Find the distance the spacecraft travels in:
    (a) 1 hour (b) 15 minutes (c) 1 second.
13. The speed of sound is $340\,\text{m s}^{-1}$. Find the distance travelled in one minute by an aircraft flying at Mach 2 (i.e. twice the speed of sound).
14. The speed of light is $3 \times 10^8\,\text{m s}^{-1}$. If the distance from the Sun to the Earth is $1{\cdot}5 \times 10^8$ km, find how long it takes light from the Sun to reach the Earth.
15. If it takes 5 seconds for the sound of thunder to reach my ears, how far am I from the place that it actually occurred? (Speed of sound is $340\,\text{m s}^{-1}$.)
16 If an athlete runs a 1500 metre race in 3 minutes 33 seconds, find his average speed for the race.
17. A, B and C are three points lying in that order on a straight road with AB = 5 km and BC = 4 km. A man runs from A to B at $20\,\text{km h}^{-1}$ and then walks from B to C at $8\,\text{km h}^{-1}$. Find:
    (a) the total time taken to travel from A to C
    (b) the average speed of the man for the journey from A to C.
18. A man walks 150 m due north, in a time of 70 s, and then 50 m due south, in a time of 30 s. Find:
    (a) his average speed
    (b) his average velocity.

19. A car is driven from Town A to Town B, 40 km away, at an average speed of 60 km h^{-1}. The car is at B for 10 minutes and is then driven back to A.
(a) Find the average speed for the journey B → A if the average speed for the complete journey is 60 km h^{-1}.
(b) What is the average velocity of the car for the complete journey?

20. A, B and C are three points lying in that order on a straight line with AB = 60 m and AC = 80 m. A body moves from A to B at an average speed of 10 m s^{-1}, then from B to C in a time of 4 s, and then returns to B. The average speed for the whole journey is 5 m s^{-1}. Find:
(a) the average speed of the body in the second stage of the motion (i.e. B → C)
(b) the average speed of the body in moving from A to C
(c) the time taken for the third stage of the motion (i.e. C → B)
(d) the average velocity for the complete motion.

i–j notation

In the first chapter, vectors were expressed using the **i–j** notation. Both the position and the velocity of a body can be given in this vector form.

Position vector

Using a suitable origin O, the position of a body at P may be given as the vector $\overrightarrow{OP}$, where:

$$\overrightarrow{OP} = (a\mathbf{i} + b\mathbf{j})\,\text{m}.$$

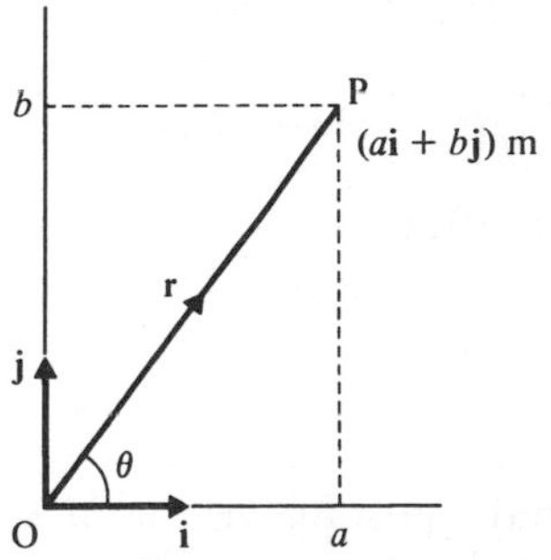

This is the position vector of the body.

As before, the vector may be denoted by a single letter, say, **r**, i.e.

$$\mathbf{r} = \overrightarrow{OP} = (a\mathbf{i} + b\mathbf{j})\,\text{m}$$

The distance of P from the origin O and the direction of $\overrightarrow{OP}$ may then be found:

$$\text{distance OP} = |\overrightarrow{OP}| \text{ or } |\mathbf{r}|$$
$$= \sqrt{(a^2 + b^2)}$$

and the direction of $\overrightarrow{OP}$ is given by $\tan\theta = \dfrac{b}{a}$

Velocity vector

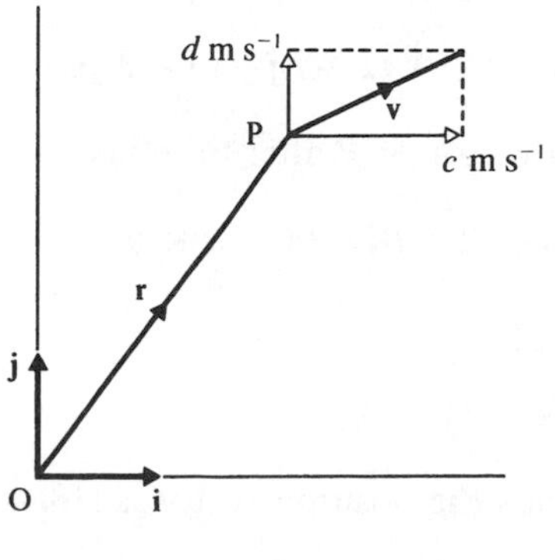

The velocity vector **v** can be expressed in the same way.
If the velocity vector of a body at P is given by:

$$\mathbf{v} = (c\mathbf{i} + d\mathbf{j})\,\text{m s}^{-1},$$

the body has a velocity of c m s^{-1} in the direction of the unit vector **i**, and d m s^{-1} in the direction of the unit vector **j**.

Example 5

The point O is the origin and the points P and Q have position vectors $(7\mathbf{i} - 24\mathbf{j})$ m and $(13\mathbf{i} - 16\mathbf{j})$ m respectively. Find:

(a) the distance OP

(b) the vector $\overrightarrow{PQ}$

(c) the distance PQ.

(a) $$\overrightarrow{OP} = (7\mathbf{i} - 24\mathbf{j})\,\text{m}$$

$$\text{distance OP} = |\overrightarrow{OP}| = \sqrt{(7^2 + (-24)^2)} = 25\,\text{m}$$

The distance OP is 25 m.

(b) Since $\overrightarrow{OP} + \overrightarrow{PQ} = \overrightarrow{OQ}$

$$\overrightarrow{PQ} = \overrightarrow{OQ} - \overrightarrow{OP}$$

$$= (13\mathbf{i} - 16\mathbf{j}) - (7\mathbf{i} - 24\mathbf{j}) = (6\mathbf{i} + 8\mathbf{j})\,\text{m}$$

The vector $\overrightarrow{PQ}$ is $(6\mathbf{i} + 8\mathbf{j})$ m.

(c) From (b) $\overrightarrow{PQ} = (6\mathbf{i} + 8\mathbf{j})$ m

$$\text{distance PQ} = |\overrightarrow{PQ}|$$

$$= \sqrt{(6^2 + 8^2)} = 10\,\text{m}$$

The distance PQ is 10 m.

Example 6

A particle P has an initial position vector $(3\mathbf{i} + 2\mathbf{j} + 4\mathbf{k})$ m.
If the particle moves with a constant velocity of $(5\mathbf{i} + \mathbf{j} - 3\mathbf{k})\,\text{m s}^{-1}$, find:

(a) the position vector of P after time t,

(b) the position vector of P after 3 seconds.

(a) Position vector after time t is given by:

$$\mathbf{r}(t) = (3\mathbf{i} + 2\mathbf{j} + 4\mathbf{k}) + t(5\mathbf{i} + \mathbf{j} - 3\mathbf{k})$$

$$\mathbf{r}(t) = (3 + 5t)\mathbf{i} + (2 + t)\mathbf{j} + (4 - 3t)\mathbf{k}.$$

The position vector of P after time t is:

$$[(3 + 5t)\mathbf{i} + (2 + t)\mathbf{j} + (4 - 3t)\mathbf{k}]\,\text{m}.$$

(b) After 3 seconds:

$$\mathbf{r}(3) = 18\mathbf{i} + 5\mathbf{j} - 5\mathbf{k}.$$

After 3 seconds the position vector is $(18\mathbf{i} + 5\mathbf{j} - 5\mathbf{k})$ m.

Exercise 2B

1. The point A has position vector $(7\mathbf{i} + 24\mathbf{j})$ m. Find how far A is from the origin.
2. The points B and C have position vectors $(8\mathbf{i} - 15\mathbf{j})$ m and $(5\mathbf{i} - 12\mathbf{j})$ m respectively. Find:
   (a) how far B is from the origin
   (b) how far C is from the origin
   (c) $\overrightarrow{BC}$ in vector form (i.e. **i**–**j** notation)
   (d) $|\overrightarrow{BC}|$.
3. The point O is the origin and points A, B and C have position vectors $(3\mathbf{i} - 4\mathbf{j} + 5\mathbf{k})$ m, $(8\mathbf{i} + 8\mathbf{j} - 3\mathbf{k})$ m, and $(4\mathbf{i} + 3\mathbf{k})$ m respectively. Find:
   (a) the distance OA
   (b) the distance OB
   (c) the distance OC
   (d) the vector $\overrightarrow{AB}$
   (e) the vector $\overrightarrow{BC}$
   (f) the vector $\overrightarrow{CB}$
   (g) the distance AB
   (h) the distance BC.
4. Find the speed of a body moving with velocity $(6\mathbf{i} - 8\mathbf{j})$ m s^{-1}.
5. Find the speed of a body moving with velocity $(7\mathbf{i} - 24\mathbf{j})$ m s^{-1}.
6. Find the speed of a body moving with velocity $(-4\mathbf{i} + \mathbf{j})$ m s^{-1}.
7. Find the speed of a body moving with velocity $(4\mathbf{i} - 10\mathbf{j} + \mathbf{k})$ m s^{-1}.
8. Find the speed of a body moving with velocity $(3\mathbf{i} - \mathbf{j} - 7\mathbf{k})$ m s^{-1}.
9. Particle A has velocity $(5\mathbf{i} + 2\mathbf{j})$ m s^{-1} and particle B has velocity $(-4\mathbf{i} + 4\mathbf{j})$ m s^{-1}. Which particle has the greater speed?
10. A body moving with a velocity $(2\mathbf{i} + a\mathbf{j})$ m s^{-1} has a speed of 5·2 m s^{-1}. Find the two possible values of a.
11. A body moving with a velocity of $[b\mathbf{i} + (b + 7)\mathbf{j}]$ m s^{-1} has a speed of 17 m s^{-1}. Find the two possible values of b.
12. A particle has an initial position vector of $(5\mathbf{i} + 3\mathbf{j})$ m.
    If the particle moves with a constant velocity of $(2\mathbf{i} + 4\mathbf{j})$ m s^{-1} find its position vector after:
    (a) 1 second
    (b) 2 seconds.
13. A particle has an initial position vector of $(5\mathbf{i} + 4\mathbf{j})$ m.
    If the particle moves with a constant velocity of $(2\mathbf{i} - \mathbf{j})$ m s^{-1}, find its position vector after:
    (a) 3 seconds (b) 5 seconds
14. A particle has an initial position vector of $(7\mathbf{i} + 5\mathbf{j})$ m. The particle moves with a constant velocity of $(a\mathbf{i} + b\mathbf{j})$ m s^{-1} and after 3 seconds has a position vector of $(10\mathbf{i} - \mathbf{j})$ m.
    Find the values of a and b.
15. Find the speed of a body which is moving with a constant velocity of $(5\mathbf{i} - 12\mathbf{j})$ m s^{-1}. If the body is initially at a point with position vector $(\mathbf{i} + 6\mathbf{j})$ m, find the position vector of the body 3 seconds later and its distance from the origin at that time.
16. A particle has an initial position vector $(4\mathbf{i} + 3\mathbf{j} + 9\mathbf{k})$ m. The particle moves with a constant velocity of $(3\mathbf{i} - 2\mathbf{j} - 5\mathbf{k})$ m s^{-1}. Find:
    (a) the position vector of the particle at time t
    (b) the position vector of the particle after 5 seconds.
    How far is the particle from the origin after 5 seconds?
17. A particle has an initial position vector $(a\mathbf{i} + b\mathbf{j} + c\mathbf{k})$ m. The particle moves with a constant velocity of $(3\mathbf{i} + \mathbf{j} + 4\mathbf{k})$ m s^{-1} and after 2 seconds has a position vector $(7\mathbf{i} + \mathbf{j} + 11\mathbf{k})$ m.
    Find the values of a, b and c. How far is the particle from the origin after 3 seconds?
18. A particle has an initial position vector $(7\mathbf{i} - 6\mathbf{j} + 3\mathbf{k})$ m with respect to an origin O. For the next two seconds the particle moves with a constant velocity of $(4\mathbf{i} - 6\mathbf{k})$ m s^{-1}. The particle then moves with a constant velocity $(a\mathbf{i} + b\mathbf{j} + c\mathbf{k})$ m s^{-1}, reaching O after a further three seconds.
    Find the values of a, b and c.
19. At time $t = 0$ two particles A and B have position vectors $(2\mathbf{i} + 3\mathbf{j} - 4\mathbf{k})$ m and $(8\mathbf{i} + 6\mathbf{k})$ m respectively. A moves with constant velocity $(-\mathbf{i} + 3\mathbf{j} + 5\mathbf{k})$ m s^{-1} and B with constant velocity $\mathbf{v}$ m s^{-1}. Given that when $t = 5$ seconds B passes through the point that A passed through one second earlier, find v.

Uniform acceleration formulae

When the motion of a body is being considered, the letters u, v, a, t and s usually have the following meanings:

u = initial velocity — v = final velocity
a = acceleration — t = time interval or time taken
s = displacement

Consider a car travelling in a straight line. If initially its velocity is $5\,\text{m s}^{-1}$ and 3 seconds later its velocity is $11\,\text{m s}^{-1}$, the car is said to be accelerating. Acceleration is a measure of the rate at which velocity is changing. In this example, the velocity increases by $6\,\text{m s}^{-1}$ in 3 s. If the acceleration a is assumed to be uniform, then it is $6\,\text{m s}^{-1}$ in 3 s, or $2\,\text{m s}^{-1}$ each second, which is written $2\,\text{m s}^{-2}$.
In general:

$$\text{acceleration} = \frac{\text{change in velocity}}{\text{time interval}}$$

$$\therefore \quad a = \frac{v-u}{t}$$

or $$at = v - u$$

Hence $$v = u + at \qquad \ldots [1]$$

If the acceleration is uniform, then the average velocity is the average of the initial and final velocities, i.e.

$$\text{average velocity} = \frac{u+v}{2}$$

But $$\text{average velocity} = \frac{\text{displacement}}{\text{time taken}} = \frac{s}{t}$$

$$\therefore \quad \frac{s}{t} = \frac{u+v}{2}$$

or $$s = \frac{(u+v)t}{2} \qquad \ldots [2]$$

Substituting the value of v from equation [1] into equation [2] gives:

$$s = \frac{(u+u+at)t}{2}$$

$$\therefore \quad s = ut + \tfrac{1}{2}at^2 \qquad \ldots [3]$$

Substitute for t from equation [1] into equation [2].

Equation [1] is rewritten as: $$t = \frac{v-u}{a}$$

and substituted into equation [2]: $$s = \frac{(u+v)}{2}\frac{(v-u)}{a}$$

giving $$2as = v^2 - u^2$$

Hence $$v^2 = u^2 + 2as \qquad \ldots [4]$$

These four formulae are very important and should be committed to memory:

$$\left.\begin{aligned} v &= u + at \\ s &= \frac{(u+v)t}{2} \\ s &= ut + \tfrac{1}{2}at^2 \\ v^2 &= u^2 + 2as \end{aligned}\right\}$$

Remember these only apply to motion involving *uniform* acceleration.

Distance and displacement

In the above formulae, s represents displacement. In practice, s is also used to denote distance because distance and displacement are often equal. There need be no confusion provided that care is taken in any particular question. When the direction of motion of a body remains unchanged, then the distance travelled and the displacement are equal.
If the direction of motion changes part way through the motion, then the distance travelled and the displacement will not be equal.

Suppose a body moves 15 km due east and then 10 km due west:

$$\text{distance moved} = 15\,\text{km} + 10\,\text{km} = 25\,\text{km}$$

$$\text{displacement from initial position} = 15\,\text{km E} + 10\,\text{km W} = 5\,\text{km E}$$

Example 7

A body moves along a straight line from A to B with uniform acceleration $\frac{2}{3}\,\text{m s}^{-2}$. The time taken is 12 s and the velocity at B is $25\,\text{m s}^{-1}$. Find:
(a) the velocity at A
(b) the distance AB.

Given $$\left.\begin{aligned} a &= \tfrac{2}{3}\,\text{m s}^{-2} \\ t &= 12\,\text{s} \\ v &= 25\,\text{m s}^{-1} \end{aligned}\right\} \text{ we need to find: (a) } u \quad \text{(b) } s$$

(a) Use
$$v = u + at$$
$$25 = u + \left(\tfrac{2}{3}\right)(12)$$
$$\therefore \quad u = 17\,\text{m s}^{-1}$$

The velocity at A is $17\,\text{m s}^{-1}$.

(b) Use $s = ut + \frac{1}{2}at^2$ or $v^2 = u^2 + 2as$ or $s = \frac{(u+v)t}{2}$

$= (17)(12) + \frac{1}{2}\left(\frac{2}{3}\right)(12)^2$

$= 204 + 48$

$s = 252\,\text{m}$

$(25^2) = (17)^2 + 2\left(\frac{2}{3}\right)s$

$625 = 289 + \frac{4}{3}s$

$s = 252\,\text{m}$

$= \frac{(17+25)12}{2}$

$s = 252\,\text{m}$

The distance AB is 252 m.

Example 8

A cyclist travelling downhill accelerates uniformly at $1\frac{1}{2}\,\text{m}\,\text{s}^{-2}$. If his initial velocity at the top of the hill is $3\,\text{m}\,\text{s}^{-1}$, find:
(a) how far he travels in 8 s
(b) how far he travels before reaching a velocity of $7\,\text{m}\,\text{s}^{-1}$.

(a) Given $a = 1{\cdot}5\,\text{m}\,\text{s}^{-2}$, $u = 3\,\text{m}\,\text{s}^{-1}$, $t = 8\,\text{s}$ — we need to find s

Use $s = ut + \frac{1}{2}at^2$
$= 3(8) + \frac{1}{2}(1{\cdot}5)(8)^2$
$= 24 + 48$
$\therefore\quad s = 72\,\text{m}$ — In 8 s the cyclist travels 72 m.

(b) Given $a = 1{\cdot}5\,\text{m}\,\text{s}^{-2}$, $u = 3\,\text{m}\,\text{s}^{-1}$, $v = 7\,\text{m}\,\text{s}^{-1}$ — we need to find s

Use $v^2 = u^2 + 2as$
$(7)^2 = (3)^2 + 2(1{\cdot}5)s$
or $49 - 9 = 3s$
$\therefore\quad s = 13\frac{1}{3}\,\text{m}$ — The distance travelled is $13\frac{1}{3}$ m.

Retardation

If a body moving at $12\,\text{m}\,\text{s}^{-1}$ is subsequently moving at $2\,\text{m}\,\text{s}^{-1}$, the body is said to be subject to a retardation, i.e. a negative acceleration. If the change in velocity takes place over a period of 4 s, the retardation is $10\,\text{m}\,\text{s}^{-1}$ in 4 s or $2\frac{1}{2}\,\text{m}\,\text{s}^{-2}$ and the acceleration is $-2\frac{1}{2}\,\text{m}\,\text{s}^{-2}$.

Example 9

A stone slides in a straight line across a horizontal sheet of ice. It passes a point A with velocity $14\,\text{m}\,\text{s}^{-1}$, and the point B $2\frac{1}{2}$ s later.
Assuming the retardation is uniform and that AB = 30 m, find:
(a) the retardation
(b) the velocity at B
(c) how long after passing A the stone comes to rest.

(a) Given $u = 14\,\text{m s}^{-1}$, $t = 2{\cdot}5\,\text{s}$, $s = 30\,\text{m}$ — we need to find retardation

Let acceleration $= a$

Use $s = ut + \frac{1}{2}at^2$

$$30 = (14)(2{\cdot}5) + \tfrac{1}{2}(a)(2{\cdot}5)^2$$

$$\therefore \quad 30 = 35 + \frac{25a}{8}$$

or $a = -1{\cdot}6\,\text{m s}^{-2}$, i.e. a retardation

The retardation is $1{\cdot}6\,\text{m s}^{-2}$.

(b) Let velocity at B $= v$

From part (a) retardation $= 1.6\,\text{m s}^{-2}$ or $a = -1{\cdot}6\,\text{m s}^{-2}$

Use $v^2 = u^2 + 2as$

$$= (14)^2 + 2(-1{\cdot}6)(30)$$
$$= 196 - 96$$
$$\therefore \quad v = 10\,\text{m s}^{-1}$$

The velocity of the stone at B is $10\,\text{m s}^{-1}$.

(c) Given $u = 14\,\text{m s}^{-1}$, $a = -1{\cdot}6\,\text{m s}^{-2}$, $v = 0$ when the stone comes to rest — we need to find t

Use $v = u + at$

$$0 = 14 + (-1{\cdot}6)t$$
$$\therefore \quad t = 8{\cdot}75\,\text{s}$$

The stone is at rest 8·75 s after passing A.

Exercise 2C

Questions **1** to **20** involve a body moving with uniform acceleration along a straight line from point A to point B.

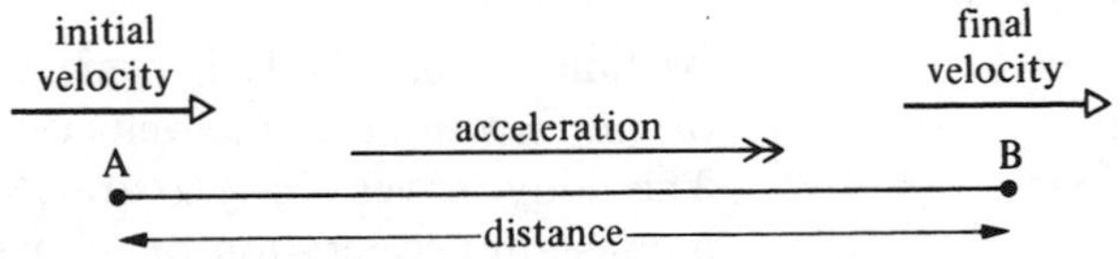

1. Initially at rest, acceleration $= 4\,\text{m s}^{-2}$, time taken $= 8\,\text{s}$. Find the distance.
2. Initial velocity $= 3\,\text{m s}^{-1}$, acceleration $= 2\,\text{m s}^{-2}$, time taken $= 6\,\text{s}$. Find the final velocity.
3. Initially at rest, acceleration $= 2\,\text{m s}^{-2}$, time taken $= 4\,\text{s}$. Find the distance.
4. Initial velocity $= 3\,\text{m s}^{-1}$, final velocity $= 5\,\text{m s}^{-1}$, time taken $= 10\,\text{s}$. Find the distance.
5. Initial velocity $= 3\,\text{m s}^{-1}$, final velocity $= 5\,\text{m s}^{-1}$, distance $= 2\,\text{m}$. Find the acceleration.

6. Final velocity $= 27\,\text{m s}^{-1}$, acceleration $= 8\,\text{m s}^{-2}$, time taken $= 2\,\text{s}$. Find the initial velocity.

7. Initial velocity $= 7\,\text{m s}^{-1}$, final velocity $= 3\,\text{m s}^{-1}$, distance $= 5\,\text{m}$. Find the acceleration.

8. Distance $= 28\,\text{m}$, acceleration $= 1\,\text{m s}^{-2}$, time taken $= 4\,\text{s}$. Find the initial velocity.

9. Distance $= 20\,\text{m}$, initial velocity $= 3\,\text{m s}^{-1}$, final velocity $= 7\,\text{m s}^{-1}$. Find the time taken.

10. Initial velocity $= 6\,\text{m s}^{-1}$, final velocity $= 8\,\text{m s}^{-1}$, acceleration $= 0{\cdot}5\,\text{m s}^{-2}$. Find the distance.

11. Initial velocity $= 2\,\text{m s}^{-1}$, final velocity $= 50\,\text{m s}^{-1}$, time taken $= 16\,\text{s}$. Find the acceleration.

12. Distance $= 500\,\text{m}$, initial velocity $= 1\,\text{m s}^{-1}$, time taken $= 10\,\text{s}$. Find the acceleration.

13. Initial velocity $= 10\,\text{m s}^{-1}$, final velocity $= 2\,\text{m s}^{-1}$, acceleration $= -4\,\text{m s}^{-2}$. Find the distance.

14. Initial velocity $= 30\,\text{m s}^{-1}$, final velocity $= 10\,\text{m s}^{-1}$, acceleration $= -4\,\text{m s}^{-2}$. Find the time taken.

15. Initial velocity $= 5\,\text{m s}^{-1}$, acceleration $= 1\,\text{m s}^{-2}$, distance $= 12\,\text{m}$. Find the time taken.

16. Distance $= 60\,\text{m}$, final velocity $= 8\,\text{m s}^{-1}$, time taken $= 12\,\text{s}$. Find the initial velocity.

17. Initial velocity $= 5\,\text{m s}^{-1}$, final velocity $= 36\,\text{km h}^{-1}$, acceleration $= 1\frac{1}{4}\,\text{m s}^{-2}$. Find the distance.

18. Acceleration $= 0{\cdot}5\,\text{m s}^{-2}$, final velocity $= 162\,\text{km h}^{-1}$, time taken $= 1$ minute. Find the initial velocity.

19. Acceleration $= 2\,\text{m s}^{-2}$, final velocity $= 10\,\text{m s}^{-1}$, time taken $= 2\,\text{s}$. Find the distance.

20. Distance $= 132\,\text{m}$, time taken $= 12\,\text{s}$, acceleration $= -1\,\text{m s}^{-2}$. Find the final velocity.

21. A train starts from rest and accelerates uniformly, at $1{\cdot}5\,\text{m s}^{-2}$, until it attains a speed of $30\,\text{m s}^{-1}$. Find the distance the train travels during this motion and the time taken.

22. A cheetah can accelerate from rest to $30\,\text{m s}^{-1}$ in a distance of 25 m. Find the acceleration (assumed constant).

23. The manufacturer of a new car claims that it can accelerate from rest to $90\,\text{km h}^{-1}$ in 10 seconds. Find the acceleration (assumed constant).

24. In travelling the 70 cm along a rifle barrel, a bullet uniformly accelerates from its initial state of rest to a muzzle velocity of $210\,\text{m s}^{-1}$. Find the acceleration involved and the time for which the bullet is in the barrel.

25. According to the Highway Code, a car travelling at $20\,\text{m s}^{-1}$ requires a minimum braking distance of 30 m. What retardation is this and what length of time will it take?

26. A car is initially at rest at a point O. The car moves away from O in a straight line, accelerating at $4\,\text{m s}^{-2}$. Find how far the car is from O after:
(a) 2 seconds
(b) 3 seconds.
How far does the car travel in the third second?

27. A body moves along a straight line uniformly increasing its velocity from $2\,\text{m s}^{-1}$ to $18\,\text{m s}^{-1}$ in a time interval of 10 s. Find the acceleration of the body during this time and the distance travelled.

28. A particle is projected away from an origin O with initial velocity of $0{\cdot}25\,\text{m s}^{-1}$. The particle travels in a straight line and accelerates at $1{\cdot}5\,\text{m s}^{-2}$. Find:
(a) how far the particle is from O after 3 seconds
(b) the distance travelled by the particle during the fourth second after projection.

29. At time $t = 0$, a body is projected from an origin O with an initial velocity of $10\,\text{m s}^{-1}$. The body moves along a straight line with a constant acceleration of $-2\,\text{m s}^{-2}$.
(a) Find the displacement of the body from O when t equals 7 seconds.
(b) How far from O does the body come to instantaneous rest and what is the value of t then?
(c) Find the distance travelled by the body during the time interval $t = 0$ to $t = 7$ seconds.

30. A, B and C are three points lying in that order on a straight line. A body is projected from B towards A with speed $3\,\text{m}\,\text{s}^{-1}$. The body experiences an acceleration of $1\,\text{m}\,\text{s}^{-2}$ towards C. If BC = 20 m, find the time taken to reach C and the distance travelled by the body from the moment of projection until it reaches C.

31. A car is being driven along a road at a steady $25\,\text{m}\,\text{s}^{-1}$ when the driver suddenly notices that there is a fallen tree blocking the road 65 metres ahead. The driver immediately applies the brakes giving the car a constant retardation of $5\,\text{m}\,\text{s}^{-2}$. How far in front of the tree does the car come to rest? If the driver had not reacted immediately and the brakes were applied one second later, with what speed would the car have hit the tree?

32. A train travels along a straight piece of track between two stations A and B. The train starts from rest at A and accelerates at $1{\cdot}25\,\text{m}\,\text{s}^{-2}$ until it reaches a speed of $20\,\text{m}\,\text{s}^{-1}$. It then travels at this steady speed for a distance of 1·56 km and then decelerates at $2\,\text{m}\,\text{s}^{-2}$ to come to rest at B. Find:
(a) the distance from A to B
(b) the total time taken for the journey
(c) the average speed for the journey.

33. A particle travels in a straight line with uniform acceleration. The particle passes through three points A, B and C lying in that order on the line, at times $t = 0$, $t = 2$ s and $t = 5$ s respectively. If BC = 30 m and the speed of the particle when at B is $7\,\text{m}\,\text{s}^{-1}$, find the acceleration of the particle and its speed when at A.

34. A, B and C are three points which lie in that order on a straight road with AB = 95 m and BC = 80 m. A car is travelling along the road in the direction ABC with constant acceleration $a\,\text{m}\,\text{s}^{-2}$. The car passes through A with speed $u\,\text{m}\,\text{s}^{-1}$, reaches B five seconds later, and C two seconds after that. Find the values of u and a.

35. A car A, travelling at a constant velocity of $25\,\text{m}\,\text{s}^{-1}$, overtakes a stationary car B. Two seconds later car B sets off in pursuit, accelerating at a uniform $6\,\text{m}\,\text{s}^{-2}$. How far does B travel before catching up with A?

Free fall under gravity

The uniform acceleration formulae developed in the last section may be used when considering the motion of a body falling under gravity. In such cases the acceleration of the body is $9{\cdot}8\,\text{m}\,\text{s}^{-2}$ and this is commonly referred to as g, the acceleration due to gravity. If the motion is vertically upward, the body will be subject to a retardation of $9{\cdot}8\,\text{m}\,\text{s}^{-2}$.

In fact the magnitude of g varies slightly at different places on the Earth's surface, but for our purposes it can be taken as having the constant value of $9{\cdot}8\,\text{m}\,\text{s}^{-2}$.

Arrow convention

In any particular example, care is needed to ensure that the directions of the vectors involved are all the same.

$u = 25\,\text{m}\,\text{s}^{-1} \uparrow$ implies that the initial velocity is $25\,\text{m}\,\text{s}^{-1}$ upwards

$a = 9{\cdot}8\,\text{m}\,\text{s}^{-2} \downarrow$ implies a downward acceleration of $9{\cdot}8\,\text{m}\,\text{s}^{-2}$

$= -9{\cdot}8\,\text{m}\,\text{s}^{-2} \uparrow$

Before substituting numerical values in the uniform acceleration formulae, the arrows of the vectors involved must all be in the same direction.

Example 10

A brick is thrown vertically downwards from the top of a building and has an initial velocity of $1{\cdot}5\,\mathrm{m\,s^{-1}}$. If the height of the building is $19\frac{2}{7}$ m, find:
(a) the velocity with which the brick hits the ground
(b) the time taken for the brick to fall.

(a) Given $u = 1{\cdot}5\,\mathrm{m\,s^{-1}}\downarrow$, $s = 19\frac{2}{7}\,\mathrm{m}\downarrow$, $a = 9{\cdot}8\,\mathrm{m\,s^{-2}}\downarrow$ — we need to find: (a) v (b) t

Use $v^2 = u^2 + 2as$

$$v^2 = (1{\cdot}5)^2 + 2(9.8)\frac{(135)}{7}$$
$$= 380{\cdot}25$$

or $v = 19{\cdot}5\,\mathrm{m\,s^{-1}}$

The brick hits the ground with a downward velocity of $19{\cdot}5\,\mathrm{m\,s^{-1}}$.

(b) Use $v = u + at$

$$19{\cdot}5 = 1{\cdot}5 + 9{\cdot}8t$$

$$\therefore \quad t = \frac{18}{9{\cdot}8} = 1{\cdot}84\,\mathrm{s}$$

The brick hits the ground after $1{\cdot}84$ s.

Example 11

A ball is thrown vertically upwards with a velocity of $14{\cdot}7\,\mathrm{m\,s^{-1}}$ from a platform $19{\cdot}6$ m above ground level. Find:
(a) the time taken for the ball to reach the ground
(b) the velocity of the ball when it hits the ground.

Given $u = 14{\cdot}7\,\mathrm{m\,s^{-1}}\uparrow$, $a = 9{\cdot}8\,\mathrm{m\,s^{-2}}\downarrow = -9{\cdot}8\,\mathrm{m\,s^{-2}}\uparrow$, $s = 19{\cdot}6\,\mathrm{m}\downarrow = -19{\cdot}6\,\mathrm{m}\uparrow$ — we need to find (a) t (b) v

(a) Use

$$s = ut + \tfrac{1}{2}at^2$$
$$-19.6 = 14.7t + \tfrac{1}{2}(-9{\cdot}8)t^2$$
$$-4 = 3t - t^2$$
$$\therefore \quad t^2 - 3t - 4 = 0$$
$$(t-4)(t+1) = 0 \quad \text{i.e.} \quad t = 4\,\mathrm{s} \quad \text{or} \quad -1\,\mathrm{s}$$

The ball reaches the ground after 4 s.

(b) Use $v = u + at$

$$v = 14{\cdot}7 + (-9{\cdot}8)4 = 14{\cdot}7 - 39{\cdot}2$$
$$v = -24{\cdot}5\,\mathrm{m\,s^{-1}}, \quad \text{i.e.} \quad 24{\cdot}5\,\mathrm{m\,s^{-1}}\downarrow$$

The ball hits the ground with a downward velocity of $24{\cdot}5\,\mathrm{m\,s^{-1}}$.

Example 12

A particle is projected vertically upwards with a velocity of $34{\cdot}3\,\mathrm{m\,s^{-1}}$. Find how long after projection the particle is at a height of 49 m above the point of projection for:
(a) the first time
(b) the second time.

Given
$$\left.\begin{aligned} u &= 34{\cdot}3\,\mathrm{m\,s^{-1}}\uparrow \\ a &= 9{\cdot}8\,\mathrm{m\,s^{-2}}\downarrow = -9{\cdot}8\,\mathrm{m\,s^{-2}}\uparrow \\ s &= 49\,\mathrm{m}\uparrow \end{aligned}\right\}\ \text{we need to find } t.$$

Use
$$s = ut + \tfrac{1}{2}at^2$$
$$49 = 34{\cdot}3t - \tfrac{1}{2}(9{\cdot}8)t^2$$
$$\therefore \quad t^2 - 7t + 10 = 0$$
$$(t-5)(t-2) = 0, \quad \text{i.e.} \quad t = 5\,\mathrm{s} \quad \text{or} \quad 2\,\mathrm{s}$$

The particle is 49 m above the point of projection:
(a) after 2 s
(b) after 5 s.

Exercise 2D

1. A book falls from a shelf 160 cm above the floor.
   Find the speed with which the book strikes the floor.
2. A stone is dropped from a position which is 40 metres above the ground.
   Find the time taken for the stone to reach the ground.
3. A stone is dropped from the top of a tower and falls to the ground below.
   If the stone hits the ground with a speed of $14\,\mathrm{m\,s^{-1}}$, find the height of the tower.
4. A ball is thrown vertically downwards from the top of a tower and has an initial speed of $4\,\mathrm{m\,s^{-1}}$. If the ball hits the ground 2 seconds later, find:
   (a) the height of the tower
   (b) the speed with which the ball strikes the ground.
5. A stone is projected vertically upwards from ground level with a speed of $21\,\mathrm{m\,s^{-1}}$. Find the height of the stone above ground:
   (a) 1 second after projection
   (b) 2 seconds after projection
   (c) 3 seconds after projection.
6. A ball is thrown vertically upwards with speed $28\,\mathrm{m\,s^{-1}}$ from a point which is 1 metre above ground level. Find:
   (a) the speed the ball will have when it returns to the level from which it was projected
   (b) the height above ground level of the highest point reached.
7. A ball is thrown vertically upwards from a point A, with initial speed of $21\,\mathrm{m\,s^{-1}}$, and is later caught again at A. Find the length of time for which the ball was in the air.
8. A ball is kicked vertically upwards from ground level with an initial speed of $14\,\mathrm{m\,s^{-1}}$. Find the height above ground level of the highest point reached and the total time for which the ball is in the air.
9. A stone is thrown vertically upwards with a speed of $20\,\mathrm{m\,s^{-1}}$ from a point at a height h metres above ground level. If the stone hits the ground 5 seconds later, find h.
10. A stone is projected vertically upwards from ground level at a speed of $24{\cdot}5\,\mathrm{m\,s^{-1}}$. Find how long after projection the stone is at a height of $19{\cdot}6$ m above the ground:
    (a) for the first time
    (b) for the second time.
    For how long is the stone at least $19{\cdot}6$ m above ground level?

11. A ball is held 1·6 m above a concrete floor and released. The ball hits the floor and rebounds with half the speed it had just prior to impact. Find the greatest height the ball reaches after:
(a) the first bounce
(b) the second bounce.

12. A body is projected vertically upwards from ground level at a speed of $49\,\text{m}\,\text{s}^{-1}$. Find the length of time for which the body is at least 78·4 m above the ground.

13. A bullet is fired vertically upwards at a speed of $147\,\text{m}\,\text{s}^{-1}$. Find the length of time for which the bullet is at least 980 m above the level of projection.

14. A body is projected vertically upwards with a speed of $14\,\text{m}\,\text{s}^{-1}$. Find the height of the body above the level of projection after:
(a) 1 second of motion
(b) 2 seconds of motion.
Find the distance travelled by the body in the 2nd second of motion.

15. Two stones are thrown from the same point at the same time, one vertically upwards with speed $30\,\text{m}\,\text{s}^{-1}$, and the other vertically downwards at $30\,\text{m}\,\text{s}^{-1}$. Find how far apart the stones are after 3 seconds.

Graphical representation

Consider the motion of a body which accelerates uniformly from a speed u to a speed v in time t and then maintains constant speed v. Plotting velocity on the vertical axis and time on the horizontal axis, we can draw a velocity–time graph .

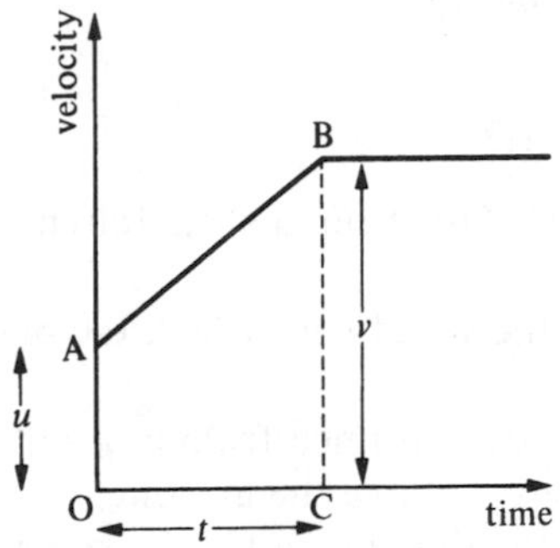

The acceleration of the body is defined as the rate of change of velocity,

i.e. $a = \dfrac{v - u}{t}$ and so the acceleration during the time interval $0 \rightarrow t$ will be the gradient or slope of the line AB.

From $s = \dfrac{(u + v)}{2}t$ it can be seen that the distance travelled by the body during the time interval $0 \rightarrow t$ is represented by the area OABC, i.e. the area 'under' the graph for that part of the motion.

Example 13

The velocity–time graph shown is for a body which starts from rest, accelerates uniformly to a velocity of $8\,\text{m}\,\text{s}^{-1}$ in 2 seconds, maintains that velocity for a further 5 seconds, and then retards uniformly to rest. The entire journey takes 11 seconds.

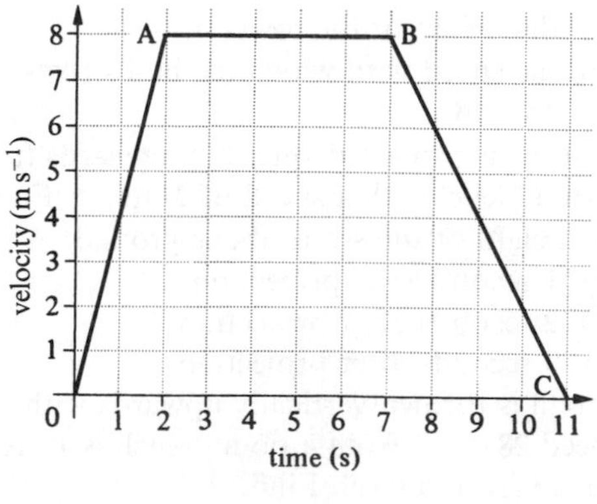

Find:
(a) the acceleration of the body during the initial part of the motion
(b) the retardation of the body during the final part of the motion
(c) the total distance travelled by the body.

(a) The initial acceleration is given by the gradient of the line OA:

$$\text{gradient of OA} = \frac{\text{vertical increase from O to A}}{\text{horizontal increase from O to A}}$$

$$= \frac{8}{2} = 4$$

The initial acceleration is $4\,\mathrm{m\,s^{-2}}$.

(b) The acceleration during the final part of the motion is given by the gradient of the line BC:

$$\text{gradient of BC} = \frac{\text{vertical increase from B to C}}{\text{horizontal increase from B to C}}$$

$$= \frac{-8}{4} = -2$$

The final retardation is $2\,\mathrm{m\,s^{-2}}$.

(c) The total distance travelled is given by the area OABC. This is a trapezium, and so:

$$\text{area OABC} = \frac{(5 + 11)8}{2}$$

$$= 64$$

The total distance travelled is 64 m.

Setting up a mathematical model of a real situation

In an earlier part of this chapter we considered free fall under gravity. In that section we used the value of $9{\cdot}8\,\mathrm{m\,s^{-2}}$ for the acceleration due to gravity, but pointed out that in fact g varied slightly dependent upon location. We simplified the real situation, in which the acceleration due to gravity may not quite equal $9{\cdot}8\,\mathrm{m\,s^{-2}}$, by adopting a more convenient and sufficiently accurate *model* that assumed g to be $9{\cdot}8\,\mathrm{m\,s^{-2}}$ everywhere on and close to the Earth's surface.

Similarly when we say that a car travels with constant speed we are choosing to neglect the small variations in speed that will probably occur in reality. Our simplified *mathematical model* chooses to neglect these small variations.

Likewise when we choose to neglect wind resistance we are again adopting a simplified mathematical model of the situation. In this way we avoid complications without seriously affecting the acceptability of the answer.

If we draw a velocity–time graph for the motion of a cyclist, we are choosing to display our mathematical model graphically. Certain assumptions may be made in our model. For example, we may draw a horizontal line on our graph to show the cyclist travelling with constant velocity, whereas in reality the cyclist's speed probably varied slightly during this time.

A good mathematical model of a real situation will make any assumptions necessary to allow appropriate mathematics to be used while at the same time not making the model so different from the real situation as to make it useless.
As with most models, these mathematical models simplify the real situation, making it more manageable, while still retaining those features of the real situation that are considered to be most significant. The model will lose some small details that exist in the real situation but can still allow useful calculations and predictions to be made. The validity of the model can then be checked by comparing the outcomes as predicted by the model with the real-life outcomes.

Note. As the above paragraphs point out, we have been using mathematical models of the real world already in this book in order to solve problems. In such cases we did not state all of the assumptions we were making in order to solve each question. However, if a question specifically asks you to "set up the model" you should clearly state any assumptions you are going to make, as the next example demonstrates.

Example 14

A road engineer for a local council needs to find a safe distance between road humps on a 30 miles per hour stretch of road. The engineer knows that the maximum safe speed to travel over the humps is 5 miles per hour and must allow for the motorist who will reach a speed of 30 miles per hour between humps. From previous research the engineer knows that the average family car has an average acceleration of $3\,\text{m}\,\text{s}^{-2}$ and a deceleration of $6{\cdot}5\,\text{m}\,\text{s}^{-2}$. Set up a model for the above situation and use the model to estimate a safe distance between humps.

Step 1 Set up the model

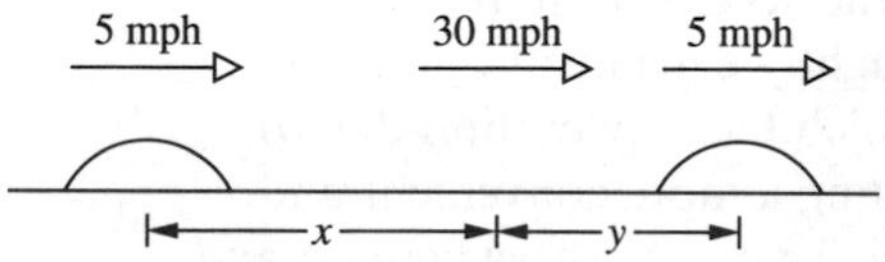

- Assume that the car is a particle.
- Assume that there are no resistive forces such as air resistance and friction.
- Assume that the car will travel throughout the motion with constant acceleration and then constant deceleration.
- Assume the road surface is horizontal.

Step 2 Apply the mathematics

Using the conversion $1\text{ mph} \approx \frac{4}{9}\,\text{m}\,\text{s}^{-1}$ gives:

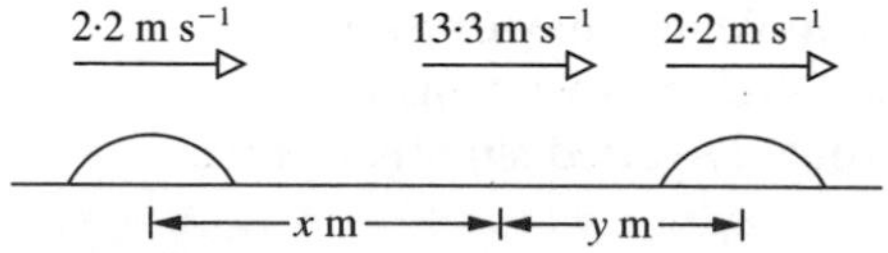

where x m is the distance travelled during acceleration and y m the distance travelled during deceleration.

Given
$$\left.\begin{aligned} u &= 2{\cdot}2\,\mathrm{m\,s^{-1}} \\ v &= 13{\cdot}3\,\mathrm{m\,s^{-1}} \\ s &= x\,\mathrm{m} \\ a &= 3\,\mathrm{m\,s^{-2}} \end{aligned}\right\}\ \text{we need to find } x$$

Use
$$\begin{aligned} v^2 &= u^2 + 2as \\ (13{\cdot}3)^2 &= (2{\cdot}2)^2 + 2(3)x \\ \therefore\quad x &\approx 28{\cdot}7 \end{aligned}$$

Given
$$\left.\begin{aligned} u &= 13{\cdot}3\,\mathrm{m\,s^{-1}} \\ v &= 2{\cdot}2\,\mathrm{m\,s^{-1}} \\ s &= y\,\mathrm{m} \\ a &= -6{\cdot}5\,\mathrm{m\,s^{-2}} \end{aligned}\right\}\ \text{we need to find } y$$

Use
$$\begin{aligned} v^2 &= u^2 + 2as \\ (2{\cdot}2)^2 &= (13{\cdot}3)^2 + 2(-6{\cdot}5)y \\ \therefore\quad y &\approx 13{\cdot}2 \end{aligned}$$

Therefore the total distance is $x\,\mathrm{m} + y\,\mathrm{m} \approx 41{\cdot}9\,\mathrm{m}$.
An estimate for the safe distance between the humps is 42 m.

Example 15

An express train of length 100 metres accelerates through Doom station. The diagram below shows the layout of the platform, which is of length 90 metres.

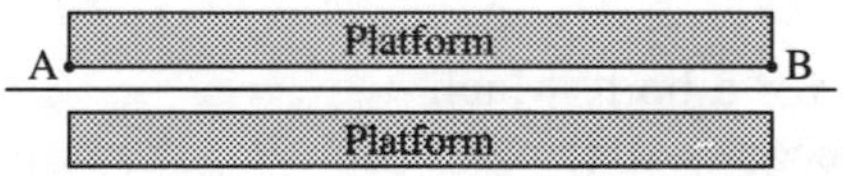

James, a keen train-spotter, knows that when the front of the train reaches point A it has a speed of $22\,\mathrm{m\,s^{-1}}$, and when it reaches point B its speed is $27\,\mathrm{m\,s^{-1}}$. Set up a model to calculate the time taken for the entire train to pass through the station.

Step 1 Set up the model

(The question specifically asks us to "set up a model" so we should clearly state any assumptions we are going to make.)

- Assume that the acceleration is constant for the period of time for which the train passes through the station.
- Assume that "pass through the station" means from when the front of the train reaches A to when the end of the last carriage passes B.

Step 2 Apply the mathematics

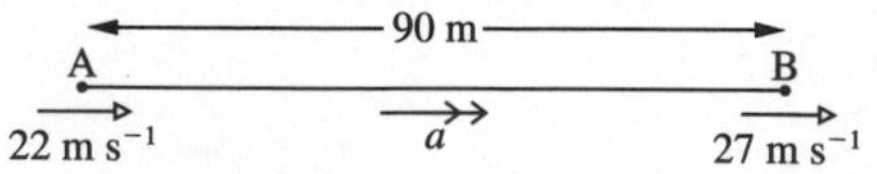

Given $u = 22\,\text{m s}^{-1}$, $v = 27\,\text{m s}^{-1}$, $s = 90\,\text{m}$ — we need to find a

Use $v^2 = u^2 + 2as$

$$27^2 = 22^2 + 2(a)90$$

$\therefore$ $a \approx 1{\cdot}36\,\text{m s}^{-2}$

For the last carriage of the train to reach B the front of the train needs to have travelled (100 m + 90 m) from A.

Given $u = 22\,\text{m s}^{-1}$, $s = 190\,\text{m}$, $a = 1{\cdot}36\,\text{m s}^{-2}$ — we need to find t

Use $s = ut + \frac{1}{2}at^2$

$$190 = 22t + \tfrac{1}{2}(1{\cdot}36)t^2$$

Thus, using the quadratic formula gives:

$$t = 7{\cdot}08\,\text{s}$$

The total time for the entire train to pass through the station is approximately 7 seconds.

Exercise 2E

1. Each of the following velocity–time graphs is for a body which accelerates uniformly for a time period of 4 seconds after which time it maintains its final velocity. In each case find:
- (i) the initial velocity of the body
- (ii) the final velocity of the body
- (iii) the acceleration of the body during the 4 seconds
- (iv) the distance travelled by the body during the 4 seconds.

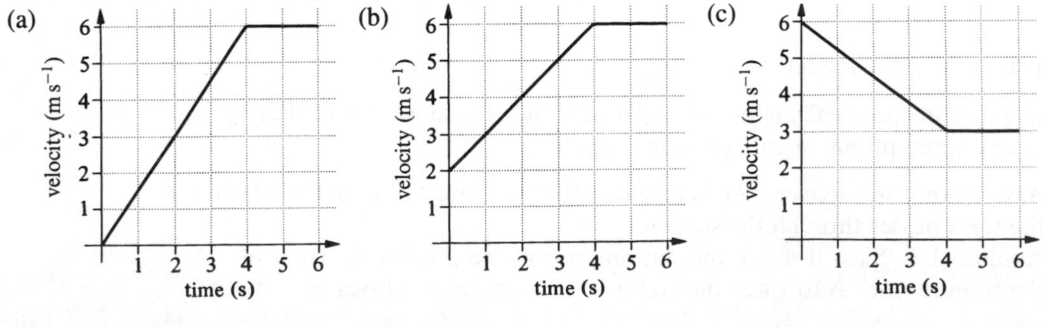

2. Each of the following velocity–time graphs are for a body which starts from rest, accelerates uniformly to a particular velocity, maintains that velocity for a period of time and then uniformly retards to rest. In each case find:
   (i) the acceleration during the initial part of the motion
   (ii) the retardation during the final part of the motion
   (iii) the total distance travelled by the body during the motion.

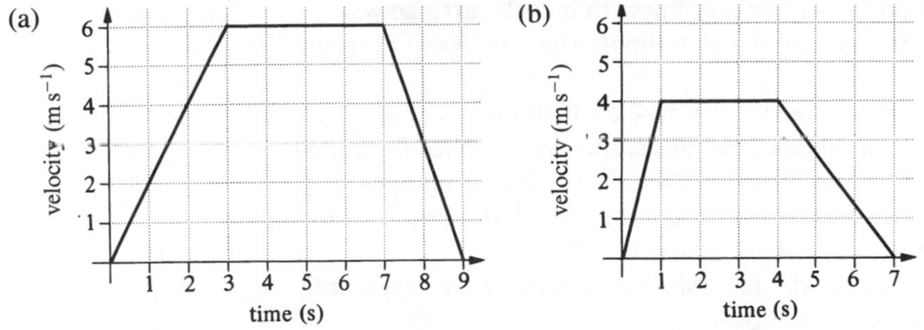

3. A cyclist rides along a straight road from a point A to a point B. He starts from rest at A and accelerates uniformly to reach a speed of $12\,\mathrm{m\,s^{-1}}$ in 8 seconds. He maintains this speed for a further 20 seconds and then uniformly retards to rest at B. If the whole journey takes 34 seconds, draw a velocity–time graph for the motion and from it find:
   (a) his acceleration for the first part of the motion
   (b) his retardation for the last part of the motion
   (c) the total distance travelled.

4. A particle is initially at rest at a point A on a straight line ABCD. The particle moves from A to B with uniform acceleration, reaching B with a speed of $12\,\mathrm{m\,s^{-1}}$ after 2 seconds. The acceleration then alters to a constant $1\,\mathrm{m\,s^{-2}}$ and 8 seconds after leaving B the particle reaches C. The particle then retards uniformly to come to rest at D after a further 10 seconds. Draw a velocity–time graph for the motion, and from it find:
   (a) the acceleration of the particle when travelling from A to B
   (b) the speed of the particle on reaching C
   (c) the retardation of the particle when travelling from C to D
   (d) the total distance from A to D.

5. A and B are two points on a straight road. A car travelling along the road passes through A when $t = 0$ and maintains a constant speed until $t = 30$ seconds and in this time covers three-fifths of the distance from A to B. The car then retards uniformly to rest at B. Sketch a velocity–time graph for the motion and find the total time taken for the car to travel from A to B.

6. Two stations A and B are a distance of $6x$ m apart along a straight track. A train starts from rest at A and accelerates uniformly to a speed $v\,\mathrm{m\,s^{-1}}$, covering a distance of x m. The train then maintains this speed until it has travelled a further $3x$ m, it then retards uniformly to rest at B. Make a sketch of the velocity–time graph for the motion and show that if T is the time taken for the train to travel from A to B, then $T = \dfrac{9x}{v}$ seconds.

7. James, the keen train-spotter of Example 15, was holidaying in the USA with his family. He noticed that some of the inter-state freight trains pulled an enormous number of carriages – so many so that when attempting to count them as the train passed by he always lost count. One day, while waiting at a level crossing for one of these trains to pass by he decided to estimate the number of carriages it had. He estimated that the train was travelling at 60 miles per hour. He noted that it took one minute for the train to pass by and knew that each carriage was about 8 metres long. Set up a model to estimate the number of carriages the train was pulling.

8. As John drove his car past three lamp posts, each 30 metres apart, on a road with a speed limit of 30 miles per hour, the traffic police registered the time taken for him to travel between the posts. The equipment registered John's time as 3 seconds to travel between lamp posts 1 and 2, and 2·2 seconds to travel between lamp posts 2 and 3.
Set up a model to determine whether John was breaking the speed limit as he drove past the middle lamp post.

9. A car approaches a set of traffic lights at 40 miles per hour. When the car is 15 metres from the stop line the lights change from green and the driver applies the brakes. Set up a model to determine the necessary deceleration of the car.
It takes 1·5 seconds in total for the lights to pass through the phase "end green – amber – start red" and the width of the junction is 20 metres (i.e. any vehicle passing over the stop line needs to travel a further 20 metres before it can be considered as having completed the crossing). If the driver of the car mentioned at the beginning of the question does not brake when the lights change from green, could the car complete the crossing without accelerating before the lights turn red? If your answer is no, determine the necessary acceleration for the crossing to be just completed in time.

Exercise 2F Examination questions

(Unless otherwise indicated take $g = 9{\cdot}8\,\text{m s}^{-2}$ in this exercise.)

1. A car is moving along a straight road with uniform acceleration. The car passes a check-point A with a speed of $12\,\text{m s}^{-1}$ and another check-point C with a speed of $32\,\text{m s}^{-1}$. The distance between A and C is 1100 m.
(a) Find the time, in s, taken by the car to move from A to C.
Given that B is the mid-point of AC,
(b) find, in m s^{-1} to 1 decimal place, the speed with which the car passes B.
(ULEAC)

2. A, B and C are three points on a straight line, in that order, and the distances AB and AC are 45 m and 77 m respectively. A particle moves along the straight line with constant acceleration $2\,\text{m s}^{-2}$. Given that it takes 5 seconds to travel from A to B, find the time taken to travel from A to C.
(UCLES)

3.

The diagram shows part of a racing circuit with two bends and a straight AB. A car comes out of the bend at A with a speed of $12\,\text{m s}^{-1}$ and accelerates uniformly in the direction shown, reaching a top speed of $48\,\text{m s}^{-1}$ in 6 s. Find
(i) the acceleration

(ii) the distance travelled from A to reach top speed.
The driver maintains a speed of $48\,\mathrm{m\,s^{-1}}$ for 2 s. He then decelerates uniformly at $7{\cdot}5\,\mathrm{m\,s^{-2}}$ until B is reached. Given that the distance from A to B is 408 m, find
(iii) the speed of the car at B
(iv) the time taken from A to B.
Sketch the velocity–time diagram for the motion from A to B. (UCLES)

4. A particle moves with constant acceleration $0{\cdot}5\,\mathrm{m\,s^{-2}}$ along a straight line passing through the points P and Q. It passes the point Q with velocity $1\,\mathrm{m\,s^{-1}}$ greater than its velocity at P. Given that the distance PQ is 25 m, calculate the velocity with which the particle passes the point P.
How long after passing the point P does it take for the velocity of the particle to reach $20\,\mathrm{m\,s^{-1}}$? (UCLES)

5. A cyclist travels on a straight road with a constant acceleration of $0{\cdot}6\,\mathrm{m\,s^{-2}}$. P and Q are two points on the road, 120 m apart. Given that the cyclist increases speed by $6\,\mathrm{m\,s^{-1}}$ as he travels from P to Q, find
(i) the speed of the cyclist at P
(ii) the time taken to travel from P to Q. (UCLES)

6. A car is moving with speed $u\,\mathrm{m\,s^{-1}}$. The brakes of the car can produce a constant retardation of $6\,\mathrm{m\,s^{-2}}$ but it is known that, when the driver decides to stop, a period of $\frac{2}{3}$ second elapses before the brakes are applied. As the car passes a point O the driver decides to stop. Find, in terms of u, an expression for the minimum distance of the car from O when the car comes to rest.
The driver is approaching traffic signals and is 95 m away from the signals when the light changes from green to amber. The light remains amber for 3 seconds before changing to red.
Show that
(a) when $u < 30$, the driver can stop before reaching the signals
(b) when $3u > 95$, the driver can pass the signal before the light turns red. (AEB 1994)

7. (Take $g = 10\,\mathrm{m\,s^{-2}}$ in this question.)
A stone is dropped vertically from the top of an overhanging cliff, and it hits the sea 3 seconds later. Assuming there is no air resistance, find the speed of the stone when it hits the sea, and the height of the cliff.
State briefly, with a reason, the effect on the estimate of height of ignoring air resistance. (UCLES)

8. (Take $g = 10\,\mathrm{m\,s^{-2}}$ in this question.)
A balloon is moving vertically upwards with a steady speed of $3\,\mathrm{m\,s^{-1}}$. When it reaches a height of 36 m above the ground an object is released from the balloon. The balloon then accelerates upwards at a constant rate of $2\,\mathrm{m\,s^{-2}}$. Find
(i) the greatest height of the object above the ground
(ii) the speed of the object as it strikes the ground
(iii) the time taken by the object from leaving the balloon to striking the ground
(iv) the speed of the balloon as the object strikes the ground.
Sketch, on the same diagram, velocity–time graphs to illustrate the motion of the object and of the balloon during the interval from the object leaving the balloon to striking the ground. (UCLES)

9. (Take $g = 10\,\mathrm{m\,s^{-2}}$ in this question.)
A ball moves in a vertical straight line under gravity. Air resistance is negligible. The ball is projected from a point 2 m above the ground with an upward speed of $3\,\mathrm{m\,s^{-1}}$.
(a) (i) Find the time taken for the ball to reach its greatest height above the ground.
(ii) Show that this maximum height above the ground is 2·45 m.
(iii) Hence, or otherwise, find the speed of the ball when it first strikes the ground.

(b) The following is a velocity–time graph for the motion of the ball. Time $t = 0$ is the moment of projection, and velocity is measured positively in the upward direction.

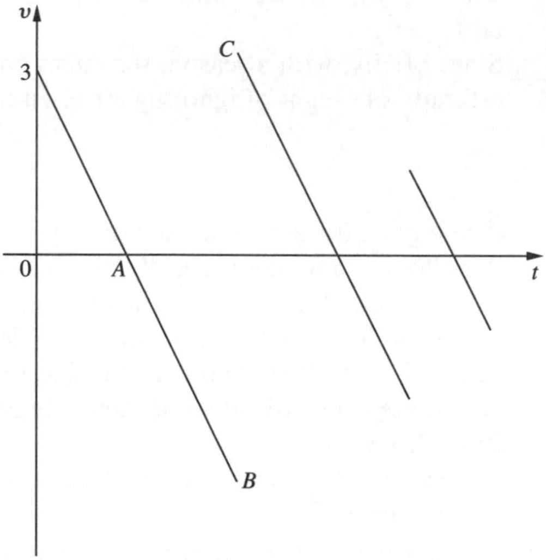

(Not drawn to scale)

(i) Describe the significance of the point A on the graph.
(ii) Explain why there is a discontinuity (i.e., a break) in the graph between B and C. (UODLE)

10. A car is travelling along a straight motorway at a constant speed $V\,\mathrm{m\,s^{-1}}$. Ten seconds after passing a speed-limit sign, the driver brakes and the car decelerates uniformly for 5 seconds, reducing its speed to $30\,\mathrm{m\,s^{-1}}$.
(a) Sketch a speed–time graph to illustrate this information.
Given that the car covers a distance of 600 m in the 15 second period, find
(b) the value of V
(c) the deceleration of the car. (ULEAC)

11. A tram travelling along a straight track starts from rest and accelerates uniformly for 15 seconds. During this time it travels 135 metres. The tram now maintains a constant speed for a further minute. It is finally brought to rest decelerating uniformly over a distance of 90 metres. Calculate the tram's acceleration and deceleration during the first and last stages of the journey. Also find the time taken and the distance travelled for the whole journey. (AEB 1989)

12. A car travels along a straight horizontal road, passing two garages, A and B. The car passes A at $u\,\mathrm{m\,s^{-1}}$ and maintains this speed for 60 s, during which time it travels 900 m.
Approaching a junction, the car then slows at a uniform rate of $a\,\mathrm{m\,s^{-2}}$ over the next 125 m to reach a speed of $10\,\mathrm{m\,s^{-1}}$, at which instant, with the road clear, the car accelerates at a uniform rate of $0{\cdot}75\,\mathrm{m\,s^{-2}}$. This acceleration is maintained for 20 s by which time the car has reached a speed of $v\,\mathrm{m\,s^{-1}}$, which is then maintained. The car passes B 45 s after its speed reaches $v\,\mathrm{m\,s^{-1}}$.
(i) Calculate the value of u, of a and of v.
(ii) Sketch a velocity–time graph for the motion of the car between A and B.
(iii) Find the distance between garages A and B and the time taken by the car to travel this distance. (UCLES)

13. A vehicle travelling on a straight horizontal track joining two points A and B accelerates at a constant rate of $0{\cdot}25\,\mathrm{m\,s^{-2}}$ and decelerates at a constant rate of $1\,\mathrm{m\,s^{-2}}$. It covers a distance of 2·0 km from A to B by accelerating from rest to a speed of $v\,\mathrm{m\,s^{-1}}$ and travelling at that speed until it starts to decelerate to rest. Express in terms of v the times taken for acceleration and deceleration.
Given that the total time for the journey is 2·5 minutes find a quadratic equation for v and determine v, explaining clearly the reason for your choice of the value of v.
(AEB 1992)

3 Force and Newton's laws

In the last chapter we considered bodies which changed their velocities. For this to occur, a *force* must act on the body.

Newton's First Law

A change in the state of motion of a body is caused by a force. The unit of force is the newton, abbreviated to N.

A body at rest

If forces act on a body and it does not move, the forces must balance. Hence, if a number of forces act on a body and it remains at rest, the resultant force in any direction must be zero.

Example 1

A body is at rest when subjected to the forces shown in the diagram. Find X and Y.

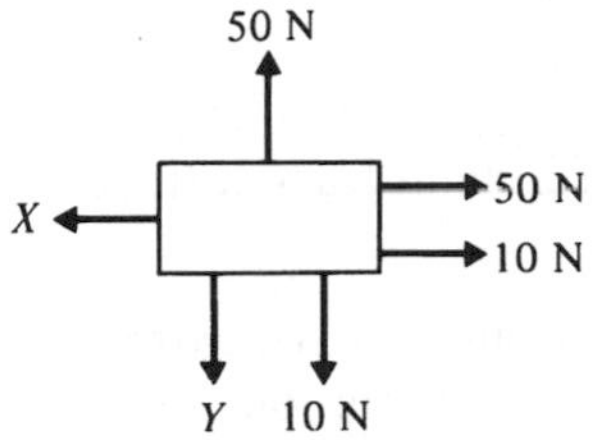

The horizontal forces balance.

$$\therefore \quad X = 50 + 10 = 60\,\text{N}$$

The vertical forces balance.

$$Y + 10 = 50$$
$$\therefore \quad Y = 40\,\text{N}$$

A body in motion

A body can only change its velocity, i.e. increase its speed, slow down or change direction, if a resultant force acts upon it. Thus, if a body is moving with constant velocity, there can be no resultant force acting on it.

Example 2

A body moves horizontally at a constant $5\,\text{m s}^{-1}$ subject to the forces shown. Find P and S.

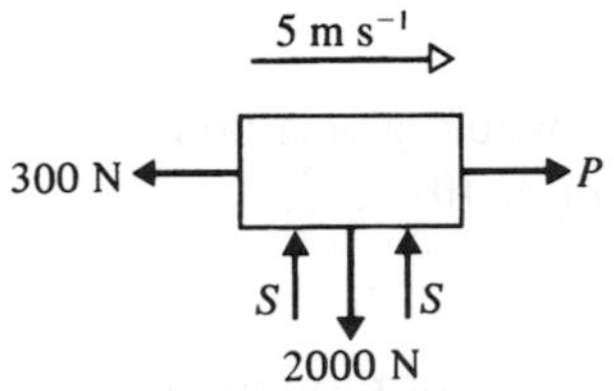

There is no vertical motion.

$$\therefore \quad S + S = 2000$$
$$\therefore \quad S = 1000\,\text{N}$$

The horizontal velocity is constant.

$$\therefore \quad P = 300\,\text{N}$$

Newton's First Law of Motion states that:

> A body will remain at rest, or will continue to move with constant velocity, unless external forces cause it to do otherwise.

Only when these external forces have a non-zero resultant will the body change from its previous state of rest or of constant velocity.

Newton's Second Law deals with such situations.

Newton's Second Law

When a resultant force acts on a body, it causes acceleration. The acceleration is proportional to the force. The same force will not produce the same acceleration in all bodies. The force which would give a cyclist, say, an acceleration of $\frac{1}{2}\,\text{m s}^{-2}$, would, when applied to a car, produce a very much smaller acceleration.
The acceleration produced by a force depends upon the *mass* of the body on which it acts. The unit of mass is the kilogram, abbreviated to kg.
A force of 1 N produces an acceleration of $1\,\text{m s}^{-2}$ in a body of mass 1 kg. In general terms, a force of **F** newtons acting on a body of mass m kg produces an acceleration of $\mathbf{a}\,\text{m s}^{-2}$, where:

$$\mathbf{F} = m\mathbf{a}$$

This is a vector equation, so the acceleration produced is in the direction of the applied force, or of the resultant force if there is more than one force acting.

Newton's Second Law can be summarized by the equation $F = ma$, which is often referred to as the equation of motion.

Example 3

A body of mass 8 kg is acted upon by a force of 10 N. Find the acceleration.

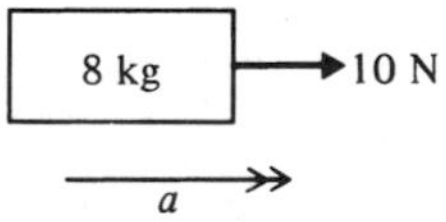

Using $F = ma$ gives the equation of motion as:

$$10 = 8 \times a$$
$$\therefore \quad a = 1\tfrac{1}{4}$$

The acceleration is $1\frac{1}{4}\,\text{m s}^{-2}$.

Example 4

Find the resultant force that would give a body of mass 200 g an acceleration of $10\,\text{m s}^{-2}$.

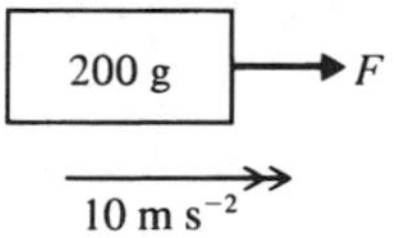

The mass is $200\,\text{g} = \frac{1}{5}\,\text{kg}$.

Using $F = ma$, gives the equation of motion as:

$$F = \tfrac{1}{5} \times 10 = 2$$

The force is 2 N.

Example 5

A body of mass 2 kg, subject to forces as shown in the diagram, accelerates uniformly in the direction indicated.
Find the acceleration and the value of P.

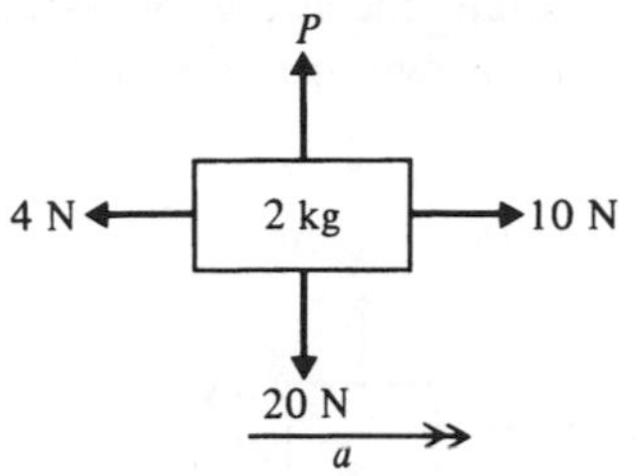

There is no vertical motion.

$$\therefore \qquad P = 20\,\text{N}$$

Horizontally, there are two forces acting, 10 N and 4 N, in opposite directions. Using $F = ma$ gives the equation of motion as:

$$10 - 4 = 2 \times a$$
$$\therefore \qquad a = 3$$

The horizontal acceleration is $3\,\text{m s}^{-2}$ and P is 20 N.

Example 6

Find, in vector form, the acceleration produced in a body of mass 5 kg subject to forces $(4\mathbf{i} + \mathbf{j})$ N and $(-\mathbf{i} + \mathbf{j})$ N. Also state the magnitude and the direction of the acceleration.

The resultant force acting is:

$$(4\mathbf{i} + \mathbf{j}) + (-\mathbf{i} + \mathbf{j})\,\text{N}$$
$$= (3\mathbf{i} + 2\mathbf{j})\,\text{N}$$

Using $\mathbf{F} = m\mathbf{a}$ gives the equation of motion as:

$$3\mathbf{i} + 2\mathbf{j} = 5 \times \mathbf{a}$$
$$\therefore \qquad \mathbf{a} = \tfrac{3}{5}\mathbf{i} + \tfrac{2}{5}\mathbf{j}$$

The magnitude of $\mathbf{a} = |\mathbf{a}| = \sqrt{\left(\tfrac{3}{5}\right)^2 + \left(\tfrac{2}{5}\right)^2}$

$$= \tfrac{1}{5}\sqrt{(3^2 + 2^2)} = \tfrac{1}{5}\sqrt{13}$$
$$= 0{\cdot}72\,\text{m s}^{-2}$$

The acceleration produced is $\left(\tfrac{3}{5}\mathbf{i} + \tfrac{2}{5}\mathbf{j}\right)\,\text{m s}^{-2}$; its magnitude is $0{\cdot}72\,\text{m s}^{-2}$ and its direction makes an angle $\tan^{-1}\tfrac{2}{3}$ with the unit vector $\mathbf{i}$.

Exercise 3A

1. In each of the following situations a body is shown at rest under the action of certain forces.
Find the magnitudes of the unknown forces X and Y.

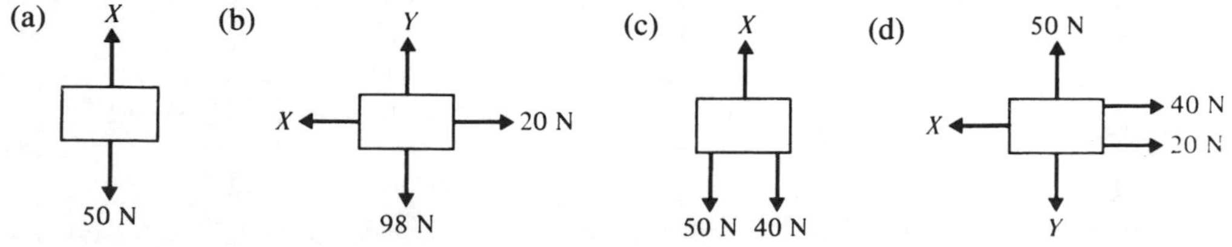

2. In each of the following situations a body is shown moving with constant velocity v under the action of certain forces. Find the magnitudes of the unknown forces X and Y.

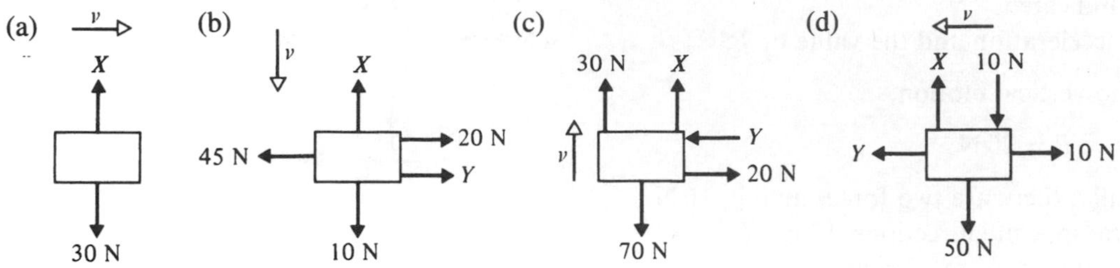

3. Find the acceleration produced when a body of mass 5 kg experiences a resultant force of 10 N.

4. Find the resultant force that would give a body of mass 3 kg an acceleration of $2\,\text{m}\,\text{s}^{-2}$.

5. A resultant force of 24 N causes a body to accelerate at $3\,\text{m}\,\text{s}^{-2}$. Find the mass of the body.

6. Find the acceleration produced when a body of mass 100 g experiences a resultant force of 5 N.

7. Find, in vector form, the resultant force required to make a body of mass 2 kg accelerate at $(5\mathbf{i} + 2\mathbf{j})\,\text{m}\,\text{s}^{-2}$.

8. Find, in vector form, the acceleration produced in a body of mass 500 g subject to forces of $(4\mathbf{i} + 2\mathbf{j})$ N and $(-\mathbf{i} + \mathbf{j})$ N.

9. Find, in vector form, the acceleration produced in a body of mass 2 kg subject to forces of $(2\mathbf{i} - 3\mathbf{j} + 4\mathbf{k})$ N and $(\mathbf{i} + 5\mathbf{j} + 2\mathbf{k})$ N.

10. A car travels a distance of 24 m whilst uniformly accelerating from rest to $12\,\text{m}\,\text{s}^{-1}$. Find the acceleration of the car.
If the car has a mass of 600 kg find the magnitude of the accelerating force.

11. A body of mass 500 g experiences a resultant force of 3 N. Find:
(a) the acceleration produced
(b) the distance travelled by the body whilst increasing its speed from $1\,\text{m}\,\text{s}^{-1}$ to $7\,\text{m}\,\text{s}^{-1}$.

12. In each of the following situations the forces acting on the body cause it to accelerate as indicated. Find the magnitude of the unknown forces X and Y.

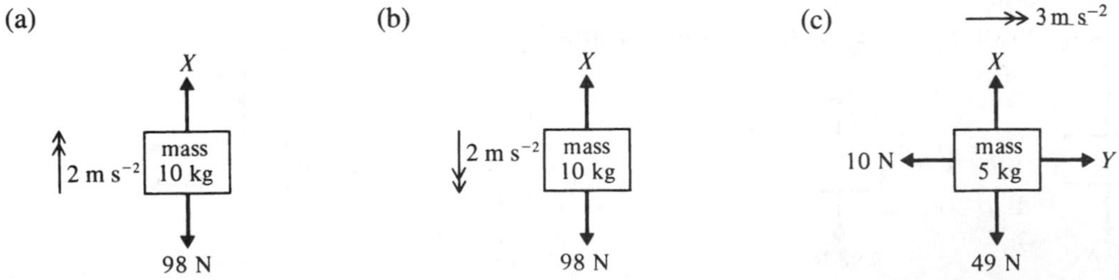

(d)

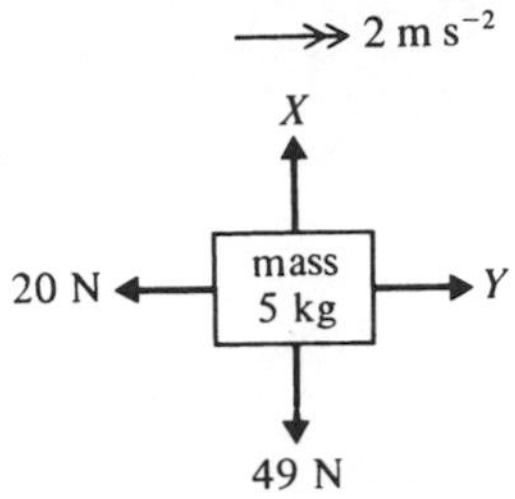

(e)

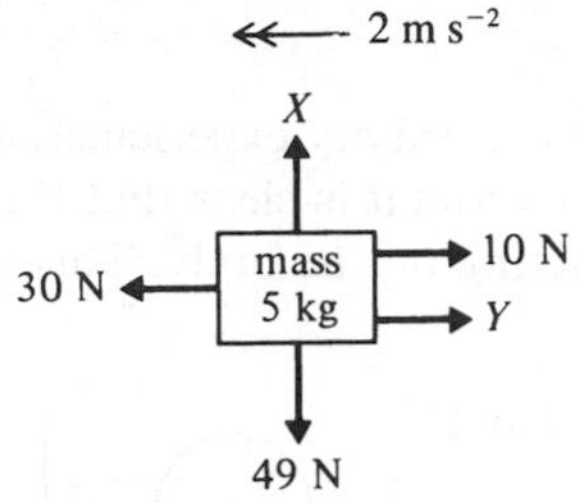

(f)

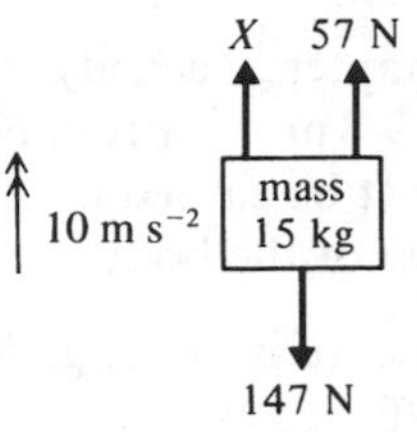

13. A car moves along a level road at a constant velocity of $22\,\text{m s}^{-1}$.
If its engine is exerting a forward force of 500 N, what resistance is the car experiencing?

14. A car of mass 500 kg moves along a level road with an acceleration of $2\,\text{m s}^{-2}$.
If its engine is exerting a forward force of 1100 N, what resistance is the car experiencing?

15. A van of mass 2 tonnes moves along a level road against resistances of 700 N.
If its engine is exerting a forward force of 2200 N, find the acceleration of the van.

16. Find the magnitude of the resultant force required to give a body of mass 2 kg an acceleration of $(\mathbf{i} - 3\mathbf{j})\,\text{m s}^{-2}$.

17. Find, in vector form, the acceleration produced in a body of mass 500 g when forces of $(5\mathbf{i} + 3\mathbf{j})$ N, $(6\mathbf{i} + 4\mathbf{j})$ N and $(-7\mathbf{i} - 7\mathbf{j})$ N act on the body.

18. Forces of $(10\mathbf{i} + 2\mathbf{j})$ N and $(a\mathbf{i} + b\mathbf{j})$ N acting on a body of mass 500 g cause it to accelerate at $(24\mathbf{i} + 3\mathbf{j})\,\text{m s}^{-2}$. Find the constants a and b.

19. Forces of $(a\mathbf{i} + b\mathbf{j} + c\mathbf{k})$ N and $(2\mathbf{i} - 3\mathbf{j} + \mathbf{k})$ N acting on a body of mass 2 kg cause it to accelerate at $(4\mathbf{i} + \mathbf{k})\,\text{m s}^{-2}$.
Find the constants a, b and c.

20. Find the constant force necessary to accelerate a car of mass 600 kg from rest to $25\,\text{m s}^{-1}$ in 12 s if the resistance to motion is (a) zero, (b) 350 N.

21. Find the constant force necessary to accelerate a car of mass 1000 kg from $15\,\text{m s}^{-1}$ to $20\,\text{m s}^{-1}$ in 10 s against resistances totalling 270 N.

22. A train of mass 60 tonnes is travelling at $40\,\text{m s}^{-1}$ when the brakes are applied.
If the resultant braking force is 40 kN, find the distance the train travels before coming to rest.

23. A train of mass 100 tonnes starts from rest at station A and accelerates uniformly at $1\,\text{m s}^{-2}$ until it attains a speed of $30\,\text{m s}^{-1}$. It maintains this speed for a further 90 s and then the brakes are applied, producing a resultant braking force of 50 kN.
If the train comes to rest at station B, find the distance between the two stations.

Gravity and weight

As stated in Chapter 2, a body falling under gravity experiences an acceleration of $9{\cdot}8\,\mathrm{m\,s^{-2}}$. From Newton's Laws it is clear that this acceleration must be caused by a force acting on the body. This force is called the *weight* of the body.

Consider a stone, of mass 2 kg, dropped from the top of a cliff.
It will fall with an acceleration of $9{\cdot}8\,\mathrm{m\,s^{-2}}$.
The force F which produces this acceleration is given by:

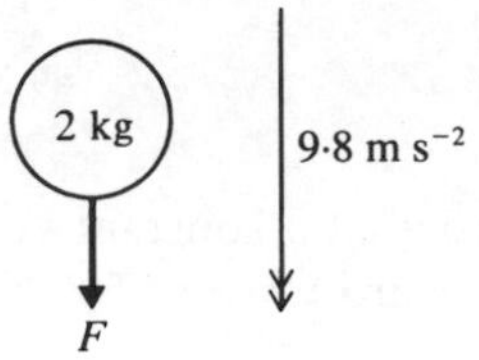

$$F = 2 \times 9{\cdot}8$$
$$= 19{\cdot}6\,\mathrm{N}$$

A body of mass m kg has a weight of mg N.

It should be remembered that, although the value of g (the acceleration due to gravity) has slightly different values at different places on the Earth's surface, it should be taken as $9{\cdot}8\,\mathrm{m\,s^{-2}}$ unless stated otherwise.

Example 7

Find:
(a) the weight in newtons of a box of mass 5 kg
(b) the mass of a stone of weight 294 N.

(a)
$$\text{mass} = 5\,\mathrm{kg}$$
$$\therefore \quad \text{weight} = 5 \times 9{\cdot}8$$
$$= 49\,\mathrm{N}$$

The weight of the box is 49 N.

(b)
$$\text{weight} = 294\,\mathrm{N}$$
$$\therefore \quad \text{mass} \times 9{\cdot}8 = 294$$
$$\therefore \quad \text{mass} = \frac{294}{9{\cdot}8} = 30\,\mathrm{kg}$$

The mass of the stone is 30 kg.

The difference between the mass and the weight of a body is well illustrated by considering a 70 kg person on the Earth and the same person on the Moon. The acceleration due to gravity on the Moon is approximately $1{\cdot}6\,\mathrm{m\,s^{-2}}$.

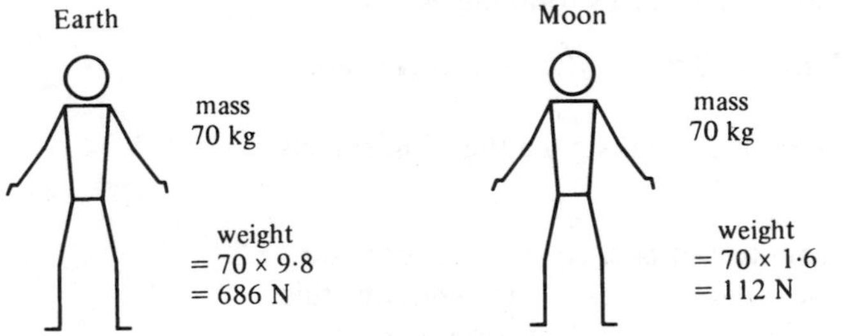

The person on the Moon has the same *mass* as they had on the Earth, but their *weight* is far less. Consequently they feel lighter when on the Moon.

Example 8

A box of mass 5 kg is lowered vertically by a rope.
Find the force in the rope when the box is lowered with an acceleration of $4\,\text{m}\,\text{s}^{-2}$.

$$\text{mass of box} = 5\,\text{kg}$$
$$\therefore \quad \text{weight of box} = 5g\,\text{N}$$

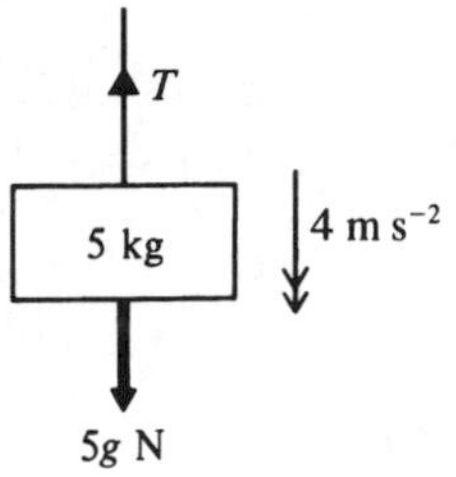

The resultant vertical force on box is $(5g - T)$ N downwards.

The downward force is required since this is the direction in which motion is taking place.

Using $F = ma$ gives the equation of motion as:

$$5g - T = 5 \times 4$$
$$\therefore \quad T = 5g - 20$$
$$= 29$$

The force in the rope is 29 N.

Example 9

A pack of bricks of mass 100 kg is hoisted up the side of a house.
Find the force in the lifting rope when the bricks are lifted with an acceleration of $\frac{1}{4}\,\text{m}\,\text{s}^{-2}$.

$$\text{mass of bricks} = 100\,\text{kg}$$
$$\therefore \quad \text{weight of bricks} = 100g\,\text{N}$$

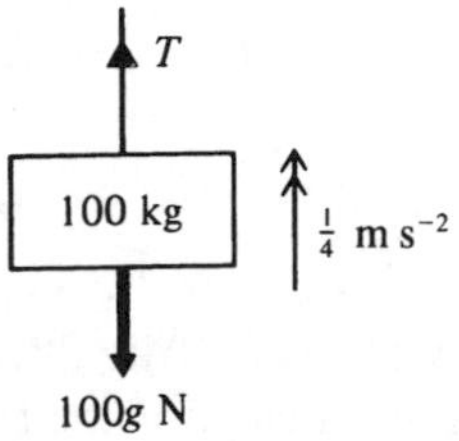

The resultant upward vertical force on the bricks is $(T - 100g)$ N.
(Upward force needed since motion is upward.)

Using $F = ma$ gives the equation of motion as:

$$T - 100g = 100 \times \tfrac{1}{4}$$
$$\therefore \quad T = 100g + 25$$
$$= 1005\,\text{N}$$

The force in the lifting rope is 1005 N.

Exercise 3B

Remember that g should be taken as $9{\cdot}8\,\text{m}\,\text{s}^{-2}$, unless otherwise stated.

1. Find the weight in newtons of a particle of mass 4 kg.
2. Find the mass of a car of weight 4900 N.
3. Find the weight, in newtons, of a particle of mass 100 g.

4. In each of the following situations, the forces acting on the body cause it to accelerate as indicated.
In (a), (b) and (c), find the magnitude of the unknown forces X and Y.

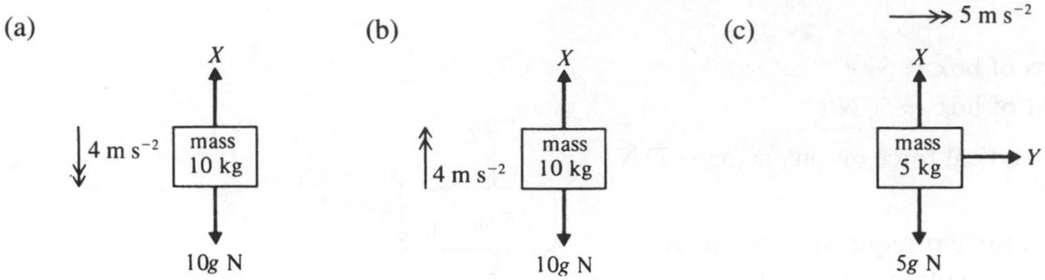

In (d), (e) and (f) find the mass m.

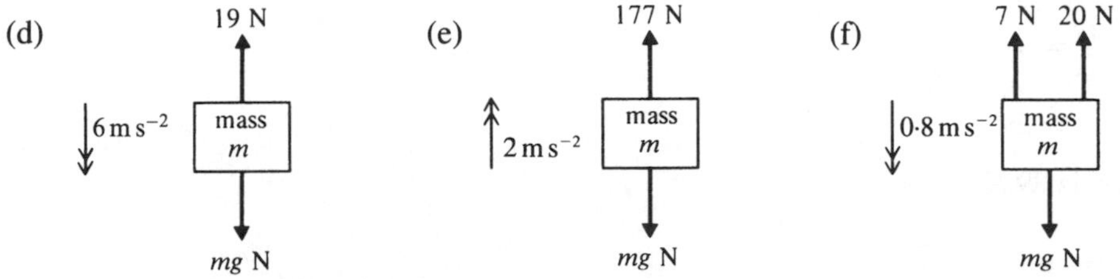

In (g), (h) and (i) find the magnitude of the acceleration a.

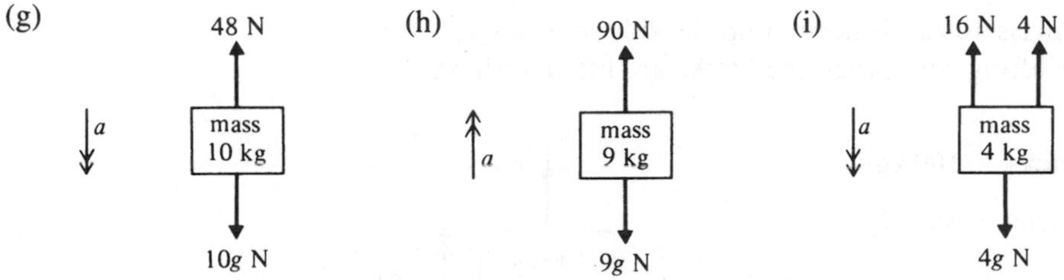

5. The diagram shows a body of mass 10 kg attached to a vertical string. Find the force T in the string in each of the following situations:
(a) the string raises the body with an acceleration of $5\,\text{m}\,\text{s}^{-2}$
(b) the string lowers the body with an acceleration of $5\,\text{m}\,\text{s}^{-2}$
(c) the string raises the body at a constant velocity of $5\,\text{m}\,\text{s}^{-1}$
(d) the string lowers the body at a constant velocity of $5\,\text{m}\,\text{s}^{-1}$.

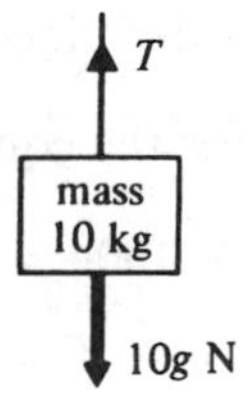

6. A particle of mass 100 g is attached to the lower end of a vertical string. Find the force in the string when it raises the particle with an acceleration of $1{\cdot}2\,\text{m}\,\text{s}^{-2}$.

7. A concrete block of mass 50 kg is hoisted up the side of a building. Find the force in the lifting rope when the block is lifted with an acceleration of $\frac{1}{5}\,\text{m}\,\text{s}^{-2}$.

8. A lift of mass 600 kg is raised or lowered by means of a cable attached to its top. When carrying passengers, whose total mass is 400 kg, the lift accelerates uniformly from rest to $2\,\text{m}\,\text{s}^{-1}$ over a distance of 5 m. Find:
(a) the magnitude of the acceleration
(b) the tension in the cable if the motion takes place vertically upwards
(c) the tension in the cable if the motion takes place vertically downwards.

9. The hot air balloon shown in the diagram, rises from the ground with uniform acceleration. After 10 s the balloon has attained a height of 25 m. If the total mass of the balloon and basket is 250 kg, find the magnitude of the lifting force F.

10. A stone of mass 50 g is dropped into some liquid and falls vertically through it with an acceleration of $5{\cdot}8\ \text{m s}^{-2}$.
Find the force of resistance acting on the stone.

11. A tile of mass 2 kg falls from the roof of a building and hits the ground, 16·6 m below, 2 s later. Assuming the resistance experienced by the tile is constant throughout the motion, find this resistance.

12. A miners' cage of mass 420 kg contains three miners of total mass 280 kg. The cage is lowered from rest by a cable. For the first 10 seconds the cage accelerates uniformly and descends a distance of 75 m. Find the force in the cable during the first 10 seconds.

13. A bucket has a mass of 5 kg when empty and 15 kg when full of water. The empty bucket is lowered into a well, at a constant acceleration of $5\ \text{m s}^{-2}$, by means of a rope. When full of water the bucket is raised at a constant velocity of $2\ \text{m s}^{-1}$.
Neglecting the weight of the rope, find the force in the rope:
(a) when lowering the empty bucket,
(b) when rasing the full bucket.

Newton's Third Law

This law states that: Action and Reaction are equal and opposite.

This means that if two bodies A and B are in contact and exert forces on each other, then the force exerted on B by A is equal in magnitude and opposite in direction to the force exerted on A by B. The following examples illustrate its application.

Example 10

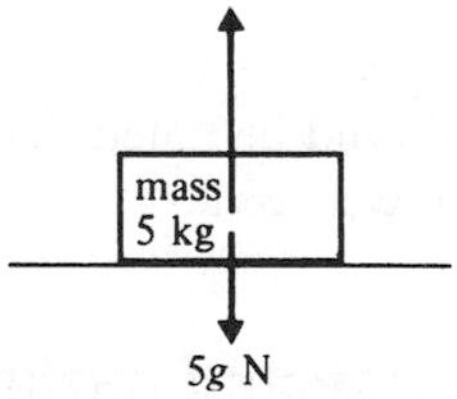

A box of mass 5 kg rests on a horizontal floor. The box exerts a force on the floor and the floor "reacts" by exerting an equal and opposite force on the box. As the box is at rest, this force of reaction R must equal the weight of the box, i.e. $R = 5g$ N. The force R is called the *normal reaction* as it acts at right angles to the surfaces in contact.

Example 11

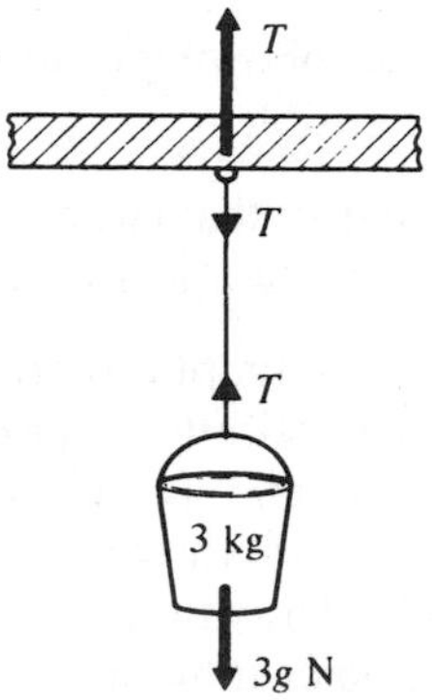

A bucket of mass 3 kg hangs on a vertical rope which is also attached to a beam. The bucket exerts a force on the rope, so the rope exerts an equal and opposite force on the bucket. As the bucket is at rest, this force T must equal the weight of the bucket, i.e. $T = 3g$ N.

At the point where the rope is attached to the beam, the rope exerts a downward force T on the beam. The beam exerts an equal and opposite force T on the rope. The rope therefore experiences a stretching force T at both ends. The force T is called the tension in the rope.

Example 12

A granite sphere of mass M kg rests on top of a pillar. The sphere exerts a force on the pillar and the pillar exerts an equal and opposite force T on the sphere. As the sphere is at rest, T must equal the weight of the sphere, i.e. $T = Mg$ N.

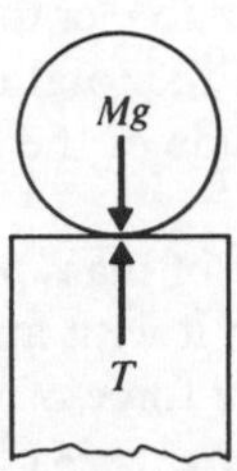

The force T is called the thrust in the pillar.

Example 13

A car pulls a trailer along a level road at a constant velocity v.
The trailer is pulled forwards by the tension T in the tow bar.
The trailer will exert an equal and opposite force T on the car.

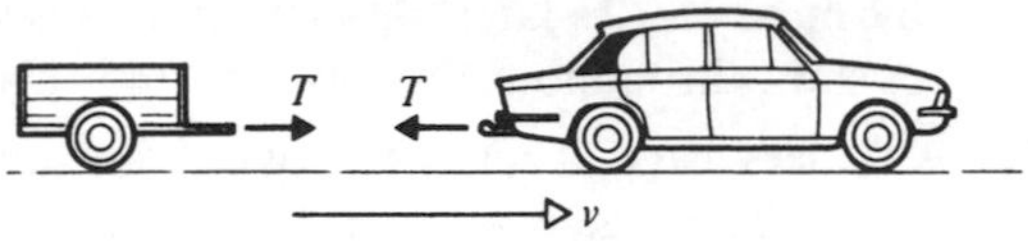

If the car is accelerating, then there must be a force acting on the trailer to produce this acceleration, and this will be provided by the tension T in the tow bar.
On the other hand, when the car is slowing down, so also is the trailer. In the absence of brakes on the trailer, some force must act in the opposite direction to the motion of the car and trailer. In this case the tow bar will exert a *thrust* on both the car and the trailer.

Example 14

A person of mass 70 kg stands on the floor of a lift which is accelerating downwards at $2\,\text{m}\,\text{s}^{-2}$. The person exerts a force on the floor and the floor exerts an equal and opposite force R on the person. Thus the resultant downward vertical force on the person is $(70g - R)$ and the equation of motion for the person is:

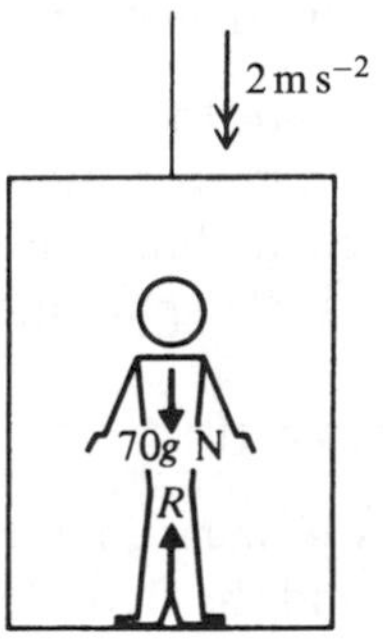

$$70g - R = 70 \times 2$$
$$\therefore \quad R = 686 - 140$$
$$= 546\text{ N}$$

It should be noted that in this case the reaction R is not equal to the weight of the person.

Connected particles

In the following examples the strings are all considered to be light and inextensible.

Note also that, when a surface is said to be smooth, it is to be assumed that the surface offers no resistance to the motion of a body across it.

These assumptions are examples of the real-life situation being *modelled* mathematically. In practice, strings do have mass, they do extend when under tension, and surfaces are not smooth. However, by making these assumptions, we simplify the situation and certain mathematical ideas can be applied more easily. The validity of the model can then be checked by comparing the results as predicted by the model with those occurring in real life.

Example 15

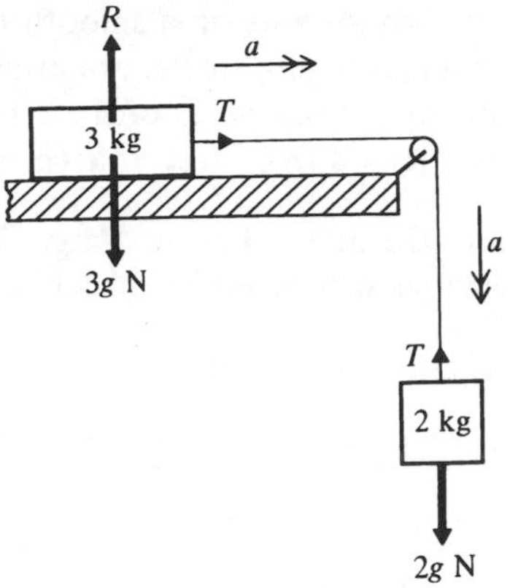

Consider a body of mass 3 kg at rest on a smooth horizontal table. This body is connected by a light string, which passes over a smooth pulley at the edge of the table, to another body of mass 2 kg hanging freely.
As the pulley is smooth, the tension in the string on both sides of the pulley will be the same.
The string is said to be light, so its weight can be ignored.
As the string is inextensible, when the system is released from rest the two bodies will have equal accelerations along the line of the string.

The 3 kg mass will not move in a vertical direction, so the vertical forces acting on it must balance.

$\therefore \quad R = 3g$

The horizontal force acting on the 3 kg mass is T.
Using $F = ma$ gives the equation of motion as:

$$T = 3 \times a \qquad \ldots [1]$$

The 2 kg mass moves vertically downwards.
Using $F = ma$ gives the equation of motion as:

$$2g - T = 2 \times a \qquad \ldots [2]$$

Solving equations [1] and [2] simultaneously gives:

$$2g = 3a + 2a$$

$$\therefore \quad a = \tfrac{2}{5}g$$

The acceleration of both bodies is $\frac{2}{5}g\,\text{m s}^{-2}$ along the line of the string; the tension in the string is $\frac{6}{5}g\,\text{N}$, obtained by substituting for a into equation [1].

Example 16

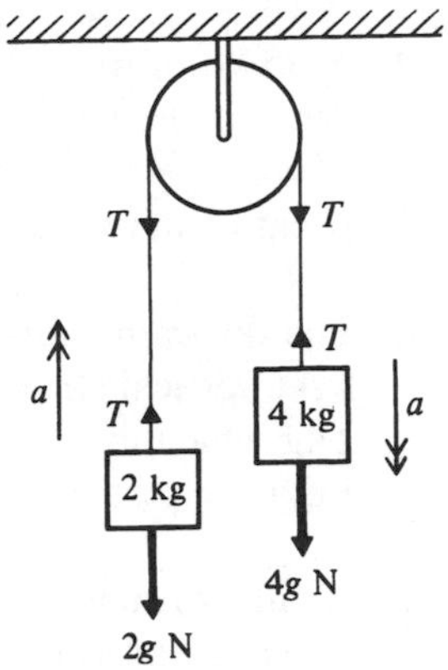

Particles of mass 4 kg and 2 kg are connected by a light string passing over a smooth fixed pulley. The particles hang freely and are released from rest. Find the acceleration of the two particles and the tension in the string. Let the acceleration be a and the tension in the string be T.

Using $F = ma$ gives:

for 2 kg mass: $T - 2g = 2 \times a \qquad \ldots [1]$

for 4 kg mass: $4g - T = 4 \times a \qquad \ldots [2]$

Adding equations [1] and [2] gives $2g = 6a$

$\therefore \quad a = \frac{1}{3}g \quad \text{and} \quad T = 2\frac{2}{3}g$

Using $g = 9{\cdot}8\,\text{m s}^{-2}$ gives the acceleration as $3{\cdot}27\,\text{m s}^{-2}$ and the tension as $26{\cdot}1$ N.

Example 17

A body A rests on a smooth horizontal table. Two bodies of mass 2 kg and 10 kg, hanging freely, are attached to A by strings which pass over smooth pulleys at the edges of the table. The two strings are taut. When the system is released from rest, it accelerates at $2\,\text{m}\,\text{s}^{-2}$. Find the mass of A.

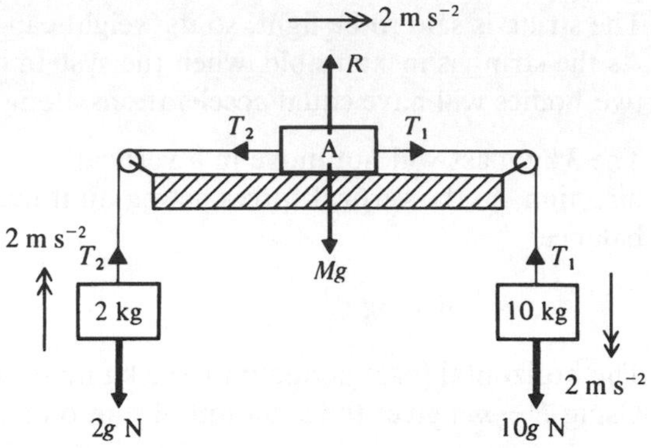

Let the mass of A be M kg. The tensions in the two strings will be different; let them be T_1 and T_2.

Using $F = ma$ gives:

for 2 kg mass: $T_2 - 2g = 2 \times 2$ ... [1]

for A: $T_1 - T_2 = M \times 2$ ... [2]

for 10 kg mass: $10g - T_1 = 10 \times 2$ ... [3]

Adding equations [1], [2] and [3] gives:

$$8g = 2M + 24$$

$$\therefore \quad M = 27{\cdot}2$$

The mass of the body A is 27·2 kg.

Force on pulley

It should be noted, in each of these examples, that there is a force acting on each of the fixed pulleys due to the tension in the string passing around the pulley. In Example 16 there is a downward force of $2T$ or 52·2 N acting on the fixed pulley, due to the string and the attached loads.

Exercise 3C

1. A box of mass 10 kg rests on a horizontal floor.
   What is the reaction that the floor exerts on the box?

2. A yo-yo of mass 200 g hangs at rest at the lower end of a vertical string.
   What is the tension in the string?

3. A cat of mass 4 kg sits on top of a vertical post. What is the thrust in the post?

4. The diagram shows a body of mass 5 kg hanging at rest at the end of a light vertical string. The other end of the string is attached to a mass of 2 kg which in turn hangs at the end of another light vertical string.
   Find the tension in each string.

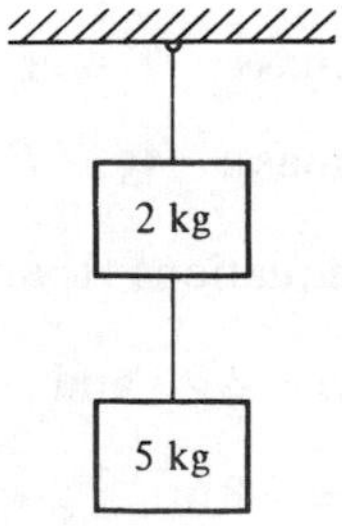

5. A cube of mass 6 kg rests on top of a horizontal table. A smaller cube of mass 2 kg is placed on top of the 6 kg cube.
   Find the reaction between the two cubes and that between the larger cube and the table.

6. The diagram shows a car of mass m_1 pulling a trailer of mass m_2 along a level road. The engine of the car exerts a forward force F, the tension in the tow bar is T and the reactions at the ground for the car and the trailer are R_1 and R_2 respectively. If the acceleration of the car is a, write down the equation of motion for:
(a) the system as a whole
(b) the car
(c) the trailer.
What can be said about R_1 and R_2?

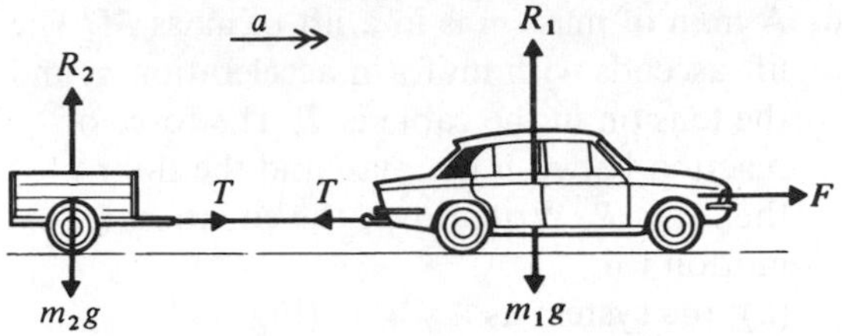

7. Masses m_1 and m_2 are connected by a light inextensible string passing over a smooth fixed pulley with $m_1 > m_2$. The masses move with acceleration a, as shown in the diagram. If the tension in the string is T, write down the equation of motion for
(a) mass m_1, (b) mass m_2.

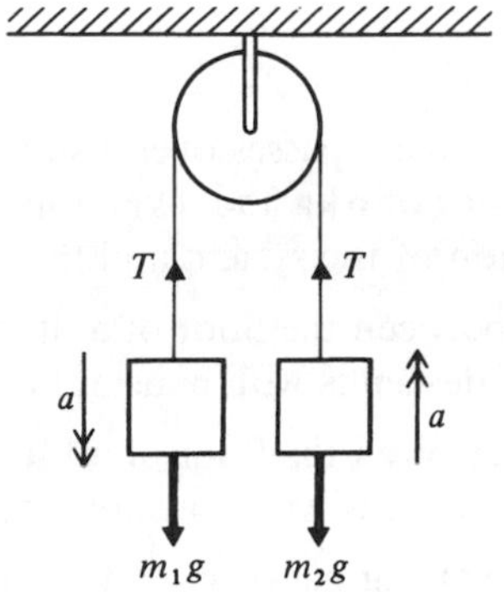

8. Mass m_1 lies on a smooth horizontal table and has one end of a light inextensible string attached to it. The string passes over a smooth fixed pulley at the edge of the table and carries a mass m_2 at its other end. T is the tension in the string, R is the reaction between m_1 and the table, and the acceleration a is as indicated in the diagram. Write down the equation of motion for
(a) mass m_1, (b) mass m_2.

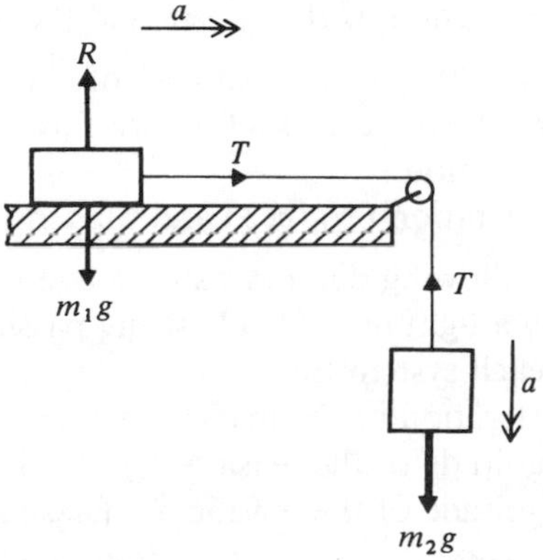

9. The diagram shows masses m_1, m_2 and m_3 connected by light inextensible strings such that m_1 and m_3 hang vertically and m_2 lies on a smooth horizontal surface. With $m_1 > m_3$, the forces and accelerations are as indicated in the diagram. Write down the equation of motion for
(a) mass m_1, (b) mass m_2, (c) mass m_3.

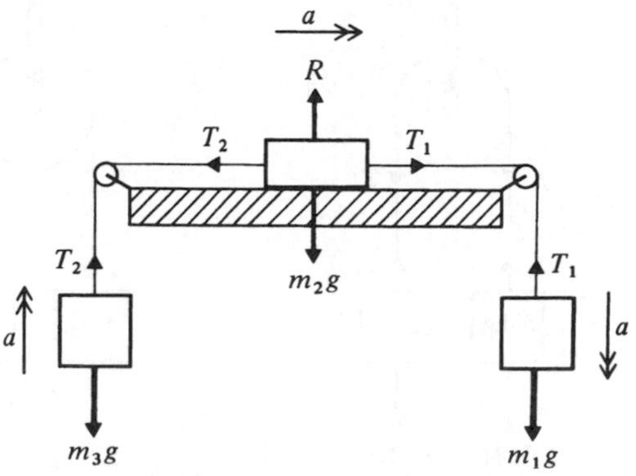

10. A man of mass m is in a lift of mass M. The lift ascends with uniform acceleration a, and the tension in the cable is T. The force of reaction between the man and the floor of the lift is R. Write down the equation of motion for
(a) the system as a whole (Fig. 1)
(b) the lift (Fig. 2)
(c) the man (Fig. 3).

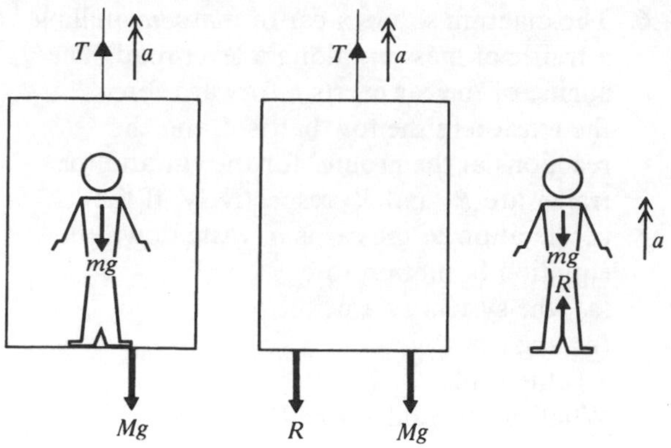

Fig. 1 Forces on man and lift **Fig. 2** Forces on lift **Fig. 3** Forces on man

11. A light inextensible string passes over a smooth fixed pulley and carries freely hanging masses of 6 kg and 4 kg at its ends.
Find the acceleration of the system and the tension in the string.

12. Find the reaction between the floor of a lift and a passenger of mass 60 kg when the lift descends with constant acceleration of $1{\cdot}3\,\mathrm{m\,s^{-2}}$.

13. Find the reaction between the floor of a lift and a passenger of mass 60 kg when the lift ascends with constant acceleration of $1{\cdot}2\,\mathrm{m\,s^{-2}}$.

14. A light inextensible string passes over a smooth fixed pulley and carries freely hanging masses of 800 g and 600 g at its ends.
Find the acceleration of the system and the force on the pulley.

15. A car of mass 900 kg tows a caravan of mass 700 kg along a level road. The engine of the car exerts a forward force of 2·4 kN and there is no resistance to motion.
Find the acceleration produced and the tension in the tow bar.

16. Each of the following diagrams shows two freely hanging masses connected by a light inextensible string passing over a smooth fixed pulley. For each system find:
(i) the acceleration of the masses,
(ii) the magnitude of the tension T_1,
(iii) the magnitude of the tension T_2. (Assume the pulley to be light.)

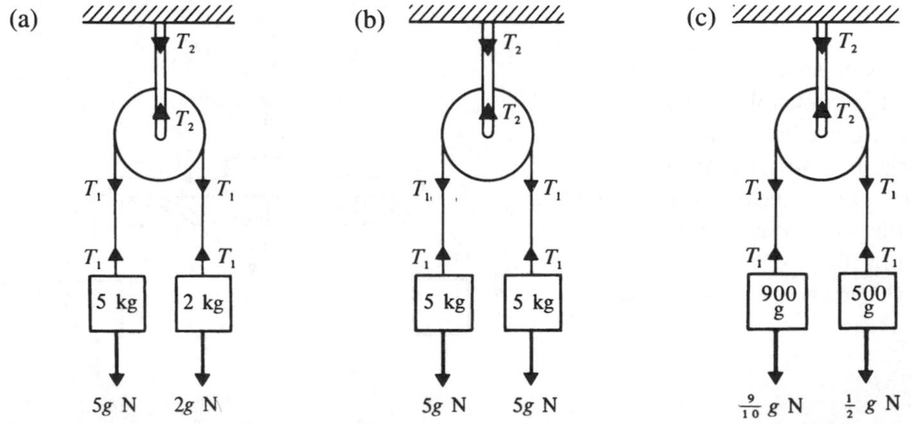

17. Bodies of mass 6 kg and 2 kg are connected by a light inextensible string which passes over a smooth fixed pulley. With the masses hanging vertically, the system is released from rest.
Find the acceleration of the system and the distance moved by the 6 kg mass in the first 2 seconds of motion.

18. Each of the following diagrams shows three bodies connected by light inextensible strings passing over smooth pulleys. One mass lies on a smooth horizontal surface and the other two masses hang freely. In each case the masses, pulleys and strings all lie in the same vertical plane. With the strings taut, each system is released from rest.
Find the ensuing accelerations and the tensions in the strings.

(a)

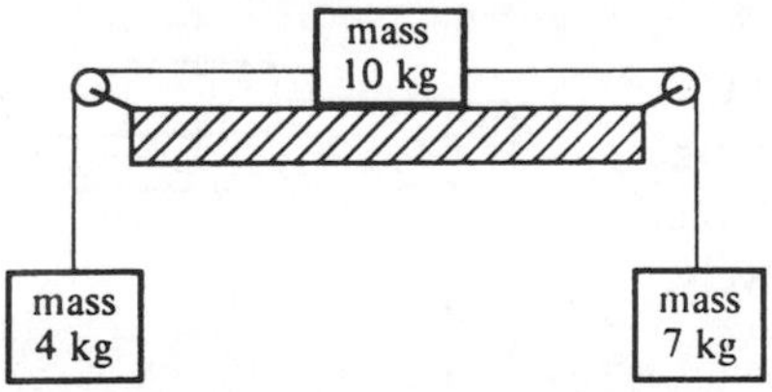

(b)

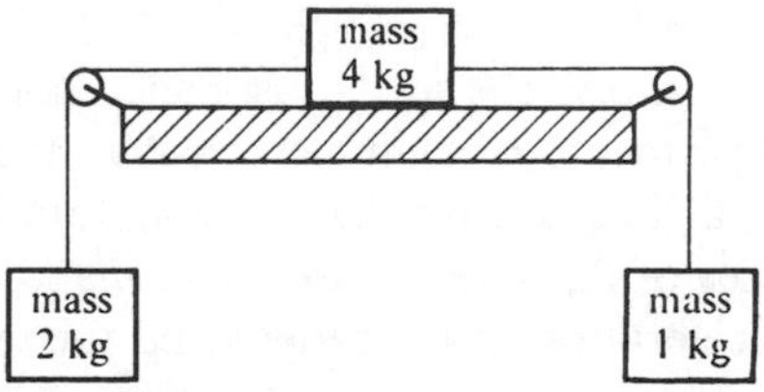

19. A body of mass 65 g lies on a smooth horizontal table. A light inextensible string runs from this body, over a smooth fixed pulley at the edge of the table, to a body of mass 5 g hanging freely. With the string taut, the system is released from rest. Find:
(a) the acceleration of the system
(b) the tension in the string
(c) the distance moved by the 5 g mass in the first 2 seconds of motion. (Assume that nothing impedes its motion in this time.)

20. The motion of a lift, when ascending from rest, is in three stages. First, it accelerates at $1\ \text{m s}^{-2}$ until it reaches a certain velocity. It then maintains this velocity for a period of time, after which it slows, with retardation $1{\cdot}2\ \text{m s}^{-2}$, until it comes to rest.
Find the reaction between the floor of the lift and a passenger, of mass 100 kg, during each of these three stages.

21. A car of mass 900 kg tows a trailer of mass 600 kg by means of a rigid tow bar. The car experiences a resistance of 200 N and the trailer a resistance of 300 N.
If the car engine exerts a forward force of 3 kN, find the tension in the tow bar and the acceleration of the system.
If the engine is switched off and the brakes now apply a retarding force of 500 N, what will be the retardation of the system, assuming the same resistances apply?
What will be the nature and magnitude of the force in the tow bar?

Exercise 3D Harder questions

1. Particles of mass m_1 and m_2 $(m_2 > m_1)$ are connected by a light inextensible string passing over a smooth fixed pulley. The particles hang vertically and are released from rest. Show that the acceleration of the system is $\dfrac{(m_2 - m_1)g}{m_1 + m_2}$ and that the tension in the string is $\dfrac{2m_1m_2g}{m_1 + m_2}$.

2. A particle of mass m_1 lies on a smooth horizontal table and is connected to a freely hanging particle of mass m_2 by a light inextensible string passing over a smooth fixed pulley situated at the edge of the table. Initially the system is at rest with m_1 a distance d from the edge of the table.

Show that the acceleration of the system is $\dfrac{m_2 g}{(m_1+m_2)}$ and that the time taken for m_1 to reach the edge of the table is $\sqrt{\dfrac{2d(m_1+m_2)}{m_2 g}}$.

3. The diagram shows the freely suspended particles A and C connected by means of light inextensible strings and smooth pulleys to particle B which lies on a smooth horizontal table. If the masses of A, B and C are $3m$, $3m$ and $4m$ respectively, find the acceleration of the system and the tensions in the strings.

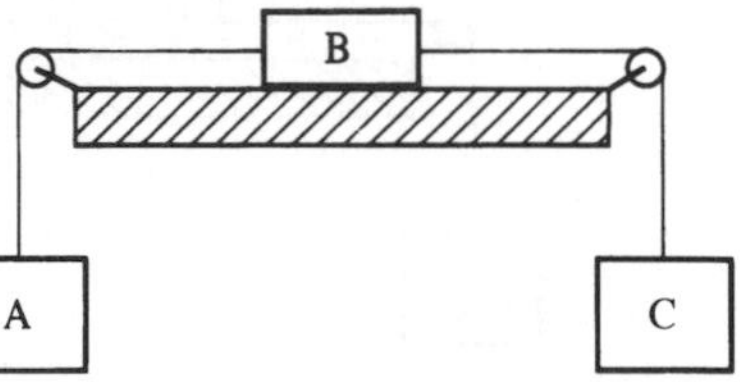

4. A car of mass 800 kg tows a trailer of mass 400 kg against resistances totalling 600 N. The separate resistances on the car and the trailer are proportional to their masses.
If the car accelerates at $1{\cdot}25\ \text{m s}^{-2}$ along a level road, find:
(a) the forward force exerted by the engine
(b) the tension in the tow bar.

5. Two particles A and B are connected by a light inextensible string passing over a smooth fixed pulley. The masses of A and B are $\frac{11}{2}m$ and $\frac{9}{2}m$ respectively. With A and B hanging vertically, the system is released from rest with particle A a distance d above the floor.
If a time t elapses before A hits the floor, show that $20d = t^2 g$.

6. A lorry of mass 3 tonnes tows a trailer of mass 1 tonne along a level road and accelerates uniformly from rest to $18\ \text{m s}^{-1}$ in 24 s. The resistances on the lorry and trailer are proportional to their masses and total 1200 N. Find:
(a) the driving force exerted by the engine of the lorry
(b) the tension in the tow bar.

7. The diagram shows a light inextensible string passing over a smooth fixed pulley, and carrying a particle A at one end and particles B and C at the other. The masses of A, B and C are $2m$, m and $2m$ respectively. Find the acceleration of the system when released from rest.
After C has travelled 50 cm it falls off and the system continues without it. Find:
(a) the velocity of B at the instant C falls off
(b) how much further B travels down before it starts to rise.

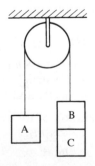

8. A car of mass 1 tonne exerts a driving force of 2·5 kN when pulling a trailer of mass 400 kg along a level road. The car and trailer start from rest and travel 18 m in the first 6 s of motion. If the resistances on the car and trailer are $1000x$ N and $400x$ N respectively, find the value of the constant x.

9. Particles of mass 600 g and 400 g are connected by a light inextensible string passing over a smooth fixed pulley. Initially both masses hang vertically, 30 cm above the ground. If the system is released from rest, find the greatest height reached above ground by the 400 g mass.

10. Particles of mass m_1 and m_2 ($m_2 > m_1$) are connected by a light inextensible string passing over a smooth fixed pulley. Initially both masses hang vertically with mass m_2 at a height x above the floor. Show that, if the system is released from rest, the mass m_2 will hit the floor with speed

$\sqrt{\dfrac{2(m_2 - m_1)gx}{m_1 + m_2}}$ and the mass m_1 will rise a further distance

$\dfrac{(m_2 - m_1)x}{m_1 + m_2}$ after this occurs.

Pulley systems

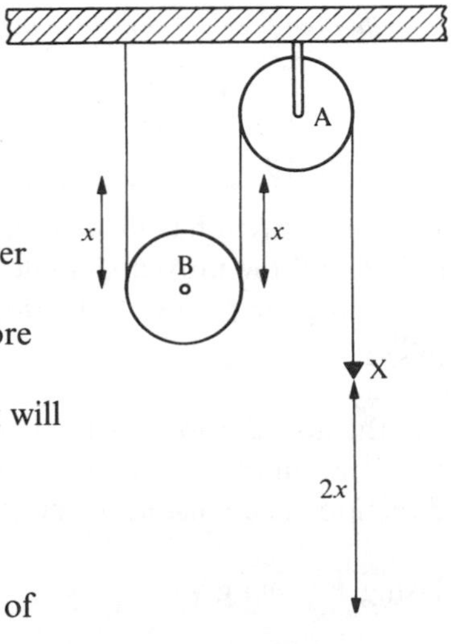

In the diagram, pulley A is fixed and pulley B may be raised by pulling down the end X of the string. All the parts of the string not in contact with the pulleys are vertical.
For B to move upwards a distance x, a length $2x$ of string must pass over the pulley A.
The distance between the pulley A and the end X of the string is therefore increased by $2x$.
Hence if B has an *upward* acceleration of a, then the end X of the string will have a *downward* acceleration of $2a$.

Example 18

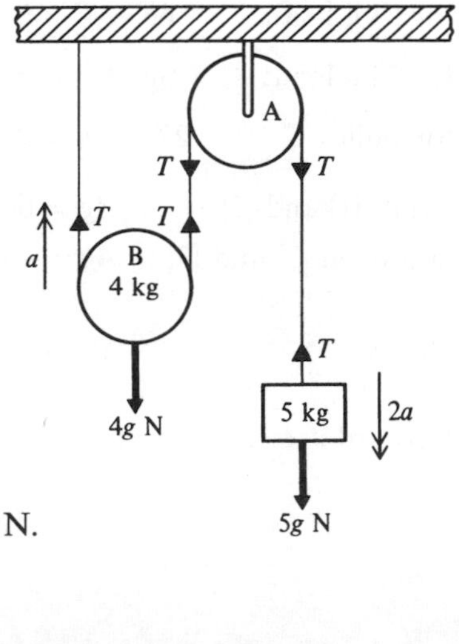

In the pulley system shown, A is a fixed pulley and pulley B has a mass of 4 kg. A load of mass 5 kg is attached to the free end of the string. Assuming the pulleys to be smooth, the tension throughout the string will be T as shown. When the system is released from rest, let the upward acceleration of B be a. The downward acceleration of the 5 kg mass will then be $2a$.

Using $F = ma$ gives:

for 5 kg load: $5g - T = 5 \times 2a$

for pulley B: $2T - 4g = 4 \times a$

Solving these equations and substituting $g = 9{\cdot}8$ m s^{-2}, we obtain:

$a = 2{\cdot}45$ and $T = 24{\cdot}5$

Pulley B has an *upward* acceleration of 2·45 m s^{-2}, the 5 kg mass has a *downward* acceleration of 4·9 m s^{-2}, and the tension in the string is 24·5 N.

In the pulley system shown, A and B are fixed pulleys and C is a moveable pulley. When the ends, X and Y, of the string move down distances x and y respectively, the length of the string between A and B is shortened by $(x + y)$. Pulley C will therefore move up a distance $\frac{1}{2}(x + y)$.

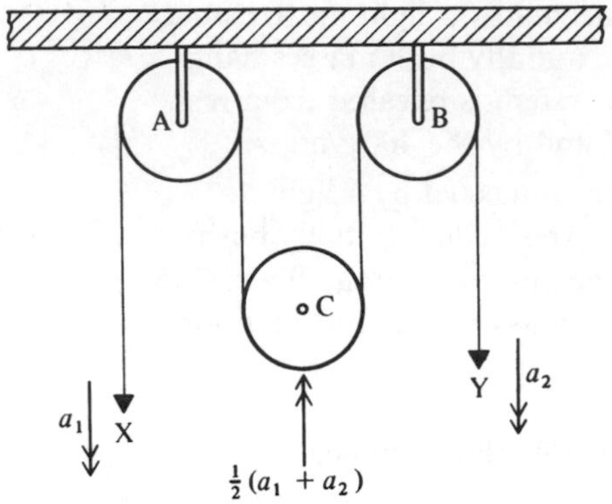

Hence, if the downward accelerations of X and Y are a_1 and a_2, the upward acceleration of C will be $\frac{1}{2}(a_1 + a_2)$.

Example 19

A pulley system has loads of 6 kg and 3 kg at the ends of the string, and the moveable pulley has a mass of 2 kg as shown. Assuming the pulleys to be smooth, find the acceleration of pulley C.

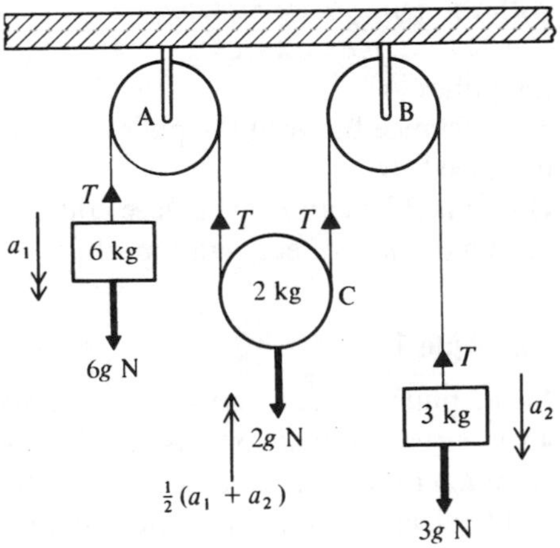

Let the accelerations of the loads be a_1 and a_2. The acceleration of the pulley C will be $\frac{1}{2}(a_1 + a_2)$ in the opposite direction. The tension of the string will be T throughout.

Using $F = ma$ gives:

for 6 kg load: $6g - T = 6 \times a_1$ ... [1]

for 3 kg load: $3g - T = 3 \times a_2$ ... [2]

for pulley C: $2T - 2g = 2 \times \frac{1}{2}(a_1 + a_2)$ ... [3]

From [1] and [2] $3g = 6a_1 - 3a_2$

and from [2] and [3] $4g = 7a_2 + a_1$

Hence $a_1 = \frac{11}{15}g$ and $a_2 = \frac{7}{15}g$

Upward acceleration of C $= \frac{1}{2}(a_1 + a_2)$

$= \frac{3}{5}g\,\text{m s}^{-2}$

Example 20

When a light pulley A is suspended from a fixed pulley, we have to consider relative accelerations. In the system shown, if the 3 kg load ascends with an acceleration a_1, the pulley A descends with an acceleration a_1.

Suppose the accelerations of the 2 kg and 6 kg loads, relative to pulley A, are a_2 upwards and a_2 downwards respectively.

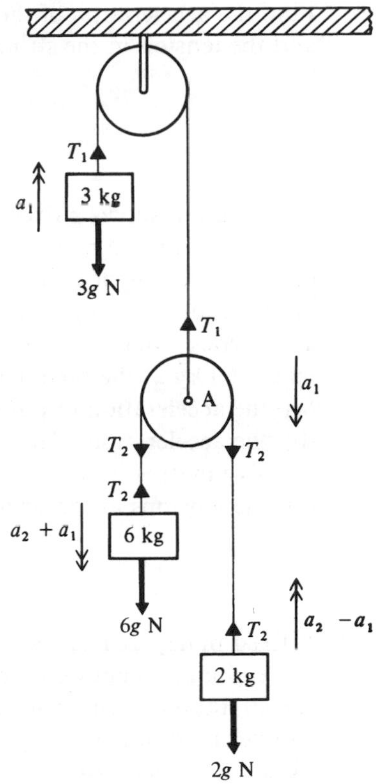

The actual accelerations of these loads will then be $(a_2 - a_1)$ upwards and $(a_2 + a_1)$ downwards respectively. Find these accelerations.

Let the tensions in the two strings be T_1 and T_2. Assuming the pulley A to be weightless and using $F = ma$, we obtain:

for 3 kg load: $T_1 - 3g = 3a_1$ ...[1]

for pulley A: $2T_2 - T_1 = 0 \times a_1$ ...[2]

for 6 kg load: $6g - T_2 = 6 \times (a_2 + a_1)$ ...[3]

for 2 kg load: $T_2 - 2g = 2 \times (a_2 - a_1)$ ...[4]

Now eliminate T_1 and T_2 from these equations.

[3] and [4] give $4g = 8a_2 + 4a_1$ ...[5]

[1] + [2] + 2[3] gives $9g = 12a_2 + 15a_1$ ...[6]

Solving [5] and [6] simultaneously gives:

$$a_1 = \tfrac{1}{3}g, \quad a_2 = \tfrac{1}{3}g$$

The 3 kg load accelerates upwards at $\frac{1}{3}g\,\text{m s}^{-2}$.

Acceleration of 6 kg load $= a_2 + a_1$

$= \frac{2}{3}g$ downwards.

Acceleration of 2 kg load $= a_2 - a_1$

$= 0$

The 6 kg load has a downward acceleration of $\frac{2}{3}g\,\text{m s}^{-2}$ and the 2 kg load remains stationary.

Exercise 3E

In this exercise all pulleys are smooth, all strings are light and inextensible, and all those parts of the strings not in contact with the pulleys are vertical.

1. A string, with one end fixed, passes under a moveable pulley of mass 2 kg, over a fixed pulley and carries a 5 kg mass at its other end (see diagram).
   Find the acceleration of the moveable pulley and the tension in the string.

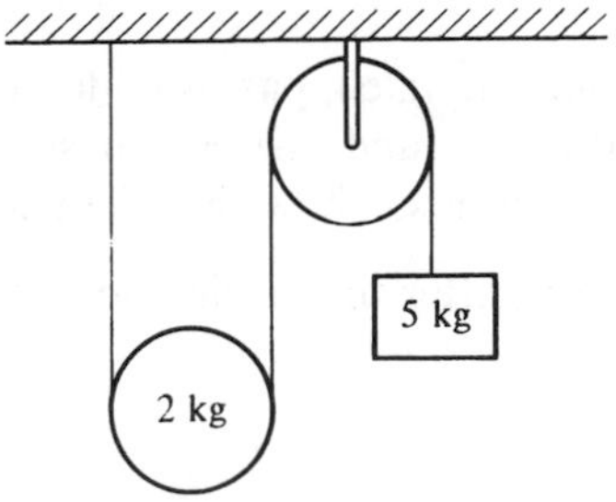

2. A string has a load of mass 2 kg attached to one end. The string passes over a fixed pulley, under a moveable pulley of mass 6 kg, over another fixed pulley and has a load of mass 3 kg attached to its other end. Find the acceleration of the moveable pulley and the tension in the string.

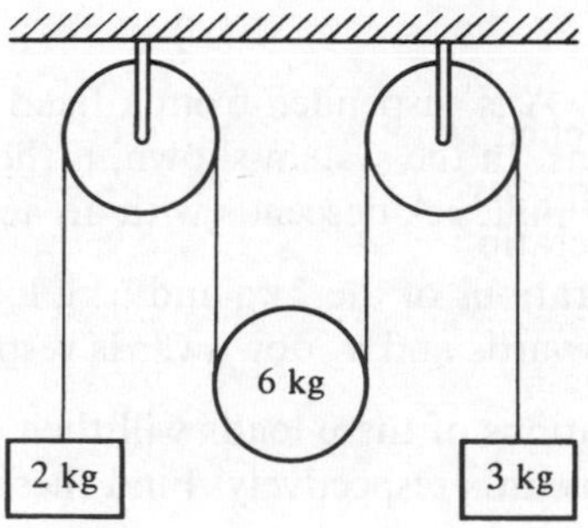

3. The diagram shows a fixed pulley carrying a string which has a mass of 4 kg attached at one end and a light pulley A attached at the other. Another string passes over pulley A and carries a mass of 3 kg at one end and a mass of 1 kg at the other end. Find:
   (a) the acceleration of pulley A
   (b) the acceleration of the 1 kg, 3 kg and 4 kg masses
   (c) the tensions in the strings.

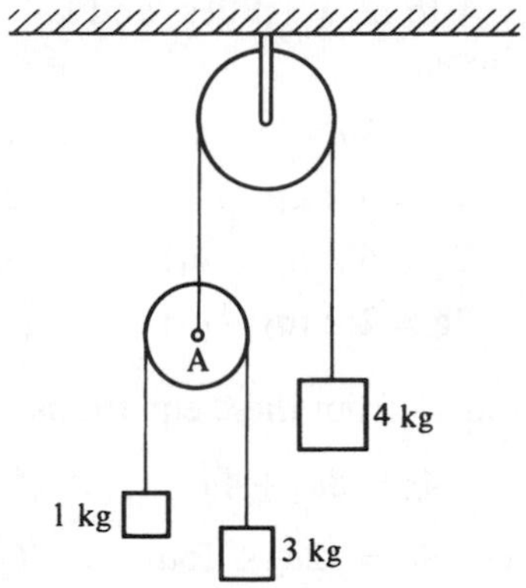

4. A fixed pulley carries a string which has a load of mass 7 kg attached to one end and a light pulley attached to the other end. This light pulley carries another string which has a load of mass 4 kg at one end, and another load of mass 2 kg at the other end.
   Find the acceleration of the 4 kg mass and the tensions in the strings.

5. A string, with one end fixed, passes under a moveable pulley of mass 8 kg, and over a fixed pulley; the string carries a 5 kg mass at its other end.
   Find the acceleration of the 5 kg mass and the tension in the string.

6. A string, carrying a particle A at one end, passes over a fixed pulley and has a light pulley attached to its other end. Over this light pulley runs another string carrying particle B at one end and particle C at the other. The masses of A, B and C are $3m$, $2m$ and m respectively.
   Find the acceleration of A and the tensions in the strings.

7. A string, with one end fixed, passes under a moveable pulley of mass m_1, over a fixed pulley, and carries a mass m_2 at its other end. With the system released from rest, show that the tension in the string is $\dfrac{3m_1m_2g}{4m_2+m_1}$ and that, after time t, the moveable pulley has moved a distance $\dfrac{gt^2(2m_2-m_1)}{2(4m_2+m_1)}$.

8. A string, with a particle A attached to one end passes over a fixed pulley, under a moveable pulley B, over another fixed pulley, and has a particle C attached to its other end. The masses of A, B and C are $3m$, $4m$ and $4m$ respectively.
Find the acceleration of A and the tension in the string.

9. In the pulley system shown in the diagram, A is a heavy pulley which is free to move. Find the mass of pulley A if it does not move upwards or downwards when the system is released from rest.

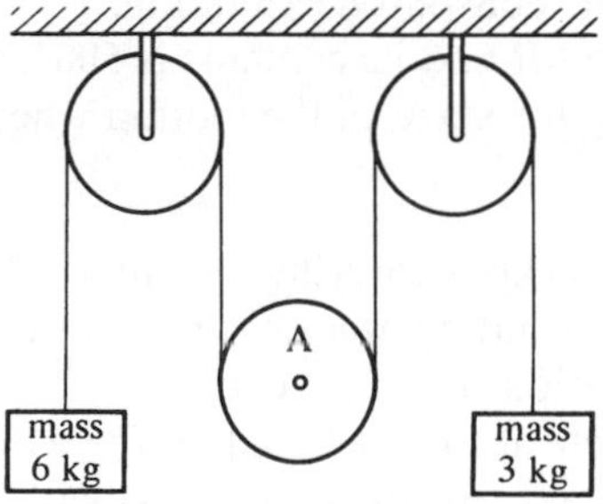

10. In the pulley system shown in the diagram, the pulley A is free to move.
Find the mass of the load B if, when the system is released from rest, pulley A does not move upwards or downwards.

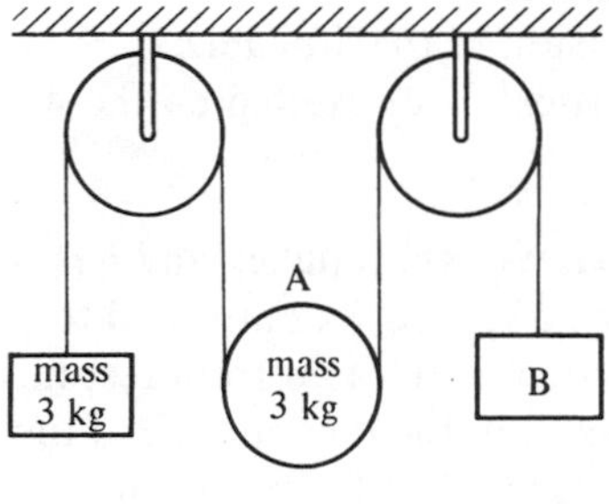

Exercise 3F Examination questions

(Unless otherwse indicated take $g = 9{\cdot}8\,\text{m s}^{-2}$ in this exercise.)

1. A ship of mass 10^7 kg is travelling at $2\,\text{m s}^{-1}$ when its engines are switched off. As a consequence the ship's speed is reduced to $1{\cdot}5\,\text{m s}^{-1}$ in a distance of 100 m. Assuming that the resistance to the ship's motion is uniform, calculate the magnitude of this resistance. (UCLES)

2. A particle of mass 2 kg moves under the action of a constant force $(2\mathbf{i} + 4\mathbf{j})$ N. At time $t = 0$ the particle is stationary and at the point with position vector $(2\mathbf{i} + 5\mathbf{j})$ m. Find the position vector of the particle at time $t = 3$ seconds. (AEB 1994)

3. Forces $(\mathbf{i} - 2\mathbf{j})$ N and $(3\mathbf{i} + 4\mathbf{j})$ N are applied to a small body of mass 2 kg. Unit vectors **i** and **j** are mutually perpendicular. All other forces on the body are in equilibrium.

(a) Find in terms of **i** and **j**:

(i) the resultant force acting on the body;

(ii) the acceleration of the body;

(b) Initially the position vector of the body is $(2\mathbf{i} - \mathbf{j})$ m and its initial velocity is $(4\mathbf{i} + 3\mathbf{j})\,\text{m s}^{-1}$.

(i) Show that after t seconds, the position vector of the body is $[(t^2 + 4t + 2)\mathbf{i} + (\frac{1}{2}t^2 + 3t - 1)]\,\mathbf{j}$ metres.

(ii) Find the value of t when the body's position vector is in the same direction as its acceleration. (UODLE)

4. A lift travels vertically upwards from rest at floor A to rest at floor B, which is 20 m above A, in three stages as follows. At first the lift accelerates from rest at A at $2\,\text{m s}^{-2}$ for 2 s. It then travels at a constant speed and finally it decelerates uniformly, coming to rest at B after a total time of $6\frac{1}{2}$ s. Sketch the (t, v) graph for this motion, and find the magnitude of the constant deceleration.
The mass of the lift and its contents is 500 kg. Find the tension in the lift cable during the stage of the motion when the lift is accelerating upwards. (UCLES Spec)

5. A lift of mass 950 kg is carrying a woman of mass 50 kg.
(a) The lift is ascending at a uniform speed. Calculate:
(i) the tension in the lift cable;
(ii) the vertical force exerted on the woman by the floor of the lift.
(b) Sometime later the lift is ascending with a downward acceleration of $2\,\text{m s}^{-2}$. Calculate:
(i) the tension in the lift cable;
(ii) the vertical force exerted on the woman by the floor of the lift.
(UODLE)

6. A light string passes over a pulley and has bodies of masses 2 kg and 3 kg attached to its ends. The system is released from rest in the position shown in the diagram. Find the tension in the string while the bodies are moving.
State any two assumptions necessary for your method to be valid.

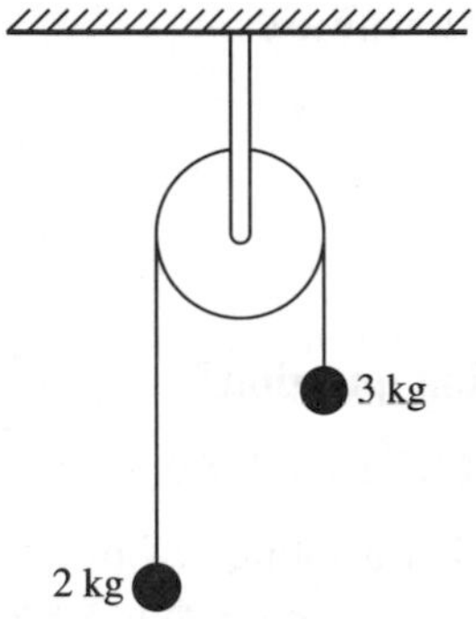

(UCLES)

7. Two particles of masses $4m$ and $6m$ respectively are attached one to each end of a light inextensible string. The string passes over a small smooth pulley and the particles are released from rest with the string vertical and taut. Find, in terms of m and g, the tension in the string during the subsequent motion. (WJEC)

8. (Take $g = 10\,\text{m s}^{-2}$ in this question.)
The diagram shows a particle P of mass 0·5 kg on a smooth horizontal table. P is connected to another particle Q, of mass 1·5 kg, by a taut light inextensible string which passes over a small fixed smooth pulley at the edge of the table, Q hanging vertically below the pulley. A horizontal force of magnitude X N acts on P as shown.

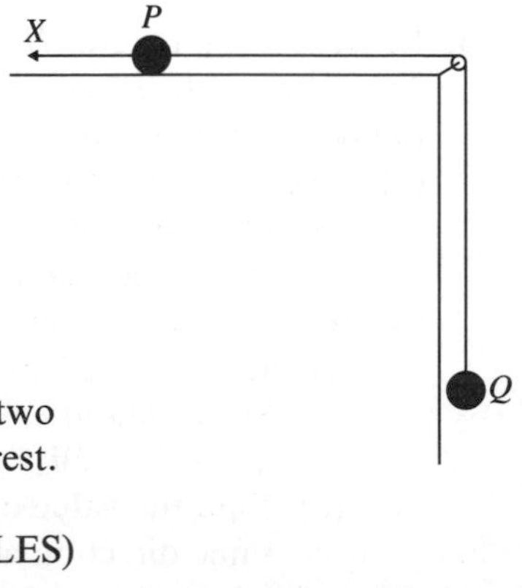

(i) Given that the system is in equilibrium, find X.
(ii) Given that $X = 12$, find the distance travelled by Q in the first two seconds of its motion following the release of the system from rest. You may assume that P does not reach the pulley in this time.
(UCLES)

9.

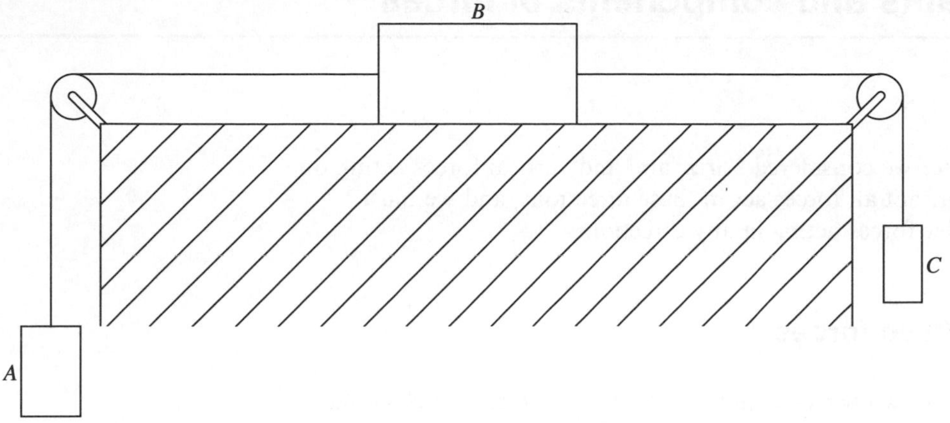

The figure shows a block B of mass 5 kg lying on a smooth table. It is connected to blocks A of mass 6 kg and C of mass 3 kg, which are hanging over the edges of the table, by light inextensible strings running over smooth pulleys. Initially the system is held at rest. After being released, the tension (in newtons) in the string joining A and B is P and the tension (in newtons) in the string joining B and C is Q. If the magnitude of the acceleration of each block is a m s^{-2}, write down three equations connecting a with P or Q (or both) and hence calculate a, P and Q.
Calculate the time taken by A to descend 60 cm from rest, assuming that B remains on the table and C remains hanging during this time.
(OCSEB)

10. A car, of mass M kilograms, is pulling a trailer, of mass λM kilograms, along a straight horizontal road. The tow-bar connecting the car and the trailer is horizontal and of negligible mass. The resistive forces acting on the car and trailer are constant and of magnitude 300 N and 200 N respectively. At the instant when the car has an acceleration of magnitude 0·3 m s^{-2}, the tractive force has magnitude 2000 N.
Show that

$$M(\lambda + 1) = 5000.$$

Given that the tension in the tow-bar is 500 N at this same instant, find the value of M and the value of λ. (ULEAC)

4 Resultants and components of forces

In the last chapter we considered horizontal and vertical forces acting on bodies. However, not all forces act in these directions, and we must therefore consider forces acting in any direction.

Resultant of two forces

The resultant R of two forces P and Q is that single force which could completely take the place of the two forces. The resultant R must have the same effect as the two forces P and Q.

When only parallel forces are involved, it is easy to find the resultant. For example:

the forces 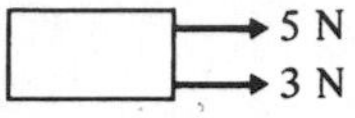

could be replaced by 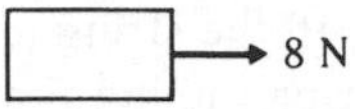

the forces 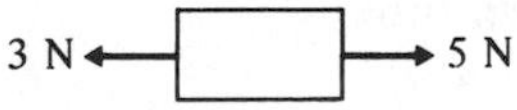

could be replaced by 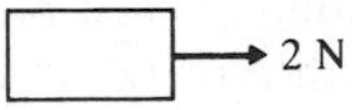

Parallelogram of forces

Two forces P and Q are represented by the line segments AB and AD.

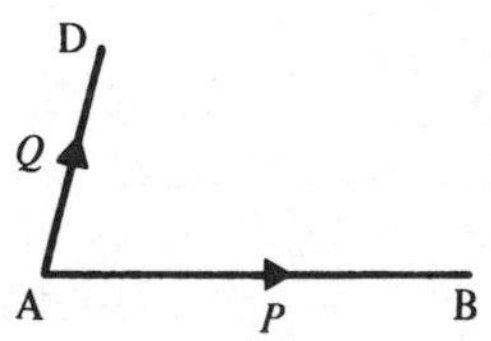

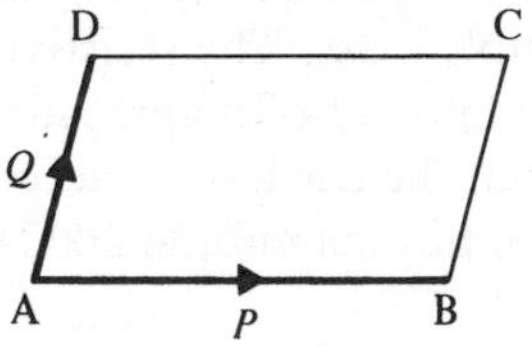

The parallelogram ABCD is completed by drawing BC and DC.

To find the resultant of the forces P and Q, we have to consider:

$$\overrightarrow{AB} + \overrightarrow{AD}$$

But $\overrightarrow{AD} = \overrightarrow{BC}$, as these are equivalent vectors.

$$\therefore \quad \overrightarrow{AB} + \overrightarrow{AD} = \overrightarrow{AB} + \overrightarrow{BC}$$
$$= \overrightarrow{AC}$$

Hence the resultant of the two forces P and Q, which are represented by the line segments AB and AD, is fully represented by the line segment AC. This is the diagonal AC of the parallelogram ABCD, which is therefore referred to as a parallelogram of forces.

Example 1

Find, by scale drawing, the magnitude of the resultant of the two forces shown in the sketch. Find also the angle that the resultant makes with the larger force.

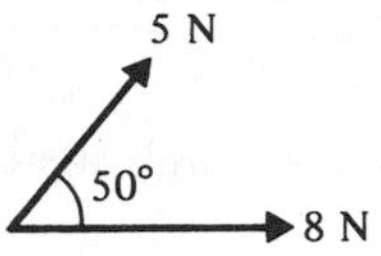

Construct parallelogram ABCD with AB = 8 cm, AD = 5 cm and angle DAB = 50°.

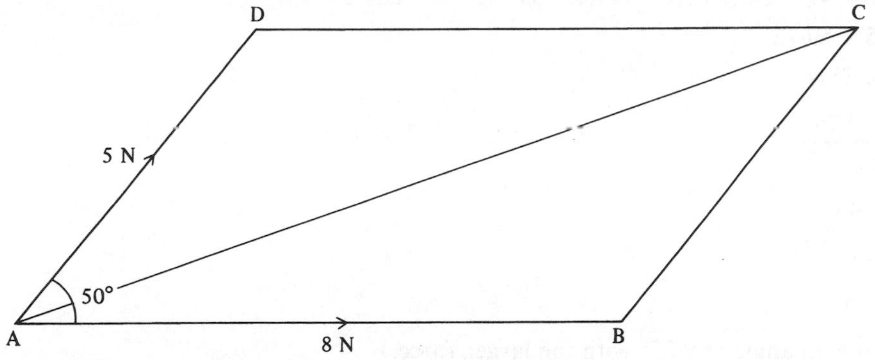

By measurement, AC = 11·9 cm and angle CAB = 19°.
The resultant is 11·9 N and makes an angle of 19° with the larger force.

It should be noted that, in the last example, the magnitude and direction of the resultant could have been obtained by considering the triangle ABC, rather than the whole parallelogram. Thus the resultant of two forces, which are represented in magnitude and direction by the sides AB and BC of the triangle ABC, is fully represented by the side AC of the triangle.

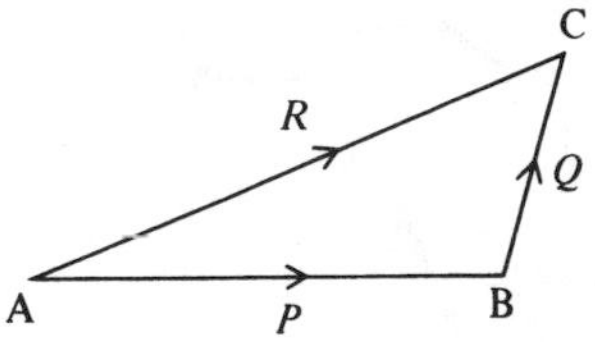

$\overrightarrow{AB} + \overrightarrow{BC} = \overrightarrow{AC}$

Note that the forces to be added are in the same sense around the triangle, in this case, anticlockwise.

From this triangle we can find the resultant, either by a scale drawing or by calculation.

Example 2

Find the magnitude of the resultant of the forces shown in the sketch, and the angle that the resultant makes with the larger force.

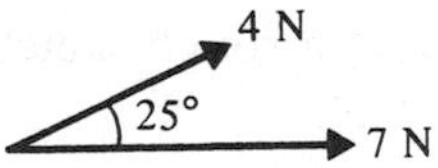

By scale drawing

Construct triangle ABC with AB = 7 cm, BC = 4 cm and angle ABC = 180° − 25° = 155°

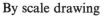

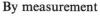

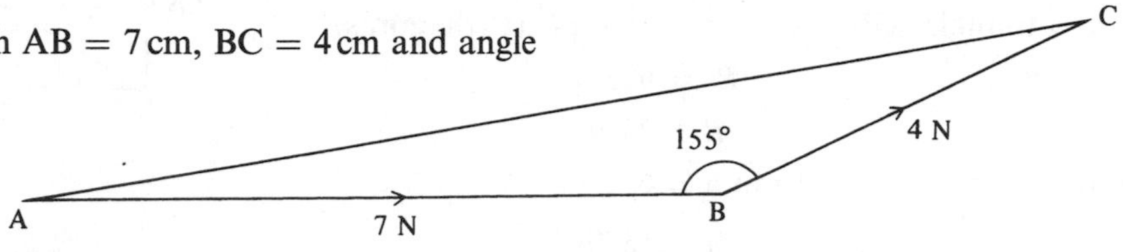

By measurement

AC = 10·8 cm and angle CAB = 9°.
The resultant is 10·8 N making an angle of 9° with the larger force.

By calculation

First, make a rough sketch.

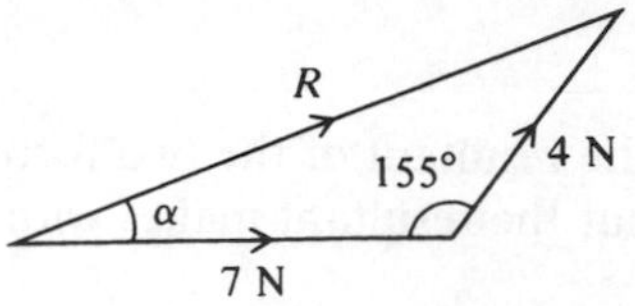

By the cosine rule $R^2 = 4^2 + 7^2 - 2 \times 4 \times 7 \cos 155°$

$= 16 + 49 + 56 \cos 25°$ since $\cos 155° = -\cos 25°$

$= 65 + 50{\cdot}75$

$\therefore \quad R = 10{\cdot}75\,\text{N}$

By the sine rule $\dfrac{4}{\sin \alpha} = \dfrac{10{\cdot}75}{\sin 155°}$

$\therefore \quad \sin \alpha = \dfrac{4 \sin 155°}{10{\cdot}75}$

$\therefore \quad \alpha = 9{\cdot}05°$

The resultant is 10·8 N making an angle of 9·05° with the larger force.

Angle between forces

In examples where a diagram is not given, it is necessary to interpret carefully the directions of the given forces.

If the angle between the forces is given as 35°, this should be interpreted as shown in the diagram.

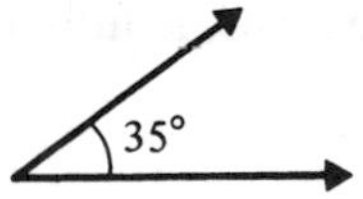

If it is stated that two forces act away from the point X and make an angle of 65° with each other, this should be interpreted as shown in the diagram.

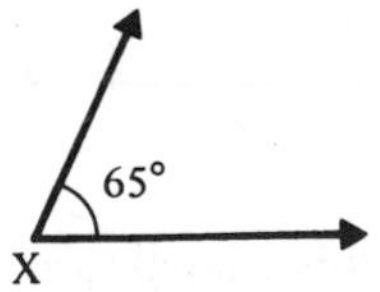

Example 3

Two forces of 7 N and 24 N act away from the point A and make an angle of 90° with each other. Find the magnitude and direction of their resultant.

First make a rough sketch.

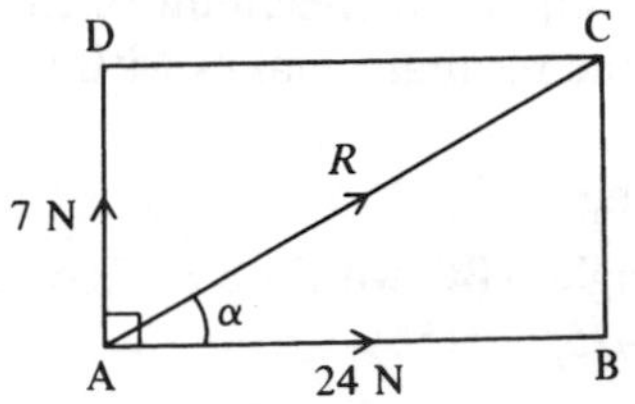

From triangle ABC: $R^2 = 7^2 + 24^2$ (Pythagoras)

$R^2 = 625$

$R = 25\,\text{N}$

and $\tan \alpha = \frac{7}{24}$

$\therefore \quad \alpha = 16{\cdot}26°$

The resultant is 25 N making an angle of 16·26° with the 24 N force.

Example 4

Find the angle between a force of 7 N and a force of 4 N if their resultant has a magnitude of 9 N.

First make a rough sketch.

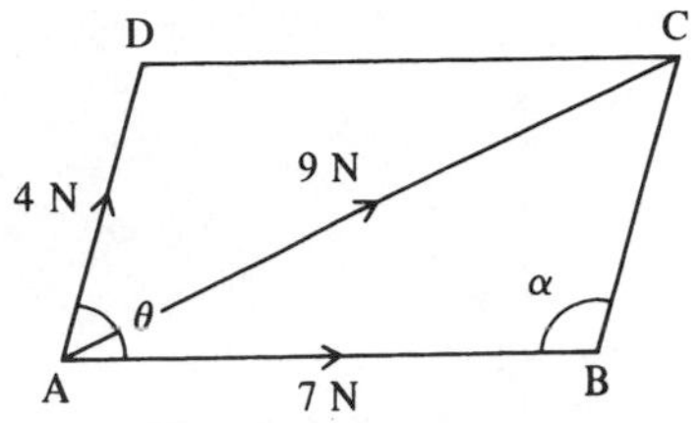

From triangle ABC, using the cosine rule, we obtain:

$$9^2 = 7^2 + 4^2 - 2 \times 7 \times 4 \cos \alpha$$

$$\therefore \quad \cos \alpha = \frac{7^2 + 4^2 - 9^2}{2 \times 7 \times 4}$$

$$= -\tfrac{2}{7}$$

Hence $\quad \alpha = 106{\cdot}60°$

$$\therefore \quad \theta = 180° - 106{\cdot}60°$$

$$= 73{\cdot}40°$$

The angle between the given forces is 73·40°.

Resultant of any number of forces

It is now known that the resultant R of any two forces P and Q can be found by constructing the triangle shown:

This method can be extended to find the resultant of any number of forces. Consider the forces P, Q, S, . . .
The forces P and Q can be combined and the resultant R_1 of these two forces can be then combined with the force S to find the resultant R_2, and so on.

Instead of drawing separate triangles, the forces can simply be added, paying due regard to their direction, and will form a polygon.

Thus to find the resultant R of the forces P, Q and S shown below, make an accurate scale drawing with the forces to be added following in the same sense around the figure:

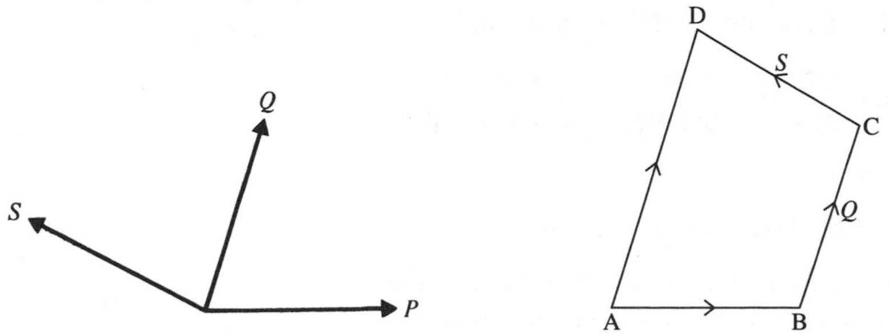

The line segment AD will then completely represent the resultant R.

Example 5

Forces of 6 N, 3 N and 4 N act as shown in the diagram. Find graphically the magnitude and the direction of the resultant of these forces.

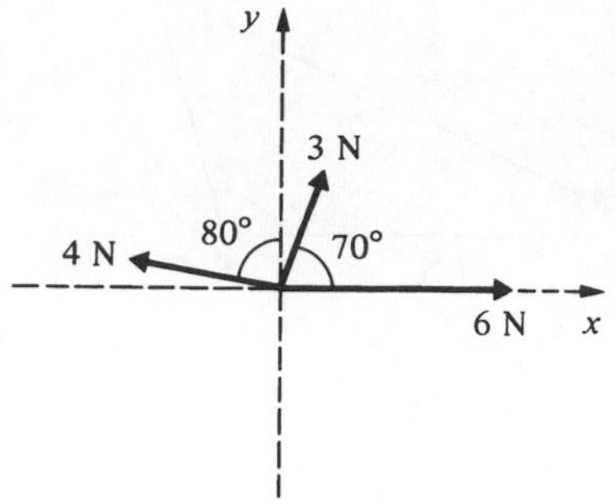

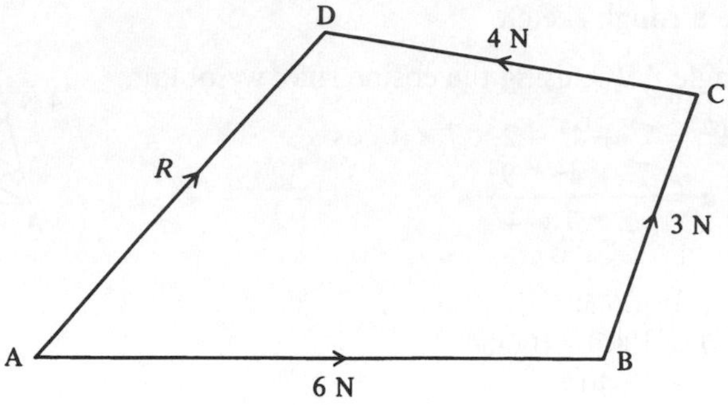

A line AB is drawn 6 cm in length and parallel to the force of 6 N.
A line BC is drawn 3 cm in length and parallel to the force of 3 N.
A line CD is drawn 4 cm in length and parallel to the force of 4 N.

Care is needed to ensure that each line is drawn in the correct direction, i.e. in the same direction as the force the line represents, and that the forces follow in the same sense around the polygon.

The polygon is completed by drawing the line AD which will represent the resultant in magnitude and direction.

By measurement, AD = 4·7 cm and angle DAB = 49°

The resultant is 4·7 N and makes an angle of 49° with the x-axis.

Example 6

Two forces of 5 N and 8 N act away from the point A and make an angle of 40° with each other. Find the angle which the resultant makes with the larger force.

Make a rough sketch and complete the parallelogram ABCD.

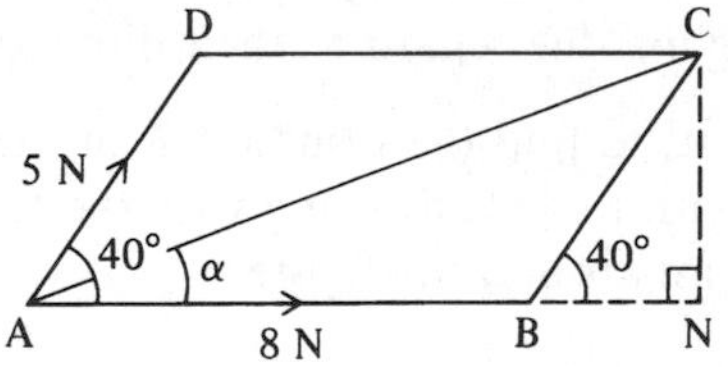

Note that, in this case, the magnitude of the resultant is not required.

Let N be the foot of the perpendicular from C to AB produced.

From the triangle BCN: $BN = 5\cos 40°$ $\quad CN = 5\sin 40°$

From the triangle ACN: $\tan\alpha = \dfrac{CN}{AN} = \dfrac{CN}{AB + BN} = \dfrac{5\sin 40°}{8 + 5\cos 40°}$

$\therefore \quad \alpha = 15{\cdot}20°$

The angle between the resultant and the larger force is 15·20°.

The angle α could have been found by first finding the magnitude of the resultant, and then using the sine rule in triangle ABC. The method of Example 6 avoids errors which could arise due to the incorrect determination of the resultant.

Exercise 4A

1. In each of the following diagrams, two forces are shown. Find, by scale drawing, the magnitude of their resultant and the angle it makes with the larger of the two forces.

(a)
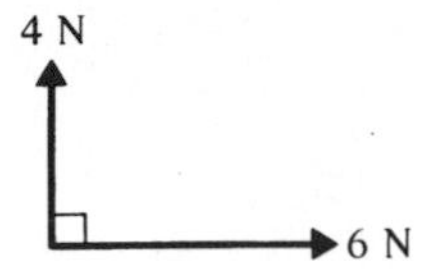

(b)
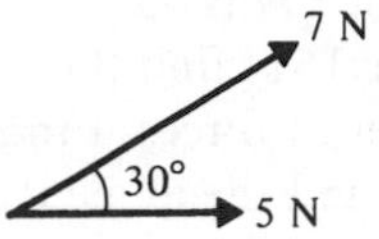

(c)
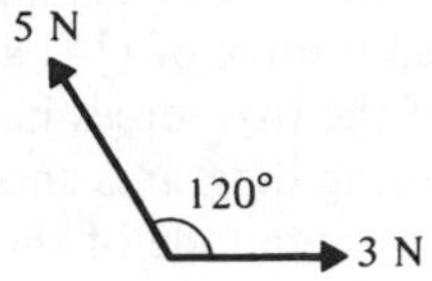

2. In each of the following diagrams, two forces are shown. Find, by calculation, the magnitude of their resultant and the angle it makes with the larger of the two forces.

(a)
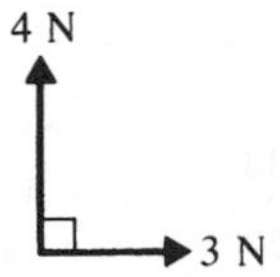

(b)
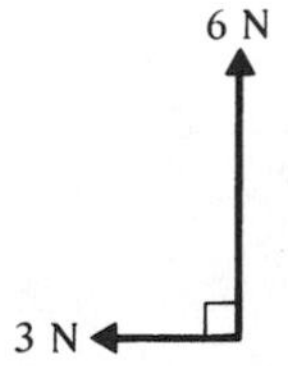

(c)
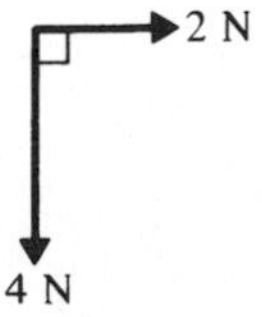

(d)
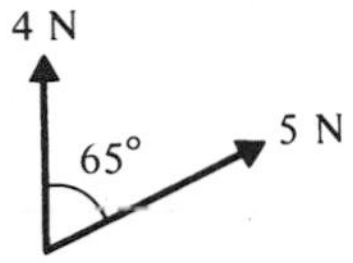

(e)
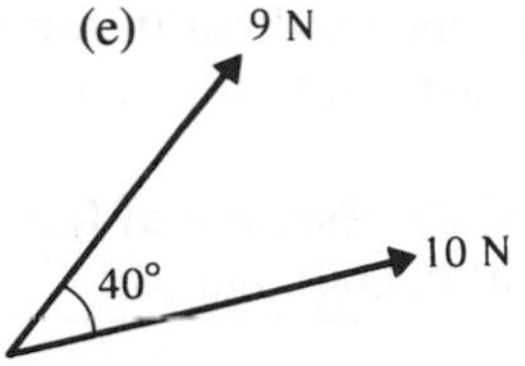

(f)
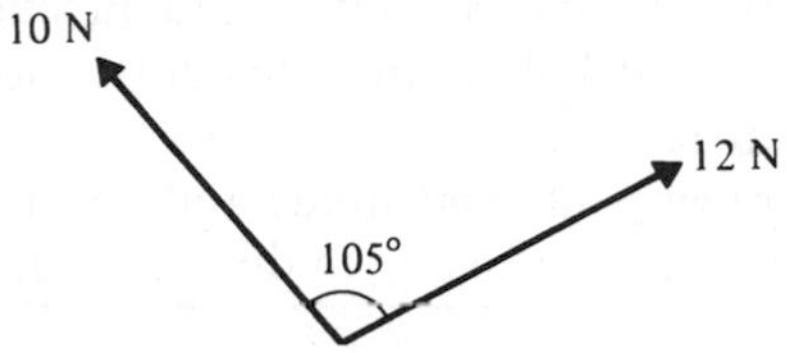

3. Find the magnitude and direction of the resultant of forces of 8 N and 3 N if the angle between the two forces is:
   (a) 60°
   (b) 50°
   (c) 160°.

4. Forces of 3 N and 2 N act along OA and OB respectively, the direction of the forces being indicated by the order of the letters. If $A\hat{O}B = 150°$, find the magnitude of the resultant of the two forces and the angle it makes with OA.

5. Forces of 6 N and 4 N act along OA and BO respectively, the direction of the forces being indicated by the order of the letters. If $A\hat{O}B = 60°$, find the magnitude of the resultant of the two forces and the angle it makes with OA.

6. Find the angle between a force of 6 N and a force of 5 N given that their resultant has magnitude 9 N.

7. Find the angle between a force of 10 N and a force of 4 N given that their resultant has magnitude 8 N.

8. The angle between a force of 6 N and a force of X N is 90°.
   If the resultant of the two forces has magnitude 8 N, find the value of X.

9. A force F N acts along $\overrightarrow{AB}$ and a force $2F$ N acts along $\overrightarrow{AC}$.
If $B\widehat{A}C = 60°$, find the magnitude of the resultant and the angle it makes with AB.

10. The angle between a force of P N and a force of 3 N is 120°.
If the resultant of the two forces has magnitude 7 N, find the value of P.

11. The angle between a force of Q N and a force of 8 N is 45°.
If the resultant of the two forces has magnitude 15 N find the value of Q.

12. Each of the following diagrams shows a number of forces. Find, by scale drawing, the magnitude of their resultant and the angle it makes with the x-axis.

(a)

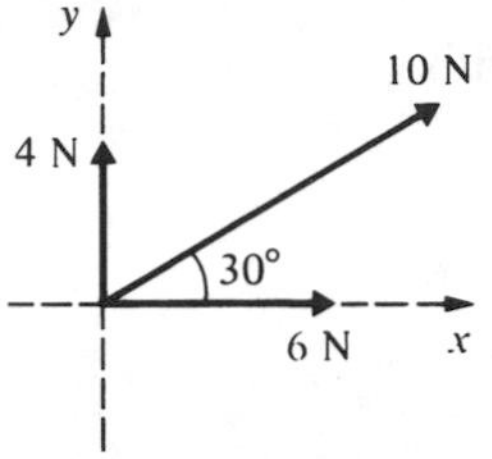

(b)

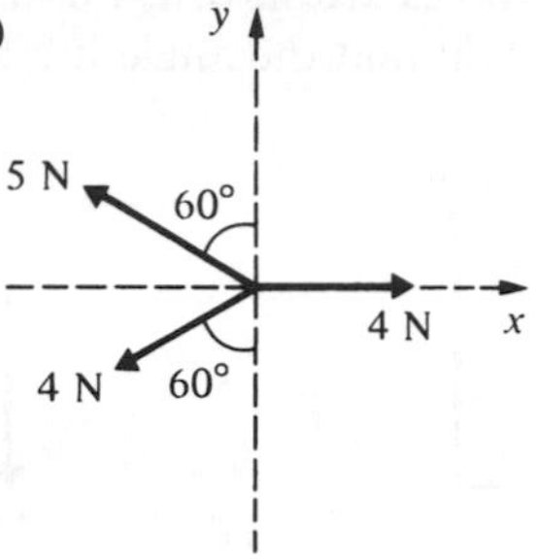

(c)

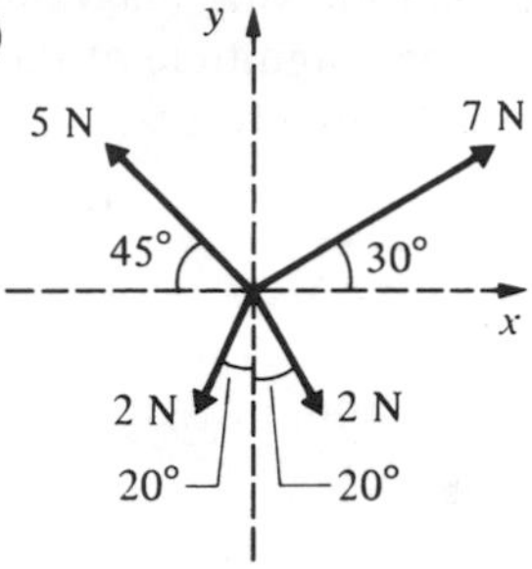

13. Find, by drawing, the magnitude and direction of the resultant of forces 5 N, 6 N, 3 N and 1 N in directions north, north-east, south-west and west respectively.

14. Find, by drawing, the magnitude and direction of the resultant of forces 5 N, 7 N, 6 N and 4 N acting in directions 050°, 100°, 200° and 310° respectively.

15. ABCD is a square. Forces of 4 N, 3 N, 2 N and 5 N act along the sides AB, BC, CD and AD respectively, in the directions indicated by the order of the letters.
Find, by drawing, the magnitude of the resultant and the angle it makes with AB.

16. ABC is an equilateral triangle. Forces of 4 N, 4 N and 6 N act along the sides AB, BC and AC respectively, in the directions indicated by the order of the letters.
Find, by drawing, the magnitude of the resultant and the angle it makes with AB.

17. A concrete block is pulled by two horizontal ropes. One rope has a tension of 500 N and is in a direction 050° and the other rope has a tension of 350 N and is in a direction 350°.
Find the magnitude and direction of the resultant pull on the block.

18. A body of mass 5 kg is being raised by forces of 75 N and 50 N as shown in the diagram. Find, by drawing, the magnitude of the resultant of the three forces acting on the body, and find the angle this resultant makes with the upward vertical.

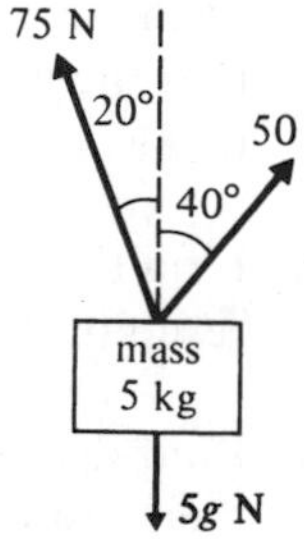

19. Two forces have magnitudes P and Q and the angle between them is θ.
If the resultant of these two forces has magnitude R, and makes an angle α with the force P, show that:

(a) $R^2 = P^2 + Q^2 + 2PQ\cos\theta$ (b) $\tan\alpha = \dfrac{Q\sin\theta}{P + Q\cos\theta}$.

If $P = Q$ and $\theta = 40°$, find α.

Components

It has been seen that two forces can be combined into a single force which is called their resultant.
There is the reverse process which consists of expressing a single force in terms of its two *components*. These components are sometimes referred to as the *resolved parts* of the force.
It is particularly useful to find two mutually perpendicular components of a force.
The direction of the two components may, for example, be horizontal and vertical, or parallel and at right angles to the surface of an inclined plane.

Definition

The component of the force F in any given direction is a measure of the effect of the force F in that direction.

Suppose the force F acts at an angle θ to the x-axis as shown in the diagram. Let ON represent the force F and the angle NXO = 90°. Then OX and OY represent the horizontal and vertical components of F, along the x and y axes.

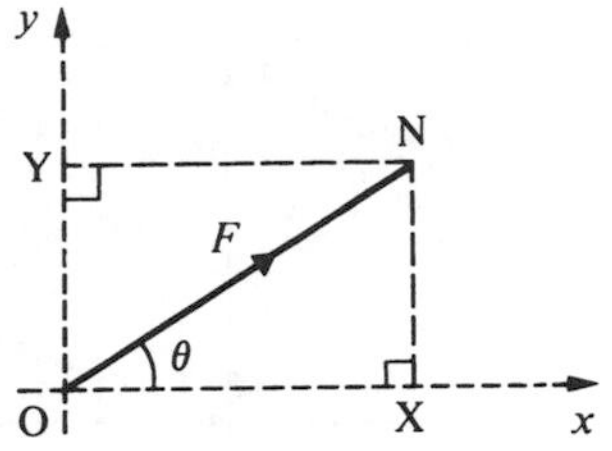

But OX = ON cos $\hat{NOX}$ and OY = ON cos $\hat{NOY}$

$$OX = ON\cos \hat{NOX} = ON\cos\theta = F\cos\theta$$

$$OY = ON\cos \hat{NOY} = ON\cos(90-\theta) = ON\sin\theta = F\sin\theta$$

Hence the components are $F\cos\theta$ and $F\sin\theta$ along the x and y axes respectively.
The rule for finding the components may be stated as:

> The component of a force in any direction is the product of the magnitude of the force and the cosine of the angle between the force and the required direction.

The components in two mutually perpendicular directions are then always $F\cos\theta$ and $F\cos(90-\theta)$ or as these are more usually written: $F\cos\theta$ and $F\sin\theta$.

It is important to remember that, when a force F has been resolved into its components in two mutually perpendicular directions, the force F is the resultant of these two components.

Example 7

Find the components of the given forces, in the direction of:
(i) the x-axis
(ii) the y-axis.

(a) (i) component along the x-axis $= 5 \times \cos 35° = 4{\cdot}10$ N

(ii) component along the y-axis $= 5 \times \cos(90° - 35°) = 5\cos 55°$ or $5\sin 35° = 2{\cdot}87$ N

(a)
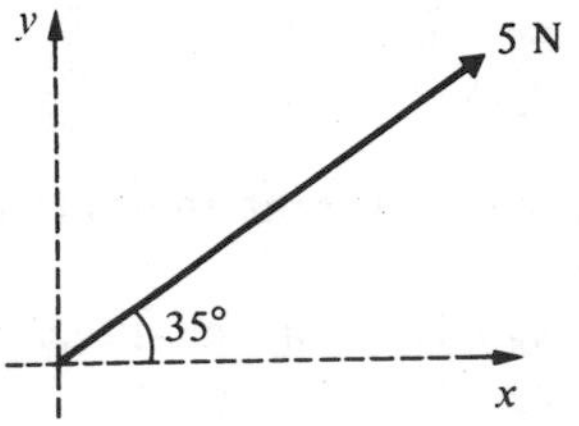

(b) (i) component along the x-axis $= 15 \times \cos 0°$
$= 15 \times 1$
$= 15\,\text{N}$

(ii) component along the y-axis $= 15 \times \cos 90°$
$= 15 \times 0$
$= 0$

(b)

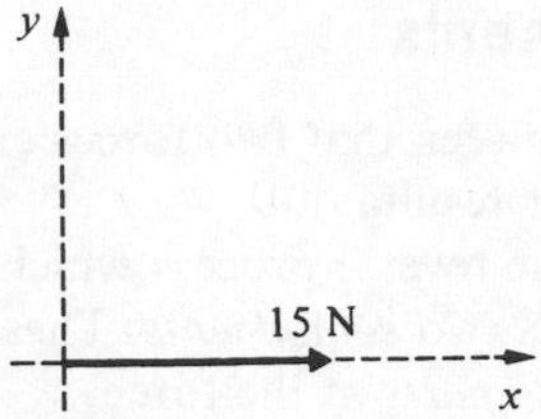

It is seen in this last case that the component in the direction of the y-axis is zero. This agrees with our experience that a force has no effect in a direction at right angles to its line of action. Since the force acts along the x-axis, its component in that direction will be equal to the whole force, 15 N.

Example 8

Express each of the following forces in the form $(a\mathbf{i} + b\mathbf{j})$.

(a)

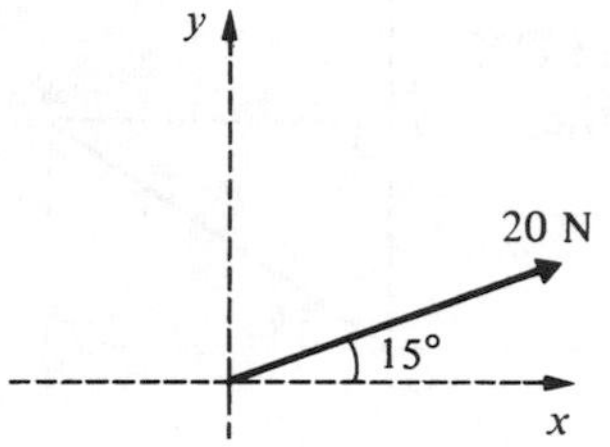

Remembering that **i** and **j** are the unit vectors in the directions of the x-axis and y-axis respectively, we obtain:

(a) component along x-axis $= 20 \cos 15° = 19{\cdot}3$
component along y-axis $= 20 \sin 15° = 5{\cdot}18$
Form required: $(19{\cdot}3\mathbf{i} + 5{\cdot}18\mathbf{j})$ N.

(b)

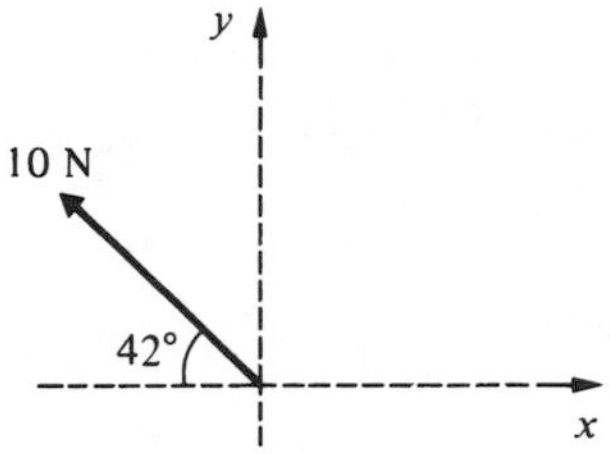

(b) component along x-axis $= -10 \cos 42° = -7{\cdot}43$
component along y-axis $= 10 \sin 42° = 6{\cdot}69$
Form required: $(-7{\cdot}43\mathbf{i} + 6{\cdot}69\mathbf{j})$ N.

Example 9

A body of mass 4 kg rests on an incline of 35°. Find the component of the weight of the body in each of the directions:
(i) down the plane
(ii) at right angles to the plane.

(i) component down the plane
$= \text{force} \times \cos\,(\text{angle between force and plane})$
$= 4g \times \cos\,(90° - 35°)$
$= 4g \times \sin 35°$
$= 22{\cdot}5\,\text{N}$

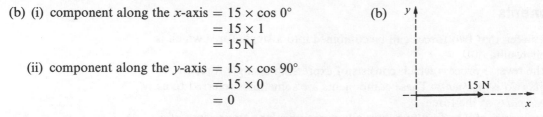

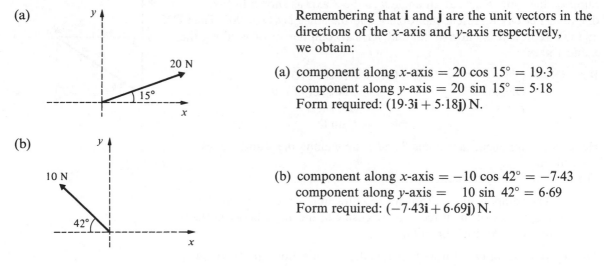

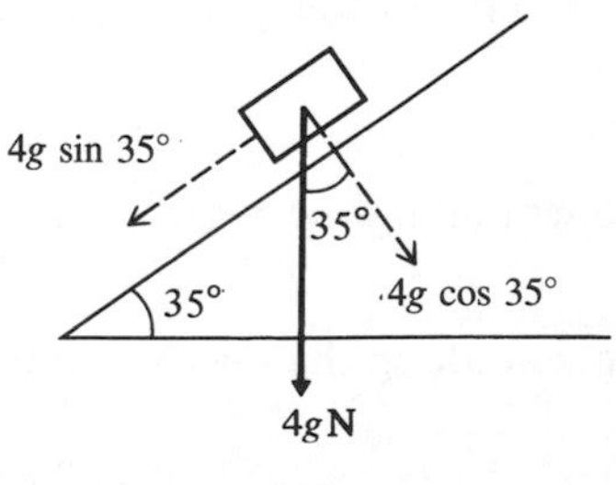

(ii) component at right angles to plane
$= 4g \times \cos 35°$
$= 4g \cos 35°$ in the direction shown in the diagram
$= 32{\cdot}1\,\text{N}$

Example 10

Find the sum of the components of the given forces in the direction of:
(i) the x-axis (ii) the y-axis.

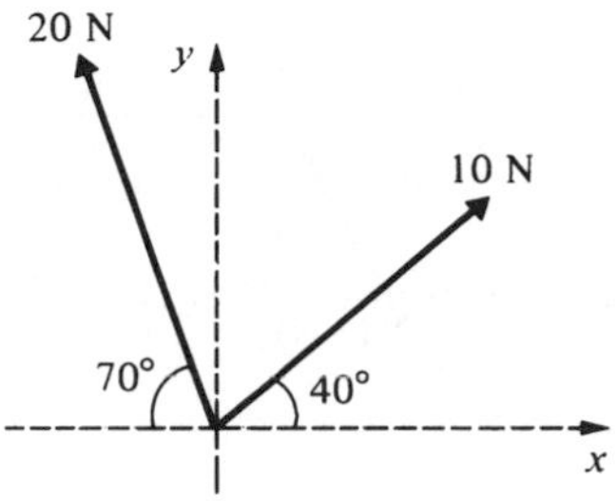

(i) Resolving along the x-axis gives us:
$10 \cos 40° - 20 \cos 70° = 7{\cdot}66 - 6{\cdot}84$
$= 0{\cdot}82\,\text{N}$

(ii) Resolving along the y-axis gives us:
$10 \sin 40° + 20 \sin 70° = 6{\cdot}43 + 18{\cdot}79$
$= 25{\cdot}22\,\text{N}$

This example illustrates that when there are a number of forces acting, their components in a particular direction can be added together, due regard being given to the directions of the components.

Exercise 4B

1. For each of the forces shown below, find the components in the direction of:
(i) the x-axis and (ii) the y-axis.

(a)
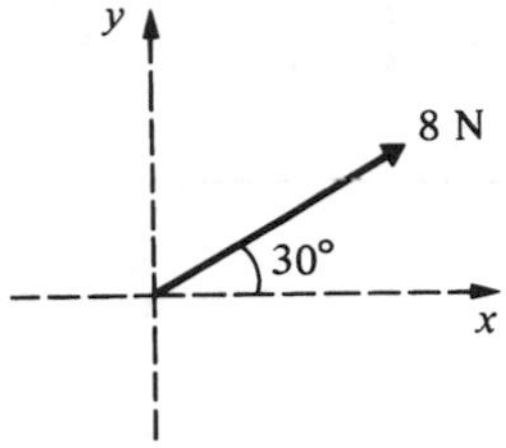

(b)
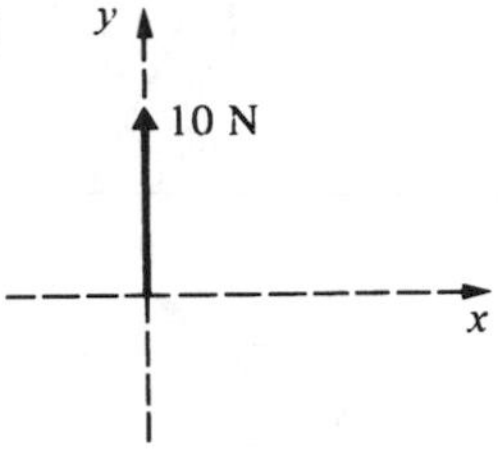

(c)
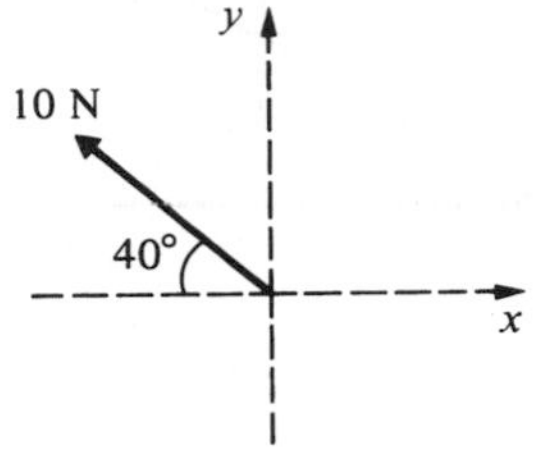

(d)
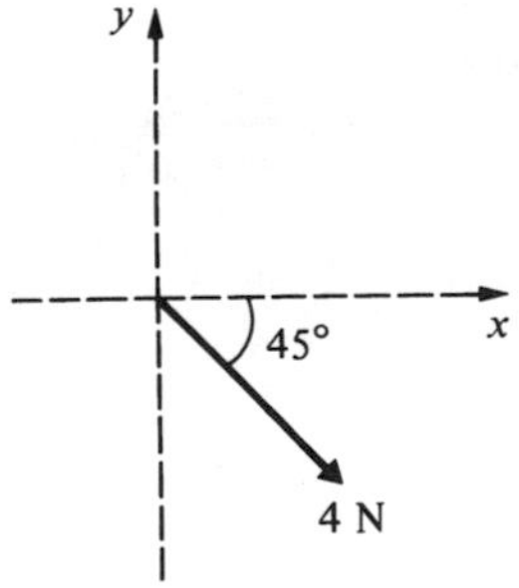

(e)
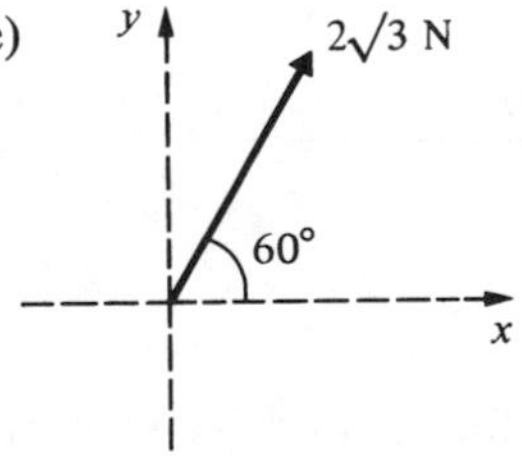

(f)
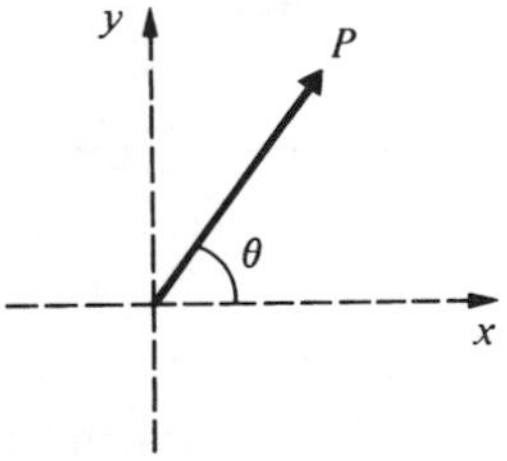

2. Express each of the following forces in the form $a\mathbf{i} + b\mathbf{j}$.

(a)
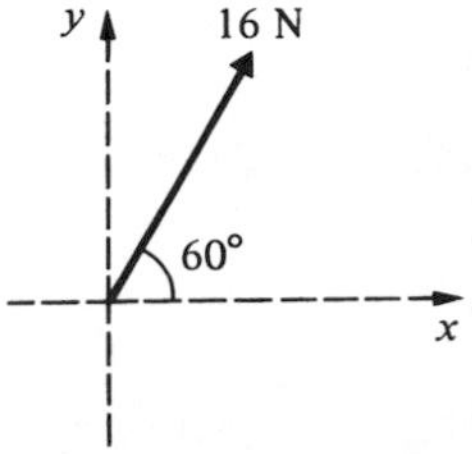

(b)
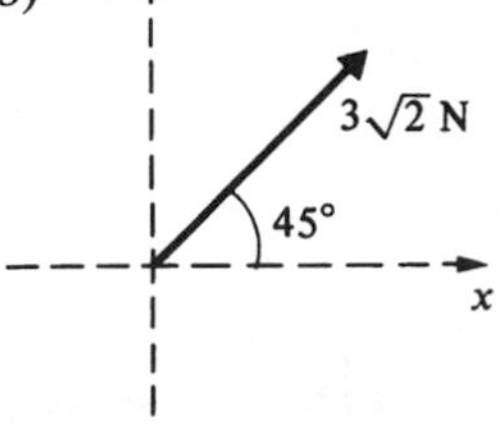

(c)
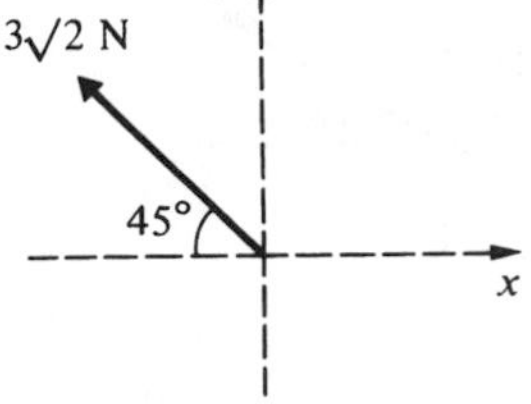

(d)

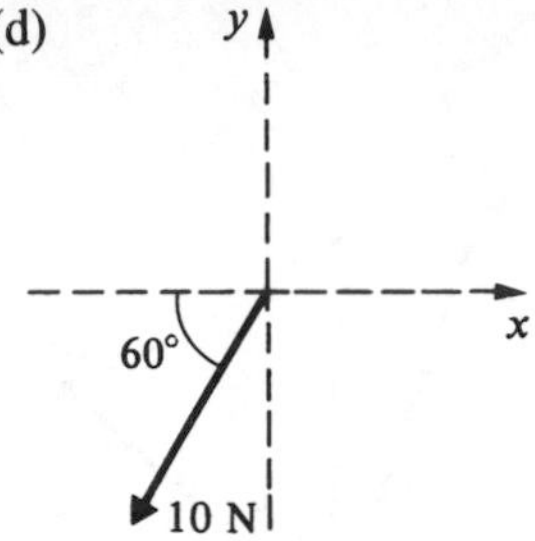

(e)

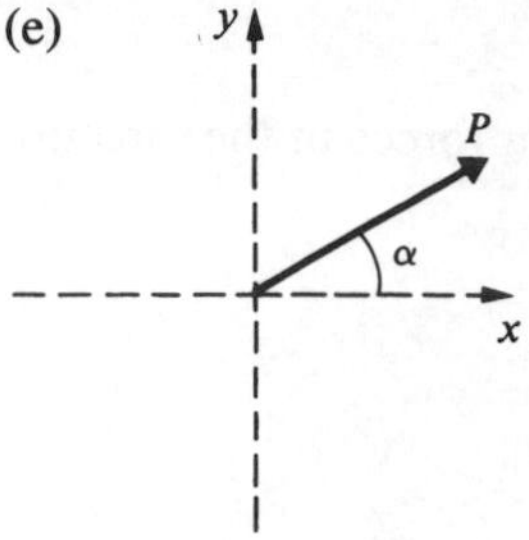

(f)

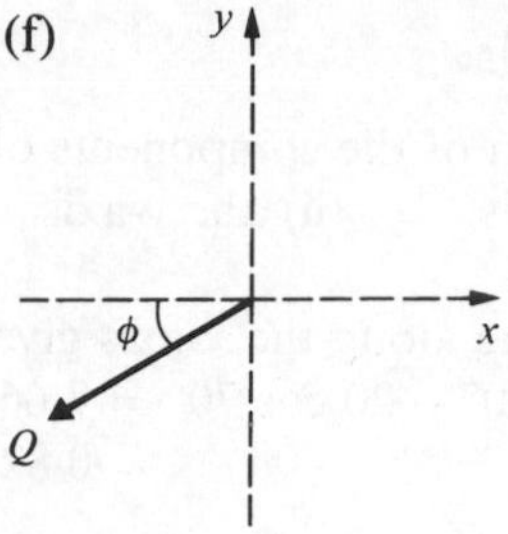

3. Each of the following diagrams shows a body of weight 10 N on an incline. In each case find the component of the weight of the body: (i) in the Ox direction and (ii) in the Oy direction.

(a)

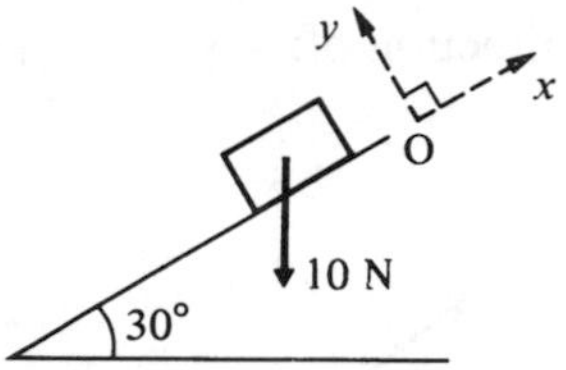

(b)

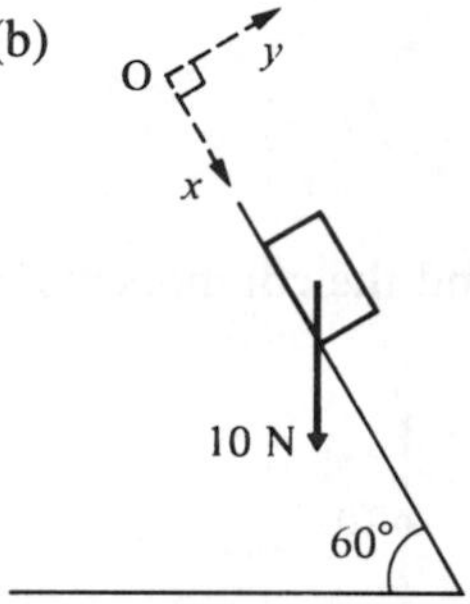

(c)

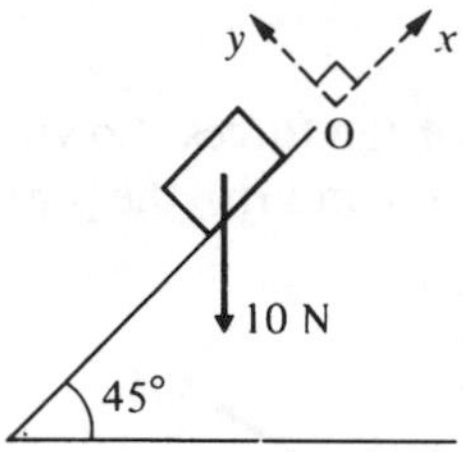

(d)

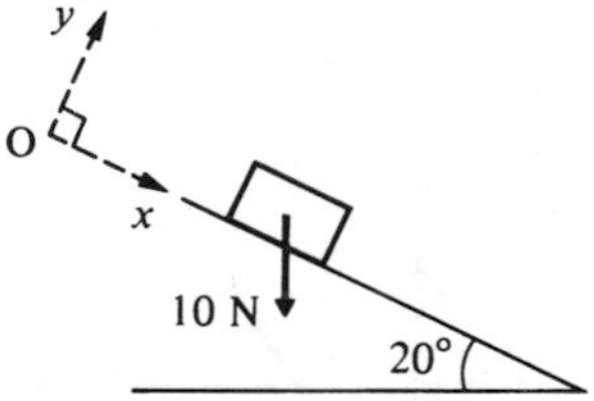

(e)

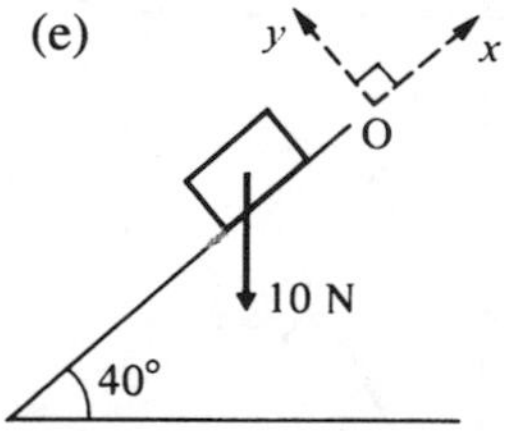

(f)

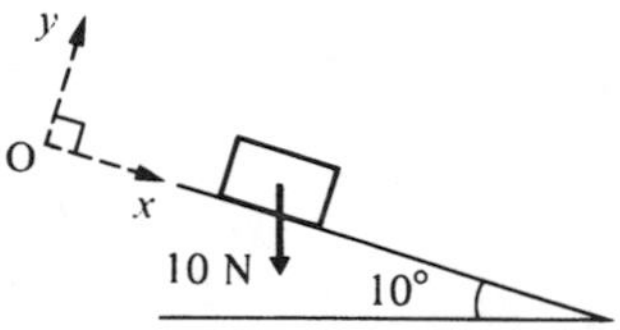

4. For each of the following systems of forces, find the sum of the components in the direction of: (i) the x-axis and (ii) the y-axis.

(a)

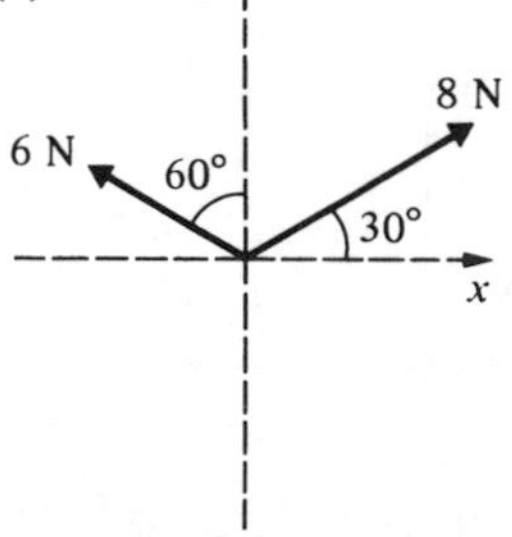

(b)

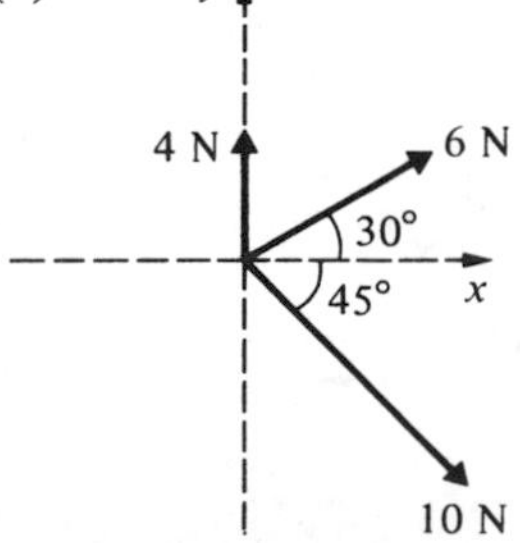

(c)

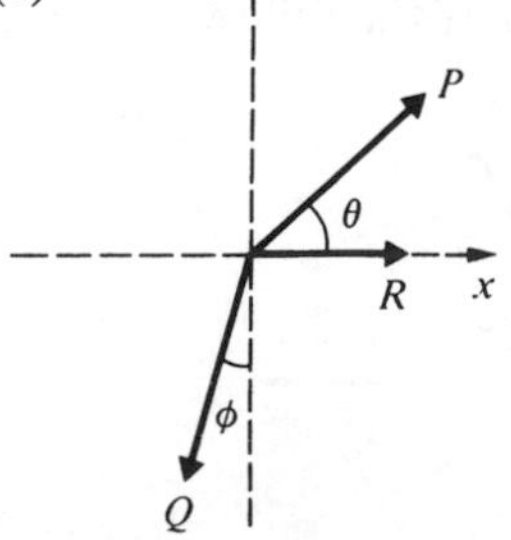

5. For each of the following systems of forces, find the sum of the components: (i) in the Ox direction and (ii) in the Oy direction.

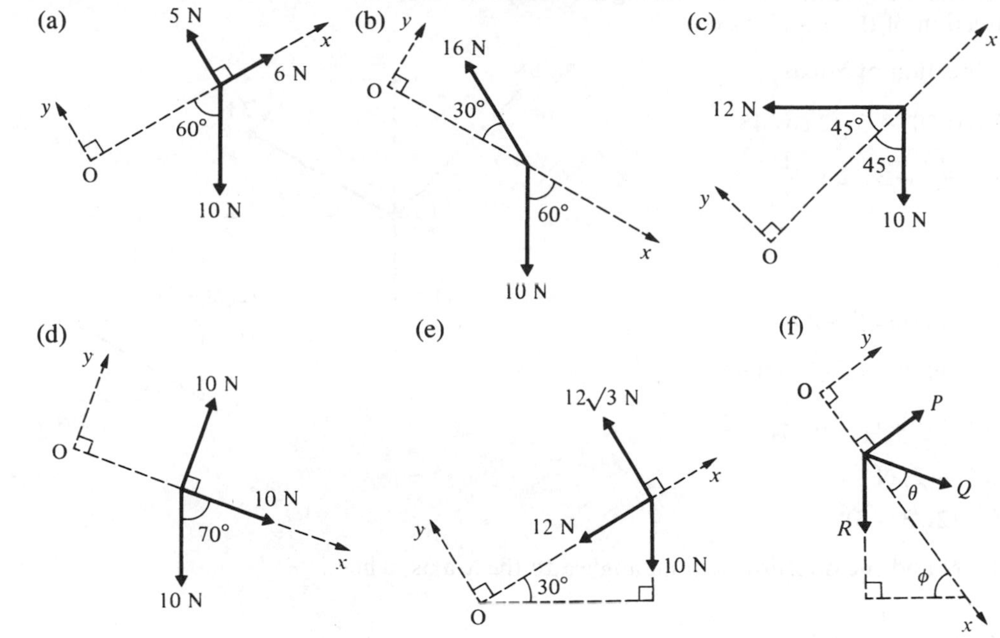

Resultant from sum of components

A number of forces, all of which lie in one plane, are said to be coplanar.

In a given system of coplanar forces, it is possible to choose two mutually perpendicular directions and find the components of all the forces in these two directions. By finding the algebraic sum of these components the resultant of the system can be found in both magnitude and direction.

If the forces are expressed in terms of the unit vectors **i** and **j**, the components of the forces are immediately known.

Example 11

Find the resultant of the following forces, giving the answer in the form $a\mathbf{i} + b\mathbf{j}$:

$$(2\mathbf{i} + 4\mathbf{j})\,\text{N},\ (5\mathbf{i} - 7\mathbf{j})\,\text{N and } (-2\mathbf{i} - \mathbf{j})\,\text{N}.$$

resultant $= 2\mathbf{i} + 4\mathbf{j} + 5\mathbf{i} - 7\mathbf{j} - 2\mathbf{i} - \mathbf{j}$

$= (5\mathbf{i} - 4\mathbf{j})\,\text{N}$

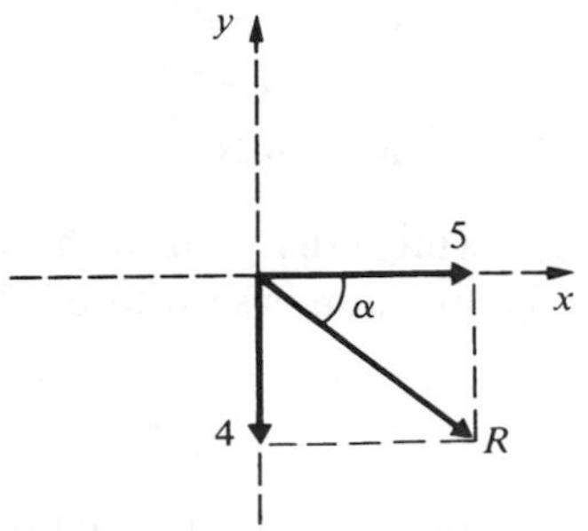

If the magnitude of the resultant is required:

magnitude $= \sqrt{(5^2 + 4^2)} = \sqrt{41} = 6{\cdot}40\,\text{N}$

The direction is given by $\tan \alpha = \frac{4}{5}$ where α is the angle below the x-axis.

Hence $\alpha = 38{\cdot}66°$, so the angle with the x-axis is $-38{\cdot}66°$

The resultant is $(5\mathbf{i} - 4\mathbf{j})$ N at an angle of 38·66° below the x-axis.

Example 12

Find the resultant of the given forces, by finding the components of the forces in the direction of the x and y axes.

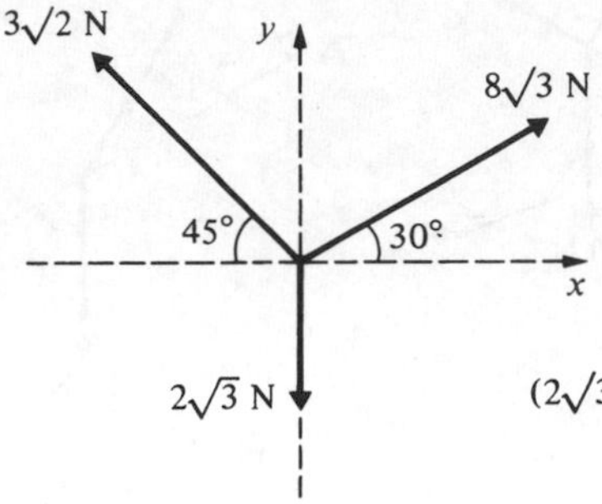

components in direction of x-axis

$$= 8\sqrt{3}\cos 30^\circ - 3\sqrt{2}\cos 45^\circ$$
$$= 8\sqrt{3} \times \frac{\sqrt{3}}{2} - 3\sqrt{2} \times \frac{1}{\sqrt{2}}$$
$$= 12 - 3$$
$$= 9\,\text{N}$$

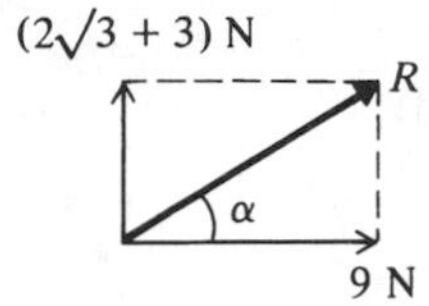

components in direction of y-axis

$$= 8\sqrt{3}\sin 30^\circ + 3\sqrt{2}\sin 45^\circ - 2\sqrt{3}$$
$$= 8\sqrt{3} \times \tfrac{1}{2} + 3\sqrt{2} \times \frac{1}{\sqrt{2}} - 2\sqrt{3}$$
$$= (2\sqrt{3} + 3)\,\text{N}$$

$\therefore \quad R^2 = 9^2 + (2\sqrt{3} + 3)^2$

$\therefore \quad R = 11{\cdot}1\,\text{N}$ and the direction is at an angle α to the x-axis, where:

$$\tan\alpha = \frac{2\sqrt{3} + 3}{9}$$

$\therefore \quad \alpha = 35{\cdot}69^\circ$

The resultant is $11{\cdot}1\,\text{N}$ at an angle of $35{\cdot}69^\circ$ above the x-axis.

Example 13

ABCD is a rectangle. Forces of 9 N, 8 N and 3 N act along the lines DC, CB and BA respectively, in the directions indicated by the order of the letters. Find the magnitude of the resultant and the angle it makes with DC.

Draw a diagram showing the forces.

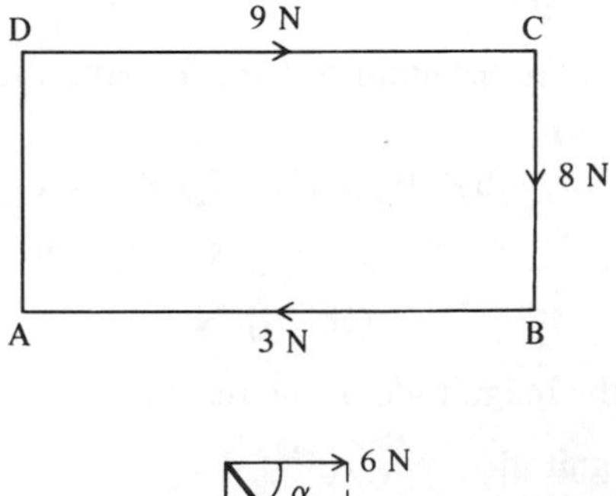

Resolving parallel to DC gives horizontal component $= 9 - 3$
$= 6\,\text{N}$

Resolving parallel to CB gives vertical component $= 8\,\text{N}$

$\therefore \quad R^2 = 6^2 + 8^2$
$\quad\quad = 36 + 64$

$\therefore \quad R = 10\,\text{N}$

Drawing a diagram to show the two components the direction is seen to be given by:

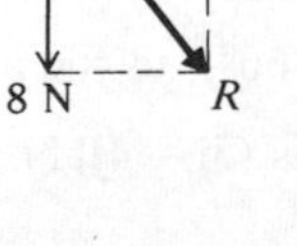

$\tan\alpha = \frac{8}{6}$

$\therefore \quad \alpha = 53.13^\circ$

The resultant is 10 N making an angle $53{\cdot}13^\circ$ with DC.

Exercise 4C

1. Find the resultant of each of the following sets of forces giving your answers in the form $a\mathbf{i} + b\mathbf{j}$:
(a) $(2\mathbf{i} + 4\mathbf{j})$ N, $(3\mathbf{i} - 4\mathbf{j})$ N
(b) $(3\mathbf{i} - 5\mathbf{j})$ N, $(2\mathbf{i} - 5\mathbf{j})$ N, $(-\mathbf{i} + 7\mathbf{j})$ N
(c) $(6\mathbf{i} + 2\mathbf{j})$ N, $(-5\mathbf{i} + \mathbf{j})$ N, $(3\mathbf{i} - 3\mathbf{j})$ N
(d) $(2\mathbf{i} + 4\mathbf{j})$ N, $(3\mathbf{i} - 5\mathbf{j})$ N, $(6\mathbf{i} + 2\mathbf{j})$ N, $(-7\mathbf{i} - 7\mathbf{j})$ N.
2. Find the resultant of each of the following sets of forces giving your answers in the form $a\mathbf{i} + b\mathbf{j} + c\mathbf{k}$:
(a) $(2\mathbf{i} + 3\mathbf{j} + 3\mathbf{k})$ N, $(2\mathbf{i} + 4\mathbf{j} - 8\mathbf{k})$ N
(b) $(7\mathbf{i} - 4\mathbf{j} + 3\mathbf{k})$ N, $(5\mathbf{i} - 2\mathbf{j} + 8\mathbf{k})$ N, $(\mathbf{i} - \mathbf{k})$ N
(c) $(2\mathbf{i} + 3\mathbf{j} - 7\mathbf{k})$ N, $(2\mathbf{i} + 5\mathbf{k})$ N, $(3\mathbf{j} + 4\mathbf{k})$ N.
3. The resultant of the forces $(5\mathbf{i} - 2\mathbf{j})$ N, $(7\mathbf{i} + 4\mathbf{j})$ N, $(a\mathbf{i} + b\mathbf{j})$ N and $(-3\mathbf{i} + 2\mathbf{j})$ N is a force $(5\mathbf{i} + 5\mathbf{j})$ N.
Find a and b.
4. The resultant of the forces $(5\mathbf{i} + 7\mathbf{j})$ N, $(a\mathbf{i} + b\mathbf{j})$ N and $(b\mathbf{i} - a\mathbf{j})$ N is a force $(11\mathbf{i} + 5\mathbf{j})$ N.
Find a and b.
5. The resultant of the forces $(\mathbf{i} - 2\mathbf{j} + 2a\mathbf{k})$ N, $(2\mathbf{i} + \mathbf{j} + 4\mathbf{k})$ N, $(b\mathbf{i} + 2\mathbf{j})$ N is $(8\mathbf{i} + c\mathbf{j} + 14\mathbf{k})$ N.
Find a, b and c.
6. Find the magnitude of the force $(4\mathbf{i} + 3\mathbf{j})$ N and the angle it makes with the direction of $\mathbf{i}$.
7. Find the magnitude of the force $(-2\mathbf{i} + 4\mathbf{j})$ N and the angle it makes with the direction of $\mathbf{i}$.
8. Find the magnitude of the force $(3\mathbf{i} + 4\mathbf{j} - 5\mathbf{k})$ N and determine the angle it makes with the direction of $\mathbf{i}$.
9. Find the magnitude of the resultant of each of the following sets of forces and state the angle that this resultant makes with the direction of $\mathbf{i}$:
(a) $(2\mathbf{i} + 3\mathbf{j})$ N, $(5\mathbf{i} - 2\mathbf{j})$ N, $(-3\mathbf{i} + 3\mathbf{j})$ N
(b) $(-2\mathbf{i} + 5\mathbf{j})$ N, $(\mathbf{i} + 2\mathbf{j})$ N
(c) $(4\mathbf{i} + 3\mathbf{j})$ N, $(-\mathbf{i} - 5\mathbf{j})$ N
(d) $(2\mathbf{i} + 4\mathbf{j})$ N, $(-6\mathbf{i} - 5\mathbf{j})$ N, $(2\mathbf{i} + \mathbf{j})$ N.
10. Find the magnitude of the resultant of the following set of forces and determine the angle this resultant makes with the direction of $\mathbf{k}$:
$(3\mathbf{i} + \mathbf{j} - \mathbf{k})$ N $(-4\mathbf{i} + \mathbf{j} + 4\mathbf{k})$ N $(3\mathbf{i} + \mathbf{j} + 3\mathbf{k})$ N.
11. For each of the following systems of forces find the resultant in the form $a\mathbf{i} + b\mathbf{j}$. Hence find the magnitude of the resultant and the angle it makes with the x-axis.

(a)
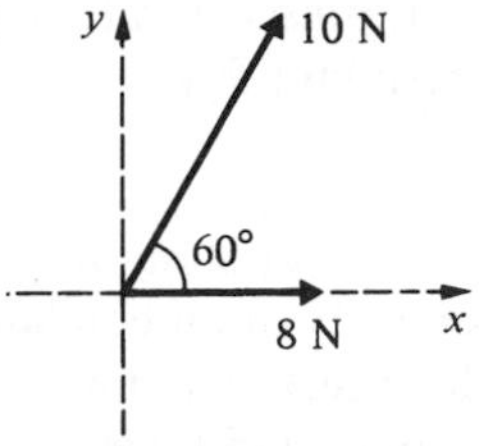

(b)
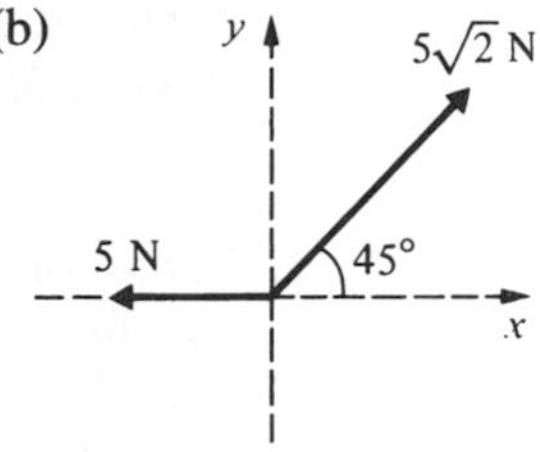

(c)
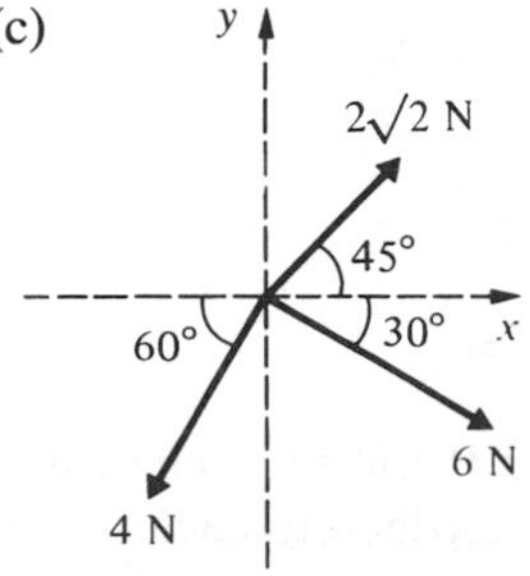

12. A sledge is being pulled across a horizontal surface by forces of $(6\mathbf{i} + 2\mathbf{j})$ N and $(4\mathbf{i} - 3\mathbf{j})$ N. What is the magnitude of the resultant pull on the sledge and what angle does this resultant make with the direction of $\mathbf{i}$?

13. Find, by calculation, the resultant of forces of 5 N, 7 N, 8 N and 5 N acting in directions north, north-east, west and north-west respectively, giving your answer in the form $a\mathbf{i} + b\mathbf{j}$. (Take $\mathbf{i}$ as a unit vector due east and $\mathbf{j}$ as a unit vector due north.)

14. Find, by calculation, the magnitude and direction of the resultant of forces of 10 N, 15 N and 8 N acting in directions 030°, 150° and 225° respectively.

15. ABCD is a rectangle. Forces of 3 N, 4 N and 1 N act along AB, BC and DC respectively, in the directions indicated by the order of the letters. By resolving in two mutually perpendicular directions find the magnitude of the resultant and the angle it makes with AB.

16. ABCD is a rectangle. Forces of $6\sqrt{3}$ N, 2 N and $4\sqrt{3}$ N act along AB, CB and CD respectively, in the directions indicated by the order of the letters.
By resolving in two mutually perpendicular directions find the magnitude of the resultant and the angle it makes with AB.

17. ABCD is a rectangle. Forces of 8 N, 4 N, 10 N and 2 N act along AB, CB, CD and AD respectively, in the directions indicated by the order of the letters.
Find the magnitude and direction of the resultant.

18. ABC is an equilateral triangle. Forces of 12 N, 10 N and 10 N act along AB, BC and CA respectively, the direction of the forces being indicated by the order of the letters.
Find the magnitude and direction of the resultant.

19. ABCD is a rectangle with AB = 4 m and BC = 3 m. Forces of 3 N, 1 N and 10 N act along AB, DC and AC respectively, in the directions indicated by the order of the letters.
Find the magnitude and direction of the resultant.

20. ABC is an equilateral triangle. Forces of 10 N act along AB, BC and AC in the directions indicated by the order of the letters.
Find the magnitude of the resultant and the angle it makes with AB.

Exercise 4D Examination Questions

1.

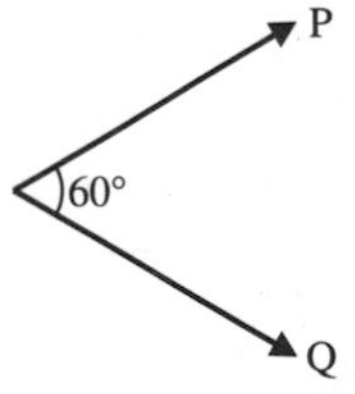

Two forces **P** and **Q**, each of magnitude 100 N, are inclined to each other at an angle of 60°, and act on an object. Find the magnitude of the resultant **P** + **Q**. (UCLES)

2. Forces of 5 N, 9 N, 7 N act along the sides AB, BC, CA respectively of an equilateral triangle ABC in the directions indicated by the order of the letters. Find their resultant in magnitude and direction. (SUJB)

3.

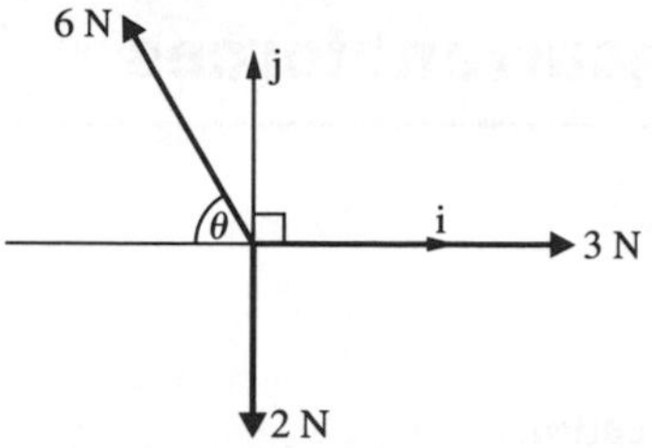

Forces of magnitude 3 N, 6 N and 2 N act at a point as shown in the above diagram. Given that $\theta = 60°$, show that the component of the resultant force in the **i** direction is zero. Calculate the magnitude of the resultant force and state its direction. (UCLES)

4. Two forces $(3\mathbf{i} + 2\mathbf{j})$ N and $(-5\mathbf{i} + \mathbf{j})$ N act at a point. Find the magnitude of the resultant of these forces and determine the angle which the resultant makes with the unit vector **i**. (AEB 1990)

5.

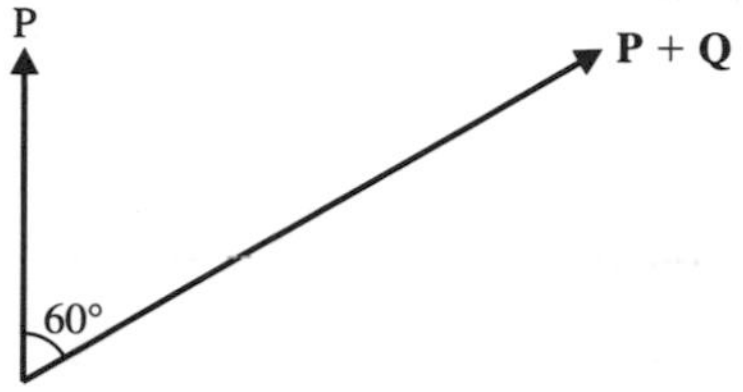

A particle is free to move on a horizontal table. It is acted on by constant forces **P** and **Q**. The force **P** has magnitude 5 N and acts due north. The resultant **P** + **Q** has magnitude 9 N and acts in the direction 060°. Calculate the magnitude and direction of **Q**. (UCLES)

6. Two forces, P and Q, are such that the sum of their magnitudes is 45 N. The resultant of P and Q is perpendicular to P and has a magnitude of 15 N. Calculate
(i) the magnitude of P and of Q,
(ii) the angle between P and Q. (UCLES)

7. A force **F** has magnitude 50 N and acts in the direction of the vector $24\mathbf{i} + 7\mathbf{j}$. Show that $\mathbf{F} = (48\mathbf{i} + 14\mathbf{j})$ N.
Two forces $\mathbf{F}_1$ and $\mathbf{F}_2$ have magnitudes α N and β N and act in the directions $\mathbf{i} - 2\mathbf{j}$ and $4\mathbf{i} + 3\mathbf{j}$ respectively. Given that the resultant of $\mathbf{F}_1$ and $\mathbf{F}_2$ is **F**, show that $\alpha = 8\sqrt{5}$ and find β. (AEB 1993)

8. Two forces, P N and Q N, are inclined at an angle θ to each other. When $P = \sqrt{12}$ and $Q = 2$, the resultant has the same magnitude, R N, as the resultant in the case when $P = \sqrt{12}$ and $Q = 4$.
Find the value of θ and of R. (UCLES)

5 Equilibrium and acceleration under concurrent forces

In Chapter 3 it was found that:

(i) if a body is not moving, then the resultant force acting in any direction must be zero,

(ii) if a body is accelerating, then the relationship $F = ma$ applies.

In this chapter these facts are now used, together with the skills of combining and resolving forces which were acquired in Chapter 4.

Terminology

Particle. A particle is that portion of matter which is so small in size that the distance between its extremities may be neglected.

Rigid body. A rigid body is, on the other hand, one in which the distances between its various parts are not negligible, and these distances remain fixed.

Equilibrium. A state of equilibrium is said to exist when two or more forces act upon a particle, or upon a rigid body, and motion does not take place.

Triangle of forces

It has already been seen in Chapter 4 that the resultant of two forces, acting at a point, can be found both graphically and by calculation.

The two forces P and Q are represented by the line segments AB and AD and the parallelogram ABCD is completed. The diagonal AC then fully represents, in magnitude and direction, the resultant R of the two forces P and Q.

Thus $\qquad \overrightarrow{AB} + \overrightarrow{AD} = \overrightarrow{AC}$

$\therefore \qquad \overrightarrow{AB} + \overrightarrow{BC} = \overrightarrow{AC}$ since $\overrightarrow{AD}$ and $\overrightarrow{BC}$ are equivalent vectors

$\therefore \qquad \overrightarrow{AB} + \overrightarrow{BC} - \overrightarrow{AC} = 0$

i.e. $\qquad P + Q - R = 0$

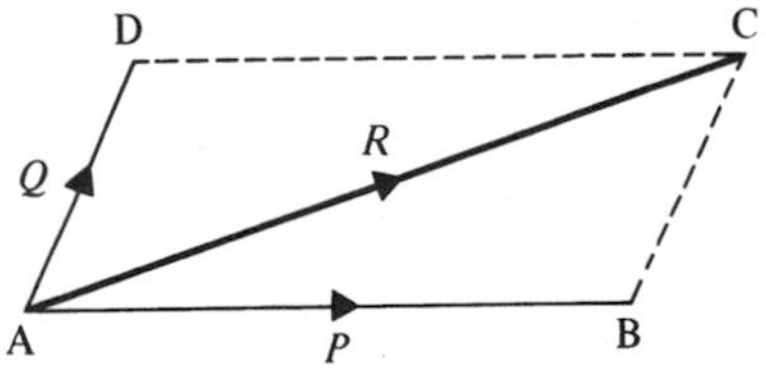

Hence, if R is the resultant of the forces P and Q, then $-R$ added to the forces P and Q will produce equilibrium.

This means that, if three forces acting at a point can be represented by the sides of a triangle, and the forces all act in the same sense around the triangle, then these forces are in equilibrium.

The triangle DEF, shown below, is said to be a *triangle of forces* for the three forces X, Y and Z.

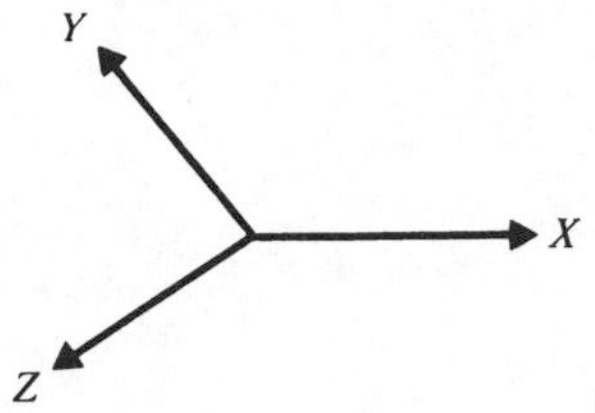

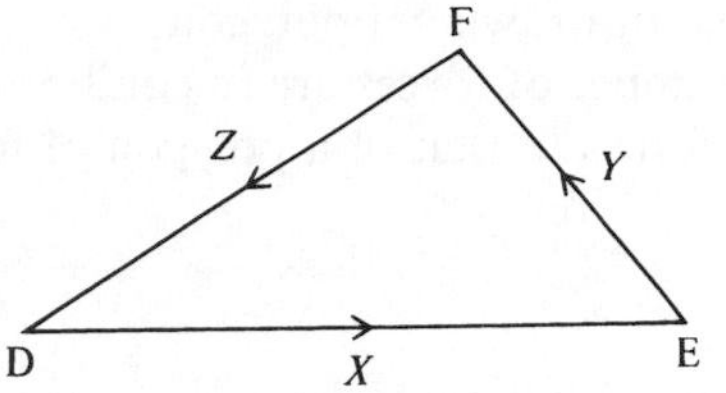

The significance of the different directions of the arrows on the side AC in the diagrams below should be carefully noted.

One force the resultant of two forces:

$$\overrightarrow{AB} + \overrightarrow{BC} = \overrightarrow{AC}$$

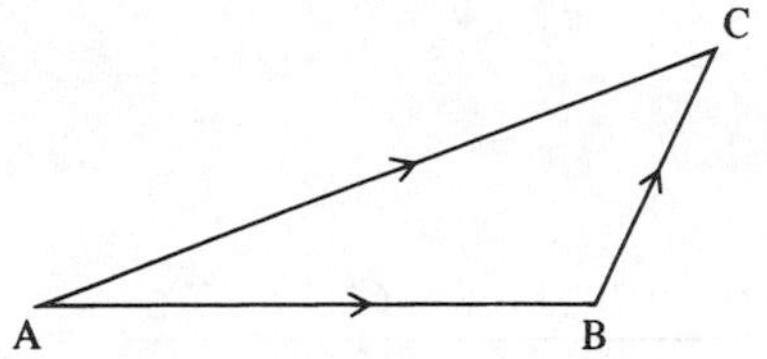

Three forces in equilibrium:

$$\overrightarrow{AB} + \overrightarrow{BC} + \overrightarrow{CA} = 0$$

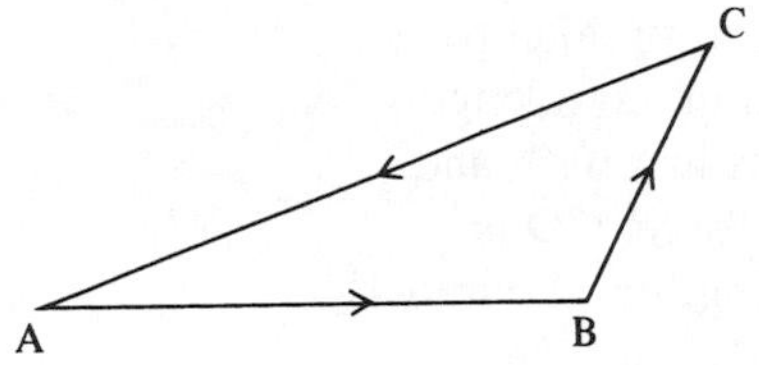

The converse of the triangle of forces is also true:

If three forces acting at a point are in equilibrium, they can be represented by the sides of a triangle.

It should be carefully noted that the directions of the forces must be parallel to the sides of the triangle, and such that the arrows on the sides of the triangle indicating the directions of the forces are all in the same sense.

Example 1

Given that the three forces shown in the diagram are in equilibrium, find, by scale drawing, the magnitude of S and θ.

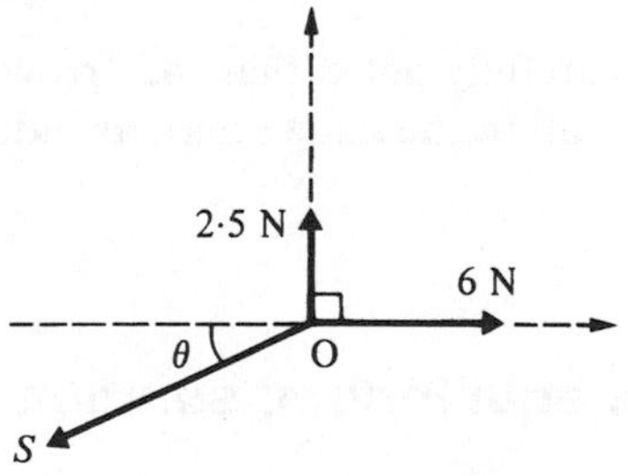

Draw a line OA, 6 cm in length, parallel to the 6 N force.
Draw a line AB, 2·5 cm in length, parallel to the 2·5 N force. Join BO.

By measurement BO = 6·5 cm and angle AOB = 23°.

The force S is 6·5 N and $\theta = 23°$.

It will be noted that this involves the same process as was used in Chapter 4 to find the *resultant* of two forces.

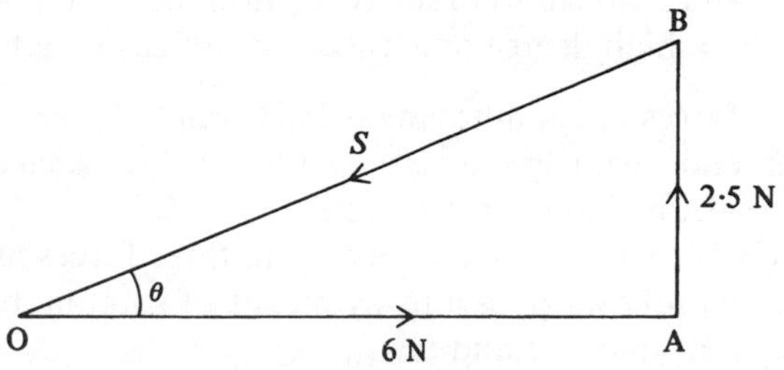

Polygon of forces

The resultant of a number of forces has already been found by extending the idea of drawing a triangle to that of drawing a polygon.
In the same way, given that a number of forces are in equilibrium, we can extend the idea of a triangle of forces to that of a polygon of forces.

Example 2

The forces shown in the diagram are known to be in equilibrium. By drawing a polygon of forces, find the magnitudes of S and θ.

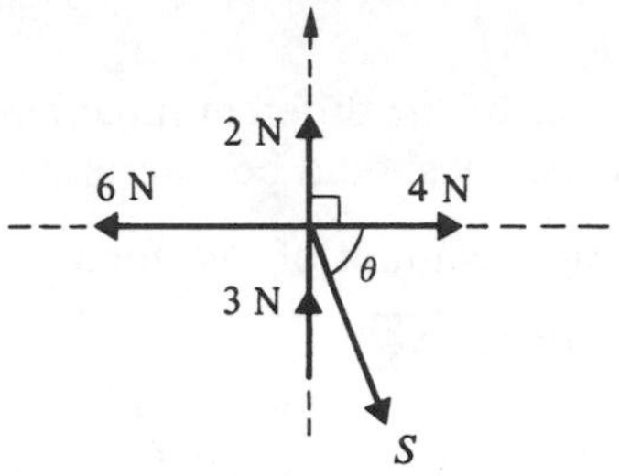

The polygon can be constructed in various ways.

The line AB is drawn parallel to the 2 N force and 2 units in length; BC is drawn 4 units in length and parallel to the 4 N force; CD is drawn 3 units in length and parallel to the 3 N force; DE is drawn 6 units in length and parallel to the 6 N force. The line EA then represents the force needed to produce equilibrium, i.e. the force S. Alternatively, a different polygon is obtained by drawing AW parallel to the 6 N force and 6 units in length, WX 3 units in length, XY 2 units in length and YE 4 units in length. The unknown force is again represented by the line EA required to complete the polygon.

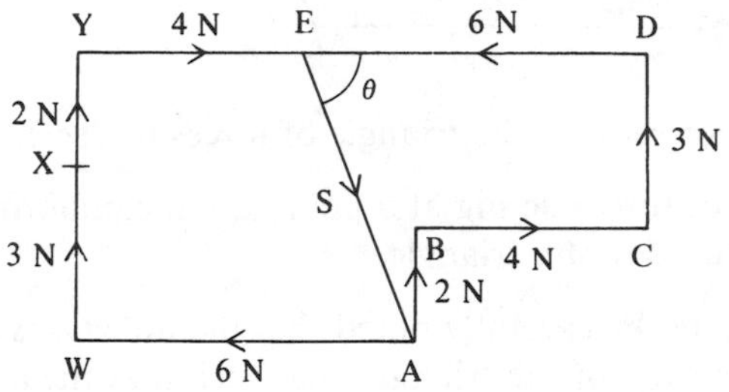

By measurement the force S is 5·4 N and the angle $\theta = 68°$.

Again, it should be carefully noted that the arrows on the sides of the polygon ABCDE are all in the same sense, as indeed they are in polygon AWXYE.

Three forces in equilibrium: solution by calculation

Examples 1 and 2 above were solved by graphical methods and, consequently, a high degree of accuracy is not easily obtained.

Given three forces in equilibrium, as in Example 1, the triangle of forces can be sketched and then trigonometry can be used to calculate the unknown force and angle, as shown in Examples 3 and 4.
Alternatively for some problems involving three forces in equilibrium, the theorem which follows gives a ready means of solution by calculation, as illustrated by Examples 5 and 6.

Lami's Theorem

Suppose the forces, X, Y and Z acting at a point are in equilibrium. The forces can therefore be represented by the sides of a triangle ABC.

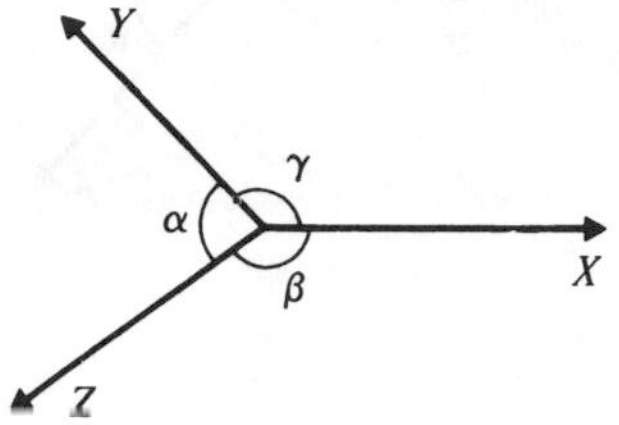

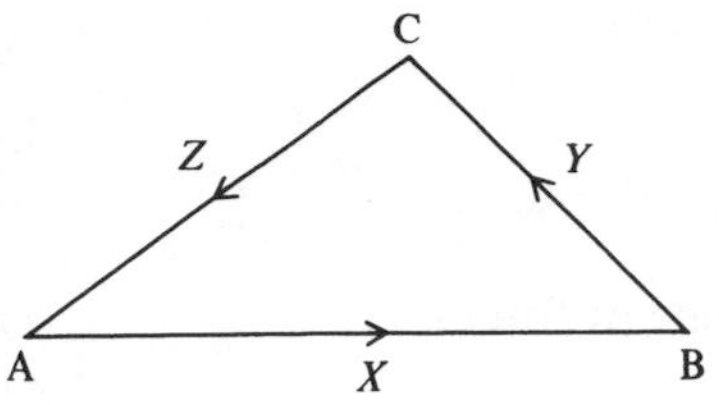

Applying the sine rule to the triangle ABC gives:

$$\frac{AB}{\sin B\hat{C}A} = \frac{BC}{\sin C\hat{A}B} = \frac{CA}{\sin A\hat{B}C}$$

But $\quad B\hat{C}A = 180° - \alpha, \quad C\hat{A}B = 180° - \beta \quad$ and $\quad A\hat{B}C = 180° - \gamma$

Hence $\quad \dfrac{AB}{\sin(180° - \alpha)} = \dfrac{BC}{\sin(180° - \beta)} = \dfrac{CA}{\sin(180° - \gamma)}$

$\therefore \quad \dfrac{AB}{\sin\alpha} = \dfrac{BC}{\sin\beta} = \dfrac{CA}{\sin\gamma}$

or $\quad \dfrac{X}{\sin\alpha} = \dfrac{Y}{\sin\beta} = \dfrac{Z}{\sin\gamma}$

This result, which only applies to three forces acting at a point, is known as Lami's Theorem.

Example 3

Given that the system of forces shown in the diagram is in equilibrium, sketch the triangle of forces and hence calculate the magnitude of the force X and the angle α.

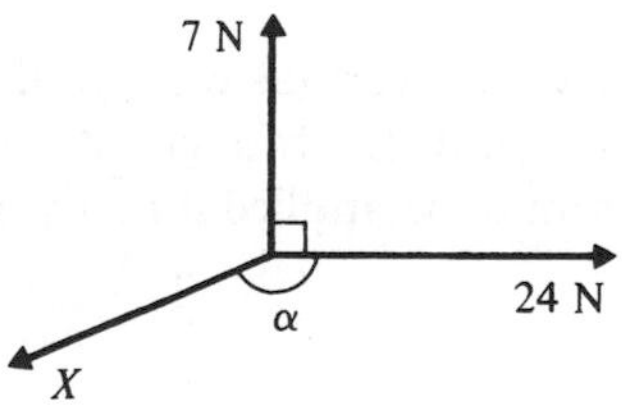

Sketch the triangle of forces.

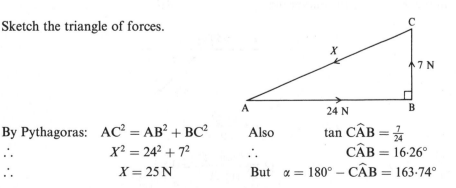

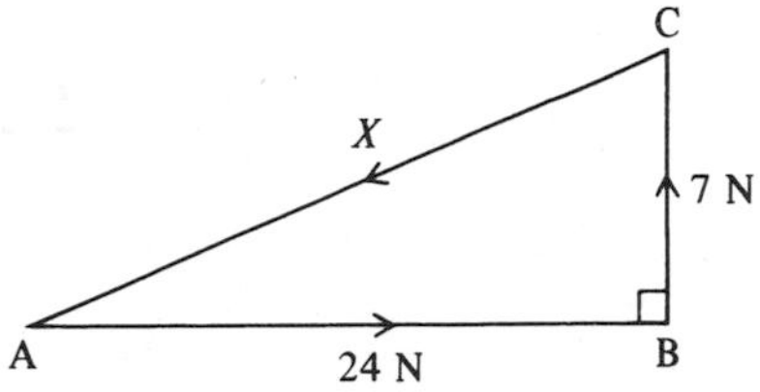

By Pythagoras: $\quad AC^2 = AB^2 + BC^2$

$\therefore \quad X^2 = 24^2 + 7^2$

$\therefore \quad X = 25\,\text{N}$

Also $\quad \tan C\hat{A}B = \frac{7}{24}$

$\therefore \quad C\hat{A}B = 16{\cdot}26°$

But $\quad \alpha = 180° - C\hat{A}B = 163{\cdot}74°$

The force X is 25 N and the angle α is 163·74°.

Example 4

Sketch the triangle of forces for the given system of forces which is in equilibrium. Calculate the magnitude of P and θ.

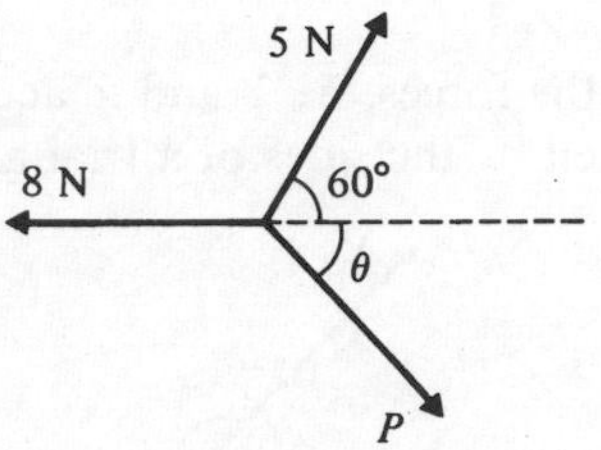

Sketch the triangle of forces.

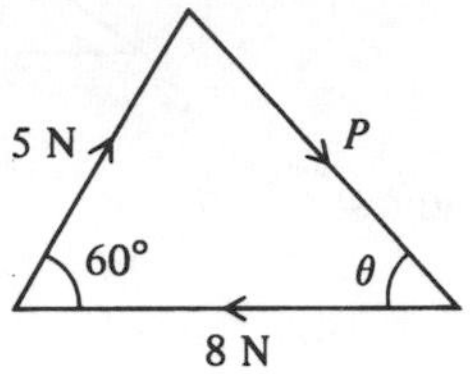

By the cosine rule from this triangle:

$$P^2 = 5^2 + 8^2 - 2 \times 5 \times 8 \times \cos 60°$$
$$= 49$$
$$\therefore \quad P = 7\,\text{N}$$

By the sine rule:

$$\frac{7}{\sin 60°} = \frac{5}{\sin \theta}$$

$$\therefore \quad \sin \theta = \frac{5 \sin 60°}{7}$$

$$\therefore \quad \theta = 38{\cdot}21°$$

The force P is 7·0 N and the angle θ is 38·21°.

It should be noted that various ways of determining the unknowns may be used, once the sketch of the triangle of forces has been made. Alternatively, Lami's Theorem may be applied directly as shown in the following example.

Example 5

The force system shown in the diagram is in equilibrium. Calculate P and Q.

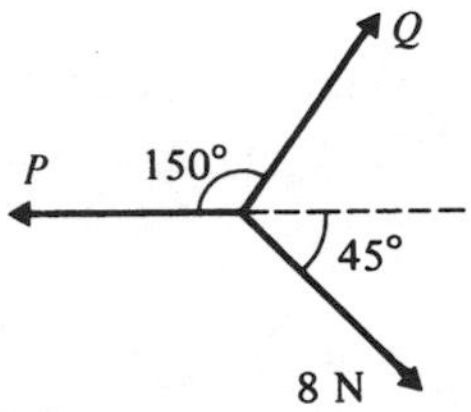

By Lami's Theorem:

$$\frac{8}{\sin 150°} = \frac{P}{\sin 75°} = \frac{Q}{\sin 135°}$$

Thus $\dfrac{8}{\sin 30°} = \dfrac{P}{\sin 75°}$ and $\dfrac{8}{\sin 30°} = \dfrac{Q}{\sin 45°}$

$\therefore \quad P = \dfrac{8 \sin 75°}{\sin 30°} = 15{\cdot}5 \qquad \therefore \quad Q = \dfrac{8 \sin 45°}{\sin 30°} = 11{\cdot}3$

The force P is 15·5 N and the force Q is 11·3 N.

Example 6

A mass of 5 kg is suspended, in equilibrium, by two light inextensible strings which make angles of 30° and 45° with the horizontal.
Calculate the tensions in the strings.

First, draw a diagram showing the position of equilibrium; let the tensions in the strings be S and T newtons.
There are seen to be three forces, acting at a point, producing equilibrium, so by Lami's Theorem:

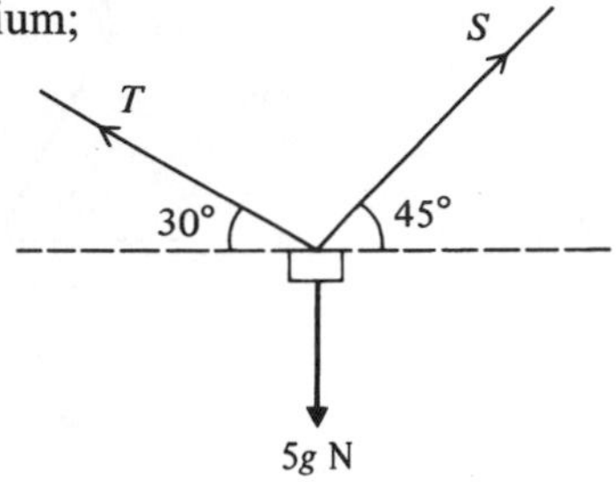

$$\frac{T}{\sin(90° + 45°)} = \frac{5g}{\sin(180° - 30° - 45°)}$$

$$\therefore \qquad T = \frac{5g \sin 135°}{\sin 105°} = 35{\cdot}87$$

and $$\frac{S}{\sin(90° + 30°)} = \frac{5g}{\sin(180° - 30° - 45°)} \qquad \therefore \quad S = \frac{5g \sin 120°}{\sin 105°} = 43{\cdot}93$$

The tensions in the strings are 43·9 N and 35·9 N.

It should be noted that an equation involving T could have been used to determine the tension S, once the value of T had been found. It is better to avoid the use of a previously determined value if it is possible, since that value may be incorrect.

Exercise 5A

1. Each of the following systems of forces is in equilibrium. By making an accurate scale drawing of the triangle of forces, find the magnitude of forces P and Q and the size of angle θ.

(a)

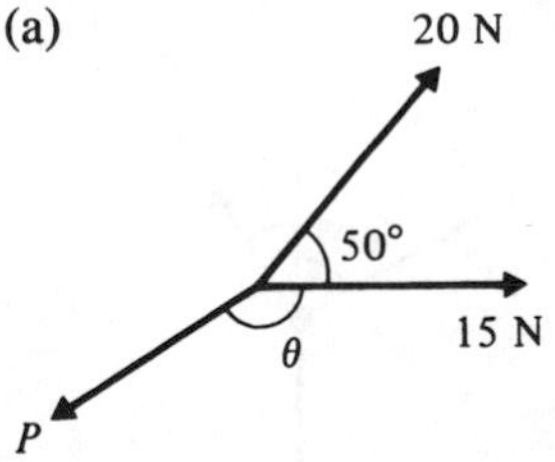

(b)

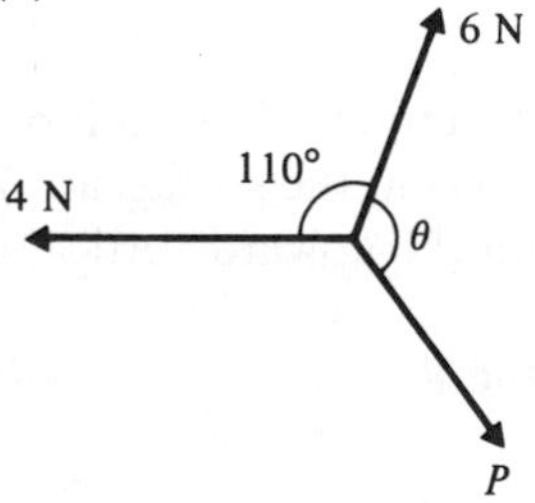

(c)

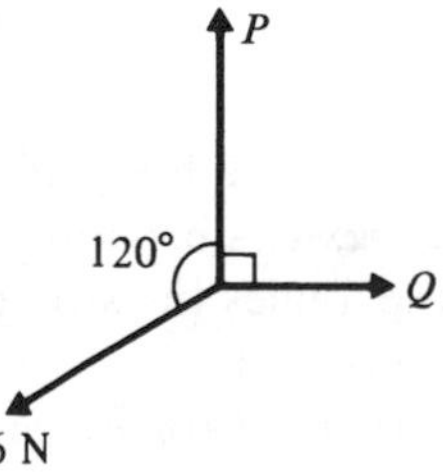

2. Each of the following systems of forces is in equilibrium. By making an accurate scale drawing of the polygon of forces, find the magnitude of forces P and Q and the size of angle θ.

(a)

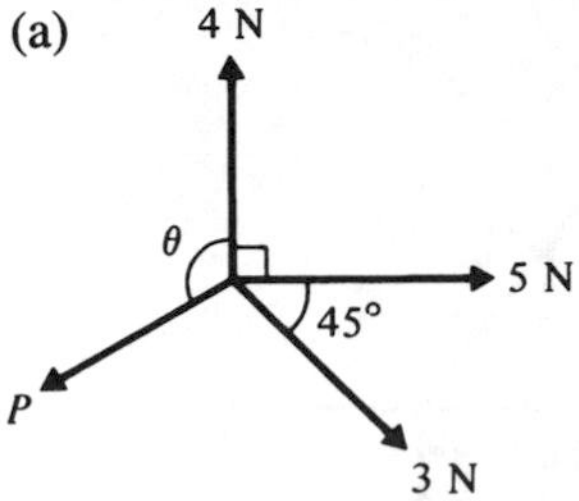

(b)

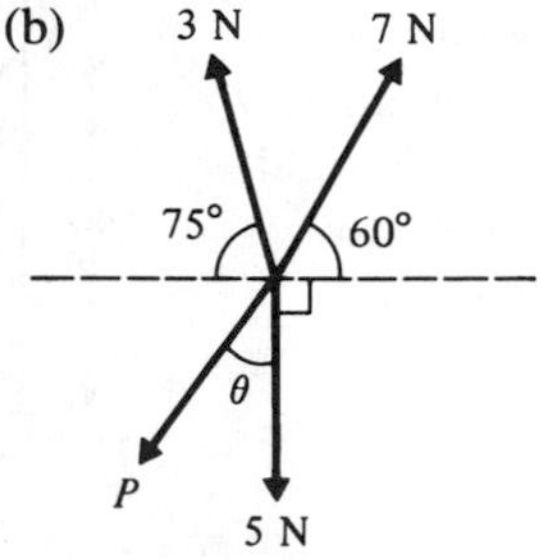

(c)

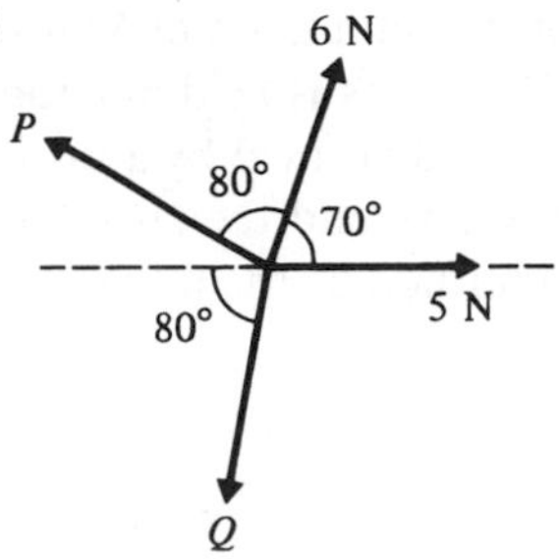

3. Each of the following systems of forces is in equilibrium. Make a sketch of the triangle of forces and hence calculate the magnitude of force P and the size of angle θ.

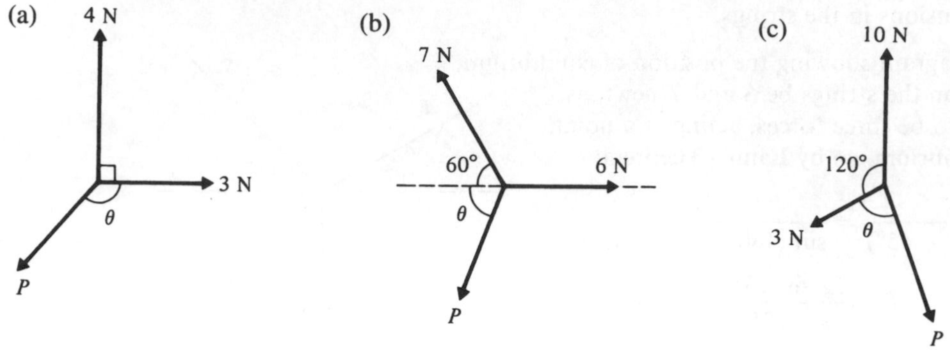

4. Each of the following systems of forces is in equilibrium. Use Lami's Theorem to find the magnitude of forces P and Q.

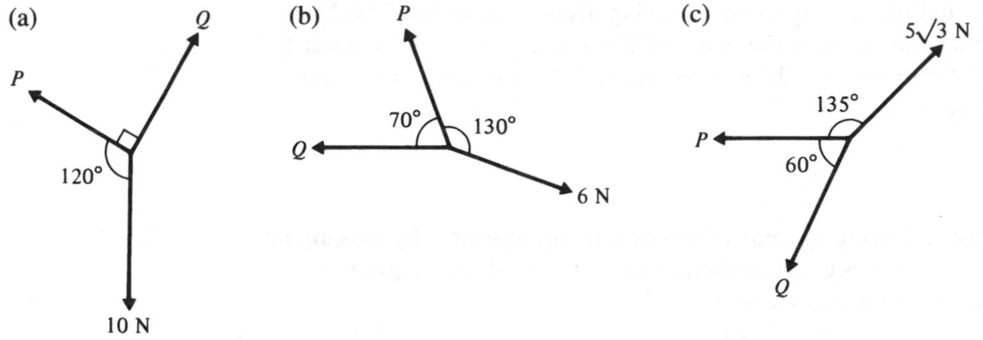

5. The diagram shows a body of weight 10 N supported in equilibrium by two light inextensible strings. The tensions in the strings are 7 N and T and the angles the strings make with the upward vertical are 60° and θ respectively.
Using the triangle of forces, calculate T and θ.

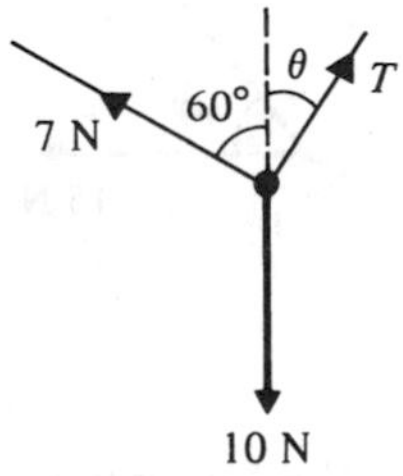

6. The diagram shows a light inextensible string with one end fixed at A and a mass of 5 kg suspended at the other end. The mass is held in equilibrium at an angle θ to the downward vertical by a horizontal force P. Find θ by trigonometry and then use Lami's Theorem to find the magnitude of the force P and the tension T.

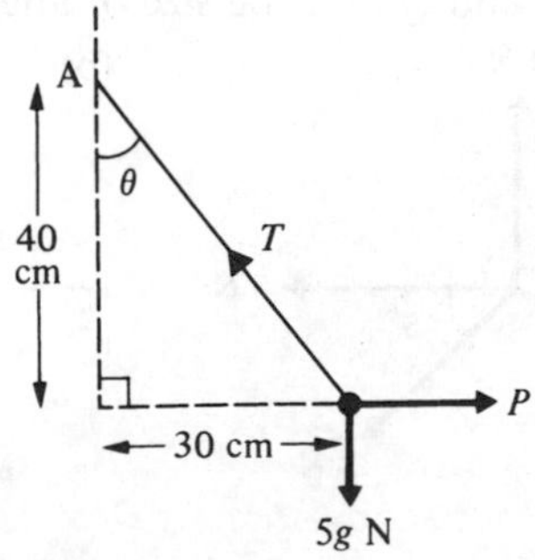

7. Four horizontal forces, all emanating from some point O, are in equilibrium. Three of the four forces have magnitudes 10 N, 20 N and 30 N in directions north, east and south-west respectively. Find the magnitude and direction of the fourth force by scale drawing.

8. A mass of 2 kg is suspended by two light inextensible strings, one making an angle of 60° with the upward vertical and the other 30° with the upward vertical. Find the tension in each string.

9. A light inextensible string of length 40 cm has its upper end fixed at a point A, and carries a mass of 2 kg at its lower end. A horizontal force applied to the mass keeps it in equilibrium, 20 cm from the vertical through A. Find the magnitude of this horizontal force and the tension in the string.

10. The diagram shows a body of mass 5 kg supported by two light inextensible strings, the other ends of which are attached to two points A and B on the same level as each other and 7 m apart. The body rests in equilibrium at C, 3 m vertically below AB. If $\hat{CBA} = 45°$, find T_1 and T_2, the tension in the strings.

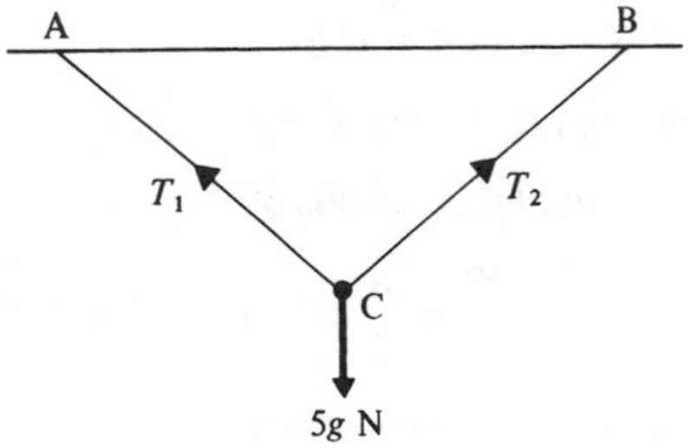

Particle in equilibrium under more than three forces: solution by calculation

When more than three forces act upon a particle and a state of equilibrium exists, it has been seen that a polygon of forces can be drawn and the magnitude and direction of an unknown force determined. Graphical methods such as this have only a limited degree of accuracy. Lami's Theorem cannot be used when there are more than three forces involved.

For a system of forces in equilibrium, the resultant force acting, in any direction, is zero. Thus the sum of the components (or resolved parts) of the forces in any and every direction must be zero. This result applies to a system of any number of forces which are in equilibrium, and gives a method of solving such problems.

For coplanar forces, we can choose two mutually perpendicular directions; by finding the components of all the forces, two equations will be obtained. Two unknown quantities can then be determined. Any other equation which may be obtained by resolving in some other direction will be a combination of the previous two equations, and it will not therefore enable more unknowns to be found.

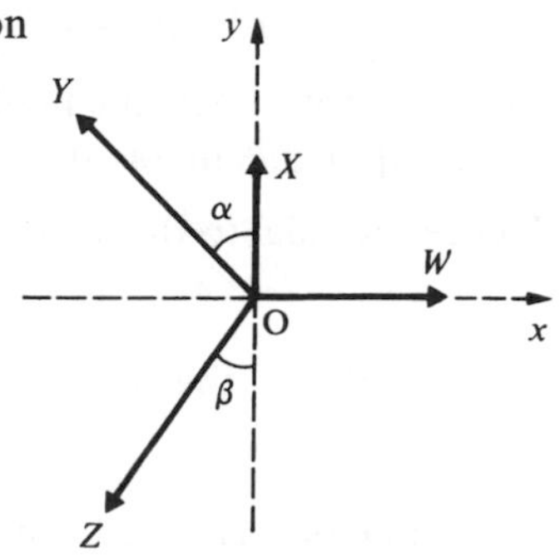

Suppose the forces W, X, Y and Z act at the point O and are in equilibrium with the direction of the forces as shown in the diagram. Since the resultant parallel to the x-axis is zero, the sum of the components of the forces in this direction is zero.
Alternatively, the sum of the resolved parts in the direction of the positive x-axis must balance those in the opposite direction, and this will give the same equation.

Resolve parallel to the x-axis:

$$W - Y\sin\alpha - Z\sin\beta = 0 \quad \text{or} \quad W = Y\sin\alpha + Z\sin\beta \qquad \ldots[1]$$

Resolve parallel to the y-axis:

$$X + Y\cos\alpha - Z\cos\beta = 0 \quad \text{or} \quad X + Y\cos\alpha = Z\cos\beta \qquad \ldots[2]$$

Hence two equations are obtained and from these equations two unknowns can then be determined.

It should be noted that the alternative way of writing down the equations does give the same equations.

Example 7

The given forces act on a particle at O which is in equilibrium. By resolving in two directions, find P and S.

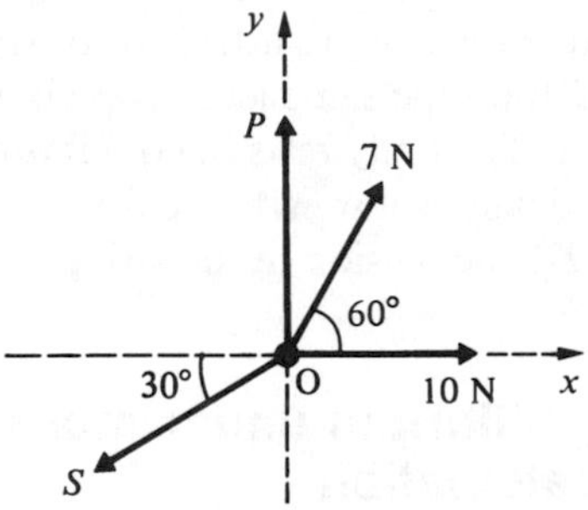

Resolving parallel to x-axis gives:

$$10 + 7\cos 60° = S\cos 30°$$

$$\therefore \qquad \frac{27}{2} = \frac{S}{2}\sqrt{3} \quad \text{or} \quad S = 9\sqrt{3}\,\text{N}$$

Resolving parallel to y-axis gives:

$$P + 7\cos 30° = S\cos 60°$$

Substituting for S gives:

$$P + \tfrac{7}{2}\sqrt{3} = 9\sqrt{3} \times \tfrac{1}{2}$$

$$\therefore \qquad P = \sqrt{3}\,\text{N}$$

The force P is $\sqrt{3}$ N and the force S is $9\sqrt{3}$ N.

Example 8

The forces $(3\mathbf{i} + 5\mathbf{j})$ N, $(a\mathbf{i} + b\mathbf{j})$ N, $(8\mathbf{i} - 6\mathbf{j})$ N and $(-4\mathbf{i} - 3\mathbf{j})$ N are in equilibrium. Find the values of a and b by calculation.

In this case a diagram is not necessary.

The resultant of these forces, in vector form, is:

$$(3\mathbf{i} + 5\mathbf{j} + a\mathbf{i} + b\mathbf{j} + 8\mathbf{i} - 6\mathbf{j} - 4\mathbf{i} - 3\mathbf{j})$$

$$= (3 + a + 8 - 4)\mathbf{i} + (5 + b - 6 - 3)\mathbf{j}\,\text{N}$$

$$= (7 + a)\mathbf{i} + (b - 4)\mathbf{j}\,\text{N}$$

Since the forces are in equilibrium, the resultant is zero; therefore the sum of the components in any direction must also be zero.

Hence, parallel to the x-axis: $a + 7 = 0$ (in the direction of $\mathbf{i}$)

$$\therefore \qquad a = -7$$

and, parallel to the y-axis: $b - 4 = 0$ (in the direction of $\mathbf{j}$)

$$\therefore \qquad b = 4$$

The value of a is -7 and of b is $+4$.

Positions of equilibrium

If a position of equilibrium of a particle acted upon by a number of forces, is described, it is important to first draw a diagram showing all the forces acting on the particle. It can then be decided which are the best directions in which to resolve the forces.

Example 9

A particle of mass 2 kg is attached to the lower end of an inextensible string. The upper end of the string is fixed. A horizontal force of 21 N and an upward vertical force of $0{\cdot}5g$ N act upon the particle, which is in equilibrium with the string making an angle θ with the vertical.
Calculate the tension in the string and the angle θ.

Draw a diagram showing the forces acting on the particle.
Let the tension in the string be T.

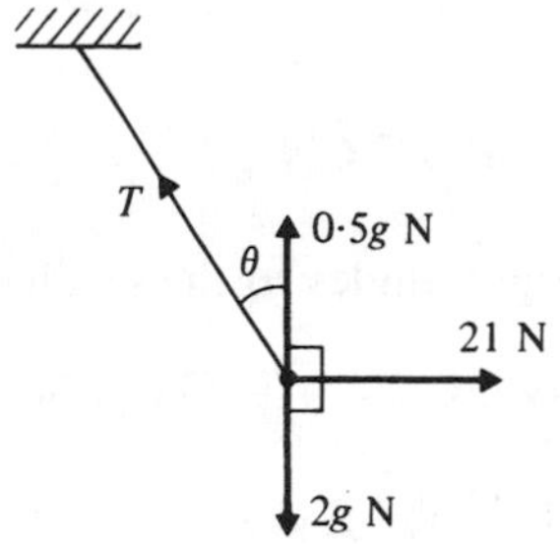

Resolving horizontally gives:

$$21 = T \sin \theta \qquad \ldots [1]$$

Resolving vertically gives:

$$2g = T \cos \theta + 0{\cdot}5g$$

$$\therefore \quad 14{\cdot}7 = T \cos \theta \qquad \ldots [2]$$

Dividing equation [1] by equation [2] gives:

$$\tan \theta = \frac{21}{14{\cdot}7} = 1{\cdot}4286$$

$$\therefore \quad \theta = 55{\cdot}01^\circ$$

Substituting into equation [1] gives:

$$T = \frac{21}{\sin 55{\cdot}01^\circ} = 25{\cdot}63\,\text{N}$$

The tension in the string is 25·63 N and the angle θ is 55·01°.

Components in other directions

Although resolving in a horizontal and vertical direction may frequently be convenient, this is not always so.
If it is possible to choose a direction, at right angles to which there is an unknown force, this may well be a sensible choice.

Inclined plane

When a mass is in equilibrium on an inclined plane, then it will usually be found expedient to resolve the forces parallel to, and at right angles to, the surface of the plane.

Example 10

A particle of mass 4 kg rests on the surface of a smooth plane which is inclined at an angle of 30° to the horizontal. When a force P acting up the plane and a horizontal force of $8\sqrt{3}$ N are applied to the particle, it rests in equilibrium.
Calculate P and the normal reaction between the particle and the plane.

Draw a diagram showing the position of equilibrium and all the forces acting on the particle. Let the normal reaction be R.

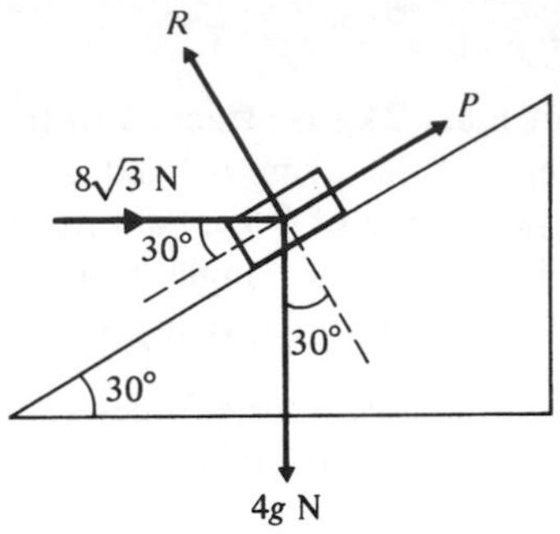

Resolving parallel to the surface of the plane gives:

$$4g \cos 60° = 8\sqrt{3} \cos 30° + P$$

$$4g \times \tfrac{1}{2} = 8\sqrt{3} \times \frac{\sqrt{3}}{2} + P$$

$$\therefore \qquad P = 7{\cdot}6\,\text{N}$$

Resolving at right angles to the surface of the plane gives:

$$R = 4g \cos 30° + 8\sqrt{3} \cos 60°$$

$$\therefore \qquad R = 4g \times \frac{\sqrt{3}}{2} + 8\sqrt{3} \times \tfrac{1}{2}$$

$$\therefore \qquad R = 40{\cdot}87\,\text{N}$$

The horizontal force is 7·6 N and the normal reaction of the plane is 40·9 N.

Systems involving more than one particle

In a more complicated system, there may be more than one particle involved in a state of equilibrium.

The diagram must show clearly all the forces acting on each particle. The equilibrium of each particle may then be considered separately. This means that the directions in which the forces are resolved for the two particles may not be the same.

Example 11

A light inextensible string passes over a smooth pulley fixed at the top of a smooth plane inclined at 30° to the horizontal. A particle of mass 2 kg is attached to one end of the string and hangs freely. A mass m is attached to the other end of the string and rests in equilibrium on the surface of the plane. Calculate the normal reaction between the mass m and the plane, the tension in the string and the value of m.

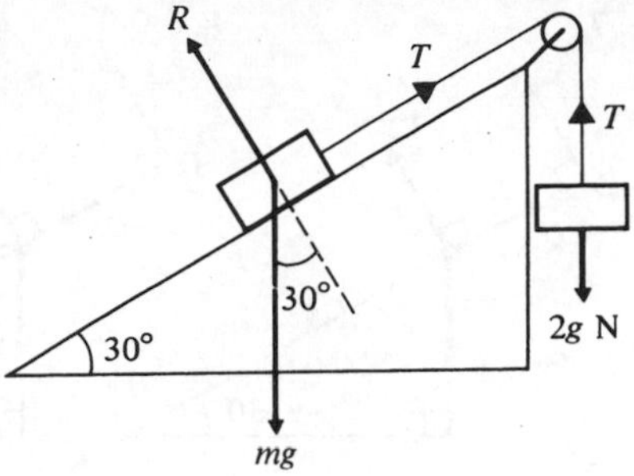

Draw a diagram showing the position of equilibrium and show the forces acting on each particle. Let the tension in the string be T.

Consider the mass of 2 kg.

Resolve vertically: $\quad T = 2g$

$\therefore \quad T = 19{\cdot}6\,\text{N}$

As no other forces act on this mass, there is only the one equation.

Consider the mass m.

Resolve parallel to the surface of the plane:

$$T = mg\cos 60°$$

Substitute for T: $\quad 19{\cdot}6 = mg \times \frac{1}{2}$

$\therefore \quad m = 4\,\text{kg}$

Resolve at right angles to the plane:

$$R = mg\cos 30°$$

Substitute for m: $\quad R = 33{\cdot}9\,\text{N}$

The normal reaction is 33·9 N, the tension in the string is 19·6 N and m is 4 kg.

Exercise 5B

Each of the diagrams in questions **1** to **9** shows a particle in equilibrium under the forces shown. In each case:

(a) obtain an equation by resolving in the direction Ox

(b) obtain an equation by resolving in the direction Oy

(c) use your equations for (a) and (b) to find the unknown forces and angles.

1.

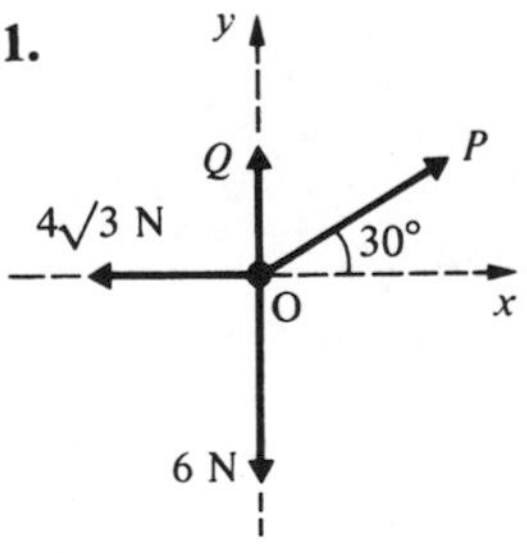

2.

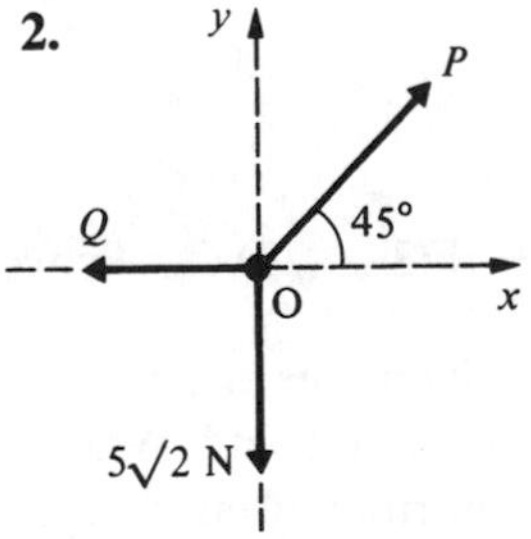

3.

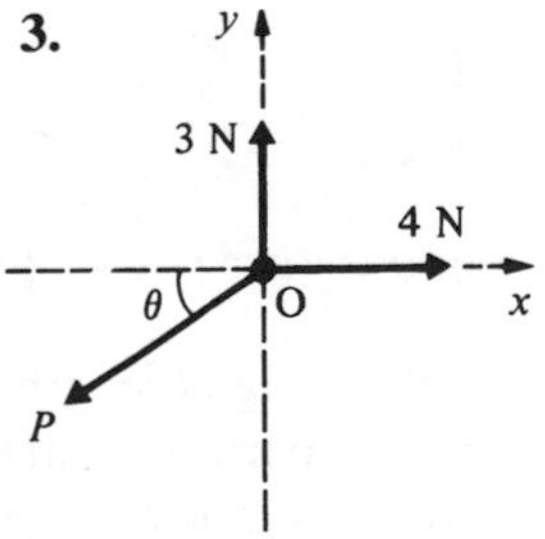

4.

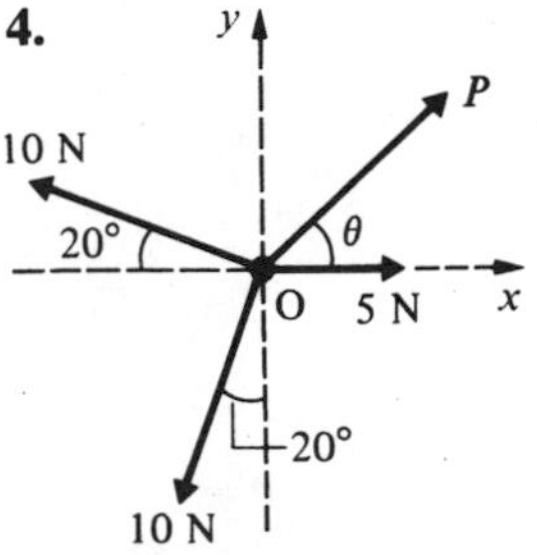

5.

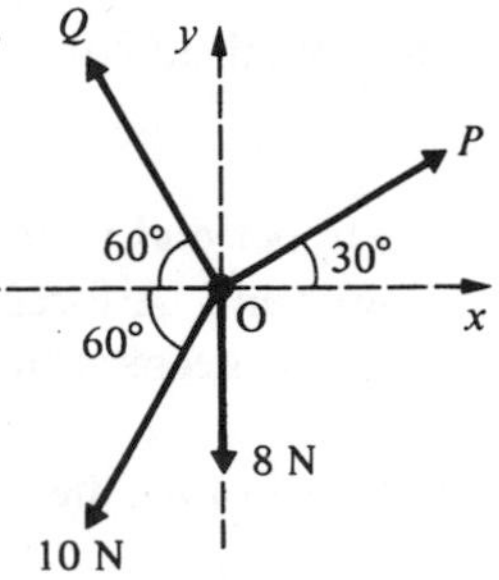

6.

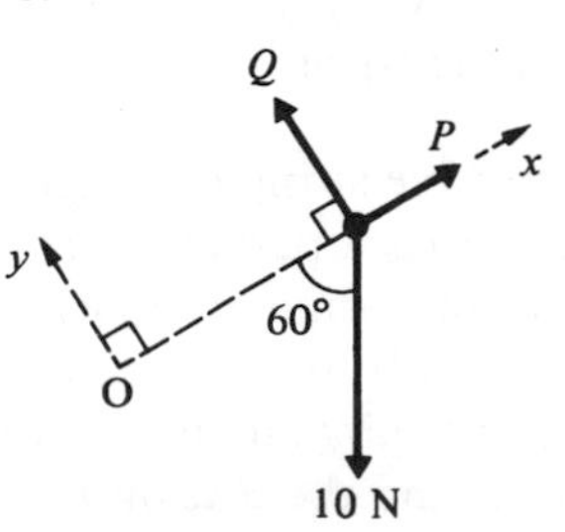

7.

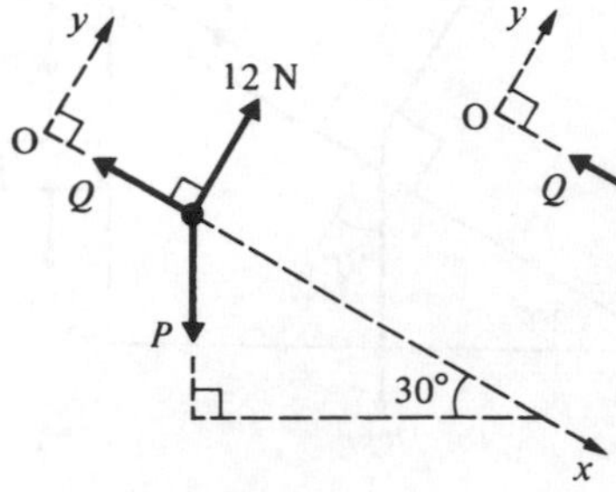

8.

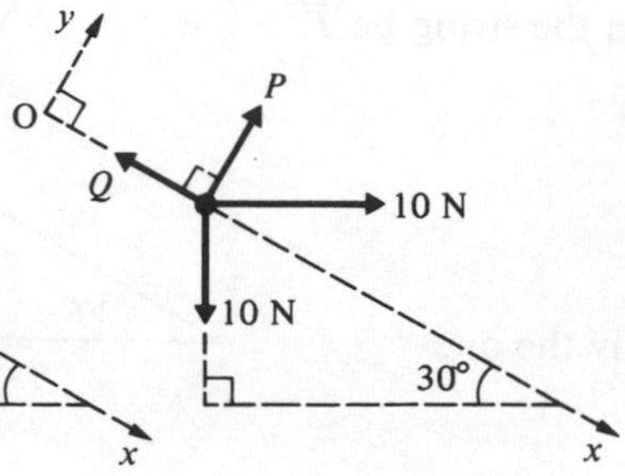

9.

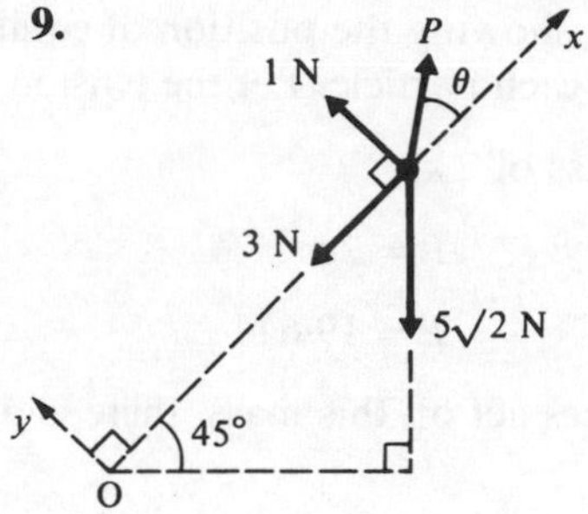

10. If each of the following sets of forces is in equilibrium, find the value of a and b in each case:

(a) $(6\mathbf{i} + 4\mathbf{j})$ N, $(-2\mathbf{i} - 5\mathbf{j})$ N, $(a\mathbf{i} + b\mathbf{j})$ N
(b) $(5\mathbf{i} + 4\mathbf{j})$ N, $(3\mathbf{i} + \mathbf{j})$ N, $(a\mathbf{i} + b\mathbf{j})$ N
(c) $(a\mathbf{i} + 3\mathbf{j})$ N, $(2\mathbf{i} - 5\mathbf{j})$ N, $(-7\mathbf{i} + b\mathbf{j})$ N
(d) $(a\mathbf{i} - 3b\mathbf{j})$ N, $(b\mathbf{i} - 2a\mathbf{j})$ N, $(-3\mathbf{i} + 8\mathbf{j})$ N
(e) $(-3\mathbf{i} + 2\mathbf{j})$ N, $(4\mathbf{i} + 7\mathbf{j})$ N, $(-8\mathbf{i} + 5\mathbf{j})$ N, $(a\mathbf{i} + b\mathbf{j})$ N.

11. If each of the following sets of forces is in equilibrium, find the value of a, b and c in each case:

(a) $(4\mathbf{i} + 3\mathbf{j} - \mathbf{k})$ N, $(\mathbf{i} - 5\mathbf{j} + 2\mathbf{k})$ N, $(a\mathbf{i} + b\mathbf{j} + c\mathbf{k})$ N
(b) $(-2\mathbf{i} + 3\mathbf{k})$ N, $(4\mathbf{j} - 7\mathbf{k})$ N, $(a\mathbf{i} + b\mathbf{j} + c\mathbf{k})$ N
(c) $(5\mathbf{i} + a\mathbf{j} + \mathbf{k})$ N, $(b\mathbf{i} - 6\mathbf{j} - \mathbf{k})$ N, $(-3\mathbf{i} + 2\mathbf{j} + c\mathbf{k})$ N.

12. Each of the following diagrams shows a particle in equilibrium under the forces shown.

(a)

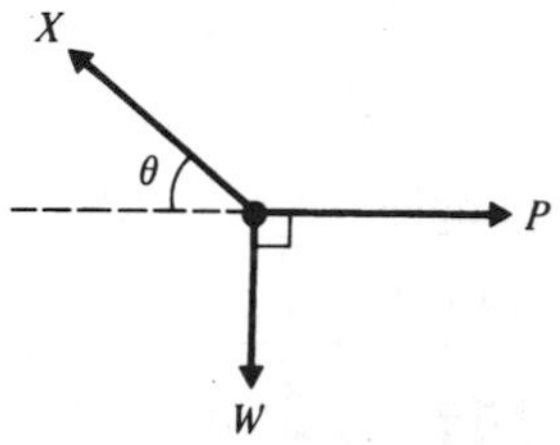

Prove: $P \tan \theta = W$

(b)

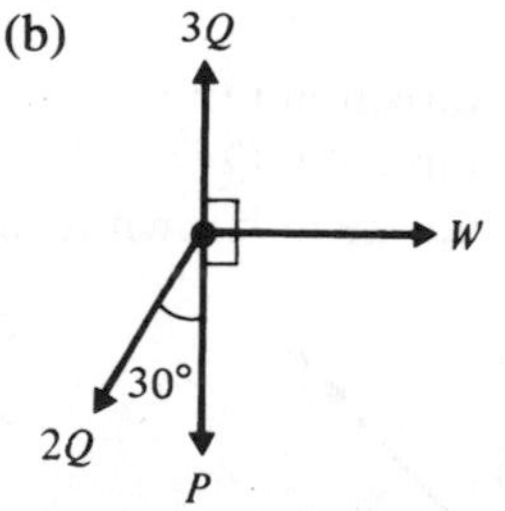

Prove: $P = W(3 - \sqrt{3})$

(c)

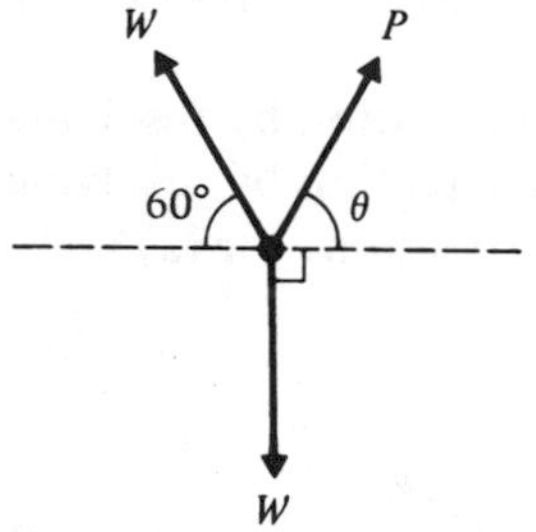

Prove: $\tan \theta = 2 - \sqrt{3}$

13. A light inextensible string of length 50 cm has its upper end fixed at point A and carries a particle of mass 8 kg at its lower end. A horizontal force P applied to the particle keeps it in equilibrium 30 cm from the vertical through A.
By resolving vertically and horizontally, find the magnitude of P and the tension in the string.

14. A light inextensible string of length 26 cm has its upper end fixed at point A and carries a particle of mass m at its lower end. A force P at right angles to the string is applied to the particle and keeps it in equilibrium 10 cm from the vertical through A.
By resolving vertically and horizontally find, in terms of m, the magnitude of P and the tension in the string.

15. A particle is in equilibrium under the action of forces 4 N due north, 8 N due west, $5\sqrt{2}$ N south-east and P. Find the magnitude and direction of P.

16. A force acting parallel to and up a line of greatest slope holds a particle of mass 10 kg in equilibrium on a smooth plane which is inclined at 30° to the horizontal. Find the magnitude of this force and of the normal reaction between the particle and the plane.

17. A horizontal force P holds a body of mass 10 kg in equilibrium on a smooth plane which is inclined at 30° to the horizontal.
Find the magnitude of P and of the normal reaction between the particle and the plane.

18. A force P holds a particle of mass m in equilibrium on a smooth plane which is inclined at 30° to the horizontal. If P makes an angle ϕ with the plane, as shown in the diagram, find ϕ when R, the normal reaction between particle and plane, is $1{\cdot}5\,mg$.

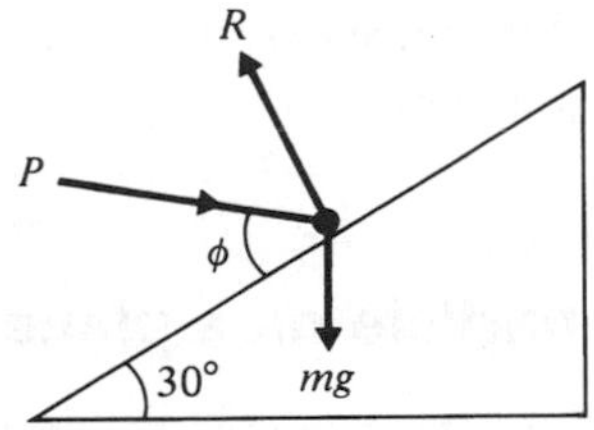

19. A particle of mass 3 kg lying on a smooth surface which is inclined at θ to the horizontal is attached to a light inextensible string which passes up the plane, along the line of greatest slope, over a smooth pulley at the top and carries a 1 kg mass freely suspended at its other end.
If the system rests in equilibrium, find:
(a) the value of θ (b) the tension in the string
(c) the normal reaction between the particle and the plane.

20. A 5 kg mass lies on a smooth horizontal table. A light inextensible string attached to this mass passes up over a smooth pulley and carries a freely suspended mass of 5 kg at its other end. The part of the string between the mass on the table and the pulley makes an angle of 25° with the horizontal. The system is kept in equilibrium by a horizontal force applied to the mass on the table. Find the magnitude of this horizontal force, the tension in the string and the normal reaction between the table and the mass resting on it.

21. The diagram shows masses of 8 kg and 6 kg lying on smooth planes of inclination θ and ϕ respectively.

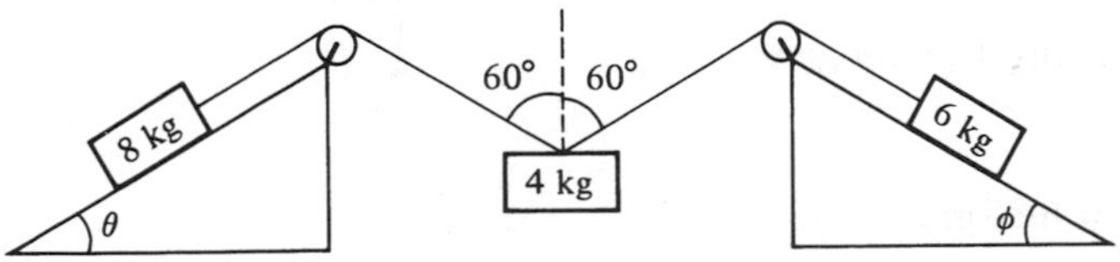

Light inextensible strings attached to these masses pass along the lines of greatest slope, over smooth pulleys and are connected to a 4 kg mass hanging freely. The strings both make an angle of 60° with the upward vertical as shown. If the system rests in equilibrium, find θ and ϕ.

22. The diagram shows masses A and B each lying on smooth planes of inclination 30°.

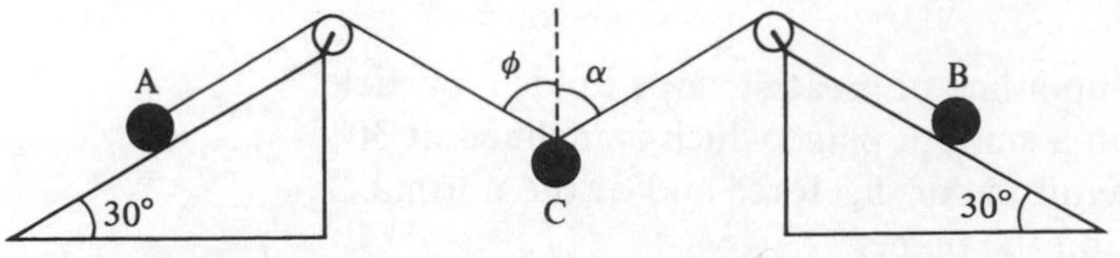

Light inextensible strings attached to A and B pass along the lines of greatest slope, over smooth pulleys and are connected to a third mass C hanging freely. The strings make angles of ϕ and α with the upward vertical as shown.
If A, B and C have masses $2m$, m and m respectively and the system rests in equilibrium, show that $\sin\alpha = 2\sin\phi$ and $\cos\alpha + 2\cos\phi = 2$.
Hence find ϕ and α.

Motion of a particle on a plane

In the examples so far considered in this chapter, the particle has been in equilibrium under the action of a number of forces.
The motion of a particle (or a rigid body) can now be considered using Newton's Laws as in Chapter 3 and the idea of resolving a force in a particular direction.

Example 12

A body of mass 4 kg has an acceleration a when it is acted upon by a force of $25\sqrt{2}$ N which is inclined at 45° to a smooth horizontal surface on which the body rests, as shown. Resolve the forces acting on the body at right angles to the surface. Calculate the normal reaction between the body and the surface and the acceleration a of the body.

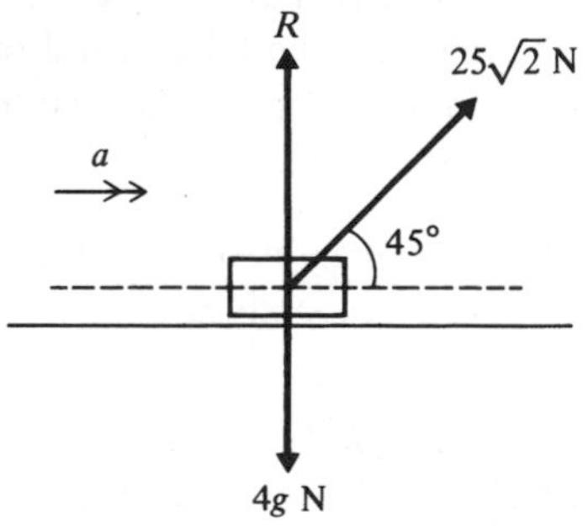

Resolving at right angles to the surface gives:

$$R + 25\sqrt{2}\cos 45° = 4g$$

$$\therefore \quad R + 25\sqrt{2} \times \frac{1}{\sqrt{2}} = 39{\cdot}2$$

$$\therefore \quad R = 14{\cdot}2\,\text{N}$$

Resolving along the surface and applying $F = ma$ gives:

$$25\sqrt{2}\cos 45° = 4 \times a$$

$$\therefore \quad 25 = 4a \quad \text{or} \quad a = 6{\cdot}25\,\text{m s}^{-2}$$

The body has an acceleration of $6{\cdot}25\,\text{m s}^{-2}$ along the surface, and the normal reaction between the body and the surface is 14·2 N.

Since the reaction R is found to be positive, i.e. 14·2 N, this implies that the body and the surface remain in contact.

Motion on an inclined plane

The same principles apply to motion on an inclined plane as were used in considering motion on a horizontal plane. The directions in which resolving takes place will necessarily be different.

Example 13

A body of mass $3\sqrt{3}$ kg on the surface of an inclined plane is acted upon by a horizontal force of $15g$ N, as shown in the diagram. Calculate the normal reaction of the plane on the body, and the acceleration of the body up the surface of the smooth inclined plane.

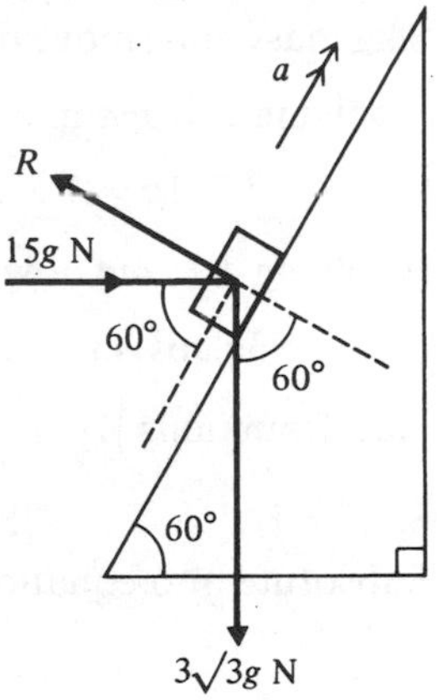

Let the acceleration of the body be a and the normal reaction R.

Resolving at right angles to the plane gives:

$$R = 3\sqrt{3}g \cos 60° + 15g \cos 30°$$

$$\therefore \quad R = 9\sqrt{3}g = 152{\cdot}8\,\text{N}$$

Resolving up the plane and applying $F = ma$ gives:

$$15g \cos 60° - 3\sqrt{3}g \cos 30° = 3\sqrt{3} \times a$$

$$\therefore \quad a = 5{\cdot}658\,\text{m}\,\text{s}^{-2}$$

The normal reaction is 153 N and the acceleration of the body is $5{\cdot}66\,\text{m}\,\text{s}^{-2}$.

Connected particles

In a more complicated system involving the motion of more than one particle, the particles may again be considered separately. Care is needed to ensure that all the forces acting on each particle are considered. Any relationship which may exist between the accelerations of the various parts of the system must also be taken into account.
At this stage, strings will normally be inextensible and pulleys over which strings pass will be treated as smooth pulleys.

Of course in real life strings are not inextensible and pulleys are not smooth. We make these assumptions to allow a *mathematical model* of the situation to be created. The real situation is complicated if every aspect is taken into account, but by concentrating on the more important features, and ignoring some small details, we can apply certain mathematical ideas more easily. The validity of the model can then be checked by comparing the results as predicted by the model with those occurring in real life. If the answers supplied by our model do not agree with reality, we must question the wisdom of the assumptions made and alter them as necessary. For example, the reader will see in Example 15, and in the next chapter, that the roughness of a surface can be taken into account.

Example 14

The bodies shown are connected by a light string which passes over a smooth pulley. Calculate the tension T, the normal reaction R and the acceleration a.

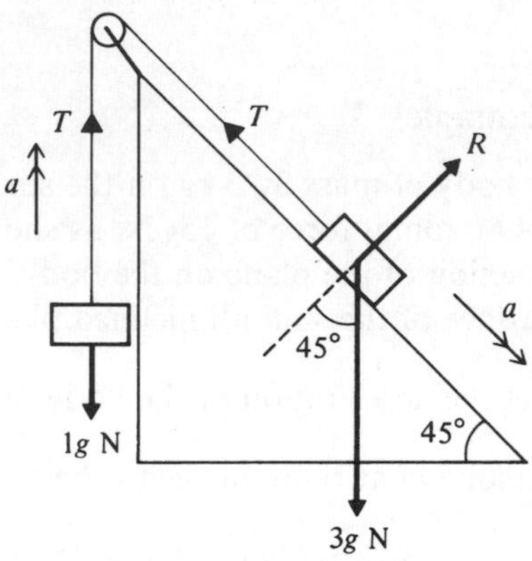

The diagram shows all the forces acting on the two bodies. Since the bodies are connected by an inextensible string, the accelerations must be equal in magnitude. If the 1 kg mass moves upwards, the 3 kg mass must move down the surface of the plane.

Applying $F = ma$ in a vertical direction for the 1 kg mass gives:

$$T - 1g = 1 \times a \qquad \ldots [1]$$

Applying $F = ma$ down the plane for the 3 kg mass gives:

$$3g \cos 45° - T = 3 \times a \qquad \ldots [2]$$

Add Equations [1] and [2]: $\frac{3}{2}g\sqrt{2} - g = 4a$

$\therefore \qquad a = 2{\cdot}747 \text{ m s}^{-2}$

Substitute into equation [1]: $T = g + 2{\cdot}747$

$= 12{\cdot}547 \text{ N}$

Resolve at right angles to the surface of the plane, for the 3 kg mass (note that in this direction there is no acceleration):

$$R = 3g \cos 45°$$
$$= 20{\cdot}79 \text{ N}$$

The tension in the string is 12·5 N, the normal reaction is 20·8 N and the acceleration of both particles is $2{\cdot}75 \text{ m s}^{-2}$.

Rough surfaces

In the foregoing examples, motion has been taking place on smooth surfaces. In practice this does not happen; all surfaces tend to impede motion. The resistance to motion is an external force acting upon the body, parallel to the surfaces in contact. It will be considered to be a constant force.

Example 15

A body of mass 5 kg is released from rest on the surface of a rough plane which is inclined at 30° to the horizontal. If the body takes $2\frac{1}{2}$ seconds to acquire a speed of 4 m s^{-1} from rest, find the resistance to motion which the body must be experiencing.

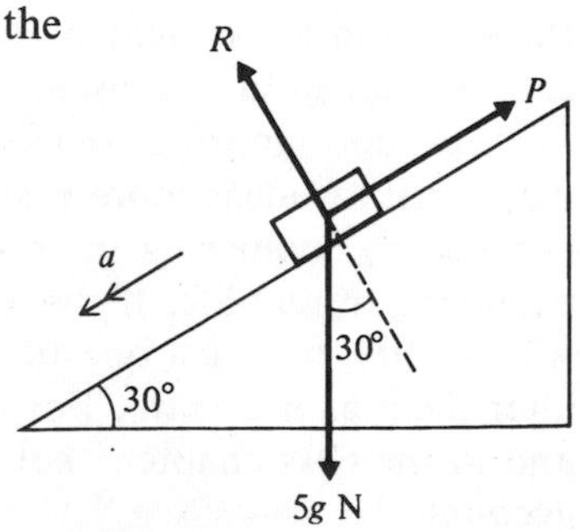

Assume that the force of resistance acting upon the body is P up the plane, and that the acceleration of the body is a down the plane.

Use $v = u + at$ for the motion of the body:

$$4 = 0 + a \times 2\tfrac{1}{2}$$

$\therefore \qquad a = 1{\cdot}6 \text{ m s}^{-2}$

Apply $F = ma$ to the motion of the body down the plane:

$$5g \cos 60° - P = 5 \times a$$

Substitute for a:

$$5g \times \tfrac{1}{2} - P = 5 \times 1{\cdot}6$$

$$\therefore \quad P = 24{\cdot}5 - 8 = 16{\cdot}5\,\text{N}$$

The resistance to motion is a force of 16·5 N.

Inclination of a plane

In the examples so far considered, the inclination of the plane to the horizontal has been given in degrees.
The gradient of a hill or of a piece of railway track is usually relatively small. The inclination is frequently given in the form 1 in 8 (or $12\tfrac{1}{2}\%$), and this means a rise of 1 unit for every 8 units measured horizontally.

Hence if the angle of inclination is θ, then

$$\tan\theta = \tfrac{1}{8} \quad \text{and} \quad \therefore \quad \theta = 7° \text{ approximately.}$$

For angles of this size, the sine and tangent are equal within 1%, so it is usual to take the sine of the angle as the gradient.

For example: an incline of 1 in 80 is taken to mean that $\sin\theta = \frac{1}{80}$.

Example 16

A body of mass 8 kg is released from rest on the surface of a plane. If the resistance to motion is 1 N acting up the plane and the slope of the plane is 1 in 40, calculate the acceleration of the body down the plane and the speed acquired 6 seconds after release.

Draw a diagram showing the forces acting on the body.

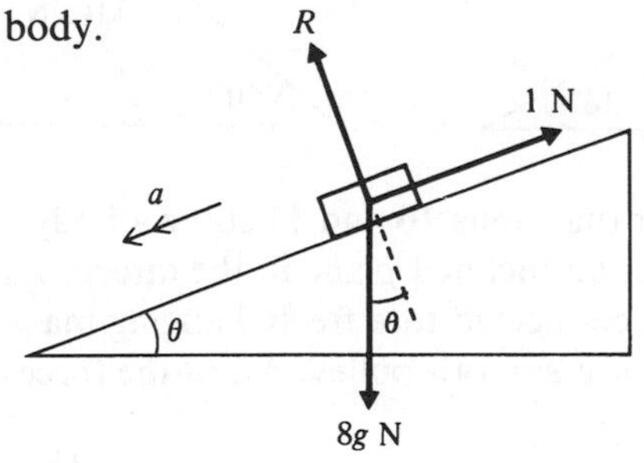

There is a component of the weight of the body which acts down the plane.

Apply $F = ma$ down the plane:

$$8g \sin\theta - 1 = 8 \times a$$

$$\therefore \quad 8g \times \tfrac{1}{40} - 1 = 8a$$

$$\therefore \quad a = \frac{0{\cdot}96}{8} = 0{\cdot}12\,\text{m s}^{-2}$$

Use $v = u + at$ and the value of a obtained to give:

$$v = 0 + 0{\cdot}12 \times 6$$

$$\therefore \quad v = 0{\cdot}72\,\text{m s}^{-1}$$

The acceleration of the body is $0{\cdot}12\,\text{m s}^{-2}$ down the plane and the speed acquired in 6 seconds is $0{\cdot}72\,\text{m s}^{-1}$.

Exercise 5C

Each of the diagrams in questions **1** to **9** shows a body of mass 10 kg accelerating along a surface in the direction indicated. All of the forces acting are as shown. In each case:

(a) obtain an equation by resolving perpendicular to the direction of motion

(b) obtain an equation by applying $F = ma$ parallel to the direction of motion

(c) use your equations to (a) and (b) to find the unknown forces, accelerations and angles.

1.

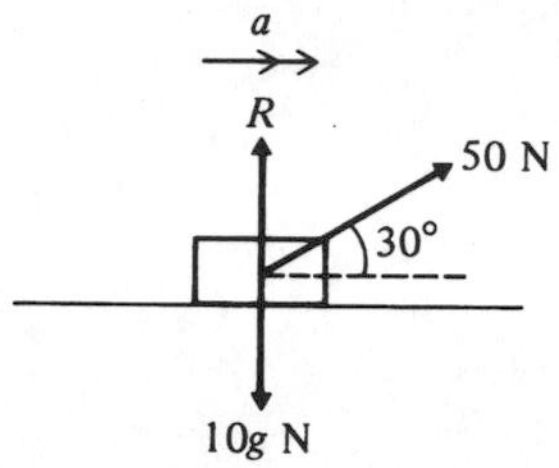

2.

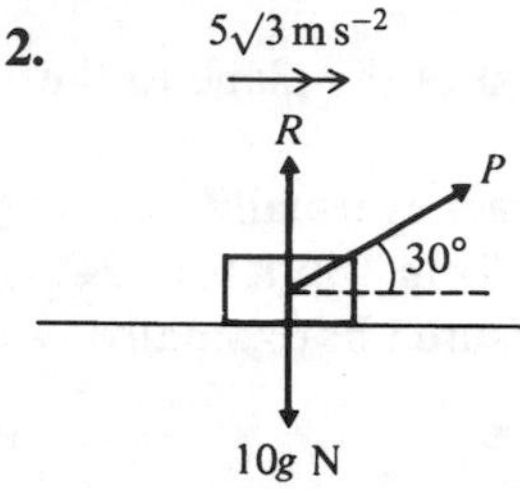

3.

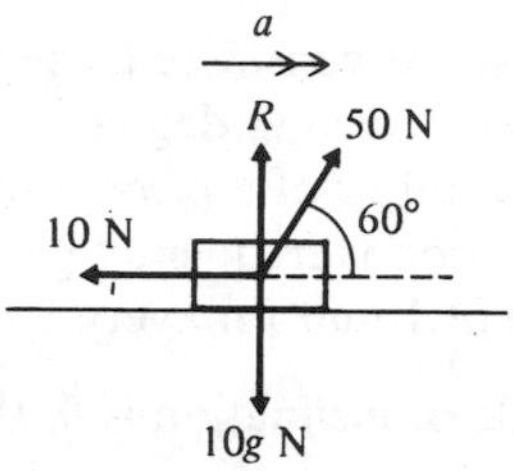

4.

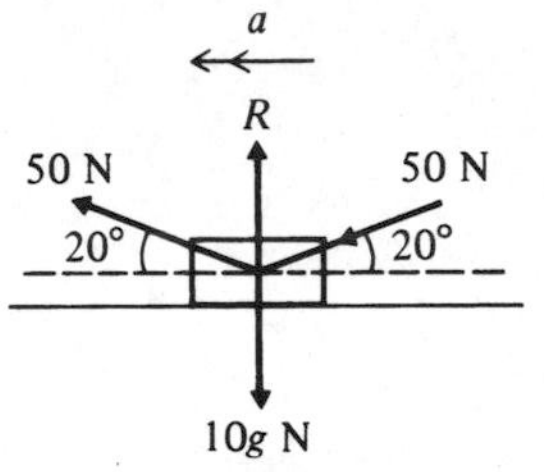

5.

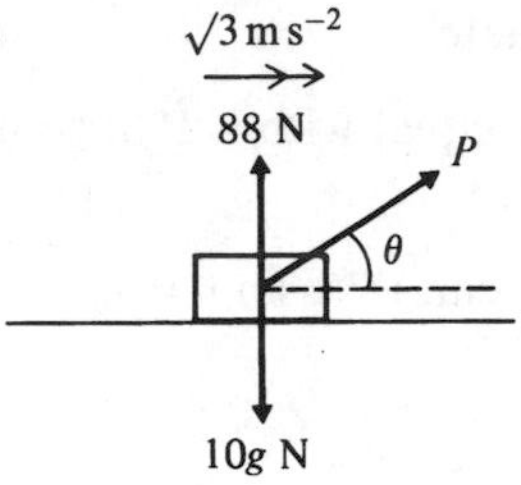

6.

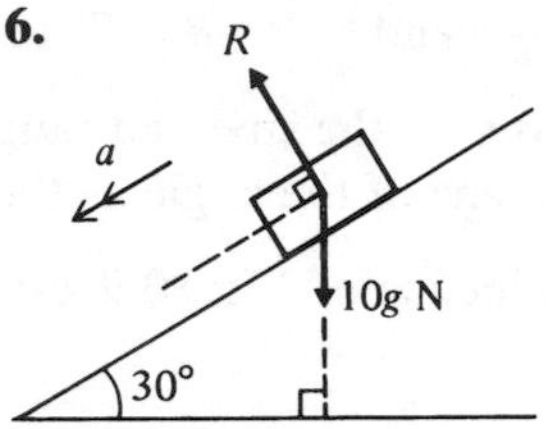

7.

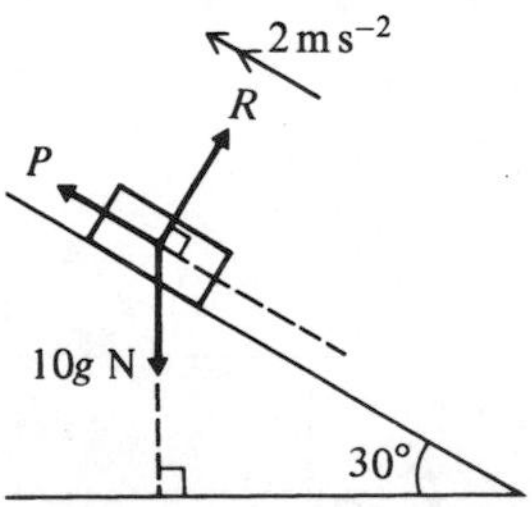

8.

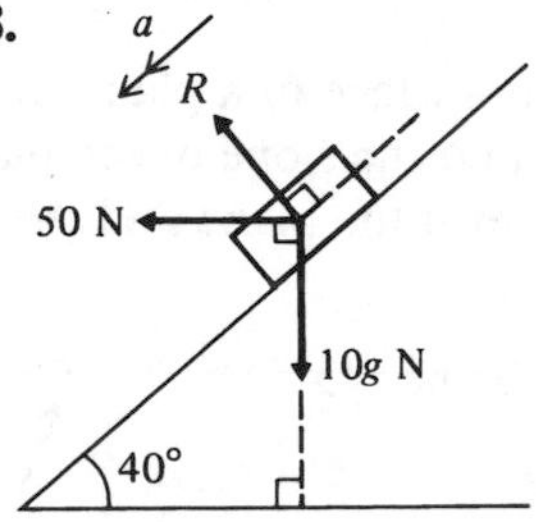

9.

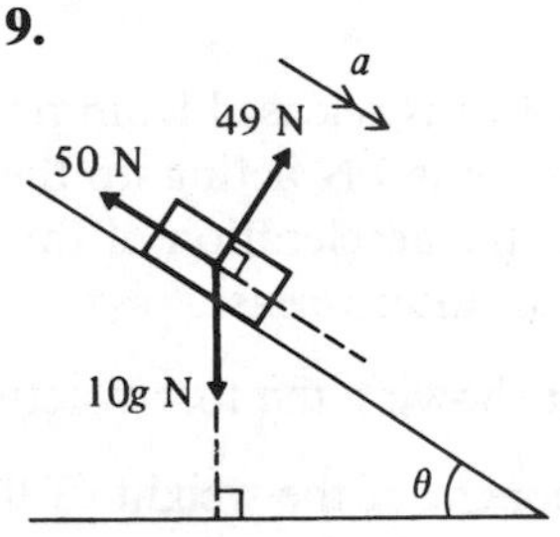

The diagrams for questions **10** and **11** show a body of mass 10 kg accelerating along an inclined plane in the direction indicated. In each case the 10 kg mass is connected to a freely hanging mass by a light inextensible string passing over a smooth pulley. All of the forces acting are as shown.

10.

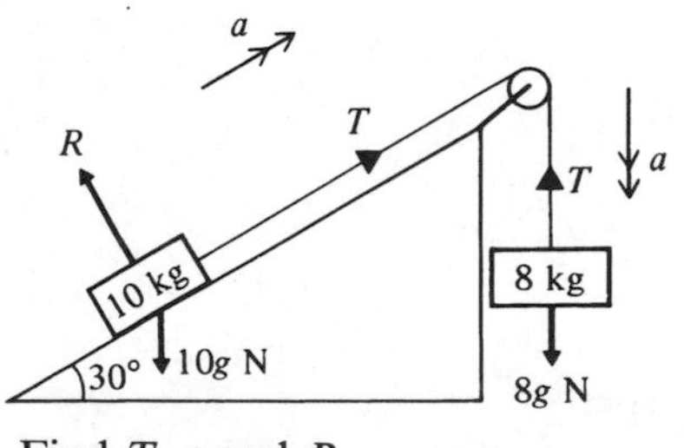

Find T, a and R

11.

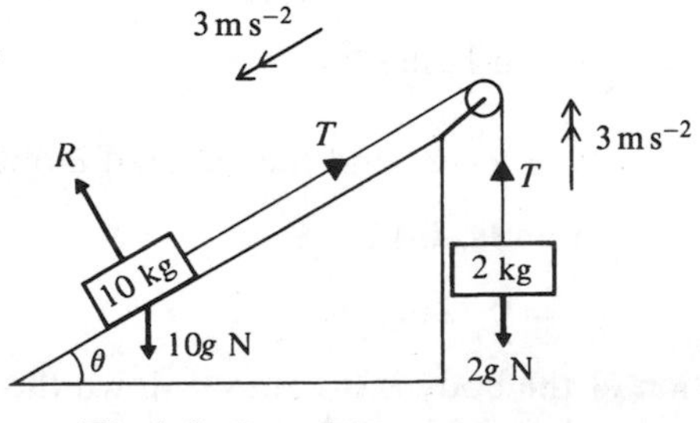

Find θ, R and T

12. A body of mass 10 kg is initially at rest on a rough horizontal surface. It is pulled along the surface by a constant force of 60 N inclined at 60° above the horizontal.
If the resistance to motion totals 10 N, find the acceleration of the body and the distance travelled in the first 3 s.

13. A body of mass 5 kg, initially at rest on a smooth horizontal surface, is pulled along the surface by a constant force P inclined at 45° above the horizontal. In the first 5 seconds of motion the body moves a distance of 10 m along the surface.
Find the acceleration of the body, the magnitude of P and the normal reaction between the body and the surface.

14. A mass of 5 kg is initially at rest at the bottom of a smooth slope which is inclined at $\sin^{-1} \frac{3}{5}$ to the horizontal. The mass is pushed up the slope by a horizontal force of 50 N.
Find the normal reaction between the mass and the plane and the acceleration up the slope. How far up the slope will the mass travel in the first 4 s?

15. A body of mass 100 kg is released from rest at the top of a smooth plane which is inclined at 30° to the horizontal.
Find the velocity of the body when it has travelled 20 m down the slope. What would your answer be if the mass had been 50 kg?

16. A body of mass 20 kg is released from rest at the top of a rough slope which is inclined at 30° to the horizontal.
If the body accelerates down the slope at $3\,\text{m}\,\text{s}^{-2}$, find the resistance to motion experienced by the body. (Assume this resistance to be constant throughout.)

17. A body of mass 20 kg is released from rest at the top of a rough slope which is inclined at 30° to the horizontal. Six seconds later the body has a velocity of $21\,\text{m}\,\text{s}^{-1}$ down the slope. Find the resistance to motion experienced by the body. (Assume this resistance to be constant throughout.)

18. Find the time interval between a particle reaching the bottom of a smooth slope of length 5 m and inclination 1 in 98, and another particle reaching the bottom of a smooth slope of length 6 m and incline 1 in 70. Both particles are released from rest at the top of their respective slopes at the same time.

19. A mass of 15 kg lies on a smooth plane of inclination 1 in 49. One end of a light inextensible string is attached to this mass and the string passes up the line of greatest slope, over a smooth pulley fixed at the top of the plane and has a freely suspended mass of 10 kg at its other end.
If the system is released from rest, find the acceleration of the masses and the distance each travels in the first 2 s. Assume that nothing impedes the motion of either mass.

20. A mass of 2 kg lies on a rough plane which is inclined at 30° to the horizontal. One end of a light inextensible string is attached to this mass and the string passes up the line of greatest slope and over a smooth pulley fixed at the top of the slope; a freely suspended mass of 5 kg is attached to its other end. The system is released from rest; as the 2 kg mass accelerates up the slope, it experiences a constant resistance to motion of 14 N down the slope owing to the rough nature of the surface. Find the tension in the string.

21. A mass of 10 kg lies on a smooth plane which is inclined at θ to the horizontal. The mass is 5 m from the top, measured along the plane. One end of a light inextensible string is attached to this mass; the string passes up the line of greatest slope and over a smooth pulley fixed at the top of the slope. The other end is attached to a freely suspended mass of 15 kg. This 15 kg mass is 4 m above the floor. The system is released from rest and the string first goes slack $1\frac{3}{7}$ s later.
Find the value of θ.

22. One of two identical masses lies on a smooth plane, which is inclined at $\sin^{-1}\frac{1}{14}$ to the horizontal, and is 2 m from the top. A light inextensible string attached to this mass passes along the line of greatest slope, over a smooth pulley fixed at the top of the incline; the other end carries the other mass hanging freely 1 m above the floor. If the system is released from rest, find the time taken for the hanging mass to reach the floor.

Exercise 5D Harder questions

Each of the diagrams in questions **1** to **9** shows a mass, or masses, accelerating in the directions indicated. In each case the forces acting are as shown and R is the normal reaction between the mass and the surface it is on.

1.

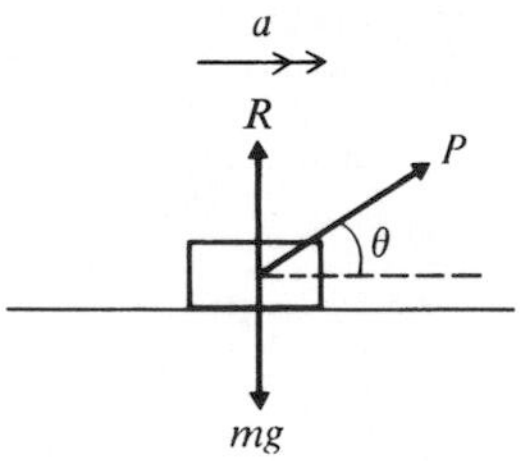

Prove: $\tan\theta = \dfrac{mg - R}{ma}$

2.

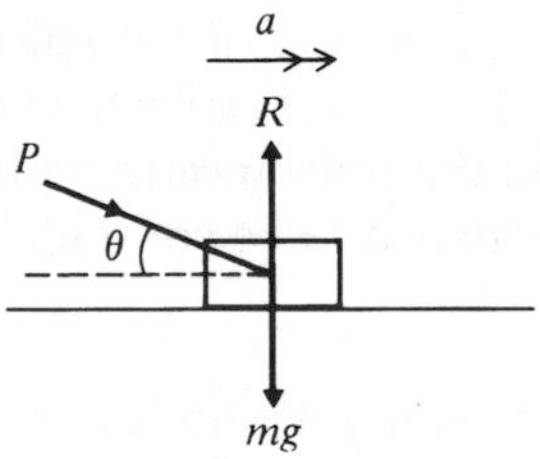

Prove: $\tan\theta = \dfrac{R - mg}{ma}$

3.

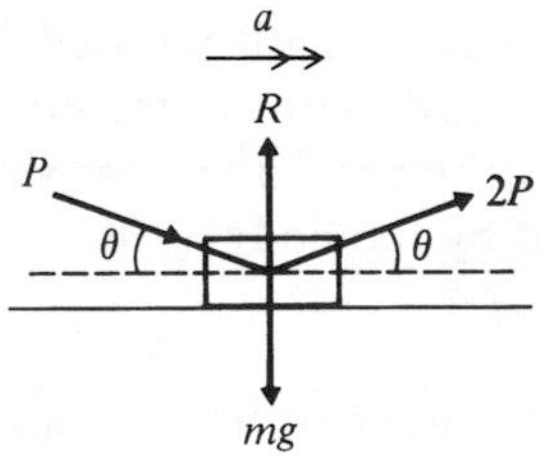

Prove: $3R = m(3g - a\tan\theta)$

4.

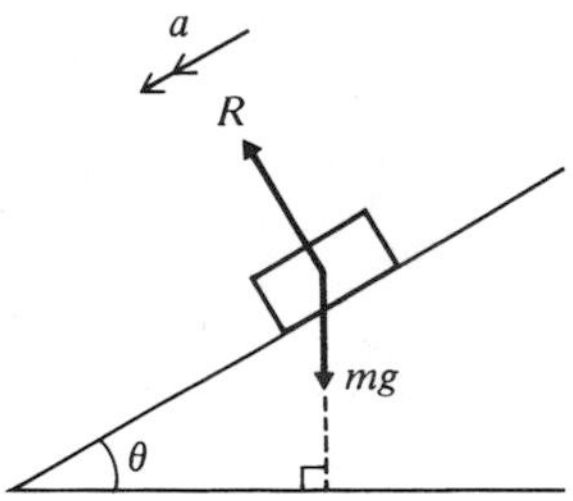

Prove: $R\tan\theta = ma$

5.

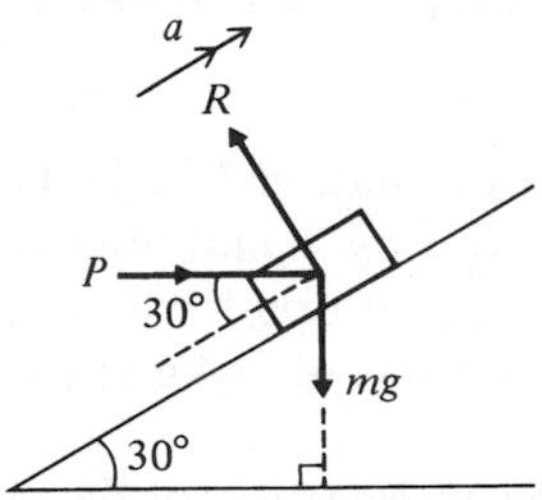

Prove: $\sqrt{3}\,P = m(2a + g)$

6.

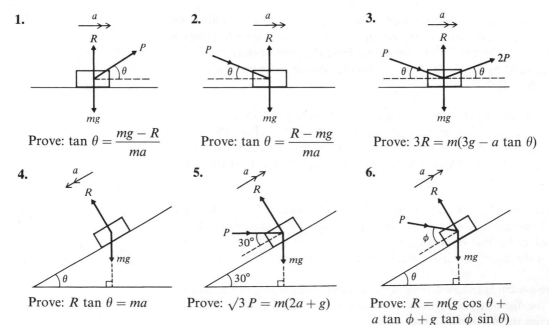

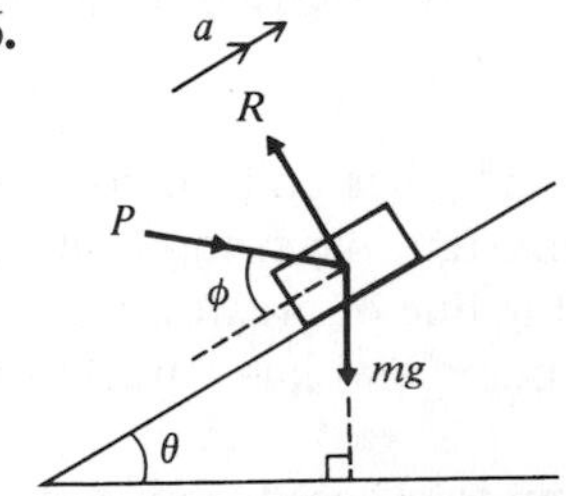

Prove: $R = m(g\cos\theta + a\tan\phi + g\tan\phi\sin\theta)$

7.

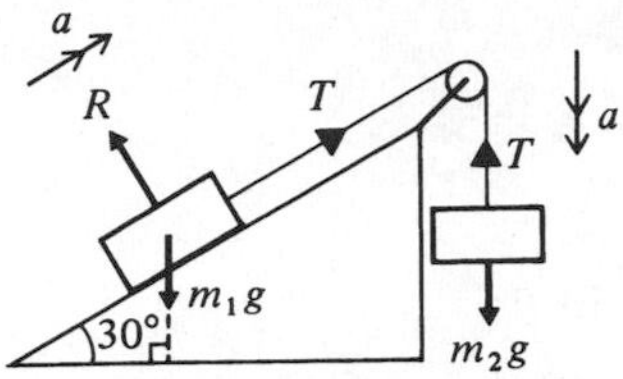

Prove: $a = \dfrac{(2m_2 - m_1)g}{2(m_1 + m_2)}$

8.

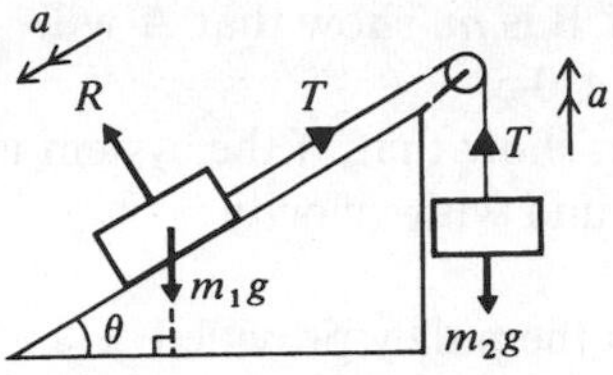

Prove: $T = \dfrac{m_1m_2g(1 + \sin\theta)}{m_1 + m_2}$

9.

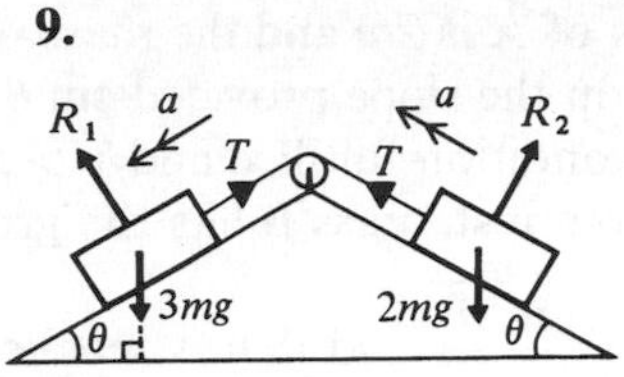

Prove: $g\sin\theta = 5a$

10. A body of mass m is pulled along a smooth horizontal surface by a force P inclined at θ above the horizontal.
If the mass starts from rest, show that the distance moved in time t is given by $\dfrac{Pt^2\cos\theta}{2m}$.

11. A body of mass m is pulled along a rough horizontal surface by a force P inclined at θ above the horizontal.
If the mass accelerates from rest to velocity v in a distance d, show that the resistance to motion (assumed constant throughout) is
$P\cos\theta - \dfrac{mv^2}{2d}$.

12. A mass of 5 kg is pulled along a rough horizontal surface by a force of 50 N inclined at 60° above the horizontal. The mass starts from rest and after 4 seconds the pulling force ceases.
If the resistance to motion is 20 N throughout, find the total distance travelled before the mass comes to rest again.

13. A body of mass m is released from rest at the top of a smooth slope which is inclined at θ to the horizontal.
Show that its velocity, when it has travelled a distance s down the slope, is given by $\sqrt{(2gs\sin\theta)}$.

14. A body of mass m is released from rest at the top of a rough plane which is inclined at θ to the horizontal. After time t the mass has travelled a distance d down the slope.

Show that the resistance to motion experienced by the body is $\dfrac{m}{t^2}(gt^2\sin\theta - 2d)$.
(Assume the resistance to be constant throughout.)

Questions **15**, **16** and **17** refer to the situation shown.
The body A lies on a smooth slope and body B is freely suspended. The pulley is smooth and the string light and inextensible.

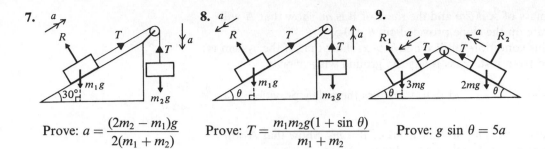

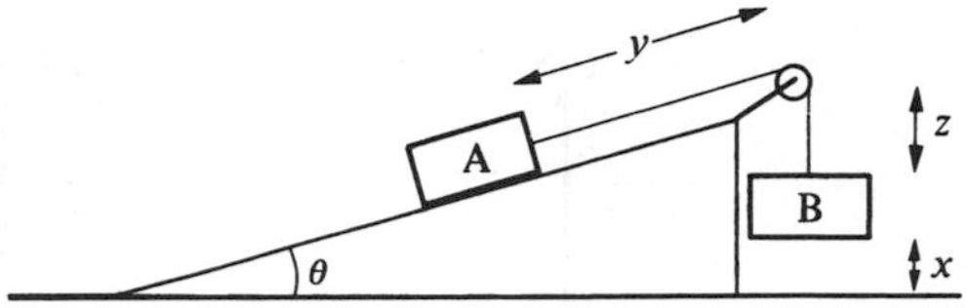

15. The mass of A is 4 kg and the mass of B is 3 kg. With $\theta = 30°$, body A will accelerate up the slope. If $y = 3$ m and $x = 2{\cdot}8$ m, find the velocity with which A hits the pulley.

16. If the mass of A is $2m$ and the mass of B is m, show that A will accelerate up the slope provided $\sin\theta < 0{\cdot}5$.
With this condition fulfilled and $y > x$, show that, if the system is released from rest, mass B hits the ground with velocity

$\sqrt{\dfrac{2gx(1-2\sin\theta)}{3}}$ and that A reaches the pulley provided $x \geqslant \dfrac{3y\sin\theta}{1+\sin\theta}$.

17. The mass of A is m_1 and the mass of B is m_2. Show that A will accelerate down the slope provided $m_1 \sin\theta > m_2$.
With this condition fulfilled, the system is released from rest and when A has travelled a distance d down the slope ($d < z$), the string connecting the two masses is cut. Show that the greatest height reached by B above the floor is $\dfrac{x(m_2+m_1)+dm_1(1+\sin\theta)}{(m_1+m_2)}$.

18. Masses m_1 and m_2 are held at rest on inclined surfaces in the positions shown in the diagram ($s > d$). They are connected by a light inextensible string passing over a smooth pulley. Show that, when the system is released, m_1 will accelerate towards the pulley provided $\sqrt{3} > \dfrac{m_1}{m_2}$.

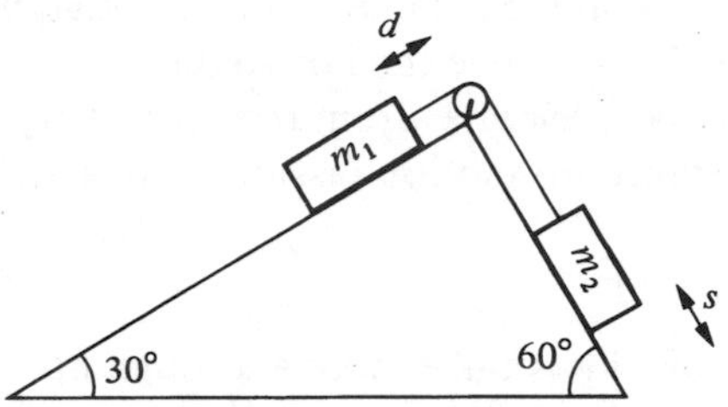

With this condition fulfilled show that m_1 hits the pulley with speed $\sqrt{\dfrac{dg(\sqrt{3}\,m_2-m_1)}{m_1+m_2}}$.

Exercise 5E Examination questions

(Unless otherwise indicated take $g = 9{\cdot}8\ \text{m s}^{-2}$ in this exercise.)

1. Fig. 1 shows three coplanar forces of magnitude 2 N, 3 N and P N all acting at a point O in the directions shown.

Given that the forces are in equilibrium obtain the numerical values of $P\cos\theta$ and $P\sin\theta$ and hence, or otherwise, find $\tan\theta$ and P.

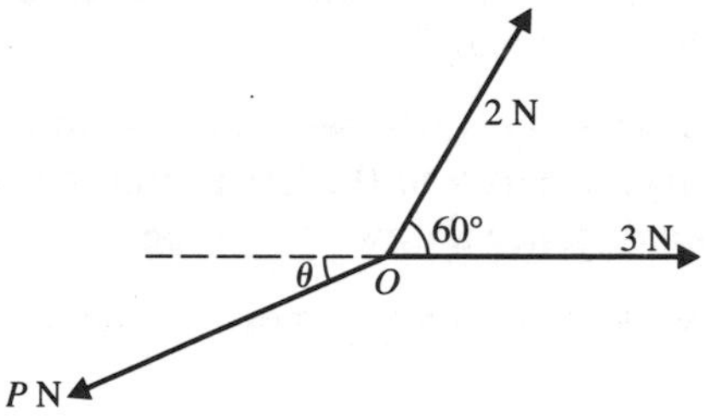

Fig.1

(AEB 1992)

2.

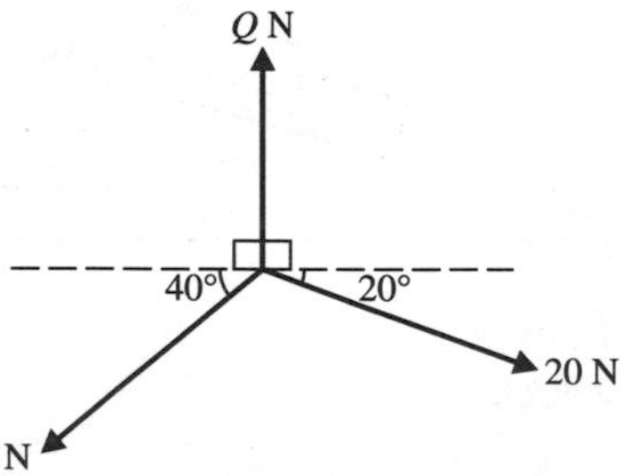

(i) The diagram shows three coplanar forces in equilibrium.
Find the value of P and of Q.

(ii) If the direction of Q is now reversed, find the magnitude and direction of the resultant of the three forces.

(UCLES)

3. Three forces $\mathbf{F}_1$, $\mathbf{F}_2$ and $\mathbf{F}_3$ act on a particle and $\mathbf{F}_1 = (-3\mathbf{i} + 7\mathbf{j})$ newtons, $\mathbf{F}_2 = (\mathbf{i} - \mathbf{j})$ newtons, $\mathbf{F}_3 = (p\mathbf{i} + q\mathbf{j})$ newtons.
(a) Given that this particle is in equilibrium, determine the value of p and the value of q.
The resultant of the forces $\mathbf{F}_1$ and $\mathbf{F}_2$ is $\mathbf{R}$.
(b) Calculate, in N, the magnitude of $\mathbf{R}$.
(c) Calculate, to the nearest degree, the angle between the line of action of $\mathbf{R}$ and the vector $\mathbf{j}$. (ULEAC)

4. (Take the acceleration due to gravity to be $10\,\mathrm{m\,s^{-2}}$ in this question, and give your answers correct to 2 significant figures.)

A particle P, of mass 0·2 kg, is suspended from a fixed point O by means of a light inextensible string. The string is taut and makes an angle of 30° with the downward vertical through O, and the particle is held in equilibrium by means of a force of magnitude F acting on the particle. Using a triangle of forces, or otherwise, find
(i) the value of F when the force acts horizontally,
(ii) the tension in the string when F takes its least possible value. (UCLES)

5. A particle of mass 0·3 kg lies on a smooth plane inclined at an angle α to the horizontal, where $\tan\alpha = \frac{3}{4}$. The particle is held in equilibrium by a horizontal force of magnitude Q newtons. The line of action of this force is in the same vertical plane as a line of greatest slope of the inclined plane. Calculate the value of Q, to one decimal place. (ULEAC)

6. A child is attempting to take two dogs for a walk. The dogs exert horizontal forces of 40 newtons and 50 newtons in directions making 120° with each other. Find the magnitude and direction of the horizontal force which the child must exert to maintain equilibrium. (NICCEA)

7. Three cables exert forces that act in a horizontal plane on the top of a telegraph pole.

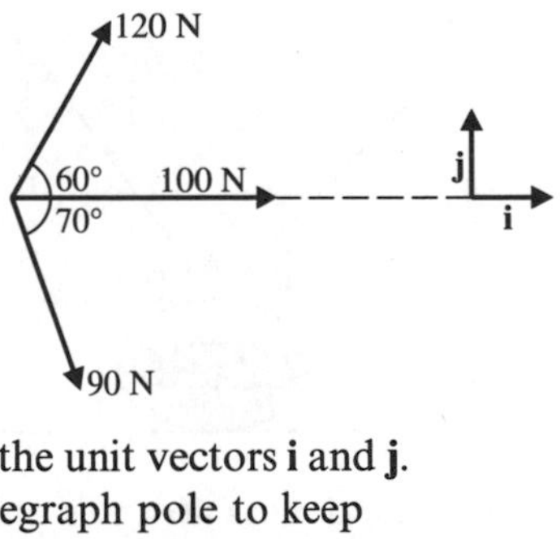

(a) Find the resultant of these 3 forces, in terms of the unit vectors $\mathbf{i}$ and $\mathbf{j}$.
(b) A fourth cable is attached to the top of the telegraph pole to keep the pole in equilibrium. Find the force, exerted by this fourth cable, in terms of $\mathbf{i}$ and $\mathbf{j}$.
(c) Show that the magnitude of the fourth force is 192 N, correct to 3 significant figures.
(d) On a diagram show clearly the direction in which the fourth force acts.
(e) The fourth cable does not lie in the same horizontal plane as the other 3 cables and the tension in this cable is in fact 200 N. Find the angle between this cable and the horizontal plane. (AEB Spec)

8. (Take $g = 10\,\mathrm{m\,s^{-2}}$ in this question.)

The diagram shows a smooth bead B, of mass M kg, which is threaded on a light inextensible string. The ends of the string are attached to two points A and C, which are on the same horizontal level. A horizontal force of 4 N maintains B vertically below C, where $BC = 1{\cdot}5$ m, with the string taut. Given that the string is of length 4 m, find

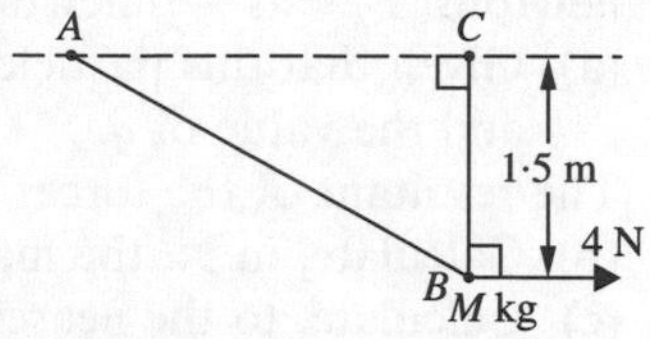

(i) the tension in the string,
(ii) the value of M. (UCLES)

9. (Take $g = 10\,\mathrm{m\,s^{-2}}$ in this question.)

Two strings AB and BC are tied to a particle of mass 0·3 kg at B, and the end A is fixed. A second particle of mass 0·4 kg is attached at C. A horizontal force of magnitude P newtons at C maintains the system in equilibrium with the string BC making an angle of 60° with the downward vertical, and with the string AB making an angle θ with the downward vertical. The forces in the diagram are in newtons.

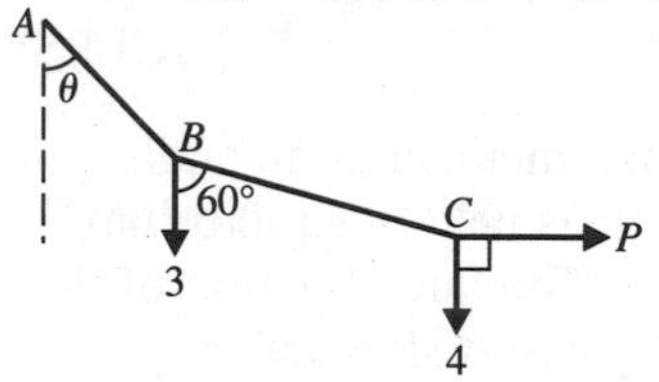

(a) Calculate the magnitude of the tension in BC, and determine the value of P.

(b) Show that $\tan\theta = \dfrac{4\sqrt{3}}{7}$ and calculate the magnitude of the tension in AB. (UODLE)

10. (Take $g = 10\,\mathrm{m\,s^{-2}}$ in this question.)

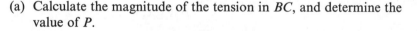

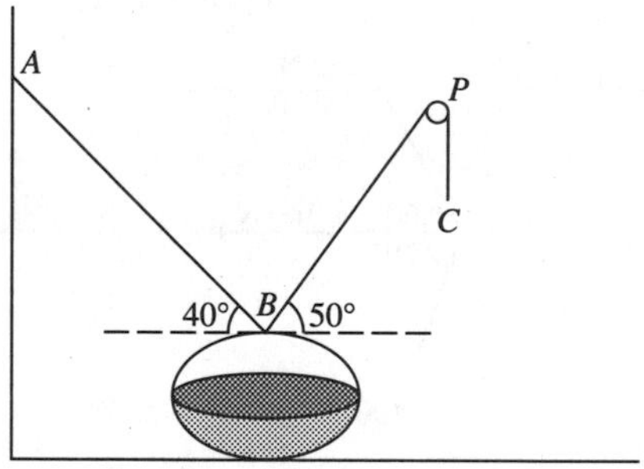

A basket of earth, of total mass 40 kg, is attached to two light inextensible ropes BA and BC. The end A is attached to a fixed point and BC passes over a fixed smooth pulley P. A downward pull is applied to the end C. The parts AB and BP of the rope make angles 40° and 50° respectively with the horizontal (see diagram). The basket is at rest on the ground; the force exerted on the basket by the ground is vertical and has magnitude 150 N. Find the tensions in AB and BP and find the magnitude and direction of the resultant force on the pulley due to the rope. (UCLES)

11. A car of mass 1000 kg, including its driver, is being pushed along a horizontal road by three people as indicated in the diagram. The car is moving in the direction PQ.

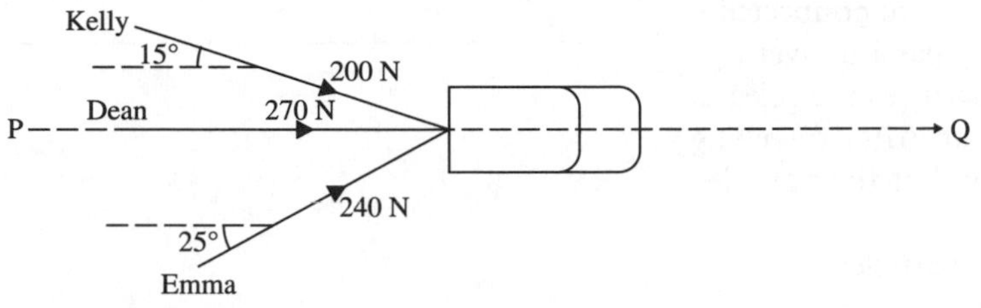

(i) Calculate the total force exerted by the three people in the direction PQ.

(ii) Calculate the force exerted overall by the three people in the direction perpendicular to PQ.

(iii) Explain briefly why the car does not move in the direction perpendicular to PQ.

Initially the car is stationary and 5 seconds later it has a speed of $2\,\text{m s}^{-1}$ in the direction PQ.

(iv) Calculate the force of resistance to the car's movement in the direction PQ, assuming the three people continue to push as described above.

(v) The car comes to a steady downhill slope. The three people stop pushing but the car maintains the same acceleration. Assuming the resistance to motion is unchanged and remains constant, what angle does the slope make with the horizontal? (OCSEB)

12. (Take $g = 10\,\text{m s}^{-2}$ in this question.)
A particle travels along a line of greatest slope of a smooth plane inclined at an angle θ to the horizontal, where $\sin\theta = \frac{3}{5}$. The particle starts from a point P where it is given an initial speed of $15\,\text{m s}^{-1}$ up the plane towards a point Q. Given that PQ is 12 m, calculate the speed of the particle when it first passes Q.
Calculate also the time for which the particle is above the level of Q.
(UCLES)

13. A particle of mass 2 kg on a smooth plane inclined at 30° to the horizontal is attached by means of a light inextensible string passing over a smooth pulley at the top edge of the plane to a particle of mass 3 kg which hangs freely. If the system is released from rest with both parts of the string taut, find the speed acquired by the particles when both have moved a distance of 1 m.

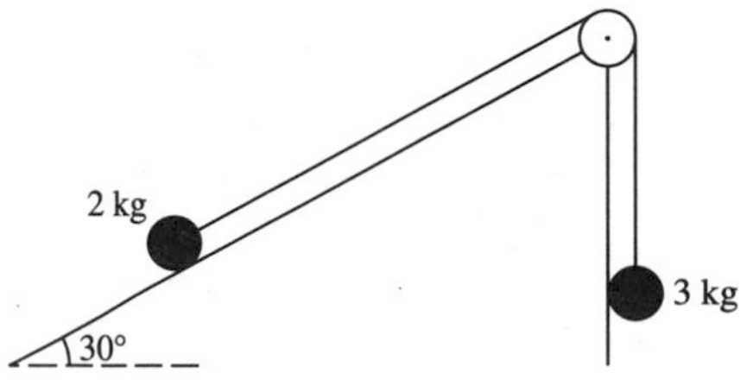

(UODLE)

14. (Take $g = 10\,\mathrm{m\,s^{-2}}$ in this question.)
The diagram shows two smooth fixed slopes each inclined at an angle α to the horizontal, where $\sin\alpha = 0{\cdot}6$. Two particles of mass 3 kg and M kg, where $M < 3$, are connected by a light inextensible string passing over a smooth fixed pulley. The particles are released from rest with the string taut. After travelling a distance of 1·08 m the speed of the particles is $1{\cdot}8\,\mathrm{m\,s^{-1}}$. Calculate

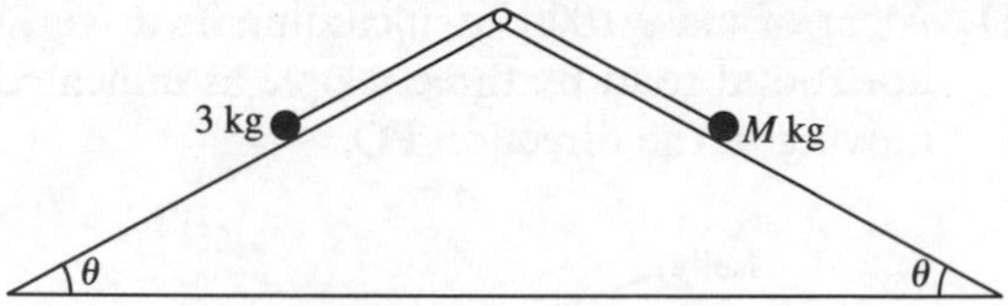

(i) the acceleration of the particles
(ii) the tension in the string
(iii) the value of M. (UCLES)

15. (Take $g = 10\,\mathrm{m\,s^{-2}}$ in this question.)
The diagram shows two particles, A of mass 2 kg and B of mass 1·5 kg, connected by a light inextensible string passing over a smooth pulley. The system is released from rest with A at a height of 3·6 m above horizontal ground and B at the foot of a smooth slope inclined at an angle θ to the horizontal where $\sin\theta = \frac{1}{6}$.

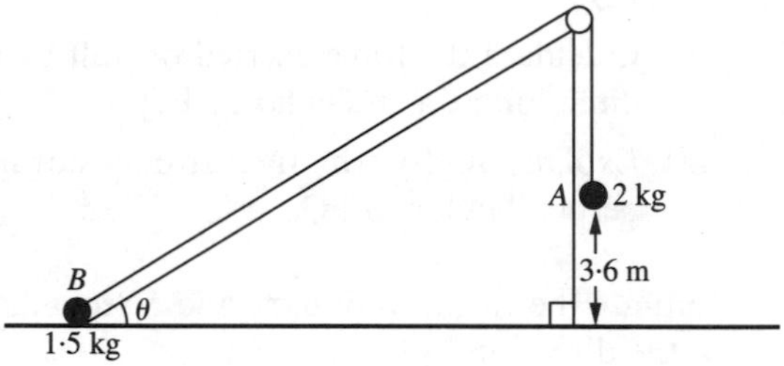

Calculate
(i) the magnitude of the acceleration of the particles,
(ii) the speed with which A reaches the ground,
(iii) the distance B moves up the slope before coming to instantaneous rest. (UCLES)

6 Friction

Rough and smooth surfaces

A block of mass M kg rests on a horizontal table and a horizontal force of P newtons is applied to the block.
From Newton's Third Law, it is known that equal and opposite forces act on the block and on the plane at right angles to the surfaces in contact.

$\therefore \qquad R = Mg$

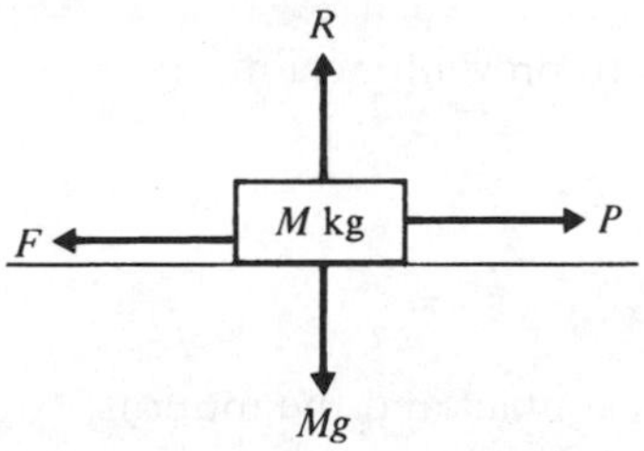

It is known from experience that, if the surfaces in contact are highly polished, it is easier to move the block than if both the underside of the block and the surface of the table are covered with sandpaper.
The force F which opposes the motion of the block is called the frictional force, and it acts in a direction to oppose the motion and is parallel to the surfaces in contact. If the surfaces were perfectly smooth, there would be no frictional force; hence F would be 0 and motion would take place however small the applied force P might be. In practice, it is not possible to have perfectly smooth surfaces, although in particular instances we may consider the surfaces to be smooth.
When the surfaces are rough, the block will only move if P is greater than the frictional force F. The magnitude of the frictional force depends upon the roughness of the surfaces in contact and also upon the force P which is trying to move the block.

Limiting equilibrium

The frictional force F for a particular block and surface is not constant, but increases as the applied force P increases until the force F reaches a value F_{max} beyond which it cannot increase. The block is then on the point of moving and is said to be in a state of *limiting equilibrium.*

Suppose $\qquad F_{max} = P_1$

If the applied force P is increased still further to a value P_2, the frictional force cannot increase as it has already reached its maximum value, and the block will therefore move.

Since force = mass × acceleration, the equation of motion is:

$$P_2 - F_{max} = M \times a$$

Coefficient of friction

The magnitude of the maximum frictional force is a fraction of the normal reaction R. This fraction is called the coefficient of friction μ for the two surfaces in contact:

$$F_{max} = \mu R$$

For a perfectly smooth surface, $\mu = 0$.

It should be noted that the maximum frictional force will only act if
(a) there is a state of limiting equilibrium, or
(b) motion is taking place.

The frictional force F is only as large as is necessary to prevent motion.

Laws of friction

The laws governing the equilibrium of two bodies in contact and the motion of one body on another, may be summarized as follows.

The frictional force:
(i) acts parallel to the surfaces in contact and in a direction so as to oppose the motion of one body across the other
(ii) will not be larger than is necessary to prevent this motion
(iii) has a maximum value μR, where R is the normal reaction between the surfaces in contact
(iv) can be assumed to have its maximum value μR when motion occurs
(v) depends upon the nature of the surfaces in contact and not upon the contact area.

Example 1

Calculate the maximum frictional force which can act when a block of mass 2 kg rests on a rough horizontal surface, the coefficient of friction between the surfaces being
(a) 0·7 (b) 0·2

(a) There is no motion perpendicular to the plane.

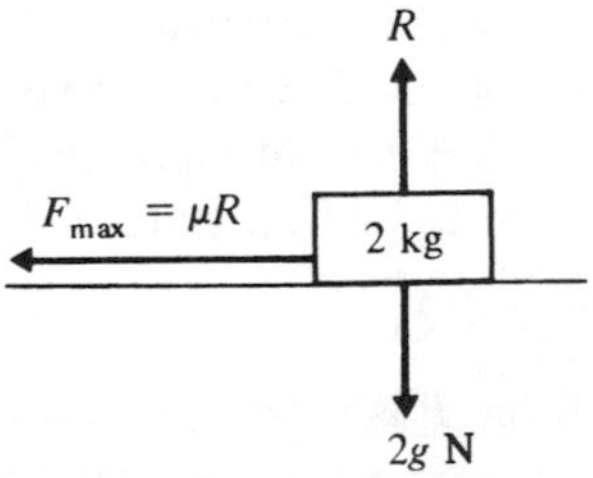

Resolve vertically: $R = 2g$

$\therefore$ $R = 19{\cdot}6\,\text{N}$

maximum frictional force $F_{max} = \mu R$

$= (0{\cdot}7) \times 19{\cdot}6$

$= 13{\cdot}72\,\text{N}$

(b) As before $R = 19{\cdot}6\,\text{N}$

maximum frictional force $= \mu R$

$= (0{\cdot}2) \times 19{\cdot}6$

$= 3{\cdot}92\,\text{N}$

Example 2

A block of mass 5 kg rests on a rough horizontal plane, the coefficient of friction between the block and the plane being 0·6. Calculate the frictional force acting on the block when a horizontal force P is applied to the block and the magnitude of P is: (a) 12 N (b) 28 N (c) 36 N.
Also calculate the magnitude of any acceleration that may occur.

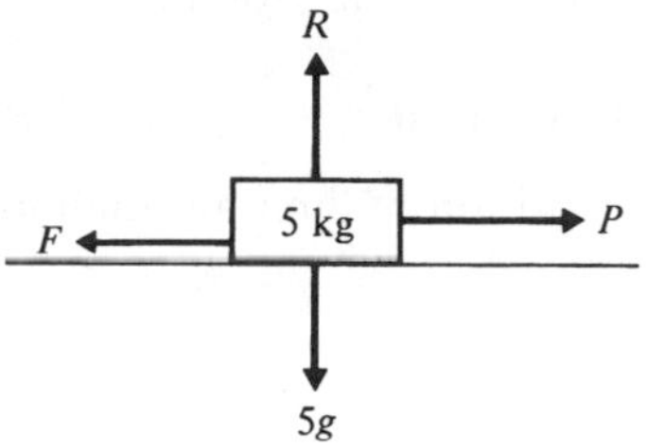

There is no motion perpendicular to the plane.

Resolve vertically: $R = 5g$

$\therefore$ $R = 49\text{ N}$

The frictional force will act in the direction opposite to that in which the force P acts. The maximum value of the frictional force is μR.

$$\mu R = 0{\cdot}6 \times 49 = 29{\cdot}4\text{ N}$$

(a) If $P = 12\text{ N}$, then P is less than μR, so there is no motion:

frictional force $F = P$
$F = 12\text{ N}$

(b) If $P = 28\text{ N}$, then again P is less than μR and there is no motion:

frictional force $F = P$
$F = 28\text{ N}$

(c) If $P = 36\text{ N}$, then P is greater than the maximum value of the frictional force, which is 29·4 N

frictional force acting = 29·4 N, which does not prevent motion

The block will move and the maximum value μR of the frictional force will be maintained.

Using $F = ma$, the equation of motion is:

$$P - \mu R = m \times a$$
$$36 - 29{\cdot}4 = 5 \times a$$
$$\therefore \quad a = 1{\cdot}32\text{ m s}^{-2}$$

Applied force not horizontal

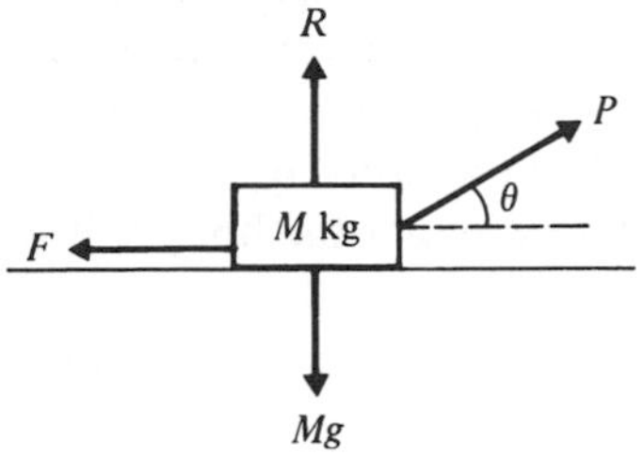

When the force P acting on the block of mass M is inclined at an angle θ above the horizontal, this has two effects:

(i) the component of P in a vertical direction decreases the magnitude of the normal reaction R,
(ii) only the component of P in a horizontal direction is tending to move the block.

Hence the value of R is less than it would have been if the force P had been applied horizontally. The maximum frictional force μR is also therefore reduced. In addition, a smaller force is tending to cause motion, a component of P rather than P.

Example 3

A 10 kg trunk lies on a horizontal rough floor. The coefficient of friction between the trunk and the floor is $\frac{\sqrt{3}}{4}$. Calculate the magnitude of the force P which is necessary to pull the trunk horizontally if P is applied:
(a) horizontally
(b) at 30° above the horizontal.

(a) Resolve vertically: $R = 10g$

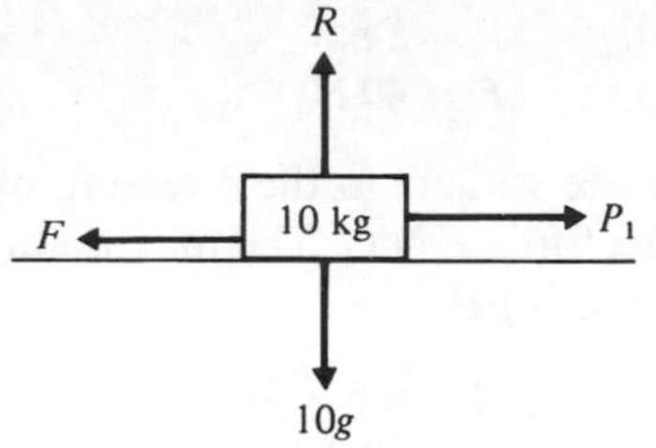

In the position of limiting equilibrium:

$$P_1 = \mu R = \frac{\sqrt{3}}{4} \times 10g = 42{\cdot}43\text{ N}$$

For motion to take place the applied force must exceed 42·43 N.

(b) Resolve vertically: $R + P_2 \cos 60° = 10g$

$$R = 98 - \frac{P_2}{2}$$

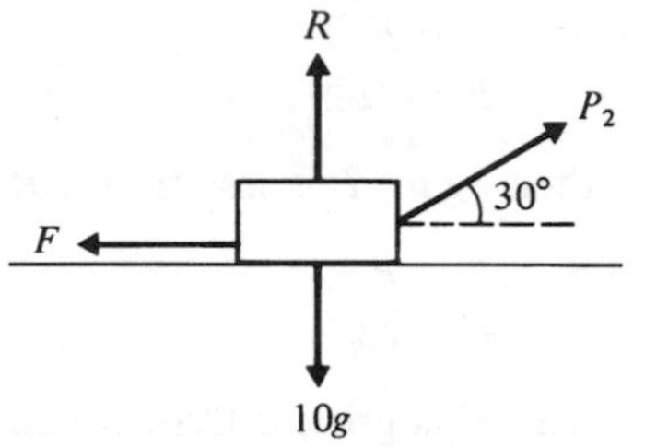

In the position of limiting equilibrium:

$$P_2 \cos 30° = \mu R$$

$$\therefore \quad P_2 \cos 30° = \mu\left(98 - \frac{P_2}{2}\right)$$

Hence $\quad P_2 \times \frac{\sqrt{3}}{2} + \frac{\sqrt{3}}{4} \times \frac{P_2}{2} = \frac{\sqrt{3}}{4} \times 98$

or $\quad P_2 = 39{\cdot}2\text{ N}$

For motion to take place the applied force must exceed 39·2 N.

Therefore it is easier to move such a trunk if the pulling force is inclined upwards as this reduces the frictional force opposing the motion.

Exercise 6A

1. Each of the following diagrams shows a body of mass 10 kg initially at rest on a rough horizontal plane. The coefficient of friction between the body and the plane is $\frac{1}{7}$. In each case, R is the normal reaction and F the frictional force exerted on the body, by the plane. Any other forces applied to the body are as shown. In each case, find the magnitude of F and state whether the body will remain at rest or will accelerate along the plane.

(a)

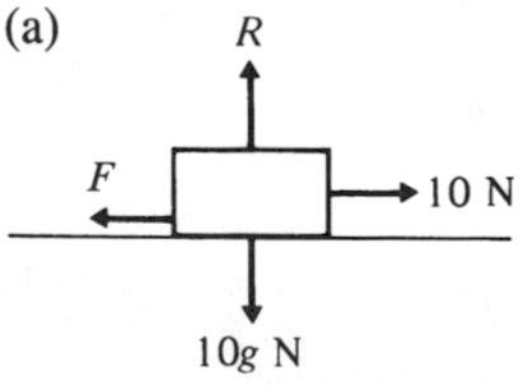

(b)

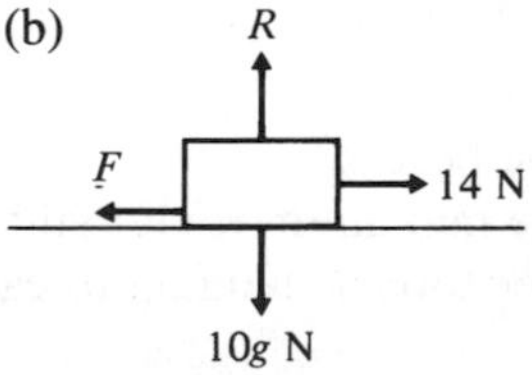

(c)

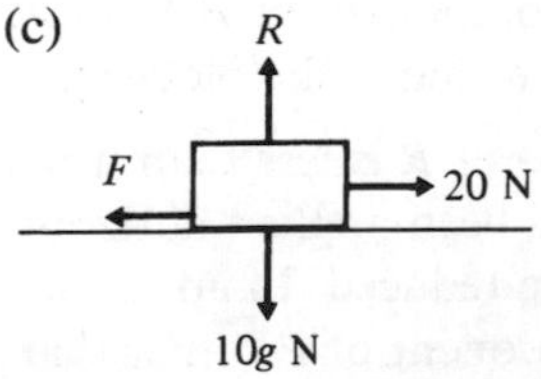

(d)

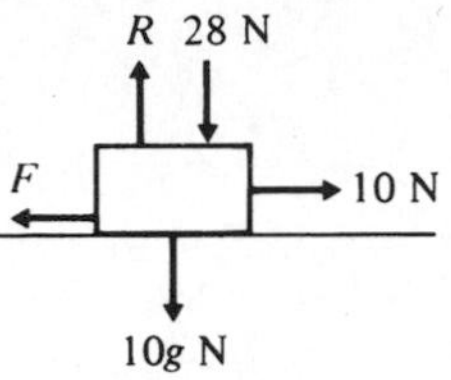

(e)

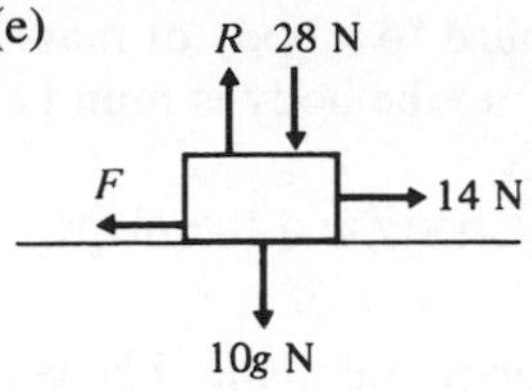

(f)

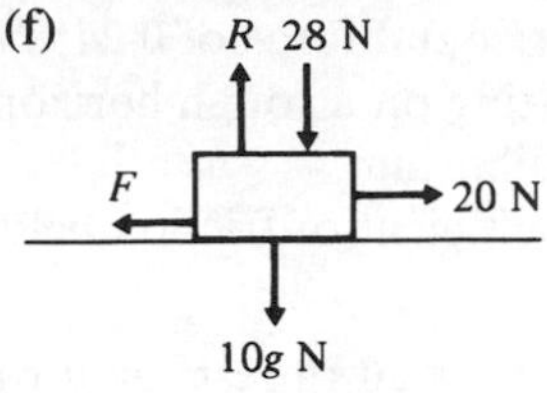

(g)

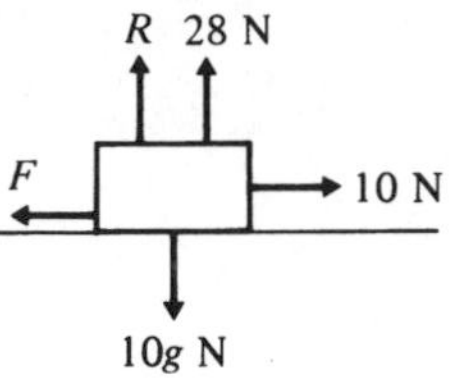

(h)

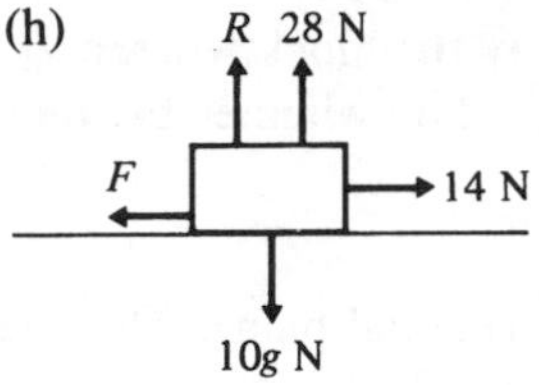

(i)

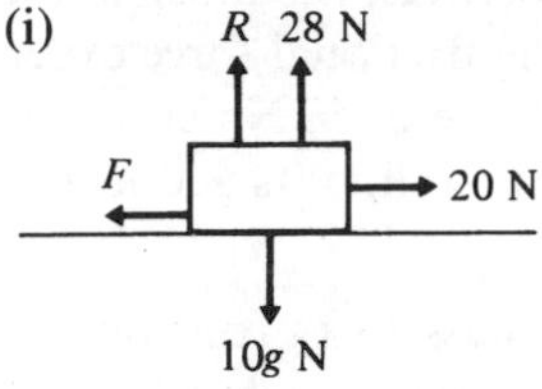

(j)

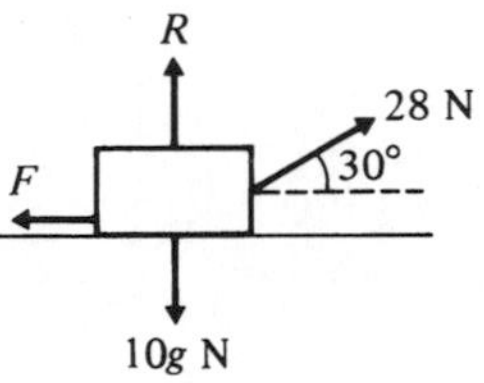

(k)

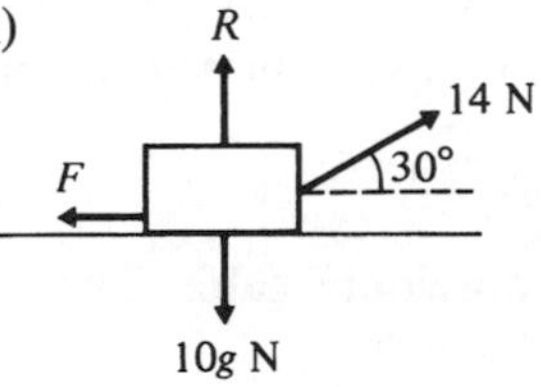

(l)

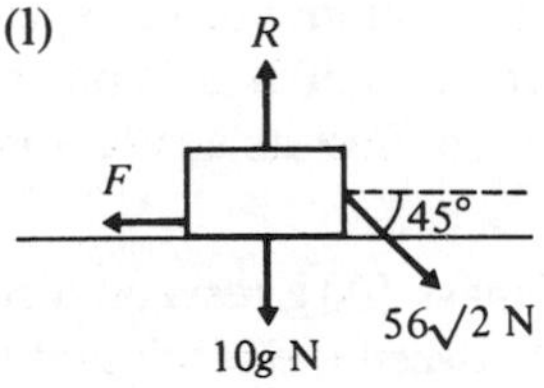

2. In each of the following situations, the forces shown cause the body of mass 5 kg to accelerate along the rough horizontal plane. The direction and magnitude of each acceleration is as indicated; R is the normal reaction and F the frictional force exerted on the body by the plane. For each case, find the coefficient of friction between the body and the plane.

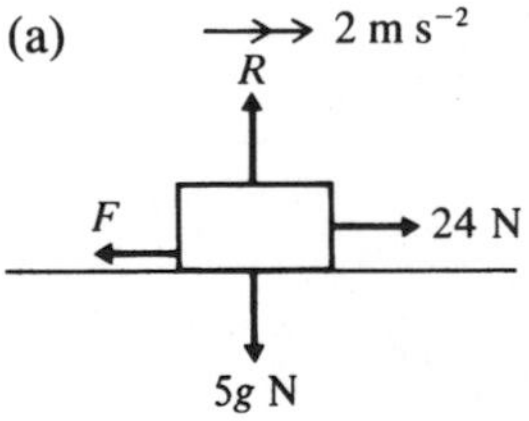

(b)

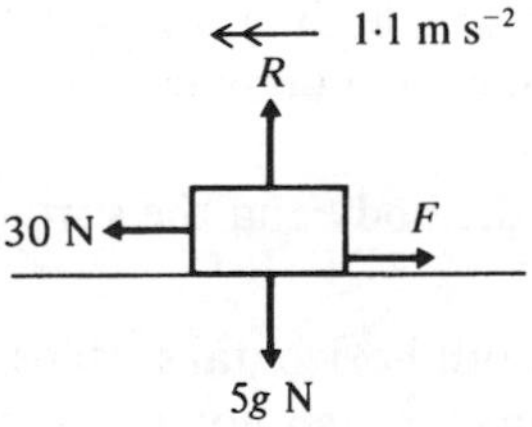

(c)

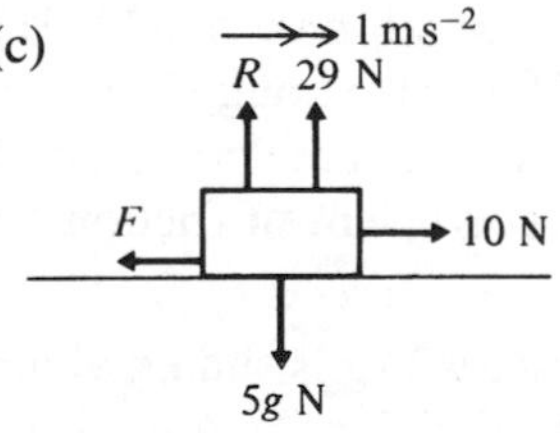

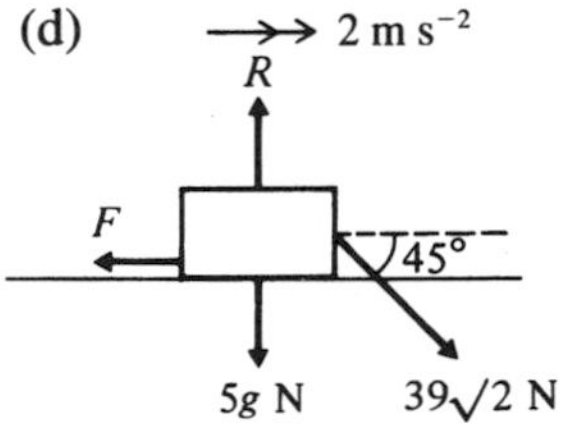

(e)

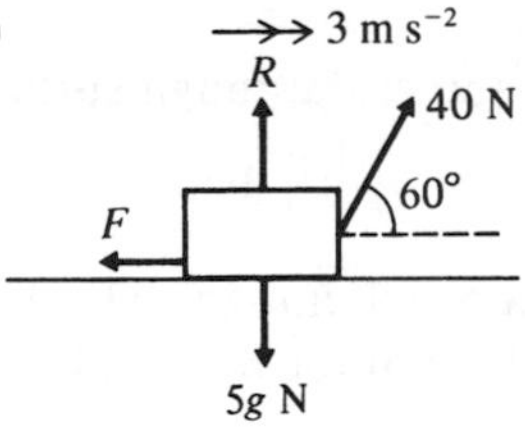

(f)

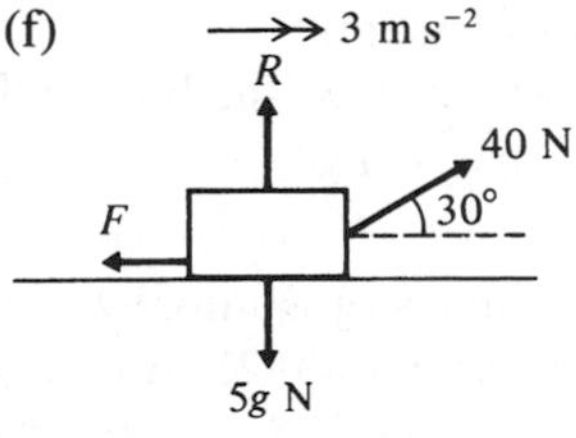

3. When a horizontal force of 28 N is applied to a body of mass 5 kg which is resting on a rough horizontal plane, the body is found to be in limiting equilibrium.
Find the coefficient of friction between the body and the plane.

4. When a horizontal force of 0·245 N is applied to a body of mass 250 g which is resting on a rough horizontal plane, the body is found to be in limiting equilibrium.
Find the coefficient of friction between the body and the plane.

5. A block of mass 20 kg rests on a rough horizontal plane. The coefficient of friction between the block and the plane is 0·25.
Calculate the frictional force experienced by the block when a horizontal force of 50 N acts on the block. State whether the block will move and, if so, find its acceleration.

6. A block of mass 15 kg rests on a rough horizontal plane. The coefficient of friction between the block and the plane is 0·35.
Calculate the frictional force experienced by the block when a horizontal force of 50 N acts on the block. State whether the block will move and, if so, find its acceleration.

7. A block of mass 500 g rests on a rough horizontal table. The coefficient of friction between the block and the table is 0·1.
Calculate the frictional force experienced by the block when a horizontal force of 1 N acts on the block. State whether the block will move and, if so, find its acceleration.

8. A block of mass 2 kg is initially at rest on a rough horizontal table. The coefficient of friction between the block and the table is 0·5.
Find the horizontal force that must be applied to the block to cause it to accelerate along the surface at: (a) $5\,\text{m}\,\text{s}^{-2}$ (b) $0{\cdot}1\,\text{m}\,\text{s}^{-2}$.

9. When a horizontal force of 37 N is applied to a body of mass 10 kg which is resting on a rough horizontal surface, the body moves along the surface with an acceleration of $1{\cdot}25\,\text{m}\,\text{s}^{-2}$.
Find μ, the coefficient of friction between the body and the surface.

10. A body of mass 2 kg is sliding along a smooth horizontal surface at a constant speed of $2\,\text{m}\,\text{s}^{-1}$ when it encounters a rough horizontal surface, coefficient of friction 0·2.
Find the distance that the body will move across the rough surface before it comes to rest.

11. A body of mass 1 kg is initially at rest on a rough horizontal surface, coefficient of friction 0·25. A constant horizontal force is applied to the body for 5 seconds and is then removed.
Given that when the force is removed, the body has a velocity of $3{\cdot}5\,\text{m}\,\text{s}^{-1}$ along the surface, find:
(a) the acceleration of the body when experiencing the applied force
(b) the magnitude of the applied force
(c) the retardation of the body when the force is removed
(d) the total distance travelled by the body.

12. A box of mass 2 kg lies on a rough horizontal floor, coefficient of friction 0·5. A light string is attached to the box in order to pull the box across the floor.
If the tension in the string is T N, find the value that T must exceed for motion to occur if the string is:
(a) horizontal
(b) 30° above the horizontal
(c) 30° below the horizontal.

13. A box of mass 2 kg lies on a rough horizontal floor, coefficient of friction 0·2. A light string is attached to the box in order to pull the box across the floor. If the tension in the string is T N, find the value that T must exceed for the motion to occur if the string is:
(a) horizontal
(b) 45° above the horizontal
(c) 45° below the horizontal.

14. A body of mass 100 g rests on a rough horizontal surface and has a light string, inclined at 20° above the horizontal, attached to it. When the tension in the string is 5×10^{-1} N, the body is found to be in limiting equilibrium.
Find the coefficient of friction between the body and the surface.
What would the tension in the string have to be for the body to accelerate along the surface at $1{\cdot}5\,\text{m s}^{-2}$?

15. A mass of 3 kg lies on a rough horizontal surface, coefficient of friction $\frac{1}{7}$. State whether or not the mass will slide along the surface when the surface is moved horizontally with an acceleration of:
(a) $1\,\text{m s}^{-2}$
(b) $1{\cdot}4\,\text{m s}^{-2}$
(c) $2\,\text{m s}^{-2}$.

16. A mass of 6 kg lies on a rough horizontal surface, coefficient of friction 0·25.
State whether or not the mass will slide along the surface when the surface is moved horizontally with an acceleration of:
(a) $2{\cdot}4\,\text{m s}^{-2}$
(b) $2{\cdot}6\,\text{m s}^{-2}$
(c) $3\,\text{m s}^{-2}$

17. A parcel is placed on the tail-board of a stationary lorry. The tail-board is horizontal and the coefficient of friction between the parcel and the tail-board is 0·2. State whether or not the parcel will slide along the surface of the tail-board when the lorry moves off horizontally with an acceleration of:
(a) $1\,\text{m s}^{-2}$
(b) $1{\cdot}5\,\text{m s}^{-2}$
(c) $2{\cdot}5\,\text{m s}^{-2}$.

18. A cake of mass 500 g lies on the horizontal surface of a plate. The coefficient of friction between the cake and the plate is 0·1. Will the cake slide across the plate when the plate is moved horizontally with an acceleration of $1{\cdot}1\,\text{m s}^{-2}$?

Questions **19** to **22** refer to the system shown in the diagram. Body A lies on a rough horizontal table and is connected to the freely hanging body B by a light inextensible string passing over a smooth pulley.

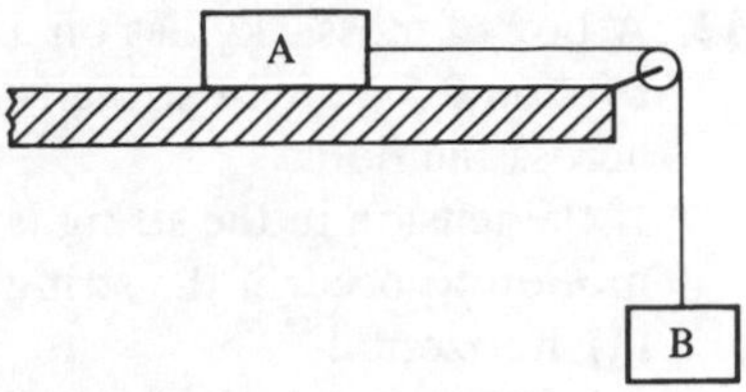

19. The masses of A and B are 6 kg and 1 kg respectively, and the coefficient of friction between body A and the table is 0·2. If the system is released from rest, find the frictional force experienced by A and state whether motion will occur.

20. The masses of A and B are 90 g and 50 g respectively, and the coefficient of friction between body A and the table is $\frac{1}{3}$.
If the system is released from rest, find:
(a) the acceleration of the system
(b) the tension in the string
(c) the distance moved by A in the first second of motion, assuming that nothing impedes the motion of either mass.

21. The masses of A and B are 1 kg and 500 g respectively and the coefficient of friction between body A and the table is $\frac{1}{3}$. The system is released from rest with A 3 metres from the pulley and B 2·5 metres above the floor.
Find:
(a) the initial acceleration of the system
(b) the speed with which B hits the floor
(c) the speed with which A hits the pulley.

22. The masses of A and B are m_1 and m_2 respectively and the coefficient of friction between body A and the table is μ. Show that, if the system is released from rest, motion will occur if $m_2 > m_1 \mu$. If this condition is fulfilled show that the resulting acceleration will be $\dfrac{g(m_2 - \mu m_1)}{m_1 + m_2}$.

Questions **23** to **25** refer to the system shown in the diagram. Body A lies on a rough horizontal table and is connected to freely hanging bodies B and C by light inextensible strings passing over smooth pulleys. The masses, pulleys and strings all lie in the same vertical plane.

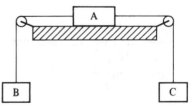

23. The masses of A, B and C are $4m$, m and $5m$ respectively, and the coefficient of friction between body A and the table is $\frac{1}{4}$. Find the acceleration of the system when released from rest.

24. The masses of A, B and C are 5 kg, 3 kg and 2 kg respectively. When the system is released from rest, body B descends witn an acceleration of $0{\cdot}28\ \text{m s}^{-2}$. Find the coefficient of friction between body A and the table.

25. The masses of A, B and C are m_1, m_2 and m_3 respectively, and the coefficient of friction between body A and the table is μ. Show that, if the system is released from rest, body B will move downwards, provided $m_2 > \mu m_1 + m_3$. If this condition is fulfilled, show that the resulting acceleration will be $\dfrac{g(m_2 - \mu m_1 - m_3)}{m_1 + m_2 + m_3}$.

Rough inclined plane

A body of mass M kg rests on a plane which is inclined at θ to the horizontal.
The vertical force Mg can be resolved into two components, parallel to and perpendicular to the surface of the plane. The plane exerts a normal reaction R on the body; since there is no motion at right angles to the plane, the normal reaction balances the component of Mg acting in this direction.

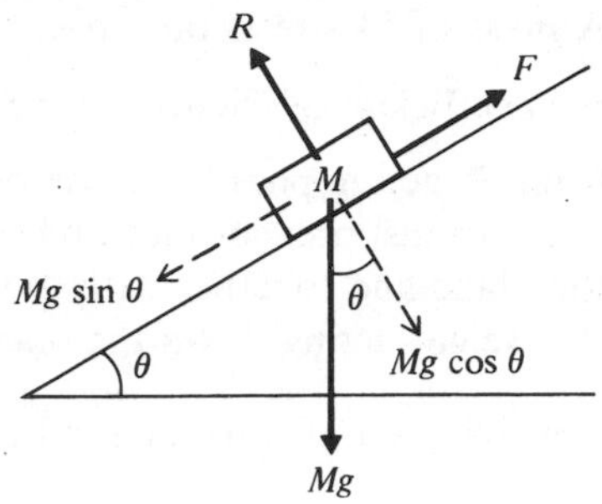

Resolve at right angles to the plane:

$$R = Mg \cos \theta$$

The component $Mg \sin \theta$ acting down the plane will cause motion unless the frictional force F, acting up the plane, balances it.

For equilibrium $\quad F = Mg \sin \theta$

The maximum value of F is, as before, μR:

$$F_{max} = \mu R = \mu Mg \cos \theta$$

In the position of limiting equilibrium, the maximum frictional force must balance the force tending to produce motion:

$$\therefore \quad \mu Mg \cos \theta = Mg \sin \theta$$

For motion to take place down the plane, $Mg \sin \theta$ must exceed $\mu Mg \cos \theta$

$$\therefore \quad Mg \sin \theta > \mu Mg \cos \theta$$
$$\therefore \quad \tan \theta > \mu$$

Example 4

A mass of 6 kg rests in limiting equilibrium on a rough plane inclined at 30° to the horizontal. Find the coefficient of friction between the mass and the plane.

The mass is on the point of moving down the plane, so the frictional force F acts up the plane and has its maximum value μR.

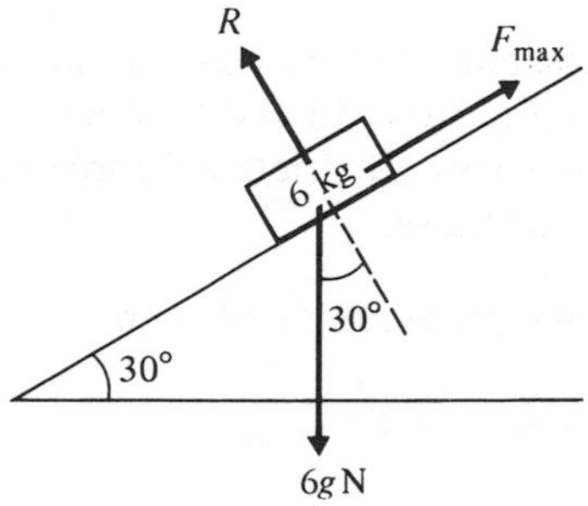

Resolving at right angles to the plane gives:

$$R = 6g \cos 30°$$
$$\therefore \quad R = 3g\sqrt{3}$$

Resolving parallel to the surface of the plane gives:

$$6g \sin 30° = \mu R$$
$$\therefore \quad 6g \sin 30° = \mu \times 3g\sqrt{3}$$
$$\therefore \quad \mu = \frac{1}{\sqrt{3}}$$

Example 5

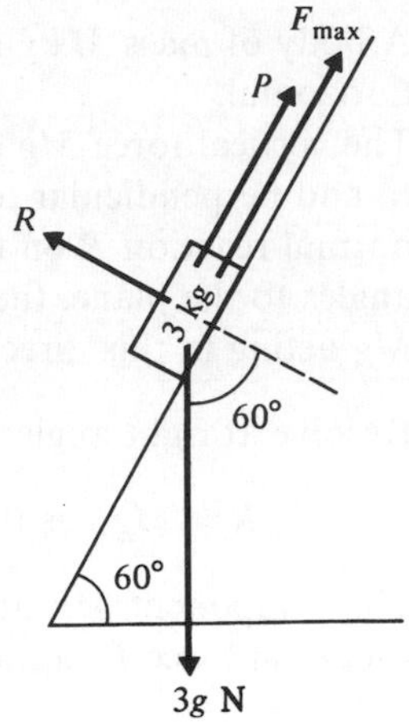

A mass of 3 kg rests on a rough plane inclined at 60° to the horizontal and the coefficient of friction between the mass and the plane is $\frac{\sqrt{3}}{5}$. Find the force P, acting parallel to the plane, which must be applied to the mass in order to just prevent motion down the plane. The frictional force F acts up the plane and together with the applied force P balances the component of the weight acting down the plane.

Resolving parallel to the surface of the plane gives:

$$P + F = 3g \sin 60° \qquad \ldots [1]$$

There is no motion perpendicular to the plane, and so resolving at right angles to the plane gives:

$$R = 3g \cos 60°$$

$$\therefore \quad R = \frac{3g}{2} \text{ N}$$

Since the mass is in limiting equilibrium (motion is *just* prevented)

$$F = \mu R$$

$$= \frac{\sqrt{3}}{5} \times \frac{3g}{2}$$

Substituting for F in equation [1] gives:

$$P + \frac{3\sqrt{3}}{10} g = \frac{3g\sqrt{3}}{2}$$

$$\therefore \quad P = \frac{6\sqrt{3}g}{5} \text{ N}$$

Motion up the plane

If the force P applied to the mass is larger than the component of Mg resolved down the plane, then the tendency will be for the mass to move *up* the plane. In this case the frictional force will act *down* the plane, opposing the motion of the mass.

Motion will take place up the plane if:

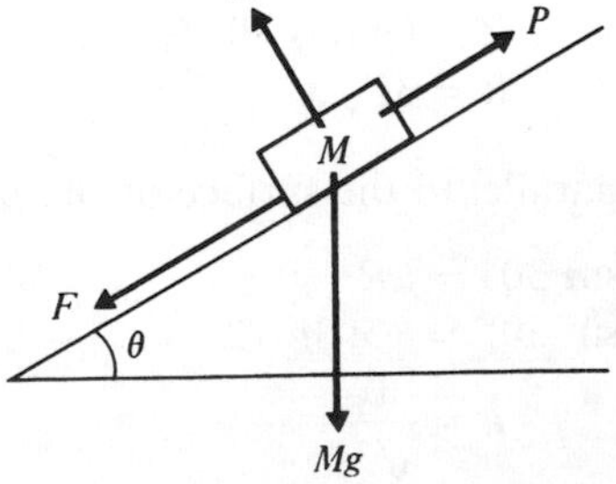

$$P > Mg \sin \theta + F_{max}$$

but $\quad F_{max} = \mu R$

$$= \mu Mg \cos \theta$$

So for motion up the plane $P > Mg \sin \theta + \mu Mg \cos \theta$.

Example 6

A mass of 0·5 kg rests on a rough plane. The coefficient of friction between the mass and the plane is $\frac{1}{\sqrt{2}}$, and the plane is inclined at angle θ to the horizontal such that $\sin\theta = \frac{1}{3}$.

Investigate the motion of the mass when it experiences a force of 6 N applied up the plane along a line of greatest slope.

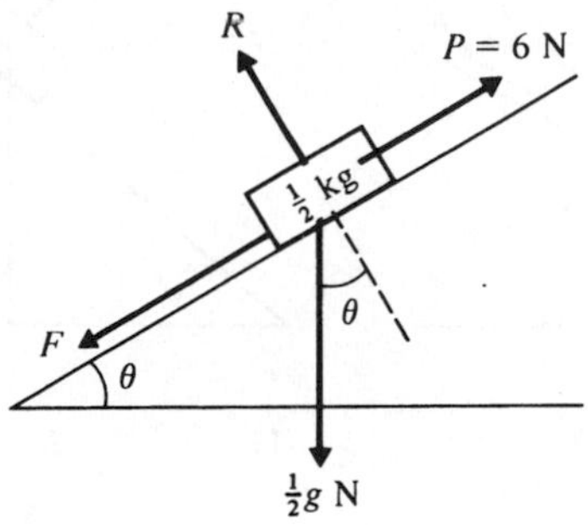

Since $\quad \sin\theta = \frac{1}{3}, \quad \cos\theta = \frac{2\sqrt{2}}{3}$

Resolve perpendicular to the plane:

$$R = \tfrac{1}{2}g\cos\theta$$

$$\therefore \qquad R = \tfrac{1}{2}g \times \frac{2\sqrt{2}}{3} = \frac{g\sqrt{2}}{3}$$

The forces acting down the plane are the component of the weight and the frictional force F if we assume the mass is tending to move up the plane. The component of the weight down the plane is $\frac{1}{2}g\sin\theta$ or $\frac{1}{2}g \times \frac{1}{3}$ or $\frac{1}{6}g$ N.

The magnitude of $F_{max} = \mu R$

$$= \frac{\mu g\sqrt{2}}{3} = \frac{1}{\sqrt{2}}\,\frac{g\sqrt{2}}{3} = \frac{g}{3}\text{N.}$$

Hence the applied force of 6 N is greater than the sum of the forces acting down the plane. Motion takes place up the plane and the equation of motion is

$$6 - \tfrac{1}{2}g\sin\theta - F_{max} = \text{mass} \times \text{acceleration}$$

$$\therefore \qquad 6 - \frac{g}{6} - \frac{g}{3} = \tfrac{1}{2} \times a$$

$$\therefore \qquad a = 2{\cdot}2\,\text{m s}^{-2}$$

Exercise 6B

1. Each of the following diagrams shows a body of mass 10 kg released from rest on a rough inclined plane. R is the normal reaction and F the frictional force exerted on the body by the plane. In each case, find the magnitude of F and state whether the body will remain at rest or will begin to slip down the plane. For (a), (b) and (c), $\mu = \frac{1}{2}$ and for (d), (e) and (f), $\mu = \frac{1}{4}$.

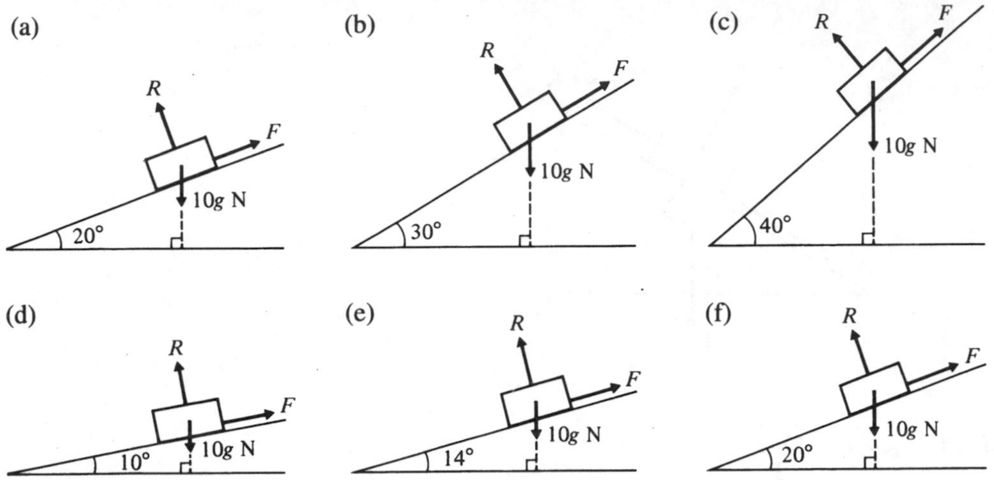

2. Each of the following diagrams shows a body of mass 5 kg on a rough inclined plane, coefficient of friction $\frac{1}{7}$; R is the normal reaction and F the frictional force exerted on the body, by the plane. In each case, find the magnitude of the force X if it just prevents the body from slipping down the plane.

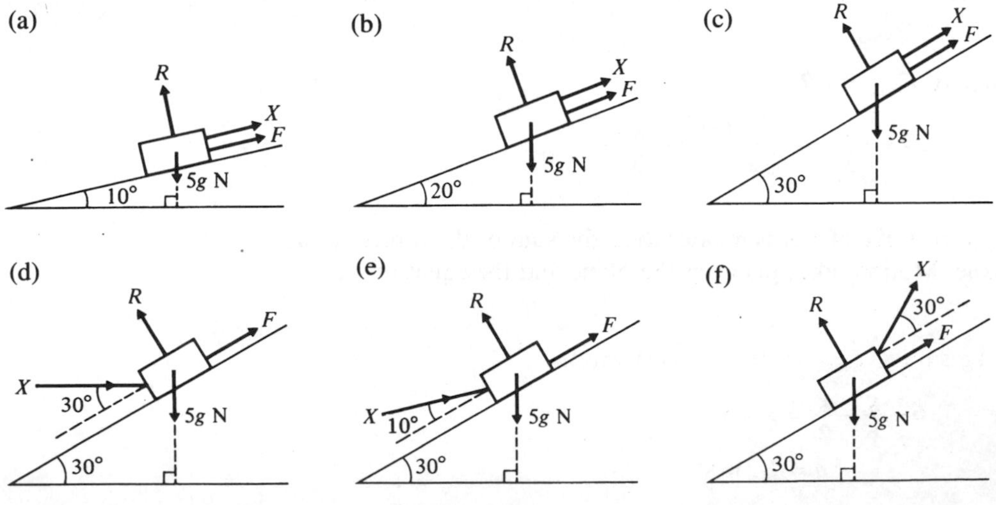

3. Each of the following diagrams shows a body of mass 3 kg on a rough inclined plane, coefficient of friction $\frac{1}{3}$; R is the normal reaction and F the frictional force exerted on the body by the plane. In each case, find the magnitude of the force X if the body is just on the point of moving up the plane.

(a)

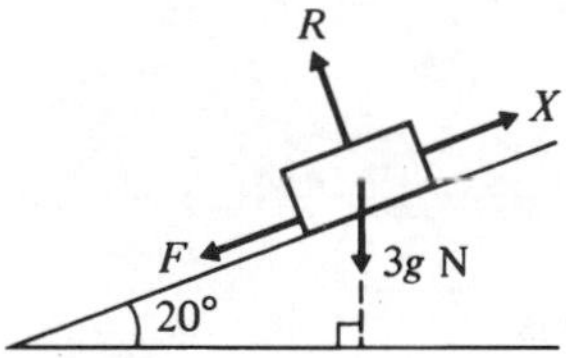

(b)

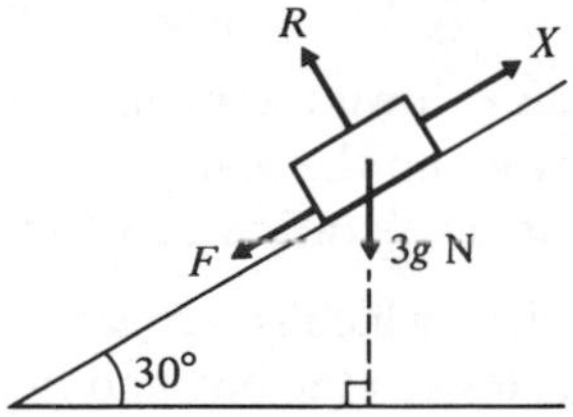

(c)

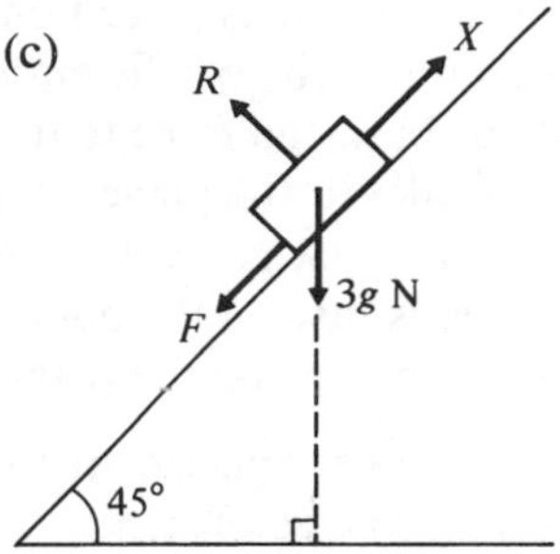

(d)

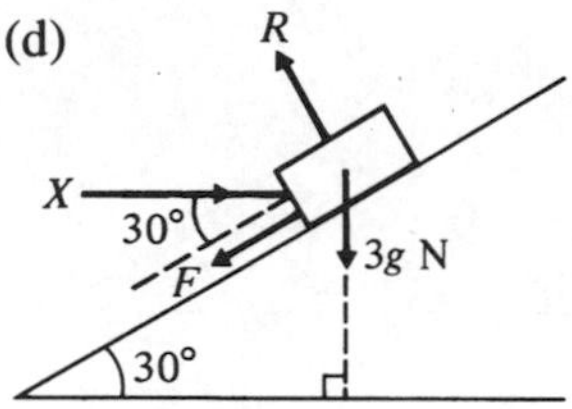

(e)

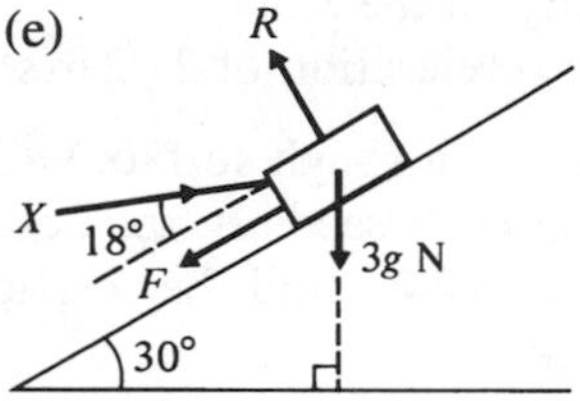

(f)

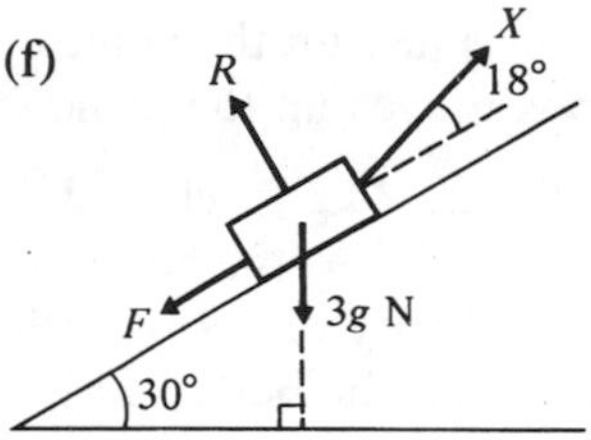

4. In each of the following situations, the forces acting on the body of mass 2 kg cause it to accelerate along the rough inclined plane as indicated. R is the normal reaction and F the frictional force exerted on the body by the plane. Find the value of μ for each situation.

(a)

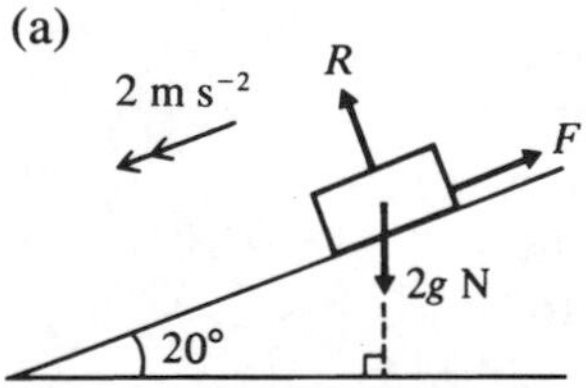

(b)

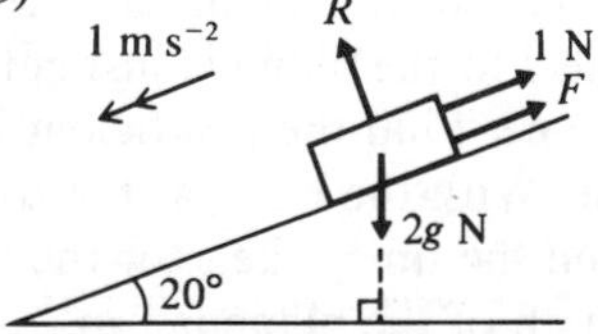

(c)

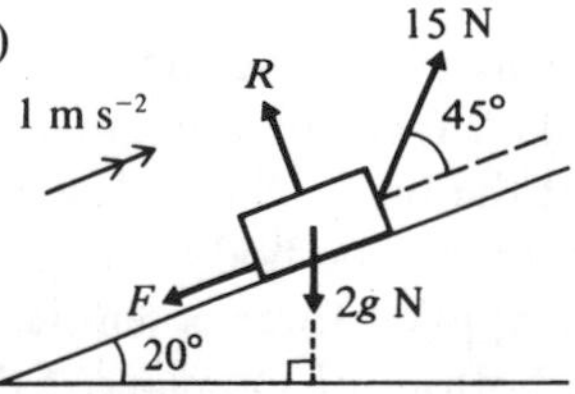

5. A body of mass 500 g is placed on a rough plane which is inclined at 40° to the horizontal. If the coefficient of friction between the body and the plane is 0·6, find the frictional force acting and state whether motion will occur.

6. A body of mass 5 kg lies on a rough plane which is inclined at 35° to the horizontal. When a force of 20 N is applied to the body, parallel to and up the plane, the body is found to be on the point of moving down the plane, i.e. in limiting equilibrium. Find μ, the coefficient of friction between the body and the plane.

7. A body of mass 2 kg lies on a rough plane which is inclined at 30° to the horizontal. When a horizontal force of 20 N is applied to the body in an attempt to push it up the plane, the body is found to be on the point of moving up the plane, i.e. in limiting equilibrium.
Find μ, the coefficient of friction between the body and the plane.

8. A body of mass 2 kg lies on a rough plane which is inclined at $\sin^{-1}\frac{5}{13}$ to the horizontal. A force of 20 N is applied to the body, parallel to and up the plane. If the body accelerates up the plane at $1{\cdot}5\,\text{m s}^{-2}$, find μ, the coefficient of friction between the body and the plane.

9. A parcel of mass 1 kg is placed on a rough plane which is inclined at 30° to the horizontal. The coefficient of friction between the parcel and the plane is 0·25. Find the force that must be applied to the parcel in a direction parallel to the plane so that:
(a) the parcel is just prevented from sliding down the plane
(b) the parcel is just on the point of moving up the plane
(c) the parcel moves up the plane with an acceleration of $1{\cdot}5\,\text{m s}^{-2}$.

10. A box of mass 6 kg is placed on a rough plane which is inclined at 45° to the horizontal. The coefficient of friction between the box and the plane is 0·5. Find the horizontal force that must be applied to the box so that:
(a) the box is just prevented from sliding down the plane
(b) the box is just on the point of moving up the plane
(c) the box moves up the plane with an acceleration of $2\sqrt{2}\,\text{m s}^{-2}$.

11. A body of mass 3 kg is released from rest on a rough surface which is inclined at $\sin^{-1}\frac{3}{5}$ to the horizontal. If, after $2\frac{1}{2}$ seconds, the body has acquired a velocity of $4{\cdot}9\,\text{m s}^{-1}$ down the surface, find the coefficient of friction between the body and the surface.

12. A particle of mass 250 g is released from rest at the top of a rough plane which is inclined at $\sin^{-1}\frac{3}{5}$ to the horizontal. The coefficient of friction between the particle and the plane is $\frac{11}{18}$ and the plane is of length 2·5 m. Find whether the particle will slide down the plane and, if it does, find its speed on reaching the bottom.

13. A body of mass 4 kg lies on a rough plane which is inclined at 16° to the horizontal. A force of 1 N applied parallel to the plane is just sufficient to prevent the body sliding down the plane. Find the coefficient of friction between the body and the plane. With the body at the top of the plane, the applied force is removed. Find the time taken for the body to reach the bottom of the plane if the length of the plane is 2 m.

14. A horizontal force of 1 N is just sufficient to prevent a brick of mass 600 g sliding down a rough plane which is inclined at $\sin^{-1}\frac{5}{13}$ to the horizontal. Find the coefficient of friction between the brick and the plane.

15. A body of mass 5 kg is initially at rest at the bottom of a rough inclined plane of length 6·3 m. The plane is inclined at 30° to the horizontal and the coefficient of friction between the body and the plane is $\dfrac{1}{2\sqrt{3}}$.
A constant horizontal force of $35\sqrt{3}$ N is applied to the body causing it to accelerate up the plane. Find the time taken for the body to reach the top and its speed on arrival.

16. (a) A mass m lies on a rough plane which is inclined at angle θ to the horizontal. The coefficient of friction between the mass and the plane is μ. Show that slipping will occur if $\tan\theta > \mu$.
(b) Will slipping occur when a body is placed on a rough plane ($\mu = 0{\cdot}5$) which is inclined at 40° to the horizontal?

(c) Will slipping occur when a body is placed on a rough plane ($\mu = 0{\cdot}25$) which is inclined at 10° to the horizontal?
(d) When a body is placed on a rough plane which is inclined at 30° to the horizontal, the body is found to be in limiting equilibrium. Find μ, the coefficient of friction between the body and the plane.

17. A mass of 4 kg lies on a rough plane which is inclined at 30° to the horizontal. A light string has one end attached to this mass, passes up the line of greatest slope, over a smooth pulley fixed at the top of the plane and carries a freely hanging mass of 1 kg at its other end. The tension in the string is just sufficient to prevent the 4 kg mass from sliding down the slope. Find the coefficient of friction between the 4 kg mass and the plane.

18. The diagram shows a mass of 1 kg lying on a rough inclined plane ($\mu = \frac{1}{4}$). From this mass, a light inextensible string passes up the line of greatest slope and over a smooth fixed pulley to a mass of 4 kg hanging freely. The plane makes an angle θ with the horizontal where $\sin\theta = \frac{3}{5}$. Show that the 1 kg mass will slide up the plane and find the velocity with which the 4 kg mass hits the floor.

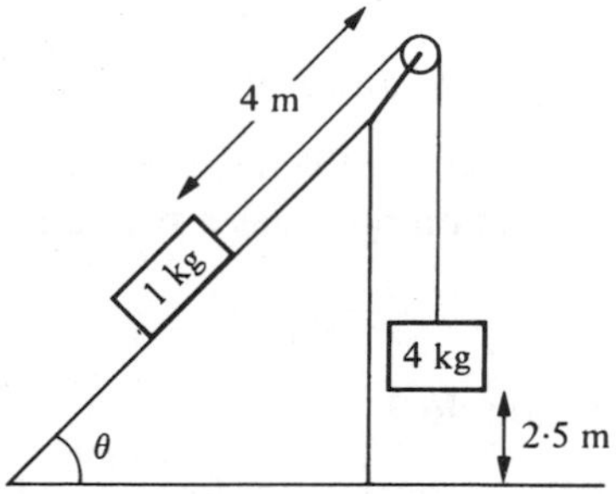

19. The diagram shows a body A of mass 13 kg lying on a rough inclined plane, coefficient of friction μ. From A, a light inextensible string passes up the line of greatest slope and over a smooth fixed pulley to a body B of mass m kg hanging freely. The plane makes an angle θ with the horizontal where $\sin\theta = \frac{5}{13}$. When $m = 1$ kg and the system is released from rest, B has an upward acceleration of $a\,\text{m s}^{-2}$; when $m = 11$ kg and the system is released from rest, B has a downward acceleration of $a\,\text{m s}^{-2}$. Find a and μ.

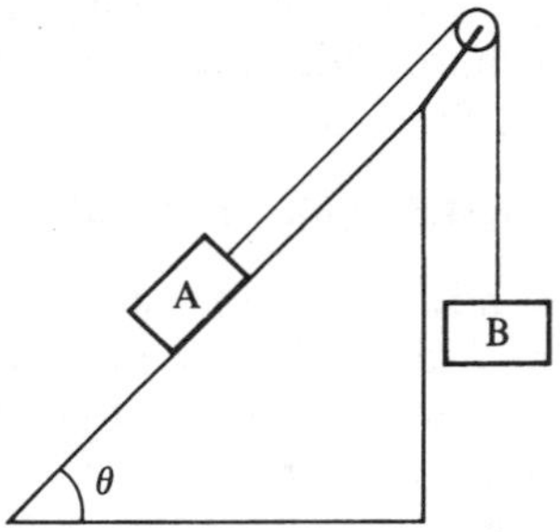

20. Masses of 5 kg and 15 kg are held at rest on inclined surfaces as shown in the diagram. The masses are connected by a light, taut, inextensible string passing over a smooth fixed pulley. The coefficient of friction between each mass and the surface with which it is in contact is 0·25. The inclination of the plane is such that $\sin\theta = \frac{3}{5}$. When the system is released from rest, the 15 kg mass accelerates down the slope. Find the magnitude of this acceleration and the tension in the string.

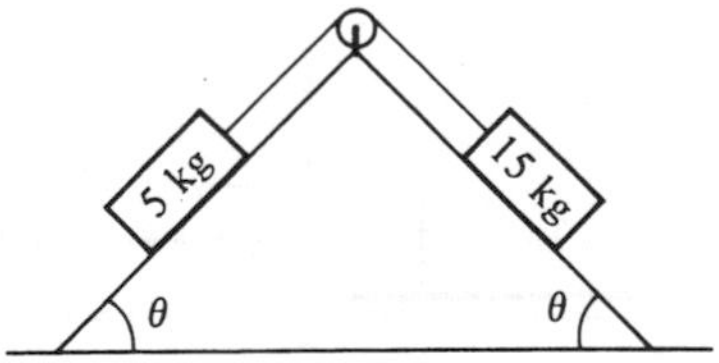

21. A force F acting parallel to and up a rough plane of inclination θ, is just sufficient to prevent a body of mass m from sliding down the plane. A force $4F$ acting parallel to and up the same rough plane causes the mass m to be on the point of moving up the plane. If μ is the coefficient of friction between the mass and the plane, show that $5\mu = 3\tan\theta$.

22. A horizontal force X is just sufficient to prevent a body of mass m from sliding down a rough plane of inclination θ. A horizontal force $4X$ applied to the same mass on the same rough plane, causes the mass to be on the point of moving up the plane. If μ is the coefficient of friction between the mass and the plane, show that $5\mu\tan^2\theta - 3(\mu^2 + 1)\tan\theta + 5\mu = 0$.

Angle of friction

When a state of limiting equilibrium exists, the frictional force F has its maximum value $F_{max} = \mu R$, where R is the normal reaction between the surfaces. If this frictional force F_{max} and the normal reaction are compounded into a single force called the resultant reaction, then the angle between the normal reaction and the resultant reaction is called the angle of friction λ.

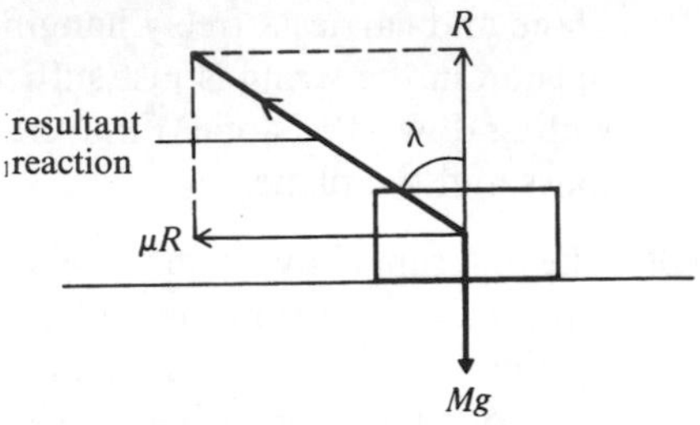

It can be seen that:

$$\tan \lambda = \frac{F_{max}}{R} = \frac{\mu R}{R} = \mu$$

Hence the angle of friction $\lambda = \tan^{-1} \mu$.

In certain problems, use of the resultant reaction and the angle of friction provides a neat and sometimes shorter solution.

Example 7

When a horizontal force of 14·7 N is applied to a body of mass 4 kg which is resting on a rough horizontal plane, the body is found to be in limiting equilibrium. Calculate the resultant reaction acting on the body and the angle of friction.

This question can be solved in several different ways.

Fig. 1 shows the forces acting on the body, including the normal reaction R and the frictional force μR, and this is used in Method 1 as explained below. Fig. 2 shows the resultant reaction P and the angle of friction λ, and this is used in Method 2.

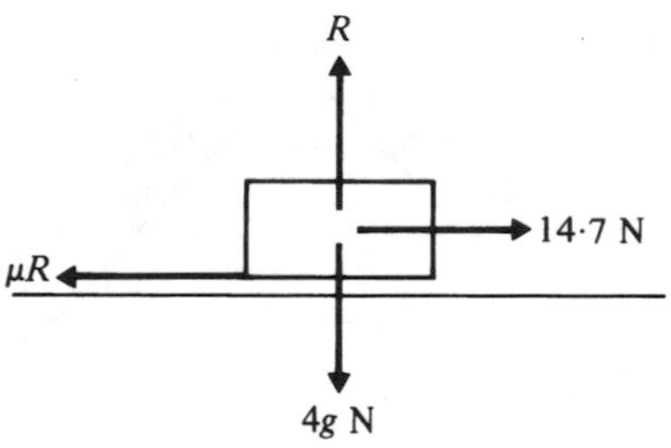

Fig. 1

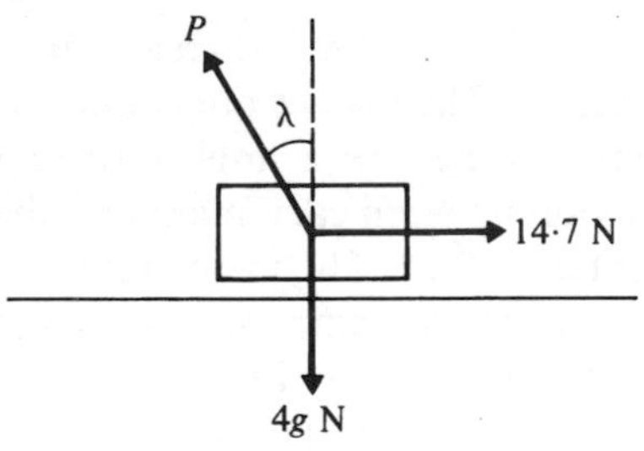

Fig. 2

Method 1, using Fig. 1

Since the body is in limiting equilibrium:

Resolve vertically: $R = 4g$

Resolve horizontally: $\mu R = 14{\cdot}7$

$\therefore \quad \mu \times 4g = 14{\cdot}7$

$\therefore \quad \mu = \frac{14{\cdot}7}{4g} = \frac{3}{8}$

If $\mu = \frac{3}{8}$ then $\tan \lambda = \frac{3}{8}$ and $\lambda = 20{\cdot}55°$.

Resultant reaction $= \sqrt{(R^2 + \mu^2 R^2)}$

$= R\sqrt{(1+\mu^2)}$

$= 4g\sqrt{(1+\frac{9}{64})} = \frac{g}{2}\sqrt{73}$ N

The resultant reaction is $\frac{g}{2}\sqrt{73}$ N and the angle of friction is 20·55°.

Method 2, using Fig. 2

Resolve vertically: $P \cos \lambda = 4g$ ...[1]
Resole horizontally: $P \sin \lambda = 14{\cdot}7$ [2]

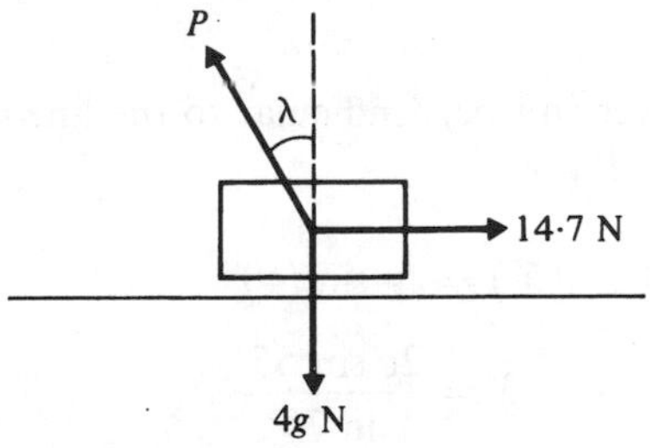

Fig. 2

Divide equation [2] by equation [1]:

$$\tan \lambda = \frac{14{\cdot}7}{4g} = \frac{3}{8}$$

$$\therefore \quad \lambda = 20{\cdot}55°$$

Substitute in equation [1]:

$$P \cos 20{\cdot}55° = 4g$$

$$\therefore \quad P = 41{\cdot}9 \text{ N}$$

The resultant reaction is 41·9 N and the angle of friction is 20·55°.

Since the situation as shown in Figure 2 is one of three forces acting on a body and producing a state of equilibrium, this problem could also be solved by the use of Lami's Theorem.

Method 3, using Fig. 2

Apply Lami's Theorem:

$$\frac{P}{\sin 90°} = \frac{14{\cdot}7}{\sin(180° - \lambda)} = \frac{4g}{\sin(90° + \lambda)}$$

$$\therefore \quad \frac{14{\cdot}7}{\sin \lambda} = \frac{4g}{\cos \lambda}$$

$$\therefore \quad \tan \lambda = \frac{14{\cdot}7}{4g} \quad \text{and} \quad \lambda = 20{\cdot}55°$$

and also $\dfrac{P}{\sin 90°} = \dfrac{14{\cdot}7}{\sin \lambda}$ $\therefore$ $P = \dfrac{14{\cdot}7}{\sin 20{\cdot}55°} = 41{\cdot}9$ N

It should also be noted that, if the question had only required λ to be found, then resolving perpendicular to the resultant reaction gives a neat solution:

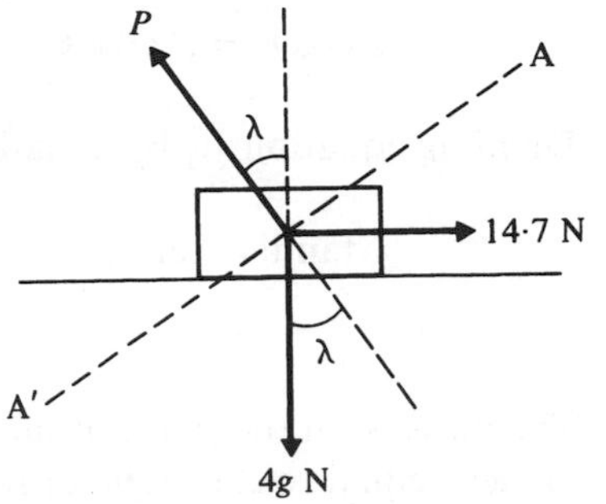

Resolving along AA′, perpendicular to P gives:

$$4g \sin \lambda = 14{\cdot}7 \cos \lambda$$

$$\therefore \quad \tan \lambda = \frac{14{\cdot}7}{4g}$$

$$\therefore \quad \lambda = 20{\cdot}55°$$

The reader should be aware of all these possible methods of solution and select the most suitable method for each question.

Example 8

A body of mass 2 kg lies on a rough plane which is inclined at 40° to the horizontal. The angle of friction between the plane and the body is 15°. Find the greatest force which can be applied to the body, parallel to and up the plane, without motion occurring.

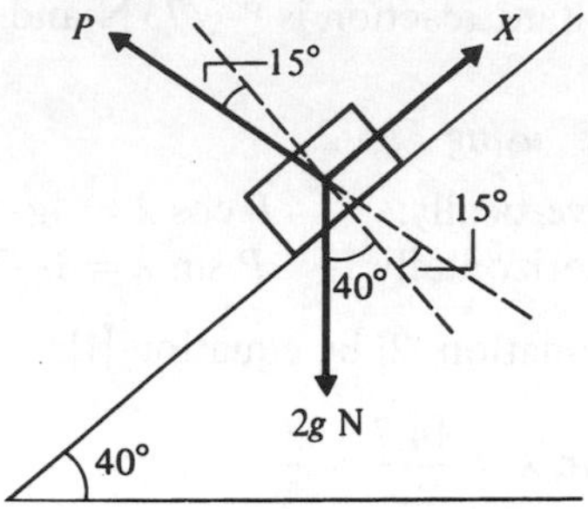

Let the applied force be X.
X will have its maximum value when the body is on the point of moving up the plane, i.e. when it is in limiting equilibrium.

Resolving in a direction perpendicular to the line of the resultant reaction P gives:

$$X \sin(90° - 15°) = 2g \sin(40° + 15°)$$

$$\therefore \quad X = \frac{2g \sin 55°}{\sin 75°}$$

$$\therefore \quad X = 16{\cdot}6\,\text{N}$$

The greatest force which can be applied to the body without motion occurring is 16·6 N.

It should be noted that this result could also have been obtained by the use of Lami's Theorem.

Example 9

A mass m rests in limiting equilibrium on a rough plane inclined at θ to the horizontal. If the angle of friction is λ, show that if the mass is on the point of moving *down* the plane, then $\theta = \lambda$.

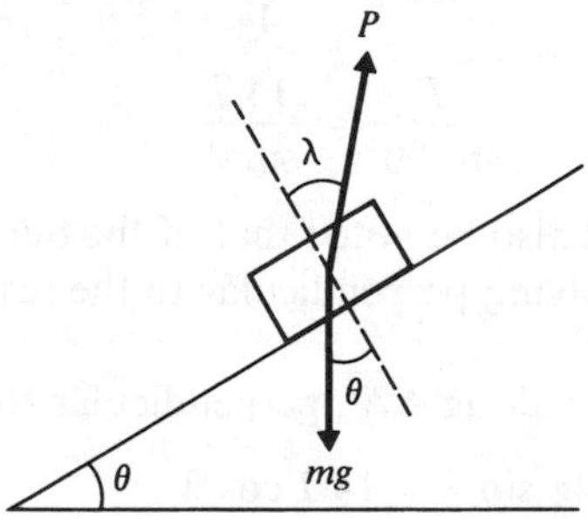

Resolving parallel to the plane gives:

$$mg \sin\theta = P \sin\lambda \qquad \ldots[1]$$

Resolving perpendicular to the plane gives:

$$mg \cos\theta = P \cos\lambda \qquad \ldots[2]$$

Dividing equation [1] by equation [2] gives:

$$\tan\theta = \tan\lambda$$

$$\therefore \quad \theta = \lambda$$

The mass is on the point of moving down the plane when the angle of the plane is equal to the angle of friction.

Example 10

If a force Q, inclined at an angle α to the surface of the plane, is applied to the mass in Example 9, find the minimum value of Q and the corresponding angle α when the mass is on the point of moving up the plane.

As the mass is on the point of moving up the plane, the resultant reaction will act as shown in the diagram.

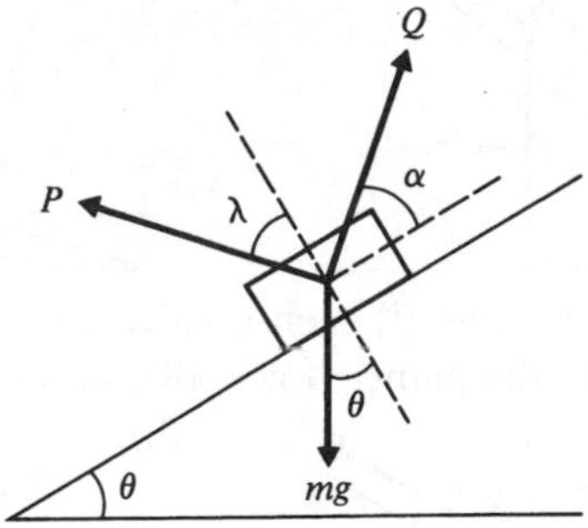

Using Lami's Theorem gives:

$$\frac{P}{\sin(90° + \theta + \alpha)} = \frac{Q}{\sin[180° - (\lambda + \theta)]} = \frac{mg}{\sin(90° - \alpha + \lambda)}$$

$$\therefore \quad \frac{Q}{\sin(\lambda + \theta)} = \frac{mg}{\cos(\alpha - \lambda)}$$

or

$$Q = \frac{mg\sin(\lambda + \theta)}{\cos(\alpha - \lambda)}$$

This expression for Q has a minimum value when $\cos(\alpha - \lambda) = 1$, or when $\alpha = \lambda$.
In this case $Q = mg\sin(\lambda + \theta)$ and the force must be applied at an angle λ to the surface of the plane.

The minimum value of Q is $mg\sin(\lambda + \theta)$ when the angle $\alpha = \lambda$.

Exercise 6C

1. The coefficient of friction for two surfaces in contact is 0·2. Find the angle of friction for the two surfaces.
2. The angle of friction for two surfaces in contact is 30°. Find the coefficient of friction for the two surfaces.
3. Each of the following diagrams shows a body on a rough plane. P is the resultant reaction of the plane on the body and λ is the angle of friction for the two surfaces in contact. Parts (a), (b) and (c) each involve a body, of mass 5 kg, on the point of moving along a horizontal surface. Find P and λ.

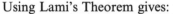

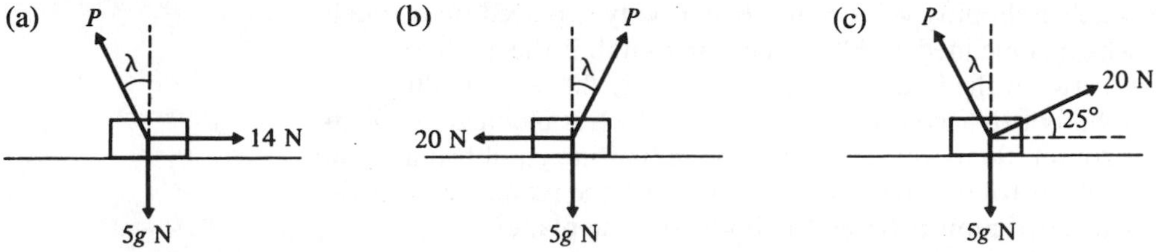

Parts (d), (e) and (f) each involve a body, of mass 2 kg, on the point of moving down a slope. If $\lambda = 25°$, find the magnitude of force X.

(d)

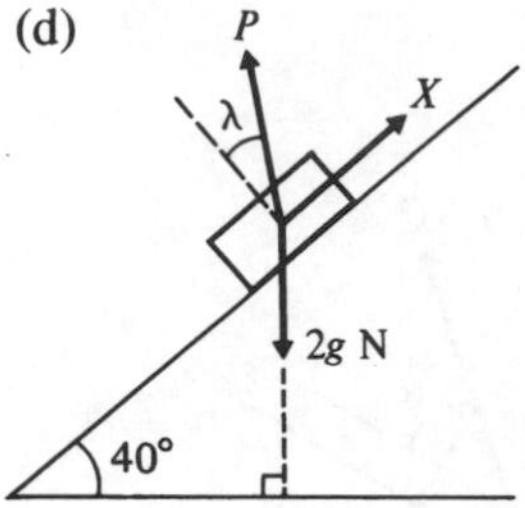

(e)

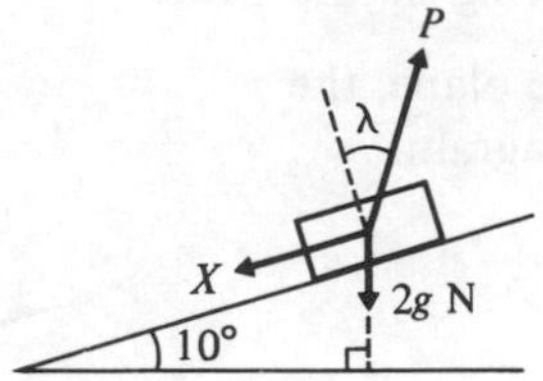

(f)

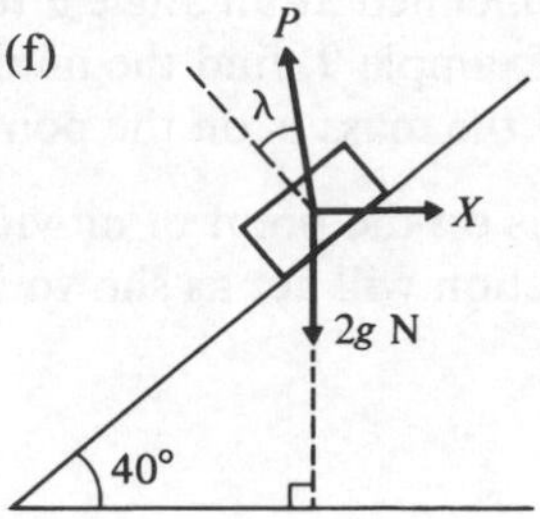

Parts (g), (h) and (i) each involve a body of mass 10 kg on the point of moving up the plane. If $\lambda = 18°$, find the magnitude of force X.

(g)

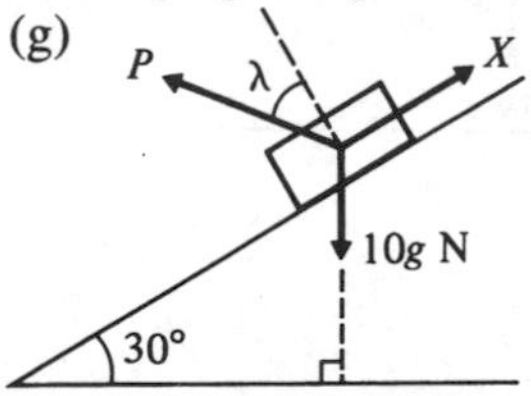

(h)

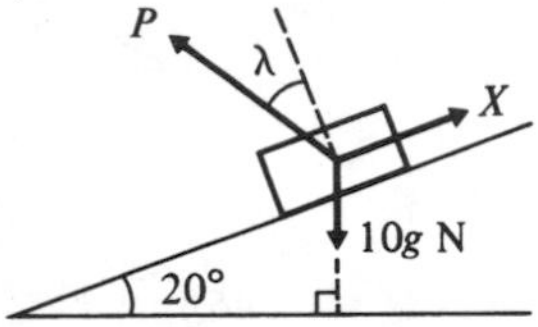

(i)

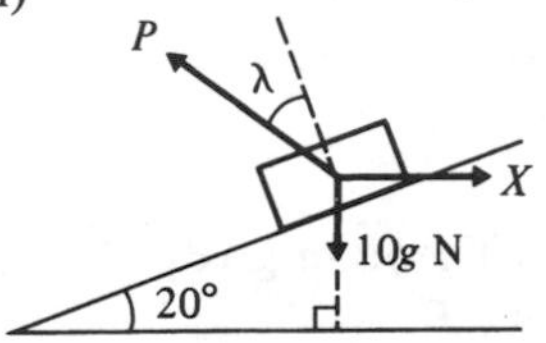

Parts (j), (k) and (l) each involve a body of mass 5 kg accelerating along the plane as indicated. Find P and λ.

(j)

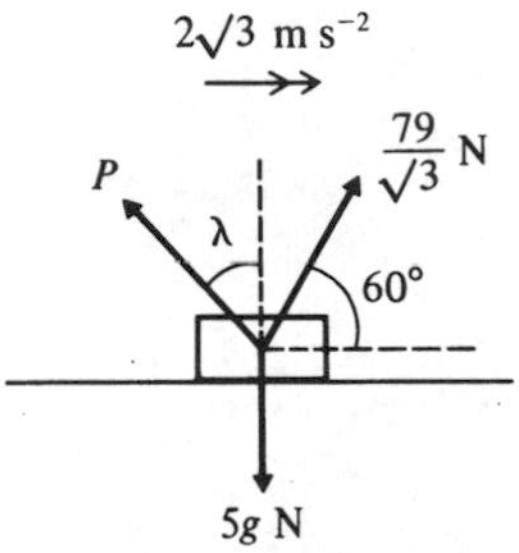

(k)

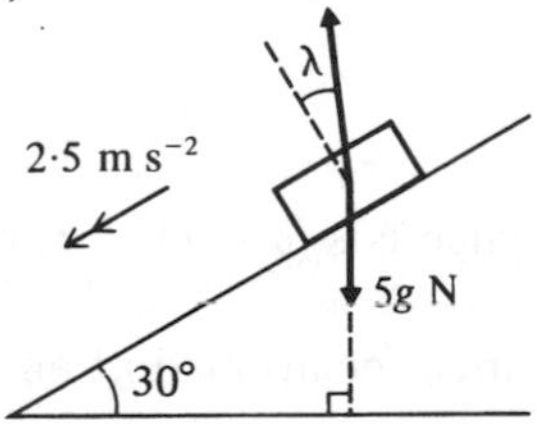

(l)

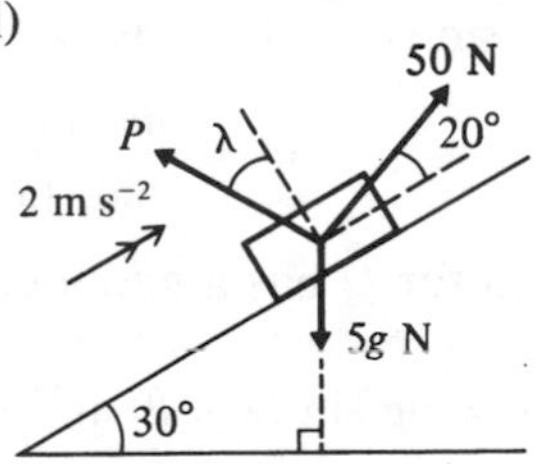

4. When a horizontal force of 4·9 N is applied to a body of mass 2 kg which is resting on a rough horizontal plane, the body is found to be in limiting equilibrium. Calculate the resultant reaction acting on the body and the angle of friction.

5. A boy pulls a sledge of mass 2 kg across a rough horizontal surface by means of a rope inclined at $\sin^{-1} \frac{3}{5}$ above the horizontal. The tension in the rope is 8 N and the angle of friction for the two surfaces is 8°. Find the resultant reaction that the surface has on the sledge and the acceleration of the sledge.

6. (a) A body of mass m is placed on a rough plane which is inclined at θ degrees to the horizontal. The angle of friction between the body and the plane is λ. Show that the body will slip down the plane if $\theta > \lambda$.

(b) State whether slipping will occur when a body is placed on a rough plane which is inclined at 30° to the horizontal, if the angle of friction between the body and the plane is: (i) 20° (ii) 40°.

(c) When a body is placed on a rough plane which is inclined at 25° to the horizontal, the body is found to be in limiting equilibrium. Find λ, the angle of friction between the body and the plane. What is the coefficient of friction between the body and the plane?

7. A body of mass 5 kg lies on a rough plane which is inclined at 35° to the horizontal. The angle of friction between the plane and the body is 20°. Find the magnitude of the least force that must be applied to the body, in a direction parallel to and up the plane, in order to prevent motion down the plane.

8. A body of mass 4 kg lies on a rough plane which is inclined at 30° to the horizontal. The angle of friction between the plane and the body is 15°. Find the magnitude of the least horizontal force that must be applied to the body to prevent motion down the plane.

9. A body of mass 2 kg lies on a rough plane which is inclined at 40° to the horizontal. The angle of friction between the plane and the body is 15°. Find the greatest horizontal force that can be applied to the body without motion occurring.

10. A body of mass 3 kg lies on a rough plane which is inclined at 20° to the horizontal. A force of 28 N applied to the body, parallel to and up the slope, causes the body to accelerate up the slope at 1·5 m s^{-2}. Find λ, the angle of friction between the body and the plane. If the applied force were subsequently removed, the body would travel on up the slope, eventually coming to rest. Would it then slip down the slope or would it remain at rest?

11. A light string is attached to a body of mass 5 kg lying on a rough horizontal surface. The angle of friction between the body and the surface is 20° and the string is pulled upwards at an angle θ to the horizontal. If the body is on the point of moving, find the tension in the string when θ is:
(a) 10° (b) 20° (c) 40°.

12. A body of mass m lies on a rough horizontal surface. The angle of friction between the body and the surface is λ. A light string is attached to the body and is pulled upwards, at an angle θ to the horizontal. If the body is on the point of moving, show that the least value for the tension in the string is $mg \sin \lambda$ and that it occurs when $\theta = \lambda$.

Questions **13** to **15** refer to Fig. 1 which shows a mass m lying on a rough plane inclined at an angle θ to the horizontal. The angle of friction between the body and the plane is λ, with $\theta > \lambda$. Force X is applied to the body and makes an angle ϕ with the plane. The body is on the point of moving down the plane.

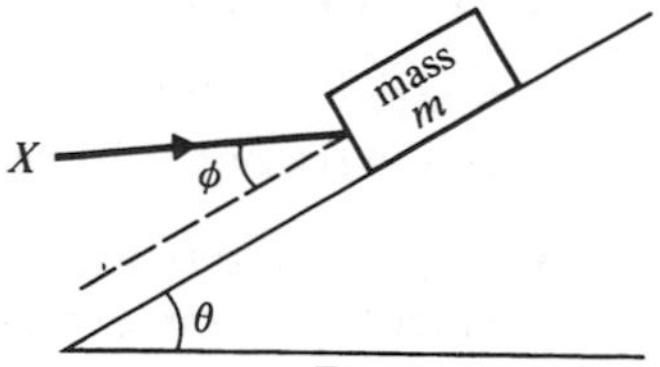

Fig. 1

13. If $\lambda = 18°$, $\theta = 30°$ and $m = 2$ kg, find the magnitude of X when ϕ is:
(a) 10° (b) 13° (c) 30°.

14. If $\lambda = 15°$, $\theta = 40°$ and $m = 5$ kg find the magnitude of X when ϕ is:
(a) 10° (b) 15° (c) 30°.

15. Show that the least force X sufficient to prevent motion down the plane is $mg \sin(\theta - \lambda)$ and that it occurs when $\phi = \lambda$.

Questions **16** to **18** refer to Fig. 2 which shows a mass m lying on a rough plane inclined at an angle θ to the horizontal. The angle of friction between the body and the plane is λ.
Force X is applied to the body and makes an angle ϕ with the plane. The body is on the point of moving up the plane.

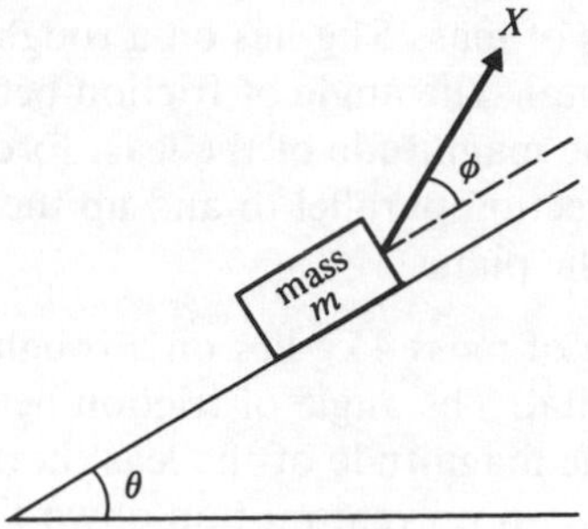

Fig. 2

16. If $\lambda = 30°$, $\theta = 20°$ and $m = 2$ kg, find the magnitude of X when ϕ is:
(a) 20° (b) 30° (c) 40°.

17. If $\lambda = 20°$, $\theta = 10°$ and $m = 5$ kg, find the magnitude of X when ϕ is:
(a) 20° (b) 30° (c) 40°.

18. Show that the least force X sufficient to ensure that the body is on the point of moving up the plane is $mg \sin(\theta + \lambda)$ and that it occurs when $\phi = \lambda$.

Exercise 6D Examination questions

(Unless otherwise indicated, take $g = 9{\cdot}8\ \text{m s}^{-2}$ in this exercise.)

1.

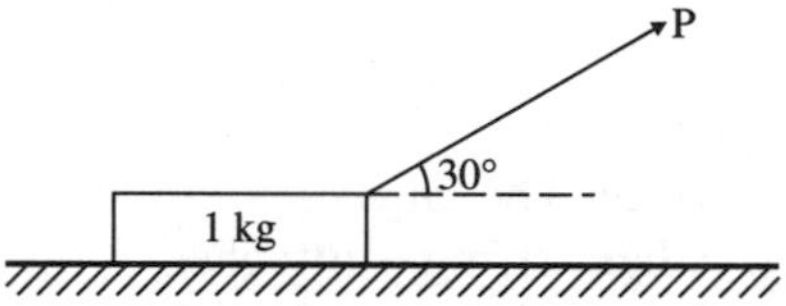

A block of mass 1 kg rests in equilibrium on a rough horizontal table under the action of a force P which acts at an angle of 30° to the horizontal, as shown in the diagram. Given that the magnitude of P is 2·53 N, calculate

(i) the normal reaction exerted by the table on the block,

(ii) the frictional force on the block.

Given that the block is about to slip, calculate the coefficient of friction, correct to 2 decimal places. (UCLES)

2. The diagram shows a particle of mass 4 kg on a rough horizontal surface. The particle is acted on by a force of P N acting at 30° to the surface and is in limiting equilibrium. Given that the normal reaction between the particle and the surface is 54 N, find

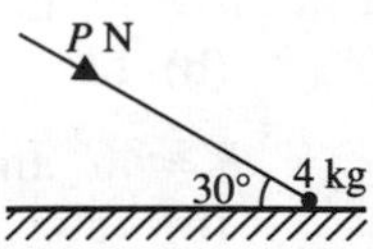

(i) the value of P

(ii) the value, correct to two decimal places, of the coefficient of friction between the particle and the surface. (UCLES)

3. (Take $g = 10\,\mathrm{m\,s^{-2}}$ in this question.)
Describe an observation that you could make which would show that the motion of a sledge on snow is subject to a friction force.
A small child of mass 20 kg sits on a sledge of mass 10 kg that rests on a horizontal surface of snow. A woman of mass 50 kg attempts to push the sledge with the child on it by applying a horizontal force.

(a) When the woman starts to push, she finds the force she applies is not sufficient to cause motion. Draw one diagram for the woman and a second for the loaded sledge, showing all of the forces that act on each.

(b) The coefficient of friction between the sledge and the snow is 0·2.
  (i) Show that the woman must apply a horizontal force to the sledge greater than 60 N so that it will move forward, assuming that she does not slip.
  (ii) Find the minimum coefficient of friction required between the woman's feet and the snow, in order that she does not slip before the sledge moves forward.

(UODLE)

4.

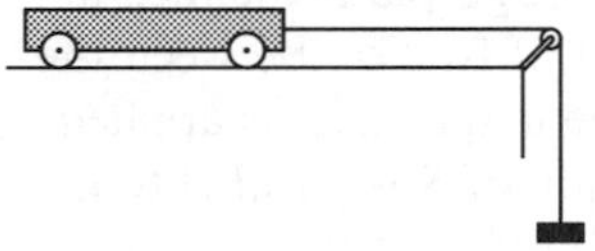

In a school physics experiment, a trolley can move on a horizontal table. A string connected to the trolley passes over a pulley fixed at the edge of the table, and a load hangs from the end of the string (see diagram). In a particular experiment, the mass of the trolley is 3 kg and the mass of the load is 1 kg. Calculate the acceleration of the trolley and the tension in the string when the system is released from rest, and state any assumptions you have made in the course of your calculation.

When the trolley is replaced by a wooden block of mass 2 kg which can slide on the table, it is found that a load of 1 kg is heavy enough to move the block while a load of 0·5 kg is not heavy enough to move the block. Use this information to find out what you can about the size of the coefficient of friction between the block and the table. (UCLES)

5. Two particles A and B of mass m and $2m$ respectively are connected by a light inextensible string which passes over a smooth pulley attached to the edge of a rough horizontal table. The particle A is held at rest on the table at a distance $2a$ from the pulley with the string taut and B hangs vertically below the pulley. The system is released from rest. The coefficient of friction between the table and A is $\frac{3}{5}$. When A is released, find the magnitude of the acceleration of each particle and the magnitude of the tension in the string.
After B has fallen a distance a it hits a horizontal floor and does not rebound. Show that A comes to rest at a distance $\frac{2}{9}a$ from the pulley.

(AEB 1992)

6. (Take $g = 10\,\mathrm{m\,s^{-2}}$ in this question.)
The diagram shows a particle of mass 0·5 kg on a rough plane inclined at an angle α to the horizontal, where $\sin\alpha = \frac{5}{13}$. The particle is acted on by a horizontal force of 8 N and is about to move up a line of greatest slope. Show that the value of the coefficient of friction between the particle and the plane is 0·71.
Determine, with working, whether or not the particle will move when the force of 8 N is removed. (UCLES)

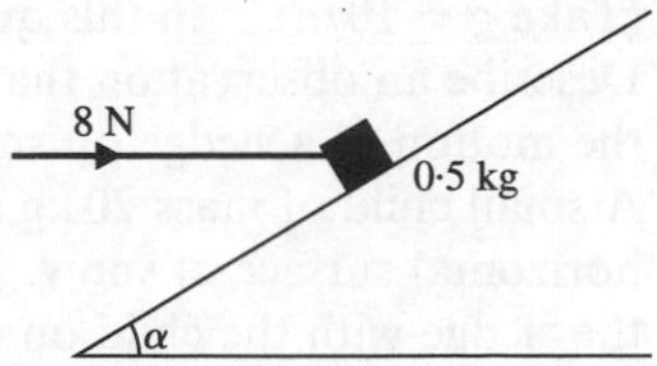

7.

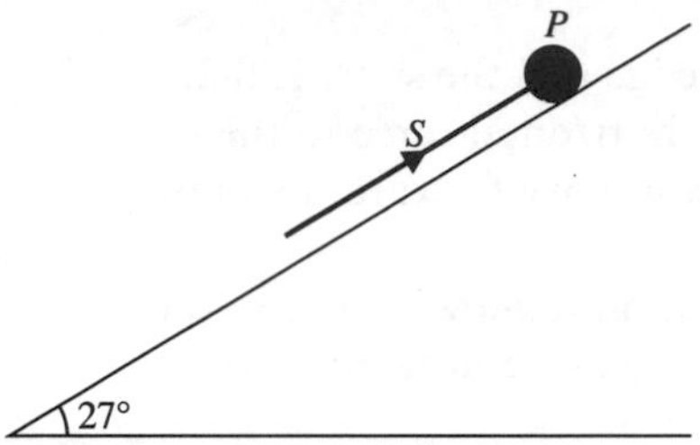

Fig. 1

A small parcel P, of mass 1·5 kg, is placed on a rough plane inclined at an angle of 27° to the horizontal. The coefficient of friction between the parcel and the plane is 0·3. A force S, of variable magnitude, is applied to the parcel as shown in Fig. 1. The line of action of S is parallel to a line of greatest slope of the inclined plane.
Determine, in N to 1 decimal place, the magnitude of S when the parcel P is in limiting equilibrium and on the point of moving

(a) down the plane

(b) up the plane. (ULEAC)

8. (Take $g = 10\,\mathrm{m\,s^{-2}}$ in this question.)
A particle of mass 0·8 kg is on a rough plane inclined at an angle α to the horizontal, where $\tan\alpha = \frac{5}{12}$. The particle is acted upon by an upward force of 4 N parallel to a line of greatest slope of the plane. Given that the particle is about to move up the plane, calculate the coefficient of friction between the particle and the plane.
Given that the force of 4 N is then removed, find the acceleration of the particle down the plane. (UCLES)

9. A particle of mass 3 kg is placed on a rough horizontal surface and a horizontal force of gradually increasing magnitude P is applied to it.

(i) Draw a diagram and show the force of friction F acting on the particle, together with all other forces.

(ii) If the coefficient of friction between the particle and the surface is μ, state the magnitude of the friction force when the particle remains in equilibrium, and determine its value when motion occurs.

(iii) Draw and label a sketch-graph of F against P. (UODLE)

10. (Take $g = 10\,\mathrm{m\,s^{-2}}$ in this question.)
After an unexpected snow fall some children take tin trays out to a local hillside and slide down the hillside sitting on the tin trays.

(a) List three factors that would affect the amount of friction between the tray and the snow.

The hill is inclined at 30° to the horizontal.

(b) (i) Show that if friction is assumed to be zero then the acceleration of a child on a tray will be $5\,\mathrm{m\,s^{-2}}$.
(ii) Find the time it takes for the child to travel 20 m down the slope, if the child starts at rest.

(c) In reality the child is given a push so that the initial speed is $2\,\mathrm{m\,s^{-1}}$ and it then takes 4 seconds to travel the 20 m. Find the actual acceleration of the child.

(d) What is the coefficient of friction between the tray and the snow?

(e) State one factor that has not been taken into account in the problem. (AEB Spec)

11. A toboggan of mass 15 kg carries a child of mass 25 kg. It starts from rest on a snow slope of inclination 10°. Given that the acceleration is $1{\cdot}2\,\mathrm{m\,s^{-2}}$ and that air resistance may be ignored, find the coefficient of friction.
Find the speed when it has moved 50 m from rest, and find the time taken.
Having reached the bottom of the slope, the toboggan and child are pulled back up the slope, at a constant speed, by a light rope which is parallel to a line of greatest slope. Find the tension in the rope.

(UCLES)

12. (Take $g = 10\,\mathrm{m\,s^{-2}}$ in this question.)
The diagram shows two slopes: the upper slope is inclined at an angle θ to the horizontal where $\sin\theta = \frac{1}{4}$, the lower slope at 30° to the horizontal. A particle A of mass 3 kg is held at rest on the upper slope. This particle is connected by a light inextensible string, passing over a smooth pulley, to a particle B of mass 2 kg on the lower slope.

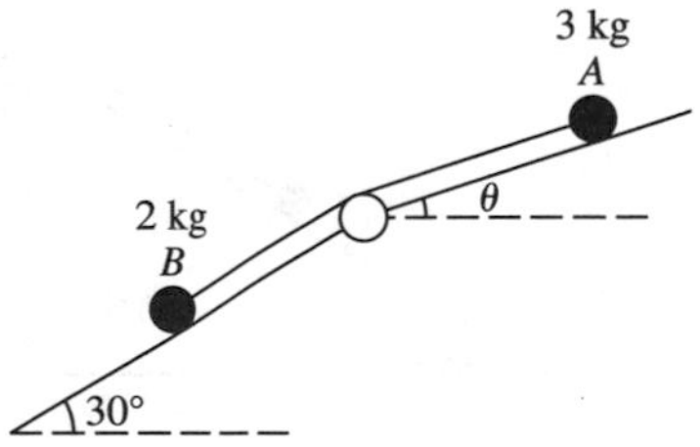

(a) In the case where both slopes are smooth and A is released from rest, find
(i) the acceleration of the particles
(ii) the tension in the string
(iii) the speed of each particle after travelling 0·63 m, assuming both particles remain on their respective slopes.

(b) In the case where the lower slope is smooth and the upper slope rough, the system is in limiting equilibrium. Find the coefficient of friction between particle A and the upper slope. (UCLES)

13. A particle of mass m kg is resting on a rough plane inclined at 30° to the horizontal and is attached to one end of a light inextensible string. The string passes over a pulley at the top of the plane and carries at its other end a mass M kg, where $M > m$. The coefficient of friction between the plane and the mass m is $\frac{1}{\sqrt{3}}$. Initially the mass M is 1 m above a horizontal plane as shown in Fig. 1.

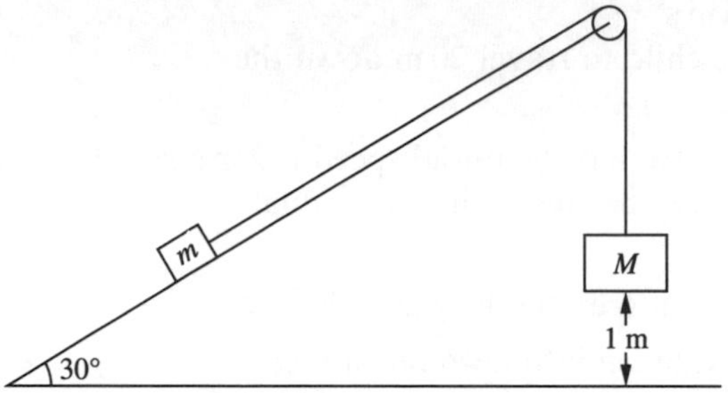

Fig. 1

(i) If the system is released from rest, show that it will move with an acceleration of

$$\frac{(M-m)g}{(M+m)}$$

(Assume that the mass m does not reach the pulley.)

(ii) Find the speed with which mass m is moving when mass M hits the horizontal plane.

(iii) For how much longer will mass m continue to move up the plane?

(iv) What will happen then? (NICCEA)

14.

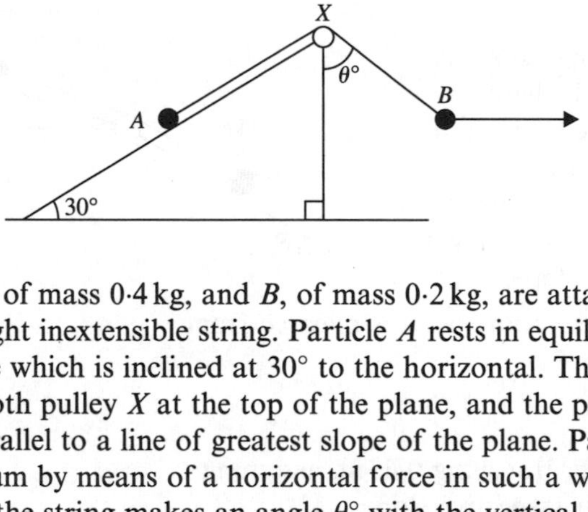

Particles A, of mass 0·4 kg, and B, of mass 0·2 kg, are attached to the ends of a light inextensible string. Particle A rests in equilibrium on a rough plane which is inclined at 30° to the horizontal. The string passes over a smooth pulley X at the top of the plane, and the part AX of the string is parallel to a line of greatest slope of the plane. Particle B is held in equilibrium by means of a horizontal force in such a way that the part XB of the string makes an angle θ° with the vertical. The points A, X, B lie in the same vertical plane (see diagram). The coefficient of friction between A and the sloping plane is 0·3. Given that A is about to slip up the plane, find the value of θ. (UCLES)

15.

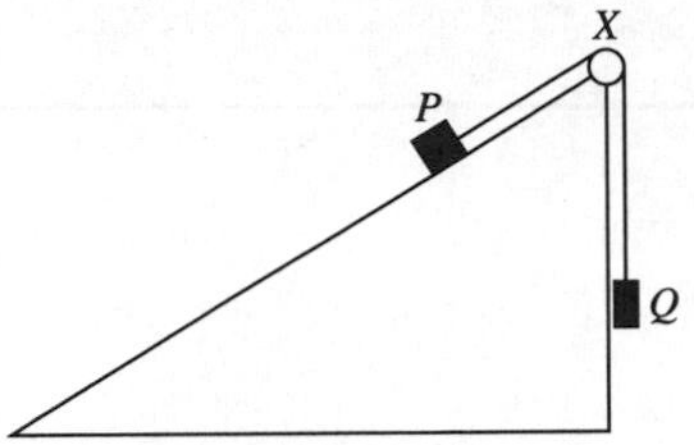

Fig. 1

Fig. 1 shows a particle P, of mass $5m$, on a rough plane, connected by a light inextensible string passing over a small smooth pulley at the top of the plane to a particle Q of mass $10m$ which is hanging freely. The string is parallel to a line of greatest slope of the plane. The coefficient of friction between P and the plane is $\frac{1}{4}$ and the plane is inclined at an angle $\sin^{-1}\frac{3}{5}$ to the horizontal. The particles are released from rest. Find the tension in the string during the subsequent motion.
After Q has dropped a distance $10d$ it hits an inelastic horizontal plane. Assuming that P does not reach the pulley, find the further distance travelled by P before first coming to instantaneous rest. (WJEC)

7 Moments

Moment of a force

From our everyday experience we know that:

(i) it is easier to undo a tight nut using a long spanner when the force is applied at the end of the spanner, rather than by using a short spanner;

(ii) if a child sits on one end of a seesaw which is pivoted at its centre, he can be balanced by a heavier adult sitting near to the centre of the seesaw;

(iii) a door is more easily closed by pushing on the edge further from the hinges than by pushing at a point part way across the door, such as Q.

(i)

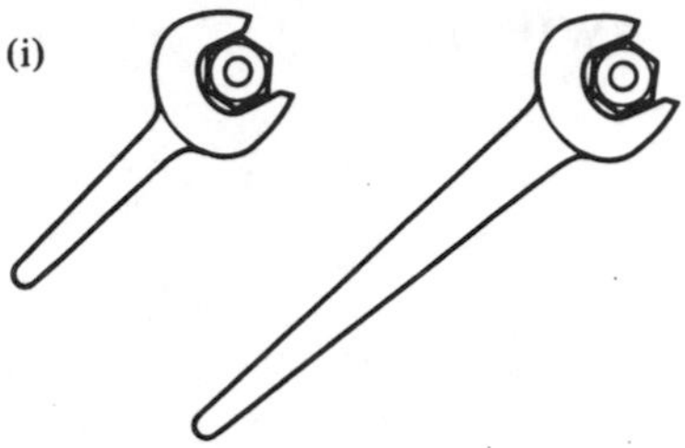

(iii)

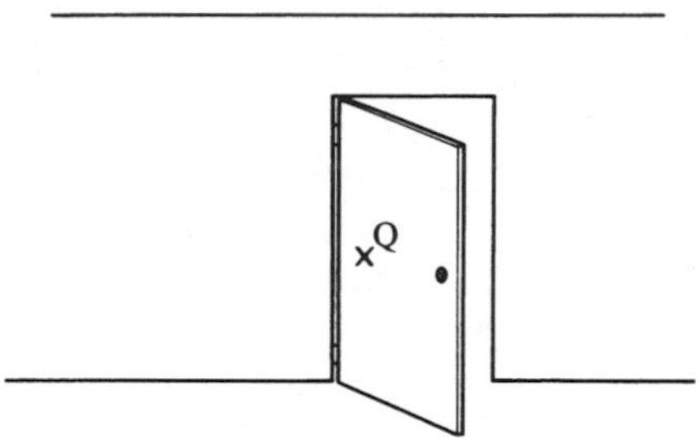

(ii)

In each of these examples, the application of the force is causing or tending to cause a body to rotate about an axis, i.e. rotational motion. Previously, only motion along a line has been considered, i.e. translational motion.

The diagram shows a rod AB pinned to a horizontal table by a pin through A. When a horizontal force of 5 N is applied at the point B, perpendicular to BA, the rod will turn about the axis through A. The turning effect of the force will be lessened if instead of being applied at B, it is applied at C or at D. This turning effect is known as the moment of the force about the point A.

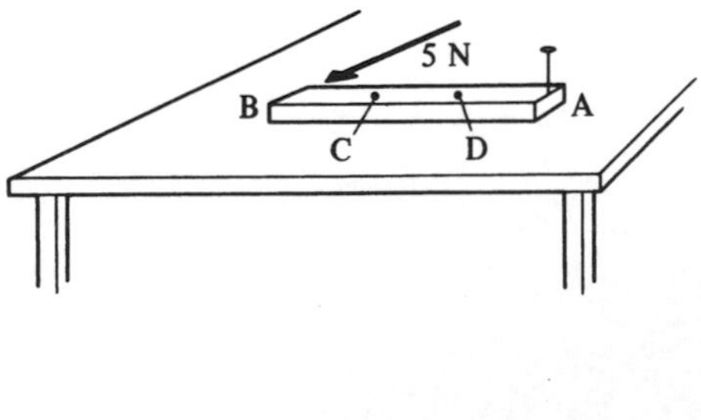

Although the phrase "the moment of the force about the point A" has just been used, strictly speaking moments are always taken about an axis, and not about a point. In the last case, the turning effect (or moment) of the force about the vertical axis through A has really been considered. However in the questions which follow, all the forces act in one plane, and so for simplicity, the phrase "moment about a point" will be used.

Definition

The moment of a force about a point is found by multiplying the magnitude of the force by the perpendicular distance from the point to the line of action of the force.

The moment of the force P about the point X is $P \times a$.

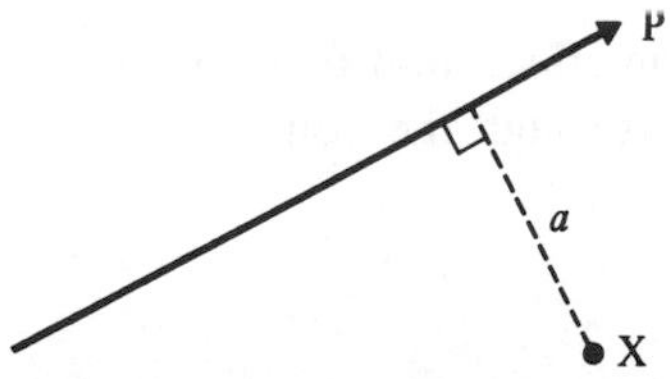

Hence a force will have no moment about a point on its line of action.

If the force is measured in newtons and the distance in metres, the moment of the force is measured in newton metres (N m).

Sense of rotation

A nut is usually rotated in an anticlockwise direction when being undone. All rotations should have their sense clearly stated: the moment of a force about a point has both magnitude and direction.

Example 1

Consider the rod AB shown in the diagram. Find the moment about A of the force of 5 N when it is applied at each of the points B, C and A.

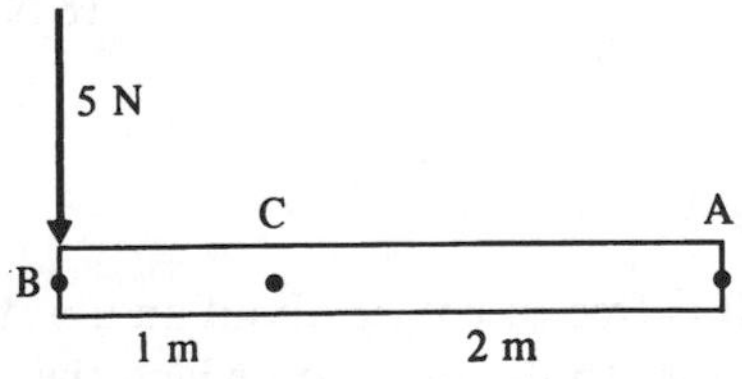

When the force of 5 N is applied at B:

$$\text{moment about A} = \text{force} \times \text{perpendicular distance from A to force}$$
$$= 5\,\text{N} \times 3\,\text{m} = 15\,\text{N m anticlockwise}$$

When the force of 5 N is applied at C:

$$\text{moment about A} = 5\,\text{N} \times 2\,\text{m} = 10\,\text{N m anticlockwise}$$

When a force of 5 N is applied at A:

$$\text{moment about A} = 5\,\text{N} \times 0\,\text{m} = 0$$

A force through A has no turning effect about the point A and therefore no moment about A: it is like trying to close a door by pressing on the hinges.

It is usual when taking moments about an axis through, say, the point K to write $\curvearrowleft\!\text{K}$ to stand for "taking moments about the point K".

Example 2

In the diagram, find the moment of the force at C about each of the points A, B and C. State the direction of the moment in each case.

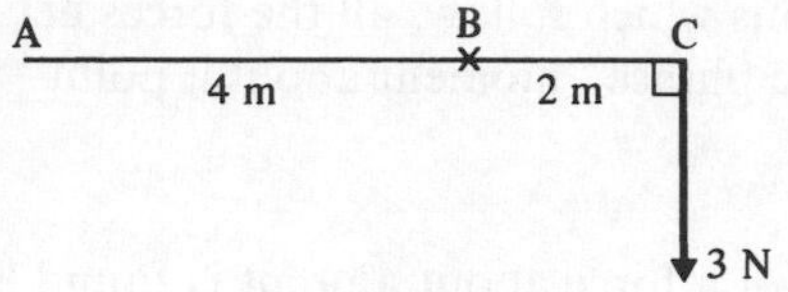

Taking moments gives:

$\overset{\frown}{A}$ $\quad 3\,\text{N} \times 6\,\text{m} = 18\,\text{N m}$ clockwise

$\overset{\frown}{B}$ $\quad 3\,\text{N} \times 2\,\text{m} = 6\,\text{N m}$ clockwise

$\overset{\frown}{C}$ $\quad 3\,\text{N} \times 0 \quad = 0$

In the third case, the force has no turning effect (and therefore no moment) about the point C since the force passes through the point C.

Algebraic sum of moments

If a number of coplanar forces act on a body, their moments about any point may be added provided due regard is given to the sense of each moment.

Example 3

Three forces acting on a body have moments of 15 N m clockwise, 10 N m anticlockwise and 13 N m clockwise, about a point X. Find the sum of these moments in magnitude and direction.

moment of 10 N m anticlockwise = moment of −10 N m clockwise

The sum of the moments = (15 N m − 10 N m + 13 N m) clockwise
= 18 N m clockwise

Example 4

A rod AB is free to rotate about an axis through the point O, perpendicular to the plane on which the rod rests. Forces of P and Q newtons act as shown. Find the combined turning effect about the point O of these forces if
(a) $P = 6$ N and $Q = 5$ N (b) $P = 7$ N and $Q = 3\frac{1}{2}$ N.

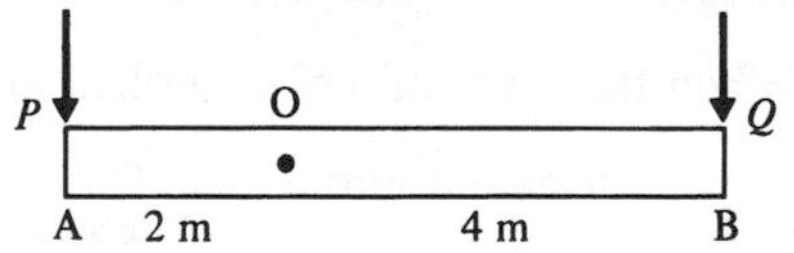

Taking moments about O gives:

(a) $\overset{\frown}{O}$ $\quad (5 \times 4)$ N m clockwise + (6×2) N m anticlockwise
= (20 N m − 12 N m) clockwise
= 8 N m clockwise

(b) $\overset{\frown}{O}$ $\quad (3\frac{1}{2} \times 4)$ N m clockwise + (7×2) N m anticlockwise
= 14 N m − 14 N m
= 0

Example 5

Find the moment, or the sum of the moments, about the point O of the forces shown in each of the following diagrams.

(a)

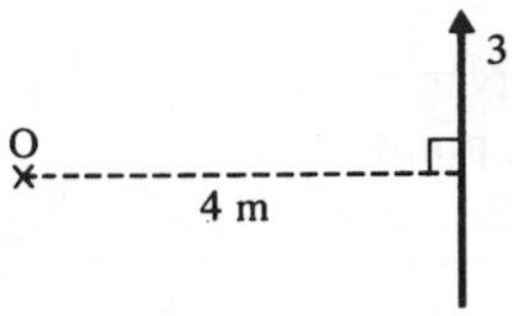

$\overset{\curvearrowright}{O}$ (3×4) N m
$= 12$ N m anticlockwise

(b)

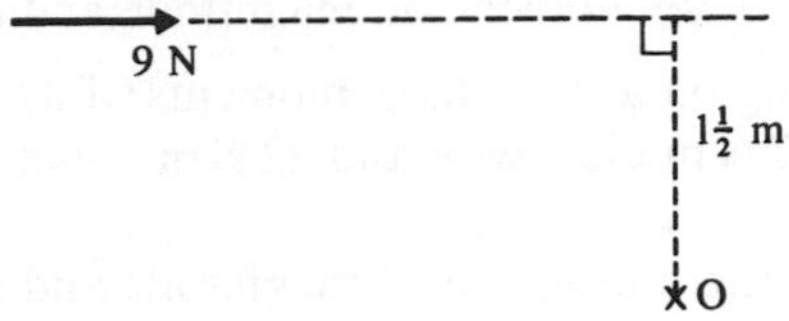

$\overset{\curvearrowright}{O}$ $(9 \times 1\frac{1}{2})$ N m
$= 13\frac{1}{2}$ N m clockwise

(c)

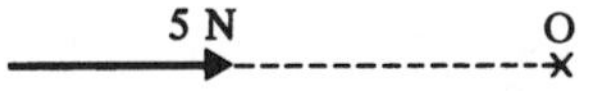

$\overset{\curvearrowright}{O}$ (5×0) N m
$= 0$

(d)

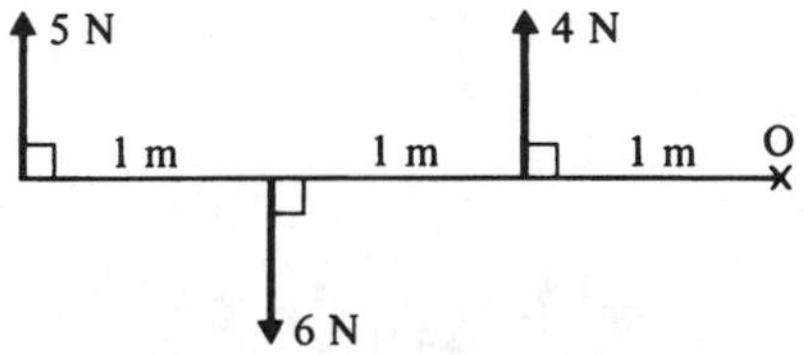

$\overset{\curvearrowright}{O}$ $(5 \times 3 + 4 \times 1)$ N m clockwise
$+ (6 \times 2)$ N m anticlockwise
$= (15 + 4 - 12)$ N m clockwise
$= 7$ N m clockwise

Example 6

Find the moment about the origin of a force of $3\mathbf{j}$ N acting at the point which has position vector $4\mathbf{i}$ m.

Drawing a diagram and taking moments about the point O gives:

$\overset{\curvearrowright}{O}$ (4×3) N m $= 12$ N m anticlockwise

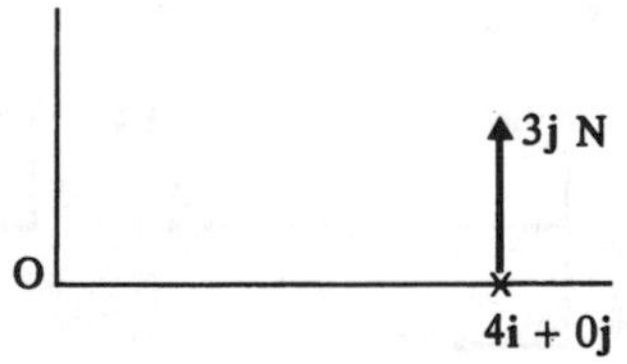

Example 7

Find the moment about the point P with position vector $(-3\mathbf{i} + 4\mathbf{j})$ m of a force $(6\mathbf{i} + 5\mathbf{j})$ N acting at a point Q which has position vector $(2\mathbf{i} + \mathbf{j})$ m.

Again drawing a diagram gives:

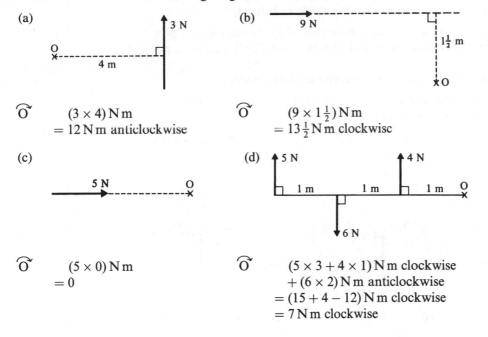

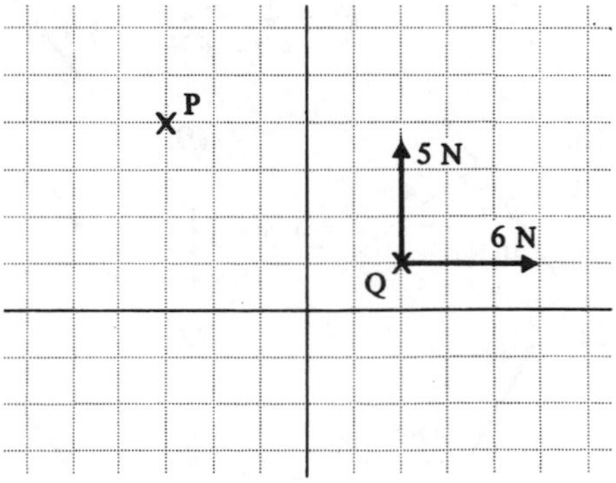

$\overset{\curvearrowright}{P}$ (5×5) N m anticlockwise $+ (6 \times 3)$ N m anticlockwise
$= (25 + 18)$ N m
$= 43$ N m anticlockwise

Exercise 7A

1. Three forces acting on a body have moments of 7 N m clockwise, 12 N m anticlockwise and 15 N m clockwise, about a point X.
   Find the sum of these moments in magnitude and direction.
2. Four forces acting on a body have moments of 8 N m clockwise, 5 N m anticlockwise, 17 N m clockwise and 22 N m anticlockwise, about a point X.
   Find the sum of these moments in magnitude and direction.

For each of the questions **3** to **23** find the moment (or the sum of the moments) about the point A of the forces shown.

3.

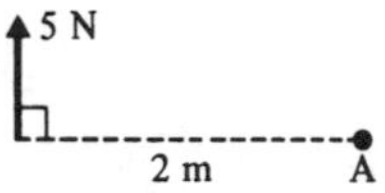

4.

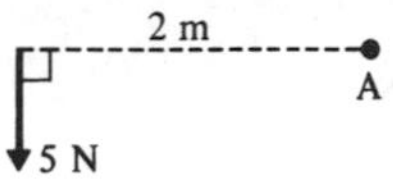

5.

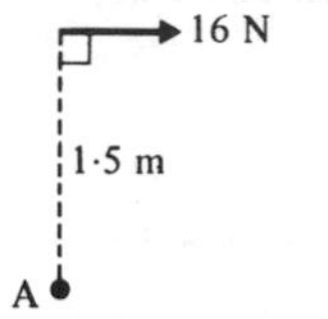

6.

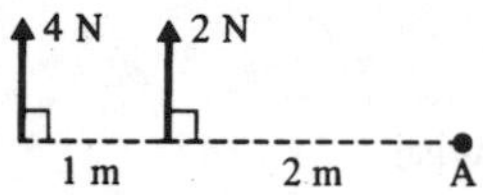

7.

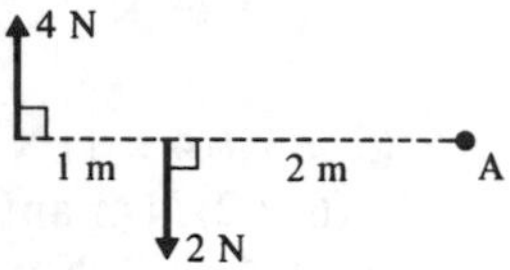

8.

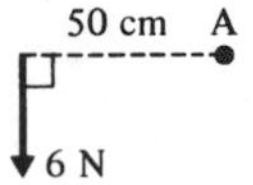

9.

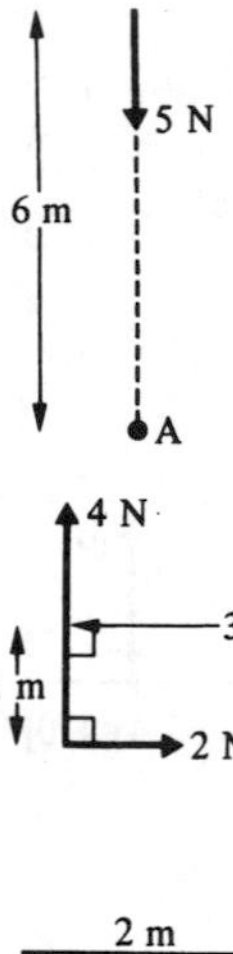

10.

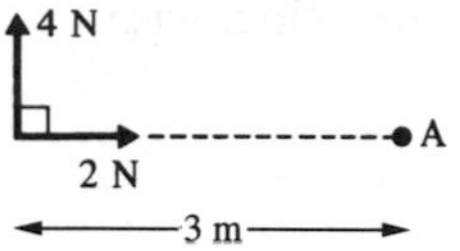

11.

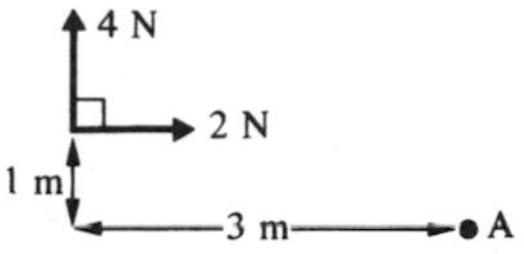

12.

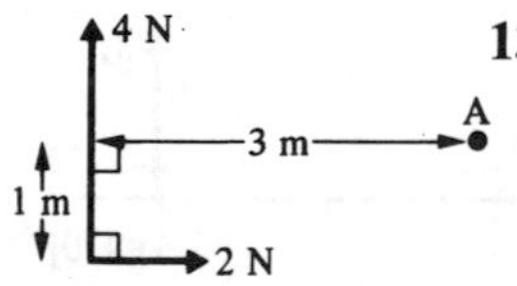

13.

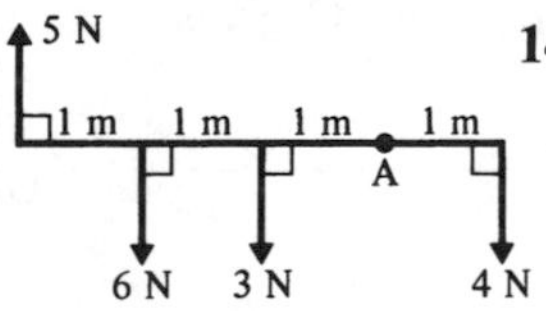

14.

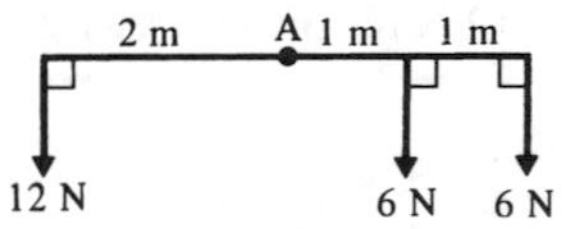

15.

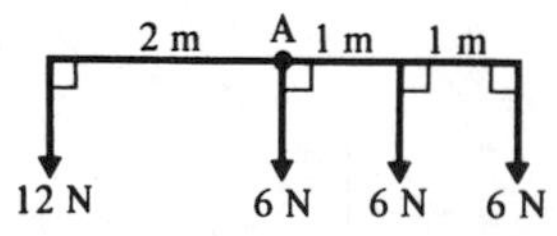

16.

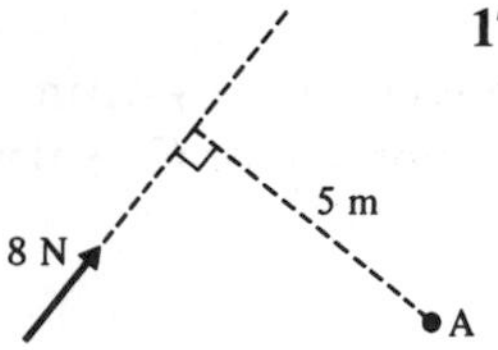

17.

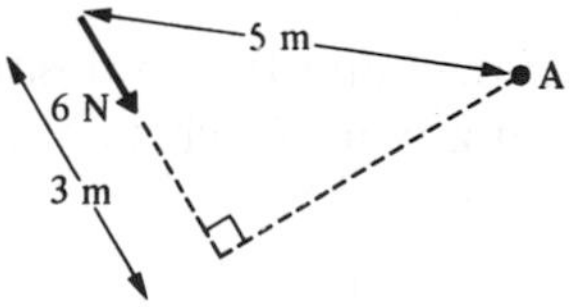

18.

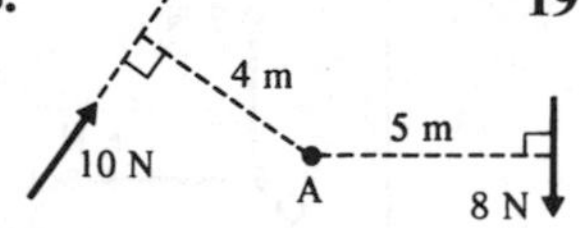

19.

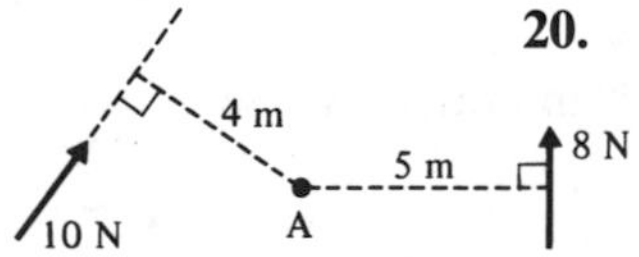

20.

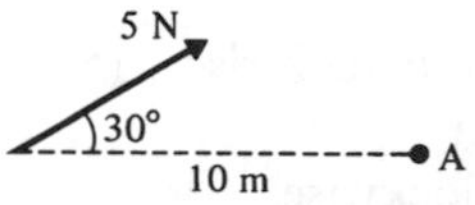

21.

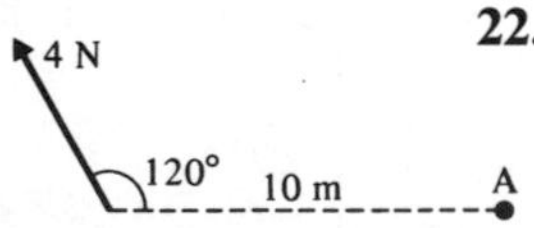

22.

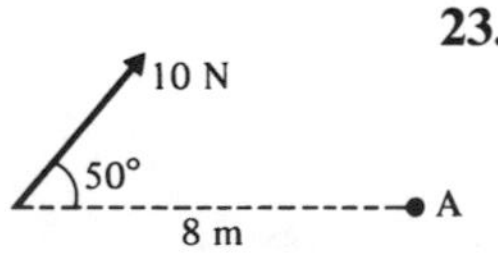

23.

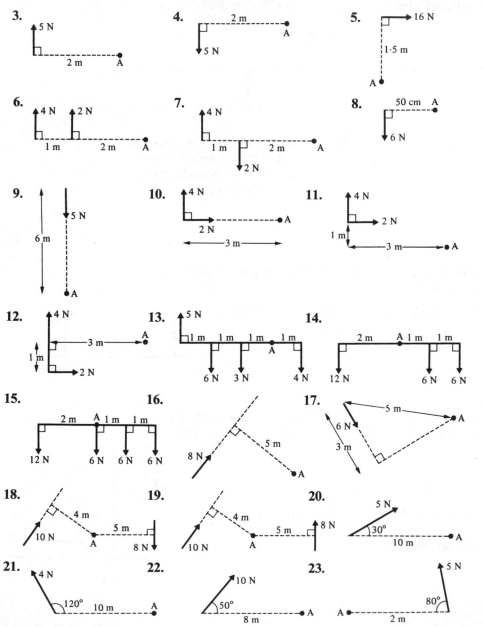

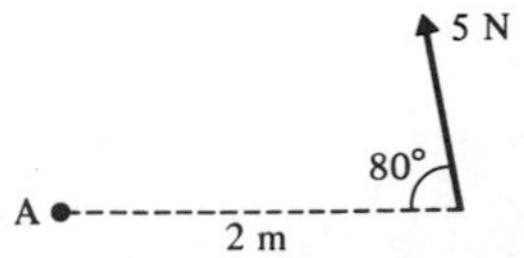

24. Find the moment about the origin of a force of 4**j** N acting at the point which has position vector 5**i** m.

25. Find the moment about the origin of a force of 4**j** N acting at the point which has position vector −5**i** m.

26. Find the moment about the origin of a force of 3**i** N acting at the point which has position vector (2**i** + 3**j**) m.

27. Find the moment about the origin of a force of (4**i** + 2**j**) N acting at the point which has position vector (3**i** + 2**j**) m.

28. A force of (3**i** − 2**j**) N acts at the point which has position vector (5**i** + **j**) m. Find the moment of this force about the point which has position vector (**i** + 2**j**) m.

29. A force of (2**i** + **j**) N acts at the point which has position vector (2**i** + 2**j**) m and a force of 5**i** N acts at the point which has position vector (−2**i** + **j**) m.
Find the sum of the moments of these forces about the origin.

30. A force of (3**i** + 2**j**) N acts at the point which has position vector (5**i** + **j**) m and a force of (**i** + **j**) N acts at the point which has position vector (2**i** + **j**) m.
Find the sum of the moments of these forces about the point which has position vector (**i** + 3**j**) m.

31. If a line AB represents the force P, both in magnitude and direction, show that the moment of force P about some point O is represented in magnitude by twice the area of the triangle AOB.

Parallel forces and couples

Example 8

Two forces, each of 5 N, are applied at the ends A and B of a rod, 3 m in length. The forces are parallel but act in opposite directions.
Find the sum of the moments of the forces about each of the points A, B and P shown in the diagram.

Since the forces are equal in magnitude but act in opposite directions, they have no *translational* effect. However, the forces will have a *turning* effect. Taking moments gives:

$\overset{\curvearrowright}{A}$ $(5 \times 3) + (5 \times 0)$ N m $=$ 15 N m clockwise

$\overset{\curvearrowright}{B}$ $(5 \times 3) + (5 \times 0)$ N m $=$ 15 N m clockwise

$\overset{\curvearrowright}{P}$ $5 \times (y + 3)$ N m clockwise $+ (5 \times y)$ N m anticlockwise

$= (5y + 15 - 5y)$ N m clockwise

$=$ 15 N m clockwise

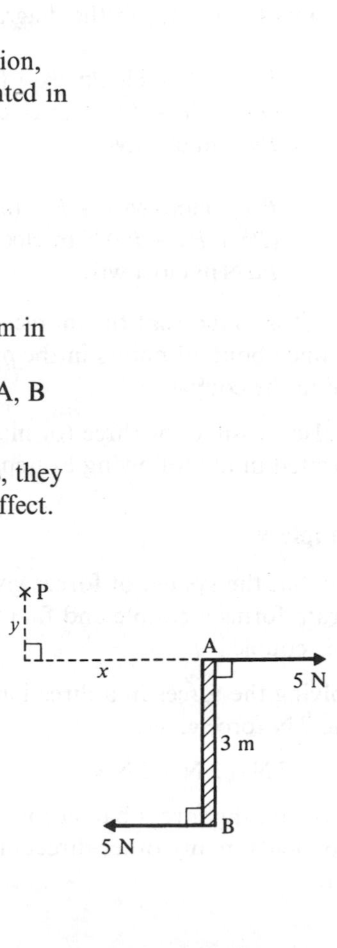

Note that the moments about each of the points is the same.

Like and unlike forces

Forces which are parallel and act in the *same* direction are said to be *like* forces.

Forces which are parallel and act in *opposite* directions are said to be *unlike* forces.

Definition of a couple

Two unlike forces of equal magnitude, not acting along the same line, are said to form a *couple*. A couple has a turning effect but cannot produce a translatory effect.

Moment of a couple

If the magnitude of each force forming a couple is P newtons and the perpendicular distance between their lines of action is a metres, the magnitude of the moment of the couple is

$$P \times a \,\text{N m}$$

The turning effect of a couple is *independent* of the point about which the turning is taking place.

This statement, which was illustrated in Example 8, can be verified more generally by considering the sum of the moments of the forces about each of the points O_1 and O_2 in the diagram below.

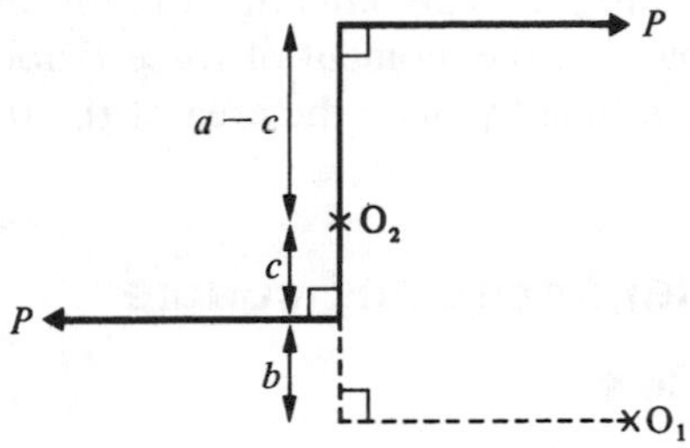

$\overset{\frown}{O_1}$ $\quad P \times (a+b)$ clockwise $+ (P \times b)$ anticlockwise
$= (Pa + Pb - Pb)\,\text{N m}$ clockwise
$= Pa\,\text{N m}$ clockwise

$\overset{\frown}{O_2}$ $\quad P \times c$ clockwise $+ P \times (a-c)$ clockwise
$= (Pc + Pa - Pc)\,\text{N m}$ clockwise
$= Pa\,\text{N m}$ clockwise

Hence it is found that the moment of this couple is the same about all points in the plane of the forces forming the couple.

It is also possible for three (or more) parallel forces to form a couple, as is illustrated in the following example.

Example 9

Show that the system of forces given in the diagram forms a couple and find the moment of this couple.

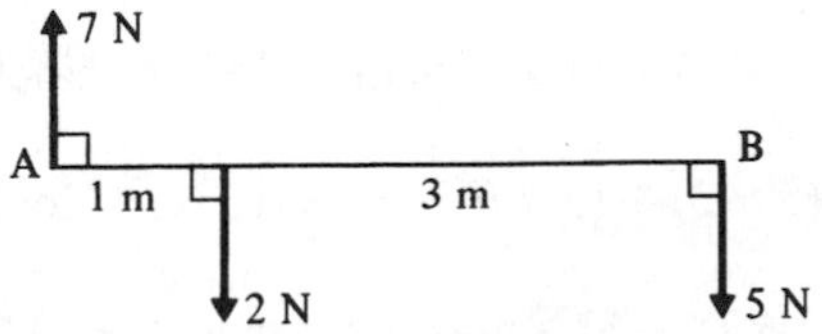

Resolving the forces in a direction parallel to the 7 N force gives:

$$7\,\text{N} - 2\,\text{N} - 5\,\text{N} = 0$$

In this case the forces balance in this direction and they have no components in any other direction; consequently they have no translatory effect.

Hence, either:

(a) their moments about any point also balance and therefore the forces have no turning effect,

or

(b) the forces have a turning effect, that is they form a couple.

Taking moments gives:

$\overset{\curvearrowright}{A}$ $\quad (7 \times 0) + (2 \times 1) + (5 \times 4)$ N m clockwise
$= 22$ N m clockwise

Therefore these forces form a couple with a moment of 22 N m in a clockwise sense.

Example 10

Forces of 5 N, 2 N, 5 N and 2 N act along the sides BA, BC, DC and DA respectively, of the square ABCD, in the directions indicated by the order of the letters. The side of the square is 3 m. Show that the forces form a couple and find the moment of this couple by taking moments about
(a) the centre of the square and (b) the point A.
Find the moment of the couple which must be applied to the system in order to produce equilibrium.

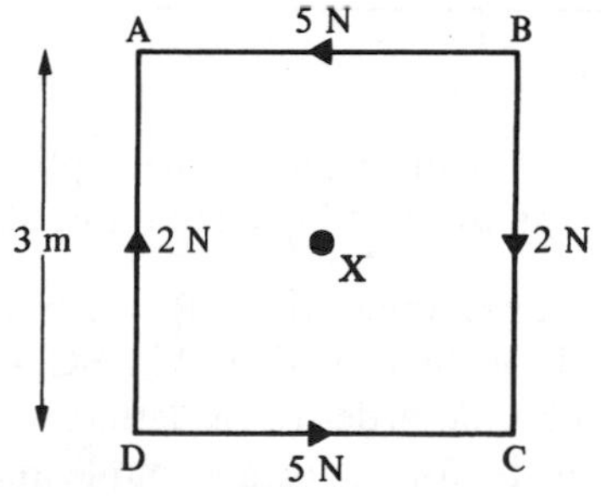

Resolving parallel to AB gives:

$$-5\,\text{N} + 5\,\text{N} = 0$$

Resolving parallel to AD gives:

$$-2\,\text{N} + 2\,\text{N} = 0$$

Hence the system has no translatory effect, and is therefore either in equilibrium or reduces to a couple.

Taking moments gives:

(a) $\overset{\curvearrowright}{X}$ $\quad (2 \times 1\tfrac{1}{2}) - (5 \times 1\tfrac{1}{2}) + (2 \times 1\tfrac{1}{2}) - (5 \times 1\tfrac{1}{2})$ N m clockwise
$= (3 - 7\tfrac{1}{2} + 3 - 7\tfrac{1}{2})$ N m
$= -9$ N m clockwise
$= 9$ N m anticlockwise

(b) $\overset{\curvearrowright}{A}$ $\quad (2 \times 3) - (5 \times 3)$ N m clockwise
$= 6 - 15$ N m clockwise
$= 9$ N m anticlockwise

As was to be expected, the moments about the points X and A are the same, since the system is equivalent to a couple. The moment of the couple is 9 N m in an anticlockwise sense.
In order to produce a state of equilibrium, a couple of the same magnitude, but opposite in sense is required:

couple required $= 9$ N m clockwise

Exercise 7B

1. Find which of the following systems will reduce to a couple and, in these cases, find the moment of the couple:

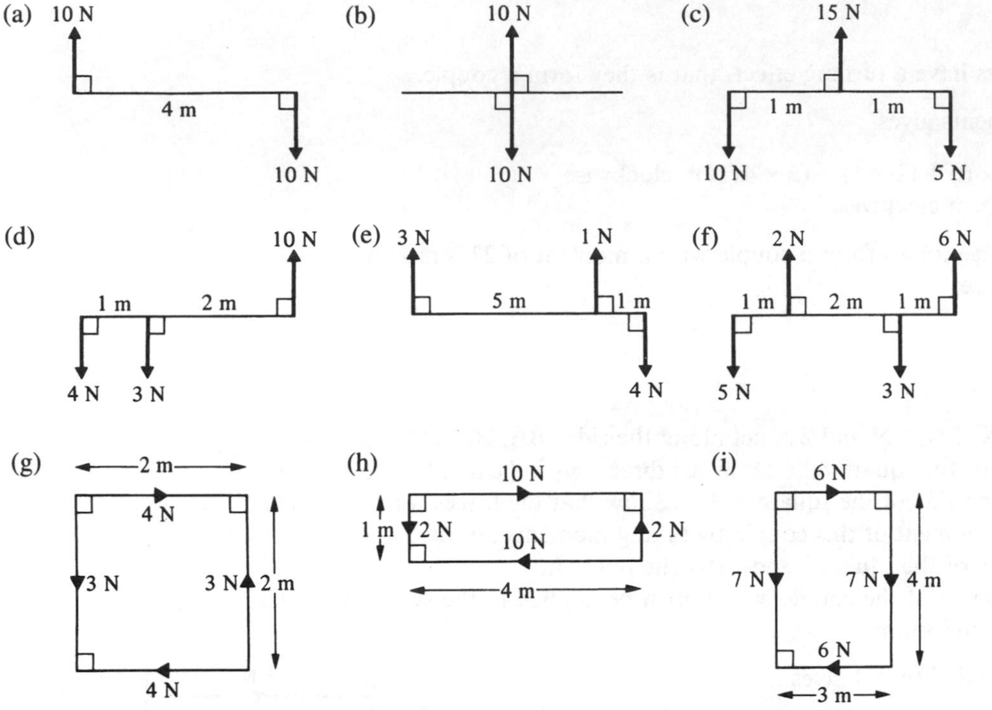

2. Find the moment of the couple applied to a corkscrew by two equal and opposite forces of 25 N acting on it, along lines 7 cm apart.

3. ABCD is a rectangle with AB = 5 m and BC = 2 m. A force of 3 N acts along each of the four sides AB, BC, CD and DA in the directions indicated by the order of the letters.
Show that the forces form a couple and find its moment.

4. ABCD is a rectangle with AB = 6 m and BC = 2 m. Forces of 5 N, 5 N, X N and X N act along CB, AD, AB and CD respectively. The directions of the forces are given by the order of the letters.
If the system is in equilibrium, find X.

5. ABCD is a square of side 40 cm. Forces of 20 N, 15 N and 20 N act along the sides AB, BC and CD respectively and a force Y acts along DA. The directions of the forces are given by the order of the letters.
If the system is equivalent to a couple, find the magnitude of Y and the moment of the couple.

6. ABCD is a square of side 60 cm. Forces of 6 N, 2 N, 6 N and 2 N act along the sides AB, CB, CD and AD respectively, in the directions indicated by the order of the letters.
Show that the forces form a couple and find the moment of the couple that must be applied to the system in order to produce equilibrium.

7. ABCD is a rectangle with AB = 8 m and BC = 3 m. Forces, each of 4 N, act along AB and CD in the directions indicated by the order of the letters. Show that the system reduces to a couple and find the moment of the couple. It is now required to reduce the system to equilibrium by applying a force P N along AD and another force P N along CB. Find the value of P.
8. A light rod of length 50 cm lies on a horizontal table. A man holds the ends of the rod. The rod is subjected to a couple, of moment 40 N m, causing it to rotate upon the table. What force perpendicular to the rod must the man exert through each hand in order to prevent the rotation?
9. A force of $(3\mathbf{i} - 5\mathbf{j})$ N acts at the point which has position vector $(6\mathbf{i} + \mathbf{j})$ m and a force of $(-3\mathbf{i} + 5\mathbf{j})$ N acts at the point which has position vector $(4\mathbf{i} + \mathbf{j})$ m.
Show that these forces reduce to a couple and find the moment of the couple.
10. A force of $(4\mathbf{i} + 3\mathbf{j})$ N acts at the point which has position vector $(6\mathbf{i} + 3\mathbf{j})$ m and a force of $(-4\mathbf{i} - 3\mathbf{j})$ N acts at the point which has position vector $(3\mathbf{i} - \mathbf{j})$ m.
Show that these forces reduce to a couple and find the moment of the couple.
11. Forces of $(\mathbf{i} + \mathbf{j})$ N, $(-4\mathbf{i} + \mathbf{j})$ N and $(3\mathbf{i} - 2\mathbf{j})$ N act at the points having position vectors $(2\mathbf{i} + 2\mathbf{j})$ m, $(-\mathbf{i} + 4\mathbf{j})$ m and $(4\mathbf{i} - 2\mathbf{j})$ m respectively.
Show that these forces reduce to a couple and find the moment of the couple.
12. Forces of $(a\mathbf{i} + b\mathbf{j})$ N and $(6\mathbf{i} - 4\mathbf{j})$ N act at the points having position vectors $(-2\mathbf{i} - 2\mathbf{j})$ m and $(3\mathbf{i} - \mathbf{j})$ m respectively.
If these forces reduce to a couple, find a and b and the moment of the couple.
13. Forces of $6\mathbf{j}$ N and $-6\mathbf{j}$ N act at the origin and at position vector $2\mathbf{i}$ m respectively.
Show that the moment of the forces about any point P(x, y) is independent of x and y.

Replacement of parallel forces by a single force

In the last section, systems of parallel forces which reduced to couples were considered. These systems produced a turning effect but did not have a translatory effect.
Systems of parallel forces which do not reduce to a couple must now be considered.

Any number of parallel forces, which are not equivalent to a couple, may be replaced by a single force. This resultant force will be parallel to the given forces, and it must have the same translational and rotational effects as the given forces.

The same translational effect is ensured if the magnitude of the resultant is obtained by resolving in the direction of the given forces.

To ensure the same rotational effect, use is made of the very important Principle of Moments, which states that:

> When a number of coplanar forces act on a body, the algebraic sum of the moments of these forces, about any point in their plane, is equal to the moment of the resultant of these forces about that point.

When finding the resultant of a system of forces, it is advisable to draw two diagrams, one showing the given forces and a second to show the equivalent resultant force.

Example 11

Two unlike parallel forces of 1 N and 3 N are 4 m apart. Find the magnitude, direction and line of action of the resultant of these forces.

First diagram (given forces)

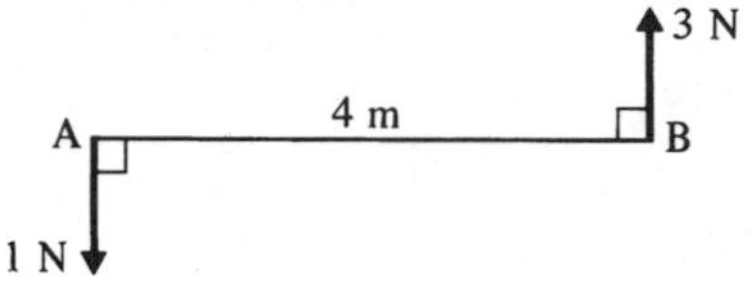

Second diagram (resultant force)

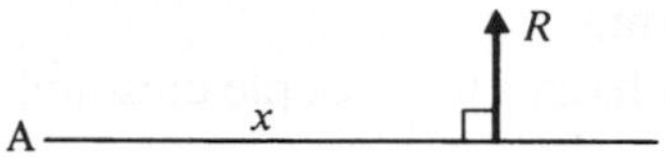

Let A and B be the points in which a line drawn at right angles to both forces meets their lines of action. Let R act at a distance of x metres from A. The resultant R of the given forces will act in the same direction as the larger force.

R must have the same translational effect. Thus, resolving at right angles to AB gives:

$$R = 3 - 1 \qquad \therefore \qquad R = 2\,\text{N}$$

R must have the same rotational effect. Thus, taking moments gives:

$$\overset{\curvearrowright}{\text{A}} \quad 3 \times 4 = R \times x \qquad \therefore \qquad x = 6\,\text{m}$$

The resultant acts at a distance of 6 m from A, is 2 N in magnitude and acts in the same direction as the 3 N force.

Example 12

Two like parallel forces of 2 N and 5 N are 21 m apart. Find the magnitude, direction and line of action of the resultant of these forces.

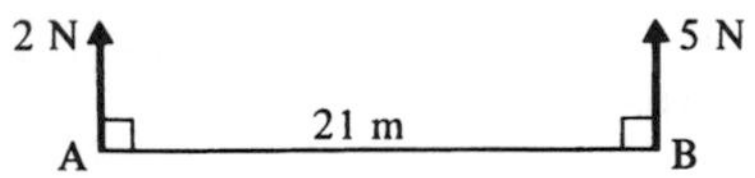

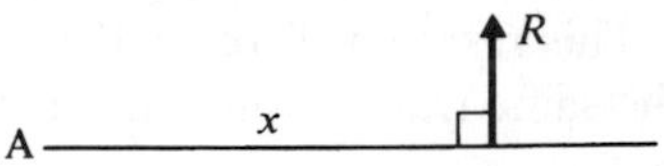

Draw two diagrams as before. The resultant R will act in the same direction as the given forces.

For R to have the same translational effect:

$$R = 2 + 5$$

$\therefore$ $R = 7\,\text{N}$ in the same direction as the given forces

For R to have the same rotational effect:

$$\overset{\curvearrowright}{\text{A}} \quad 5 \times 21 = R \times x$$

$\therefore$ $105 = 7x$ or $x = 15\,\text{m}$

The magnitude of the resultant is 7 N, acting in the same direction as the given forces, and its line of action is 15 m from the 2 N force.

Position of resultant

It should be carefully noted that the resultant of two like forces will have its line of action between the two forces; if the forces are unlike, the resultant will lie outside the given forces.

Exercise 7C

1. In each of the following cases, find the magnitude of the resultant of the forces shown and the distance of its line of action from A.

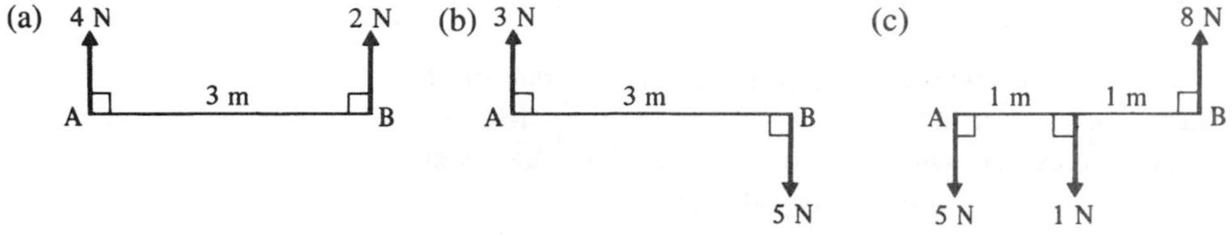

2. The lines of action of two like parallel forces of 3 N and 2 N are 10 m apart. Find the magnitude of the resultant of the forces and the distance between its line of action and that of the 3 N force.
3. The lines of action of two like parallel forces of 8 N and 12 N are 5 m apart. Find the magnitude of the resultant of the forces and the distance between its line of action and that of the 8 N force.
4. The lines of action of two unlike parallel forces of 5 N and 3 N are 4 m apart.
   Find the magnitude of the resultant of the forces and the distance between its line of action and that of the 3 N force.
5. The lines of action of two unlike parallel forces of 8 N and 12 N are 5 m apart.
   Find the magnitude of the resultant of the forces and the distance between its line of action and that of the 8 N force.
6. Two like vertical forces of 4 N and Q N act at points A and B respectively where AB is horizontal and of length 6 m. Their resultant is a force of P N and acts at a point X, between A and B, where AX = 2 m. Find P and Q.
7. Two unlike vertical forces of 10 N and Q N act at points A and B respectively where AB is horizontal and of length 3 m. The resultant of the two forces is a force of P N which acts at a point X on the line BA produced such that XA = 4·5 m. Find P and Q.

8. A, B and C are three points on a horizontal line with B between A and C. AB = 1 m and BC = 3 m. Forces of 5 N, 3 N and 2 N act at A, B and C respectively in a direction vertically downwards. Find the magnitude of the single force that could replace these forces and the distance of its line of action from A.

9. A, B and C are three points on a horizontal line with B between A and C; AB = 1 m and AC = 3 m. Forces of 2 N and 0·5 N act vertically downwards at A and C respectively and a force of 4 N acts vertically upwards at B. Find the magnitude of the resultant of these three forces and the distance of its line of action from A.

10. A, B and C are three points on a horizontal line with AC = 4 m and point B situated between A and C. Vertical forces of 4 N, 1 N and 3 N act at A, B and C respectively. The resultant of these forces is a force of 6 N, vertically upwards, acting through the point X on AB such that AX = $1\frac{1}{2}$ m. State the direction of the forces at A, B and C and find the distance AB.

11. Show that the resultant of two like parallel forces, of magnitudes P N and Q N, is a force in the same direction as the two forces and whose line of action divides the distance between the two forces internally, in the ratio $Q:P$.

12. Show that the resultant of two unlike parallel forces, of magnitudes P N and Q N, where P is greater than Q, is a force in the direction of the larger of the two forces and whose line of action divides the distance between the two forces externally, in the ratio $Q:P$.

Parallel forces in equilibrium

A system of parallel forces is either equivalent to a couple or it can be replaced by a single force or resultant.
Consider a system of parallel forces in equilibrium. Such forces cannot be equivalent to a couple because, if they were, they would have a turning effect and would not therefore be in equilibrium. The second possibility is that the forces could be replaced by a resultant. However, as the forces are in equilibrium, this resultant force must be zero. Furthermore, if the resultant force is zero, then the moment of this resultant force about any point must also be zero and so, by the Principle of Moments, the algebraic sum of the moments of the forces about any point must be zero.

Thus for parallel forces in equilibrium:

(i) the resultant force in any direction is zero

and

(ii) the algebraic sum of the moments of the forces about any point is zero (i.e. we can equate clockwise and anticlockwise moments).

Example 13

A uniform beam, of length 2 m and mass 4 kg, has a mass of 3 kg attached at one end and a mass of 1 kg attached at the other end. Find the position of the support if the beam rests in a horizontal position.

Suppose the beam is supported at the point C, s metres from the 3 kg mass, and that the support exerts an upward force of P newtons on the beam. The forces due to the two masses are shown on the diagram. Since the beam is uniform, it is assumed that its weight will act at its centre.

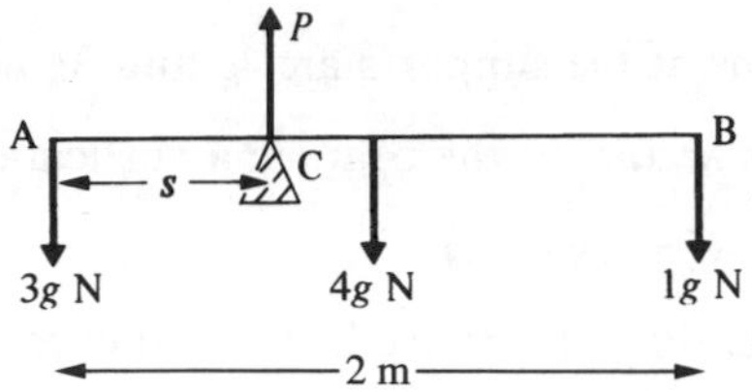

Method 1

Equate the clockwise and anticlockwise moments:

$$\overset{\curvearrowright}{C} \qquad 3g \times s = 4g \times (1 - s) + 1g \times (2 - s)$$

$$\therefore \qquad 8gs = 4g + 2g$$

$$\therefore \qquad s = \tfrac{3}{4}\text{ m}$$

Method 2

Equate the forces in a vertical direction:

$$3g + 4g + 1g = P$$

$$\therefore \qquad P = 8g$$

$$\overset{\curvearrowright}{A} \qquad 4g \times 1 + 1g \times 2 = P \times s$$

$$\therefore \qquad 6g = 8gs$$

$$\therefore \qquad s = \tfrac{3}{4}\text{ m}$$

Thus the beam must be supported at a point $\frac{3}{4}$ m from the 3 kg mass.

The first method has the advantage that the unknown force at the support is not needed, since the moments are taken about this point.

Example 14

A light horizontal beam of length 2 m rests with its ends A and B on smooth supports. The beam carries masses of 5 kg and 2 kg at distances of 60 cm and 150 cm respectively, from A.
Find the reaction at each support.

Since the beam is light, its mass can be ignored.
Suppose the reactions are P and Q newtons.

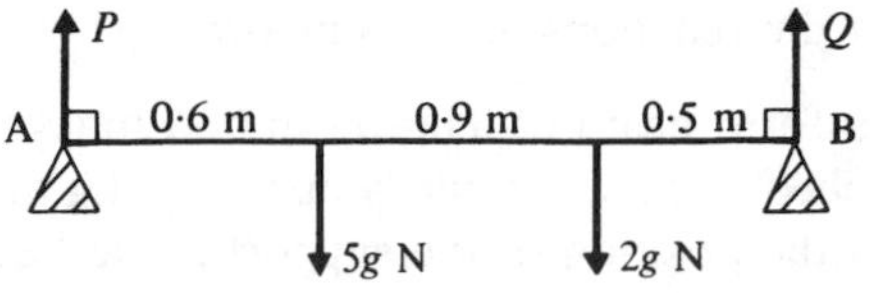

Equate the clockwise and anticlockwise moments:

$$\overset{\curvearrowright}{A} \qquad 5g \times 0{\cdot}6 + 2g \times 1{\cdot}5 = Q \times 2$$

$$\therefore \qquad Q = 3g$$

The force P does not appear in this equation since it has no moment about the point A.

$\overset{\frown}{\text{B}}$ $\quad 2g \times 0{\cdot}5 + 5g \times 1{\cdot}4 = P \times 2$

$\therefore \quad P = 4g$

The reactions at the supports are $4g$ and $3g$ newtons.

If the forces acting on the beam in a vertical direction are considered:

$$P + Q = 5g + 2g$$

and this equation can be used, either to check the values of P and Q, or in place of one of the moment equations.

Exercise 7D

Questions **1** to **8** involve light horizontal rods in equilibrium. Each diagram shows the forces acting on the rods. Find the magnitudes of the forces P and Q and the distance x, as applicable.

1.
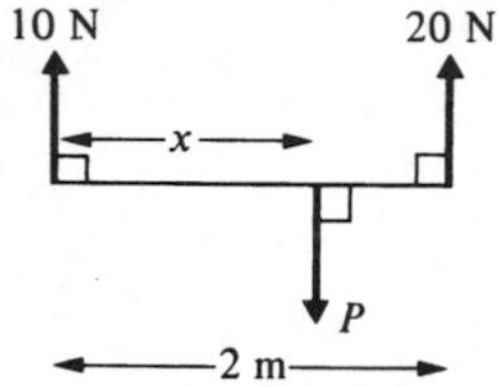

2.
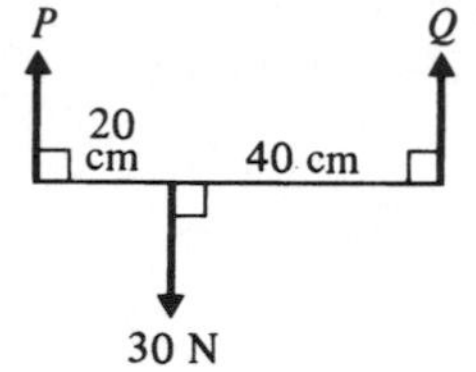

3.
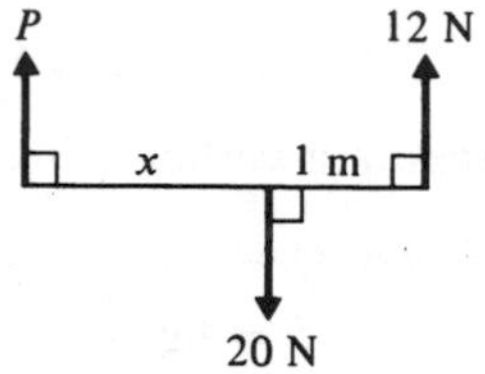

4.
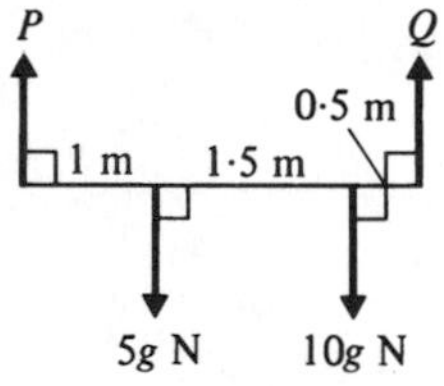

5.
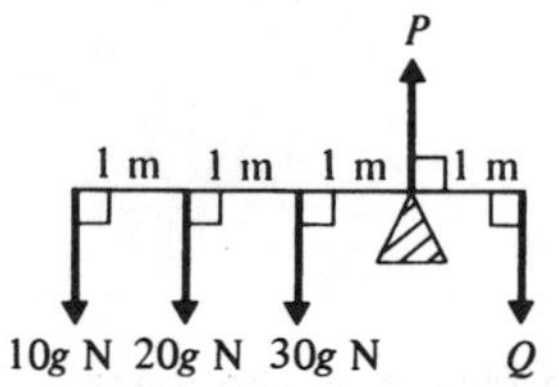

6.
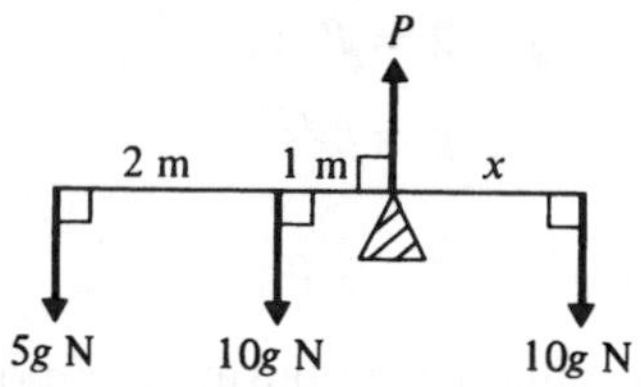

7.
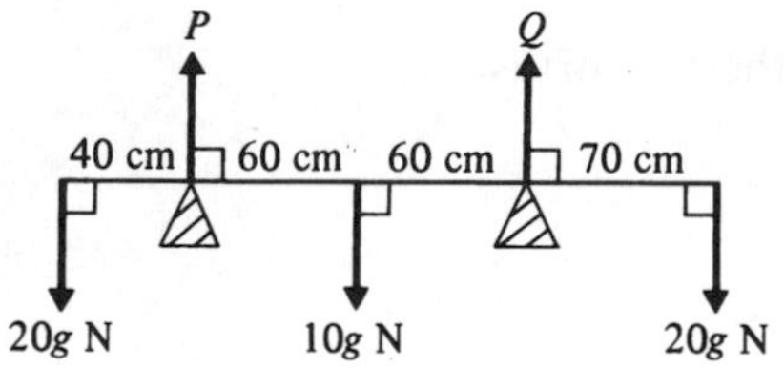

8.
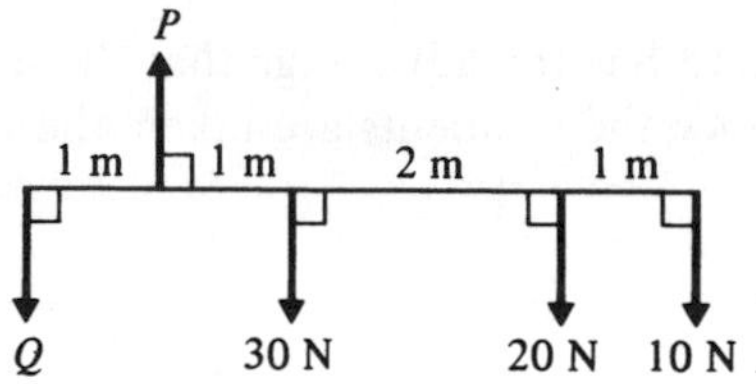

Some of the following questions involve *uniform* beams. This means that the weight of the beam can be taken as acting at the centre point of the beam.

9. A uniform beam of length 4 m and mass 50 kg rests horizontally, supported at each end. A mass of 20 kg is placed on the beam, 1 m from one end.
Find the reactions at the supports.

10. A uniform beam of length 6 m and mass 8 kg has a mass of 10 kg attached at one end and a mass of 3 kg attached at the other end.
Find the position of the support if the beam rests in a horizontal position.

11. A playground seesaw consists of a uniform beam of length 4 m supported at its mid-point. If a girl of mass 25 kg sits at one end of the seesaw, find where her brother of mass 40 kg must sit if the seesaw is to balance horizontally.

12. A broom consists of a uniform broomstick of length 120 cm and mass 4 kg with a broom head of mass 6 kg attached at one end. Find where a support should be placed so that the broom will balance horizontally.

13. A non-uniform beam AB is of length 4 m and its weight of 5 N can be considered to act at a point 1·8 m from the end A. The beam rests horizontally on smooth supports at A and B. Find the reactions at the supports.

14. A uniform beam AB of mass 10 kg and length 4 m rests horizontally on two supports, one at A and the other 1 m from B.
Where must a boy of mass 50 kg stand on the beam if he wishes to make the reactions at the supports equal?

15. A non-uniform rod AB of length 4 m is supported horizontally on two supports, one at A and the other at B. The reactions at these supports are $5g$ N and $3g$ N respectively.
If instead the rod were to rest horizontally on one support, find how far from end A this support would have to be placed.

16. A pole vaulter uses a uniform pole of length 4 m and mass 5 kg. He holds the pole horizontally by placing one hand at one end of the pole and the other hand at a position on the pole 80 cm away.
Find the vertical forces exerted by his hands.

17. Three uniform rods of mass 2, 4 and 8 kg and each of length 20 cm, are joined together in the order mentioned to form one long rigid rod of length 60 cm. This rod is then suspended horizontally by a vertical string attached to the rod at a point x cm from its mid-point.
Find the value of x and the tension in the string.

18. The diagram shows a spade which consists of a handle, a uniform shaft and a uniform rectangular blade. The handle is of mass 0·5 kg, the shaft of mass 2 kg and the blade of mass 2 kg.

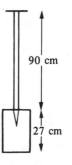

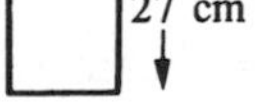

(a) If the spade is to rest horizontally on one support, where should this support be placed?

(b) If a man carries the spade horizontally with one hand on the handle and the other at a distance of 72 cm from the handle, find the vertical forces exerted by his hands when a brick of mass 6 kg is placed at the centre of the blade.

Non-parallel forces in equilibrium

The results stated for parallel forces in equilibrium also apply to non-parallel forces which are in equilibrium.

In particular, for any system of forces in equilibrium the algebraic sum of the moments of the forces about any point must be zero.

Problems involving forces which are not parallel can be solved by using this principle.

Example 15

A pendulum AM consists of a string, of length 2 m, and a bob of mass 5 kg. The pendulum is suspended from A and held in equilibrium by a horizontal force F applied at M so that the string makes an angle of 30° with the vertical. Find the force F.

Suppose the horizontal and vertical distances of M from A are x and y respectively. Taking moments about the point A for all the forces acting on the bob gives:

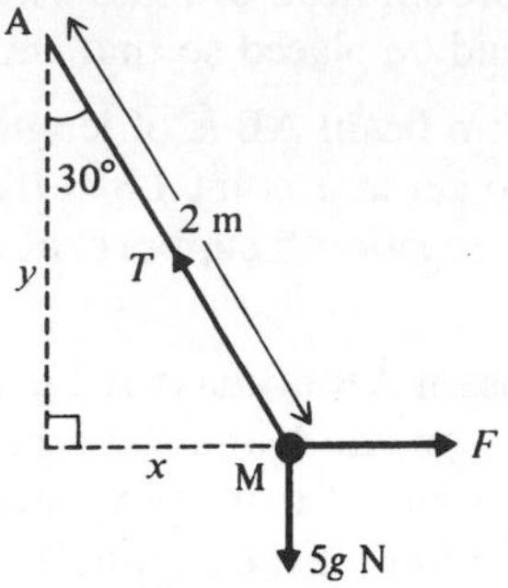

$$\overset{\curvearrowright}{A} \qquad 5g \times x = F \times y$$

$$\therefore \qquad 5g \times 2\cos 60° = F \times 2\cos 30°$$

$$\therefore \qquad F = \frac{5g\sqrt{3}}{3}\text{ N}$$

$$= 28{\cdot}3\text{ N}$$

Note that the tension T in the string does not appear in the moment equation since its line of action passes through the point A about which moments are taken. It therefore has no moment about this point.

Moment of a force as sum of moments

When finding the moment of a force about a point, it is sometimes useful first to resolve the force into its components.

Suppose the force P acts at the point N at a distance s from the point O and in a direction making an angle θ with ON produced.

Resolving the force P gives:

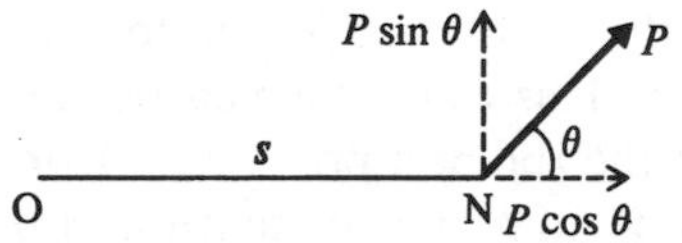

component in direction ON is $P\cos\theta$

component at right angles to ON is $P\sin\theta$

moment of P about O = sum of moments of these components about O

$$= P\cos\theta \times 0 + P\sin\theta \times s$$

$$= Ps\sin\theta$$

Example 16

A uniform beam AB of mass 100 kg and length $2l$ m has its lower end A resting on rough horizontal ground. It is held in equilibrium at an angle of 20° with the horizontal by a rope attached to the end B which makes an angle of 40° with BA.
Find the tension in the rope.

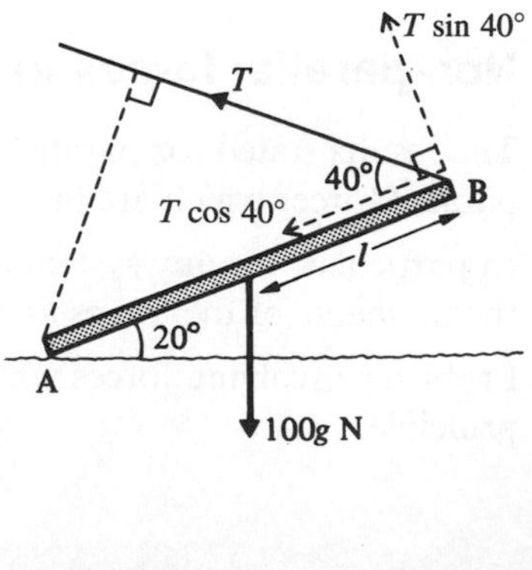

Let T be the tension in the rope. There are two forces acting on the beam, other than the action between the foot of the beam and the ground. If moments are taken about the point A, the force at this point will have no moment and will not therefore appear in the equation.

Resolve T into its components $T \cos 40°$ and $T \sin 40°$ along and at right angles to the beam respectively:

$$\overset{\frown}{A} \qquad 100g \times l \cos 20° = T \sin 40° \times 2l + T \cos 40° \times 0$$

$$\therefore \qquad T = \frac{100g \cos 20°}{2 \sin 40°} = 716\,\text{N}$$

The tension in the rope is 716 N.

It should be noted that, for these bodies in equilibrium, the algebraic sum of the forces resolved in any direction must be zero. However, in Examples 15 and 16 above, and in Exercise 7E, this fact need not be used as the required answers can be obtained by taking moments about a suitably chosen point.

Exercise 7E

1. Each of the following diagrams shows a uniform beam, of mass 5 kg and length 4 m, freely hinged at a point A and resting horizontally in equilibrium. Find the magnitude of the force T in each case.

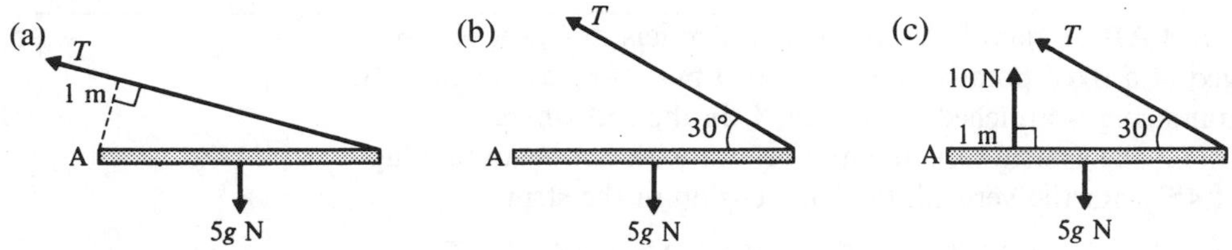

2. Each of the following diagrams shows a pendulum consisting of a light string with one end freely pivoted at A and the other end carrying a bob of mass 1 kg. The pendulum is held in equilibrium at an angle to the vertical by a force F.
Find the magnitude of F in each case.

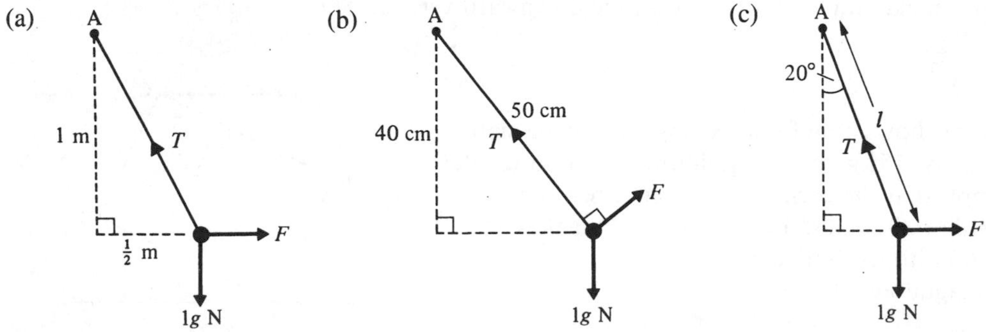

3. Each of the following diagrams shows a uniform beam of mass 5 kg and length 4 m with one end freely hinged at a point A. The beam rests in equilibrium at an angle to the horizontal.
Find the magnitude of the force T in each case.

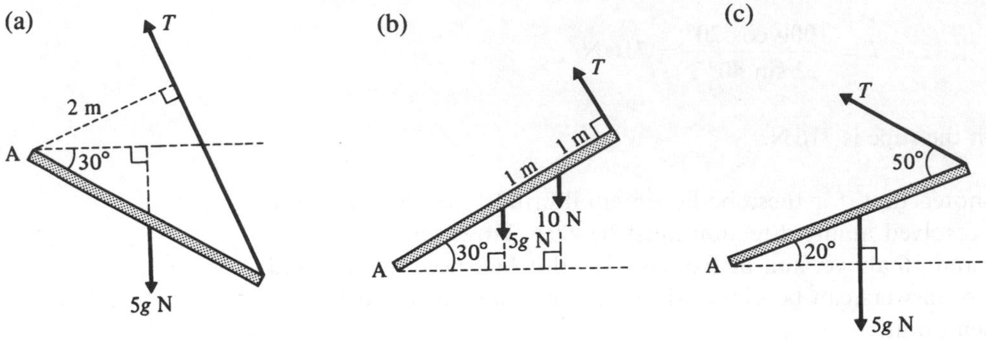

4. A pendulum consists of a light string AB of length 60 cm with end A fixed and a bob of mass 3 kg attached to B. Find the horizontal force that must be applied to the bob to keep the string at an angle of 25° to the downward vertical.

5. A light string AB of length 80 cm has end A fixed and a particle of mass 4 kg attached to B. Find the magnitude of the least force that must be applied to the particle so that it is held at a distance of 40 cm from the vertical through A with the string taut.

6. A uniform rod AB of mass 10 kg and length 2 m has its lower end A freely hinged at a fixed point and a particle of mass 4 kg attached to B. A horizontal string is attached to a point X on the rod where AX = 1·5 m. If the system rests in equilibrium, with the beam making an angle of 45° with the vertical, find the tension in the string.

7. A uniform horizontal shelf of mass 5 kg is freely hinged to a vertical wall and is supported by a chain CD as shown in the diagram:
The tension in the chain is 98 N, AD = 15 cm and angle CDA = 50°.
Find the length AB.

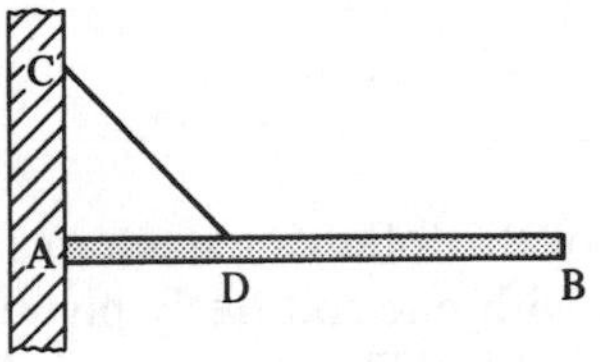

8. A uniform rod AB of mass m hangs vertically with end A freely hinged to a fixed point. The rod is pulled aside by a horizontal force F, applied at B, until it makes an angle of 30° with the downward vertical. Show that $F = \dfrac{mg}{2\sqrt{3}}$.

9. The diagram shows a uniform rectangular paving slab ABCD of mass 15 kg held in equilibrium by a horizontal force F applied at the corner C. Corner A rests on the ground; AB = 30 cm and BC = 40 cm. AD makes an angle of 30° with the horizontal.
Find the magnitude of F.

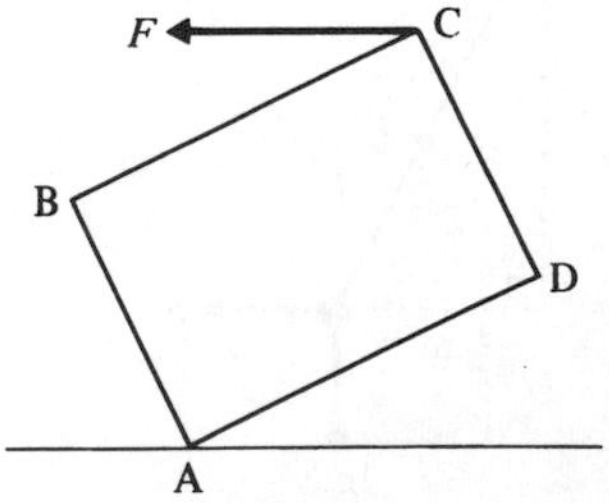

10. A non-uniform beam AB is of mass 100 kg and length 12 m. X and Y are points on the beam such that AX = 4 m and AY = 7 m. The weight of the beam can be considered to act at point X. The beam has its lower end A resting on rough horizontal ground and is kept in equilibrium, at an angle of 26° to the horizontal, by a rope attached to point Y and making an angle of 55° with YA. Find the tension in the rope.

11. A uniform beam AB of mass m and length $2l$ has its lower end A resting on rough horizontal ground and is kept in equilibrium, at an angle of 45° to the horizontal, by a rope attached to end B. If the rope makes an angle of 60° with BA and T is the tension in the rope, show that $T = \dfrac{mg}{\sqrt{6}}$.

Equivalent systems of forces

It has already been seen that a system of parallel forces is either equivalent to a couple or can be replaced by a single force or resultant. In fact *any* system of coplanar forces is either equivalent to a couple or can be replaced by a single force or resultant. In the latter case the magnitude, direction and line of action of the resultant can be found by resolving and taking moments.

Example 17

The forces 4, 2, 3 and 5 N act along the sides AB, BC, CD and DA respectively of the rectangle ABCD in which AB = 6 cm and BC = 3 cm. The forces act in the directions indicated by the order of the letters. Forces X and Y, acting as shown in the second diagram, are equivalent to the given system of forces.
Find the magnitude of X and Y and also the distance x. Hence find the magnitude and the direction of the single force which is equivalent to the given system.

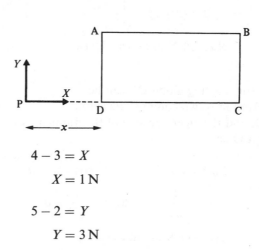

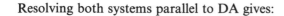

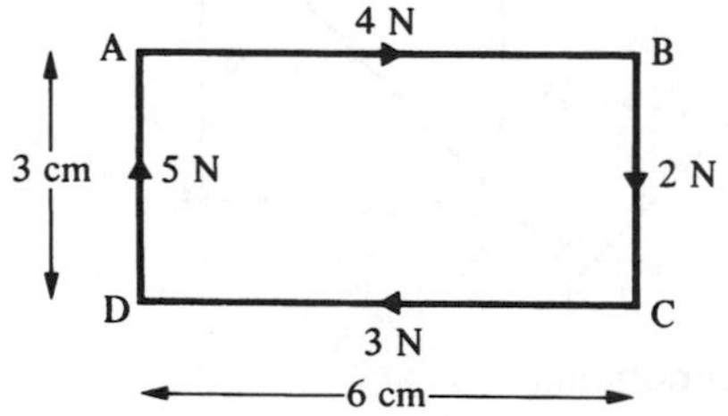

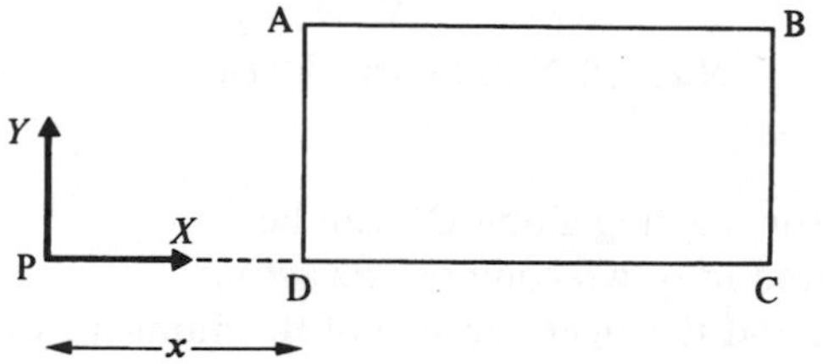

Resolving both systems parallel to AB gives: $4 - 3 = X$

$\therefore \quad X = 1\text{ N}$

Resolving both systems parallel to DA gives: $5 - 2 = Y$

$\therefore \quad Y = 3\text{ N}$

Equating the moments of the two systems gives:

$\overset{\frown}{D}$ $\quad 2(6) + 4(3) = Y \times x$

$\therefore \quad 24 = 3x$

$\therefore \quad x = 8\,\text{cm}$

$X = 1\,\text{N}$, $Y = 3\,\text{N}$ and the distance $x = 8\,\text{cm}$.

The single force R equivalent to the system is given by:

$R^2 = X^2 + Y^2$

$\therefore \quad R^2 = 1^2 + 3^2$

$\therefore \quad R = 3{\cdot}16\,\text{N}$

and $\quad \tan\theta = \frac{3}{1}$

$\therefore \quad \theta = 71{\cdot}57°$

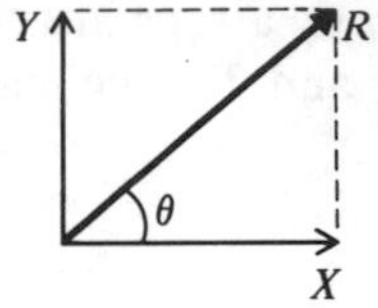

The single force is 3·16 N acting at an angle of 71·57° to the line DC.

Use of components when taking moments

The following example illustrates the use of the components of a force when the moment of a force about a point is required.

Example 18

The square ABCD has sides of length a metres. Find the moment of the force $3\sqrt{2}\,\text{N}$, acting along DB, about the point O as shown in the diagram.

(i) The moment, by definition, is:

$\overset{\frown}{O}$ $\quad 3\sqrt{2} \times \text{OP}$

$= 3\sqrt{2} \times (a + y)\cos 45°$

$= 3\sqrt{2} \times (a + y) \times \dfrac{1}{\sqrt{2}}$

$= 3(a + y)\,\text{N m anticlockwise}$

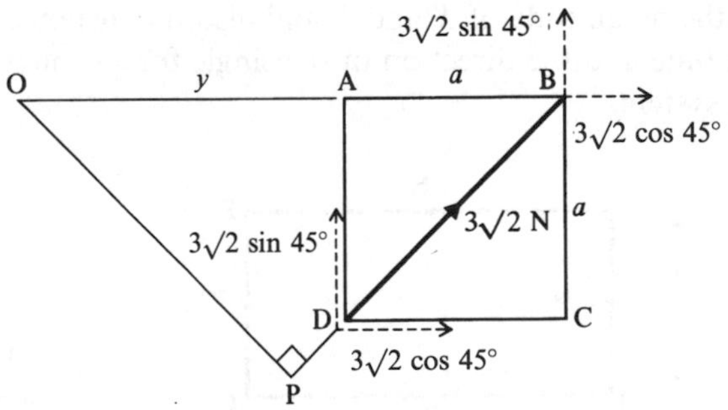

(ii) The force acting along DB can be resolved into two components acting at D, and the algebraic sum of the moments of these components about O is:

$\overset{\frown}{O}$ $\quad 3\sqrt{2}\sin 45° \times y + 3\sqrt{2}\cos 45° \times a$

$= 3\sqrt{2} \times \dfrac{1}{\sqrt{2}} \times y + 3\sqrt{2} \times \dfrac{1}{\sqrt{2}} \times a$

$= 3(a + y)\,\text{N m anticlockwise}$

(iii) Alternatively the force acting along DB can be resolved into two components at B:

$$\overset{\frown}{O} \qquad 3\surd 2 \sin 45° \times (a+y) + 3\surd 2 \cos 45° \times 0$$
$$= 3\surd 2 \frac{1}{\surd 2} \times (a+y)$$
$$= 3(a+y) \text{ N m anticlockwise}$$

It is seen that the same result is obtained each time, but care is necessary in distinguishing between case (ii) and case (iii).

Example 19

The forces 2, 4, 2, 5 and $3\surd 2$ N act along the sides AB, BC, CD, DA and DB respectively of the square ABCD in the directions indicated by the order of the letters. Find the magnitude and direction of the force R which is equivalent to this system of forces. If the line of action of the force R cuts BA produced at a point P, x cm from A, and the length of each side of the square is 10 cm, find x.

Draw the two diagrams showing the given system of forces and the required force R.

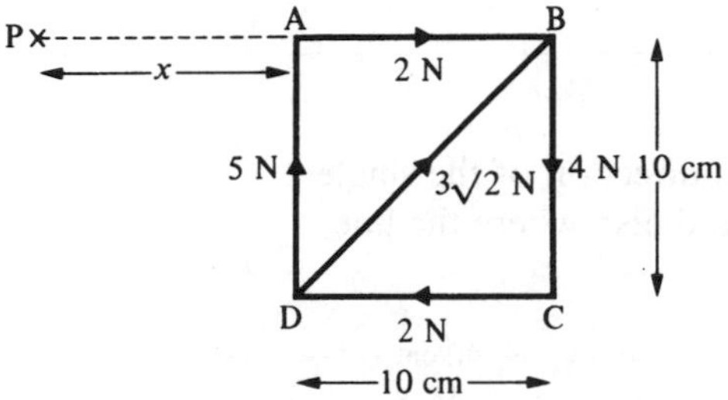

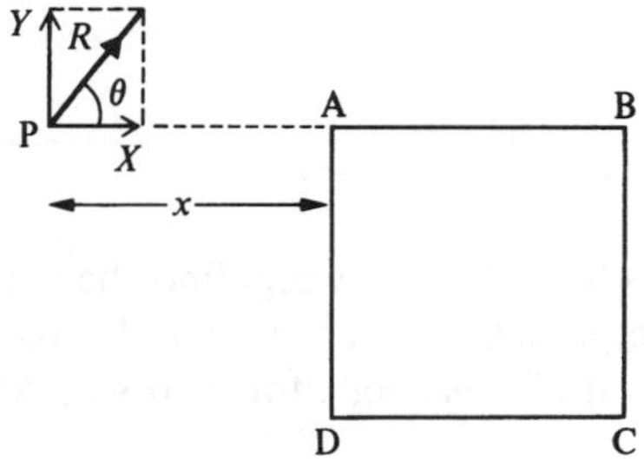

Resolving both systems parallel to AB gives:

$$2 + 3\surd 2 \cos 45° - 2 = X$$
$$\therefore \qquad 3\text{ N} = X \qquad \ldots[1]$$

Resolving both systems parallel to DA gives:

$$5 + 3\surd 2 \sin 45° - 4 = Y$$
$$\therefore \qquad 4\text{ N} = Y \qquad \ldots[2]$$

Squaring and adding equations [1] and [2] gives:

$$R^2 = X^2 + Y^2$$
$$= 9 + 16$$
$$= 25$$
$$\therefore \qquad R = 5\text{ N}$$

also $\qquad \tan\theta = \dfrac{Y}{X} = \frac{4}{3}$

$$\therefore \qquad \theta = 53{\cdot}13°$$

Equating the moments of the two systems about the point B gives:

$\overset{\frown}{B}$ $(5 \times 10) + (2 \times 10) = Y(x + 10)$

Now substituting $Y = 4$ gives:

$$x = 7\tfrac{1}{2} \text{ cm}$$

The equivalent force is 5 N acting at an angle of 53·13° to the line AB and cutting BA produced $7\frac{1}{2}$ cm from A.

Note that alternatively moments can be taken about the point P. In this case the moment of the system consisting of the resultant R will be zero.

Hence, equating moments of the two systems about the point P gives:

$\overset{\frown}{P}$ $(2 \times 10) + 4(10 + x) - 5x - 3\sqrt{2} \sin 45°(x + 10) = R \times 0$

$\therefore$ $20 + 40 + 4x - 5x - 3x - 30 = 0$

and again $x = 7\frac{1}{2}$ cm is obtained.

Exercise 7F

1. For each of the following, find the magnitude and direction of the single force equivalent to the system of forces shown. Find also where the line of action of this single force crosses AB.

(a)

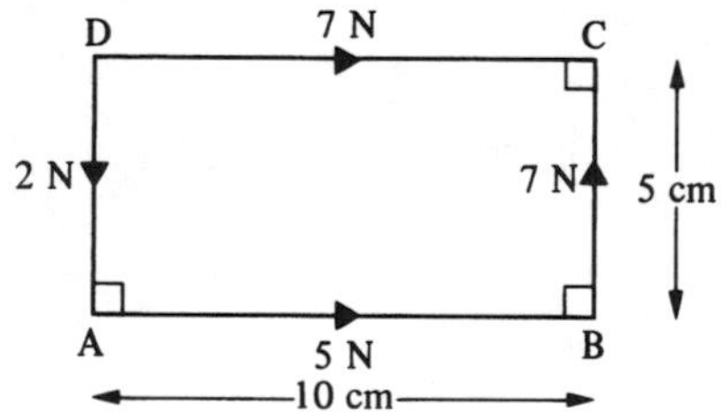

(b)

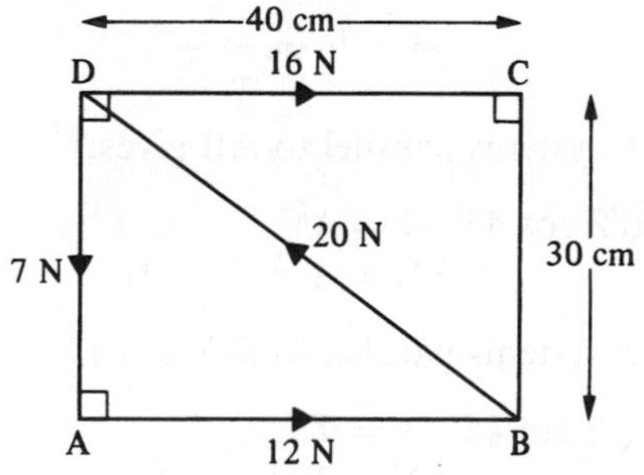

(c)

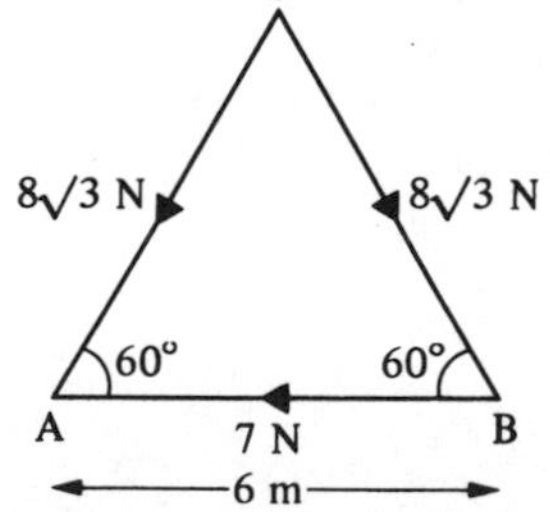

(d)

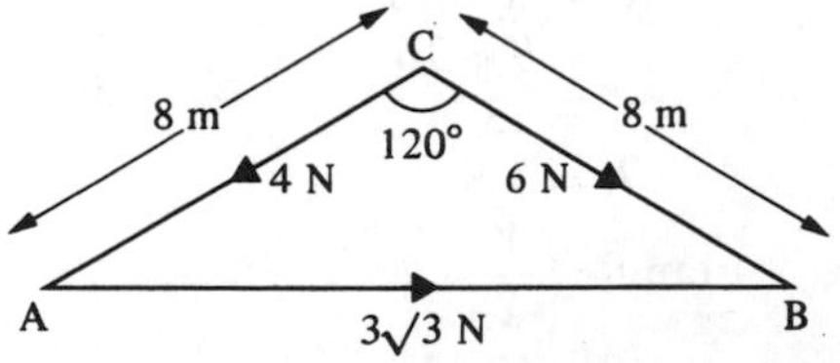

2. Show that each of the following systems of forces is equivalent to a couple and in each case find the moment of the couple.

(a)

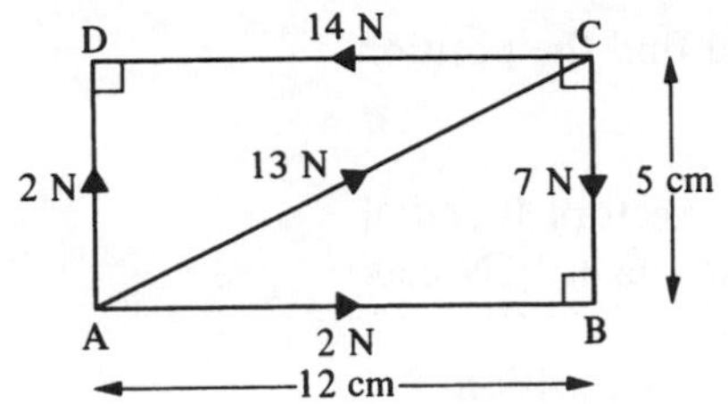

(b)

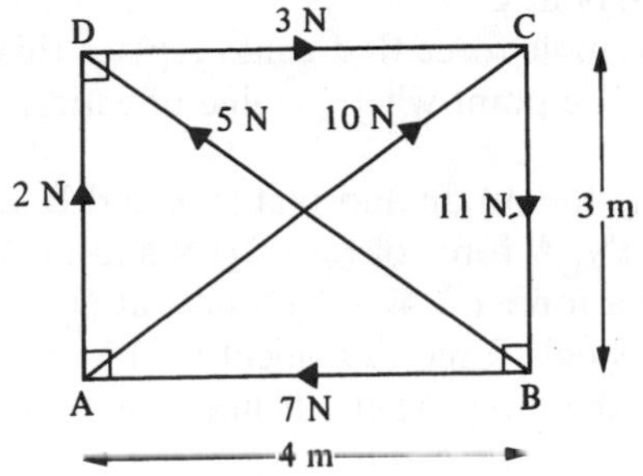

(c)

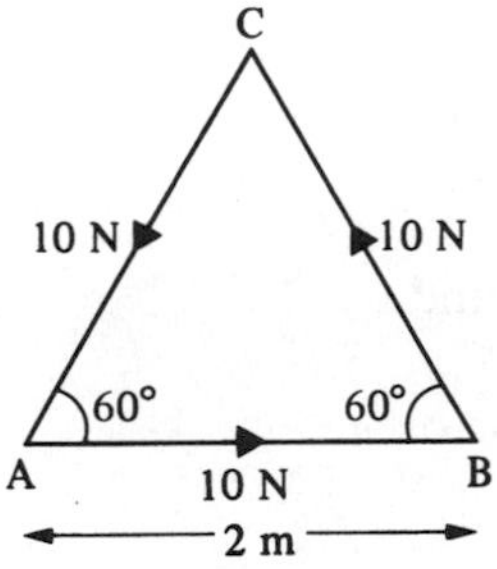

(d)

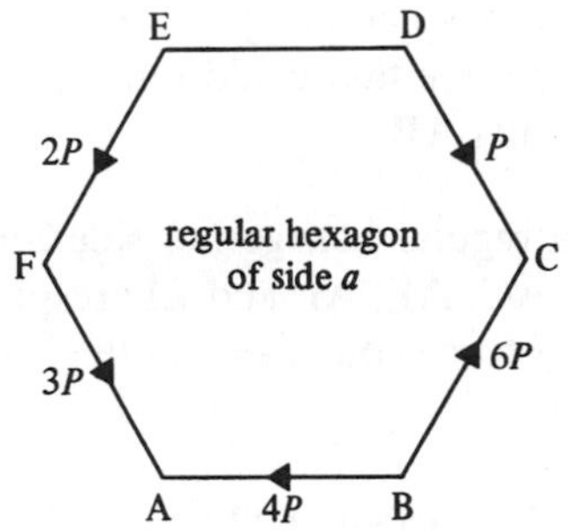

In the following questions involving geometrical figures, the forces act along the given lines in the directions indicated by the order of the letters.

3. ABCD is a rectangle with AB = 1·5 m and AD = 1 m. Forces of 2 N, 1 N, 1 N and 3 N act along AB, BC, DC and AD respectively. Calculate the magnitude and direction of the single force that could replace this system of forces and find where its line of action cuts AB.

4. ABCD is a square of side 5 m. Forces of 4 N, 6 N, 8 N and 10 N act along BD, DC, CA and CB respectively. When a force P acts along AD and a force Q acts along AB, the whole system is equivalent to a couple. Find the magnitudes of P and Q and the moment of the couple.

5. ABCD is a square of side a metres. Forces of 1 N, 4 N, 3 N and 6 N act along AB, CB, DC and AD respectively. Calculate the magnitude and direction of the single force that could replace this system of forces and find where its line of action cuts AB.

6. ABCD is a rectangle with AB = 3 m and $\hat{CAB}$ = 30°. Forces of 10 N, 20 N and 20 N act along AC, AD and DB respectively. Calculate the magnitude and direction of the single force that could replace this system of forces and find where its line of action cuts AB.

7. ABC is an equilateral triangle of side a metres. Forces of 10 N, 6 N and 10 N act along AB, CB and AC respectively. Find the magnitude and direction of the single force equivalent to this system and find where its line of action cuts AB.

8. Point O is the origin and points A, B and C have position vectors $3\mathbf{i}$, $3\mathbf{i} + 2\mathbf{j}$ and $2\mathbf{j}$ respectively. A force of $(\mathbf{i} + 4\mathbf{j})$ N acts at A, $5\mathbf{i}$ N at B and $(-2\mathbf{i} + 2\mathbf{j})$ N at C.
Find the single force that could replace this system and find the position vector of the point where its line of action cuts OA.

9. Point O is the origin and points A and B have position vectors $4\mathbf{i}$ and $4\mathbf{j}$ respectively. A force of $(5\mathbf{i} + 7\mathbf{j})$ N acts at A, a force of $(-6\mathbf{i} + 3\mathbf{j})$ N acts at B and a force of $(4\mathbf{i} - 6\mathbf{j})$ N acts at O.
Find the single force equivalent to this system and find the position vector of the point where its line of action cuts OA.

10. ABC is an isosceles triangle, right-angled at A with AB = 1 m. Forces of 8 N, 4 N and 6 N act along BA, BC and CA respectively.
Find the single force that could replace this system and find where its line of action cuts AB.

11. ABCDEF is a regular hexagon of side 2 m. Forces of 2 N, 3 N, 4 N and 5 N act along AC, AE, AF and ED respectively.
Find the single force equivalent to this system and find where its line of action cuts AB.

12. ABCDE is a regular pentagon of side 2 m. Forces of 5 N act along AB, BC and AD.
Find the single force that could replace these three forces and find where its line of action cuts AB.

13. ABCD is a rectangle with AB = 10 cm and $C\hat{A}B = 20°$. Forces of 5 N act along BA, CD and AD and forces of 10 N and 20 N act along DB and CA respectively.
Find the single force equivalent to this system and find where its line of action cuts AB.
If the 5 N force along AD were along BC instead, how would this affect your answers?

14. ABCD is a rectangle with AB = 70 cm and AD = 20 cm. Forces of 5 N, 2 N, 3 N and 6 N act along AB, BC, CD and AD respectively.
Find the single force that could replace this system and find where its line of action cuts AB.
The force along AB is now replaced by a force through A which reduces the system to a couple. Find the magnitude and direction of this force and the moment of the couple.

15. ABCD is a rectangle with AB = 4 m and BC = 3 m. Forces of 4 N, 5 N and 10 N act along CB, DC and DB respectively.
Find the single force equivalent to this system and find where its line of action cuts AB.
A couple of moment 15 N m, in the sense ABCD, is now introduced to the system.
Find the magnitude and direction of the single force that will replace this new system and show that its line of action passes through B.

16. Point O is the origin and points A, B, C and D have position vectors $(3\mathbf{i} + \mathbf{j})$ m, $(\mathbf{i} + 3\mathbf{j})$ m, $(-2\mathbf{i} + \mathbf{j})$ m and $(-2\mathbf{i} - 2\mathbf{j})$ m respectively. Forces of $(3\mathbf{i} + 3\mathbf{j})$ N, $(4\mathbf{i} - 5\mathbf{j})$ N, $(-5\mathbf{i} + 2\mathbf{j})$ N and $(2\mathbf{i} + 3\mathbf{j})$ N act at points A, B, C and D respectively.
Find the single force that could replace this system and find where its line of action cuts the horizontal axis through O.
A couple of moment a N m anticlockwise and a force $(b\mathbf{i} + c\mathbf{j})$ N acting through the point which has position vector $(2\mathbf{i} + \mathbf{j})$ m are now added to the system.
If these reduce the system to equilibrium, find a, b and c.

Exercise 7G Examination questions

(Take $g = 9{\cdot}8\ \mathrm{m\,s^{-2}}$ throughout this exercise.)

1. In this question the unit of distance is the metre and the unit of force is the newton.
The force $\mathbf{P} = 2\mathbf{i}$ acts through the point with position vector $5\mathbf{j}$. The force $\mathbf{Q} = 6\mathbf{i}$ acts through the point with position vector $\mathbf{j}$. The resultant of $\mathbf{P}$ and $\mathbf{Q}$ acts through the point with position vector $n\mathbf{j}$. Find the value of n and the magnitude of the resultant of $\mathbf{P}$ and $\mathbf{Q}$. (ULEAC)

2. Three forces are represented by the vectors $-2\mathbf{i} - 3\mathbf{j}$, $3\mathbf{i} + 4\mathbf{j}$ and $-\mathbf{i} - \mathbf{j}$. The forces act at the points (2, 0), (0, 3) and (1, 1) respectively. Show that the three forces combine to form a couple, and find the magnitude of the couple. (AEB)

3. A force $\mathbf{P}$ in the direction $4\mathbf{i} + 3\mathbf{j}$ and of magnitude 10 N acts through the point with position vector $(2\mathbf{j})$ m relative to an origin O. Another force $\mathbf{Q}$ in the direction $\mathbf{j}$ and of magnitude 2 N acts through the point with position vector $(4\mathbf{i})$ m relative to O. Find
(a) $\mathbf{P}$ and $\mathbf{Q}$ in the form $\alpha\mathbf{i} + \beta\mathbf{j}$ where α, β are scalars to be determined in each case,
(b) the magnitude of the resultant of $\mathbf{P}$ and $\mathbf{Q}$,
(c) the magnitude of the total moment of $\mathbf{P}$ and $\mathbf{Q}$ about O. State whether the moment is clockwise or anticlockwise. (AEB 1991)

4. (a) The resultant of forces $(p\mathbf{i} + \mathbf{j})$, $(2q\mathbf{i} + 3p\mathbf{j})$ and $(\mathbf{i} + q\mathbf{j})$ is $-6\mathbf{i}$. Find the values of p and q.
(b) The force $5\mathbf{j}$ acts at the point (2, 3) and the force $\mathbf{F}$ acts at (6, 1). If the system forms a couple,
(i) state the force $\mathbf{F}$,
(ii) find the magnitude of the couple. (NICCEA)

5. A thin non-uniform beam AB, of length 6 m and mass 50 kg, is in equilibrium resting horizontally on two smooth supports which are respectively 2 m and 3·5 m from A. The thrusts on the two supports are equal. Find the position of the centre of gravity of the beam. The original supports are removed and a load of 10 kg is attached to the beam at B. The loaded beam rests horizontally on two new smooth supports at A and C, where C is a point on the beam 1 m from B. Calculate the thrusts on each of the new supports. (AEB)

6. A light rod AB has length 2 m. A body of mass 12 kg hangs from A and a body of mass 8 kg hangs from B. The system is suspended from a point P of the rod, where P is x m from A, and is in equilibrium with the rod horizontal. Find x. (UCLES)

7.

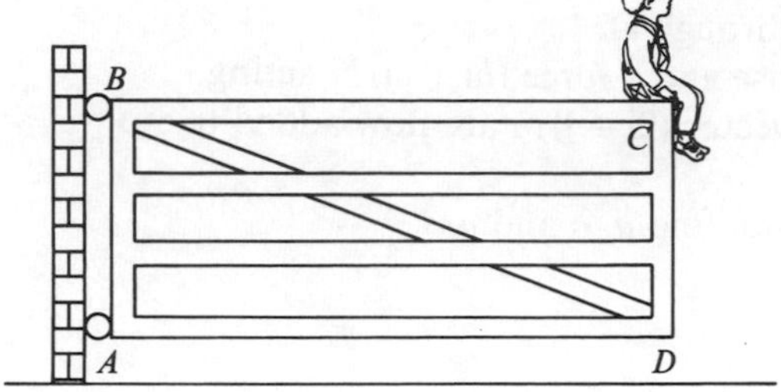

A rectangular gate $ABCD$, where $AB = 1$ m and $AD = 3$ m, is supported by smooth pins at A and B, where B is vertically above A. The pins are located in such a way that the force at B is always horizontal. The gate has mass 120 kg and it can be modelled by a uniform rectangular lamina*. A boy, of mass 45 kg, sits on the gate with his centre of mass vertically above C (see diagram). Find the magnitudes of the forces on the gate at B and at A. (UCLES)

*This means that you should consider the gate as a uniform rectangle and consider its weight to act from the intersection of the diagonals. In this book the word *lamina* is introduced in the next chapter (p.168).

8.

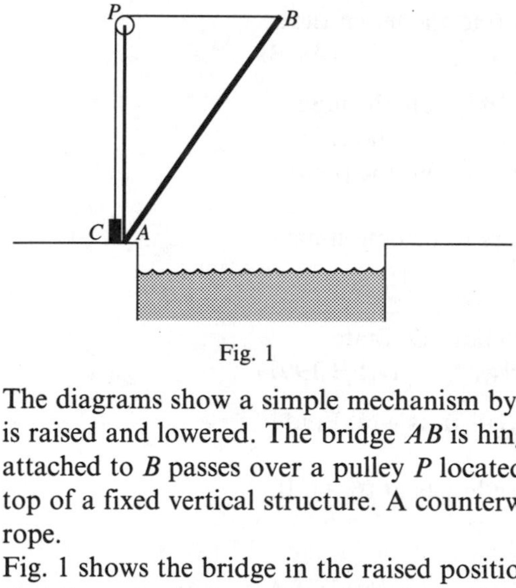

Fig. 1

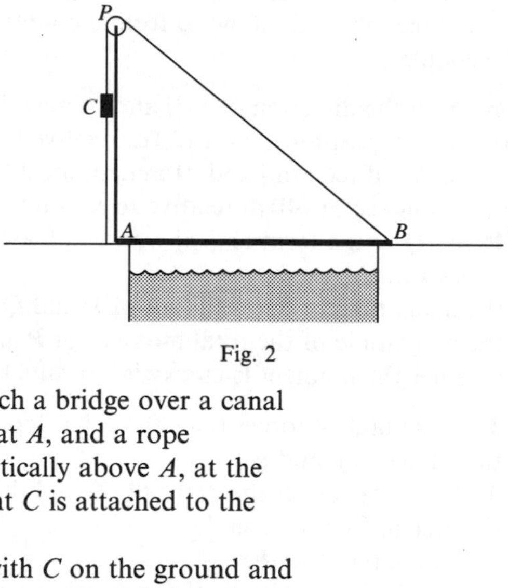

Fig. 2

The diagrams show a simple mechanism by which a bridge over a canal is raised and lowered. The bridge AB is hinged at A, and a rope attached to B passes over a pulley P located vertically above A, at the top of a fixed vertical structure. A counterweight C is attached to the rope.

Fig. 1 shows the bridge in the raised position, with C on the ground and B at the same horizontal level as P, and Fig. 2 shows the bridge lowered to its horizontal position. The mass of the bridge AB is 300 kg, and when raised (as in Fig. 1) the angle PAB is 30°.

Making suitable assumptions, which should be stated, find

(i) the least mass needed for the counterweight C if it is to be capable of holding the bridge in the raised position,

(ii) the extra force that needs to be applied to start raising the bridge from the horizontal position by pulling on the rope, if the mass of C is the minimum value found in (i). (UCLES)

9.

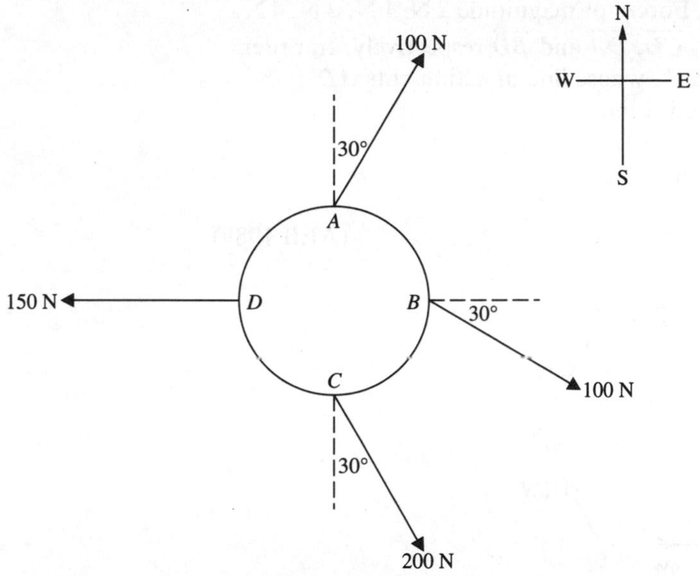

The diagram shows the circular cross-section of a vertical telegraph pole to which horizontal wires are attached at points A, B, C, D on its north, east, south, west sides respectively. The bearings of the four wires are N 30° E, S 60° E, S 30° E, due W respectively, and the tensions in the wires are 100 N, 100 N, 200 N, 150 N respectively. By considering the easterly and northerly components of these tensions, calculate the magnitude of the resultant force exerted by the wires on the pole, and the bearing of this resultant (correct to the nearest degree). Show that the resultant force passes through the centre of the cross-section. (OCSEB)

10.

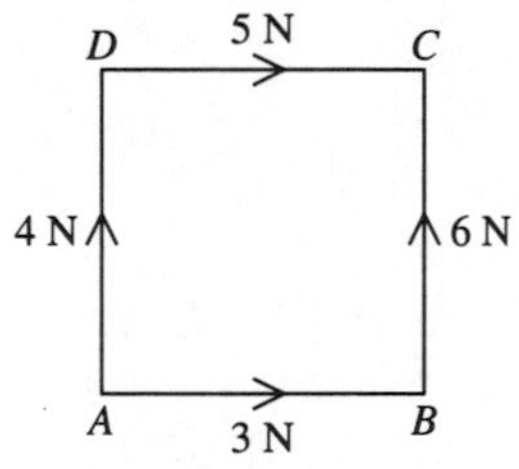

The diagram shows a square $ABCD$ of side 1 metre with forces of magnitude 3 N, 6 N, 5 N and 4 N acting along the sides AB, BC, DC and AD, the sense of each force being indicated by the order of the letters. Find

(i) the magnitude of the resultant of the forces
(ii) the total moment of the forces about A
(iii) the perpendicular distance from A to the line of action of the resultant. (WJEC)

11. *ABCD* is a square of side 2 m. Forces of magnitude 2 N, 1 N, 3 N, 4 N and $2\sqrt{2}$ N act along $\overrightarrow{AB}$, $\overrightarrow{BC}$, $\overrightarrow{CD}$, $\overrightarrow{DA}$ and $\overrightarrow{BD}$ respectively. In order to maintain equilibrium a force **F**, whose line of action cuts *AD* produced at *E*, has to be applied. Find

(a) the magnitude of **F**,

(b) the angle **F** makes with *AD*,

and

(c) the length *AE*. (AEB 1989)

12.

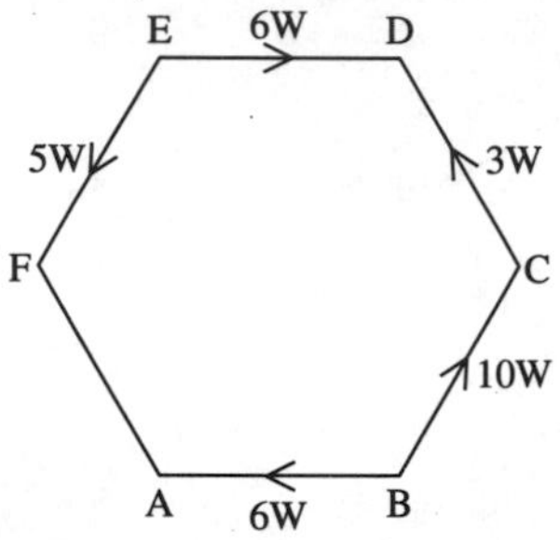

The diagram shows a regular plane hexagon with sides of length *a*. Forces having magnitudes $6W$, $10W$, $3W$, $6W$ and $5W$ act along five of the sides of the hexagon as shown. Prove that the resultant of this system of forces intersects AB produced at a point X distant $\frac{1}{8}a$ from B. The magnitude of the resultant is R and its direction makes an angle θ with AB. Find R and θ.

(OCSEB)

8 Centre of gravity

Attraction of the Earth

Every particle is attracted towards the centre of the Earth, and this force of attraction is the force previously referred to as the weight of the particle. As was explained in Chapter 3, a particle of mass m kg has a vertical force of mg newtons acting upon it, i.e. its weight is mg newtons.

For a number of particles m_1, m_2 and m_3, these forces may be considered to be parallel, all being directed towards the centre of the earth.
If the relative positions of these particles are fixed and known, then the resultant of these parallel forces can be determined.

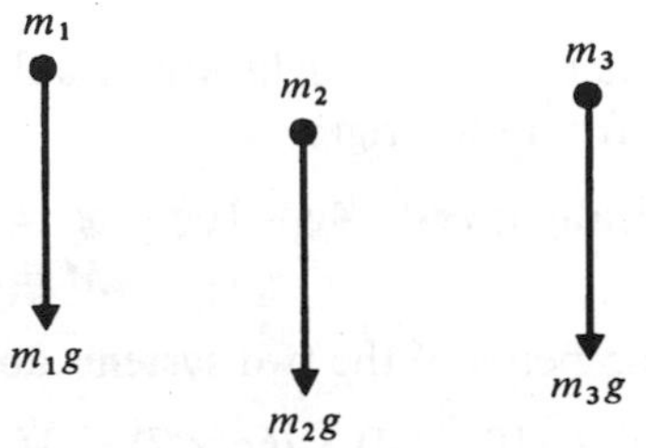

Centre of gravity of a system of particles

The centre of gravity of a number of particles is the point through which the line of action of the resultant of these parallel forces always passes, i.e. it is the point through which the resultant weight of the system acts.
In particular, if two equal particles are at the points A and B, then the centre of gravity of the system made up of these two particles will be the mid-point of the line AB.
Whatever the positions of the two particles, the resultant of the two forces acting on the particles is known to pass through the point G, where AG = GB.
Hence G is the centre of gravity.

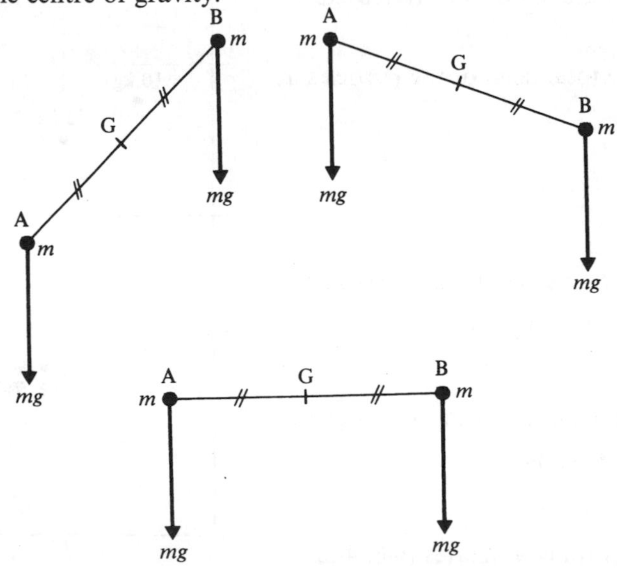

Example 1

Find the position of the centre of gravity of three particles of mass 4 kg, 10 kg and 6 kg, which lie on the x-axis at the points (3, 0), (4, 0) and (7,0) respectively.

Draw two diagrams, the first showing the forces due to the masses of the particles and the second showing the resultant force Mg due to the total mass M.

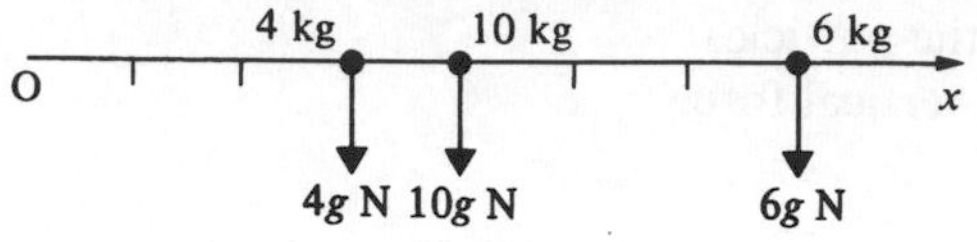

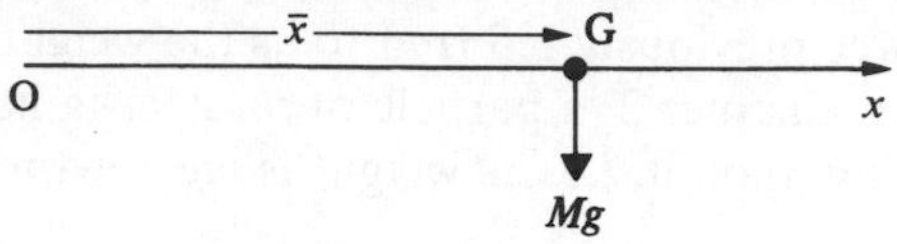

Suppose that the resultant weight Mg acts through a point G on the x-axis, at a distance $\bar{x}$ from the origin.

Resolving vertically gives: $4g + 10g + 6g = Mg$

$\therefore$ $M = 20\,\text{kg}$

Equating the moments of the two systems about the point O gives:

$$(4g \times 3) + (10g \times 4) + (6g \times 7) = Mg \times \bar{x}$$

and substituting for M gives: $\bar{x} = 4{\cdot}7$

The centre of gravity is at a point on the x-axis, 4·7 units from the origin.

Note that, in this example, all the particles lie on one line, the x-axis. It is therefore apparent that the centre of gravity is also on this line.

When a system of particles is such that the particles are not at collinear points, then the position of the centre of gravity relative to two axes must be considered.

Example 2

Find the coordinates of the centre of gravity of the given system of particles (Fig. 1).

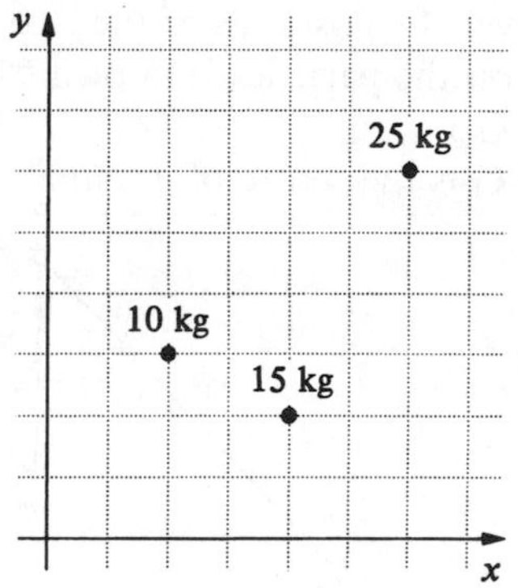

Fig. 1

Draw a second diagram showing the total mass of the particles at the point $(\bar{x}, \bar{y})$ (Fig. 2).

Resolving vertically gives:

$$10g + 15g + 25g = Mg$$

$\therefore$ $M = 50\,\text{kg}$

Equating the moments of the two systems about the y-axis gives:

$$(10g \times 2) + (15g \times 4) + (25g \times 6) = Mg \times \bar{x}$$

and substituting for M gives: $\bar{x} = 4{\cdot}6$

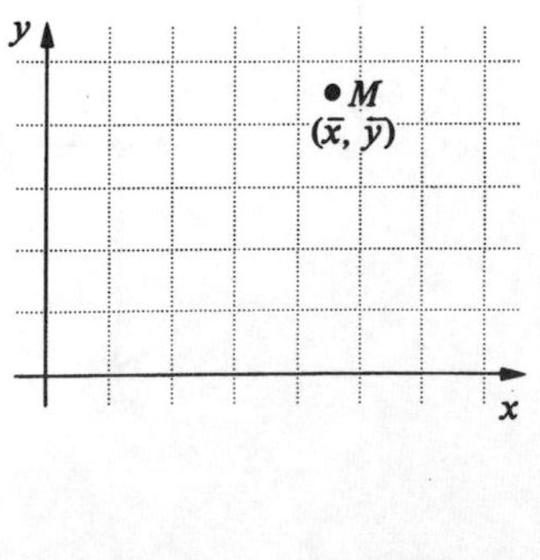

Fig. 2

Equating the moments of the two systems about the x-axis gives:

$$(10g \times 3) + (15g \times 2) + (25g \times 6) = Mg \times \bar{y}$$

and substituting for M gives: $\bar{y} = 4{\cdot}2$

The centre of gravity is at the point with coordinates (4·6, 4·2).

Example 3

The rectangle ABCD has AB = 20 cm and AD = 30 cm. Particles of mass 4 kg, 4 kg, 5 kg and 2 kg are placed at the points A, B, C and D respectively. Find the position of the centre of gravity of the system of particles.

Draw two diagrams.

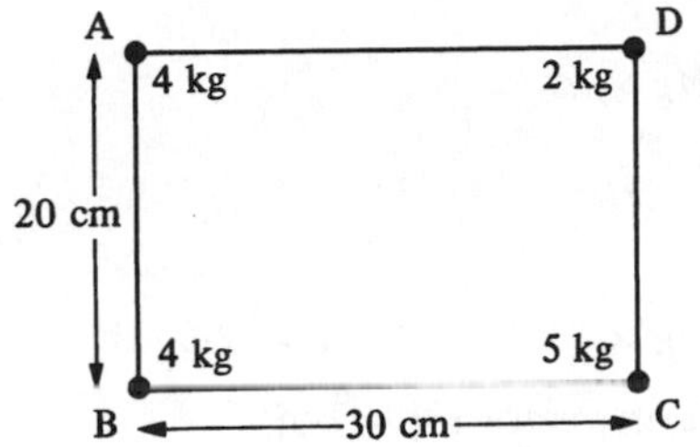

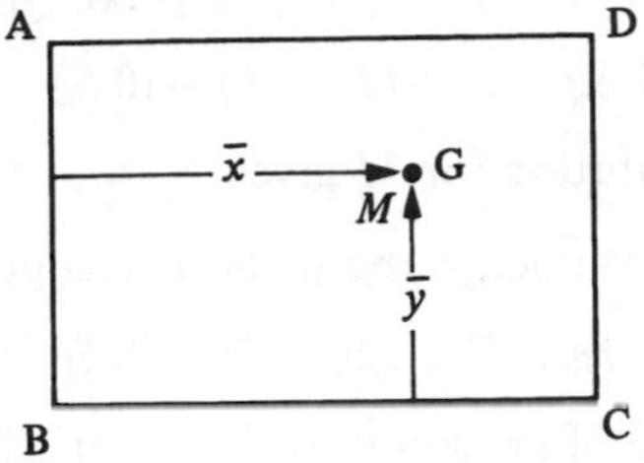

Suppose that the centre of gravity of the particles is at the point G, where $\bar{x}$ and $\bar{y}$ are the distances of G from the sides AB and BC respectively. Hence the resultant weight of the particles acts through the point G, perpendicular to the plane of the rectangle ABCD.

Resolving perpendicular to the plane ABCD gives:

$$4g + 4g + 5g + 2g = Mg$$

$$\therefore \quad M = 15\,\text{kg}$$

Equating moments about the axis AB gives:

$$(4g \times 0) + (4g \times 0) + (5g \times 30) + (2g \times 30) = Mg \times \bar{x}$$

and substituting for M gives: $\bar{x} = 14\,\text{cm}$

Equating moments about the axis BC gives:

$$(4g \times 0) + (4g \times 20) + (5g \times 0) + (2g \times 20) = Mg \times \bar{y}$$

and substituting for M gives: $\bar{y} = 8\,\text{cm}$

The centre of gravity is at a point 14 cm from AB and 8 cm from BC, and lies within the rectangle.

Example 4

Find the position vector of the centre of gravity of particles of mass 0·5 kg, 1·5 kg and 2 kg which are at the points with position vectors (6**i** − 3**j**), (2**i** + 5**j**) and (3**i** + 2**j**) respectively.

Draw two diagrams:

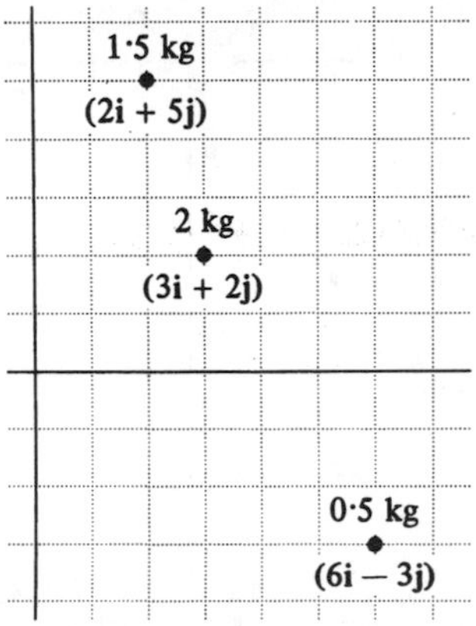

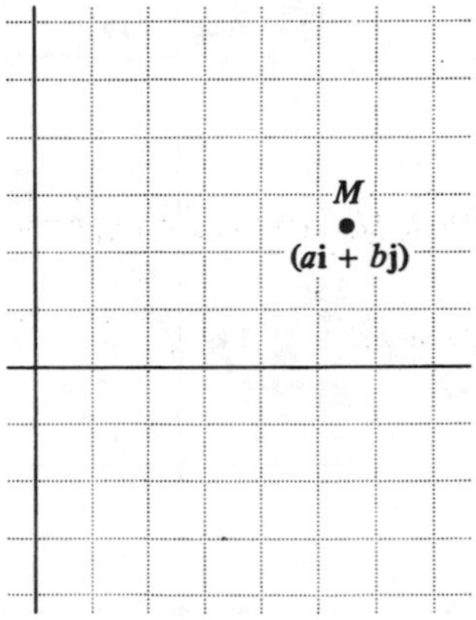

Suppose that the resultant weight acts at the point with position vector $(a\mathbf{i} + b\mathbf{j})$.

Resolving perpendicular to the plane of the axes gives:

$$0{\cdot}5g + 1{\cdot}5g + 2g = Mg \qquad \therefore \quad M = 4\,\text{kg}$$

Equating moments about the y-axis gives:

$$(1{\cdot}5g \times 2) + (2g \times 3) + (0{\cdot}5g \times 6) = Mg \times a$$

and substituting for M gives: $a = 3$

Equating moments about the x-axis gives:

$$(1{\cdot}5g \times 5) + (2g \times 2) - (0{\cdot}5g \times 3) = Mg \times b \qquad \therefore \quad b = 2{\cdot}5$$

The centre of gravity is at the point with position vector $(3\mathbf{i} + 2{\cdot}5\mathbf{j})$.

The effect of the negative sign in the position vector of one of the particles should be carefully noted.

General result

Although each example should, at this stage, be considered from first principles, it is possible to state a general result for the position of the centre of gravity of a system of particles $m_1, m_2, m_3, \ldots, m_n$ at the points with coordinates $(x_1\, y_1), (x_2\, y_2), (x_3\, y_3), \ldots, (x_n, y_n)$.

The centre of gravity is at the point $(\bar{x}, \bar{y})$ given by:

$$m_1x_1 + m_2x_2 + m_3x_3 + \ldots, m_nx_n = (m_1 + m_2 + m_3 + \ldots\ m_n)\,\bar{x}$$

or

$$\bar{x} = \frac{\Sigma(m_\text{i}x_\text{i})}{\Sigma m_\text{i}}$$

and similarly for $\bar{y}$.

Exercise 8A

1. Find the coordinates of the centre of gravity of each of the following systems of particles. The grid squares on each graph are unit squares.

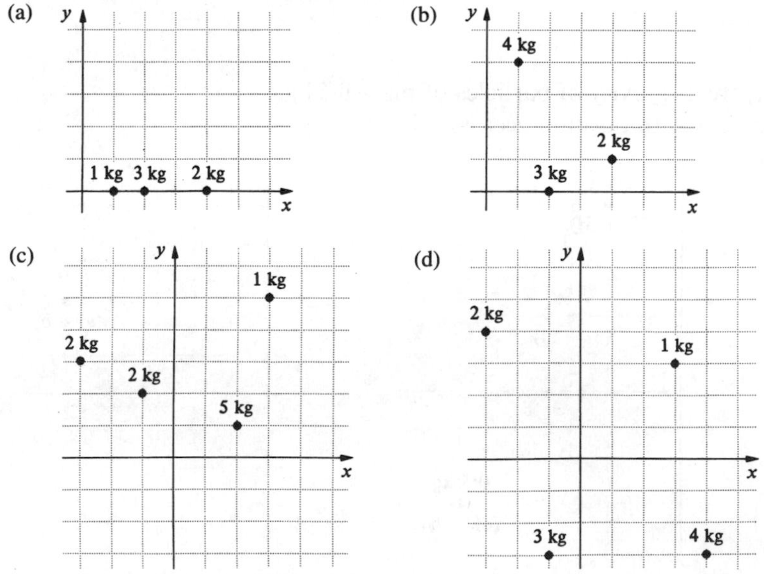

2. Find the position of the centre of gravity of three particles of mass 1 kg, 5 kg and 2 kg which lie on the y-axis at the points (0, 2), (0, 4) and (0, 5) respectively.
3. Find the coordinates of the centre of gravity of four particles of mass 5 kg, 2 kg, 2 kg and 3 kg situated at (3, 1), (4, 3), (5, 2) and (−3, 1) respectively.
4. Find the coordinates of the centre of gravity of four particles of mass 60 g, 30 g, 70 g and 40 g situated at (4, 3), (6, 5), (−6, 5) and (−5, −2) respectively.
5. Three particles of mass 2 kg, 1 kg and 3 kg are situated at (4, 3), (1, 0) and (a, b) respectively.
If the centre of gravity of the system lies at (0, 2), find the values of a and b.
6. The rectangle ABCD has AB = 4 cm and AD = 2 cm. Particles of mass 3 kg, 5 kg, 1 kg and 7 kg are placed at the points A, B, C and D respectively.
Find the distance of the centre of gravity of the system from each of the lines AB and AD.
7. The rectangle EFGH has EF = 3 m and EH = 2 m. Particles of mass 2 g, 3 g, 6 g and 1 g are placed at the mid-points of the sides EF, FG, GH and EH respectively.
Find the distance of the centre of gravity of the system from each of the lines EF and EH.
8. Find the position vector of the centre of gravity of particles of mass 2 kg, 1 kg, 3 kg and 2 kg which are at the points with position vectors $(6\mathbf{i} + 6\mathbf{j})$, $(3\mathbf{i} + 5\mathbf{j})$, $(7\mathbf{i} + 3\mathbf{j})$ and $(2\mathbf{i} - \mathbf{j})$ respectively.
9. Find the position vector of the centre of gravity of particles of mass 50 g, 60 g, 20 g and 20 g which are at the points with position vectors $(5\mathbf{i} - 7\mathbf{j})$, $(-3\mathbf{i} + 2\mathbf{j})$, $(3\mathbf{i} - 5\mathbf{j})$ and $(\mathbf{i} - 6\mathbf{j})$ respectively.
10. Particles of mass 1 kg, 2 kg, 3 kg and 4 kg lie at the points with position vectors $6\mathbf{i}$, $(\mathbf{i} - 5\mathbf{j})$, $(3\mathbf{i} + 2\mathbf{j})$ and $(a\mathbf{i} + b\mathbf{j})$ respectively.
If the centre of gravity of this system lies at the point with position vector $\left(2\tfrac{1}{2}\mathbf{i} - 2\mathbf{j}\right)$, find the values of a and b.
11. Particles of mass $2m$, m, $5m$ and $2m$ are situated at (4, −5), (1, 2), (3, −6) and (0, 3) respectively. Find the coordinates of the centre of gravity of the system.
12. Three particles of mass 1 kg, 2 kg and m kg are situated at (5, 2), (1, 5) and (1, −2) respectively.
If the centre of gravity of the system lies at $(2, \bar{y})$, find the values of m and $\bar{y}$.
13. Particles of mass 2 kg, 1 kg and 3 kg lie on the y-axis at the points (0, 7), (0, 4) and (0, −2) respectively. Where must a 6 kg mass be placed to ensure that the centre of gravity of the entire system lies at the origin?
14. Particles of mass $5m$, $4m$ and $3m$ are placed at the points (−5, 0), $\left(4, \tfrac{1}{2}\right)$ and (−4, −3) respectively. Where must a particle of mass $7m$ be placed to ensure that the centre of gravity of the entire system lies at the origin?
15. PQRS is a rectangle with PQ = 8 cm and PS = 6 cm. Particles of mass 2 g, 2 g and 3 g are placed at points P, Q and R respectively.
Find the mass that must be placed at S for the centre of gravity of the entire system to lie 3 cm from the line PQ.
With this mass in place, find the distance of the centre of gravity of the system from the line PS.

Centre of gravity of a rigid body

A rigid body is made up of a large number of particles. The position of its centre of gravity may be found by considering the constituent particles. If the rigid body has an axis of symmetry, then the centre of gravity will lie on that axis, since there will be an equal amount of matter, i.e. number of particles, in similar positions on either side of the line of symmetry. The position of the centre of gravity of some rigid bodies can be determined by considering their symmetry.

Uniform

A *uniform* body is one in which equal volumes have the same masses. It is of uniform or constant density throughout. A uniform rod is therefore such that its length is proportional to its mass. Once again this is an example of an assumption being made to allow the real situation to be modelled mathematically. In practice, most bodies are not perfectly uniform. Small differences in density will exist within the body, but our model chooses to neglect these differences on the assumption that they will not significantly effect the validity of our answers. Should we find that the answers supplied by our model do not agree with reality, we would question the wisdom of assuming the material to be uniform. Later in this chapter the reader will encounter questions in which a body is not uniform but instead consists of two or three parts that are of different densities.

Lamina

A *lamina* is a flat body, the thickness of which is negligible compared with its other two dimensions, its length and breadth. Thus a piece of card, a sheet of paper, or a thin metal sheet may be taken as examples of laminac. A uniform lamina is therefore one in which equal areas of the lamina have equal masses, i.e. the whole of the lamina is ofthe same material.

Uniform rod

The centre of gravity lies at its middle point, G mid-way between its ends A and B.

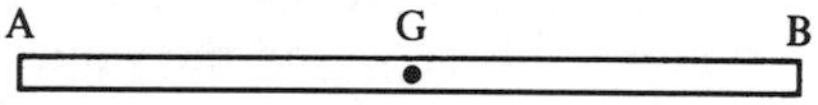

Uniform rectangular lamina

The line XX′ which bisects AD and BC, is an axis of symmetry, as also is the line YY′ which bisects the lines AB and DC.
The centre of gravity must lie on each of these lines of symmetry, and is therefore at their intersection G.

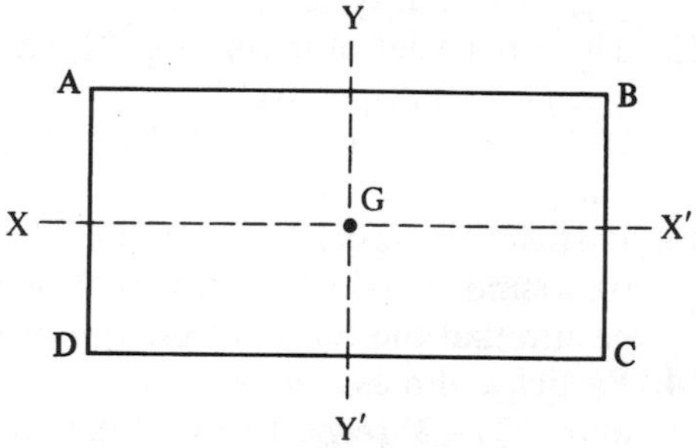

Uniform circular lamina

Since all diameters of a circle are lines of symmetry, the centre of gravity lies at the centre of the circular lamina.

The position of the centre of gravity of a solid can also be determined by consideration of its symmetry.

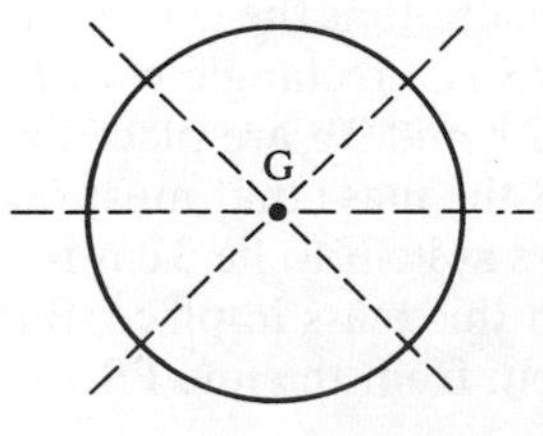

Uniform sphere

Since this solid has an infinite number of planes of symmetry, all of which contain the centre of the sphere, this is therefore its centre of gravity.

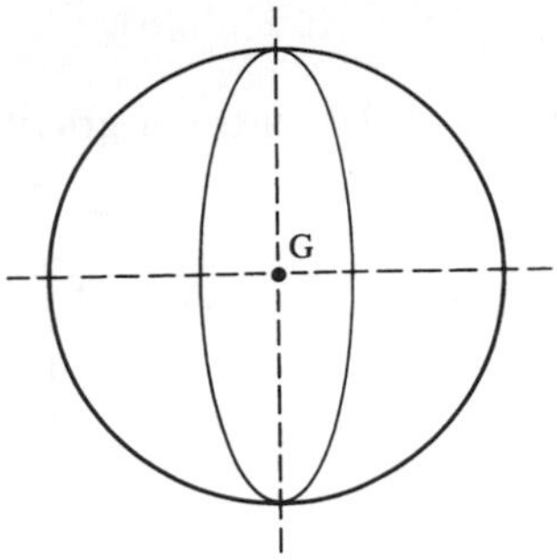

Uniform right circular cylinder

The cross-section is a circle, and this means that the cylinder has an infinite number of planes of symmetry, all of which contain the axis of the cylinder. Hence the centre of gravity must lie on this axis. The cylinder also has a plane of symmetry which bisects the axis of the cylinder and is parallel to the plane ends of the cylinder. This plane of symmetry intersects the other planes of symmetry, on the axis of the cylinder, at the point G mid-way between the ends of the cylinder. Point G is therefore the centre of gravity of the cylinder.

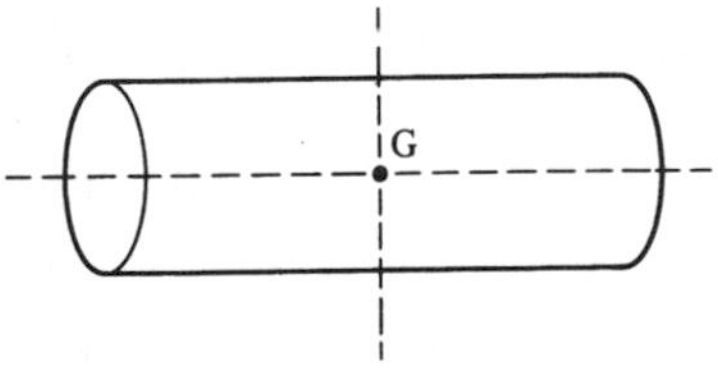

Uniform triangular lamina

This does not, in general, have an axis of symmetry, although an equilateral triangular lamina will indeed have three. These three axes are the medians of the triangle, and their point of intersection is the centre of gravity of the triangular lamina; it lies on the median and two-thirds of the distance from the vertex to the other side of the triangle.

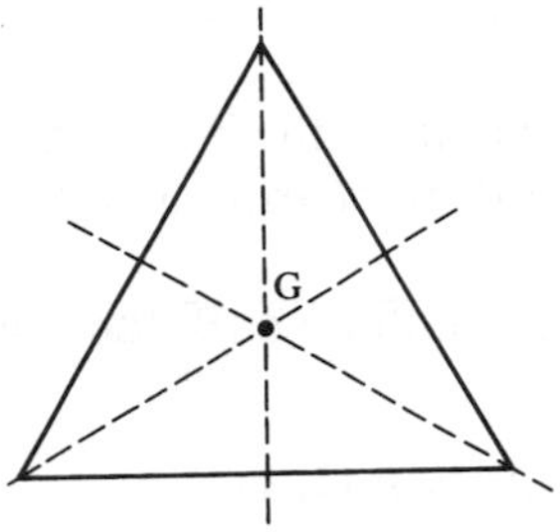

An isosceles triangular lamina has one axis of symmetry and the centre of gravity lies on it. The centre of gravity will again lie at the point of intersection of the medians of the triangle, and this result is true for all triangles as the following reasoning shows.

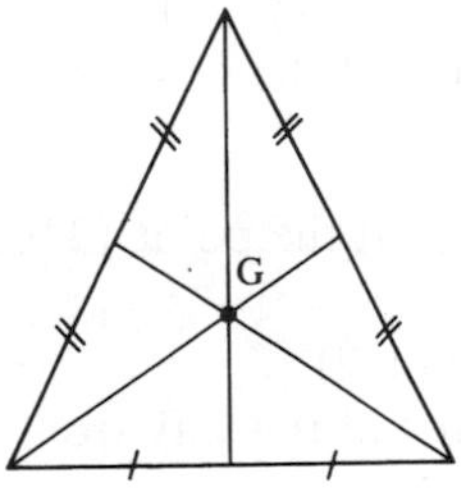

Suppose the triangle ABC is divided into a large number of thin parallel strips, all parallel to the side BC of the triangle. The centre of gravity of each of these strips will be at its centre point $X_1, X_2, \ldots$. But these points X all lie on the median drawn from the point A. Hence the centre of gravity of the triangle must also lie on this median.

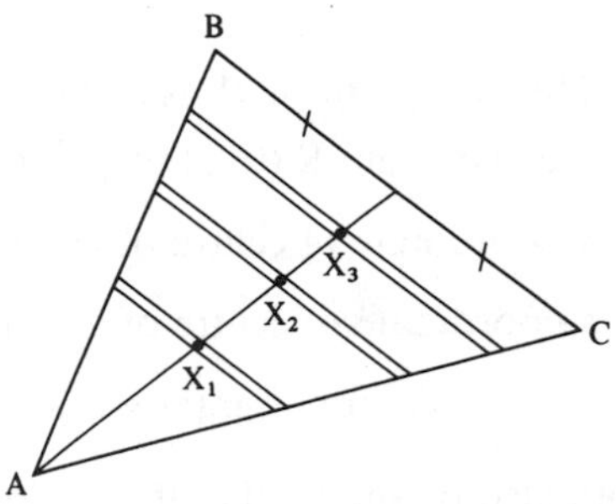

In a similar way the centre of gravity can be shown to lie on each of the other two medians.
Therefore the centre of gravity is at the intersection of the medians, i.e. two-thirds of the distance from the vertex to the mid-point of the opposite side.

Example 5

Write down the coordinates of the centre of gravity of each of the following triangular laminae.

(a)

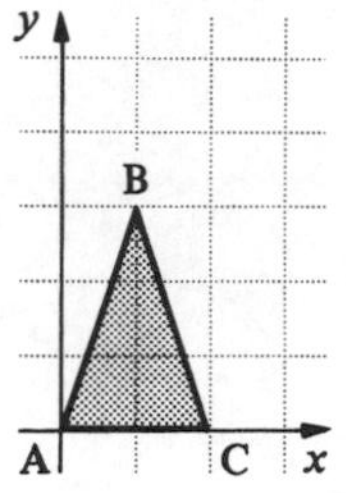

(b)

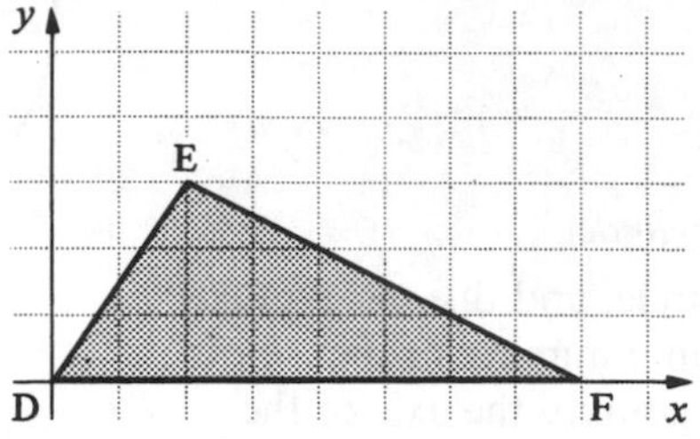

(c)

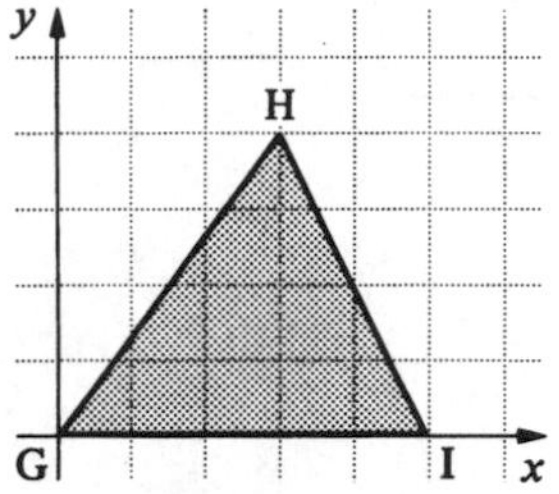

(d)

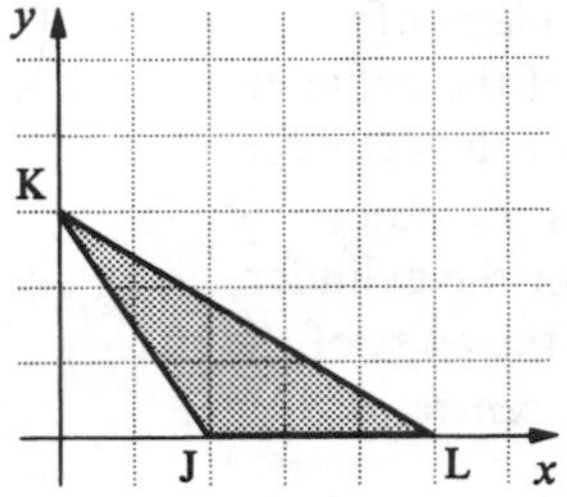

(a) The centre of gravity will lie at the intersection of the medians, i.e. at a point two-thirds of the distance from a vertex to the mid-point of the opposite side.
By considering the median through B, the centre of gravity is seen to be at the point (1, 1).

$\therefore$ centre of gravity is at (1, 1)

(b) The mid-point of EF is at $\left(5, 1\frac{1}{2}\right)$. Call this point M.
The point that is two-thirds of the way from D (0, 0) to M $\left(5, 1\frac{1}{2}\right)$ is $\left(3\frac{1}{3}, 1\right)$.

$\therefore$ centre of gravity is at the point $\left(3\frac{1}{3}, 1\right)$

(c) The mid-point of HI is at (4, 2)
The point that is two-thirds of the way from (0, 0) to (4, 2) is $\left(2\frac{2}{3}, 1\frac{1}{3}\right)$.

$\therefore$ centre of gravity is at the point $\left(2\frac{2}{3}, 1\frac{1}{3}\right)$

(d) The mid-point of JL is at $\left(3\frac{1}{2}, 0\right)$. The centre of gravity is two-thirds of the way from K (0, 3) to $\left(3\frac{1}{2}, 0\right)$.

x-coordinate of centre of gravity is $\frac{2}{3} \times 3\frac{1}{2} = 2\frac{1}{3}$

y-coordinate of centre of gravity is $3 - \left(\frac{2}{3} \times 3\right) = 1$

$\therefore$ centre of gravity is at the point $\left(2\frac{1}{3}, 1\right)$

It can also be shown that if three equal particles are placed at the vertices A, B, C of a triangle, then the centre of gravity of these particles is at the same point as the centre of gravity of the uniform triangular lamina ABC.

Example 6

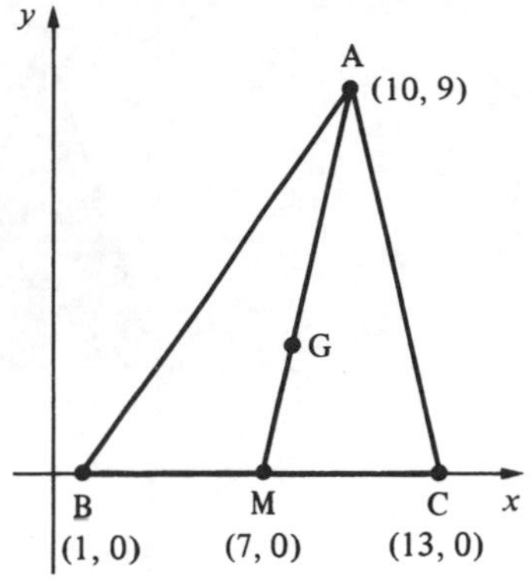

The triangle ABC has its vertices at the points with coordinates (10, 9), (1, 0) and (13, 0). Particles, each of mass m, are placed at the points A, B and C.
Suppose M is the mid-point of BC.

The centre of gravity of the lamina lies on the median AM. If G is such that $AG = \frac{2}{3}AM$, then G is the centre of gravity.

The centre of gravity of the particles is found by finding the position of the resultant of the forces due to their masses.

The resultant of mg at (1, 0) and mg at (13, 0) will be a force of $2mg$ at the mid-point, M, of BC, i.e. at the point (7, 0).

The resultant of $2mg$ at M and mg at A will be at a point P such that:

$$2mg \times MP = mg \times AP$$

or $$AP = 2MP, \qquad \text{i.e. } AP = \tfrac{2}{3}AM$$

Hence the point P and the point G are the same point, and the centre of gravity of the lamina and the three particles are at the same point.

The coordinates of the centre of gravity can be determined as follows.

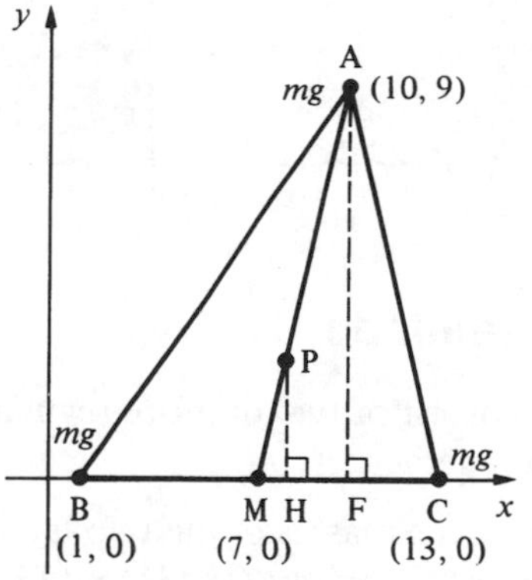

Through the point P, draw lines PH and AF parallel to the y-axis.

Since $MP = \frac{1}{3}(MA)$

then $HP = \frac{1}{3}(FA)$

$= \frac{1}{3}(9) = 3$

$MH = \frac{1}{3}(MF)$

$= \frac{1}{3}(10 - 7) = 1$

The coordinates of the centre of gravity are (8, 3).

Exercise 8B

Write down the coordinates of the centre of gravity of each of the following uniform laminae.
Each grid consists of unit squares.

1.

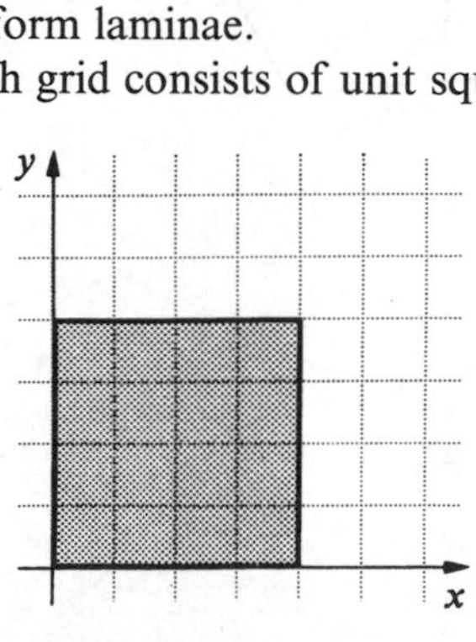

2.

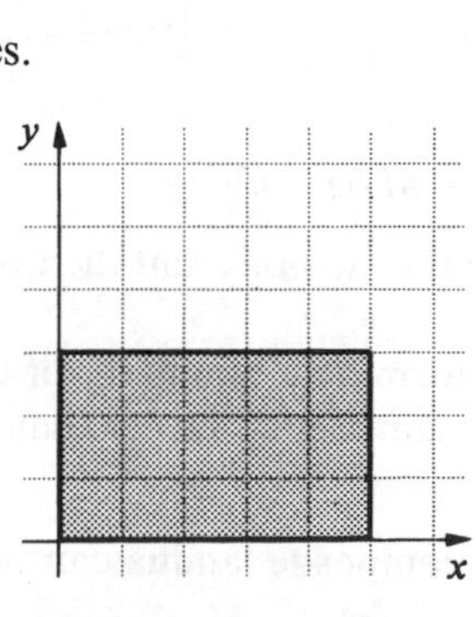

3.

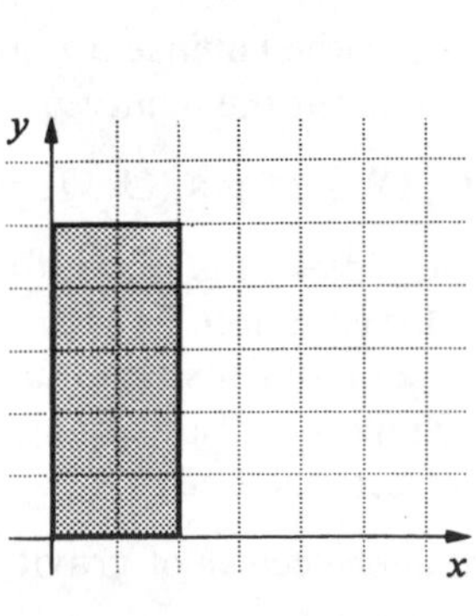

4.
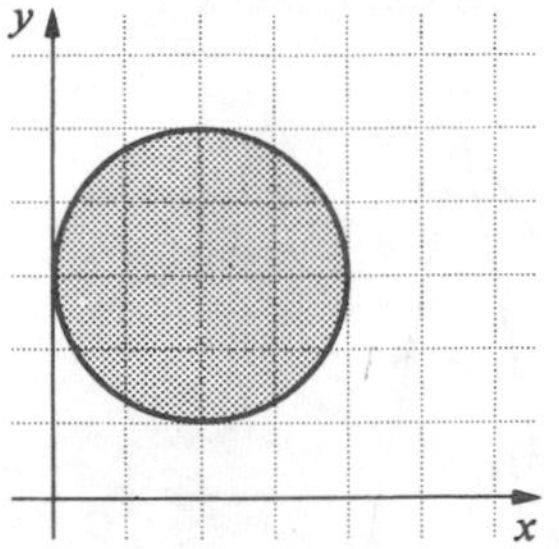

5.
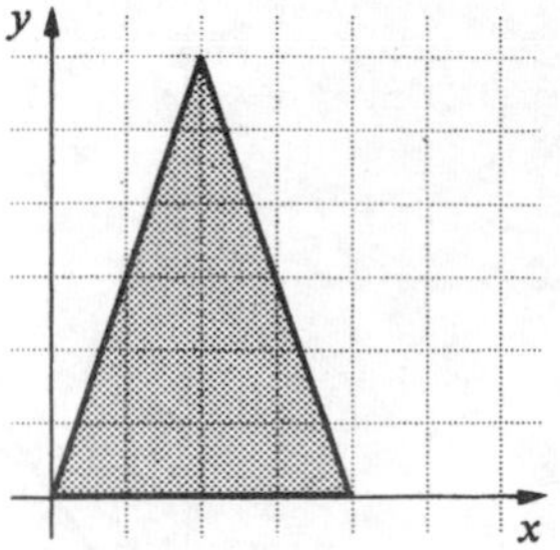

6.
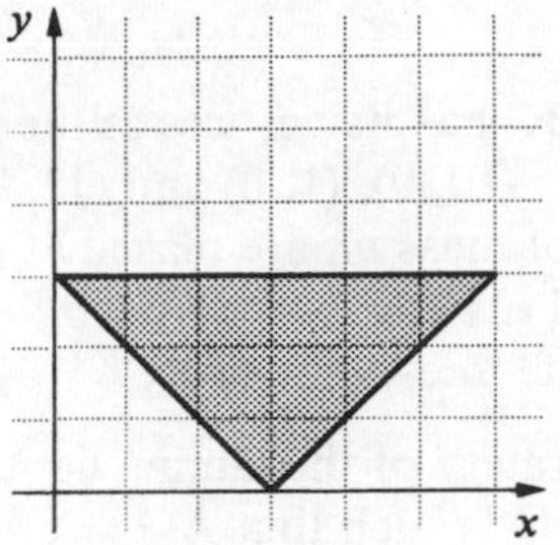

7.
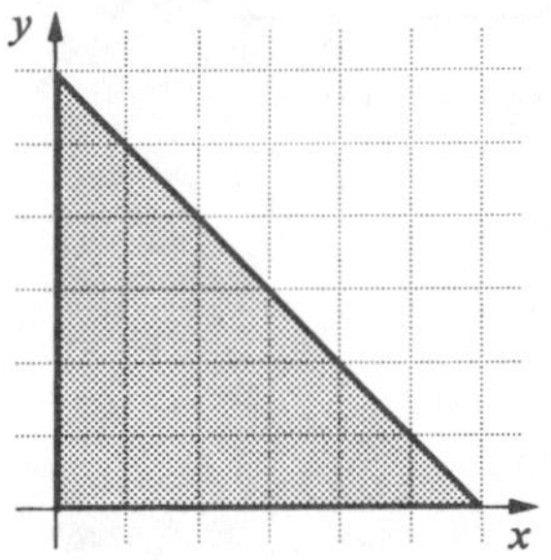

8.
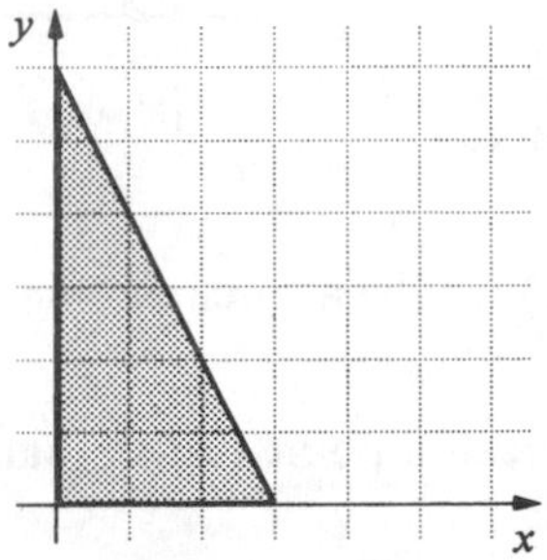

9.
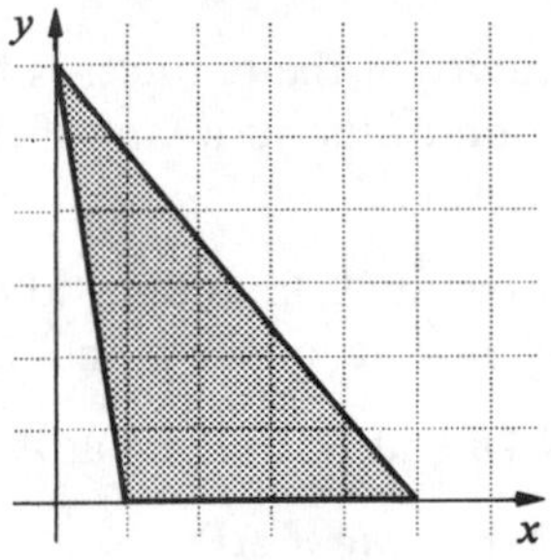

10.
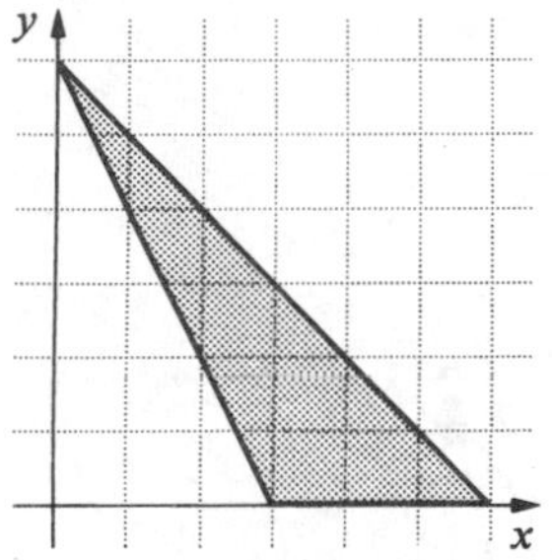

11.
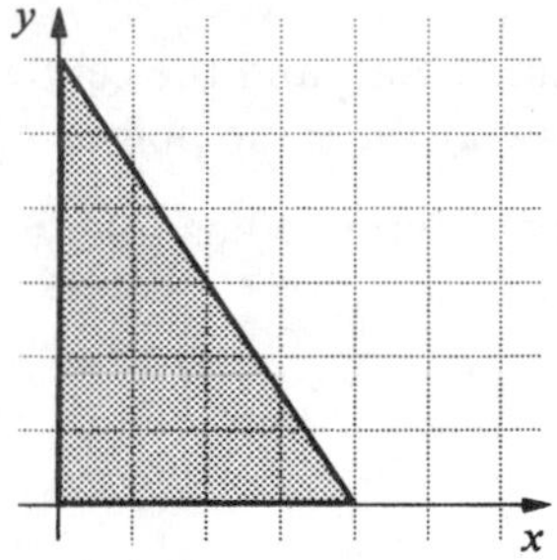

12.
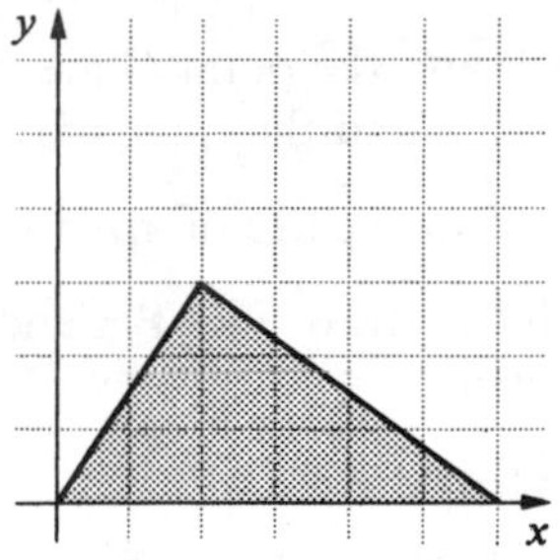

Composite laminae

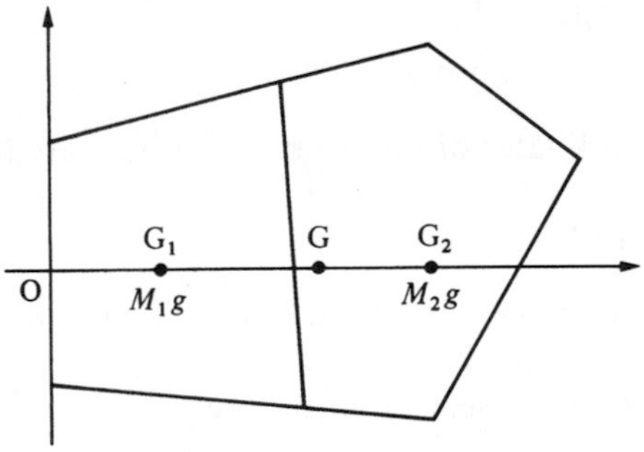

A lamina may consist of two or more regular laminae joined together.

The forces due to the masses of the two laminae will act at their centres of gravity G_1 and G_2. The resultant of these two forces will act through a point, G, on the line $G_1 G_2$.

Suppose the masses of the laminae are M_1 and M_2. Take moments about the point O:

$$M_1g \times (G_1 O) + M_2g \times (G_2 O) = (M_1 + M_2)g \times GO$$

The position of the centre of gravity of the composite body can then be determined from this equation.
It is advisable to take moments about an axis forming a boundary of the body so that all the moments are acting in the same sense, rather than using an axis which intersects the body.

Hence it is seen that the centre of gravity of a composite lamina can be found in an exactly similar way to the centre of gravity of a system of particles. It is only necessary to know the masses, and the positions of the centre of gravity, of each of the constituent parts.

Example 7

Two uniform rectangular laminae, of mass per unit area m, are joined together as is shown in the diagram. Find the distance of the centre of gravity of the complete lamina from the edge AB.

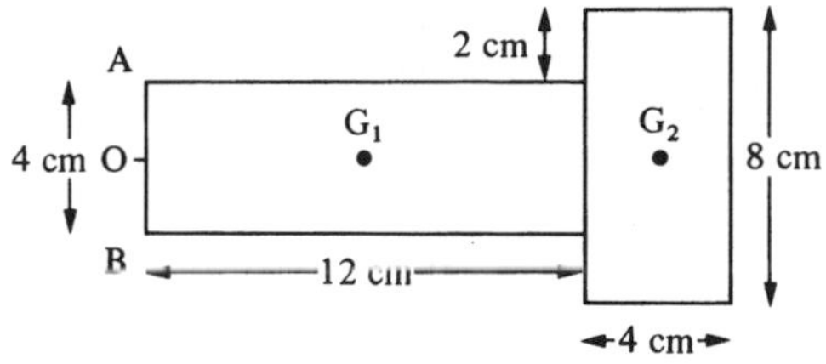

The centres of gravity of the rectangles are at G_1 and G_2 and lie on the line of symmetry of the lamina. Suppose O is the mid-point of AB.

Draw two diagrams.

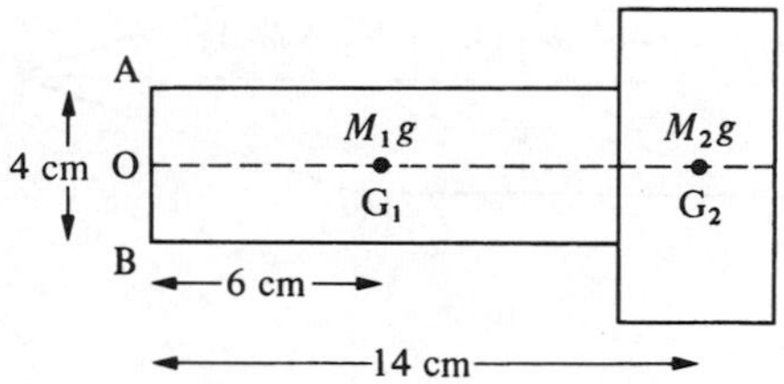

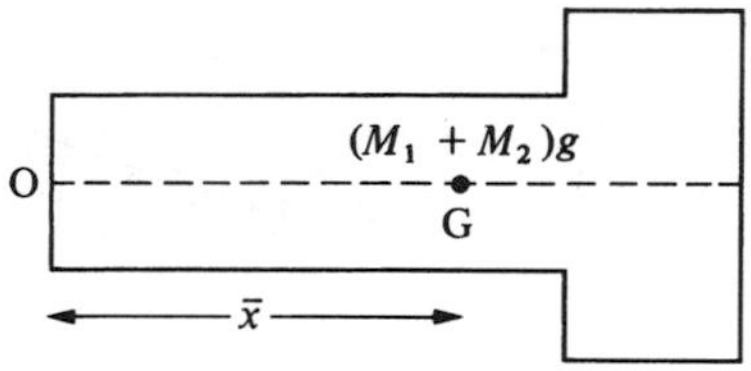

In this example the composite lamina has a line of symmetry and the point G will lie on it. Hence G lies on the line G_1 G_2 at a distance $\bar{x}$ from O. Take moments about O:

$\curvearrowright O$ $\quad M_1g \times 6 + M_2g \times 14 = (M_1 + M_2)g\,\bar{x}$

but $\quad M_1 = 48m \quad$ and $\quad M_2 = 32m$

$\therefore \quad 48mg \times 6 + 32mg \times 14 = 80mg\,\bar{x}$

$\therefore \quad 288 + 448 = 80\,\bar{x}$

$\therefore \quad \bar{x} = 9{\cdot}2\,\text{cm}$

The centre of gravity of the complete lamina is 9·2 cm from O, the mid-point of AB.

With a less regular composite lamina, which does not have an axis of symmetry, it is necessary to take moments about two axes so as to find both coordinates of the centre of gravity.

Example 8

Three uniform laminae, all made of the same material, are joined together as shown in the diagram. ABCD is a rectangle and the triangles are isosceles. If the laminae have the dimensions shown, find the position of the centre of gravity of the composite lamina.

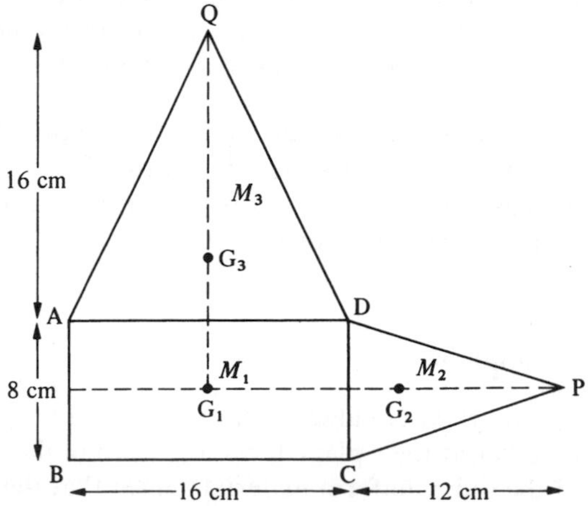

The masses and the positions of the centre of gravity of each of the constituent laminae are known. Suppose the mass per unit area of each lamina is m.

Draw two diagrams.

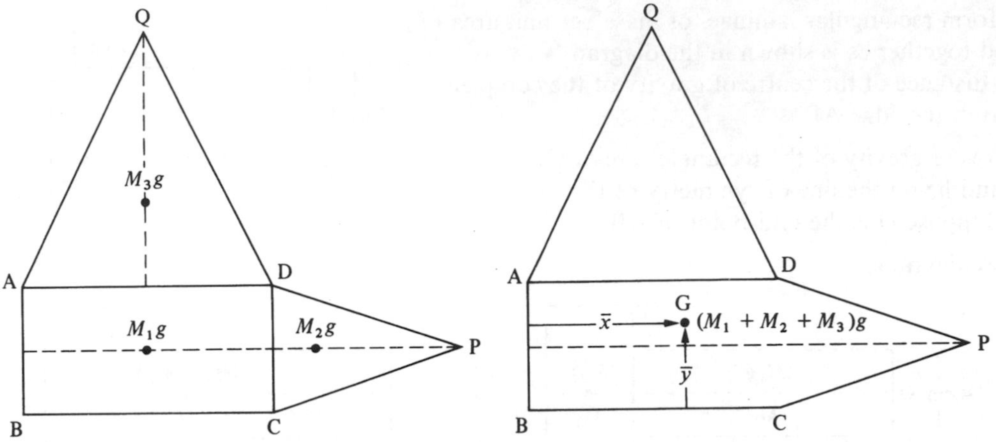

Taking moments about the axis AB and using

$$M_1 = 16 \times 8m, \quad M_2 = \tfrac{1}{2} \times 8 \times 12m \quad \text{and} \quad M_3 = \tfrac{1}{2} \times 16 \times 16m$$

gives:

$$(16 \times 8mg) \times (8) + \left(\tfrac{1}{2} \times 8 \times 12mg\right) \times (16 + 4) + \left(\tfrac{1}{2} \times 16 \times 16mg\right) \times (8) = (304mg)\,\bar{x}$$

$$\therefore \qquad \bar{x} = 9{\cdot}89 \text{ cm}$$

Taking moments about the axis BC gives:

$$(16 \times 8mg) \times (4) + \left(\tfrac{1}{2} \times 8 \times 12mg\right) \times (4) + \left(\tfrac{1}{2} \times 16 \times 16mg\right) \times \left(8 + 5\tfrac{1}{3}\right) = (304mg)\,\bar{y}$$

$$\therefore \qquad \bar{y} = 7{\cdot}93 \text{ cm}$$

The centre of gravity is 9·89 cm from AB and 7·93 cm from BC.

It should be noted that, in the case of a uniform lamina, the constants m and g will cancel from each term of the moments equation; nevertheless it is essential that they should be written in the original equation.

The method used to find the centre of gravity of a composite lamina may also be used to determine the centre of gravity of a composite three-dimensional body, provided that the centres of gravity and the masses of the various parts are known.

Example 9

A solid right circular cylinder and a solid right circular cone are joined together at their plane faces as shown in the diagram. The solids are made of the same uniform material. Given that the centre of gravity of a uniform solid cone is at a distance $\frac{h}{4}$ from its plane base, where h is the height of the cone, find the position of the centre of gravity of the composite body.

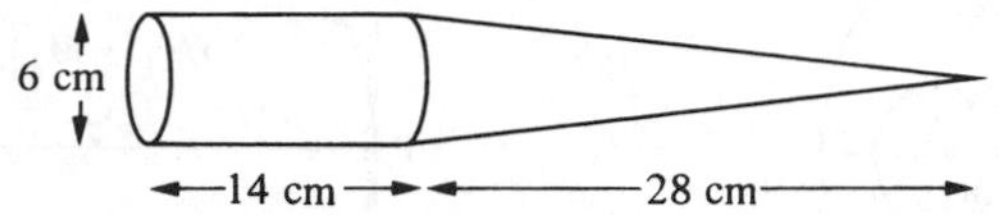

Draw two diagrams.

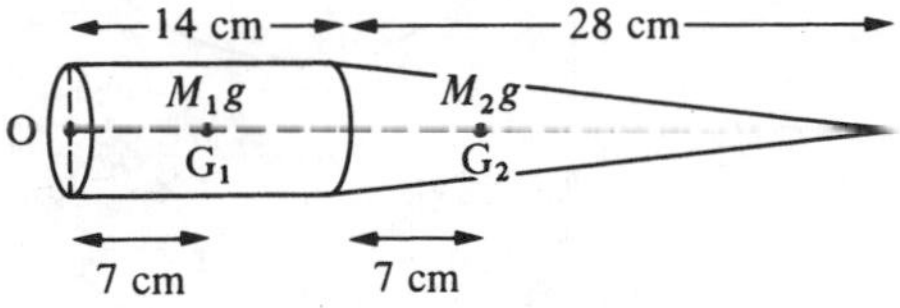

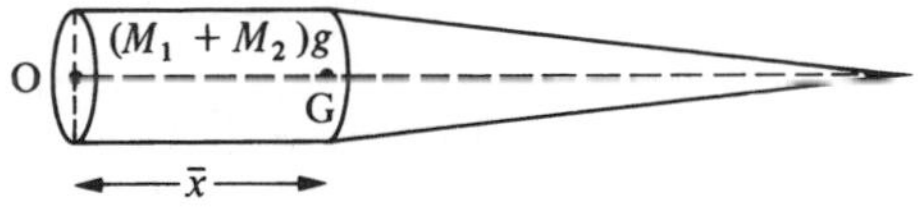

Let the mass per unit volume, i.e. the density, of the cylinder and the cone be σ.

Then $\quad M_1g\,(\text{cylinder}) = \pi \times 3^2 \times 14\sigma g \quad$ and $\quad M_2g\,(\text{cone}) = \frac{1}{3}\pi \times 3^2 \times 28\sigma g$

$$= 126\pi\sigma g \qquad\qquad = 84\pi\sigma g$$

The centre of gravity of the body will lie on the axis of symmetry through the points O, G_1 and G_2.

Equating moments about the vertical axis through O gives:

$$126\pi\sigma g \times 7 + 84\pi\sigma g \times (14 + 7) = \bar{x}(126\pi\sigma g + 84\pi\sigma g)$$

$$\therefore \qquad \bar{x} = 12{\cdot}6\,\text{cm}$$

The centre of gravity is on the axis of symmetry and 12·6 cm from the base of the cylinder.

It is sometimes required to find the position of the centre of gravity of a lamina from which a portion has been removed – for example, a perforated disc or a rectangle from which a triangle has been removed.
The method of the previous examples can still be used. The portion which is removed and the part remaining can be considered as the constituent parts, which together make up the whole lamina.
The essential difference to the calculation is that the unknown centre of gravity will now be that of one of the parts of the whole lamina, rather than that of the composite lamina itself.

Example 10

A uniform circular disc, of mass per unit area m and radius 8 cm, has a circular hole of radius 2 cm made in it. The centre of this hole lies 4 cm from the centre of the disc.

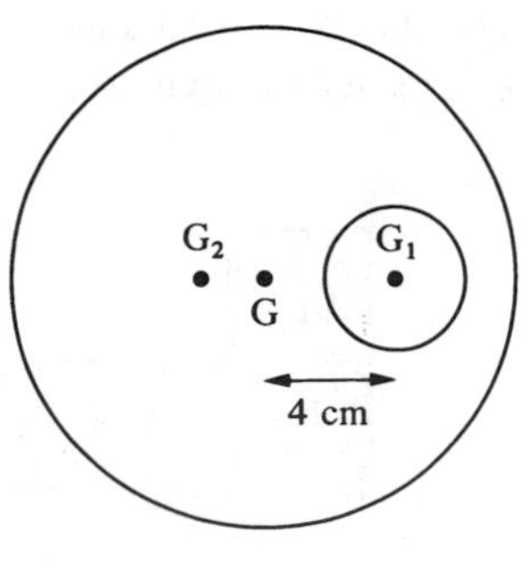

Let G be the centre of gravity of the whole disc, and G_1 and G_2 be the centres of gravity of the disc which is removed and the perforated disc respectively.
G_2 will lie on the diameter passing through G_1. M_1 and M_2 are the masses of the part removed and of the perforated disc respectively.

Draw two diagrams.

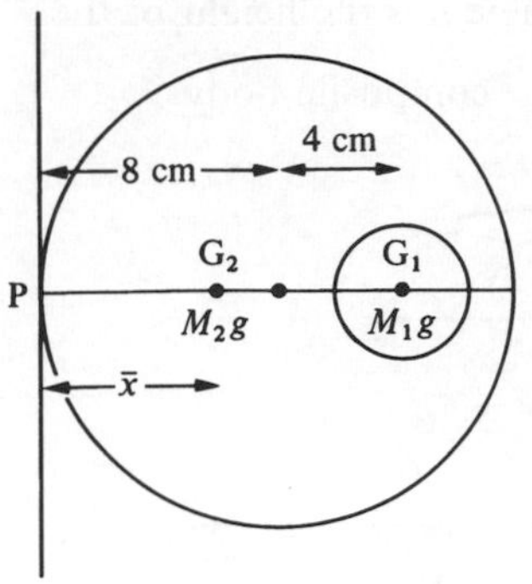

$M_1 = (\pi \times 2^2)m$

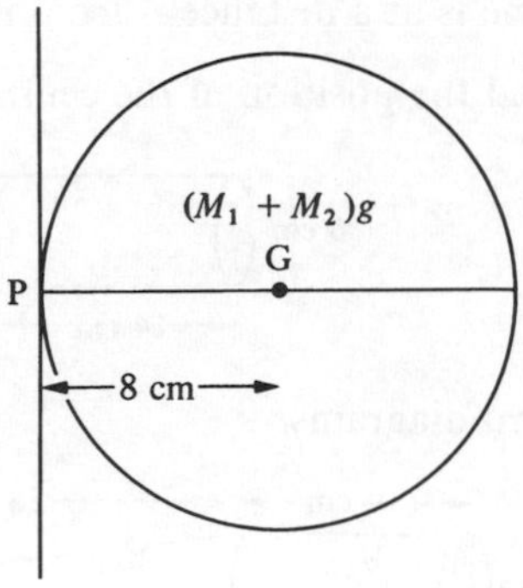

$M_2 = (\pi \times 8^2 - \pi \times 2^2)m$

Taking moments about the axis through P gives:

$$(\pi \times 2^2 mg) \times (8 + 4) + (\pi \times 8^2 - \pi \times 2^2)mg \times \bar{x} = (\pi \times 8^2 mg) \times 8$$

$$\therefore \qquad \bar{x} = 7{\cdot}73\,\text{cm}$$

The centre of gravity is 0·27 cm from the centre of the original disc and lies on the diameter passing through the centre of the disc and the centre of the hole.

Standard results

The following standard results will be required for some of the questions of Exercises 8C and 8D.

Uniform semicircular lamina of radius r:

Centre of gravity lies on the axis of symmetry at a distance $\frac{4r}{3\pi}$ from the straight edge.

Uniform solid right circular cone of height h:

Centre of gravity lies on the axis of symmetry at a distance $\frac{h}{4}$ from the plane base.

Uniform solid hemisphere of radius r:

Centre of gravity lies on the axis of symmetry at a distance $\frac{3r}{8}$ from the plane surface.

Exercise 8C

Questions **1** to **13** show uniform laminae. Find the coordinates of the centre of gravity of each one. Each grid consists of unit squares.

1. 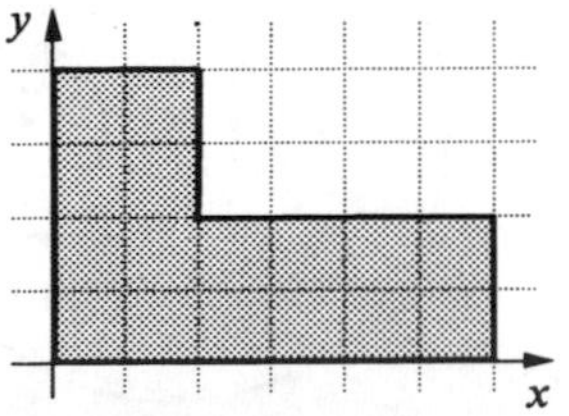

2. 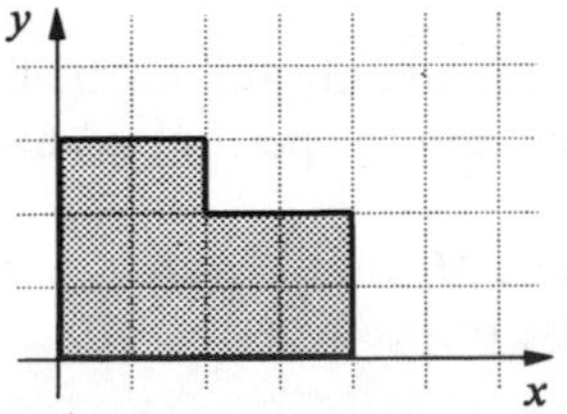

3. 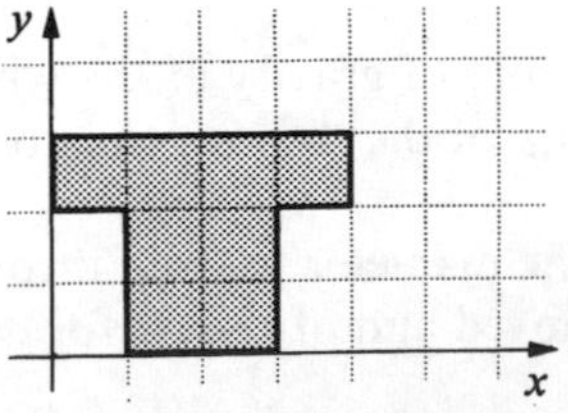

4.

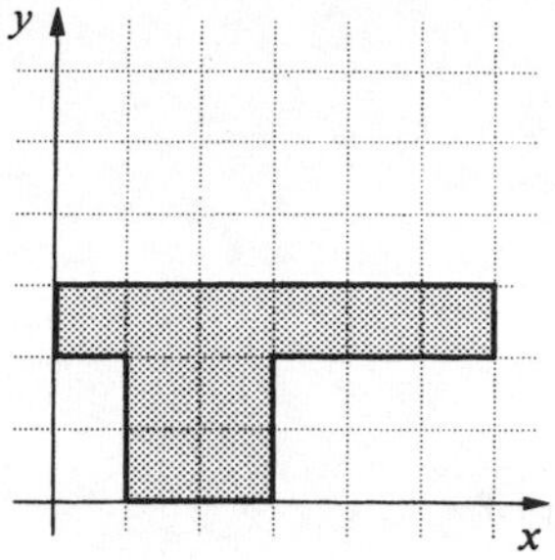

5.

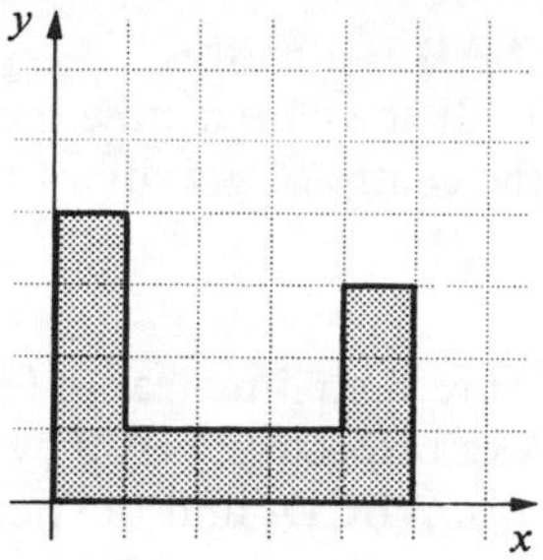

6.

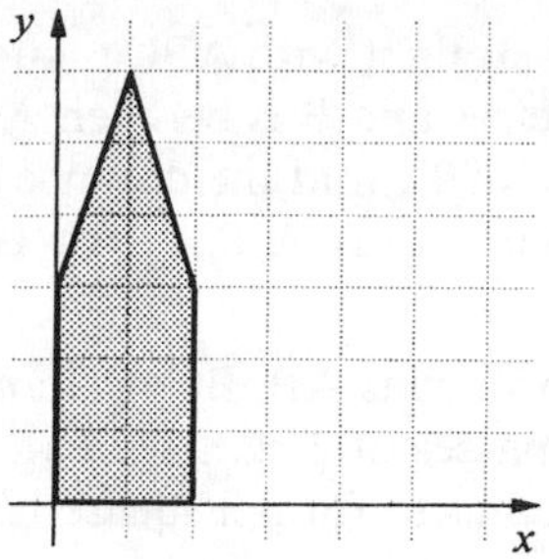

7.

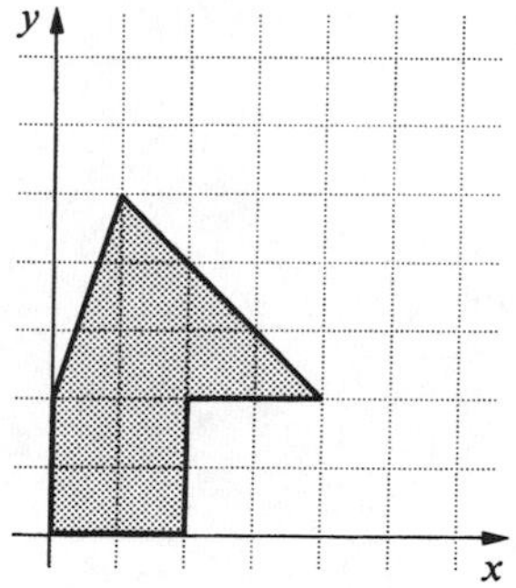

8.

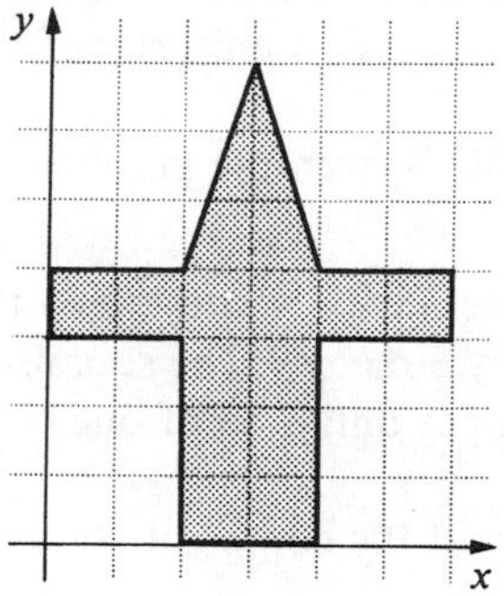

9.

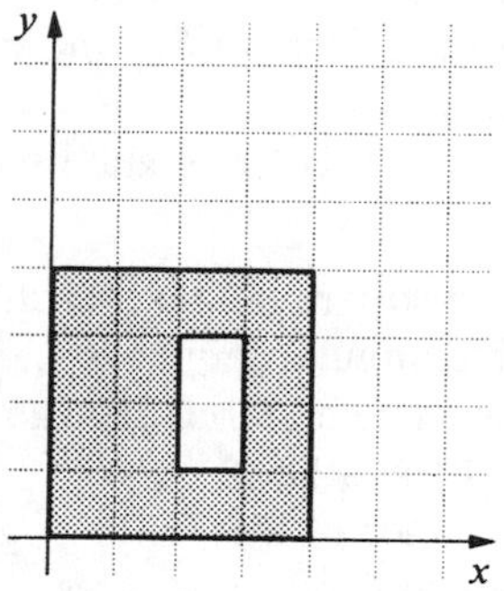

10.

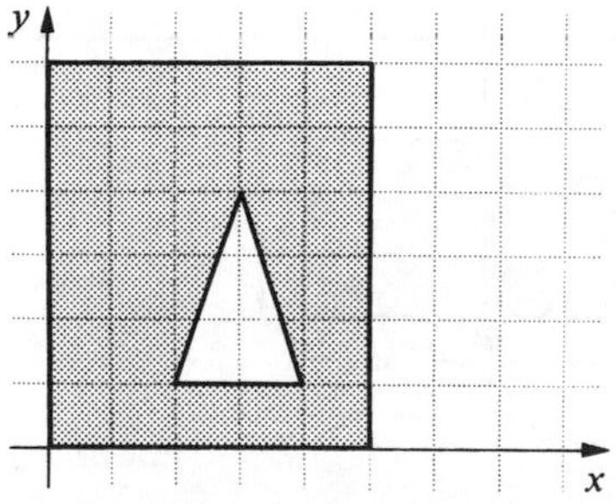

11.

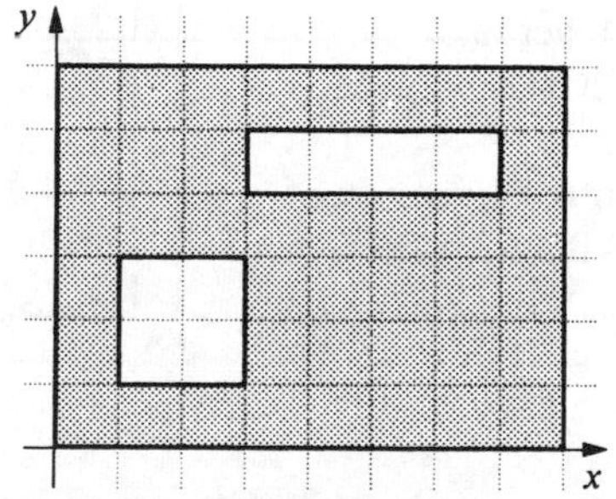

12.

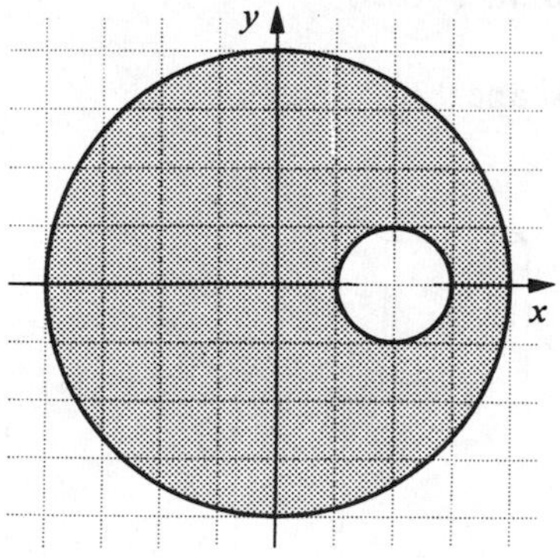

13.

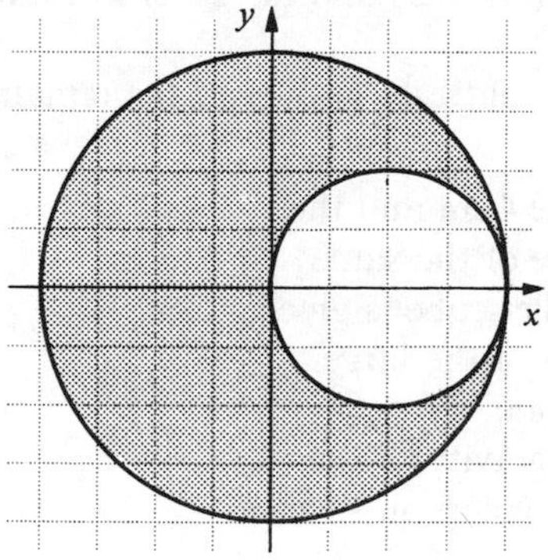

14. A uniform straight wire ABC is such that AB = 4 metres, AC = 6 metres and B is between A and C. If the wire is now bent at B until $A\hat{B}C = 90°$, find the distance that the centre of gravity of this bent wire is from: (a) AB, (b) BC.

15. Four uniform rods AB, BC, CD and DA are each 4 metres in length and have masses of 2 kg, 3 kg, 1 kg and 4 kg respectively. If they are joined together to form a square framework ABCD, find the position of its centre of gravity.

16. Two rods AB and BC are joined together at B such that $A\hat{B}C = 60°$. AB is uniform, of length 6 m and mass 4 kg. BC is uniform, of length 4 m and mass 4 kg.
Find how far the centre of gravity of ABC is from the point B.

17. Two uniform square laminae, each of side 3 metres, are joined together to form a rectangular lamina, 6 metres by 3 metres. The squares are not made of the same material and the mass per unit area of one of them is twice that of the other.
Find the distance of the centre of gravity of the composite body from the common edge of the squares.

18. The diagram shows two uniform squares, ABFG and BCDE, joined together.
The mass per unit area of BCDE is twice that of ABFG.
Find the distance of the centre of gravity of the composite body from AB and AG.

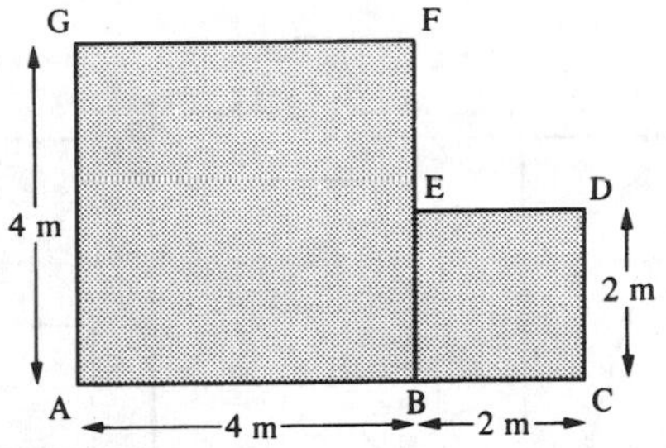

19. A circular lamina, made of uniform material, has its centre at the origin and a radius of 6 units. Two smaller circles are cut from this circle, one of radius 1 unit and centre (−1, −3) and the other of radius 3 units and centre (1, 2).
Find the coordinates of the centre of gravity of the remaining shape.

20. Two solid cubes, one of side 4 cm and the other of side 2 cm, are made of the same uniform material. The smaller cube is glued centrally to one of the faces of the larger cube, as shown in the diagram.
Find how far the centre of gravity of the composite body is from the common surface of the cubes.

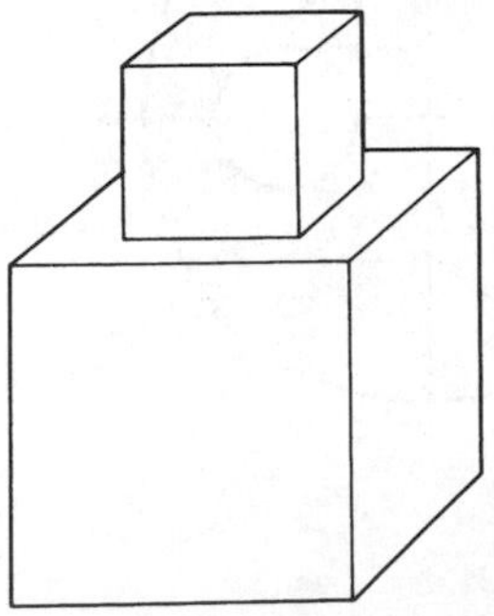

21. A solid cube of side 4 cm is made from uniform material. From this cube a smaller cube of side 2 cm is removed, as shown in the diagram.
Find the position of the centre of gravity of the remaining body.

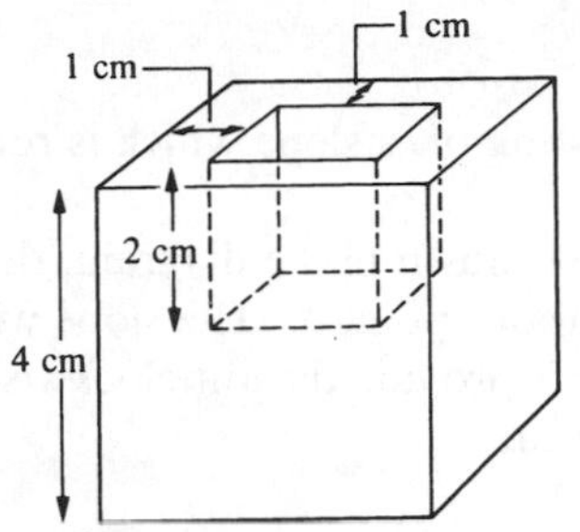

22. The diagram shows a uniform semicircular lamina of radius 6 m, with a semicircular portion of radius 3 m missing.
Find the distance of the centre of gravity of the remaining shape from the point A shown in the diagram.

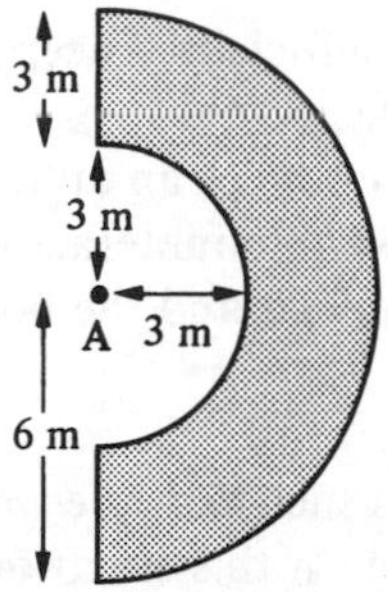

23. A uniform semicircular lamina of radius 6 cm is joined to another uniform semicircular lamina of radius 3 cm. The centres of the straight edges of each lamina coincide, but the laminae do not overlap. If the two laminae are made of the same material, find the position of the centre of gravity of the composite lamina so formed.

24. Repeat question **23** but now with the smaller semicircular lamina having a mass per unit area equal to twice that of the larger one.

25. A solid right circular cylinder has a base radius of 3 cm and a height of 6 cm. The cylinder's circular top forms the base of a solid right circular cone of base radius 3 cm and perpendicular height 4 cm. The cylinder and the cone are made from the same uniform material.
Find the position of the centre of gravity of the composite body.

26. A conical hole is made in one end of a right circular cylinder (see diagram). The axis of symmetry of the cone is the same as that of the cylinder. The cylinder is of radius 2 cm and length 6 cm. The conical hole penetrates 4 cm into the cylinder and the circular hole at the end of the cylinder is of radius $1\frac{1}{2}$ cm.
Find the position of the centre of gravity of the remaining body.

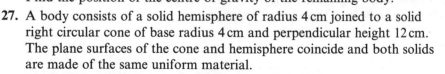

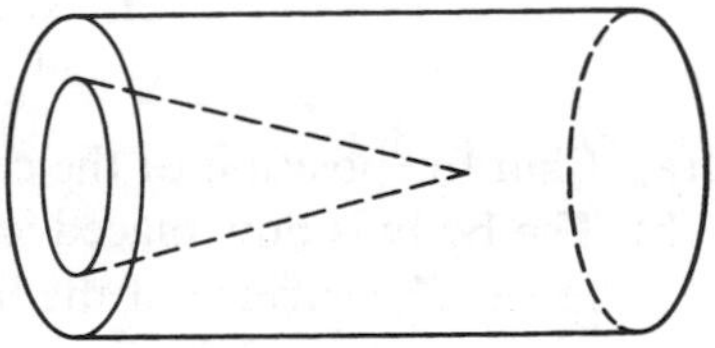

27. A body consists of a solid hemisphere of radius 4 cm joined to a solid right circular cone of base radius 4 cm and perpendicular height 12 cm. The plane surfaces of the cone and hemisphere coincide and both solids are made of the same uniform material.
Find the position of the centre of gravity of the body.

28. A body consists of a solid hemisphere of radius r joined to a solid right circular cone of base radius r and perpendicular height h. The plane surfaces of the cone and hemisphere coincide and both solids are made of the same uniform material.
Show that the centre of gravity of the body lies on the axis of symmetry at a distance $\dfrac{3r^2 - h^2}{4(h + 2r)}$ from the base of the cone.

Toppling

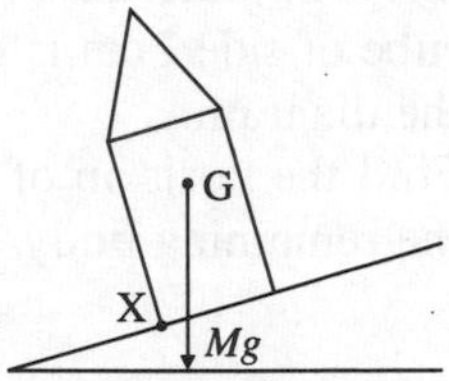

Consider a body resting on a slope which is rough enough to prevent slipping.
With the situation as shown in the diagram, the weight Mg will produce a clockwise moment about point X. The slope will exert a normal reaction on the body and this will provide the anticlockwise moment about X necessary to maintain equilibrium.

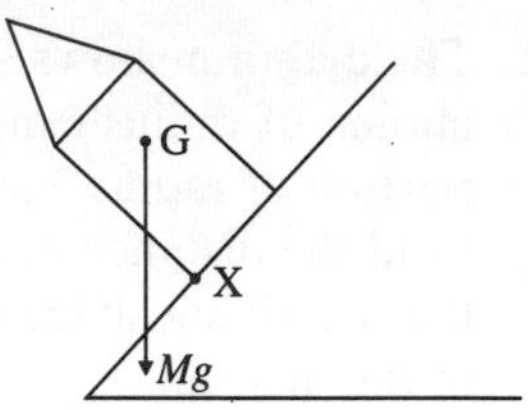

If the angle of the slope were sufficiently steep, the situation shown in the diagram on the right could occur.
The weight of the body now produces an anticlockwise moment about X.
The normal reaction is not able to counteract this anticlockwise moment.
Thus equilibrium is not maintained and the body will topple about X.

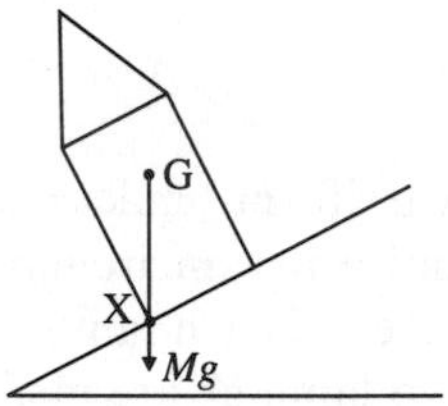

When the angle of the slope is such that the weight acts through X the body will be *on the point of* toppling. In this situation the centre of gravity of the body, G, is vertically above point X.

Example 11

A body consists of a uniform solid cylinder of mass $6m$, base radius r and height $2r$, attached at a plane face to the plane face of a uniform solid cone of mass $4m$, base radius r and height r.

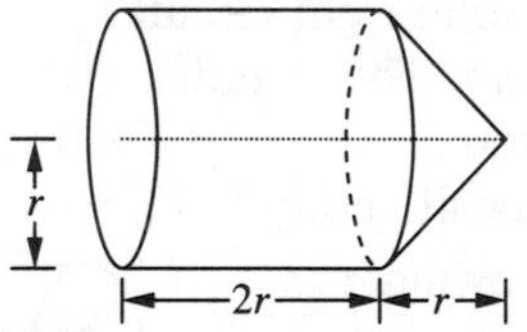

(a) Find the position of the centre of gravity of the body.
(b) The body is now placed with its plane face in contact with a horizontal table. The surface of the table is rough enough to prevent the body slipping as the table is slowly tilted. Find the angle through which the table has been tilted when the body is on the point of toppling.

(a) Draw two diagrams.

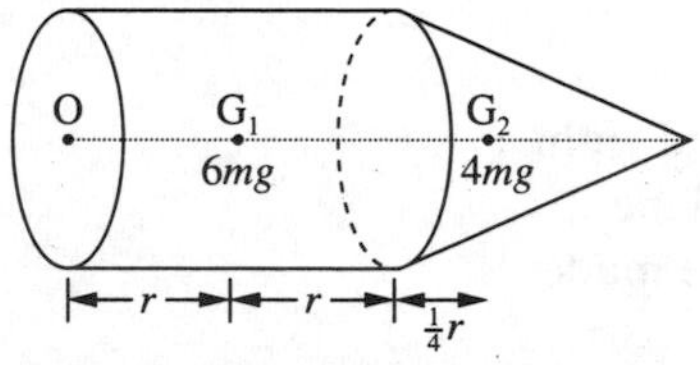

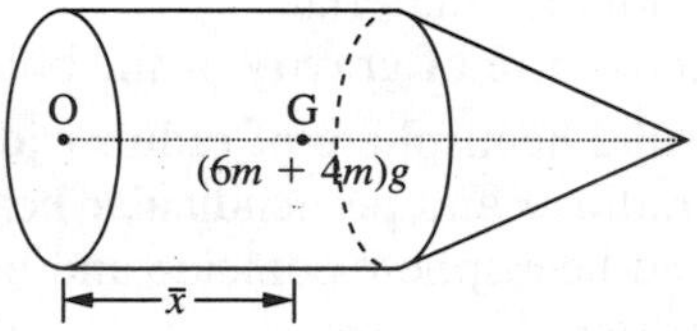

Point G, the centre of gravity of the combined body, will lie on the axis of symmetry passing through G_1 and G_2, the centres of gravity of the cylinder and the cone respectively.

Suppose OG $= \bar{x}$.
Taking moments about the vertical axis through O gives:

$$6mgr + 4mg\,(2r + \tfrac{1}{4}r) = 10mg\bar{x}$$

$$\therefore \qquad \bar{x} = 1{\cdot}5r$$

The centre of gravity of the body is on the axis of symmetry and $1{\cdot}5r$ above the plane face.

(b) The diagram on the right shows the body on the point of toppling. Point G, the centre of gravity of the body, will be vertically above point X (see diagram).

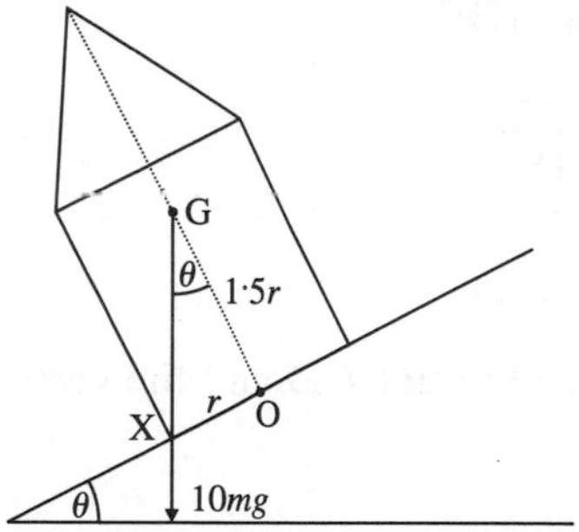

From ΔXOG:

$$\tan\theta = \frac{XO}{GO} = \frac{r}{1{\cdot}5r} = \frac{2}{3}$$

$$\theta = 33{\cdot}7^\circ \quad \text{(correct to 1 d.p.)}$$

When the table has been tilted through $33{\cdot}7^\circ$ the body is on the point of toppling.

Equilibrium of a suspended lamina

Consider a rectangular lamina, freely suspended by a string attached at the vertex A.

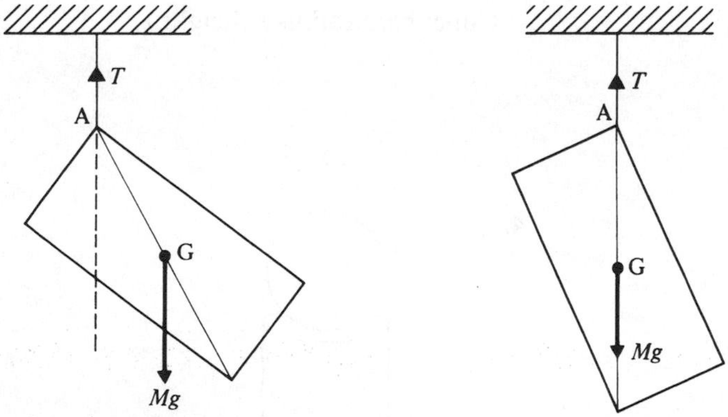

As there is a resultant moment about the point A, the position shown in the first diagram is not a possible position of equilibrium. Since there are only two forces acting on the lamina and their lines of action are parallel, they can only produce equilibrium if they act along the same line. Hence, equilibrium is only possible if the rectangle hangs so that its centre of gravity G lies vertically below the point A, as shown in the second diagram.

Example 12

A rectangular lamina ABCD is freely suspended by one vertex C. If CD = 15 cm, BC = 5 cm and the lamina is uniform, find the angle θ between the side CD and the vertical in the position of equilibrium.

The centre of gravity G of the rectangle ABCD lies on the vertical through the point C, and G is also the mid-point of the diagonal CA.

From the triangle ADC,

$$\tan\theta = \frac{AD}{DC}$$

$$= \frac{5}{15}$$

$$\therefore \quad \theta = 18{\cdot}43°$$

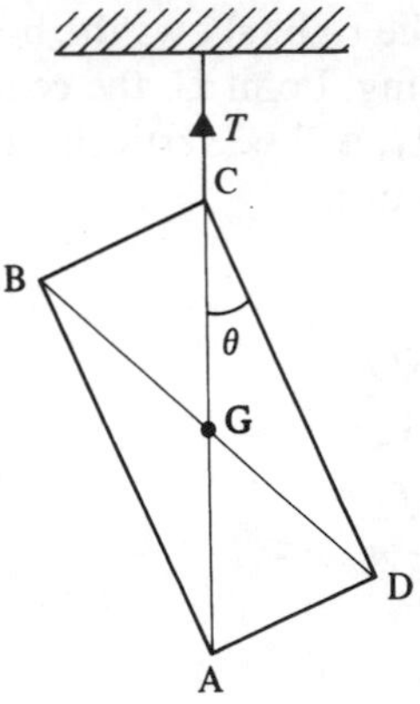

The angle between the side CD and the vertical is 18·43°.

Exercise 8D

In questions 1-7 each body consists of either one uniform solid or two uniform solids stuck together, and each has a vertical axis of symmetry. Each body is placed upright on a rough horizontal plane which is slowly tilted until the body is about to topple.
Find, in each case, the angle through which the plane is tilted.

1.

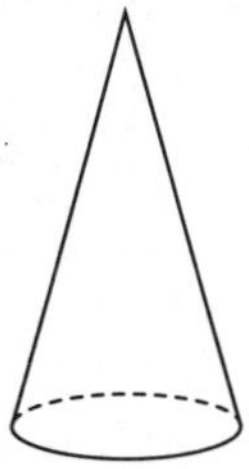

Cone: base radius r, height $4r$.

2.

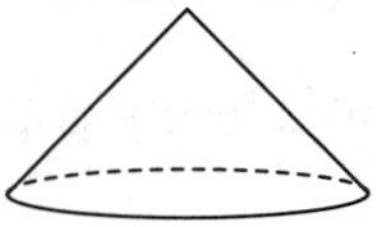

Cone: base radius r, height r.

3.

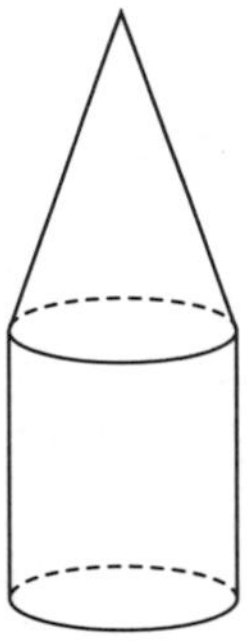

Cone: mass $2m$, base radius r, height $2r$.
Cylinder: mass $12m$, base radius r, height $3r$.

4.

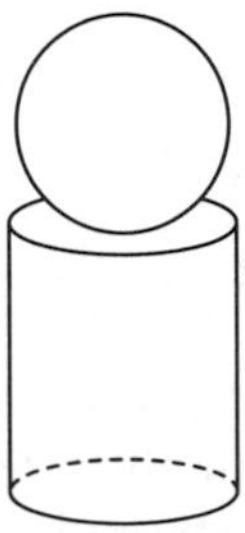

Sphere: mass m, radius r.
Cylinder: mass $4m$, base radius r, height $4r$.

5.

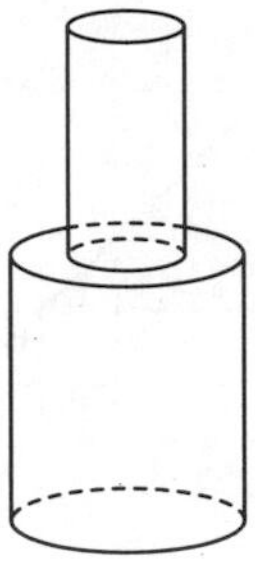

Cylinder: mass $3m$, base radius r, height $2r$
Cylinder: mass $6m$, base radius $2r$, height $3r$.

6.

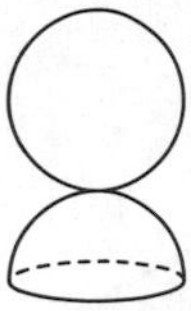

Sphere: mass $16m$, radius r.
Hemisphere: mass $10m$, radius r.

7.

Sphere: mass $6m$, radius r.
Cone: mass $8m$, base radius r, height $2r$.

In questions 8–16 the uniform laminae are freely suspended from point A. In each case find the angle that the line AB makes with the vertical. Each grid consists of unit squares.

8.

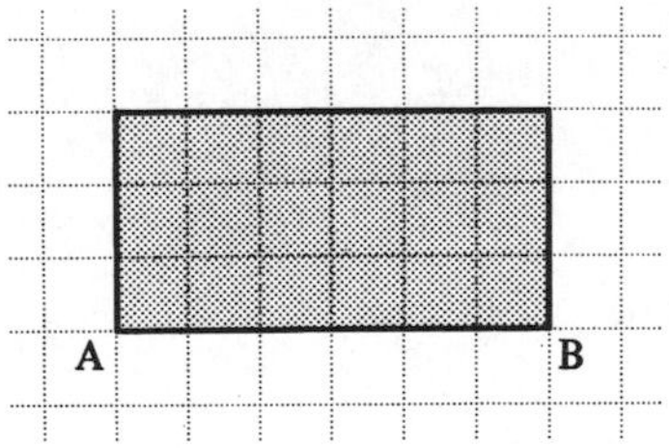

9.

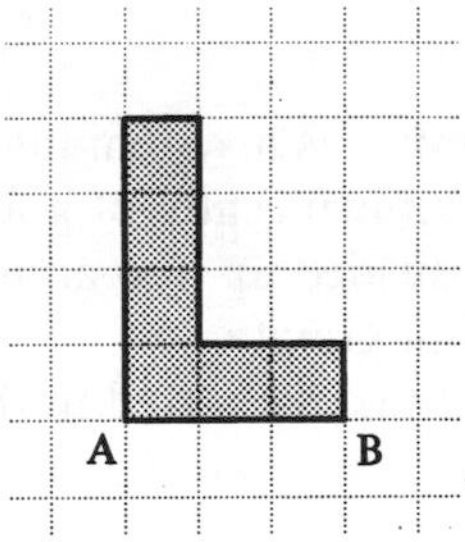

10.

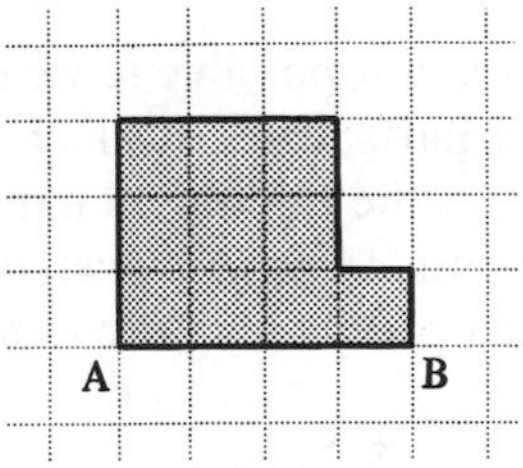

11.

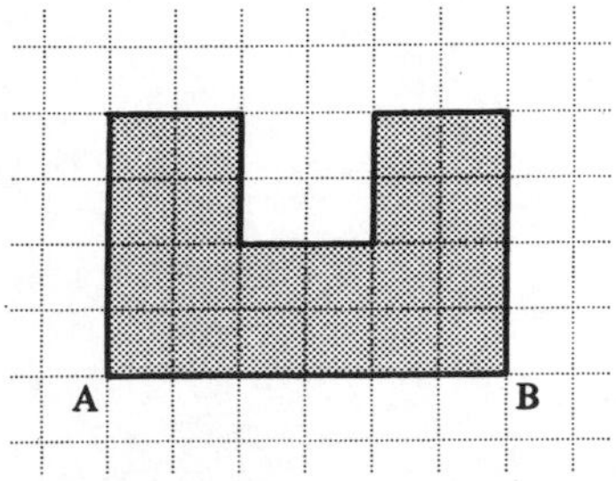

12.

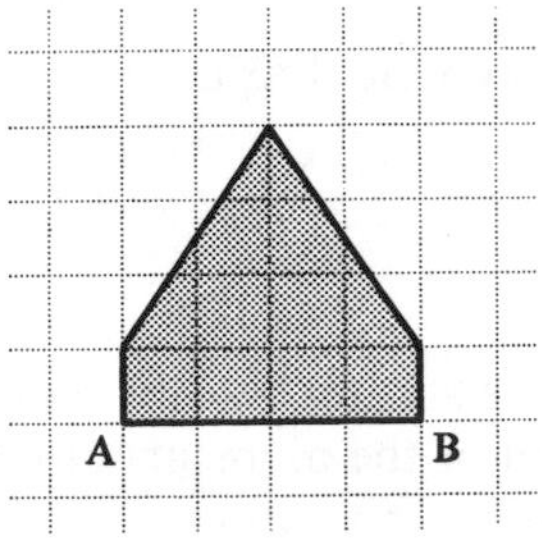

13.

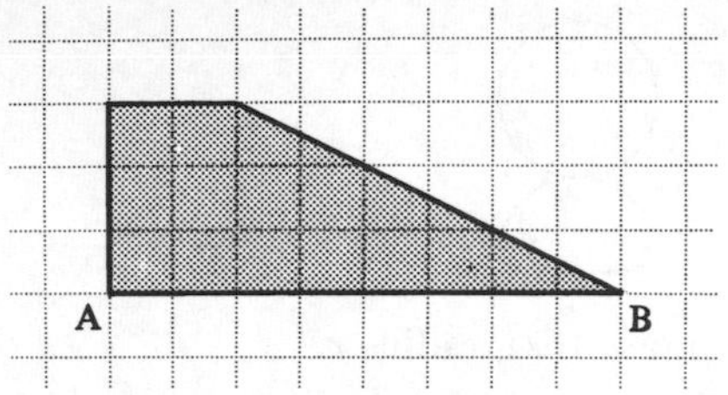

14.

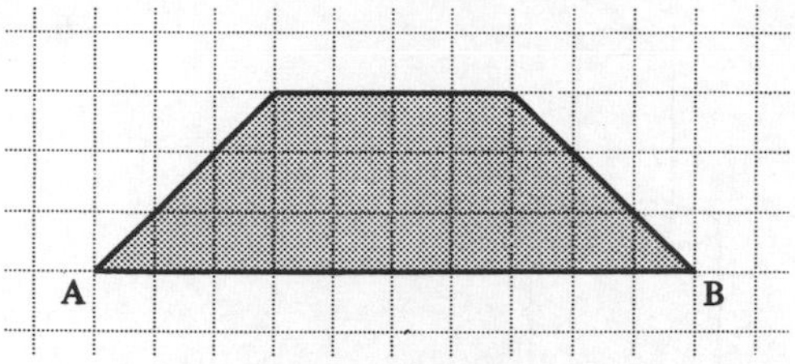

15.

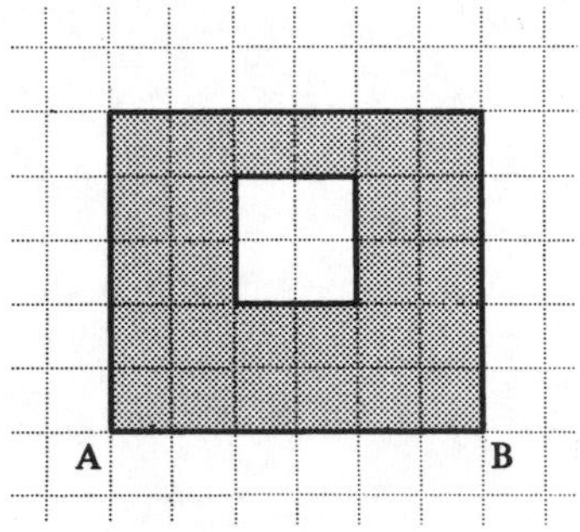

16.

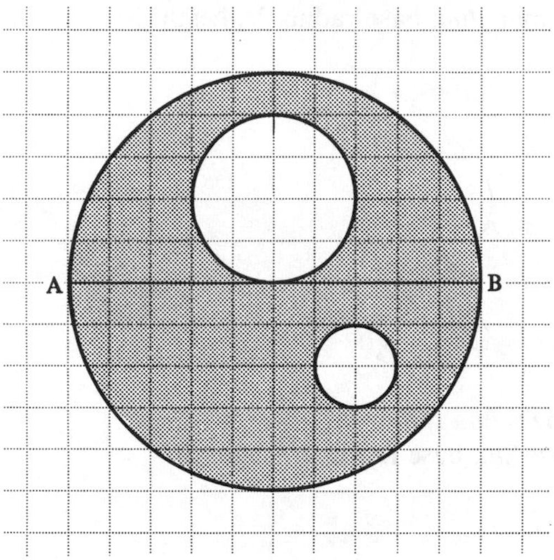

Use of calculus

In some cases it is not possible to treat a lamina as a system involving a finite number of parts, each of which has a known centre of gravity. By the methods of the calculus, a lamina is divided up into an *infinite* number of elements, all with known centres of gravity.
As was explained previously, a general result may be stated in the form:

$$\bar{x}\sum m_i g = \sum m_i g\, x_i$$

Using the calculus notation, we may write this as:

$$\bar{x}\int mg\,\mathrm{d}x = \int mg\,x\,\mathrm{d}x$$

In words, this may be interpreted as stating that the moment of the whole is equal to the algebraic sum of the moments of the parts about a given axis.

Example 13

A uniform semicircular lamina has a radius of 6 cm and mass per unit area m.
Find the position of the centre of gravity.

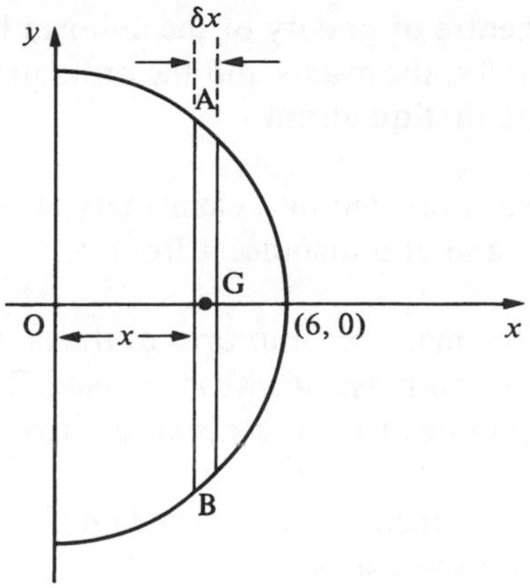

Consider an elementary strip AB of thickness δx parallel to the y-axis and a distance x from it.
The centre of gravity of this strip will be on the x-axis, as will the centres of gravity of all similar parallel strips.
Hence the centre of gravity of the lamina will also lie on the x-axis.

$$AG^2 + x^2 = 6^2 \qquad \text{or} \qquad AG = \sqrt{(6^2 - x^2)}$$

$$\text{area of strip} = 2\sqrt{(36 - x^2)} \times \delta x \quad \text{and mass} = 2\sqrt{(36 - x^2)} \times \delta x \times m$$

$$\text{area of whole lamina} = \tfrac{1}{2}\pi 6^2 = 18\pi \qquad \text{and mass} = 18\pi m$$

Equating the moments of the whole lamina and the sum of the moments of the strips gives:

$$18\pi m\bar{x}g = \sum_{x=0}^{6} 2\sqrt{(36 - x^2)} \times m\delta x \times xg$$

Using calculus gives: $\quad 18\pi m\bar{x}g = \int_0^6 2mx(36 - x^2)^{\frac{1}{2}}g\,dx$

$$= 2mg\left[-\tfrac{2}{6}(36 - x^2)^{\frac{3}{2}}\right]_0^6$$

$$= 2mg \times 72$$

$$\therefore \qquad \bar{x} = \frac{8}{\pi}\text{cm}$$

The centre of gravity is at a distance of $\frac{8}{\pi}$ cm from the point O, on the axis of symmetry of the lamina.

In cases where the lamina does not have an axis of symmetry, two coordinates will have to be found. Also, in many cases an integral giving the mass of the whole lamina will have to be evaluated rather than as in the above example where the area of the lamina (and hence its mass) was known.

Example 14

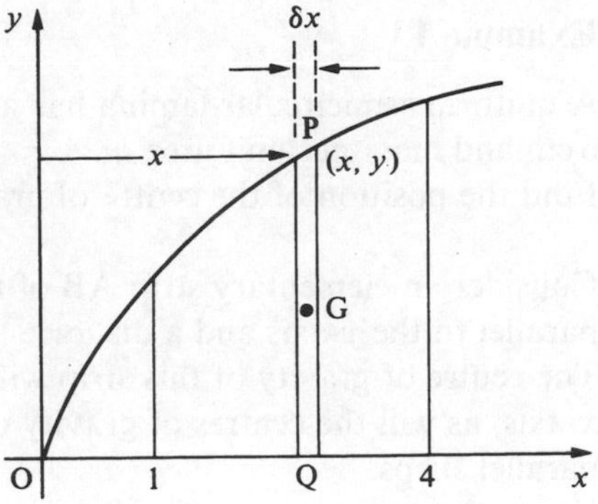

Find the centre of gravity of the uniform lamina bounded by the curve $y^2 = 9x$, the x-axis and the ordinates $x = 1$ and $x = 4$, and lying in the first quadrant.

The lamina is divided into elementary strips like PQ parallel to the y-axis, and at a distance x from it.

Let m be the mass per unit area of the lamina, then area of a strip $= y\delta x$ and mass of a strip $= ym\delta x$. The centre of gravity of this strip will be at G, at a distance x from the y-axis.

Equate the moment of the whole lamina and the sum of the moments of the strips, about the y-axis:

$$\bar{x}\int_1^4 ymg\,\mathrm{d}x = \int_1^4 xymg\,\mathrm{d}x$$

and substitute for y as $3x^{\frac{1}{2}}$ from the equation of the curve:

$$\bar{x}\int_1^4 3x^{\frac{1}{2}}\,mg\,\mathrm{d}x = \int_1^4 x\,3x^{\frac{1}{2}}\,mg\,\mathrm{d}x$$

Hence $$\bar{x}\int_1^4 3x^{\frac{1}{2}}\,\mathrm{d}x = \int_1^4 3x^{\frac{3}{2}}\,\mathrm{d}x$$

$\therefore$ $$\bar{x} = 2{\cdot}66$$

In a similar way, the y-coordinate of the centre of gravity is found by taking moments about the x-axis. The distance of the centre of gravity of each strip from the x-axis will be $\dfrac{y}{2}$.

Equate moments about the x-axis:

$$\bar{y}\int_1^4 ymg\,\mathrm{d}x = \int_1^4 \frac{y}{2}\times ymg\,\mathrm{d}x$$

and substitute for y as $3x^{\frac{1}{2}}$, as before, to give:

$$\bar{y} = 2{\cdot}41$$

The coordinates of the centre of gravity are (2·66, 2·41).

Solids of revolution

When a plane area is revolved about, say, the x-axis, a solid body is formed. Due to the way in which it is formed, it is sometimes referred to as a solid of revolution. Such bodies will necessarily have an axis of symmetry (the x-axis), and consequently the centre of gravity of the body must lie on the x-axis. There is therefore only one coordinate to determine.

It will usually be possible to divide up the solid into an infinite number of elementary discs, each with its centre of gravity on the x-axis. Then, by taking moments about the y-axis, the x-coordinate of the centre of gravity can be readily determined.

Example 15

A uniform solid cone of height 12 cm and base radius 3 cm is formed by rotating a line about the x-axis, as shown in the diagram. Find the distance of the centre of gravity of the cone from the origin.

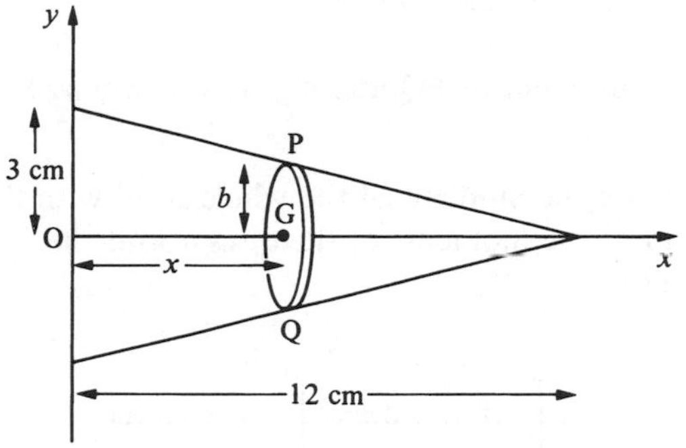

Suppose the density of the cone is σ.
Consider an elementary disc PQ, thickness δx and radius b, with its plane at right angles to the x-axis. If the distance of PQ from the y-axis is x, then by geometry:

$$\frac{b}{12-x} = \frac{3}{12} \quad \text{or} \quad b = \frac{12-x}{4}$$

$$\text{mass of elementary disc PQ} = \pi b^2 \sigma \,\delta x$$

$$\text{moment of PQ about } y\text{-axis} = \pi b^2 \sigma g \delta x \times x$$

$$\text{mass of whole cone} = \pi \frac{3^2 \times 12}{3} = 36\pi$$

$$\text{moment of whole cone about } y\text{-axis} = 36\pi\sigma g \bar{x}$$

Equate the moment of the whole solid with the sum of the moments of the discs about the y-axis:

$$36\pi\sigma g \bar{x} = \int_0^{12} \frac{\pi}{4^2} (12-x)^2 \sigma g x \,\mathrm{d}x$$

$$\therefore \quad 36\bar{x} = \tfrac{1}{16}\int_0^{12} (144x - 24x^2 + x^3)\,\mathrm{d}x$$

$$36\bar{x} = 108, \text{ giving } \bar{x} = 3\,\text{cm}$$

The centre of gravity of the whole cone is 3 cm from the origin.

In the above example, the solid was a cone for which the volume was known. In many cases, the solid of revolution will be such that calculus is needed to find its volume.

Example 16

An area is enclosed by the curve $y^2 = 5x$, the x-axis, the lines $x = 1$, $x = 3$, and it lies in the first quadrant. The area is rotated about the x-axis through one revolution.
Find the coordinates of the centre of gravity of the uniform solid so formed.

The centre of gravity will lie on the axis of symmetry, the x-axis.

Suppose the density of the solid is σ.

Consider an elementary disc PQ, thickness δx and radius y, with its plane at right angles to the x-axis.

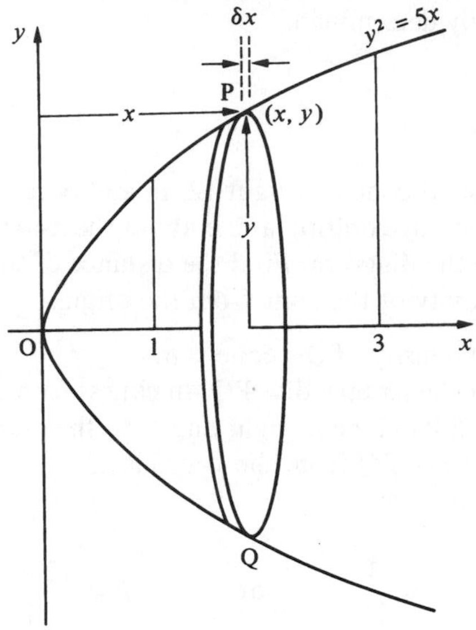

$$\text{mass of elementary disc PQ} = \pi y^2 \sigma \, \delta x$$

$$\text{moment of PQ about } y\text{-axis} = \pi y^2 \sigma g \delta x \times x$$

Equate the moment of the whole solid with the sum of the moments of the discs about the y-axis:

$$\bar{x} \int_1^3 \pi y^2 \sigma g \, \mathrm{d}x = \int_1^3 \pi y^2 \sigma g x \, \mathrm{d}x$$

Substitute for y^2 as $5x$ from the equation of the curve:

$$\bar{x} \int_1^3 \pi 5x\sigma g \, \mathrm{d}x = \int_1^3 \pi 5x^2 \sigma g \, \mathrm{d}x$$

$$\therefore \qquad 20\,\bar{x} = \frac{130}{3}$$

$$\therefore \qquad \bar{x} = \frac{13}{6}$$

The centre of gravity of the solid of revolution has coordinates $\left(2\frac{1}{6}, 0\right)$.

It should be noted that the answers to Examples 13 and 15 agree with the standard results stated prior to Exercise 8C. The proof of these and other standard results are covered in questions **1** to **6** of Exercise 8E.

Exercise 8E

1. Uniform semicircular lamina
   The diagram shows a uniform semicircular lamina of radius r; AB is an elementary strip of the lamina and lies parallel to the y-axis and at a distance x from it.
   The strip is of thickness δx.
   By considering all such strips, show that the centre of gravity of the lamina lies on its axis of symmetry at a distance of $\frac{4r}{3\pi}$ from O.

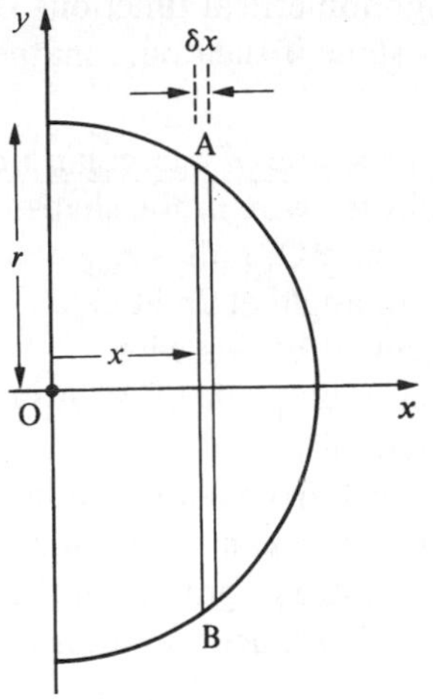

2. Uniform solid right circular cone
   The diagram shows a uniform solid right circular cone of height h and base radius r; AB is an elementary disc of thickness δx which has its plane parallel to the base of the cone and its centre at a distance x from the y-axis.
   By considering all such discs, show that the centre of gravity of the cone lies on its axis of symmetry at a distance of $\frac{3h}{4}$ from O.

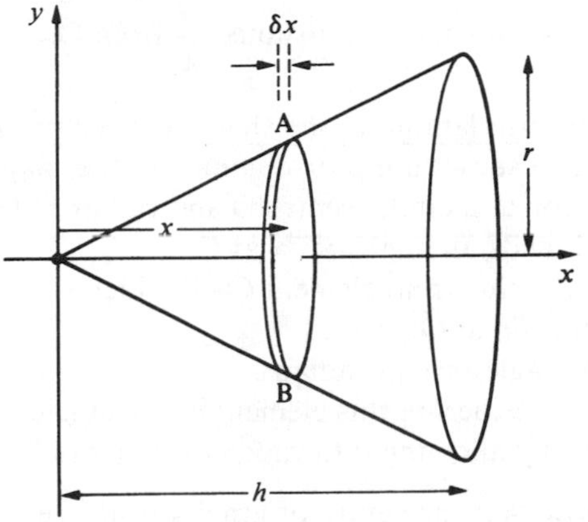

3. Uniform solid hemisphere
   The diagram shows a uniform solid hemisphere of radius r. AB is an elementary disc of the hemisphere which has its plane parallel to the plane surface of the hemisphere, lies at a distance x from this plane surface and is of thickness δx.
   By considering all such discs, show that the centre of gravity of the hemisphere lies on its axis of symmetry at a distance of $\frac{3r}{8}$ from O.

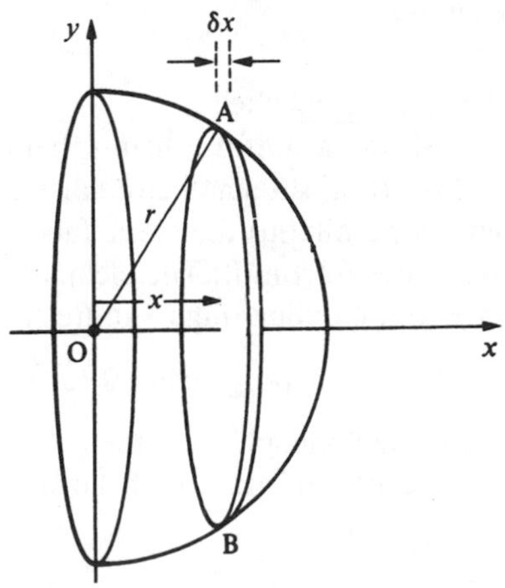

Note Questions **4**, **5** and **6** require an understanding of radians and an ability to integrate trigonometrical functions. The reader may wish to omit these questions at this stage if such pure mathematics topics have not yet been covered.

4. Uniform wire in the shape of a circular arc
Fig. 1 shows a uniform wire in the shape of an arc of a circle centre O, radius r.
The arc subtends an angle of 2α at O.
Fig 2 shows an element of this wire.
The element subtends an angle of $\delta\theta$ at O.
All angles are in radians.
By allowing θ to range from $+\alpha$ to $-\alpha$, all such elements of the wire can be considered.

(a) Show that the centre of gravity of the wire lies on its axis of symmetry at a distance of $\dfrac{r \sin \alpha}{\alpha}$ from O.

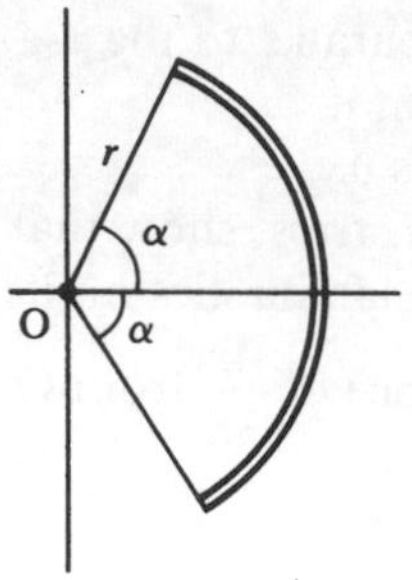

Fig.1

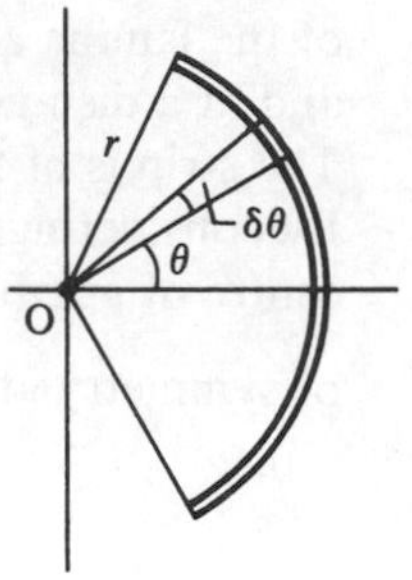

Fig. 2

(b) Show that if the wire were in the form of a semicircular arc of radius r, centre at O, the centre of gravity would lie on the axis of symmetry at a distance $\dfrac{2r}{\pi}$ from O.

5. Uniform lamina in the shape of a sector of a circle
Fig. 3 shows a uniform lamina in the shape of a sector of a circle, centre O and radius r. The sector subtends an angle of 2α at O.
Fig. 4 shows an element OAB which subtends an angle $\delta\theta$ at O.
All angles are in radians.
By considering this element as a triangle and by allowing θ to range from $+\alpha$ to $-\alpha$,

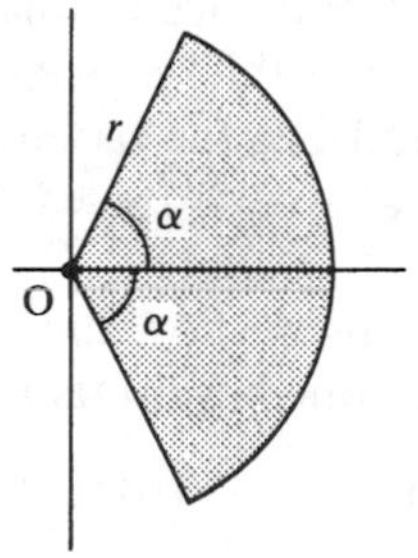

Fig.3

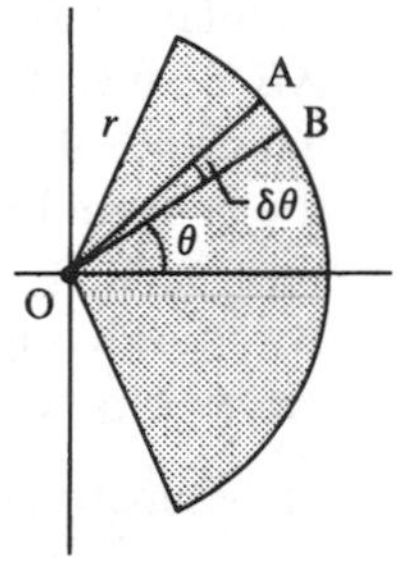

Fig. 4

show that the centre of gravity of the sector lies on its axis of symmetry at a point which is $\dfrac{2r \sin \alpha}{3\alpha}$ from O.
Show that this result is compatible with that of question **1** in this exercise.

6. Uniform hemispherical shell
The diagram shows a uniform hemispherical shell of radius r.
The shaded portion is a small circular element of the shell.
This element is parallel to the plane face of the hemisphere and at a distance $r \cos \theta$ from it. The element may be considered to approximate to a circular ring of radius $r \sin \theta$ and thickness $r\delta\theta$. If θ .is allowed to range from 0 to $\dfrac{\pi}{2}$, these elements together form the hemispherical shell.
Show that the centre of gravity of the shell lies on the x-axis at a distance $\dfrac{r}{2}$ from O.

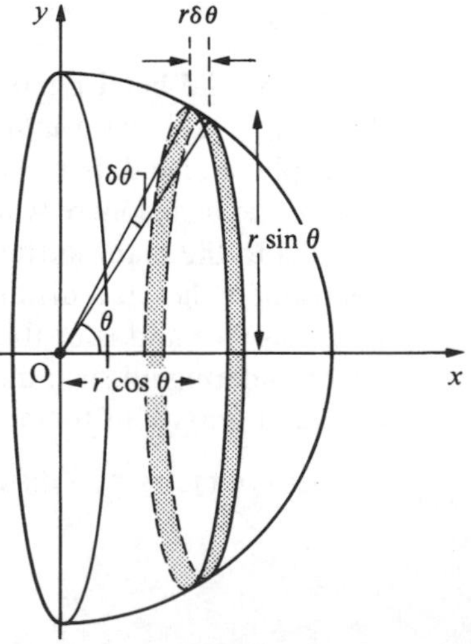

7. Find the coordinates of the centre of gravity of the uniform lamina enclosed by the curve $y = x^2$, the x-axis and the line $x = 2$.

8. Find the coordinates of the centre of gravity of the uniform lamina enclosed between the curve $y = 2x - x^2$, and the x-axis.

9. Find the coordinates of the centre of gravity of the uniform lamina enclosed by the curve $y = x^2 + 2$, the x-axis and the lines $x = 1$ and $x = 2$.

10. Find the coordinates of the centre of gravity of the uniform lamina lying in the first quadrant and enclosed by the curve $y^2 = 8x$, the x-axis and the lines $x = 2$ and $x = 8$.

11. Find the coordinates of the centre of gravity of the uniform lamina enclosed between the line $y = 3x$ and the curve $y = x^2$.

12. Find the coordinates of the centre of gravity of the uniform lamina which lies in the first quadrant and is enclosed by the curves $y = 3x^2$, $y = 4 - x^2$ and the y-axis.

13. The area enclosed by the curve $y^2 = x$, the x-axis, the line $x = 4$ and lying in the first quadrant, is rotated about the x-axis through one revolution.
    Find the coordinates of the centre of gravity of the uniform solid so formed.

14. The area enclosed by the curve $y^2 = x$, the x-axis, the lines $x = 2$, $x = 4$ and lying in the first quadrant, is rotated about the x-axis through one revolution.
    Find the coordinates of the centre of gravity of the uniform solid so formed.

15. The area lying in the first quadrant and enclosed by the curve $y = x^2$, and the lines $y = 0$, $x = 2$ and $x = 4$, is rotated about the x-axis through one revolution.
    Find the coordinates of the centre of gravity of the uniform solid so formed.

16. The area enclosed by the curve $y = x^2 + 3$, the x-axis, the y-axis and the line $x = 2$ is rotated about the x-axis through one revolution.
    Find the coordinates of the centre of gravity of the uniform solid so formed.

17. Find the coordinates of the centre of gravity of the uniform lamina enclosed between the curve $y = x^3$, the x-axis and the line $x = 3$.
    If this lamina is rotated about the x-axis through one revolution, find the coordinates of the centre of gravity of the uniform solid so formed.

Exercise 8F Examination questions

1. A straight piece of uniform wire of length $6(2+\sqrt{2})$ cm is bent so as to form a right-angled triangle ABC, with $AB = BC = 6$ cm and angle $ABC = 90°$. Find the perpendicular distances of the centre of mass of the triangle from AB and BC. (WJEC)

2.

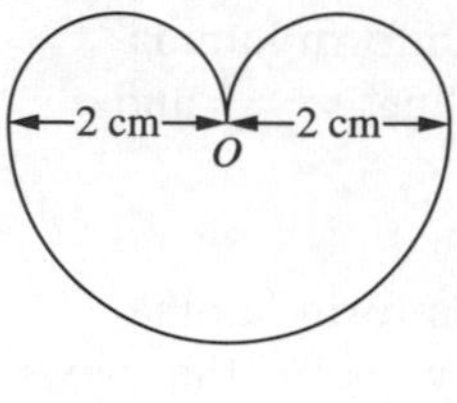

A badge is cut from a uniform thin sheet of metal. The badge is formed by joining the diameters of two semicircles, each of radius 1 cm, to the diameter of a semicircle of radius 2 cm, as shown in the diagram. The point of contact of the two smaller semicircles is O. Determine, in terms of π, the distance from O of the centre of mass of the badge. (ULEAC)

3. A uniform rectangular plate $OABC$ has mass $4m$, $OA = BC = 2d$ and $OC = AB = d$. Particles of mass $2m$, m and $3m$ are attached at A, B, and C respectively on the plate. Find, in terms of d, the distance of the centre of mass of the loaded plate
   (a) from OA,
   (b) from OC.
   The corner O of the loaded plate, is freely hinged to a fixed point and the plate hangs at rest in equilibrium.
   (c) Calculate, to the nearest degree, the angle between OC and the downward vertical. (ULEAC)

4.

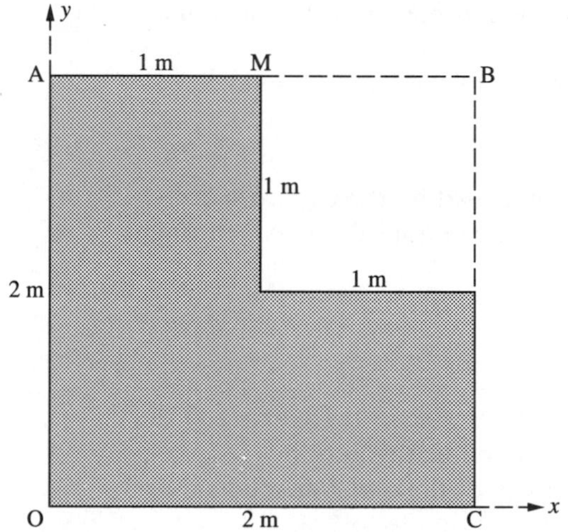

A thin uniform square plate originally of mass 12 kg has a square part cut off as shown in the diagram, leaving the shaded shape.

(i) Referred to the axes shown in the diagram, find the centre of mass of the plate.

(ii) What angle will AC make with the vertical if the plate is freely suspended at A?

A particle of mass m is added to the shape at M, the mid-point of AB, so that AC is vertical when the plate is freely suspended at A.

(iii) Show that $m = 3$ and find the coordinates of the centre of mass of the plate with the mass added, giving your answer referred to the axes shown in the diagram. (OCSEB)

5. A uniform rectangular sheet of metal $ABCD$, of mass 10 kg, is suspended from A. In equilibrium AB, which has length 0·3 m, is inclined at 20° to the vertical. Find the length of AD.
A metal ball, of mass 5 kg is now suspended from D. Find the angle between AB and the vertical in the new position of equilibrium. (UCLES)

6.

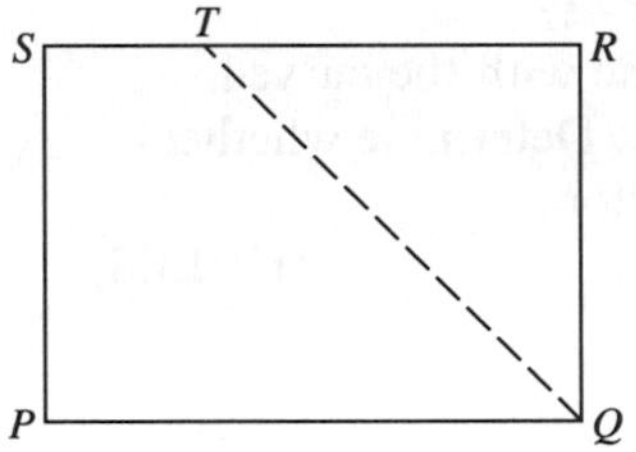

The diagram shows a rectangular sheet $PQRS$ of uniform thin metal with $PQ = 4$ m and $QR = 3$ m. T is a point on RS such that $RT = 3$ m. The sheet is folded about the line QT until R lies on PQ.

(a) Find the distances from PQ and PS of the centre of mass of the folded sheet.

The folded sheet is freely suspended from the point S and hangs in equilibrium.

(b) Calculate the angle of inclination of the edge PS to the vertical. (AEB 1994)

7.

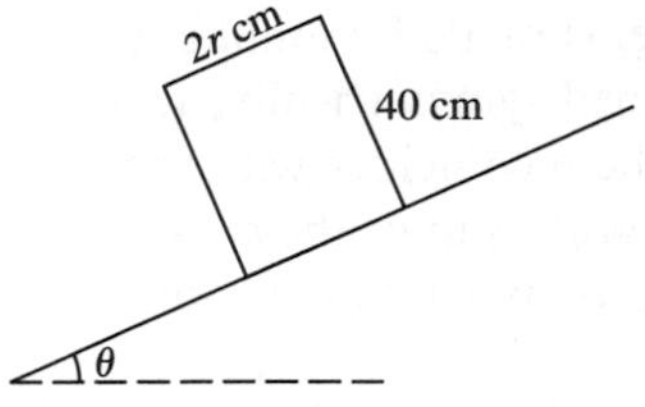

A uniform right cylinder has height 40 cm and base radius r cm. It is placed with its axis vertical on a rough horizontal plane. The plane is slowly tilted, and the cylinder topples when the angle of inclination θ (see diagram) is 20°. Find r.
What can be said about the coefficient of friction between the cylinder and the plane? (UCLES)

8. The base of a uniform solid hemisphere has radius $2a$ and its centre is at O. A uniform solid S is formed by removing, from the hemisphere, the solid hemisphere of radius a and centre O. Determine the position of the centre of mass of S. (The relevant result for a solid hemisphere may be assumed without proof). (AEB 1990)

9. A uniform body consists of a right circular cylinder of base radius r and height $2r$ and a right circular cone of base radius r and height r, fixed together so that the base of the cone coincides with one of the plane faces of the cylinder. Show that the centre of mass of the body is at a distance $\frac{51}{28}r$ from the vertex of the cone.

(i) The body is placed on a rough plane, which is inclined at an angle α to the horizontal, with the base of the cylinder in contact with the plane. The plane is rough enough to prevent sliding. Show that if the body remains in equilibrium then $\tan\alpha \leqslant \frac{28}{33}$.

(ii) The body is now placed on a horizontal plane with the curved surface of the cone in contact with the plane. Determine whether the body remains in equilibrium in this position.

(UCLES)

10.

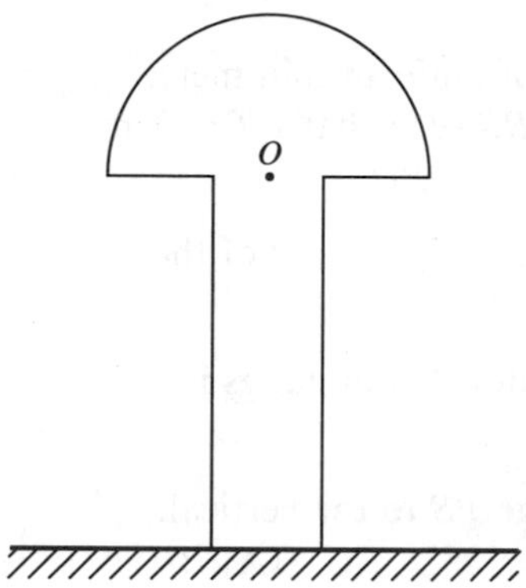

A uniform wooden "mushroom", used in a game, is made by joining a solid cylinder to a solid hemisphere. They are joined symmetrically, such that the centre O of the plane face of the hemisphere coincides with the centre of one of the ends of the cylinder. The diagram shows the cross-section through a plane of symmetry of the mushroom, as it stands on a horizontal table.

The radius of the cylinder is r, the radius of the hemisphere is $3r$, and the centre of mass of the mushroom is at the point O.

(a) Show that the height of the cylinder is $r\sqrt{(\frac{81}{2})}$.

The table top, which is rough enough to prevent the mushroom from sliding, is slowly tilted until the mushroom is about to topple.

(b) Find, to the nearest degree, the angle with the horizontal through which the table top has been tilted. (ULEAC)

11.

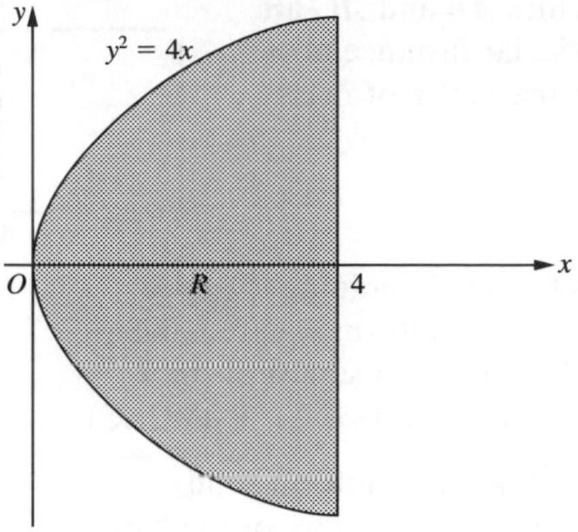

The diagram shows a sketch of the region R bounded by the curve with equation $y^2 = 4x$ and the line with equation $x = 4$. The unit of length on both the x-axis and the y-axis is the centimetre. The region R is rotated through π radians about the x-axis to form a solid S.

(a) Show that the volume of S is $32\pi\,\text{cm}^3$.

Given that the solid S is uniform,

(b) find the distance of the centre of mass of S from O. (ULEAC)

12. A solid uniform cylindrical piece of metal, of height h and radius r, has a cone shape removed from it as shown in the diagram. The base of the cone is of radius r and its height is h.

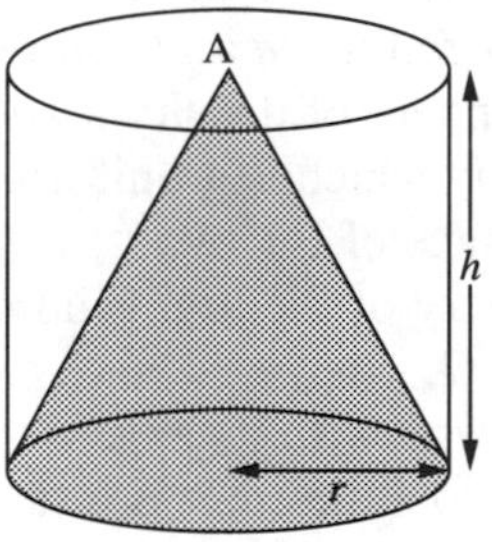

(i) Show that the centre of gravity of the resulting solid is at a distance of $\dfrac{3h}{8}$ from the point A measured along the axis of symmetry.

The solid is placed on an inclined plane with the open end in contact with the surface of the plane.

(ii) Given that $h = 4r$ and the coefficient of friction between the solid and the plane is $0{\cdot}75$, show that, as the inclination of the plane increases from zero degrees, the solid will topple before it slides.

(NICCEA)

13. In a uniform triangular lamina ABC the sides AB and AC are equal, BC is of length $2h$ and the perpendicular distance of A from BC is $3h$. Show, by integration, that the centre of mass of the lamina is at a distance $2h$ from A.

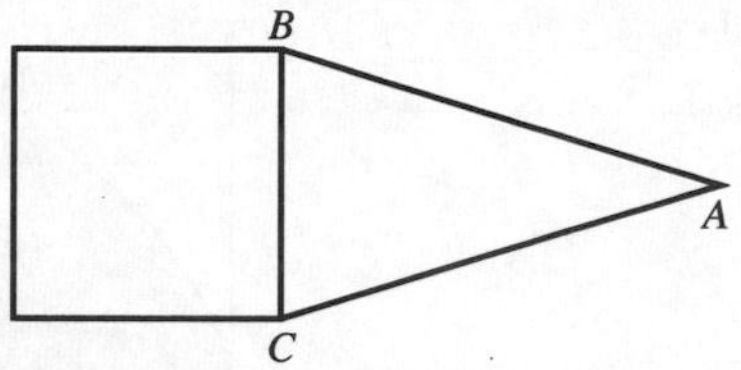

The diagram shows the composite lamina formed by joining the edge BC of the above lamina to one of the edges of a uniform square lamina, made of the same material, and of edge $2h$. Find the distance of the centre of mass of the composite lamina from the point A. (WJEC)

14. Use integration to show that the centre of mass of a uniform solid hemisphere of base radius a is a distance $\frac{3}{8}a$ from the centre of the base of the hemisphere.
The diagram shows a child's toy which consists of a head and a body.

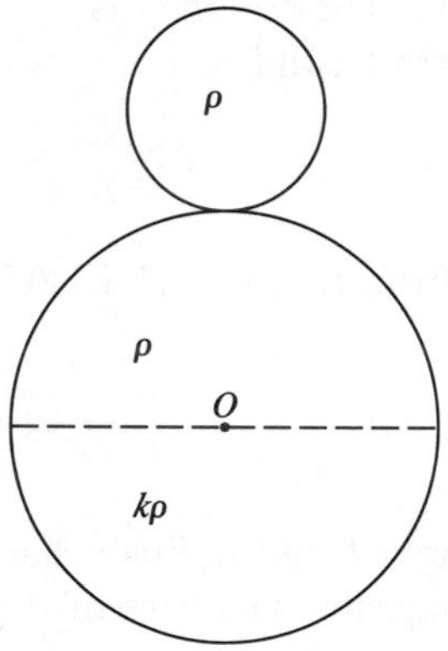

The head is a uniform sphere of radius a and density ρ. The head is rigidly attached to the body, which is spherical, of radius $2a$ and centre O. The upper half of the body is a uniform hemisphere of density ρ and this is rigidly attached to the lower half of the body which is a uniform hemisphere of density $k\rho$. The common plane surface of the two hemispheres is perpendicular to the axis of symmetry of the toy. Find the distance of the centre of mass of the toy from O.
The toy is now placed on a horizontal plane and rests with its axis of symmetry vertical. Show that when it is displaced to a different position, it will always return to the vertical position provided that $k > 2$.

(AEB 1993)

9 General equilibrium of a rigid body

In Chapter 5, forces acting on a particle were considered, and by the definition of a particle this ensured that the forces were concurrent. Thus for equilibrium it was only necessary to show that there was no resultant force acting in any direction.
A rigid body, on the other hand, has size; the forces acting on the body may not be concurrent and so rotation could occur. Thus, for equilibrium we must ensure that there is no resultant force acting *and* that the forces have no turning effect.

Three forces

If a rigid body is in equilibrium under the action of only three forces, these forces must be either concurrent or parallel.

Suppose the forces are P, Q and S and that the lines of action of any two of the three forces, for example P and Q, intersect at the point A. The forces P and Q will have no turning effect about the point A; but if the line of action of the third force S does not pass through A, then the force S will have a turning effect about A and the forces cannot be in equilibrium. Hence, if the forces are in equilibrium, the line of action of S must pass through A: the forces are then concurrent.

Again, suppose that any two of the three forces (say P and Q) have parallel lines of action, then the resultant of these two forces will be parallel to P and Q. Equilibrium is then only possible if the third force S is equal and opposite to the resultant of P and Q, and hence S must be parallel to the other two forces.
It is sometimes possible to use these facts in determining the direction of an unknown third force which is maintaining equilibrium.

When three forces are maintaining equilibrium, it is possible to solve the problem either by the use of the triangle of forces or of Lami's Theorem.

The next section considers the method of resolving in two directions and taking moments as a general method of solution for a rigid body in equilibrium under the action of any number of forces. This general method can also be used for three forces in equilibrium and is often simpler to use than the specific three-force properties mentioned above.

General method

In all problems concerning the equilibrium of a rigid body, the following procedure should be adopted:
(i) interpret the information given and draw a diagram
(ii) show on the diagram all the forces acting on the body, indicating clearly the directions of these forces

(iii) equate the clockwise and anticlockwise moments of the forces acting on the body, about any convenient point
(iv) in each of two perpendicular directions, equate the resolved parts (or components) of the forces acting in one direction to those resolved parts acting in the opposite direction.

Note. Careful choice of the point about which moments are taken may well simplify the solution of a particular problem. Usually the best directions in which to resolve the forces are:
(a) horizontally and vertically, or
(b) parallel and at right angles to the surface of an inclined plane.
However, there are examples which are more quickly solved by choosing other directions.

Example 1

A uniform rod AB of mass 4 kg and length 80 cm is freely hinged to a vertical wall.
A force P, as shown in the diagram, is applied at the point B and keeps the rod horizontal and in equilibrium. The forces X and Y are the horizontal and vertical components of the reaction at the hinge.
Find the magnitudes of the forces X, Y and P.

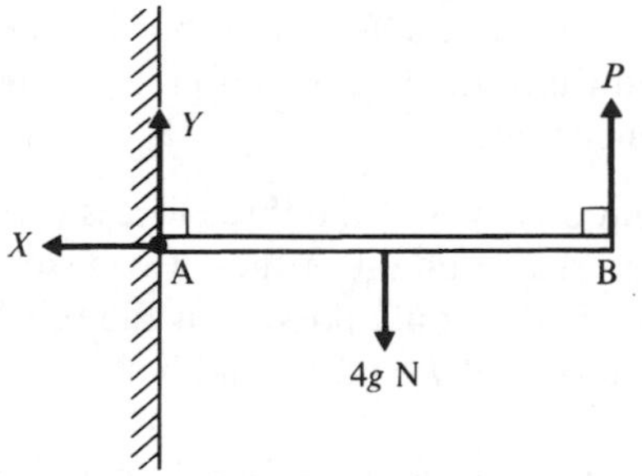

There are four forces acting on the rod.

Take moments about A:

A↻ $P \times 80 = 4g \times 40$

$\therefore \quad P = 2g$

or $\quad P = 19{\cdot}6\,\text{N}$

Resolve vertically:

$P + Y = 4g$

$\therefore \quad Y = 4g - 2g$

$Y = 19{\cdot}6\,\text{N}$

Resolve horizontally:

$X = 0$

The force $X = 0$, $Y = 19{\cdot}6\,\text{N}$ and $P = 19{\cdot}6\,\text{N}$.

Reaction at hinge

Instead of considering the horizontal and vertical components of the reaction at the hinge A, the force on the rod due to the hinge may be represented by a single force R acting at an angle θ to the vertical. The following example illustrates this method.

Example 2

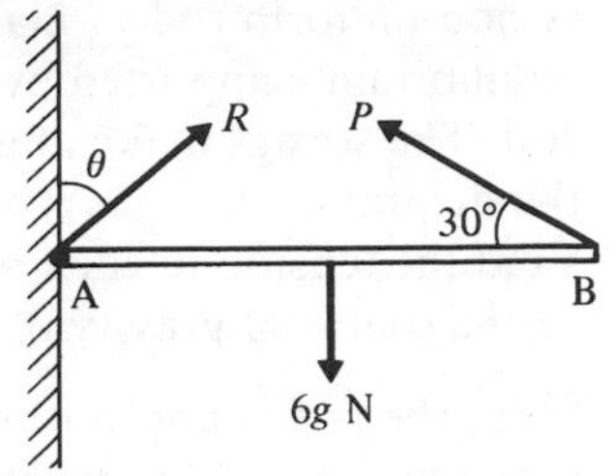

A uniform rod AB of mass 6 kg and length 4 m is freely hinged at A to a vertical wall. The force P applied at B as shown in the diagram, keeps the rod horizontal and in equilibrium; R is the force of reaction at the hinge and θ is the angle that the line of action of this force makes with the vertical. Find the magnitude of the forces P and R and the angle θ.

Take moments about A:

$\overset{\frown}{A}$ $\quad 6g \times 2 = P \times 4 \sin 30°$

$\therefore \quad P = 6g$

Resolve vertically:

$$R \cos \theta + P \cos 60° = 6g$$

Substitute for P:

$$R \cos \theta = 3g \qquad \ldots [1]$$

Resolve horizontally:

$$R \sin \theta = P \cos 30°$$

Substitute for P:

$$R \sin \theta = 6g \cos 30° \qquad \ldots [2]$$

Divide equation [2] by equation [1]:

$$\frac{R \sin \theta}{R \cos \theta} = \frac{6g \cos 30°}{3g}$$

$\therefore \quad \tan \theta = \sqrt{3} \quad$ or $\quad \theta = 60°$

Substitute in equation [1]:

$$R \cos 60° = 3g$$

$\therefore \quad R = 6g$

The force $P = 6g$ N, $R = 6g$ N and the angle $\theta = 60°$.

Alternative method

Since there are only three forces acting on the rod in Example 2, the force R must pass through the point of intersection O of the forces P and $6g$, as shown in the diagram below.

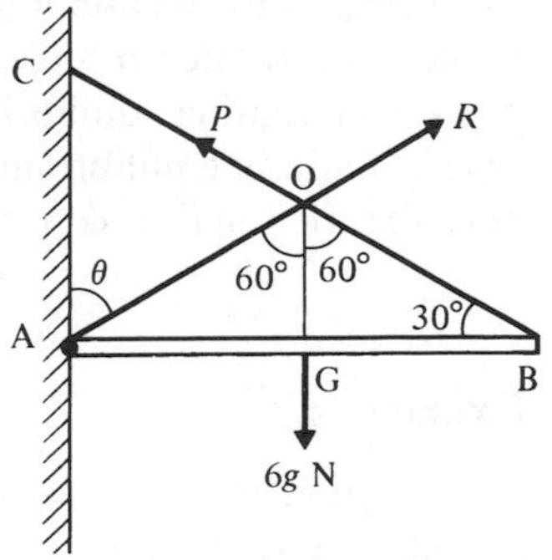

G is the mid-point of AB, and OG is perpendicular to AB. Hence the triangle AOB is isosceles and AO = OB.

$\therefore \quad$ angle OAB = 30°

or $\quad \theta = 60°$

The angles between the forces are each 120°.

Apply Lami's Theorem at the point O:

$$\frac{6g}{\sin 120°} = \frac{P}{\sin 120°} = \frac{R}{\sin 120°}$$

$\therefore \quad P = 6g$ and $R = 6g$ as before

Alternatively from the same diagram, if BO is produced to meet the wall at C, then triangle AOC may be used as a triangle of forces in order to solve the problem.

Example 3

A non-uniform rod of mass 3 kg and length 40 cm rests horizontally in equilibrium, supported by two strings attached at the ends A and B of the rod. The strings make angles of 45° and 60° with the horizontal, as shown in the diagram.
Find the tension in each of the strings and the position of the centre of gravity of the rod.

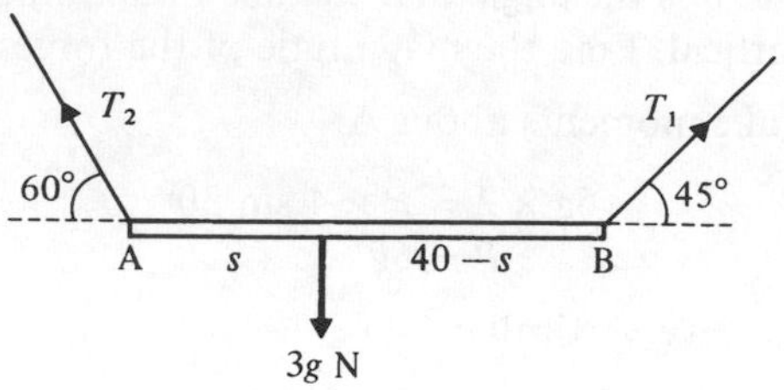

Since the rod is not uniform, the force of $3g$ N is shown acting at a distance s cm from the end A.

Resolve horizontally:

$$T_2 \cos 60° = T_1 \cos 45°$$

$$\therefore \quad T_2 = T_1\sqrt{2} \qquad \ldots [1]$$

Resolve vertically:

$$T_2 \sin 60° + T_1 \sin 45° = 3g \qquad \ldots [2]$$

Take moments about A:

$$\overset{\curvearrowright}{A} \quad 3g \times s = (T_1 \sin 45°) \times 40 \qquad \ldots [3]$$

From equation [1] and [2]: $T_2 = \dfrac{6g}{\sqrt{3}+1}$ and $T_1 = \dfrac{3\sqrt{2}g}{\sqrt{3}+1}$

Substitute in equation [3]:

$$s = \frac{40}{\sqrt{3}+1}$$

The tensions in the strings are 21·5 N and 15·2 N and the centre of gravity is 14·6 cm from end A.

Limiting equilibrium

When a rigid body is in equilibrium under the action of any number of forces, three equations may be obtained by resolving in two directions and by taking moments about a point.
If there is a frictional force acting, it is necessary to be clear whether the body is in limiting equilibrium. It should be remembered that only in the case of limiting equilibrium, when motion is on the point of taking place, does the frictional force F have its maximum value μR.

Example 4

The diagram shows a uniform rod AB of mass 4 kg with its lower end A resting on a rough horizontal floor, coefficient of friction μ. A string attached to the end B keeps the rod in equilibrium. T is the tension in the string, F is the frictional force at A, and R is the normal reaction at A. Find the magnitudes of the forces T, F and R, and also the least possible value of μ for equilibrium to be possible.

Let the rod be of length $2l$. If motion were to take place, the end A of the rod would tend to move to the left so the frictional force F acts in the opposite direction.

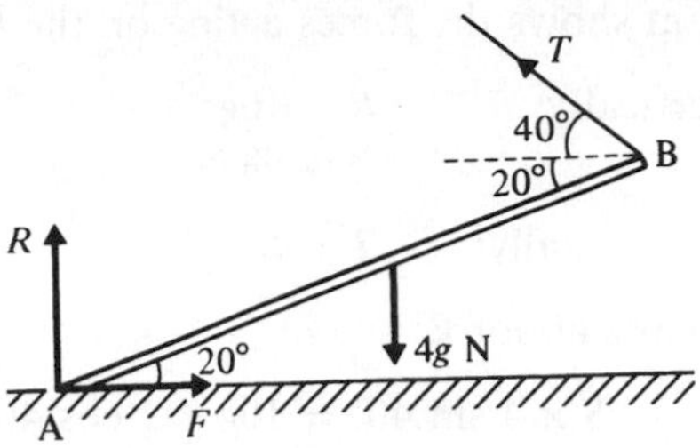

Take moments about A:

$$\overset{\frown}{A} \quad 4g \times l \cos 20° = T \times 2l \sin 60°$$

$$\therefore \quad T = \frac{2g \cos 20°}{\sin 60°}$$

$$= 21{\cdot}3\,\text{N}$$

Resolve horizontally: $T \cos 40° = F$

$\therefore$ $F = 16{\cdot}3\,\text{N}$

Resolve vertically: $R + T \sin 40° = 4g$

Substitute for T: $R = 25{\cdot}5\,\text{N}$

Since $\mu = \dfrac{F_{max}}{R}$ the least value of μ necessary is $\dfrac{16{\cdot}3}{25{\cdot}5} = 0{\cdot}64$ and the rod would then be in limiting equilibrium.

The force $T = 21{\cdot}3$ N, $F = 16{\cdot}3$ N, and $R = 25{\cdot}5$ N, and the least possible value of μ is $0{\cdot}64$.

Ladder problems

The situation of a ladder resting against a wall, with the foot of the ladder on the ground, gives rise to a variety of problems. The wall may be rough or smooth, as also may the ground. The ground may, or may not, be horizontal.
It should be remembered that where the ladder rests against a smooth surface, there will only be a normal reaction R at that point.
When the surfaces in contact are rough, there is also a frictional force F which acts parallel to the surfaces in contact, and in a direction opposite to that in which the ladder would move.

Example 5

A uniform ladder AB, of mass 10 kg and length 4 m, rests with its upper end A against a smooth vertical wall and end B on smooth horizontal ground. A light horizontal string, which has one end attached to B and the other end attached to the wall, keeps the ladder in equilibrium inclined at 40° to the horizontal. The vertical plane containing the ladder and the string is at right angles to the wall.
Find the tension T in the string and the normal reactions at the points A and B.

Suppose R and S are the normal reaction at the ground and the wall respectively.

The diagram shows the forces acting on the ladder.

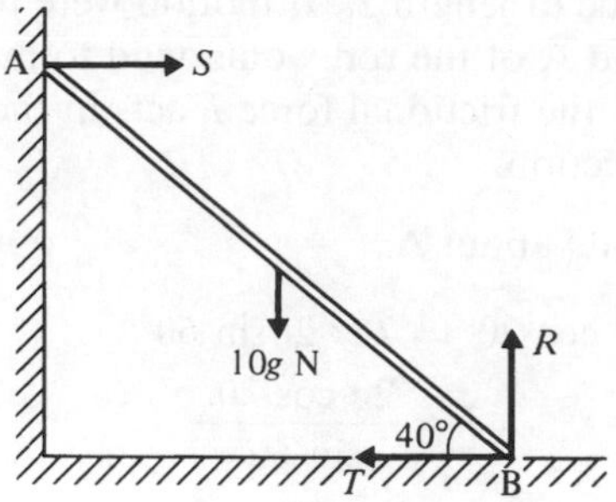

Resolve vertically: $R = 10g$

$= 98\text{ N}$

Resolve horizontally: $T = S$

Take moments about B:

$\overset{\curvearrowright}{B}$ $\quad S \times 4 \sin 40° = 10g \times 2 \cos 40°$

$\therefore \quad S = \dfrac{5g \cos 40°}{\sin 40°}$

$\therefore \quad S = 58{\cdot}4\text{ N}$

It follows that: $T = 58{\cdot}4\text{ N}$

The tension in the string is 58·4 N and the normal reactions at the top and foot of the ladder are 58·4 N and 98 N respectively.

It should be noted that whether equilibrium is possible or not will depend upon whether the string can take a tension of 58·4 N without breaking.

Rough contact at foot of ladder

If the ladder rests on ground which is rough, then there will be a frictional force F acting on the ladder at this point. The effect of this force is similar to that of the tension in the string in Example 5. The maximum value of this frictional force depends upon the roughness of the contact between the ladder and the ground.

Example 6

The diagram shows a ladder AB of mass 8 kg and length 6 m resting in equilibrium at an angle of 50° to the horizontal with its upper end A against a smooth vertical wall and its lower end B on rough horizontal ground, coefficient of friction μ. Find the forces S, F and R and the least possible value of μ if the centre of gravity G of the ladder is 2 m from B.

Take moments about B:

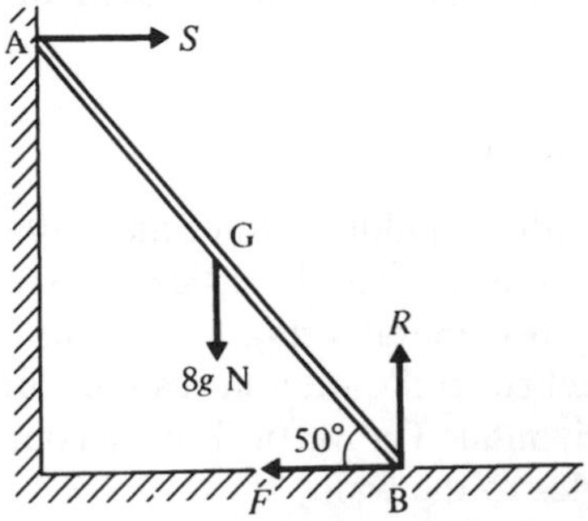

$\overset{\curvearrowright}{B}$ $\quad S \times 6 \sin 50° = 8g \times 2 \cos 50°$

$\therefore \quad S = \dfrac{8g \cos 50°}{3 \sin 50°}$

$\therefore \quad S = 21{\cdot}9\text{ N}$

Resolve horizontally: $F = S$

$\therefore \quad F = 21{\cdot}9\text{ N}$

Resolve vertically: $R = 8g$

$\therefore \quad R = 78{\cdot}4\text{ N}$

Since $\mu = \dfrac{F_{max}}{R}$ and $\dfrac{F}{R} = \dfrac{21{\cdot}9}{78{\cdot}4} = 0{\cdot}28$, μ must be at least 0·28

The force $S = 21{\cdot}9$ N, $F = 21{\cdot}9$ N, $R = 78{\cdot}4$ N, and μ must be at least 0·28.

Climbing a ladder

Whether or not it is safe to ascend to the top of a ladder will depend upon the magnitude of the frictional force which acts on the foot of the ladder. This will depend upon the roughness of the ground on which the ladder rests.
If the ladder is found to be in limiting equilibrium when a person is part way up a ladder, then any further ascent will cause the ladder to slip.
To determine how far a ladder may be ascended, consider the situation when the climber is at a distance s up the ladder and the ladder is in limiting equilibrium. The following example illustrates the method.

Example 7

A uniform ladder of mass 30 kg and length 5 m rests against a smooth vertical wall with its lower end on rough ground, coefficient of friction $\frac{2}{5}$. The ladder is inclined at 60° to the horizontal. Find how far a man of mass 80 kg can ascend the ladder without it slipping.

Assume the man can ascend a distance s m from the foot of the ladder, which is then in limiting equilibrium. The maximum frictional force μR will then act at the foot of the ladder.

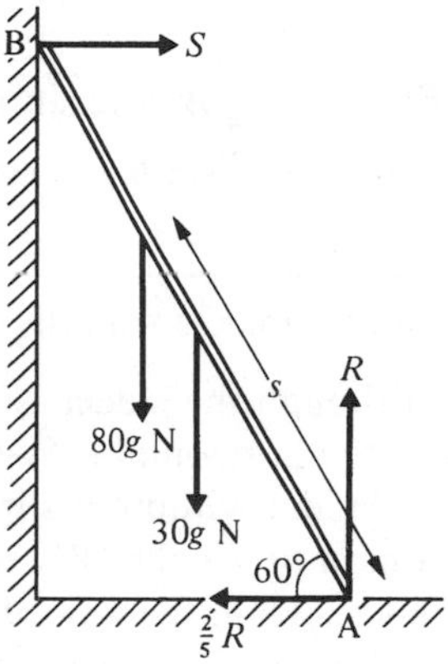

The forces acting on the ladder are then as shown.
Resolve vertically:

$$R = 30g + 80g = 110g \qquad \ldots[1]$$

Resolve horizontally:

$$S = \tfrac{2}{5}R \qquad \ldots[2]$$

Take moments about A:

$$\text{A}\curvearrowright \quad 30g \times \tfrac{5}{2}\cos 60^\circ + 80g \times s\cos 60^\circ = S \times 5\sin 60^\circ \qquad \ldots[3]$$

From equations [1] and [2]: $\quad S = \frac{2}{5}(110g) = 44g$

Substitute in equation [3]: $\quad \dfrac{75g}{2} + 40gs = 44g \times 5\sin 60^\circ$

or $\quad s = 3{\cdot}83\,\text{m}$

The man can climb 3·83 m up the ladder, at which point the ladder will be on the point of slipping.

Example 8

A uniform ladder rests in limiting equilibrium with its top end against a rough vertical wall and its lower end on a rough horizontal floor. If the coefficients of friction at the top and foot of the ladder are $\frac{2}{3}$ and $\frac{1}{4}$ respectively, find the angle which the ladder makes with the floor.

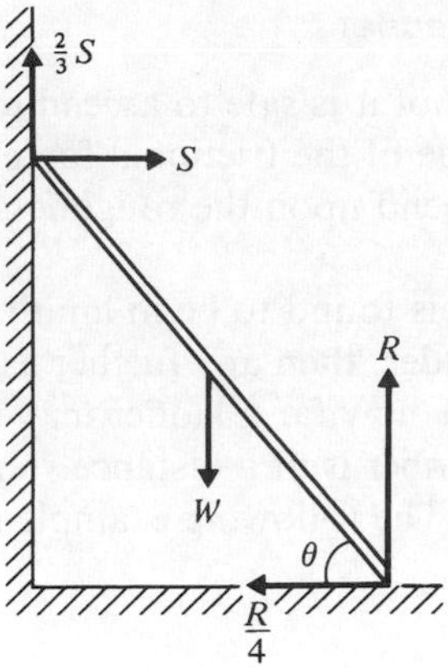

Let the ladder be of length $2l$ and weight W.
The forces acting on the ladder will be as shown in the diagram.
Since the ladder is in limiting equilibrium, both ends of the ladder will be on the point of moving, so the maximum frictional forces will act at both ends.

Resolve vertically:

$$\tfrac{2}{3}S + R = W \qquad \ldots[1]$$

Resolve horizontally:

$$S = \tfrac{1}{4}R \qquad \ldots[2]$$

Take moments about the foot of the ladder:

$$W \times l \cos\theta = S \times 2l \sin\theta + \tfrac{2}{3}S \times 2l \cos\theta \qquad \ldots[3]$$

Eliminate R from equations [1] and [2]:

$$\tfrac{2}{3}S + 4S = W$$

or $$S = \tfrac{3}{14}W$$

Substitute for S in equation [3]:

$$Wl \cos\theta = \tfrac{3}{14}W \times 2l \sin\theta + \tfrac{2}{3} \times \tfrac{3}{14}W \times 2l \cos\theta$$

$\therefore$ $$\cos\theta = \tfrac{3}{7}\sin\theta + \tfrac{2}{7}\cos\theta$$

$\therefore$ $$\tan\theta = \tfrac{5}{3} \quad \text{or} \quad \theta = 59{\cdot}04°$$

The angle the ladder makes with the floor is 59·04°.

Note that, even though the ladder is in limiting equilibrium when the angle of inclination to the horizontal is 59·04°, it is possible for a person to ascend part way up the ladder without it slipping.
Suppose that a man of weight $3W$ ascends the ladder of Example 8 to a point at a distance s from the foot of the ladder, and that the ladder is then on the point of slipping, i.e. it is in limiting equilibrium.

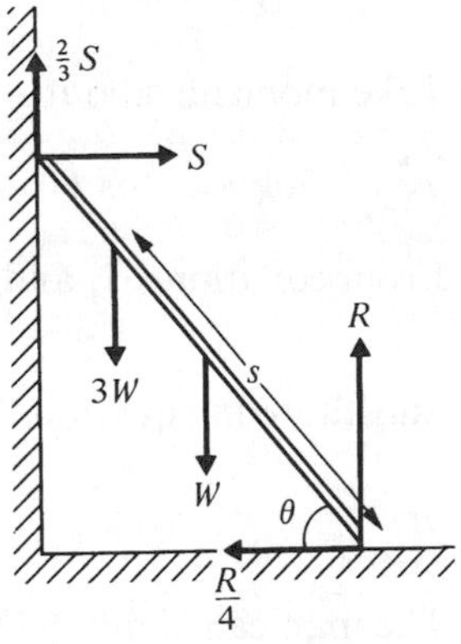

The three equations then become:

$$\tfrac{2}{3}S + R = W + 3W$$

$\therefore$ $$S = \tfrac{1}{4}R$$

and $$W \times l \cos\theta + 3W \times s \cos\theta = S \times 2l \sin\theta + \tfrac{2}{3}S \times 2l \cos\theta$$

Eliminate R as before:

$$\tfrac{2}{3}S + 4S = 4W$$

$\therefore$ $$S = \tfrac{6}{7}W$$

Then $$Wl \cos\theta + 3Ws \cos\theta = \tfrac{6}{7}W \times 2l \sin\theta + \tfrac{4}{7}W \times 2l \cos\theta$$

$\therefore$ $$3s = \left(\tfrac{12}{7}\tan\theta + \tfrac{1}{7}\right)l$$

Substitute $\tan\theta = \tfrac{5}{3}$

to give: $s = l$

The man can therefore ascend half-way up the ladder.

Practical explanation

The point to notice in this example is that as soon as the man steps on to the foot of the ladder, the normal reaction R at that point is increased. Therefore the maximum value of the frictional force ($F_{max} = \mu R$) is increased. This means that more friction is available at the foot of the ladder and it is no longer in a state of limiting equilibrium. As the man ascends the ladder, his weight has an increasing moment, anticlockwise in the diagram, about the foot of the ladder, and this together with the moment due to the weight of the ladder will eventually balance the maximum clockwise moment of S and $\frac{2}{3}S$ about the point A. This explains why it is safer to ascend a ladder when another person is standing on the foot of the ladder, or a mass is placed on the bottom rung.

Exercise 9A

For all those questions in this exercise which involve a rigid body in contact with a vertical wall, take the vertical plane through the rigid body as being perpendicular to the wall.

1. Each of the following diagrams shows a uniform rod AB of mass 10 kg and length 4 m freely hinged at A to a vertical wall. An applied force P keeps the rod in equilibrium. Forces X and Y are the horizontal and vertical components of the reaction at the hinge. By resolving vertically and horizontally and taking moments, find the magnitudes of the forces X, Y and P.

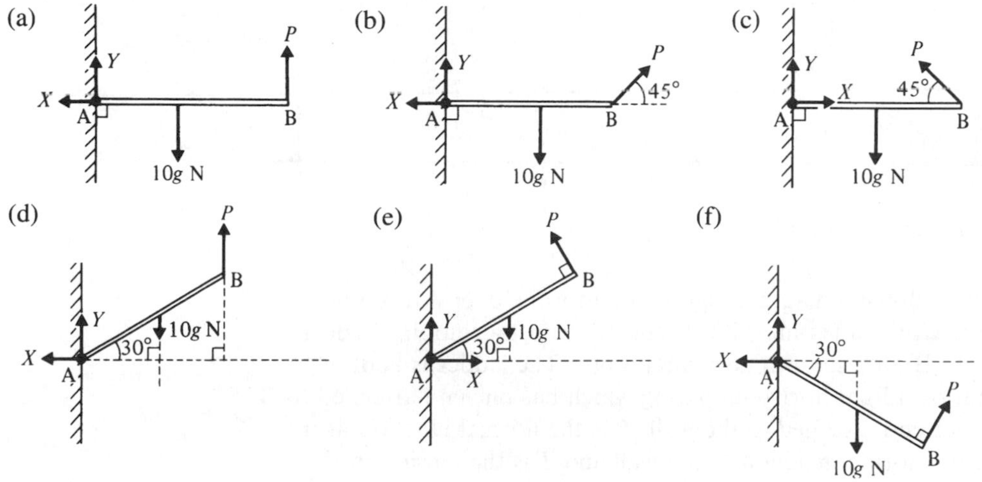

2. Each of the following diagrams shows a uniform rod AB of mass 5 kg and length 6 m freely hinged at A to a vertical wall. An applied force P keeps the rod in equilibrium. R is the force of reaction at the hinge and θ is the angle the line of action this force makes with the wall.
For each case, find the magnitudes of the forces P and R and the size of the angle θ.

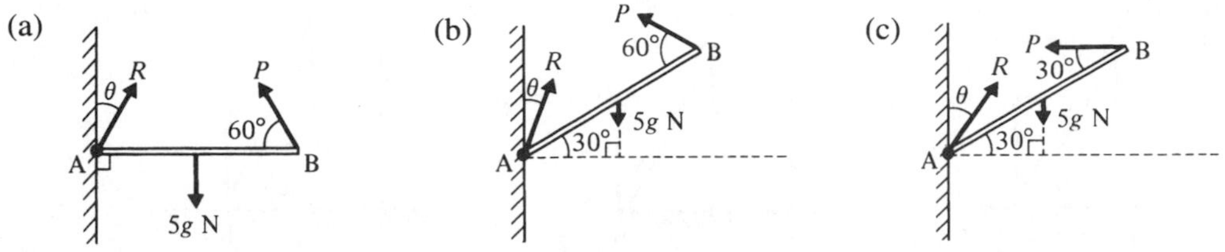

(d)

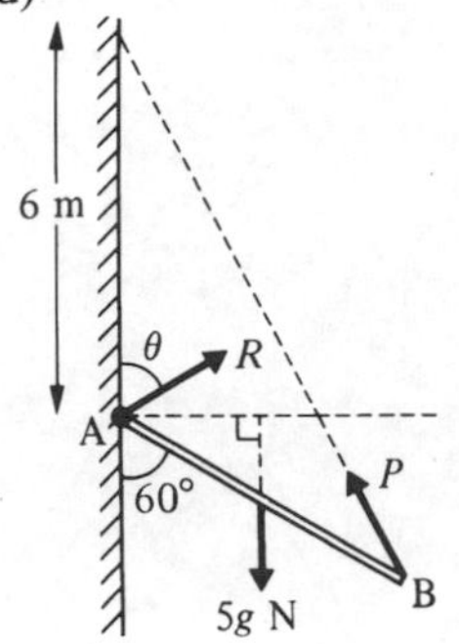

(e)

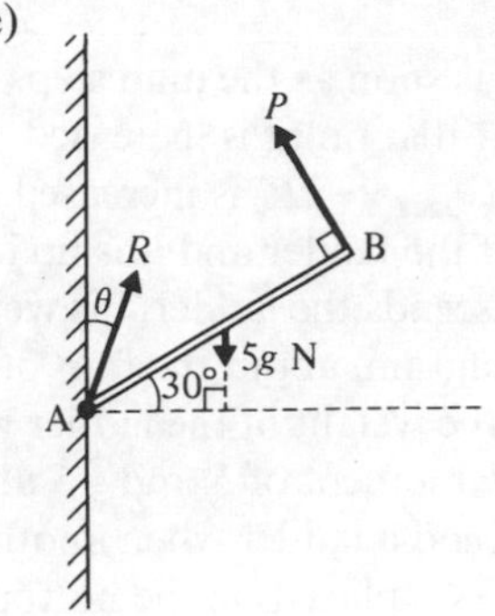

(f)

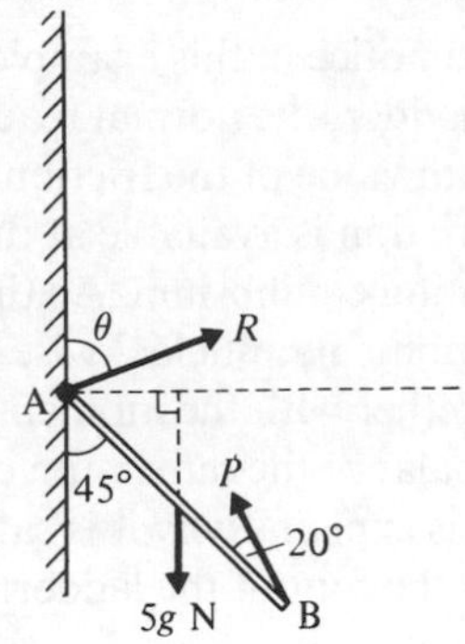

3. Each of the following diagrams shows a uniform rod AB, of mass 10 kg, with its lower end A resting on a rough horizontal floor, coefficient of friction μ. A string attached to end B keeps the rod in equilibrium. T is the tension in this string, F is the frictional force at A and R is the normal reaction at A.
Find the magnitude of T, F and R and the least possible value of μ for each situation below.

(a)

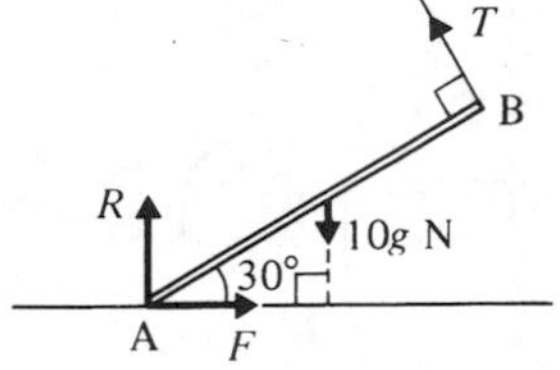

(b)

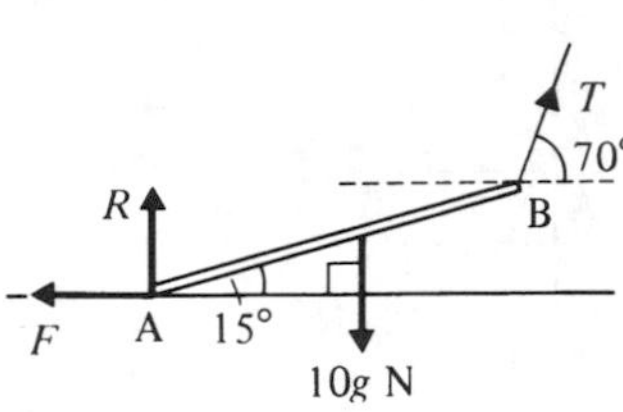

(c)

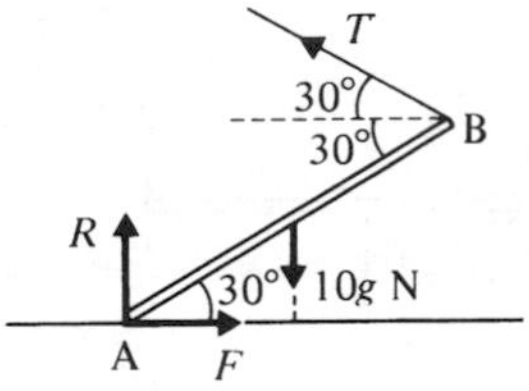

4. Each of the following diagrams shows a uniform ladder AB of mass 30 kg and length 6 m resting with its end A against a smooth vertical wall and end B on a smooth horizontal floor. The ladder is kept in equilibrium by a light horizontal string which has one end attached to B and the other end attached to the wall. R is the normal reaction at the floor, S is the normal reaction at the wall and T is the tension in the string.
Find the magnitude of T for each situation below.

(a)

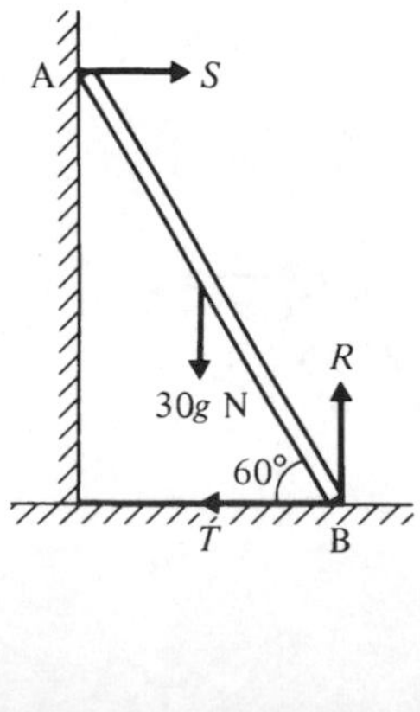

(b)

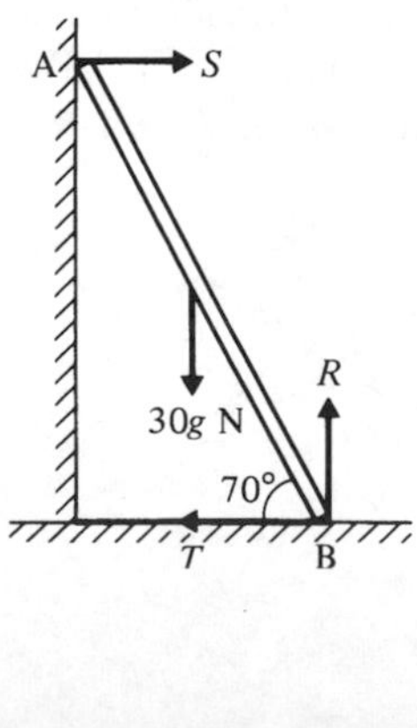

(c)

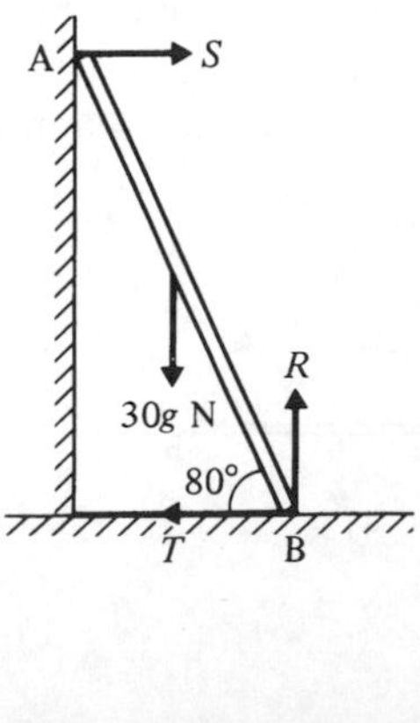

5. Each of the following diagrams shows a uniform ladder of mass 20 kg and length $2l$ resting in equilibrium with its upper end against a smooth vertical wall and its lower end on a rough horizontal floor, coefficient of friction μ. S is the normal reaction at the wall, F is the frictional force at the ground and R is the normal reaction at the ground.
Find the magnitude of S, F and R, and the least possible value of μ for each situation below.

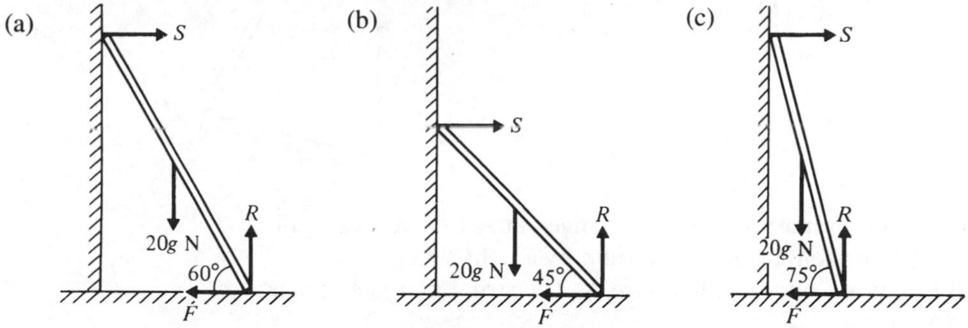

6. The diagram shows a uniform ladder AB of weight W N and length 4 m resting with its end A against a smooth vertical wall and its end B on a smooth horizontal floor. The ladder is kept in equilibrium at an angle θ to the floor by a light horizontal string attached to the wall and to a point C on the ladder.
If $\tan\theta = 2$, find the tension in the string when BC is of length:
(a) 1 m (b) 2 m (c) 3 m.

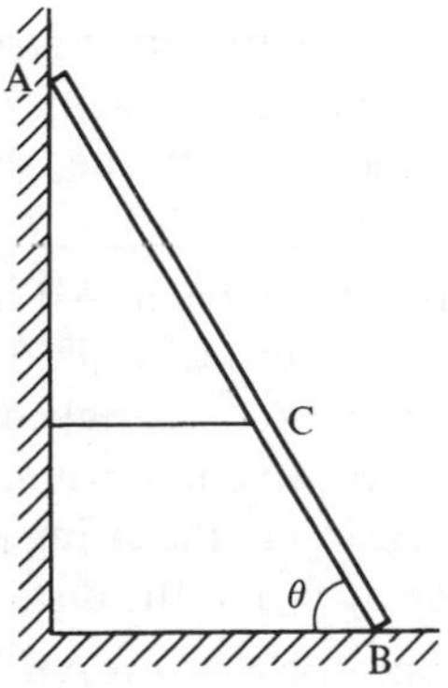

7. The diagram shows a uniform ladder of mass m and length $2l$ resting in limiting equilibrium with its upper end against a rough vertical wall (coefficient of friction μ_1) and its lower end against a rough horizontal floor (coefficient of friction μ_2).
The normal reactions at the wall and the floor are S and R respectively with $\mu_1 S$ and $\mu_2 R$ the corresponding frictional forces.
The ladder makes an angle θ with the floor.
Find θ for each of the following cases:
(a) $\mu_1 = \frac{1}{5}$ and $\mu_2 = \frac{1}{5}$
(b) $\mu_1 = \frac{1}{5}$ and $\mu_2 = \frac{1}{3}$
(c) $\mu_1 = \frac{1}{3}$ and $\mu_2 = \frac{1}{3}$.

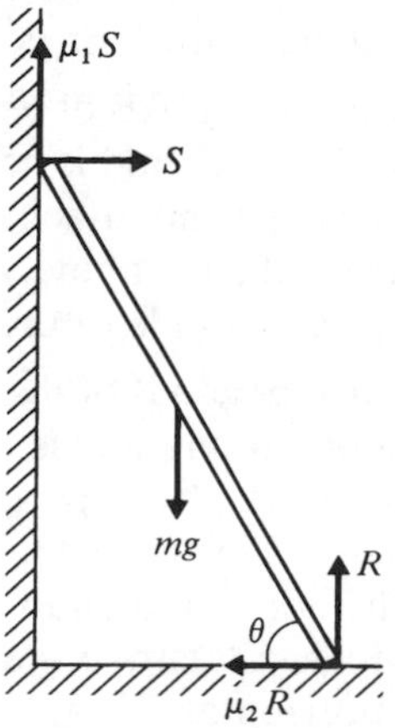

8. The diagram shows a uniform ladder AB of weight W and length $2l$ resting in equilibrium with its upper end A against a smooth vertical wall and its lower end B on a smooth inclined plane. The inclined plane makes an angle θ with the horizontal and the ladder makes an angle ϕ with the wall.
Find ϕ when θ equals:
(a) 10° (b) 20° (c) 30°.

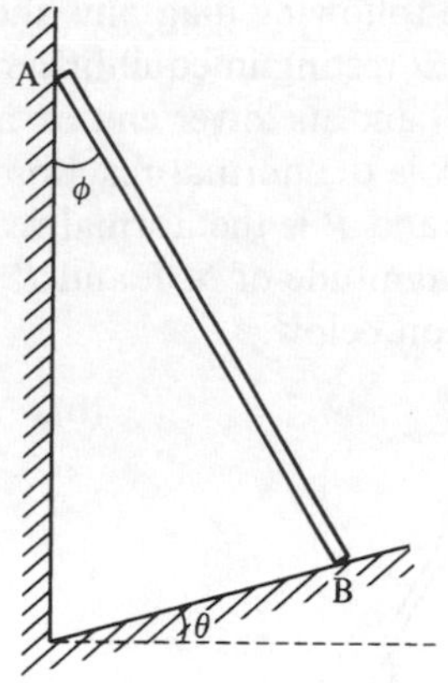

9. A uniform beam AB of mass 5 kg is freely hinged at A to a vertical wall and is maintained horizontally in equilibrium by a light string connecting B to a point on the wall, above A. The string makes an angle of 30° with BA.
Find the tension in the string and the magnitude and direction of the reaction at the hinge.

10. A non-uniform beam of mass 5 kg rests horizontally in equilibrium, supported by two light strings attached to the ends of the beam. The tensions in the strings are T_1 and T_2 and the strings make angles of 30° and 40° with the beam, as shown in the diagram.
Find the magnitudes of T_1 and T_2.

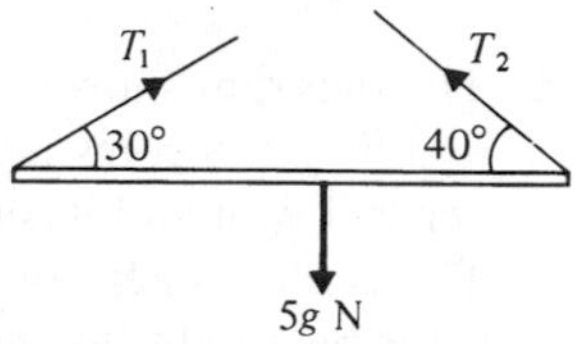

11. A non-uniform beam AB of weight 20 N and length 4 m, has end A freely hinged to a vertical wall. A light string linking B to a point on the wall above A, makes an angle of 60° with BA and allows the beam to rest horizontally in equilibrium.
If the tension in the string is 12 N, find the magnitude and direction of the reaction at A and the distance from A to the centre of gravity of the beam.

12. A non-uniform beam AB is of length 8 m and its weight of 10 N acts from a point G between A and B such that AG = 6 m. The beam is supported horizontally by strings attached at A and B. The string attached to A makes an angle of 30° with AB.
Find the angle that the string attached to B makes with BA, and find the tensions in the strings.

13. A uniform beam AB of length 4 m and weight 50 N is freely hinged at A to a vertical wall and is held horizontally, in equilibrium, by a string which has one end attached to B and the other end attached to a point C on the wall, 4 m above A.
Find the magnitude of the reaction at A.

14. A uniform pole AB of mass 100 kg has its lower end A on rough horizontal ground and is being raised into a vertical position by a rope attached to B. The rope and the pole lie in the same vertical plane and A does not slip across the ground.
Find the horizontal and vertical components of the reaction at the ground when the rope is at right angles to the pole and the pole is at 20° to the horizontal.

15. A non-uniform pole AB of mass 50 kg has its centre of gravity at the point of trisection of its length, nearer to B. The pole has its lower end A on rough horizontal ground and is being raised into a vertical position by a rope attached to B. The rope and the pole lie in the same vertical plane and A does not slip across the ground.
Find the horizontal and vertical components of the reaction at the ground when the rope is at right angles to the pole and the pole is at 30° to the horizontal.

16. A uniform beam AB of length $2l$ rests with end A in contact with rough horizontal ground. A point C on the beam rests against a smooth support. AC is of length $\frac{3l}{2}$ with C higher than A, and AC makes an angle of 60° with the horizontal. If the beam is in limiting equilibrium, find the coefficient of friction between the beam and the ground.

17. A uniform ladder of weight W and length $2l$ rests with one end on a smooth horizontal floor and the other end against a smooth vertical wall. The ladder is held in this position by a light, horizontal, inextensible string of length l, which has one end attached to the bottom of the ladder and the other end fastened to a point at the base of the wall, vertically below the top of the ladder. Show that the tension in the string is $\frac{W}{2\sqrt{3}}$.

18. A uniform ladder of mass 8 kg rests in equilibrium with its base on a smooth horizontal floor and its top against a smooth vertical wall. The base of the ladder is 1 m from the wall and the top of the ladder is 2 m from the floor. The ladder is kept in equilibrium by a light string attached to the base of the ladder and to a point on the wall, vertically below the top of the ladder and 1 m above the floor.
Find the tension in the string.

19. A uniform ladder of mass 25 kg rests in equilibrium with its base on a rough horizontal floor and its top against a smooth vertical wall. If the ladder makes an angle of 75° with the horizontal, find the magnitude of the normal reaction and of the frictional force at the floor, and state the minimum possible value of the coefficient of friction μ between the ladder and the floor.

20. A uniform ladder of mass 15 kg rests with its foot on a rough horizontal floor (angle of friction 15°) and its top against a smooth vertical wall. Find the minimum horizontal force that must be applied to the foot of the ladder to keep the ladder in equilibrium inclined at 60° to the horizontal.

21. A uniform ladder rests in limiting equilibrium with its base on rough horizontal ground (coefficient of friction μ) and its top against a rough vertical wall (coefficient of friction $\frac{1}{4}$).
If the ladder is inclined at 30° to the vertical, find the value of μ.

22. A non-uniform ladder AB of length 6 m has its centre of gravity at a point C on the ladder such that AC = 4 m. The ladder rests in limiting equilibrium with end A on rough horizontal ground (coefficient of friction $\frac{1}{3}$) and end B against a rough vertical wall (coefficient of friction $\frac{1}{4}$). If the ladder makes an acute angle θ with the ground, show that $\tan\theta = \frac{23}{12}$.

23. A non-uniform ladder AB of length 10 m has its centre of gravity at a point C. The ladder rests in limiting equilibrium with end A on a rough horizontal floor (angle of friction 17°) and end B against a smooth vertical wall.
If the ladder is inclined at an angle of 63° to the floor, find the length AC.

24. Each of the following diagrams shows a uniform rod AB of weight W, with end A freely hinged to a vertical wall. The rod is in equilibrium under the forces shown.

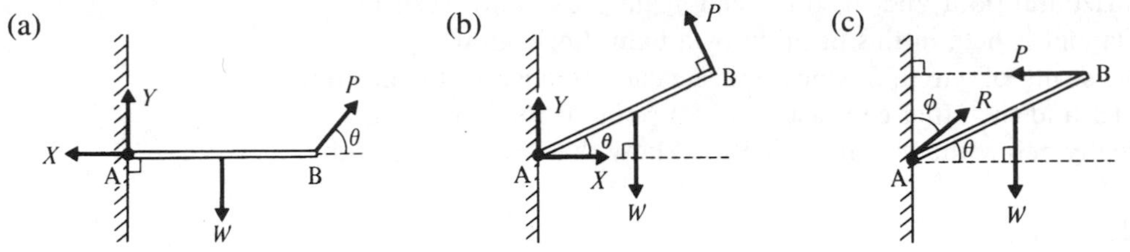

Prove: $X \tan\theta = \dfrac{W}{2}$ Prove: $2Y = W(1 + \sin^2\theta)$ Prove: $\tan\theta \tan\phi = \frac{1}{2}$

25. Each of the following diagrams shows a uniform ladder AB of weight W, with its lower end on a horizontal floor and its top against a smooth vertical wall. The ladder is in equilibrium under the forces shown.

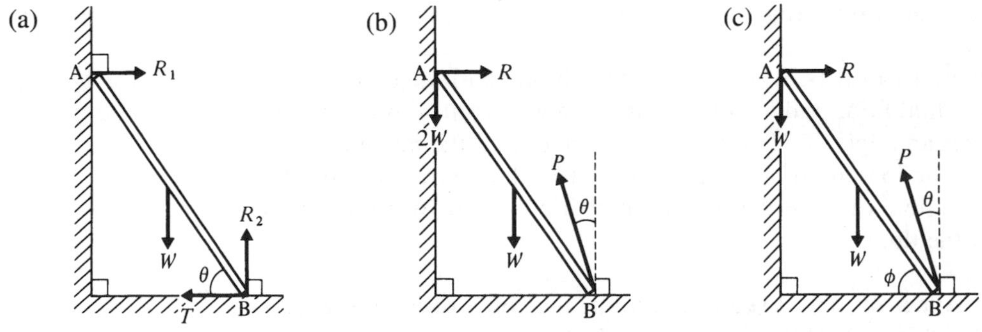

Prove: $2T \tan\theta = W$ Prove: $R = 3W \tan\theta$ Prove: (a) $R = 2W \tan\theta$
(b) $\tan\theta \tan\phi = \frac{3}{4}$

26. A uniform ladder rests in limiting equilibrium with its top end against a smooth vertical wall and its base on a rough horizontal floor (coefficient of friction μ).
If the ladder makes an angle of θ with the floor, prove that $2\mu \tan\theta = 1$.

27. A uniform ladder rests in limiting equilibrium with its top end against a rough vertical wall (coefficient of friction μ_1), and its base on a rough horizontal floor (coefficient of friction μ_2).
If the ladder makes an angle of θ with the floor, prove that $\tan\theta = \dfrac{1-\mu_1\mu_2}{2\mu_2}$.

28. The diagram shows a uniform ladder resting in equilibrium with its top end against a smooth vertical wall and its base on a smooth inclined plane. The plane makes an angle of θ with the horizontal and the ladder makes an angle of ϕ with the wall.
Prove that $\tan\phi - 2\tan\theta$.

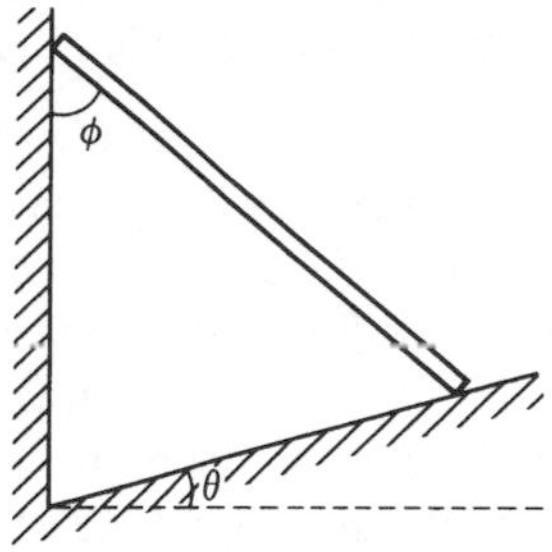

29. A uniform beam AB is supported at an angle θ to the horizontal by a light string attached to end B, and with end A resting on rough horizontal ground (angle of friction λ). The beam and the string lie in the same vertical plane and the beam rests in limiting equilibrium with the string at right angles to the beam.
Prove that $\tan\lambda = \dfrac{\sin 2\theta}{3-\cos 2\theta}$.

30. A uniform ladder of mass 30 kg is placed with its base on a rough horizontal floor (coefficient of friction $\frac{1}{4}$), and its top against a smooth vertical wall, with the ladder making an angle of 60° with the floor.
Find the magnitude of the minimum horizontal force that must be applied at the base of the ladder in order to prevent slipping.
What is the maximum horizontal force that could be applied at the base without slipping occurring?

31. A uniform ladder of mass 25 kg is placed with its base on a rough horizontal floor (coefficient of friction $\frac{1}{5}$), and its top against a rough vertical wall (coefficient of friction $\frac{1}{3}$), with the ladder making an angle of 61° with the floor.
Find the magnitude of the minimum horizontal force that must be applied at the base of the ladder in order to prevent slipping.
What is the maximum horizontal force that could be applied at the base without slipping occurring?

32. A uniform ladder of mass 10 kg and length 4 m rests with one end on a smooth horizontal floor and the other end against a smooth vertical wall. The ladder is kept in equilibrium, at an angle $\tan^{-1} 2$ to the horizontal, by a light horizontal string attached to the base of the ladder and to the base of the wall, at a point vertically below the top of the ladder. A man of mass 100 kg ascends the ladder.
If the string will break when the tension exceeds 490 N, find how far up the ladder the man can go before this occurs.
What tension must the string be capable of withstanding if the man is to reach the top of the ladder?

33. A uniform ladder of mass 30 kg and length 10 m has its base resting on rough horizontal ground and its top against a smooth vertical wall. The ladder rests in equilibrium, at 60° to the horizontal, with a man of mass 90 kg standing on the ladder at a point 7·5 m from its base. Find the magnitude of the normal reaction and of the frictional force at the ground. Find the minimum value for the coefficient of friction between the ladder and the ground that would enable the man to climb to the top of the ladder.

34. A uniform ladder of length 10 metres and weight W N rests with its base on a rough horizontal floor (coefficient of friction $\frac{1}{3}$), and its top against a smooth vertical wall. The ladder makes an angle θ with the horizontal, where $\tan\theta = 1{\cdot}7$. A man of weight $2W$ N starts to climb the ladder. How far up the ladder can the man climb before slipping occurs?
Find, in terms of W, the magnitude of the least horizontal force that must be applied to the base of the ladder to enable the man to reach the top safely.

35. A uniform ladder AB is of weight $2W$ N and length 10 metres. It rests with end A on a rough horizontal floor and end B against a rough vertical wall. The coefficient of friction at the wall and at the floor is $\frac{1}{3}$ and the ladder makes an angle θ with the horizontal, such that $\tan\theta = \frac{16}{7}$. A man of weight $5W$ N starts to climb the ladder.
How far up the ladder can the man climb before slipping occurs?
When a boy of weight X N stands on the bottom rung of the ladder, i.e. at A, the man is just able to climb to the top safely.
Find X in terms of W.

36. A non-uniform ladder AB of length 12 m and mass 30 kg has its centre of gravity at the point of trisection of its length, nearer to A. The ladder rests with end A on rough horizontal ground (coefficient of friction $\frac{1}{4}$), and end B against a rough vertical wall (coefficient of friction $\frac{1}{5}$). The ladder makes an angle θ with the horizontal such that $\tan\theta = \frac{9}{4}$. A straight horizontal string connects A to a point at the base of the wall, vertically below B. A man of mass 90 kg begins to climb the ladder.
How far up the ladder can he go without causing tension in the string?
What tension must the string be capable of withstanding if the man is to reach the top of the ladder safely?

Exercise 9B Examination questions

(Take $g = 9{\cdot}8\ \text{m s}^{-2}$ throughout this exercise.)

1. A uniform ladder of length $2l$ and mass m kg, rests with one end against a smooth vertical wall and the other end on horizontal ground. The ladder is inclined at θ to the vertical.

(i) Explain why the ladder will slip if the ground is smooth.
The bottom of the ladder is attached to a horizontal rope, the other end of the rope is attached to the wall as shown in the diagram.

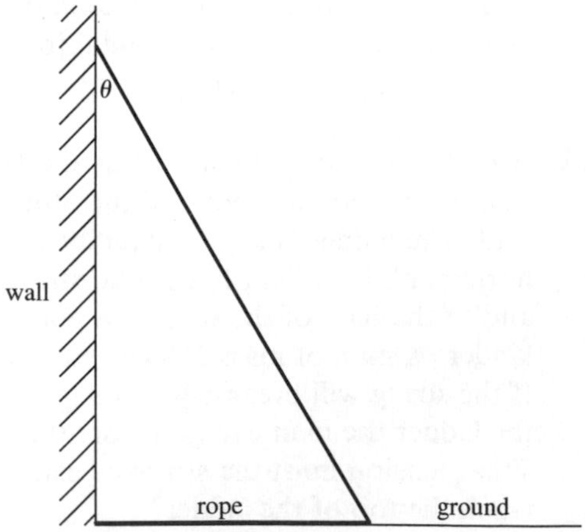

(ii) A man of mass $4m$ kg stands on the ladder at a distance of $\frac{l}{2}$ from the bottom of the ladder.

Find first the tension in the rope, and then the reaction normal to the ground at the bottom of the ladder, and the reaction normal to the wall at the top of the ladder.

(iii) If the maximum tension which the rope can bear without breaking is $4mg\tan\theta$, find how far up the ladder the man can safely climb. (NICCEA)

2. A uniform ladder of length 5 m and weight 80 newtons stands on rough level ground and rests in equilibrium against a smooth horizontal rail which is fixed 4 m vertically above the ground. If the inclination of the ladder to the *vertical* is θ, where $\tan\theta \leqslant \frac{3}{4}$, find expressions in terms of θ for the vertical reaction R of the ground, the friction F at the ground and the normal reaction N at the rail.
Given that the ladder does not slip, show that F is a maximum when $\tan\theta = \frac{1}{\sqrt{2}}$ and give this maximum value. (OCSEB)

3. The diagram shows a uniform rod AB, of length $2a$, in equilibrium in a vertical plane with the end A in contact with a vertical wall and the end B in contact with a horizontal floor. The normal reaction and frictional force at B are R and F respectively, acting in the directions shown. The corresponding forces at A, in the directions shown, are denoted by S and P, respectively. The rod is of weight W and is inclined at an angle θ to the horizontal.

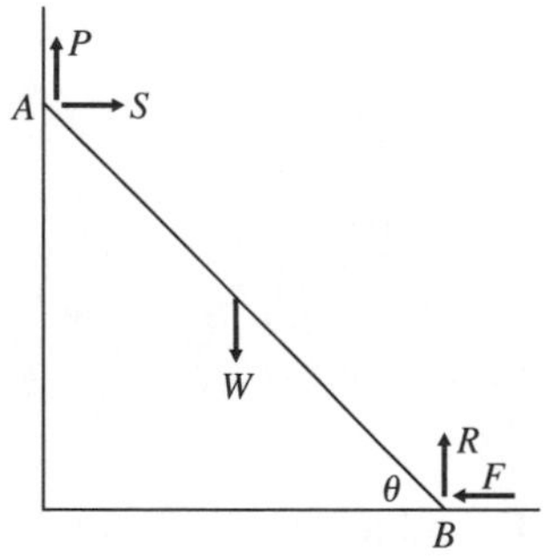

By resolving horizontally and taking moments about the centre of AB, or otherwise, express $R - P$ in terms of S and θ.

Also obtain an equation relating R, P and W and show that

$$R = \tfrac{1}{2}(W + 2S\tan\theta), \quad P = \tfrac{1}{2}(W - 2S\tan\theta).$$

(a) Find, for the case when the wall is smooth and the coefficient of friction at B is 1, the value of $\tan\theta$ for limiting equilibrium.
(b) Show that, when the wall and floor are equally rough, the coefficient of friction being 1, limiting equilibrium is not possible for $\theta \neq 0$. (AEB 1992)

4.

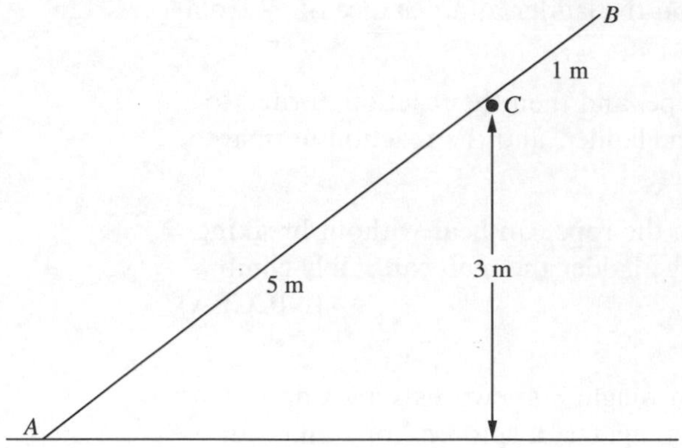

A smooth horizontal rail is fixed at a height of 3 m above a horizontal playground whose surface is rough. A straight uniform pole *AB*, of mass 20 kg and length 6 m, is placed to rest at a point *C* on the rail with the end *A* on the playground. The vertical plane containing the pole is at right angles to the rail. The distance *AC* is 5 m and the pole rests in limiting equilibrium (see diagram).
Calculate:

(a) the magnitude of the force exerted by the rail on the pole, giving your answer to the nearest N,

(b) the coefficient of friction between the pole and the playground, giving your answer to 2 decimal places,

(c) the magnitude of the force exerted by the playground on the pole, giving your answer to the nearest N. (ULEAC)

5.

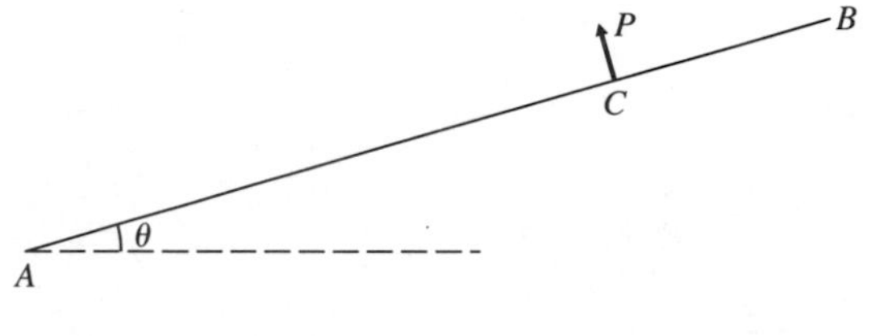

The diagram shows a uniform rod *AB* of weight W and length $2a$ freely hinged to a fixed point at *A*. When a weight $2W$ is attached at *B*, the rod is kept in equilibrium by a force of magnitude P acting at *C* perpendicular to *AB*, in the vertical plane containing *AB*. In the equilibrium position *AB* is inclined at angle $\theta(>0)$ to the horizontal and $AC = b\,(a < b < 2a)$. By taking moments about *A*, or otherwise, find *P*. Find also

(i) the vertical and horizontal components of the reaction at *A*,

(ii) the maximum value of the horizontal component of the reaction at *A* as θ varies,

(iii) the value of a/b when the reaction at *A* is along the rod,

(iv) $\cos^2\theta$ when the reaction at *A* acts at an angle θ to the upward vertical. (WJEC)

6. A uniform diving board AB, of length 4 metres and mass 40 kg, is fixed at A to a vertical wall and is maintained in a horizontal position by means of a light strut DC. D is a point on the wall 1 metre below A and C is a point on the board where AC = 1 metre. An object of mass 60 kg is placed at end B.
  (i) Draw a neat and clearly labelled diagram of the board showing all the forces acting on it.
  (ii) Find the position of the centre of mass of the 60 kg mass and the mass of the board AB combined.
  (iii) Using a triangle of forces, or otherwise find
    (a) the thrust in the strut
    (b) the magnitude of the reaction at A. (NICCEA)

7. A uniform rod AC, of mass 2 kg and length 120 cm, hangs at rest in a vertical plane with end A in contact with a vertical wall. An inelastic string, of length 70 cm, is attached to a point B on AC such that AB is 90 cm. The other end of the string is attached to the wall at a point D, 140 cm vertically above A.

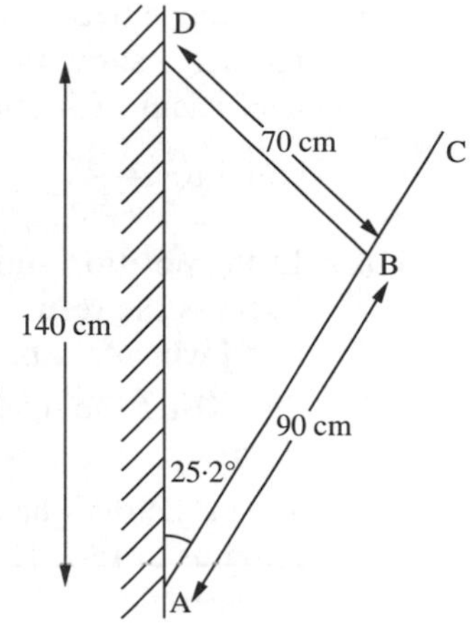

If the string is taut and the angle $D\hat{A}C$ is 25·2° find
  (i) angle $D\hat{B}A$,
  (ii) the tension in the string. (NICCEA)

8. The diagram shows a uniform ladder *AB*, of length 2*a* and mass *m*, with the end *A* resting on a rough horizontal floor.

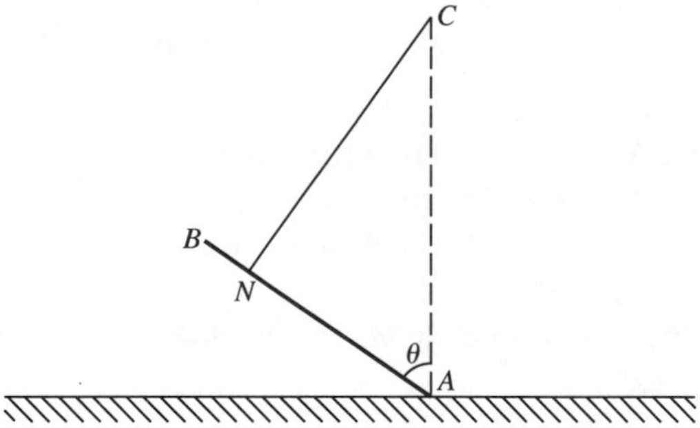

The ladder is held at an angle θ to the vertical by means of a light rope attached to the point *N*, where $AN = \frac{3}{2}a$. The other end of the rope is attached to a point C, which is at a height 3*a* vertically above the end *A* of the ladder. By taking moments about *C* show that the magnitude of the force of friction acting on the ladder at *A* is $\frac{mg}{3}\sin\theta$. Also show that the magnitude of the vertical component of the reaction at *A* is $\frac{mg}{3}(1+\cos\theta)$.
Given that the coefficient of friction between the ladder and the floor is $\frac{1}{\sqrt{3}}$, show that when the ladder is on the point of slipping at *A* its inclination to the vertical is given by $\theta = \frac{\pi}{3}$. (AEB 1993)

9. The diagram shows a man of mass 70 kg at rest while abseiling down a vertical cliff. Assume that the rope is attached to the man at his centre of mass. *You should model the man as a rod and assume that he is not holding the rope.*

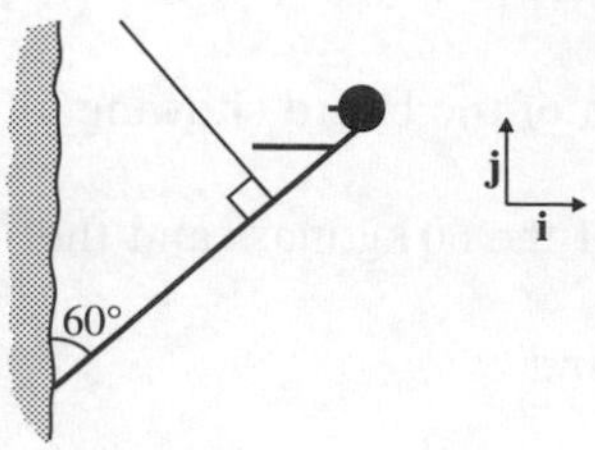

(a) Draw a diagram to show the forces acting on the man.

(b) The angle between the man's legs and the cliff is 60°. By taking moments show that the man can only remain in this position if the coefficient of friction between his feet and the wall is greater than or equal to $\frac{1}{\sqrt{3}}$.

(c) In the position shown above, the rope is at 90° to the man's body. Express the resultant force on the man in terms of the unit vectors **i** and **j** which act horizontally and vertically respectively.

(d) Show that the tension in the rope is 594 N correct to 3 significant figures.

(e) When the man has descended a further distance he again stops and remains at rest. How would the magnitude of the Normal Reaction between the man and the wall now compare with its value in the position illustrated above? (AEB Spec)

10. A smooth uniform rod AB, of length $3a$ and weight $2w$ is pivoted at A so that it can rotate in a vertical plane. A light ring is free to slide over the rod. A light inextensible string is attached to the ring and passes over a fixed smooth peg at a point C, a height $4a$ above A, and carries a particle of weight w hanging freely, as illustrated.
(*You may assume that the weight of the rod is a force acting at the mid-point of AB.*)

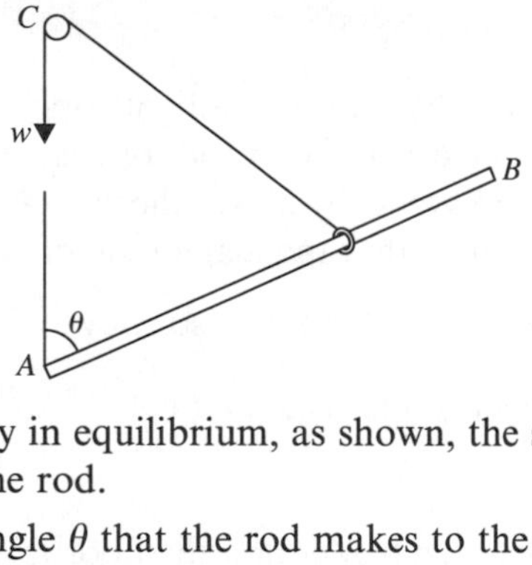

(a) Give reasons why in equilibrium, as shown, the string will be at right angles to the rod.

(b) Show that the angle θ that the rod makes to the vertical in equilibrium is given by $\tan\theta = \frac{4}{3}$

(c) Find the magnitude of the force of the pivot on the rod at A in terms of w. (UODLE)

11. A smooth uniform rod AB, of length $3a$ and weight W, is pivoted at A so that it can rotate in a vertical plane. A light ring is free to slide along the rod. A light inextensible string is attached to the ring and passes over a fixed smooth peg at a point C, a height $4a$ above A, and carries a particle of weight w hanging freely, as illustrated.

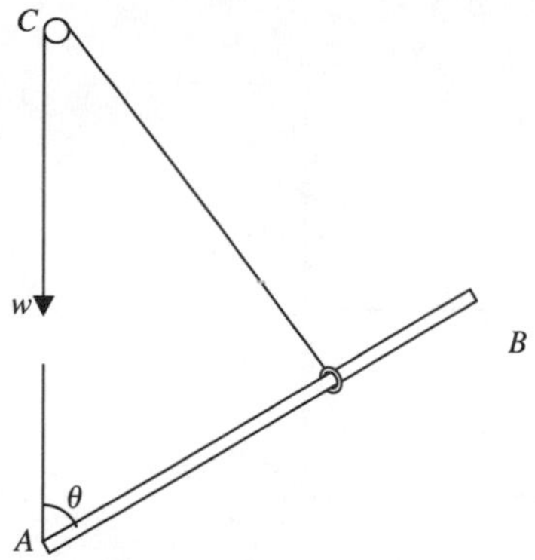

(a) Give reasons why in equilibrium, as shown, the string will be at right angles to the rod.

(b) (i) Show that the angle θ that the rod makes to the vertical in equilibrium is given by

$$\tan\theta = \frac{8w}{3W}.$$

(ii) Find the smallest value of the ratio w/W for which equilibrium is possible. (UODLE)

12. The diagram shows a heavy uniform rod of mass m and length $2a$ resting in equilibrium with its ends on two smooth inclined planes. The normal reactions at the ends of the rod have magnitudes R and S. The inclinations of the planes to the horizontal are $\frac{\pi}{6}$ and $\frac{\pi}{4}$ and the rod lies in a vertical plane containing lines of greatest slope of both planes.

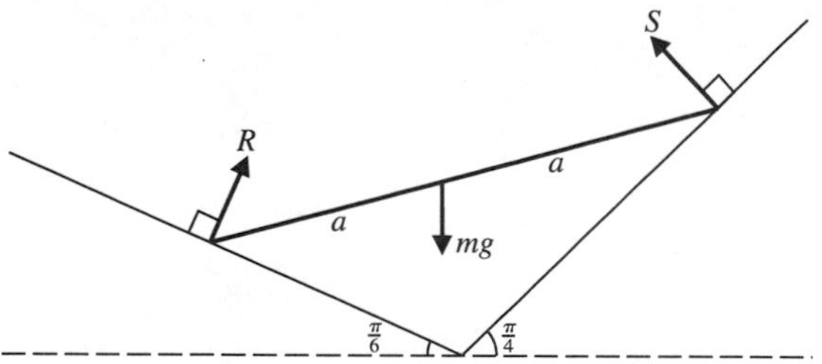

(a) Show that $R = S\sqrt{2}$.

(b)* By taking moments about the centre of the rod, prove that the inclination of the rod to the horizontal is $\cot^{-1}(1 + \sqrt{3})$.

(AEB 1993)

***Note.** The following expansions may be useful for (b):
$\cos(A + B) = \cos A \cos B - \sin A \sin B$
$\cos(A - B) = \cos A \cos B + \sin A \sin B$

13. Prove, by integration, that the centre of mass of a uniform hemispherical shell of radius a is a distance $\frac{1}{2}a$ from the centre of its plane face.
The figure shows a uniform hemispherical shell of mass m resting with its curved surface in contact with a rough horizontal floor and a rough vertical wall.

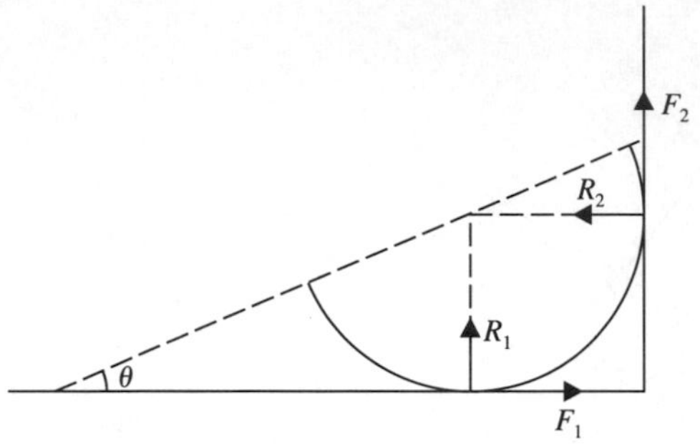

The reactions between the shell and the floor and wall have components R_1, F_1, R_2 and F_2 as indicated. Given that the coefficient of friction at both points of contact is μ and that equilibrium is limiting at both points of contact, show that

$$R_1 = \frac{mg}{1+\mu^2}.$$

Find, in terms of μ, an expression for $\sin\theta$. (AEB 1992)

10 Resultant velocity and relative velocity

Resultant velocity

Velocity is a vector quantity since it has both magnitude and direction. Two velocities can therefore be combined by the same method as used for forces in Chapter 4.

Example 1

Find, in vector form, the resultant of the following velocities: $(4\mathbf{i} - 2\mathbf{j})\,\text{m s}^{-1}$, $(-7\mathbf{i} + 5\mathbf{j})\,\text{m s}^{-1}$ and $(8\mathbf{i} - 6\mathbf{j})\,\text{m s}^{-1}$.

$$\begin{aligned}\text{Resultant velocity} &= (4\mathbf{i} - 2\mathbf{j}) + (-7\mathbf{i} + 5\mathbf{j}) + (8\mathbf{i} - 6\mathbf{j})\\ &= (5\mathbf{i} - 3\mathbf{j})\,\text{m s}^{-1}\end{aligned}$$

The resultant velocity is $(5\mathbf{i} - 3\mathbf{j})\,\text{m s}^{-1}$.

The following example shows how the magnitude and direction of the resultant velocity may be found by a scale drawing.

Example 2

Find, by scale drawing, the magnitude and direction of the resultant of the velocities $16\,\text{m s}^{-1}$ due east and $10\,\text{m s}^{-1}$ in a direction N 38° E.
Draw a rough sketch showing the given velocities.

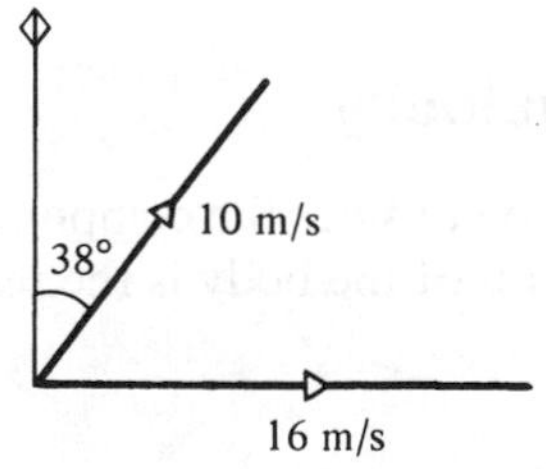

Using a scale of $1\,\text{cm} \equiv 2\,\text{m s}^{-1}$, construct a vector triangle ABC for the given velocities.

$$AB = 8\,\text{cm} \qquad BC = 5\,\text{cm}$$
$$\text{angle } ABC = 128°$$

The resultant velocity is represented by $\overrightarrow{AC}$.

By measurement:

$$AC = 11{\cdot}75\,\text{cm}$$
$$\text{angle } \alpha = 20°$$

The resultant velocity is $23{\cdot}5\,\text{m s}^{-1}$ in a direction N 70° E.

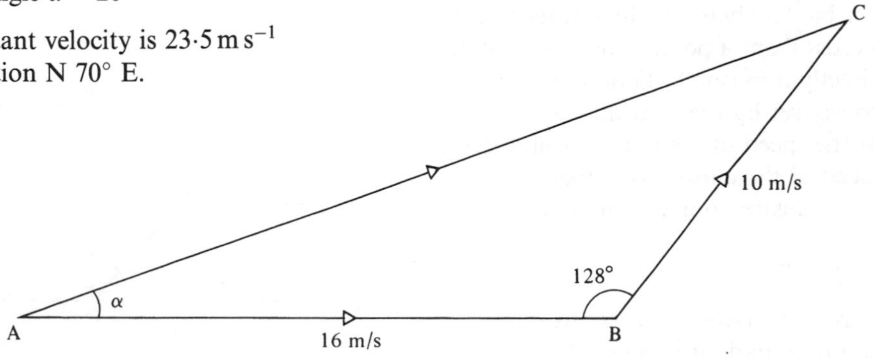

Example 3

Calculate the magnitude and the direction of the resultant of the velocities $8\,\text{km}\,\text{h}^{-1}$ in a direction N 80° W and $5\,\text{km}\,\text{h}^{-1}$ in a direction S 25° W.

Draw a rough sketch showing the given velocities.

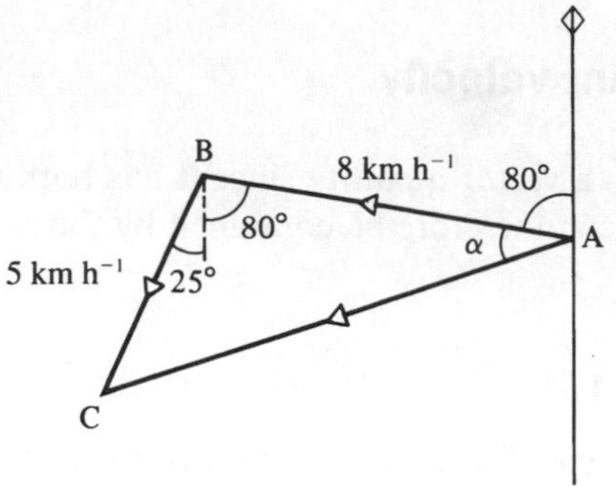

angle ABC = 105°.
Let $B\hat{A}C = \alpha$

The resultant velocity is represented by $\overrightarrow{AC}$.
From triangle ABC, by the cosine rule:

$$AC^2 = 8^2 + 5^2 - 2(5)(8)\cos 105°$$
$$= 89 + 80 \cos 75°$$
$$\therefore \quad AC = 10{\cdot}47$$

From triangle ABC, by the sine rule:

$$\frac{AC}{\sin 105°} = \frac{5}{\sin \alpha}$$

Substituting for AC gives:

$$\sin \alpha = \frac{5 \sin 105°}{10{\cdot}47}$$

$\therefore \quad \alpha = 27{\cdot}47°$ or $152{\cdot}53°$ (The obtuse angle is not applicable.)

The resultant velocity is $10{\cdot}5\,\text{km}\,\text{h}^{-1}$ in a direction S 72·53° W.

Components of velocity

It is sometimes useful to consider the components of the velocity of a body, particularly if the motion of the body is the result of the combination of two velocities.

Crossing a river by boat

Consider the problem of crossing from a point on one bank of a river to a point on the other bank. There are three cases to consider.

(i) In order to cross from a point O on one bank to a point P directly opposite to O on the other bank, the course set by the boat must be upstream. If the speed of the boat in still water is v and the speed of the current is u, then the component of v upstream must counteract u.

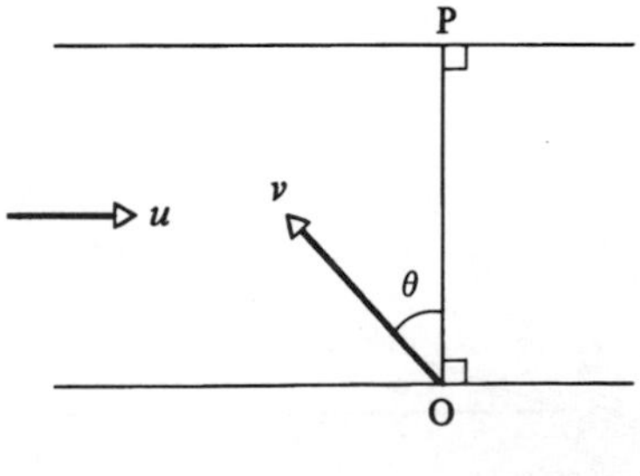

$$\therefore \quad v \sin \theta = u$$

The speed across the river is then $v \cos \theta$ and the crossing is made from O to P.

(ii) If the course set is directly across the river, then the current will carry the boat downstream. The boat has two velocities, **v** the velocity in still water, and **u** the velocity of the current downstream. The resultant velocity **V** of the boat can be found from the vector triangle. If the magnitude of **V** is written as V, then:

$$V^2 = v^2 + u^2$$

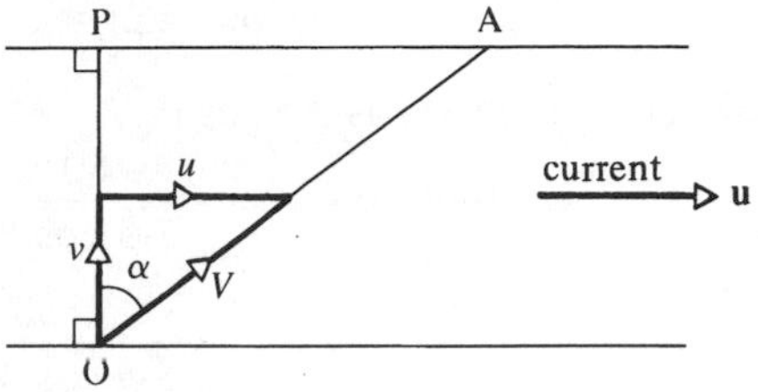

and the boat will travel at an angle α to the line OP where:

$$\tan \alpha = \frac{u}{v}$$

The time to cross the river is $t = \dfrac{\text{OP}}{v}$ and the boat is carried downstream a distance PA, where $\text{PA} = u \times t$.
This will be the quickest crossing.

(iii) In order to cross the river and reach a point B on the other bank, the course set must be in such a direction that the resultant velocity of the boat is in the direction OB.

Suppose B is upstream.

From the diagram:

$$\mathbf{V} = \mathbf{v} + \mathbf{u}$$

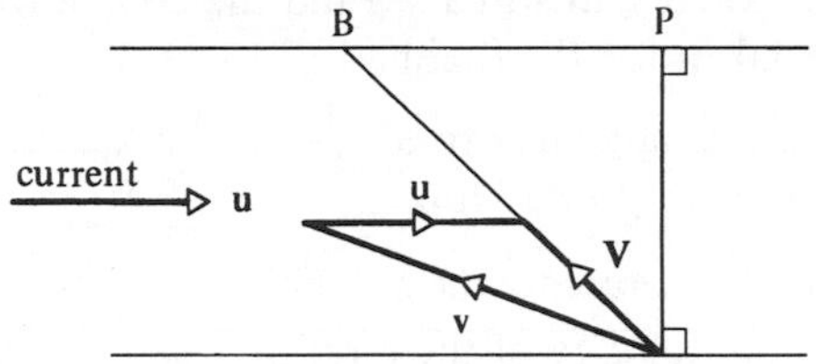

Example 4

A boat can travel at $3{\cdot}5\,\text{m}\,\text{s}^{-1}$ in still water. A river is 80 m wide and the current flows at $2\,\text{m}\,\text{s}^{-1}$. Calculate:

(a) the shortest time taken to cross the river and the distance downstream that the boat is carried

(b) the course that must be set to cross the river to a point exactly opposite the starting point and the time taken for the crossing.

(a) To cross in the shortest time, the course set is directly across the river:

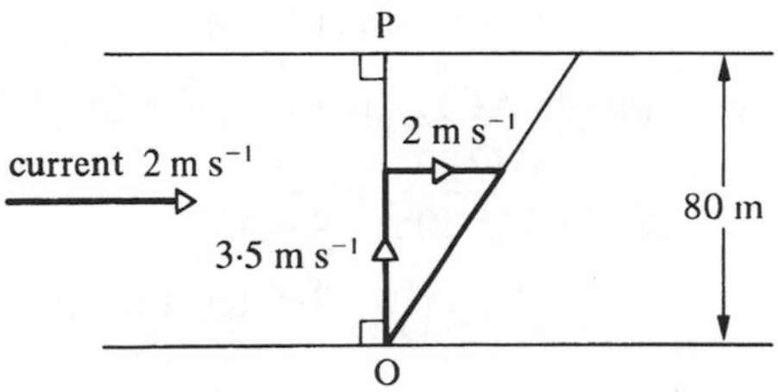

$$\text{time to cross} = \frac{\text{OP}}{3{\cdot}5} = \frac{80}{3{\cdot}5} = 22{\cdot}86\,\text{s}$$

$$\text{distance downstream} = 2 \times 22{\cdot}86 = 45{\cdot}72\,\text{m}$$

The time for the quickest crossing is 22·9 s and the distance downstream is 45·7 m.

(b) To cross the river directly, the course is set at an angle θ, where:

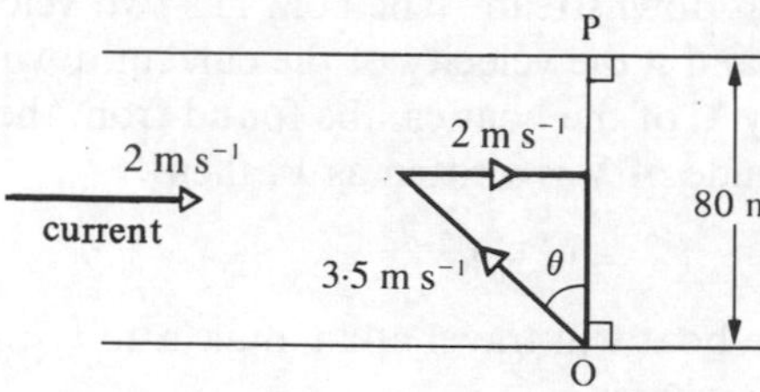

$$3{\cdot}5 \sin\theta = 2$$

$$\therefore \quad \sin\theta = \frac{4}{7}$$

i.e. $\theta = 34{\cdot}85°$

Speed across the river is $3{\cdot}5 \cos\theta$

$$\text{time to cross} = \frac{80}{3{\cdot}5\cos\theta}$$

$$= \frac{80}{3{\cdot}5\cos 34{\cdot}85°} = 27{\cdot}85\,\text{s}$$

The time taken to cross the river directly is 27·9 s and the course to be set is upstream at an angle 55·15° to the bank of the river.

Example 5

A pilot has to fly his aircraft from the point A to the point B, where B is 600 km due east of A. There is a wind of 80 km h^{-1} blowing from the north-west and the aircraft flies at 350 km h^{-1} in still air. Find the course which the pilot must set and the time taken for the flight.

This is similar to the oarsman crossing a river to a previously determined point on the opposite bank.

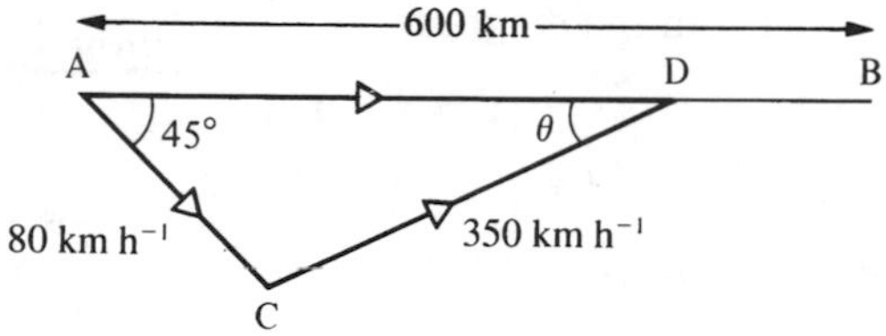

Draw a rough sketch in which $\overrightarrow{AC}$ represents the velocity of the wind and $\overrightarrow{CD}$ represents the velocity of the aircraft in still air.

The resultant $\overrightarrow{AD}$ of these two velocities must lie along the line AB since this is the line along which the aircraft is to travel. The line CD gives the direction of the course to be set by the pilot.

From triangle ACD, by the sine rule:

$$\frac{350}{\sin 45°} = \frac{80}{\sin\theta}$$

$$\therefore \quad \sin\theta = \frac{80 \sin 45°}{350}$$

$$\therefore \quad \theta = 9{\cdot}30° \text{ or } 170{\cdot}70° \text{ (The obtuse angle is not applicable.)}$$

$$\therefore \quad \text{angle ACD} = 180° - 45° - 9{\cdot}30° = 125{\cdot}70°$$

From triangle ACD, by the sine rule:

$$\frac{AD}{\sin 125{\cdot}70°} = \frac{350}{\sin 45°}$$

$$\therefore \quad AD = \frac{350 \sin 125{\cdot}70°}{\sin 45°}$$

$$\therefore \quad AD = 402{\cdot}0$$

$$\text{time of flight} = \frac{600}{402} = 1{\cdot}493\,\text{h}$$

The course to be set is 080·70° and the time taken for the flight is 1·49 h.

The modelling process

Some of the examples encountered so far in this chapter have involved such things as the flow of a river, the motion of a boat, the flight of an aircraft and the effect of wind. In each case the velocities involved were assumed to be constant. Once again this is an example of an assumption being made to allow a mathematical model to be set up and answers to be determined. Should we find that the answers supplied by our model do not agree with reality we would question the wisdom of assuming the velocities to be constant. Of course in reality, the situation is continually being reviewed. The pilot of an aircraft frequently checks the wind speed and the aircraft's position and the captain of a boat frequently monitors any changes in wind and water conditions.

Exercise 10A

1. Find in vector form the resultant of each of the following sets of velocities:
(a) $(5\mathbf{i}+2\mathbf{j})\,\mathrm{m\,s^{-1}}$, $(4\mathbf{i}-3\mathbf{j})\,\mathrm{m\,s^{-1}}$
(b) $(6\mathbf{i}+2\mathbf{j})\,\mathrm{m\,s^{-1}}$, $(2\mathbf{i}+3\mathbf{j})\,\mathrm{m\,s^{-1}}$, $(-\mathbf{i}+4\mathbf{j})\,\mathrm{m\,s^{-1}}$
(c) $(2\mathbf{i}-5\mathbf{j})\,\mathrm{m\,s^{-1}}$, $(3\mathbf{i}+7\mathbf{j})\,\mathrm{m\,s^{-1}}$, $(-6\mathbf{i}-8\mathbf{j})\,\mathrm{m\,s^{-1}}$
(d) $(18\mathbf{i}+9\mathbf{j})\,\mathrm{km\,h^{-1}}$, $(10\mathbf{i}+5\mathbf{j})\,\mathrm{m\,s^{-1}}$
(e) $(15\mathbf{i}-5\mathbf{j})\,\mathrm{m\,s^{-1}}$, $(15\mathbf{i}+3\mathbf{j})\,\mathrm{km\,h^{-1}}$, $(-6\mathbf{i}+15\mathbf{j})\,\mathrm{km\,h^{-1}}$.

2. Find by scale drawing the magnitude and direction of the resultant of each of the following pairs of velocities.
(a) $24\,\mathrm{m\,s^{-1}}$ due north, $7\,\mathrm{m\,s^{-1}}$ due east
(b) $5\,\mathrm{km\,h^{-1}}$ due north, $5\,\mathrm{km\,h^{-1}}$ N 60° E
(c) $5\,\mathrm{m\,s^{-1}}$ due north, $7\,\mathrm{m\,s^{-1}}$ S 60° E
(d) $10\,\mathrm{m\,s^{-1}}$ N 30° E, $8\,\mathrm{m\,s^{-1}}$ N 70° E
(e) $70\,\mathrm{km\,h^{-1}}$ S 35° E, $90\,\mathrm{km\,h^{-1}}$ N 25° E.

3. Find by calculation the magnitude and direction of the resultant of each of the following pairs of velocities.
(a) $9\,\mathrm{m\,s^{-1}}$ due west, $12\,\mathrm{m\,s^{-1}}$ due north
(b) $6\,\mathrm{m\,s^{-1}}$ due east, $4\,\mathrm{m\,s^{-1}}$ NW
(c) $17\,\mathrm{km\,h^{-1}}$ due north, $15\,\mathrm{km\,h^{-1}}$ N 26° E
(d) $10\,\mathrm{km\,h^{-1}}$ N 35° W, $15\,\mathrm{km\,h^{-1}}$ S 40° W
(e) $72\,\mathrm{km\,h^{-1}}$ N 65° E, $20\,\mathrm{m\,s^{-1}}$ SE.

4. If the resultant of $(3\mathbf{i}+4\mathbf{j})\,\mathrm{m\,s^{-1}}$ and $(a\mathbf{i}+b\mathbf{j})\,\mathrm{m\,s^{-1}}$ is $(7\mathbf{i}-\mathbf{j})\,\mathrm{m\,s^{-1}}$, find the values of a and b.

5. If the resultant of $(a\mathbf{i}+b\mathbf{j})\,\mathrm{km\,h^{-1}}$ and $(b\mathbf{i}-a\mathbf{j})\,\mathrm{km\,h^{-1}}$ is $(10\mathbf{i}-4\mathbf{j})\,\mathrm{km\,h^{-1}}$, find the values of a and b.

6. The resultant of two velocities is a velocity of $10\,\mathrm{km\,h^{-1}}$, N 30° W.
If one of the velocities is $10\,\mathrm{km\,h^{-1}}$ due west, find the magnitude and direction of the other velocity.

7. The resultant of two velocities is a velocity of $6\,\mathrm{m\,s^{-1}}$ due east.
If one of the velocities is $5\,\mathrm{m\,s^{-1}}$ N 30° W, find the magnitude and direction of the other velocity.

8. The resultant of two velocities is a velocity of $19\,\mathrm{m\,s^{-1}}$ S 60° E.
If one of the velocities is $10\,\mathrm{m\,s^{-1}}$ due east, find the magnitude and direction of the other velocity.

9. A man wishes to row across a river to reach a point on the far bank, exactly opposite his starting point. The river is 100 m wide and flows at $3\,\mathrm{m\,s^{-1}}$. In still water the man can row at $5\,\mathrm{m\,s^{-1}}$.
Find at what angle to the bank the man must steer the boat in order to complete the crossing, and the time it takes him.

10. A man wishes to row across a river to reach a point on the far bank, exactly opposite his starting point. The river is 125 m wide and flows at $1\,\mathrm{m\,s^{-1}}$.
If the man can row at $3\,\mathrm{m\,s^{-1}}$ in still water, find the direction the man must steer in order to complete the crossing, and the time it takes him.

11. A boy wishes to swim across a river, 100 m wide, as quickly as possible. The river flows at 3 km h^{-1} and the boy can swim at 4 km h^{-1} in still water.
Find the time that it takes the boy to cross the river and how far downstream he travels.

12. A man who can swim at 2 m s^{-1} in still water, wishes to swim across a river, 120 m wide, as quickly as possible.
If the river flows at 0·5 m s^{-1}, find the time the man takes for the crossing and how far downstream he travels.

13. A pilot has to fly his aircraft from airport A to airport B, 100 km due east of A. In still air the aircraft flies at 125 km h^{-1}.
If there is a wind of 35 km h^{-1} blowing from the north, find the course that the pilot must set in order to reach B and the time the journey takes.

14. Two airfields A and B are 500 km apart with B on a bearing 060° from A. An aircraft which can travel at 200 km h^{-1} in still air, is to be flown from A to B.
If there is a wind of 40 km h^{-1} blowing from the west, find the course that the pilot must set in order to reach B and find, to the nearest minute, the time taken.

15. An aircraft capable of flying at 250 km h^{-1} in still air, is to be flown from airport A to airport B, situated 300 km from A on a bearing 320°.
If there is a wind of 50 km h^{-1} blowing from 030°, find the course the pilot must set and find, to the nearest minute, the time taken for the journey.

16. A man swims at 5 km h^{-1} in still water. Find the time it takes the man to swim across a river 250 m wide, flowing at 3 km h^{-1}, if he swims so as to cross the river
(a) by the shortest route
(b) in the quickest time.

17. A man wishes to row a boat across a river to reach a point on the far bank that is 35 m downstream from his starting point. The man can row the boat at 2·5 m s^{-1} in still water. If the river is 50 m wide and flows at 3 m s^{-1}, find the two possible courses the man could set and find the respective crossing times.

18. Airfield A is 500 km due south of airfield B. A pilot, wishing to fly his aircraft from A to B, is told that there is a wind of 50 km h^{-1} blowing from N 60° E. In still air the aircraft flies at 300 km h^{-1}.
What course should the pilot set in order to reach B and how long will the flight take?
Assuming the wind does not change, how long would the return flight take?

19. Two heliports A and B are 150 km apart with B on a bearing 045° from A. A wind of 30 km h^{-1} is blowing from a direction 260°. Assuming this wind remains constant throughout, find the time required for a helicopter to fly from A to B and back to A again, if the helicopter can fly at 100 km h^{-1} in still air.

20. When swimming in a river a man finds that he has a maximum speed v when swimming downstream and u when swimming upstream.

(a) Find an expression for his maximum speed when swimming in still water.

(b) If the river is of width s, show that the shortest time in which the man can swim across is $\dfrac{2s}{v+u}$ and that such a crossing would take him a distance of $\dfrac{s(v-u)}{v+u}$ downstream from his starting point.

(c) If the man wishes to swim as quickly as possible from a point on one bank to a point exactly opposite on the other bank, show that he must swim in a direction that makes an angle $\cos^{-1}\left(\dfrac{v-u}{v+u}\right)$ with the bank and that the crossing will take a time $\dfrac{s}{\sqrt{(uv)}}$.

Relative velocity

Suppose A and B are two moving bodies. The velocity of A with respect to B is the velocity of A as it appears to an observer on B and this is usually denoted by ${}_A\mathbf{v}_B$. The simplest case to consider is when A and B are moving along parallel lines.

Example 6

A man M on a train travelling due west at $25\,\mathrm{m\,s^{-1}}$ notices a second train passing him on a parallel track at $40\,\mathrm{m\,s^{-1}}$. Calculate the velocity of the second train S relative to the first train if the directions of motion are
(a) the same, (b) opposite.

(a) Draw a diagram showing the velocities.

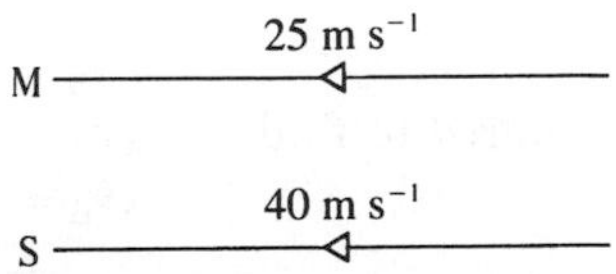

(a) The velocity of S as it appears to M is:

$$\begin{aligned} {}_S\mathbf{v}_M &= 40 - 25 \\ &= 15\,\mathrm{m\,s^{-1}} \text{ due west} \end{aligned}$$

S appears to be travelling more slowly than it is doing.

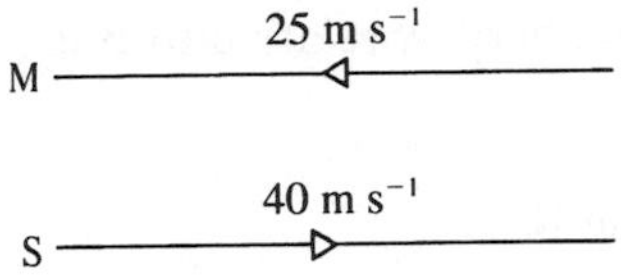

(b) The velocity of S as it appears to M is:

$$\begin{aligned} {}_S\mathbf{v}_M &= 40 + 25 \\ &= 65\,\mathrm{m\,s^{-1}} \text{ due east} \end{aligned}$$

S appears to be travelling more quickly than it is doing.

Note that the observer is only concerned with the motion of the other train *relative* to himself, i.e. as though he were not moving.

Non-parallel courses

The previous example suggests the general method. Since the observer takes no account of his own motion, the velocity of A relative to B is found by reducing the observer on B to rest.

Suppose bodies A and B have velocities $\mathbf{v}_A$ and $\mathbf{v}_B$ respectively, as shown in Fig. 1.
To find the velocity of A relative to B, i.e. ${}_A\mathbf{v}_B$, B must be reduced to rest. To do this, B is given a velocity equal and opposite to its own velocity. The same velocity must be applied to A, as shown in Fig. 2.
The resultant velocity of A will then be the velocity of A relative to B. Thus, from Fig. 2 we have the vector equation:

$${}_A\mathbf{v}_B = \mathbf{v}_A - \mathbf{v}_B \qquad \ldots[\mathbf{1}]$$

If the velocities $\mathbf{v}_A$ and $\mathbf{v}_B$ are given in vector form, ${}_A\mathbf{v}_B$ can be found very easily from equation [1].

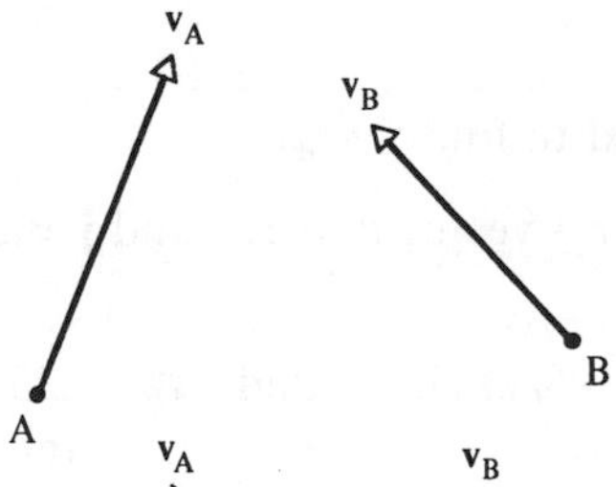

Fig. 1

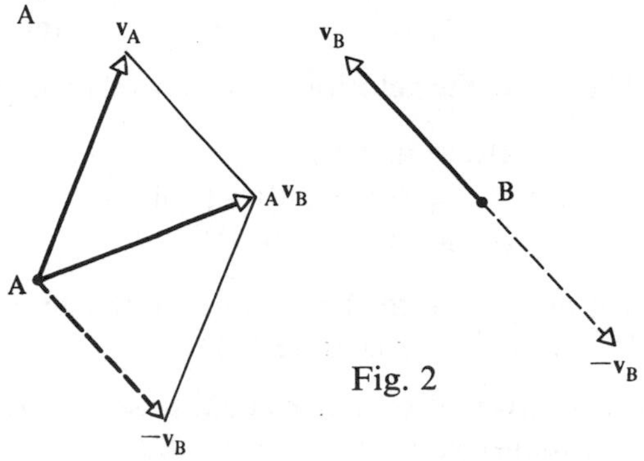

Fig. 2

If $\mathbf{v}_A$ and $\mathbf{v}_B$ are not given in vector form, then the vector equation [1] can be thought of as ${}_A\mathbf{v}_B = \mathbf{v}_A + (-\mathbf{v}_B)$ and the corresponding vector triangle can be drawn.
The magnitude and the direction of the relative velocity can be found in the usual way either by calculation, or by scale drawing.

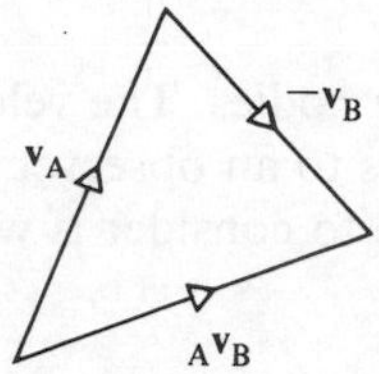

Example 7

Particle A has a velocity of $(3\mathbf{i} + 7\mathbf{j} - 3\mathbf{k})\,\mathrm{m\,s^{-1}}$ and particle B has a velocity of $(5\mathbf{i} + 2\mathbf{j} + 4\mathbf{k})\,\mathrm{m\,s^{-1}}$. Find the velocity of A relative to B.

Given $\mathbf{v}_A = (3\mathbf{i} + 7\mathbf{j} - 3\mathbf{k})\,\mathrm{m\,s^{-1}}$
and $\mathbf{v}_B = (5\mathbf{i} + 2\mathbf{j} + 4\mathbf{k})\,\mathrm{m\,s^{-1}}$
We are required to find ${}_A\mathbf{v}_B$

$$\begin{aligned} {}_A\mathbf{v}_B &= \mathbf{v}_A - \mathbf{v}_B \\ &= (3\mathbf{i} + 7\mathbf{j} - 3\mathbf{k}) - (5\mathbf{i} + 2\mathbf{j} + 4\mathbf{k}) \\ {}_A\mathbf{v}_B &= -2\mathbf{i} + 5\mathbf{j} - 7\mathbf{k} \end{aligned}$$

The velocity of A relative to B is $(-2\mathbf{i} + 5\mathbf{j} - 7\mathbf{k})\,\mathrm{m\,s^{-1}}$.

Example 8

A girl walks at $5\,\mathrm{km\,h^{-1}}$ due west and a boy runs at $12\,\mathrm{km\,h^{-1}}$ on a bearing of 150°. Find the velocity of the boy relative to the girl.

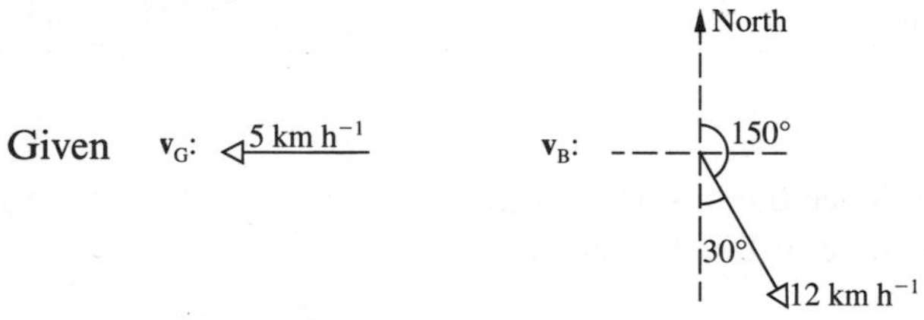

We are required to find ${}_B\mathbf{v}_G$

Taking $\mathbf{i}$ as a unit vector due east and $\mathbf{j}$ as a unit vector due north gives:

$$\mathbf{v}_G = -5\mathbf{i}\,\mathrm{km\,h^{-1}} \quad \text{and} \quad \mathbf{v}_B = (12\sin 30°\,\mathbf{i} - 12\cos 30°\,\mathbf{j})\,\mathrm{km\,h^{-1}} = (6\mathbf{i} - 6\sqrt{3}\,\mathbf{j})\,\mathrm{km\,h^{-1}}$$

Therefore, the velocity of the boy relative to the girl is given by:

$$\begin{aligned} {}_B\mathbf{v}_G &= \mathbf{v}_B - \mathbf{v}_G \\ &= (6\mathbf{i} - 6\sqrt{3}\mathbf{j}) - (-5\mathbf{i}) \\ \therefore \quad {}_B\mathbf{v}_G &= 11\mathbf{i} - 6\sqrt{3}\mathbf{j} \end{aligned}$$

The velocity of the boy relative to the girl is $(11\mathbf{i} - 6\sqrt{3}\mathbf{j})\,\mathrm{km\,h^{-1}}$, i.e. $15{\cdot}1\,\mathrm{km\,h^{-1}}$ on a bearing 133°.

Alternatively Example 8 could be solved by scale drawing or by trigonometry.

True velocity

Suppose $\mathbf{v}_A$, the true velocity of A, is known and ${}_B\mathbf{v}_A$, the velocity of B relative to A, is also known, then by using ${}_B\mathbf{v}_A = \mathbf{v}_B - \mathbf{v}_A$ the true velocity of B can be found.
Examples 9 and 10 demonstrate this idea with Example 9 using a component vector approach and Example 10 a trigonometry approach.
Note also that Example 9 reminds you that the vector $a\mathbf{i} + b\mathbf{j}$ can be written as the column matrix $\begin{pmatrix} a \\ b \end{pmatrix}$, as introduced in Chapter 1.

Example 9

To the captain of a ship S travelling with velocity $\begin{pmatrix} 12 \\ -15 \end{pmatrix}$ km h^{-1}, a second ship T appears to have a velocity of $\begin{pmatrix} 10 \\ 20 \end{pmatrix}$ km h^{-1}. Find the true velocity of T.

Given $\quad \mathbf{v}_S = \begin{pmatrix} 12 \\ -15 \end{pmatrix}$ km h^{-1} and ${}_T\mathbf{v}_S = \begin{pmatrix} 10 \\ 20 \end{pmatrix}$ km h^{-1}

we are required to find $\quad \mathbf{v}_T$

Use

$$\begin{aligned} {}_T\mathbf{v}_S &= \mathbf{v}_T - \mathbf{v}_S \\ \mathbf{v}_T &= {}_T\mathbf{v}_S + \mathbf{v}_S \\ &= \begin{pmatrix} 10 \\ 20 \end{pmatrix} + \begin{pmatrix} 12 \\ -15 \end{pmatrix} \\ &= \begin{pmatrix} 22 \\ 5 \end{pmatrix} \end{aligned}$$

The true velocity of T is $\begin{pmatrix} 22 \\ 5 \end{pmatrix}$ km h^{-1}.

Example 10

To a girl running at 6 m s^{-1} on a bearing 155°, a low flying bird appears to be moving at 7 m s^{-1} on a bearing of 250°. Find the true velocity of the bird.

Given $\quad \mathbf{v}_G$:

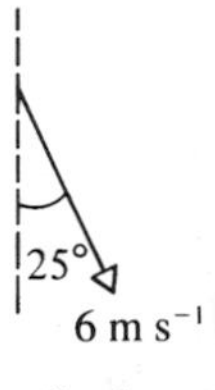

and $\quad {}_B\mathbf{v}_G$:

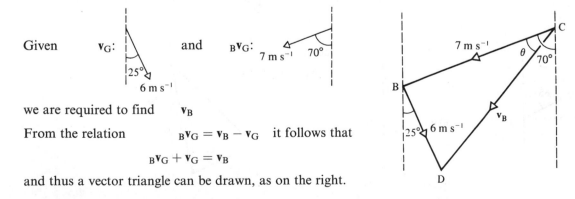

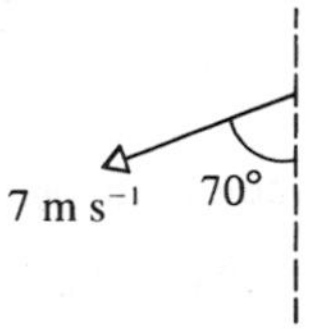

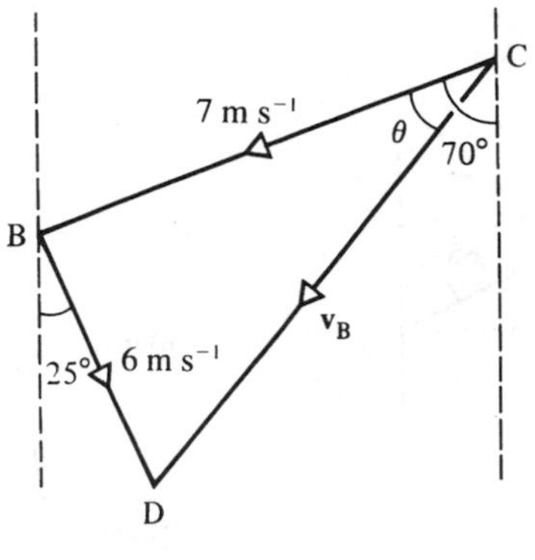

we are required to find $\quad \mathbf{v}_B$

From the relation $\quad {}_B\mathbf{v}_G = \mathbf{v}_B - \mathbf{v}_G \quad$ it follows that

$${}_B\mathbf{v}_G + \mathbf{v}_G = \mathbf{v}_B$$

and thus a vector triangle can be drawn, as on the right.

From triangle BCD, by the cosine rule:

$$CD^2 = 7^2 + 6^2 - 2(7)(6)\cos 85°$$
$$= 85 - 84 \cos 85°$$
$$\therefore \quad CD = 8{\cdot}813$$

From triangle BCD, by the sine rule:

$$\frac{6}{\sin\theta} = \frac{8{\cdot}813}{\sin 85°}$$

$$\therefore \quad \sin\theta = \frac{6 \sin 85°}{8{\cdot}813}$$

$$\therefore \quad \theta = 42{\cdot}70°$$

The true velocity of the bird is $8{\cdot}81\,\text{m}\,\text{s}^{-1}$ in a direction S 27·30° W.

Example 11

To a man rowing due south at $4\,\text{m}\,\text{s}^{-1}$, a boy in a boat appears to be moving due west. To a woman swimming at $1\frac{1}{2}\,\text{m}\,\text{s}^{-1}$ in a direction N 65° W, the boy appears to be moving in a direction S 10° W. Find the true magnitude and direction of the velocity of the boy.

Denote the man by M, the woman by W, and the boy by B.

We are given

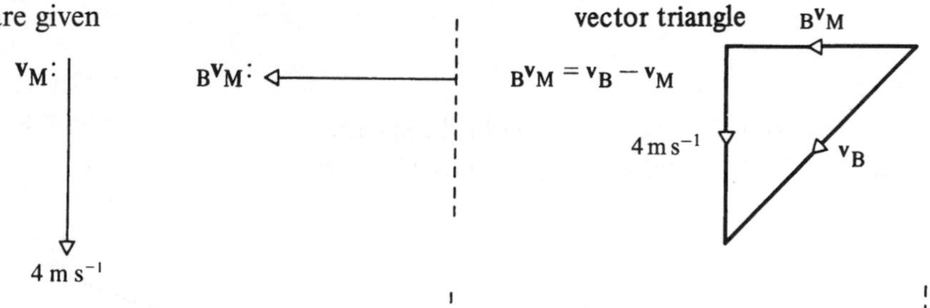

and

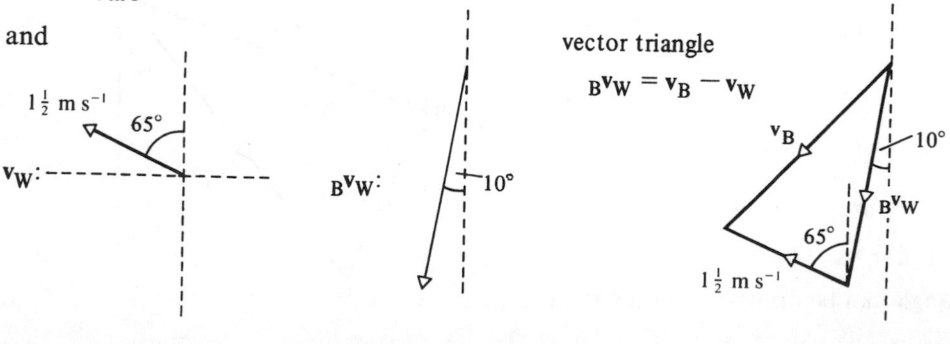

Since the two vector triangles have a common side which represents the true velocity $\mathbf{v}_B$ of the boy, they can be combined in one diagram.

From triangle ACD, right-angled at D:

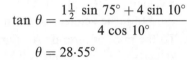

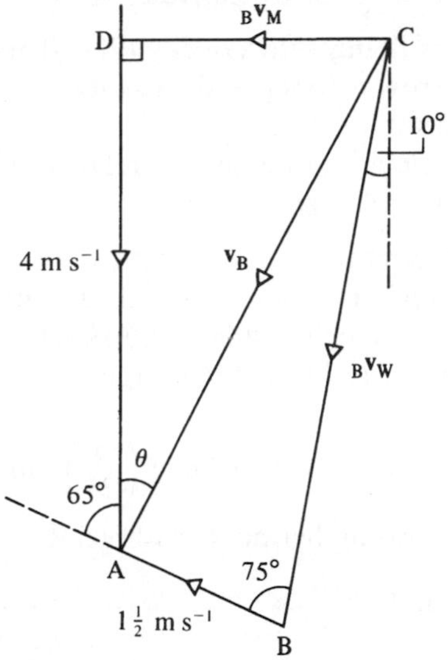

$$\cos\theta = \frac{4}{|\mathbf{v}_B|} \qquad \ldots [1]$$

From triangle ACB, by the sine rule:

$$\frac{|\mathbf{v}_B|}{\sin 75°} = \frac{1\frac{1}{2}}{\sin(\theta - 10°)} \qquad \ldots [2]$$

Substituting for $\mathbf{v}_B$ in equation [2] from [1] gives:

$$4\sin(\theta - 10°) = 1\tfrac{1}{2}\sin 75°\cos\theta$$

$$\therefore \quad 4\sin\theta\cos 10° = \left(1\tfrac{1}{2}\sin 75° + 4\sin 10°\right)\cos\theta$$

$$\therefore \quad \tan\theta = \frac{1\frac{1}{2}\sin 75° + 4\sin 10°}{4\cos 10°}$$

$$\therefore \quad \theta = 28{\cdot}55°$$

and from equation [1]:

$$|\mathbf{v}_B| = \frac{4}{\cos 28{\cdot}55°} = 4{\cdot}554\ \text{m s}^{-1}$$

The true velocity of the boy is $4{\cdot}55\ \text{m s}^{-1}$ in a direction S 28·55° E.

Alternatively, the solutions could be obtained:

- by making an accurate scale drawing of ABCD,

or
- by expressing the original data as **i**–**j** vectors, letting $\mathbf{v}_B = a\mathbf{i} + b\mathbf{j}$, determining a and b and hence the true magnitude and direction of the velocity of the boy.

Exercise 10B

1. A cruiser is moving at $30\ \text{km h}^{-1}$ due north and a battleship is moving at $20\ \text{km h}^{-1}$ due north.
Find the velocity of the cruiser relative to the battleship.

2. Particle A is moving due north at $30\ \text{m s}^{-1}$ and particle B is moving due south at $20\ \text{m s}^{-1}$.
Find the velocity of A relative to B.

3. A yacht and a trawler leave a harbour at 8 a.m. The yacht travels due west at $10\ \text{km h}^{-1}$ and the trawler due east at $20\ \text{km h}^{-1}$.
What is the velocity of the trawler relative to the yacht?
How far apart are the boats at 9.30 a.m.?

4. At 10.30 a.m. a car, travelling at $25\ \text{m s}^{-1}$ due east, overtakes a motorbike travelling at $10\ \text{m s}^{-1}$ due east.
What is the velocity of the car relative to the motorbike and how far apart are the vehicles at 10.31 a.m.?

5. Particle A has a velocity of $\begin{pmatrix}12\\5\end{pmatrix}\ \text{m s}^{-1}$ and particle B has a velocity of $\begin{pmatrix}4\\3\end{pmatrix}\ \text{m s}^{-1}$.
Find the velocity of A relative to B.

6. Particle A has a velocity of $(4\mathbf{i} + 6\mathbf{j} - 5\mathbf{k})\ \text{m s}^{-1}$ and particle B has a velocity of $(-10\mathbf{i} - 2\mathbf{j} + 6\mathbf{k})\ \text{m s}^{-1}$.
Find the velocity of A relative to B.

7. Bird A has a velocity of $(7\mathbf{i} - 3\mathbf{j} + 10\mathbf{k})\,\text{m s}^{-1}$ and bird B has a velocity of $(6\mathbf{i} - 17\mathbf{k})\,\text{m s}^{-1}$. Find the velocity of B relative to A.

8. A pigeon is flying with velocity $(7\mathbf{i} - \mathbf{j})\,\text{m s}^{-1}$ and a sparrow is flying with velocity $(5\mathbf{i} + 6\mathbf{j})\,\text{m s}^{-1}$. Find the velocity of the pigeon relative to the sparrow.

9. A bomber aircraft is moving with velocity $(300\mathbf{i} - 100\mathbf{j})\,\text{km h}^{-1}$ and a fighter aircraft is moving with velocity $(400\mathbf{i} + 500\mathbf{j})\,\text{km h}^{-1}$. Find the velocity of the fighter relative to the bomber.

10. Joe rides his horse with velocity $\begin{pmatrix} 5 \\ 24 \end{pmatrix}$ km h^{-1} while Jill is riding her horse with velocity $\begin{pmatrix} 5 \\ 12 \end{pmatrix}$ km h^{-1}.
Find Joe's velocity as seen by Jill.
What is Jill's velocity as seen by Joe?

11. Tom walks at $4\,\text{km h}^{-1}$ due north and Jane walks at $3\,\text{km h}^{-1}$ due east. Find Tom's velocity relative to Jane.

12. A and B are two yachts. A has a velocity of $8\,\text{km h}^{-1}$ due south and B has a velocity of $15\,\text{km h}^{-1}$ due west. Find the velocity of A relative to B.

13. What is the velocity of a cruiser moving at $20\,\text{km h}^{-1}$ due north as seen by an observer on a liner moving at $15\,\text{km h}^{-1}$ in a direction N 30° W.

14. A car is being driven at $20\,\text{m s}^{-1}$ on a bearing 040°. The wind is blowing from 330° with a speed of $10\,\text{m s}^{-1}$. Find the velocity of the wind as experienced by the driver of the car.

15. An aircraft is moving at $250\,\text{km h}^{-1}$ in a direction N 60° E. A second aircraft is moving at $200\,\text{km h}^{-1}$ in a direction N 20° W. Find the velocity of the first aircraft as seen by the pilot of the second aircraft.

16. Find the velocity of a crow flying at $16\,\text{m s}^{-1}$ due north as seen by a blackbird flying at $12\,\text{m s}^{-1}$ due east. What is the velocity of the blackbird as seen by the crow?

17. To a man standing on the deck of a ship which is moving with a velocity of $(-6\mathbf{i} + 8\mathbf{j})\,\text{km h}^{-1}$, the wind seems to have a velocity of $(7\mathbf{i} - 5\mathbf{j})\,\text{km h}^{-1}$. Find the true velocity of the wind.

18. To the pilot of a bomber aircraft travelling with velocity $\begin{pmatrix} 150 \\ -200 \end{pmatrix}$ km h^{-1}, a fighter aircraft appears to have a velocity $\begin{pmatrix} 150 \\ 440 \end{pmatrix}$ km h^{-1}.
Find the true velocity of the fighter.

19. To a pigeon flying with a velocity of $(-2\mathbf{i} + 3\mathbf{j} + \mathbf{k})\,\text{m s}^{-1}$, a hawk appears to have a velocity of $(\mathbf{i} - 5\mathbf{j} - 10\mathbf{k})\,\text{m s}^{-1}$. Find the true velocity of the hawk.

20. To the pilot of aircraft A, flying with a velocity of $(160\mathbf{i} + 100\mathbf{j} - 10\mathbf{k})\,\text{m s}^{-1}$, a second aircraft, B, appears to have a velocity of $(40\mathbf{i} - 40\mathbf{j})\,\text{m s}^{-1}$. Find the true velocity of B.

21. To a cyclist riding at $3\,\text{m s}^{-1}$ due east, the wind appears to come from the south with speed $3\sqrt{3}\,\text{m s}^{-1}$. Find the true speed and direction of the wind.

22. To the pilot of an aircraft A, travelling at $300\,\text{km h}^{-1}$ due south, it appears that an aircraft B is travelling at $600\,\text{km h}^{-1}$ in a direction N 60° W. Find the true speed and direction of aircraft B.

23. Jane is riding her horse at $5\,\text{km h}^{-1}$ due north and sees Sue riding her horse apparently with velocity $4\,\text{km h}^{-1}$, N 60° E. Find Sue's true velocity.

24. To a person walking due east at $3\,\text{km h}^{-1}$, the wind appears to come from the north-east at $7\,\text{km h}^{-1}$. Find the true velocity of the wind.

25. To the driver of a motorboat moving at $6\,\text{km h}^{-1}$ on a bearing 345°, a yacht appears to be moving at $18\,\text{km h}^{-1}$ on a bearing 220°. Find the true velocity of the yacht.

26. A starling, flying at $8\,\text{m s}^{-1}$ on a bearing 240°, sees a thrush apparently flying at $5\,\text{m s}^{-1}$ on bearing 300°. Find the true velocity of the thrush.

27. A train is travelling at 80 km h^{-1} in a direction N 15° E. A passenger on the train observes a plane apparently moving at 125 km h^{-1} in a direction N 50° E. Find the true velocity of the plane.

28. To a passenger on a boat which is travelling at 20 km h^{-1} on a bearing of 230°, the wind seems to be blowing from 250° at 12 km h^{-1}. Find the true velocity of the wind.

29. To a jogger jogging at 12 km h^{-1} in a direction N 10° E, the wind seems to come from a direction N 20° W at 15 km h^{-1}. Find the true velocity of the wind.

30. A, B and C are three aircraft. A has velocity $(200\mathbf{i} + 170\mathbf{j})$ m s^{-1}. To the pilot of A it appears that B has velocity $(50\mathbf{i} - 270\mathbf{j})$ m s^{-1}. To the pilot of B it appears that C has a velocity $(50\mathbf{i} + 170\mathbf{j})$ m s^{-1}. Find, in vector form, the velocities of B and C.

31. To an observer on a liner moving with velocity $(18\mathbf{i} - 17\mathbf{j})$ km h^{-1}, a yacht appears to have a velocity $(-8\mathbf{i} + 29\mathbf{j})$ km h^{-1}. To someone on the yacht the wind appears to have a velocity $(-5\mathbf{i} - 5\mathbf{j})$ km h^{-1}. Find the true velocity of the wind, giving your answer in vector form.

32. When a man cycles due north at 10 km h^{-1}, the wind appears to come from the east. When he cycles in a direction N 60° W at 8 km h^{-1} it appears to come from the south. Find the true velocity of the wind.

33. To a bird flying due east at 10 m s^{-1}, the wind seems to come from the south. When the bird alters its direction of flight to N 30° E without altering its speed, the wind seems to come from the north-west. Find the true velocity of the wind.

34. To an observer on a trawler moving at 12 km h^{-1} in a direction S 30° W, the wind appears to come from N 60° W. To an observer on a ferry moving at 15 km h^{-1} in a direction S 80° E, the wind appears to come from the north. Find the true velocity of the wind.

Interception and collision

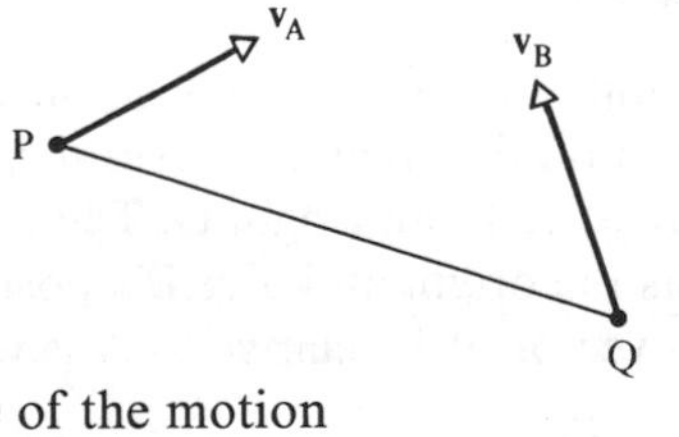

Consider two ships A and B initially at points P and Q. Suppose A is moving with uniform velocity $\mathbf{v}_A$ and B with uniform velocity $\mathbf{v}_B$.
By imposing a velocity of $-\mathbf{v}_B$ on both A and B, then B can be considered to be at rest and the velocity of A is then relative to B, i.e. ${}_A\mathbf{v}_B$.
Thus, if ${}_A\mathbf{v}_B$ is in the direction of PQ, then during the course of the motion A and B will meet. This may be by design (i.e. one ship intending to intercept the other) or by accident (i.e. one ship colliding with the other). Thus for collision or interception to occur, ${}_A\mathbf{v}_B$ must be in the direction of the line joining the original position of A to that of B.

Example 12

A speedboat A and a ship B are initially 570 m apart and B is due north of A. The ship has a constant velocity of $(7\mathbf{i} + 5\mathbf{j})$ m s^{-1} and the speed boat has a constant speed of 25 m s^{-1}.
Find, in vector form, the velocity of A if it is to intercept B, and find the time taken to do so. ($\mathbf{i}$ represents a unit vector due east and $\mathbf{j}$ a unit vector due north.)

Draw a diagram showing the initial positions of A and B.

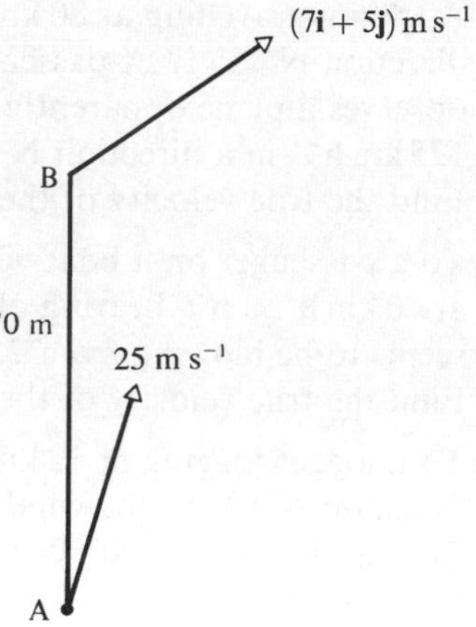

Let

$$\mathbf{v}_A = (a\mathbf{i} + b\mathbf{j})\ \text{m s}^{-1}$$

$$\begin{aligned} {}_A\mathbf{v}_B &= \mathbf{v}_A - \mathbf{v}_B \\ &= (a\mathbf{i} + b\mathbf{j}) - (7\mathbf{i} + 5\mathbf{j}) \\ &= (a - 7)\mathbf{i} + (b - 5)\mathbf{j} \end{aligned}$$

For interception, ${}_A\mathbf{v}_B$ must be in the direction due north.

Hence ${}_A\mathbf{v}_B = 0\mathbf{i} + (b - 5)\mathbf{j}$
and $(b - 5)$ must be positive

$\therefore \quad a - 7 = 0 \quad \text{or} \quad a = 7$

But the speed of A is $25\ \text{m s}^{-1}$.

$\therefore \quad \sqrt{(a^2 + b^2)} = 25$

$\therefore \quad a^2 + b^2 = 625$

Substituting for a gives:

$$b^2 = 625 - 49$$

$\therefore \quad b = 24$ (negative value not applicable as $(b - 5)$ must be positive)

$\therefore \quad \mathbf{v}_A = (7\mathbf{i} + 24\mathbf{j})\ \text{m s}^{-1}$

velocity of A relative to B $= (a - 7)\mathbf{i} + (b - 5)\mathbf{j} = 0\mathbf{i} + 19\mathbf{j}$

$\therefore \quad {}_A\mathbf{v}_B$ is $19\ \text{m s}^{-1}$ due north.

$$\text{time to intercept} = \frac{570}{|{}_A\mathbf{v}_B|} = \frac{570}{19} = 30\ \text{s}$$

Velocity of A is $(7\mathbf{i} + 24\mathbf{j})\ \text{m s}^{-1}$ and the time to intercept B is 30 s.

Relative position

The next example refers to "the position vector of B relative to A, ${}_B\mathbf{r}_A$". In the diagram on the right the position vectors $\mathbf{r}_A$ and $\mathbf{r}_B$ give the positions of A and B relative to an origin O. The position vector of B relative to A will take A as the origin and give B's position relative to A. The position vector of B relative to A is written as ${}_B\mathbf{r}_A$, where:

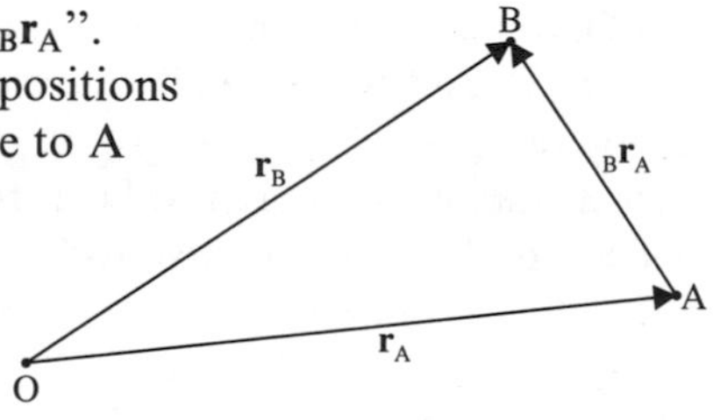

$$\begin{aligned} {}_B\mathbf{r}_A &= \overrightarrow{AB} \\ &= \overrightarrow{AO} + \overrightarrow{OB} \\ &= -\mathbf{r}_A + \mathbf{r}_B \\ &= \mathbf{r}_B - \mathbf{r}_A \end{aligned}$$

Similarly the position vector of A relative to B is written ${}_A\mathbf{r}_B$ and:

$${}_A\mathbf{r}_B = \mathbf{r}_A - \mathbf{r}_B$$

The reader will notice the similarity between this relative position result: ${}_A\mathbf{r}_B = \mathbf{r}_A - \mathbf{r}_B$

and the relative velocity result: ${}_A\mathbf{v}_B = \mathbf{v}_A - \mathbf{v}_B$.

Note in the next example that the vectors in the question are written in the form $a\mathbf{i} + b\mathbf{j}$ but the working is performed using the column matrix form.

This is quite all right, and you can use whichever style you wish but do note that any answers are stated in the style consistent with that used in the question itself.

Example 13

At 8 a.m. the position vectors **r** and velocity vectors **v** of two particles, A and B, are as follows:

$$\mathbf{r}_A = (5\mathbf{i} - 3\mathbf{j} + 4\mathbf{k})\text{ km} \qquad \mathbf{v}_A = (2\mathbf{i} + 5\mathbf{j} + 3\mathbf{k})\text{ km h}^{-1}$$
$$\mathbf{r}_B = (7\mathbf{i} + 5\mathbf{j} - 2\mathbf{k})\text{ km} \qquad \mathbf{v}_B = (-3\mathbf{i} - 15\mathbf{j} + 18\mathbf{k})\text{ km h}^{-1}$$

(a) Find ${}_B\mathbf{r}_A(t)$, the position vector of B relative to A at time t hours past 8 a.m.

(b) Show that if the velocities remain constant, a collision will occur, and find the time of the collision and the position vector of the point where it occurs.

(a) position vector of A, t hours after 8 a.m. $= \mathbf{r}_A(t)$

$$= \mathbf{r}_A + t(\mathbf{v}_A)$$

$$= \begin{pmatrix} 5 \\ -3 \\ 4 \end{pmatrix} + t\begin{pmatrix} 2 \\ 5 \\ 3 \end{pmatrix} = \begin{pmatrix} 5+2t \\ -3+5t \\ 4+3t \end{pmatrix} \qquad \ldots[1]$$

position vector of B, t hours after 8 a.m. $= \mathbf{r}_B(t)$

$$= \mathbf{r}_B + t(\mathbf{v}_B)$$

$$= \begin{pmatrix} 7 \\ 5 \\ -2 \end{pmatrix} + t\begin{pmatrix} -3 \\ -15 \\ 18 \end{pmatrix} = \begin{pmatrix} 7-3t \\ 5-15t \\ -2+18t \end{pmatrix} \qquad \ldots[2]$$

Thus the position vector of B relative to A t hours after 8 a.m. will be given by:

$${}_B\mathbf{r}_A(t) = \mathbf{r}_B(t) - \mathbf{r}_A(t)$$

$$= \begin{pmatrix} 7-3t \\ 5-15t \\ -2+18t \end{pmatrix} - \begin{pmatrix} 5+2t \\ -3+5t \\ 4+3t \end{pmatrix} = \begin{pmatrix} 2-5t \\ 8-20t \\ -6+15t \end{pmatrix}$$

The position vector of B relative to A at t hours past 8 a.m. is $(2-5t)\mathbf{i} + (8-20t)\mathbf{j} + (-6+15t)\mathbf{k}$

(b) The particles will collide if ${}_B\mathbf{r}_A(t) = 0$.

i.e. $\quad 2-5t=0 \quad$ and $\quad 8-20t=0 \quad$ and $\quad -6+15t=0$

$\quad t=\frac{2}{5} \quad$ and $\quad t=\frac{2}{5} \quad$ and $\quad t=\frac{2}{5}$

So they do collide and substituting $t = \frac{2}{5}$ into [1] or [2] gives the position vector of the particles as $\begin{pmatrix} 5{\cdot}8 \\ -1 \\ 5{\cdot}2 \end{pmatrix}$ km.

The particles will collide at 8.24 a.m. at the point with position vector $(5{\cdot}8\mathbf{i} - \mathbf{j} + 5{\cdot}2\mathbf{k})$ km. (Note that $\frac{2}{5}$h = 24 min.)

Alternatively Example 13 can be solved using the fact that for a collision to occur the velocity of A relative to B must be in the direction of the line joining the initial position of A to that of B.

Thus for collision we require ${}_A\mathbf{v}_B$ to be in the direction of $\overrightarrow{AB}$ (see diagram).

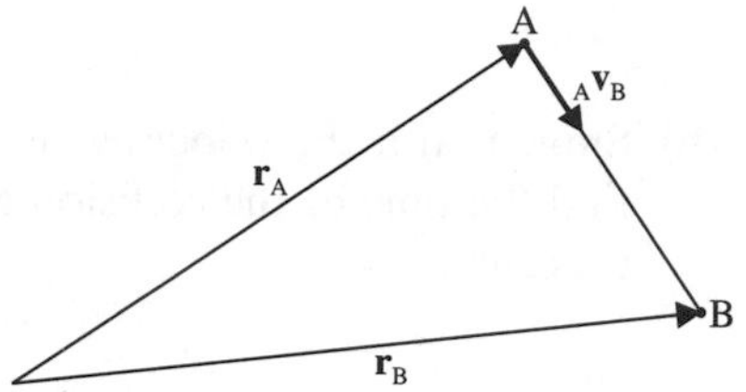

But
$$\begin{aligned} {}_A\mathbf{v}_B &= \mathbf{v}_A - \mathbf{v}_B \\ &= (2\mathbf{i} + 5\mathbf{j} + 3\mathbf{k}) - (-3\mathbf{i} - 15\mathbf{j} + 18\mathbf{k}) \\ &= 5\mathbf{i} + 20\mathbf{j} - 15\mathbf{k} \end{aligned}$$

and
$$\begin{aligned} \overrightarrow{AB} &= \mathbf{r}_B - \mathbf{r}_A \\ &= (7\mathbf{i} + 5\mathbf{j} - 2\mathbf{k}) - (5\mathbf{i} - 3\mathbf{j} + 4\mathbf{k}) \\ &= 2\mathbf{i} + 8\mathbf{j} - 6\mathbf{k} \\ &= \tfrac{2}{5}(5\mathbf{i} + 20\mathbf{j} - 15\mathbf{k}) \end{aligned}$$

Thus $\overrightarrow{AB}$ and ${}_A\mathbf{v}_B$ are parallel vectors, because one is a multiple of the other. Therefore a collision will occur.

From $\mathbf{s} = \mathbf{v}t$: $\quad 2\mathbf{i} + 8\mathbf{j} - 6\mathbf{k} = (5\mathbf{i} + 20\mathbf{j} - 15\mathbf{k})t$

giving $\quad t = \frac{2}{5}$

Thus again the collision is found to occur at 8.24 a.m. and its location may be found as before.

Velocities and positions not in vector form

Similar problems involving collisions and interceptions may be posed where the data is not given in vector form.

Example 14

At a certain instant two particles P and Q are at the points A and B which are 150 cm apart with B due east of A. P is travelling at $10\sqrt{3}$ cm s^{-1} due south and Q is travelling at 20 cm s^{-1} in a direction S 30° W. Show that if the velocities of P and Q remain unchanged, a collision will take place and find the time which elapses before it does so.

If a velocity of $-\mathbf{v}_P$ is imposed on both P and Q, then P will be at rest and ${}_Q\mathbf{v}_P$ must be in the direction BA if a collision is to occur.

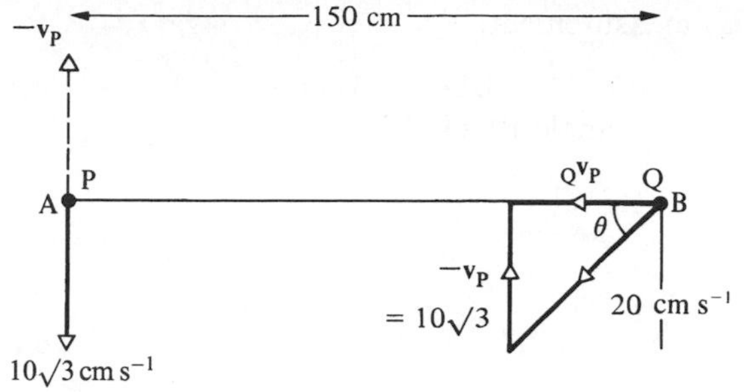

From the diagram:

$$\sin\theta = \frac{10}{20}\sqrt{3}$$

$$= \frac{\sqrt{3}}{2}$$

hence for collision $\theta = 60°$

But we know that the direction of motion of Q is S 30° W, so angle θ is 60°.

Collision occurs when Q, relative to P, is at the point A.

From the diagram: $\cos 60° = \dfrac{|{}_Q\mathbf{v}_P|}{20}$ and time taken $= \dfrac{AB}{|{}_Q\mathbf{v}_P|}$

$\therefore$ $|{}_Q\mathbf{v}_P| = 10$ $\qquad = \dfrac{150}{10} = 15$

A collision does take place after 15 s.

Note that if the actual position of the collision is required, then the position of either P or Q after 15 s has to be found.

After 15 s, P is $10\sqrt{3} \times 15$ cm or $150\sqrt{3}$ cm due south of A and this is where the collision occurs.

Example 15

At 8 a.m. two particles P and Q are at the points A and B, 12 km apart, with B on a bearing of 250° from A. P is moving at 4 km h^{-1} on a bearing of 320°. If the maximum speed of Q is 7 km h^{-1}, find the course on which Q should be set in order to intercept P as soon as possible, and find when the interception occurs.

Draw a sketch showing the initial positions of P and Q.
If a velocity of $-\mathbf{v}_P$ is imposed on both P and Q, then P will be at rest and, for interception to occur ${}_Q\mathbf{v}_P$ must be in the direction BA.
Draw a scale diagram in which D lies on BA (see below, p.236)

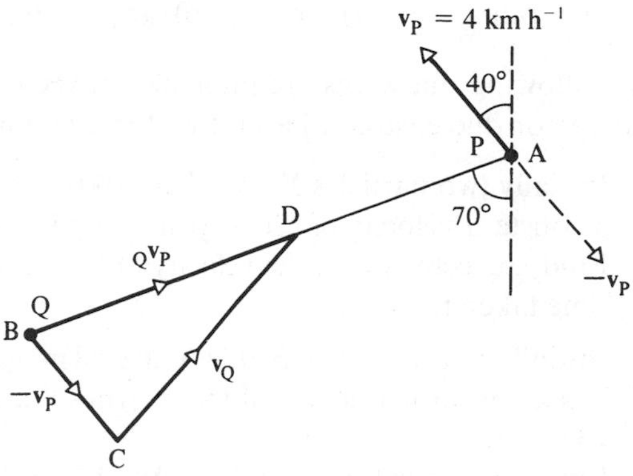

and $\qquad \overrightarrow{BC} = -\mathbf{v}_P$

$\qquad = 4$ km h^{-1} on a bearing of 140°

and $\qquad \overrightarrow{CD} = \mathbf{v}_Q$

$\qquad = 7$ km h^{-1}

By measurement:

$$BD = 7{\cdot}3\,\text{cm}$$
$$\text{angle BCD} = 78°$$

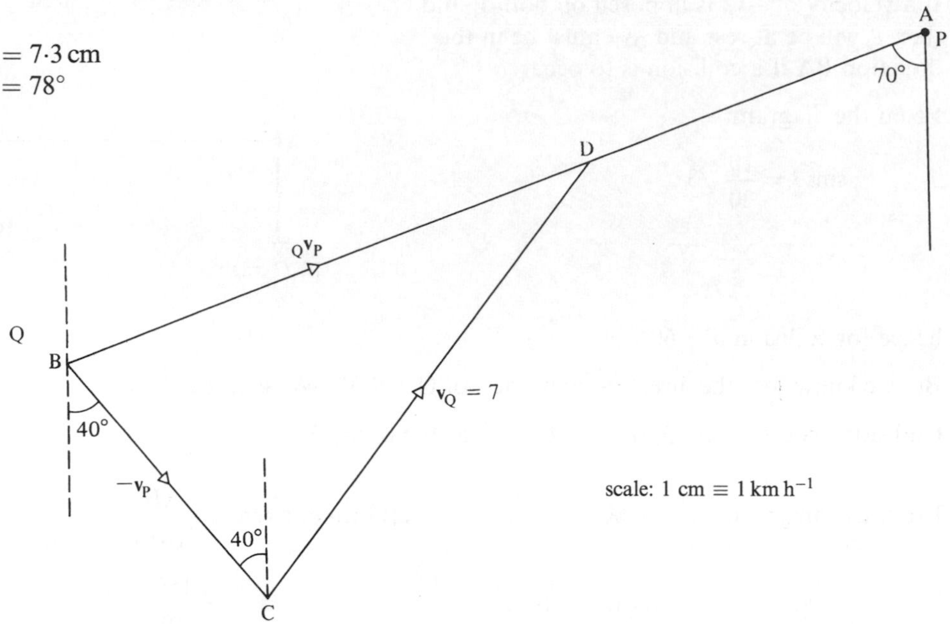

Time to intercept $\dfrac{AB}{|_Q\mathbf{v}_P|} = \dfrac{12}{7{\cdot}3} = 1{\cdot}644\,\text{h} = 1\,\text{h}\,39\,\text{min}$

The course to be set is 038° and interception takes place at 9.39 a.m.

Again, it should be noted that these answers could have been calculated from the triangle BCD using trigonometry.

Exercise 10C

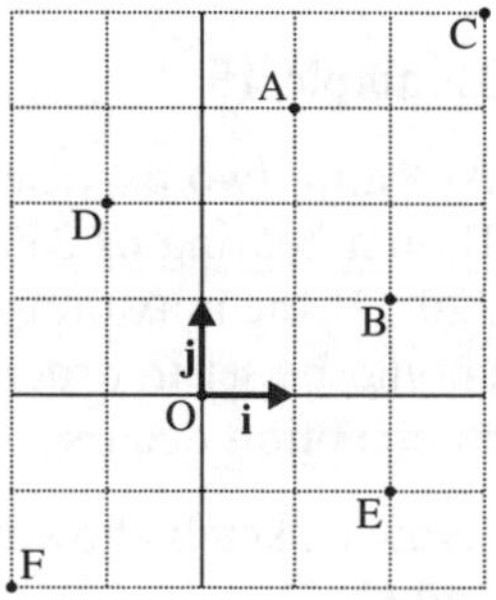

1. With O as the origin and points A, B, C, D, E and F as shown in the diagram determine:

(a) $\mathbf{r}_A$ (b) $\mathbf{r}_B$ (c) $\mathbf{r}_C$ (d) $\mathbf{r}_D$
(e) $_C\mathbf{r}_A$ (f) $_A\mathbf{r}_B$ (g) $_A\mathbf{r}_D$ (h) $_D\mathbf{r}_A$
(i) $_D\mathbf{r}_F$ (j) $_F\mathbf{r}_D$ (k) $_E\mathbf{r}_F$ (l) $_E\mathbf{r}_D$

2. Points A, B and C have position vectors of $2\mathbf{i} + 3\mathbf{j} - 5\mathbf{k}$, $6\mathbf{i} - \mathbf{j} + 2\mathbf{k}$ and $3\mathbf{j} + 4\mathbf{k}$ respectively.
Determine: (a) $_A\mathbf{r}_B$ (b) $_B\mathbf{r}_A$ (c) $_A\mathbf{r}_C$ (d) $_B\mathbf{r}_C$ (e) $_C\mathbf{r}_B$

The following questions are presented in vector form with **i** representing a unit vector due east and **j** a unit vector due north.

3. Initially two particles X and Y are 100 m apart with X due east of Y. X has a constant velocity of $(2\mathbf{i} + 3\mathbf{j})\,\text{m s}^{-1}$ and Y a constant speed of $5\,\text{m s}^{-1}$. Find, in vector form, the velocity of Y if it is to intercept X and find the time taken to do so.

4. Initially two particles X and Y are 48 m apart with Y due north of X. X has a constant velocity of $(5\mathbf{i} + 4\mathbf{j})\,\text{m s}^{-1}$ and Y a constant speed of $13\,\text{m s}^{-1}$.
Find, in vector form, the velocity of Y if it is to intercept X and find the time taken to do so.

5. At 12 noon the position vectors **r** and velocity vectors **v** of two ships A and B are as follows:

$$\mathbf{r}_A = (\mathbf{i} + 7\mathbf{j})\text{ km} \qquad \mathbf{v}_A = (6\mathbf{i} + 2\mathbf{j})\text{ km h}^{-1}$$

$$\mathbf{r}_B = (6\mathbf{i} + 4\mathbf{j})\text{ km} \qquad \mathbf{v}_B = (-4\mathbf{i} + 8\mathbf{j})\text{ km h}^{-1}$$

Show that if the ships do not alter their velocities, a collision will occur and find the time at which it occurs and the position vector of its location.

6. At 12 noon the position vectors **r** and velocity vectors **v** of two ships A and B are as follows:

$$\mathbf{r}_A = \begin{pmatrix} 5 \\ 2 \end{pmatrix}\text{ km} \qquad \mathbf{v}_A = \begin{pmatrix} 15 \\ 10 \end{pmatrix}\text{ km h}^{-1}$$

$$\mathbf{r}_B = \begin{pmatrix} 7 \\ 7 \end{pmatrix}\text{ km} \qquad \mathbf{v}_B = \begin{pmatrix} 9 \\ -5 \end{pmatrix}\text{ km h}^{-1}$$

Show that if the ships do not alter their velocities, a collision will occur and find the time at which it occurs and the position vector of its location.

7. At 11.30 a.m. a battleship is at a place with position vector $(-6\mathbf{i} + 12\mathbf{j})$ km and is moving with velocity $(16\mathbf{i} - 4\mathbf{j})$ km h^{-1}. At 12 noon a cruiser is at a place with position vector $(12\mathbf{i} - 15\mathbf{j})$ km and is moving with velocity $(8\mathbf{i} + 16\mathbf{j})$ km h^{-1}.
Show that if these velocities are maintained the two ships will collide and find when and where the collision occurs.

8. At 11.30 a.m. a jumbo jet has a position vector $(-100\mathbf{i} + 220\mathbf{j})$ km and a velocity vector $(300\mathbf{i} + 400\mathbf{j})$ km h^{-1}. At 11.45 a.m. a cargo plane has a position vector $(-60\mathbf{i} + 355\mathbf{j})$ km and a velocity vector $(400\mathbf{i} + 300\mathbf{j})$ km h^{-1}.
Show that if these velocities are maintained the planes will crash into each other and find the time and position vector of the crash.

9. At 2 p.m. the position vectors **r** and velocity vectors **v** of three ships A, B and C are as follows:

$$\mathbf{r}_A = (5\mathbf{i} + \mathbf{j})\text{ km} \qquad \mathbf{v}_A = (9\mathbf{i} + 18\mathbf{j})\text{ km h}^{-1}$$
$$\mathbf{r}_B = (12\mathbf{i} + 5\mathbf{j})\text{ km} \qquad \mathbf{v}_B = (-12\mathbf{i} + 6\mathbf{j})\text{ km h}^{-1}$$
$$\mathbf{r}_C = (13\mathbf{i} - 3\mathbf{j})\text{ km} \qquad \mathbf{v}_C = (9\mathbf{i} + 12\mathbf{j})\text{ km h}^{-1}$$

(a) Assuming that all three ships maintain these velocities, show that A and B will collide and find when and where the collision occurs.
(b) Find the position vector of C when A and B collide and find how far C is from the collision.
(c) When the collision occurs, C immediately changes its course, but not its speed, and steams direct to the scene. When does C arrive?

10. At 12 noon the position vectors **r** and velocity vectors **v** of three ships A, B and C are as follows:

$\mathbf{r}_A = (10{\cdot}5\mathbf{i} + 6\mathbf{j})\,\text{km}$ $\quad\mathbf{v}_A = (9\mathbf{i} + 18\mathbf{j})\,\text{km h}^{-1}$

$\mathbf{r}_B = (7\mathbf{i} + 20\mathbf{j})\,\text{km}$ $\quad\mathbf{v}_B = (12\mathbf{i} + 6\mathbf{j})\,\text{km h}^{-1}$

$\mathbf{r}_C = (10\mathbf{i} + 15\mathbf{j})\,\text{km}$ $\quad\mathbf{v}_C = (6\mathbf{i} + 12\mathbf{j})\,\text{km h}^{-1}$

Assuming that all three ships maintain these velocities, show that A and B will collide and find when and where the collision occurs.
When the collision occurs, C immediately changes its course but not its speed, and steams direct to the scene. When does C arrive?

11. At certain times the position vectors **r** and velocity vectors **v** of three ships A, B and C are as follows:

$\mathbf{r}_A = (6\mathbf{i} + 17\mathbf{j})\,\text{km}$ $\quad\mathbf{v}_A = (4\mathbf{i} - 20\mathbf{j})\,\text{km h}^{-1}$ at 11.30 a.m.

$\mathbf{r}_B = (5\mathbf{i} - 18\mathbf{j})\,\text{km}$ $\quad\mathbf{v}_B = (2\mathbf{i} + 14\mathbf{j})\,\text{km h}^{-1}$ at 11.45 a.m.

$\mathbf{r}_C = (2\mathbf{i} - 5\mathbf{j})\,\text{km}$ $\quad\mathbf{v}_C = (12\mathbf{i} - 4\mathbf{j})\,\text{km h}^{-1}$ at 12 noon

Assuming that all three ships maintain these velocities, show that two of them will collide and find when and where the collision will occur.
When the collision occurs the ship not involved immediately changes its course but not its speed, and steams direct to the scene. When will it arrive?

12. At certain times the position vectors **r** and velocity vectors **v** of three ships A, B and C are as follows:

$\mathbf{r}_A = (8\mathbf{i} + 26\mathbf{j})\,\text{km}$ $\quad\mathbf{v}_A = (7\mathbf{i} + 24\mathbf{j})\,\text{km h}^{-1}$ at 7.48 a.m.

$\mathbf{r}_B = (6\mathbf{i} - 10\mathbf{j})\,\text{km}$ $\quad\mathbf{v}_B = (15\mathbf{i} + 5\mathbf{j})\,\text{km h}^{-1}$ at 8.00 a.m.

$\mathbf{r}_C = (10\mathbf{i} + 6\mathbf{j})\,\text{km}$ $\quad\mathbf{v}_C = (10\mathbf{i} - 15\mathbf{j})\,\text{km h}^{-1}$ at 8.00 a.m.

Show that, if the ships maintain these velocities, two of the ships will collide and find when and where the collision will occur.
When the collision occurs the ship not involved immediately changes its course but not its speed, and steams direct to the scene. Find, in vector form, the velocity of this ship during this part of its motion and find its time of arrival.

13. A fighter pilot wishes to rendezvous with a supply aircraft for mid-air refuelling. At 0900 hours the position vectors of the two aircraft were as follows:

$$\mathbf{r}_{\text{fighter}} = \begin{pmatrix} 2300 \\ -350 \\ 2 \end{pmatrix}\text{km} \quad \mathbf{r}_{\text{supply aircraft}} = \begin{pmatrix} 2320 \\ -300 \\ 1 \end{pmatrix}\text{km}$$

The fighter pilot immediately changes the velocity of his aircraft to $\mathbf{v}_F$ and the supply plane maintains $\mathbf{v}_S$ where:

$$\mathbf{v}_F = \begin{pmatrix} 1000 \\ 900 \\ -10 \end{pmatrix}\text{km h}^{-1} \qquad \mathbf{v}_S = \begin{pmatrix} 800 \\ 400 \\ 0 \end{pmatrix}\text{km h}^{-1}$$

Show that if these velocities are maintained interception will occur. Find the position vector of its location and the time it occurs.

14. At time $t = 0$ seconds the operator at space traffic control headquarters, position vector $(0\mathbf{i} + 0\mathbf{j} + 0\mathbf{k})$ m, notifies the pilot of *Space Vehicle Alpha* that she risks collision with a piece of space debris. At this time the position and velocity vectors of the space vehicle and the debris are as follows:

$$\mathbf{r}_{\text{space vehicle}} = (-3300\mathbf{i} + 1000\mathbf{j} + 500\mathbf{k})\text{ m}$$
$$\mathbf{r}_{\text{debris}} = (2700\mathbf{i} + 4500\mathbf{k})\text{ m}$$

$$\mathbf{v}_{\text{space vehicle}} = (400\mathbf{i} + 150\mathbf{j} + 250\mathbf{k})\text{ m s}^{-1}$$
$$\mathbf{v}_{\text{debris}} = (100\mathbf{i} + 200\mathbf{j} + 50\mathbf{k})\text{ m s}^{-1}$$

(a) Show that if these velocities are maintained *Vehicle Alpha* will collide with the debris and find the time and position vector of the collision.

(b) Instead of changing her vehicle's velocity the pilot of *Vehicle Alpha* fires a "deflector missile" to hit and deflect the space debris from its path. This missile is fired when $t = 5$ and hits the debris when $t = 10$. For this period of time find:
  (i) the velocity of the missile (assumed constant)
  (ii) the velocity of the missile relative to the space vehicle.

Exercise 10D

1. At 12 noon two ships A and B are 10 km apart with B due east of A. A is travelling at 20 km h^{-1} in a direction N 60° E and B is travelling at 10 km h^{-1} due north.
Show that, if the two ships maintain these velocities, they will collide and find, to the nearest minute, when the collision occurs.

2. At 11 p.m. two ships A and B are 10 km apart with B due north of A. A is travelling north-east at 18 km h^{-1} and B is travelling due east at $9\sqrt{2}$ km h^{-1}.
Show that, if the two ships do not change their velocities, they will collide and find, to the nearest minute, when the collision occurs.

3. A coastguard vessel wishes to intercept a yacht suspected of smuggling. At 1 a.m. the yacht is 10 km due east of the coastguard vessel and is travelling due north at 15 km h^{-1}. If the coastguard vessel travels at 20 km h^{-1}, in what direction should it steer in order to intercept the yacht? When would this interception occur?

4. A lifeboat sets out from a harbour at 9.10 p.m. to go to the assistance of a yacht which is, at that time, 5 km due south of the harbour and is drifting due west at 8 km h^{-1}. If the lifeboat travels at 20 km h^{-1}, find the course it should set so as to reach the yacht as quickly as possible and the time when it arrives (to the nearest half minute).

5. At 12 noon two ships A and B are 12 km apart with B on a bearing 140° from A. Ship A has a maximum speed of 30 km h^{-1} and wishes to intercept ship B, which is travelling at 20 km h^{-1} on a bearing 340°.
Find the course A should set in order to intercept B as soon as possible and the time when interception occurs.

6. A helicopter sets off from its base and flies at 50 m s^{-1} to intercept a ship which, when the helicopter sets off, is at a distance of 5 km on a bearing of 335° from the base. The ship is travelling at 10 m s^{-1} on a bearing 095°.
Find the course that the helicopter pilot should set if he is to intercept the ship as quickly as possible and the time interval between the helicopter taking off and it reaching the ship.

7. The driver of a speed boat travelling at $75\,\text{km h}^{-1}$ wishes to intercept a yacht travelling at $20\,\text{km h}^{-1}$ in a direction N 40° E. Initially the speedboat is positioned 10 km from the yacht on a bearing S 30° E.
Find the course that the driver of the speed boat should set to intercept the yacht and how long the journey will take.

8. A batsman hits a ball at $15\,\text{m s}^{-1}$ in a direction S 80° W. A fielder, 45 m and S 65° W from the batsman, runs at $6\,\text{m s}^{-1}$ to intercept the ball. Assuming the velocities remain unchanged, find in what direction the fielder must run to intercept the ball as quickly as possible.
How long does it take him, to the nearest tenth of a second?

Closest approach

If two bodies do not collide, then there will be an instant at which they are closer to each other than they are at any other instant.

Example 16

At time $t = 0$ the position vectors and velocity vectors of two bodies A and B are as follows:

$$\mathbf{r}_A = (3\mathbf{i} + \mathbf{j} + 5\mathbf{k})\,\text{m} \qquad \mathbf{v}_A = (4\mathbf{i} + \mathbf{j} - 3\mathbf{k})\,\text{m s}^{-1}$$

$$\mathbf{r}_B = (\mathbf{i} - 3\mathbf{j} + 2\mathbf{k})\,\text{m} \qquad \mathbf{v}_B = (\mathbf{i} + 2\mathbf{j} + 2\mathbf{k})\,\text{m s}^{-1}$$

Find:
(a) the position vector of B relative to A, at time t
(b) the value of t when A and B are closest together
(c) the least distance between A and B.

(a) position vector of A at time t is given by:

$$\mathbf{r}_A(t) = (3\mathbf{i} + \mathbf{j} + 5\mathbf{k}) + t(4\mathbf{i} + \mathbf{j} - 3\mathbf{k})$$
$$= (3 + 4t)\mathbf{i} + (1 + t)\mathbf{j} + (5 - 3t)\mathbf{k}$$

position vector of B at time t is given by:

$$\mathbf{r}_B(t) = (\mathbf{i} - 3\mathbf{j} + 2\mathbf{k}) + t(\mathbf{i} + 2\mathbf{j} + 2\mathbf{k})$$
$$= (1 + t)\mathbf{i} + (-3 + 2t)\mathbf{j} + (2 + 2t)\mathbf{k}$$

Therefore

$${}_B\mathbf{r}_A(t) = \mathbf{r}_B(t) - \mathbf{r}_A(t)$$
$$= (-2 - 3t)\mathbf{i} + (-4 + t)\mathbf{j} + (-3 + 5t)\mathbf{k}$$

The position vector of B relative to A, at time t, is:

$$(-2 - 3t)\mathbf{i} + (-4 + t)\mathbf{j} + (-3 + 5t)\mathbf{k}.$$

(b) If the distance between the ships is AB, then:

$$|\overrightarrow{AB}|^2 = AB^2 = (-2 - 3t)^2 + (-4 + t)^2 + (-3 + 5t)^2$$
$$= 35t^2 - 26t + 29 \qquad \ldots [1]$$

The value of t which gives the minimum can be found using calculus.

The distance AB is not negative and so will be a minimum when AB^2 is a minimum. Any maximum or minimum value of AB^2 will occur

when $\dfrac{d(AB^2)}{dt} = 0$

i.e. $70t - 26 = 0$

giving $t = \dfrac{13}{35}$

Furthermore $\dfrac{d^2(AB^2)}{dt^2} = 70$ which, being positive, indicates a minimum value.

When $t = \dfrac{13}{35}$ the bodies A and B are closest.

(c) The minimum value of AB is now found by substituting this value of t into [1], giving

$$(AB_{min})^2 = 35\left(\frac{13}{35}\right)^2 - 26\left(\frac{13}{35}\right) + 29$$

$$\therefore \quad AB_{min} \approx 4{\cdot}92\,\text{m}$$

The least distance between A and B is 4·92 m.

Example 17

Two ships A and B are initially 20 km apart with B on a bearing of N 67° E from A. Ship A is moving at 18 km h^{-1} in a direction S 20° E and B is moving at 12 km h^{-1} due south. Assuming the velocities of A and B remain unchanged, find the least distance apart of the ships in the subsequent motion and the time at which this position is reached.

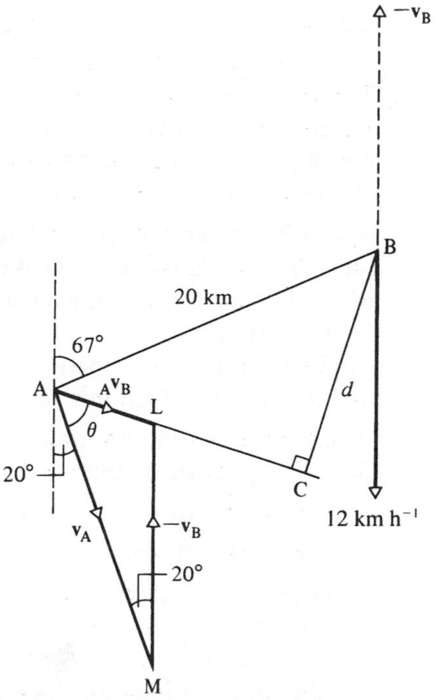

Draw a diagram showing the initial positions of the ships and consider the motion of A relative to B.

Relative to B, ship A travels along AC and

$${}_A\mathbf{v}_B = \mathbf{v}_A - \mathbf{v}_B$$

The least distance apart is d.

From triangle ALM

$$|{}_A\mathbf{v}_B|^2 = 18^2 + 12^2 - 2(18)(12)\cos 20°$$
$$= 468 - 432\cos 20°$$
$$\therefore \quad |{}_A\mathbf{v}_B| = 7{\cdot}877$$

By the sine rule:

$$\frac{|{}_A\mathbf{v}_B|}{\sin 20°} = \frac{12}{\sin\theta}$$

$$\therefore \quad \sin\theta = \frac{12\sin 20°}{7{\cdot}877}$$

$$\therefore \quad \theta = 31{\cdot}40°$$

Then angle BAC $= 180° - \theta - 20° - 67° = 61{\cdot}60°$

$$d = \text{AB} \sin \widehat{\text{BAC}} = 20 \sin 61{\cdot}60°$$
$$= 17{\cdot}59\,\text{km}$$

$$\text{time to reach this position} = \frac{\text{AC}}{|{}_{\text{A}}\mathbf{v}_{\text{B}}|}$$
$$= \frac{20 \cos 61{\cdot}60°}{7{\cdot}877} = 1{\cdot}208\,\text{h}$$

The least distance between the ships is 17·6 km and occurs after 1 hour and 12 minutes.

Exercise 10E

The following questions are presented in vector form with **i** representing a unit vector due east and **j** a unit vector due north.

1 At 12 noon the position vectors **r** and velocity vectors **v** of two ships A and B are as follows:

$$\mathbf{r}_A = (-9\mathbf{i} + 6\mathbf{j})\,\text{km} \qquad \mathbf{v}_A = (3\mathbf{i} + 12\mathbf{j})\,\text{km h}^{-1}$$
$$\mathbf{r}_B = (16\mathbf{i} + 6\mathbf{j})\,\text{km} \qquad \mathbf{v}_B = (-9\mathbf{i} + 3\mathbf{j})\,\text{km h}^{-1}$$

(a) Find how far apart the ships are at 12 noon.
(b) Assuming the velocities do not change, find the least distance between the ships in the subsequent motion.
(c) Find when this distance of closest approach occurs and the position vectors of A and B at that time.

2. At 8 a.m. two ships A and B are 11 km apart with B due west of A. A and B travel with constant velocities of $(-4\mathbf{i} + 3\mathbf{j})\,\text{km h}^{-1}$ and $(2\mathbf{i} + 4\mathbf{j})\,\text{km h}^{-1}$ respectively.
Find the least distance between the two ships in the subsequent motion and the time, to the nearest minute, at which this situation occurs.

3. At 7.30 a.m. two ships A and B are 8 km apart with B due north of A. The velocities of A and B are $12\mathbf{j}\,\text{km h}^{-1}$ and $-5\mathbf{i}\,\text{km h}^{-1}$ respectively. Assuming these velocities do not change, find the least distance between the ships in the subsequent motion and the time, to the nearest minute, at which this situation occurs.

4. A and B are two tankers, and at 1300 hours B has a position vector of $(4\mathbf{i} + 8\mathbf{j})$ km relative to A. Tanker A is moving with a constant velocity of $(6\mathbf{i} + 9\mathbf{j})\,\text{km h}^{-1}$ and tanker B is moving with a constant velocity of $(-3\mathbf{i} + 6\mathbf{j})\,\text{km h}^{-1}$.
Find the least distance between the tankers in the subsequent motion and the time at which this situation occurs.

5. At certain times the position vectors **r** and velocity vectors **v** of two ships A and B are as follows:

$\mathbf{r}_A = 20\mathbf{j}\,\text{km}$	$\mathbf{v}_A = (9\mathbf{i} - 2\mathbf{j})\,\text{km h}^{-1}$	at 1400 hours
$\mathbf{r}_B = (\mathbf{i} + 4\mathbf{j})\,\text{km}$	$\mathbf{v}_B = (4\mathbf{i} + 8\mathbf{j})\,\text{km h}^{-1}$	at 1500 hours

Assuming these velocities do not change, find:
(a) the position vector of A at 1500 hours
(b) the least distance between A and B
(c) the time at which this least separation occurs.

6. At 12 noon two ships A and B have the following position vectors **r** and velocity vectors **v**,

$$\mathbf{r}_A = \begin{pmatrix} 5 \\ 1 \end{pmatrix} \text{km} \qquad \mathbf{v}_A = \begin{pmatrix} 7 \\ 3 \end{pmatrix} \text{km h}^{-1}$$

$$\mathbf{r}_B = \begin{pmatrix} 8 \\ 7 \end{pmatrix} \text{km} \qquad \mathbf{v}_B = \begin{pmatrix} 2 \\ -1 \end{pmatrix} \text{km h}^{-1}$$

If both ships maintain these velocities, find the least distance between the ships and the time, to the nearest minute, at which this situation occurs.

7. Two ships A and B have the following position vectors **r** and velocity vectors **v** at the times indicated:

$$\mathbf{r}_A = (3\mathbf{i} + \mathbf{j})\text{ km} \qquad \mathbf{v}_A = (2\mathbf{i} + 3\mathbf{j})\text{ km h}^{-1} \qquad \text{at 11 a.m.}$$

$$\mathbf{r}_B = (2\mathbf{i} - \mathbf{j})\text{ km} \qquad \mathbf{v}_B = (3\mathbf{i} + 7\mathbf{j})\text{ km h}^{-1} \qquad \text{at 12 noon}$$

Assuming that the ships maintain these velocities, find:
(a) the position vector of ship A at 12 noon
(b) the distance between A and B at 12 noon
(c) the least distance between A and B during the motion
(d) the time, to the nearest minute, when the least separation occurs.

8. A battleship B and a cruiser C have the following position vectors **r** and velocity vectors **v** at 12 noon:

$$\mathbf{r}_B = (13\mathbf{i} + 5\mathbf{j})\text{ km} \qquad \mathbf{v}_B = (3\mathbf{i} - 10\mathbf{j})\text{ km h}^{-1}$$

$$\mathbf{r}_C = (3\mathbf{i} - 5\mathbf{j})\text{ km} \qquad \mathbf{v}_C = (15\mathbf{i} + 14\mathbf{j})\text{ km h}^{-1}$$

Assuming that the ships do not alter their velocities, find the closest distance that they come to each other.
The battleship has guns with a range of up to 5 km. Find the length of time during which the cruiser is within range of the battleship's guns.

9. Two ships A and B have the following position vectors **r** and velocity vectors **v** at the times stated:

$$\mathbf{r}_A = (-2\mathbf{i} + 3\mathbf{j})\text{ km} \qquad \mathbf{v}_A = (12\mathbf{i} - 4\mathbf{j})\text{ km h}^{-1} \qquad \text{at 11.45 a.m.}$$

$$\mathbf{r}_B = (8\mathbf{i} + 7\mathbf{j})\text{ km} \qquad \mathbf{v}_B = (2\mathbf{i} - 14\mathbf{j})\text{ km h}^{-1} \qquad \text{at 12 noon}$$

Assuming that the ships do not alter their velocities, find their least distance of separation. If ship B has guns with a range of up to 2 km, find for what length of time A is within range.

10. At time $t = 0$ the position vectors and velocity vectors of two particles A and B are as follows:

$$\mathbf{r}_A = \begin{pmatrix} 1 \\ 2 \\ 3 \end{pmatrix} \text{m} \qquad \mathbf{v}_A = \begin{pmatrix} -6 \\ 0 \\ 1 \end{pmatrix} \text{m s}^{-1}$$

$$\mathbf{r}_B = \begin{pmatrix} 4 \\ -14 \\ 1 \end{pmatrix} \text{m} \qquad \mathbf{v}_B = \begin{pmatrix} -5 \\ 1 \\ 7 \end{pmatrix} \text{m s}^{-1}$$

Find:
(a) the position vector of B relative to A at time t seconds
(b) the value of t when A and B are closest together
(c) the least distance between A and B.

11. Two particles A and B move with constant velocities $4\mathbf{i}\,\mathrm{m\,s^{-1}}$ and $(\mathbf{i} - 10\mathbf{j} + \mathbf{k})\,\mathrm{m\,s^{-1}}$ respectively. At time $t = 3$ seconds the position vectors of A and B are $(13\mathbf{i} + \mathbf{j} - \mathbf{k})\,\mathrm{m}$ and $(7\mathbf{i} - 27\mathbf{j} + 4\mathbf{k})\,\mathrm{m}$ respectively. Find the least distance between A and B.

12. At time $t = 0$ the position vectors and velocity vectors of two particles A and B are as follows:

$$\mathbf{r}_A = (\lambda\mathbf{i})\,\mathrm{m} \qquad \mathbf{v}_A = (2\mathbf{i} + \mathbf{j} - 5\mathbf{k})\,\mathrm{m\,s^{-1}}$$
$$\mathbf{r}_B = (2\lambda\mathbf{i})\,\mathrm{m} \qquad \mathbf{v}_B = (\mathbf{i} - 5\mathbf{j} + \mathbf{k})\,\mathrm{m\,s^{-1}}$$

where λ is a constant.
Find the value of t when A and B are closest together and show that the least distance between A and B is $\dfrac{6\lambda\sqrt{2}}{\sqrt{73}}$ metres.

13. Two particles A and B move with constant velocities $(\lambda\mathbf{i} + 3\mathbf{j} + 30\mathbf{k})\,\mathrm{m\,s^{-1}}$ and $(4\mathbf{i} - 2\mathbf{j} - 15\mathbf{k})\,\mathrm{m\,s^{-1}}$, respectively, where λ is a constant. At time $t = 0$ the position vectors of A and B are $(2\mathbf{i} + \mathbf{j} - 15\mathbf{k})\,\mathrm{m}$ and $(-\mathbf{i} + 4\mathbf{j} + 12\mathbf{k})\,\mathrm{m}$ respectively.
(a) Find the value of λ such that A and B will collide, and the value of t when this collision occurs.
(b) In the particular case when $\lambda = 2$ find the least distance between A and B.

14. Two particles A and B move with constant velocities $\mathbf{i}\,\mathrm{m\,s^{-1}}$ and $(5\mathbf{i} - 3\mathbf{k})\,\mathrm{m\,s^{-1}}$ respectively. At time $t = 0$ the position vectors of A and B are $(a\mathbf{i} + b\mathbf{j} + c\mathbf{k})\,\mathrm{m}$ and $(\mathbf{i} + 4\mathbf{j} - 7\mathbf{k})\,\mathrm{m}$ respectively. Show that if the particles are closest together at a time $t_1 > 0$ then $4a - 3c > 25$.
Given that the particles are closest together when $t = 2$ and at this instant the position vector of A is $(32\mathbf{i} + 7\mathbf{j} + 15\mathbf{k})\,\mathrm{m}$, find the values of a, b and c.

15. A lizard lies in wait at point A, position vector $\mathbf{r}_A = \begin{pmatrix} 65 \\ 40 \\ 0 \end{pmatrix}$ cm.

At time $t = 0$ seconds a fly has position vector $\mathbf{r}_F$ and velocity vector $\mathbf{v}_F$ as follows:

$$\mathbf{r}_F = \begin{pmatrix} 37 \\ 16 \\ 22 \end{pmatrix}\,\mathrm{cm} \qquad \mathbf{v}_F = \begin{pmatrix} 5 \\ 2 \\ -1 \end{pmatrix}\,\mathrm{cm\,s^{-1}}$$

If the fly were to continue with this velocity, find the closest distance it would come to the lizard and the value of t when this occurs.

16. With respect to a stationary bird watcher, the position vectors, $\mathbf{r}\,\mathrm{m}$, and velocity vectors, $\mathbf{v}\,\mathrm{m\,s^{-1}}$, of a pigeon and a bird of prey, at time $t = 0$ seconds, were as follows:

$$\mathbf{r}_{\text{pigeon}} = (11\mathbf{i} + 38\mathbf{j} + 11\mathbf{k})\,\mathrm{m} \qquad \mathbf{v}_{\text{pigeon}} = (3\mathbf{i} + 6\mathbf{j} - \mathbf{k})\,\mathrm{m\,s^{-1}}$$
$$\mathbf{r}_{\text{bird of prey}} = (4\mathbf{i} - 60\mathbf{j} + 88\mathbf{k})\,\mathrm{m} \qquad \mathbf{v}_{\text{bird of prey}} = (4\mathbf{i} + 20\mathbf{j} - 12\mathbf{k})\,\mathrm{m\,s^{-1}}$$

(a) Assuming the above velocities are maintained determine the least distance between each bird and the bird watcher, for $t \geqslant 0$, and the values of t for which these least distances occur.

(b) Again assuming the above velocities are maintained, show that the bird of prey will intercept the pigeon and find the value of t when this would occur.

(c) When $t = 6$ the pigeon suddenly changes its velocity to $(6\mathbf{j} - \mathbf{k})\,\text{m s}^{-1}$. If the bird of prey does not alter its velocity what is the least distance between the birds in the subsequent motion? (Give your answer to the nearest 0·1 m.)

Exercise 10F

1. Initially two ships A and B are 65 km apart with B due east of A. A is moving due east at $10\,\text{km h}^{-1}$ and B due south at $24\,\text{km h}^{-1}$. The two ships continue moving with these velocities.
Find the least distance between the ships in the subsequent motion, and the time taken to the nearest minute for such a situation to occur.

2. Two aircraft A and B are flying at the same altitude with velocities $180\,\text{m s}^{-1}$ due east and $240\,\text{m s}^{-1}$ due north respectively.
Initially B is 5 km due south of A.
Given that the aircraft do not change their velocities, find the shortest distance between the aircraft in the subsequent motion, and the time taken for such a situation to occur.

3. A road running north–south crosses a road running east–west at a junction O. Initially Paul is on the east–west road, 1·7 km west of O, and is cycling towards O at $15\,\text{km h}^{-1}$. At the same time Pat is at O cycling due north at $8\,\text{km h}^{-1}$.
If Paul and Pat do not alter their velocities, find the least distance they are apart in the subsequent motion and the time taken for that situation to occur.

4. At 7.30 a.m. two ships A and B have velocities $15\,\text{km h}^{-1}$, N 30° E and $20\,\text{km h}^{-1}$ due east respectively, with B 5 km due west of A. In the subsequent motion A and B do not alter their velocities.
Find the distance between A and B when they are closest together and the time at which this situation occurs, to the nearest minute.

5. A road running north–south crosses a road running east–west at a junction O. John cycles towards O from the west at $3\,\text{m s}^{-1}$ as Tom cycles towards O from the south at $4\,\text{m s}^{-1}$. Initially John is 600 m from O and Tom is 250 m from O.
If Tom and John do not alter their velocities, find the least distance they are apart during the motion and the time taken to reach that situation.
How far, and in what direction, are Tom and John then from O?

6. Two aircraft A and B are flying, at the same altitude, with velocities $200\,\text{m s}^{-1}$, N 30° E and $300\,\text{m s}^{-1}$, N 50° W respectively. Initially A and B are 2 km apart with B on a bearing S 70° E from A.
Given that A and B do not alter their velocities find the least distance of separation between the two aircraft in the subsequent motion and the time taken to reach such a situation.

7. At 1500 hours a trawler is 10 km due east of a launch. The trawler maintains a steady $10\,\text{km h}^{-1}$ on a bearing 180° and the launch maintains a steady $20\,\text{km h}^{-1}$ on a bearing 071°.
Find the minimum distance the boats are apart in the subsequent motion, and the time at which this occurs.
Find, to the nearest minute, the length of time for which the two boats are within 8 km of each other.

8. A battleship and a cruiser are initially 16 km apart with the battleship on a bearing N 35° E from the cruiser. The battleship travels at 14 km h^{-1} on a bearing S 29° E and the cruiser at 17 km h^{-1} on a bearing N 50° E. The guns on the battleship have a range of up to 6 km.

   Find:
   (a) the least distance between the cruiser and the battleship in the subsequent motion
   (b) the length of time for which the battleship has the cruiser within range of its guns.

Course for closest approach

The following example illustrates the method used to find the course which must be set in order that one body may pass as close as possible to a second body.

Example 18

A ship A is moving with a constant speed of 18 km h^{-1} in a direction N 55° E and is initially 6 km from a second ship B, the bearing of A from B being N 25° W. If B moves with a constant speed of 15 km h^{-1}, find the course B must set in order to pass as close as possible to A, the distance between the ships when they are closest together, and the time for this to occur, to the nearest minute.

Draw a diagram showing the initial positions of A and B, and consider the motion of B relative to A. To pass as close as possible to A, B must travel relative to A along a line BC which makes as small an angle as possible with BA.
In the vector triangle BLM, LM must be perpendicular to BC.

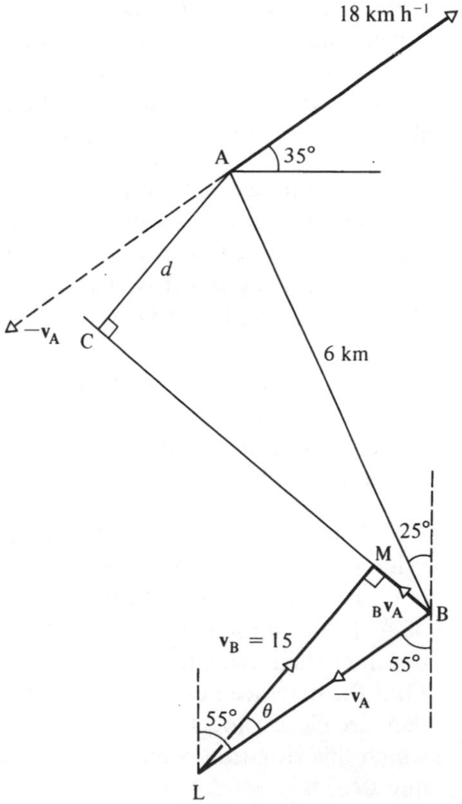

Hence $\cos\theta = \frac{15}{18}$

$\therefore \quad \theta = 33{\cdot}56°$

and course to be set is $55° - 33{\cdot}56°$
$= \text{N}\,21{\cdot}44°\,\text{E}$

$|_B\mathbf{v}_A| = \sqrt{(18^2 - 15^2)}$
$= 9{\cdot}950$

angle ABC $= 180° - 25° - 55° - (90° - \theta)$
$= 43{\cdot}56°$

The closest approach is *d*.

$$d = 6\sin 43{\cdot}56°$$

$$\therefore \quad d = 4{\cdot}135\text{ km}$$

$$\text{time taken} = \frac{\text{BC}}{|_B\mathbf{v}_A|} = \frac{6\cos 43{\cdot}56°}{9{\cdot}950} = 0{\cdot}437\text{ h} \approx 26\text{min}$$

The course to be set by B is N 21·44° E; the least distance apart is 4·14 km, and this occurs after 26 minutes.

Exercise 10G

1. Motorboat B is travelling at a constant velocity of $10\,\text{m s}^{-1}$ due east and motorboat A is travelling at a constant speed of $8\,\text{m s}^{-1}$. Initially A and B are 600 m apart with A due south of B. Find the course that A should set in order to get as close as possible to B. Find this closest distance and the time taken for the situation to occur.

2. At 8 a.m. two boats A and B are 5·2 km apart with A due west of B, and B travelling due north at a steady $13\,\text{km h}^{-1}$. If A travels with a constant speed of $12\,\text{km h}^{-1}$ show that, for A to get as close as possible to B, A should set a course of N $\theta°$ E where $\sin\theta = \frac{5}{13}$.
Find this closest distance and the time at which it occurs.

3. Two aircraft A and B are flying at the same altitude with A initially 5 km due north of B, and B flying at a steady $300\,\text{m s}^{-1}$ on a bearing 060°.
If A flies at a constant speed of $200\,\text{m s}^{-1}$, find the course that A must set in order to fly as close as possible to B, the distance between the planes when they are closest, and the time taken for this to occur.

4. A ship A is moving with a constant speed of $24\,\text{km h}^{-1}$ in a direction N 40° E and is initially 10 km from a second ship B, the bearing of A from B being N 30° W.
If B moves with a constant $22\,\text{km h}^{-1}$, find the course that B must set in order to pass as close as possible to A, the distance between the ships when they are closest, and the time taken (to the nearest minute) for this to occur.

5. At 12 noon a cruiser is 16 km due west of a destroyer. The cruiser is travelling at $40\,\text{km h}^{-1}$ on a bearing N 30° E and the destroyer is travelling at $20\,\text{km h}^{-1}$.
If the velocity of the cruiser and the speed of the destroyer do not change, find the course that the destroyer should set to get as close as possible to the cruiser, and find when this closest approach occurs. If the guns on the destroyer have a range of up to 10 km, find the length of time for which the cruiser will be within range (to the nearest minute).
If the guns on the cruiser have a range of up to 9 km, find the length of time for which the battleship will be within range (to the nearest minute).

6. A fielder is positioned 40 m from a batsman and on a bearing of S 70° W from the batsman. The fielder can run at $8\,\text{m s}^{-1}$ and the batsman hits a ball at $17\,\text{m s}^{-1}$ in a direction N 70° W.
Assuming that the fielder runs at $8\,\text{m s}^{-1}$ from the moment the ball is hit and neglecting any change in the velocity of the ball, find the closest distance that the fielder can get to the ball and how long it takes him to get there, from the time the ball is hit.

7. A battleship commander is informed that there is a lone cruiser positioned 40 km away from him on a bearing N 70° W. The guns on the battleship have a range of up to 8 km and the top speed of the battleship is $30\,\text{km h}^{-1}$. The cruiser maintains a constant velocity of $50\,\text{km h}^{-1}$, N 60° E.
Show that whatever the course the battleship sets it cannot get the cruiser within range of its guns.

Exercise 10H Examination questions

1. A river with long straight banks is 500 m wide and flows with a constant speed of $3\,\text{m s}^{-1}$. A man rowing a boat at a steady speed of $5\,\text{m s}^{-1}$, relative to the river, sets off from a point A on one bank so as to arrive at the point B directly opposite A on the other bank. Find the time taken to cross the river.
A woman also sets off at A rowing at $5\,\text{m s}^{-1}$ relative to the river and crosses in the shortest possible time. Find this time and the distance downstream of B of the point at which she lands. (AEB 1991)

2. A river flows with a constant speed of $7\,\text{m s}^{-1}$ from North to South and a man sets off from a point A on the West bank so as to cross the river. Relative to the river he sets off with speed $10\,\text{m s}^{-1}$ at an angle θ to North. Taking unit vectors $\mathbf{i}$ and $\mathbf{j}$ in the Northerly and Easterly directions, respectively, express his actual velocity in the form $a\mathbf{i} + b\mathbf{j}$. Hence, or otherwise, find the value of $\cos\theta$ such that the man lands on the Easterly bank directly opposite A. (WJEC)

3. Two towns S and T are 360 km apart and are such that T is on a bearing of 070° from S. A plane whose speed in still air is $300\,\text{km h}^{-1}$ flies directly from S to T. Given that there is a wind blowing from the north at $60\,\text{km h}^{-1}$, find

(i) the direction in which the pilot must steer the aircraft

(ii) the time taken, to the nearest minute, for the journey.

When the aircraft returns from T to S, the speed of the wind has changed, but it is still blowing from the north. Given that the pilot steers the plane on a bearing of 270°, find the new speed of the wind. (UCLES)

4. A city A is 600 km west of a city B. The time taken for an aircraft to fly directly from A to B is 2·5 hours. The aircraft steers a course of 120°. The speed of the aircraft in still air is $250\,\text{km h}^{-1}$; the velocity of the wind is constant.

(i) Sketch a vector triangle showing the velocity of the aircraft, the velocity of the aircraft relative to the wind, and the velocity of the wind.

(ii) Using the vector triangle, calculate the velocity of the wind.

(iii) By drawing the corresponding vector triangle for the return journey from B to A, determine the direction that the aircraft should steer to complete the return journey.

(iv) Would you expect the journey from B to A to take longer than that from A to B? Give a reason. (UODLE)

5. (i) State how the velocity of B relative to A is related to the true velocity of A and the true velocity of B.

(ii) Ship A, travelling due East at $15\,\text{km h}^{-1}$, is 12 km due North of ship B. Ship B, which can travel at $25\,\text{km h}^{-1}$, wishes to intercept ship A. In which direction should ship B travel for this to happen?

(iii) If ship B travels along this path, what is the position of B relative to A after 30 mins have elapsed?

(iv) Ship B now changes direction and sails in the new direction N 50° E. What is the shortest distance between the ships during the subsequent motion? (NICCEA)

6. Two long straight roads intersect at right angles at a cross-roads O. Two vehicles, one on each road, travel with constant speeds of $25\,\text{m s}^{-1}$ and $30\,\text{m s}^{-1}$ respectively. At time $t = 0$ they are both at a distance of 610 m from O and are approaching O. Write down their respective distances from O at time t seconds and find the value of t when the vehicles are closest together. (AEB 1992)

7. Two cars travelling with constant speeds $30\,\text{m s}^{-1}$ and $20\,\text{m s}^{-1}$ on perpendicular straight roads are approaching O, the point of intersection of the roads. At time $t = 0$ both are at a distance of 442 m from O. Determine their displacements from O at any subsequent time t seconds and find the value of t when the cars are closest together. (WJEC)

8. A car A is travelling with a constant velocity of $20\,\text{km h}^{-1}$ due west and a cyclist B has a constant velocity of $16\,\text{km h}^{-1}$ in the direction of the vector $(-4\mathbf{i} + 3\mathbf{j})$, where $\mathbf{i}$ and $\mathbf{j}$ are unit vectors due east and due north respectively. At noon, A is 1·2 km due north of B. Take the position of A at noon as the origin and obtain expressions for the position vectors of A and B at time t hours after noon, and hence show that the position vector of A relative to B is $\mathbf{r}$ km, where

$$5\mathbf{r} = 6\{-6t\,\mathbf{i} + (1 - 8t)\mathbf{j}\}.$$

Deduce that the distance between A and B is d km, where

$$25d^2 = 36(100t^2 - 16t + 1).$$

Hence show that the minimum separation between A and B is 720 m and find the time at which this occurs. (AEB 1994)

9. At time $t = 0$ seconds, the position vectors (in metres) of two particles A and B are $7\mathbf{i} + 9\mathbf{j}$ and $5\mathbf{i} + 3\mathbf{j}$ respectively, and the particles are moving at all times with constant velocities (in $\mathrm{m\,s^{-1}}$) of $4\mathbf{i} - 7\mathbf{j}$ and $9\mathbf{i} + 8\mathbf{j}$ respectively.
(a) Write down the velocity of B relative to A.
(b) Find the position vector of B relative to A at time t seconds.
(c) Determine whether the particles will collide.
(d) Find the cosine of the angle between their velocities. (WJEC)

10. Two joggers, A and B, are each running with constant velocity on level parkland. At a certain instant, A and B have position vectors $(-60\mathbf{i} + 210\mathbf{j})$ m and $(30\mathbf{i} - 60\mathbf{j})$ m respectively, referred to a fixed origin O. Ninety seconds later, A and B meet at the point with position vector $(210\mathbf{i} + 120\mathbf{j})$ m.
(a) Find, as a vector in terms of $\mathbf{i}$ and $\mathbf{j}$, the velocity of A relative to B.
(b) Verify that the magnitude of the velocity of A relative to B is equal to the speed of A. (ULEAC)

11. Two small boats move at constant velocity on the open sea. Boat P is travelling due West at $5\,\mathrm{m\,s^{-1}}$, and boat Q at $12\,\mathrm{m\,s^{-1}}$ due North. Taking unit vectors $\mathbf{i}$ and $\mathbf{j}$ in directions East and North respectively, write down, in terms of $\mathbf{i}$ and $\mathbf{j}$, the velocity of P, the velocity of Q, and also the velocity of Q relative to P.
Hence determine the magnitude and direction of the velocity of Q relative to P.
At time $t = 0$, Q is at a position 800 m South and 180 m West of P. Write down the distance PQ at this moment.
Show that the shortest distance between the two boats in the ensuing motion is approximately 142 m, and find the time t seconds at which they are closest to each other. (UODLE)

12. At noon, a ship, A, is travelling with a velocity of $4\mathbf{i} + 20\mathbf{j}\,\mathrm{km\,h^{-1}}$ and a second ship, B, due north of it, is travelling with a velocity of $-3\mathbf{i} - 4\mathbf{j}\,\mathrm{km\,h^{-1}}$ where $\mathbf{i}$ and $\mathbf{j}$ are unit vectors acting due East and North respectively.
(i) Find the velocity of A relative to B.
(ii) If the shortest distance between the vessels is 4·2 km find
(a) the time, to the nearest minute, at which they are closest together, and
(b) their original distance apart at noon.
(iii) If visibility at the time is 12 km show that they are within sight of each other for 54 minutes (to the nearest minute).
(iv) When the distance apart is a minimum what is the bearing of B from A? (NICCEA)

13. A particle A has speed $6\,\mathrm{m\,s^{-1}}$ in the direction $\mathbf{i} + 2\mathbf{j} - 2\mathbf{k}$ and particle B has speed $7\,\mathrm{m\,s^{-1}}$ in the direction $2\mathbf{i} + 3\mathbf{j} - 6\mathbf{k}$. Find the position vectors of A and B relative to a fixed origin O at time t seconds, given that when $t = 0$ the position vectors are $(\mathbf{i} - \mathbf{j} + 2\mathbf{k})$ m and $(\mathbf{i} + 5\mathbf{j} + 4\mathbf{k})$ m respectively. Show that in the subsequent motion the minimum distance between A and B is $2\sqrt{5}$ m and find the time at which this position of minimum separation occurs. When $t = 1$ a particle C is at the point with position vector $(\mathbf{i} + 4\mathbf{j} - 4\mathbf{k})$ m relative to O. Given that C moves with constant velocity $\mathbf{u}\,\mathrm{m\,s^{-1}}$ and that B and C collide when $t = 3$, find $\mathbf{u}$. (AEB 1993)

11 Work, energy and power

Work done by a constant force

If the point of application of a constant force of F newtons moves through a distance s metres in the direction of the force, then the work done is defined to be:

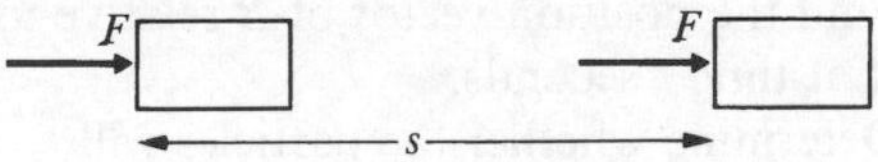

$$F \times s \text{ joules}$$

The unit of work is the joule, abbreviated J.

Work done against gravity

In order to raise a mass of m kg vertically at a constant speed, a force of mg N must be applied vertically upwards to the mass. In raising the mass a distance s metres, the work done against gravity will be:

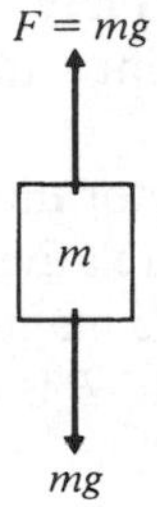

$$mgs \text{ joules}$$

Example 1

Find the work done against gravity when an object of mass 3·5 kg is raised through a vertical distance of 6 m.

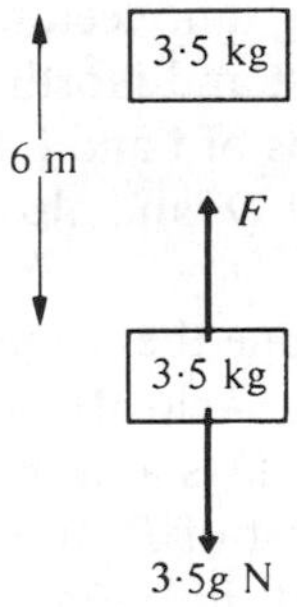

The vertical force required is F:

$$F = 3{\cdot}5g$$

work done $= F \times s$

$$= 3{\cdot}5g \times 6 = 205{\cdot}8 \text{ J}$$

The work done against gravity is 205·8 J.

General motion at constant speed

In order to move a body at a constant speed, a force equal in magnitude to the forces of resistance acting on the body has to be applied to the body.

Example 2

A block of wood is pulled a distance of 4 m across a horizontal surface against resistances totalling 7·5 N. If the block moves at a constant velocity, find the work done against the resistances.

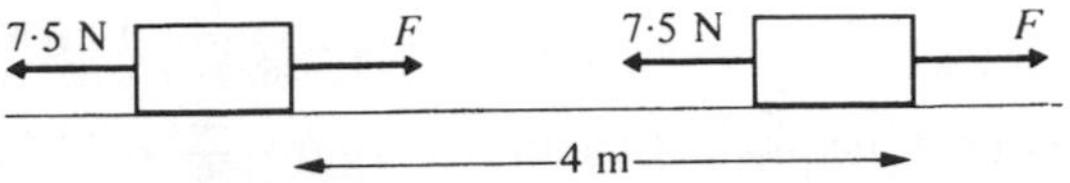

Let the pulling force be F.

Resolve horizontally: $F = 7{\cdot}5\,\text{N}$

work done against resistances

$$= 7{\cdot}5 \times \text{horizontal distance moved}$$
$$= 7{\cdot}5 \times 4 = 30\,\text{J}$$

The work done against the resistances is 30 J.

Example 3

A horizontal force pulls a body of 2·25 kg a distance of 8 m across a rough horizontal surface, coefficient of friction $\frac{1}{3}$. The body moves with constant velocity and the only resisting force is that due to friction. Find the work done against friction.

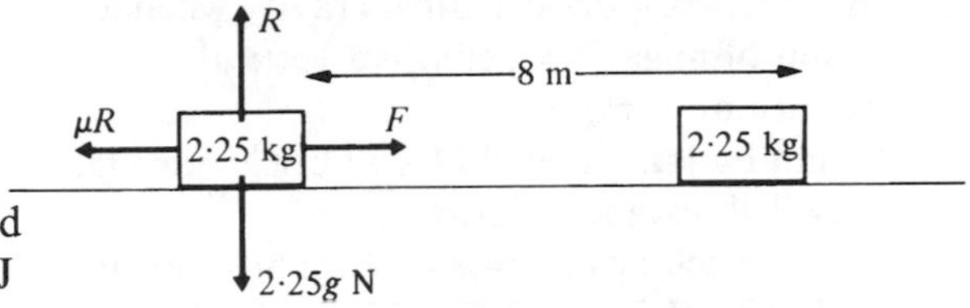

The frictional force is μR.

Resolve vertically: $R = 2{\cdot}25g$

work done against friction $= \mu R \times \text{distance moved}$
$= \frac{1}{3}(2{\cdot}25g) \times 8 = 58{\cdot}8\,\text{J}$

The work done against friction is 58·8 J.

Work done against gravity and friction

When a body is pulled at a uniform speed up the surface of a rough inclined plane, work is done both against gravity and against the frictional force which is acting on the body due to the contact with the rough surface of the plane.

Example 4

A rough surface is inclined at $\tan^{-1}\frac{7}{24}$ to the horizontal. A body of mass 5 kg lies on the surface and is pulled at a uniform speed a distance of 75 cm up the surface by a force acting along a line of greatest slope. The coefficient of friction between the body and the surface is $\frac{5}{12}$. Find:
(a) the work done against gravity
(b) the work done against friction.

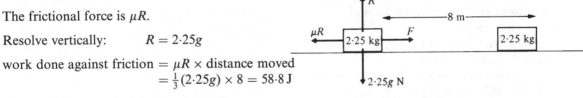

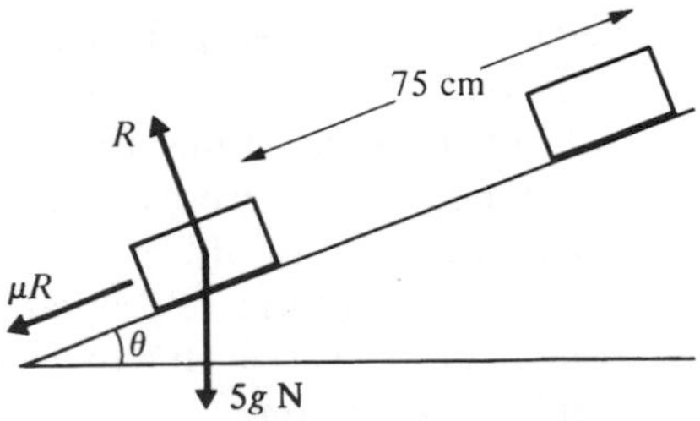

Given $\tan\theta = \frac{7}{24}$, $\sin\theta = \frac{7}{25}$, $\cos\theta = \frac{24}{25}$.

(a) work done against gravity = force × vertical distance moved
$$= 5g \times \tfrac{75}{100}\sin\theta = 5g \times \tfrac{75}{100} \times \tfrac{7}{25}$$
$$= 10{\cdot}29\,\text{J}$$

The work done against gravity is 10·3 J.

(b) Resolving perpendicular to the plane gives:

$$R = 5g \cos\theta = 5g \times \tfrac{24}{25} = \tfrac{24}{5}g$$

frictional force $= \mu R$

$= \tfrac{5}{12} \times \tfrac{24}{5}g = 19{\cdot}6\,\text{N}$

work done against friction $= 19{\cdot}6 \times \tfrac{75}{100} = 14{\cdot}7\,\text{J}$

The work done against friction is 14·7 J.

Exercise 11A

1. Find the work done against gravity when a body of mass 5 kg is raised through a vertical distance of 2 m.
2. Find the work done against gravity when a body of mass 1 kg is raised through a vertical distance of 3 m.
3. Find the work done against gravity when a person of mass 80 kg climbs a vertical distance of 25 m.
4. A man building a wall lifts 50 bricks through a vertical distance of 3 m. If each brick has a mass of 4 kg, how much work does the man do against gravity?
5. A body of mass 2 kg is moved vertically upwards at a constant speed of $5\,\text{m}\,\text{s}^{-1}$. Find how much work is done against gravity in each second.
6. A body of mass 200 g is moved vertically upwards at a constant speed of $2\,\text{m}\,\text{s}^{-1}$. Find how much work is done against gravity in each second.
7. A body of mass 10 kg is pulled a distance of 20 m across a horizontal surface against resistances totalling 40 N. If the body moves with uniform velocity, find the work done against the resistances.
8. A box is pulled a distance of 8 m across a horizontal surface against resistances totalling 7 N. If the box moves with uniform velocity, find the work done against the resistances.
9. A horizontal force pulls a body of mass 5 kg a distance of 8 m across a rough horizontal surface, coefficient of friction 0·25. The body moves with uniform velocity and the only resisting force is that due to friction. Find the work done.
10. A horizontal force pulls a body of mass 3 kg a distance of 20 m across a rough horizontal surface, coefficient of friction $\frac{2}{7}$. The body moves with uniform velocity and the only resisting force is that due to friction. Find the work done.
11. A horizontal force drags a body of mass 4 kg a distance of 10 m across a rough horizontal floor at a constant speed. The work done against friction is 49 J. Find the coefficient of friction between the body and the surface.
12. A smooth surface is inclined at 30° to the horizontal. A body of mass 15 kg lies on the surface and is pulled at a uniform speed a distance of 10 m up a line of greatest slope. Find the work done against gravity.
13. A surface is inclined at $\tan^{-1}\frac{3}{4}$ to the horizontal. A body of mass 50 kg lies on the surface and is pulled at a uniform speed a distance of 5 m up a line of greatest slope against resistances totalling 50 N. Find:
    (a) the work done against gravity
    (b) the work done against the resistances.
14. A rough surface is inclined at $\tan^{-1}\frac{5}{12}$ to the horizontal. A body of mass 130 kg is pulled at a uniform speed a distance of 50 m up the surface by a force acting along a line of greatest slope. The coefficient of friction between the body and the surface is $\frac{2}{7}$. Find:
    (a) the frictional force acting
    (b) the work done against friction
    (c) the work done against gravity.

15. A rough surface is inclined at 30° to the horizontal. A body of mass 100 kg is pulled at a uniform speed a distance of 20 m up the surface by a force acting along a line of greatest slope. The coefficient of friction between the body and the surface is 0·1. Find:
    (a) the work done against friction
    (b) the work done against gravity.
16. A rough surface is inclined at $\tan^{-1}\frac{3}{4}$ to the horizontal.
    Find the total work done when a body of mass 50 kg is pulled at a uniform speed a distance of 15 m up the surface by a force acting along a line of greatest slope. The coefficient of friction between the body and the surface is $\frac{1}{3}$ and the only resistances to motion are those due to gravity and friction.
17. A rough surface is inclined at an angle θ to the horizontal. A body of mass m is pulled at a uniform speed a distance x up the surface by a force acting along a line of greatest slope. The coefficient of friction between the body and the plane is μ.
    If the only resistances to motion are those due to gravity and friction, show that the total work done on the body is $mgx(\sin\theta + \mu\cos\theta)$.

Energy

The energy of a body is a measure of the capacity which the body has to do work.
When a force does work on a body it changes the energy of the body.
Energy can exist in a number of different forms, but we shall consider two main types: kinetic energy and potential energy.

Kinetic energy

The kinetic energy of a body is that energy which it possesses by virtue of its motion.
When a force does work on a body so as to increase its speed, then the work done is a measure of the increase in the kinetic energy of the body.

Suppose a constant force F acts on a body of mass m, which is initially at rest on a smooth horizontal surface, and after moving a distance s across the surface the body has a speed v.

$$\text{work done on the body} = F \times s$$

But $$F = \text{mass} \times \text{acceleration, and acceleration} = \frac{v^2 - 0}{2s}$$

$$\therefore \quad F = \frac{mv^2}{2s}$$

$$\therefore \quad \text{work done on the body} = \frac{mv^2}{2s} \times s = \frac{mv^2}{2}$$

The quantity $\frac{mv^2}{2}$ is defined as the kinetic energy of mass m moving with velocity v. A body at rest therefore has zero kinetic energy.

Example 5

Find the kinetic energy of a particle of mass 250 g moving with a speed of $4\sqrt{2}\,\text{m s}^{-1}$.

$$\text{kinetic energy} = \tfrac{1}{2}mv^2$$
$$= \tfrac{1}{2}(0{\cdot}25)(4\sqrt{2})^2 = 4\,\text{J}$$

The kinetic energy of the particle is 4 J.

Potential energy

The potential energy of a body is that energy it possesses by virtue of its position.
When a body of mass m kg is raised vertically a distance h metres, the work done against gravity is mgh joules. The work done against gravity is a measure of the *increase* in the potential energy of the body, i.e. the capacity of the body to do work is increased.
When a body is lowered vertically its potential energy is decreased.
There is no zero of potential energy, although an arbitrary level may be used from which *changes* in the potential energy of a body may be measured.

Example 6

Find the change in the potential energy of a child of mass 48 kg when
(a) ascending a vertical distance of 2 m
(b) descending a vertical distance of 2 m.

(a) change in potential energy = work done against gravity

$$= mgh$$
$$= 48g \times 2 = 940{\cdot}8\,\text{J}$$

The change in potential energy is 940·8 J.

(b) When descending, the child is *losing* potential energy, i.e. losing its potential to do work

$$\text{loss in potential energy} = mgh$$
$$= 48g \times 2 = 940{\cdot}8\,\text{J}$$

The loss in potential energy is 940·8 J.

Example 7

A cricket ball of mass 400 g moving at $3\,\text{m s}^{-1}$ and a golf ball of mass 100 g have equal kinetic energies. Find the speed at which the golf ball is moving.

$$\text{kinetic energy of cricket ball} = \tfrac{1}{2}mv^2$$
$$= \tfrac{1}{2}\left(\tfrac{400}{1000}\right)(3)^2 = 1{\cdot}8\,\text{J}$$
$$\text{kinetic energy of golf ball} = \tfrac{1}{2}\left(\tfrac{100}{1000}\right)v^2$$
$$\therefore \quad 1{\cdot}8 = \frac{v^2}{20}$$
$$\therefore \quad v^2 = 36 \quad \text{or} \quad v = 6\,\text{m s}^{-1}$$

The golf ball is moving at $6\,\text{m s}^{-1}$.

Example 8

A body of mass 4 kg decreases its kinetic energy by 32 J. If initially it had a speed of $5\,\text{m}\,\text{s}^{-1}$, find its final speed.

Initial kinetic energy $= \frac{1}{2}mv^2$

$= \frac{1}{2}(4)(5)^2 = 50\,\text{J}$

$\therefore$ final kinetic energy $= 50 - 32 = 18\,\text{J}$

Let final speed of body be v.

Then $18 = \frac{1}{2}(4)v^2$ or $v = 3\,\text{m}\,\text{s}^{-1}$

The final speed of the body is $3\,\text{m}\,\text{s}^{-1}$.

Exercise 11B

1. Find the kinetic energy of:
   (a) a body of mass 5 kg moving with speed $4\,\text{m}\,\text{s}^{-1}$
   (b) a body of mass 2 kg moving with speed $3\,\text{m}\,\text{s}^{-1}$
   (c) a car of mass 800 kg moving with speed $10\,\text{m}\,\text{s}^{-1}$
   (d) a particle of mass 100 g moving with speed $20\,\text{m}\,\text{s}^{-1}$
   (e) a bullet of mass 10 g moving with speed $400\,\text{m}\,\text{s}^{-1}$.
2. Find the potential energy gained by:
   (a) a body of mass 5 kg raised through a vertical distance of 10 m
   (b) a man of mass 60 kg ascending a vertical distance of 5 m
   (c) a lift of mass 1 tonne ascending a vertical distance of 20 m.
3. Find the potential energy lost by:
   (a) a body of mass 20 kg falling through a vertical distance of 2 m
   (b) a man of mass 80 kg descending a vertical distance of 10 m
   (c) a lift of mass 500 kg descending a vertical distance of 20 m.
4. Find the gain in kinetic energy when:
   (a) a car of mass 1 tonne increases its speed from $5\,\text{m}\,\text{s}^{-1}$ to $6\,\text{m}\,\text{s}^{-1}$
   (b) a body of mass 5 g increases its speed from $200\,\text{m}\,\text{s}^{-1}$ to $300\,\text{m}\,\text{s}^{-1}$.
5. Find the loss in kinetic energy when:
   (a) a body of mass 2 kg decreases its speed from $2\,\text{m}\,\text{s}^{-1}$ to $1\,\text{m}\,\text{s}^{-1}$
   (b) a car of mass 800 kg decreases its speed from $18\,\text{km}\,\text{h}^{-1}$ to rest.
6. A body of mass 5 kg, initially moving with speed $2\,\text{m}\,\text{s}^{-1}$, increases its kinetic energy by 30 J.
   Find the final speed of the body.
7. A body of mass 5 kg, initially moving with speed $3\,\text{m}\,\text{s}^{-1}$, increases its kinetic energy by 40 J.
   Find the final speed of the body.
8. Find the gain in kinetic energy when a car of mass 900 kg accelerates at $2\,\text{m}\,\text{s}^{-2}$ for 5 s if initially it is at rest.
9. A body of mass 6 kg, initially moving with speed $12\,\text{m}\,\text{s}^{-1}$, experiences a constant retarding force of 10 N for 3 s.
   Find the kinetic energy of the body at the end of this time.

The Principle of Conservation of Energy

Suppose we have a situation involving a moving body in which
(a) there is no work done against friction, and
(b) gravity is the only external force which does work on the body, (or against which the body has to do work).

The total energy possessed by the body will then be the total of its kinetic energy and its potential energy and, by the *Principle of Conservation of Energy*, this will be a constant.

i.e. total energy = kinetic energy (KE) + potential energy (PE)

= constant

Expressing this in another way, convenient for problem solving:

total energy in the initial state = total energy in the final state

Example 9

The point A is 4 metres vertically above the point B. A body of mass 0·2 kg is projected from A vertically downwards with speed $3\,\text{m}\,\text{s}^{-1}$.
Find the speed of the body when it reaches B.

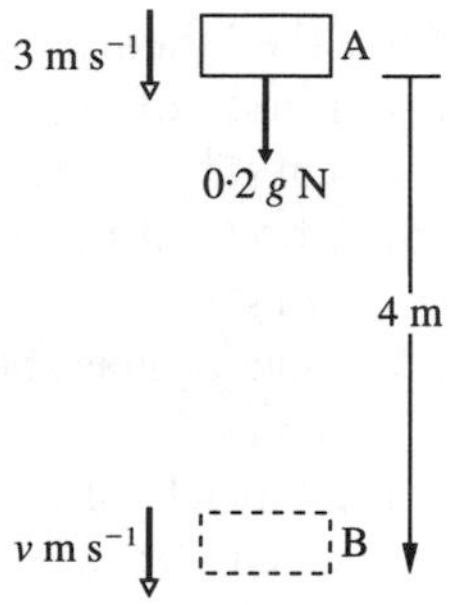

There is no friction, and gravity is the only external force.
We shall calculate the total energy at A and the total energy at B and then use the Principle of Conservation of Energy.
We shall choose to measure the potential energy from the level of B and we will let $v\,\text{m}\,\text{s}^{-1}$ be the speed of the body at B.

At A:

$$\begin{aligned} \text{KE} &= \tfrac{1}{2}mv^2 \\ &= \tfrac{1}{2}(0{\cdot}2)(3)^2 \\ &= 0{\cdot}9\,\text{J} \\ \text{PE} &= mgh \\ &= (0{\cdot}2)(9{\cdot}8)(4) \\ &= 7{\cdot}84\,\text{J} \\ \text{total energy} &= 8{\cdot}74\,\text{J} \end{aligned}$$

At B:

$$\begin{aligned} \text{KE} &= \tfrac{1}{2}mv^2 \\ &= \tfrac{1}{2}(0{\cdot}2)(v)^2 \\ &= 0{\cdot}1\,v^2\,\text{J} \\ \text{PE} &= 0\,\text{J} \\ \text{total energy} &= 0{\cdot}1\,v^2\,\text{J} \end{aligned}$$

By the Principle of Conservation of Energy:

$$\text{total energy at A} = \text{total energy at B}$$

i.e. $8{\cdot}74 = 0{\cdot}1\,v^2$

$\therefore$ $v^2 = 87{\cdot}4$

$v = 9{\cdot}35$

The speed of the body at B is $9{\cdot}35\,\text{m}\,\text{s}^{-1}$.

Example 10

The point A is vertically below the point B. A particle of mass 0·1 kg is projected from A vertically upwards with speed 21 m s^{-1} and passes through point B with speed 7 m s^{-1}. Find the distance from A to B.

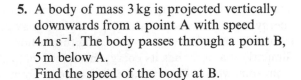

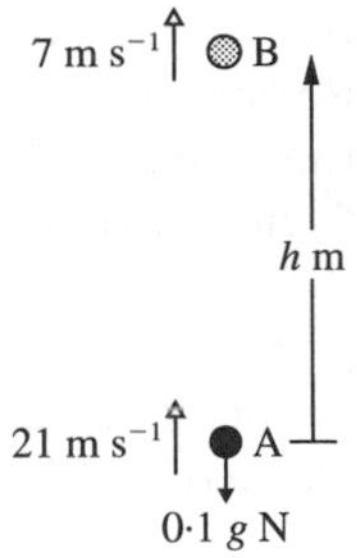

We shall choose to measure the potential energy from the level of A. Let the distance from A to B be h metres.

At A:

$$\begin{aligned} \text{KE} &= \tfrac{1}{2}mv^2 \\ &= \tfrac{1}{2}(0{\cdot}1)(21)^2 \\ &= 22{\cdot}05\text{ J} \\ \text{PE} &= 0\text{ J} \\ \text{total energy} &= 22{\cdot}05\text{ J} \end{aligned}$$

At B:

$$\begin{aligned} \text{KE} &= \tfrac{1}{2}mv^2 \\ &= \tfrac{1}{2}(0{\cdot}1)(7)^2 \\ &= 2{\cdot}45\text{ J} \\ \text{PE} &= mgh \\ &= (0{\cdot}1)\,(9{\cdot}8)\,(h) \\ &= 0{\cdot}98h\text{ J} \\ \text{total energy} &= 2{\cdot}45 + 0{\cdot}98h\text{J} \end{aligned}$$

By the Principle of Conservation of Energy:

$$\text{total energy at A} = \text{total energy at B}$$

i.e. $22{\cdot}05 = 2{\cdot}45 + 0{\cdot}98h$

$\therefore$ $h = 20$

The distance from A to B is 20 metres.

Exercise 11C

Use the Principle of Conservation of Energy to solve the following problems.

1. A body of mass 2 kg is released from rest, and falls freely under gravity.
   Find its speed when it has fallen a distance of 10 m.
2. A body of mass 6 kg is released from rest, and falls freely under gravity.
   Find its speed when it has fallen a distance of 90 m.
3. A body of mass 5 kg is released from rest, and falls freely under gravity.
   Find the distance it has fallen when its speed is 7 m s^{-1}.
4. A body of mass 20 kg is released from rest, and falls freely under gravity.
   Find the distance it has fallen when its speed is 21 m s^{-1}.
5. A body of mass 3 kg is projected vertically downwards from a point A with speed 4 m s^{-1}. The body passes through a point B, 5 m below A.
   Find the speed of the body at B.
6. A body of mass 10 kg is projected vertically upwards from a point A with speed 4 m s^{-1}. The body reaches its highest point and then falls, passing through a point B, 5 metres below A.
   Find the speed of the body at B.
   Would your answer have been different had the mass been different?
7. A and B are two points in a vertical line, with A above B. A body of mass 4 kg is released from A and falls vertically, passing through B with speed 7 m s^{-1}.
   Find the distance AB.

8. A and B are two points in a vertical line with B a distance of 4 m above A.
A body P, of mass m kg, is projected vertically upwards from A with a speed of $10\,\text{m s}^{-1}$. Find:
(a) the speed of P when it passes upwards through B
(b) the speed of P when it passes through B again, travelling downwards
(c) the speed of P when it returns to the point of projection
(d) the height above A of the highest point reached by P.

9. A smooth slope is inclined at $\tan^{-1}\frac{3}{4}$ to the horizontal. A body of mass 4 kg is released from rest at the top of the slope and reaches the bottom with speed $7\,\text{m s}^{-1}$.
Find the length of the slope.

10. Point A is situated at the base of a smooth plane which is inclined at an angle $\tan^{-1}\frac{5}{12}$ to the horizontal. A body is projected from A with a speed of $14\,\text{m s}^{-1}$ up the line of greatest slope and first comes to rest at B.
Find the distance AB.

11. Suppose that a perfect helter-skelter is invented, in which there is no friction or air resistance.

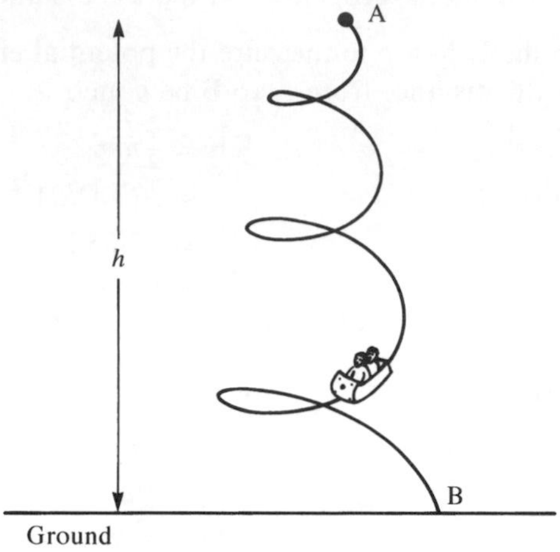

The track follows a downward path from a point A at height h above the ground to a point B at ground level. Find the speed v attained at B if the carriage starts from rest, showing that v does not depend on the shape of the path connecting A to the ground level.

Equivalence of work and energy

The principle that energy is conserved means that if a body loses energy this energy must go somewhere. For example, it may go to do work against some frictional force that is impeding the motion of the body.
Similarly if a body has its energy increased, that extra energy must come from somewhere. For example, it may come from work being done by some force acting upon the body causing it to move.
These ideas are used in the examples which follow.

Example 11

A constant force pulls a body of mass 0·5 kg in a straight line across a smooth horizontal surface. The body passes through a point A with a speed of $3\,\text{m s}^{-1}$ and then through a point B with speed of $5\,\text{m s}^{-1}$, B being 3 m from A.
Find the magnitude of the force.

We need only consider the kinetic energy in this case since the potential energy does not change.

Let the constant force be F N (see diagram).

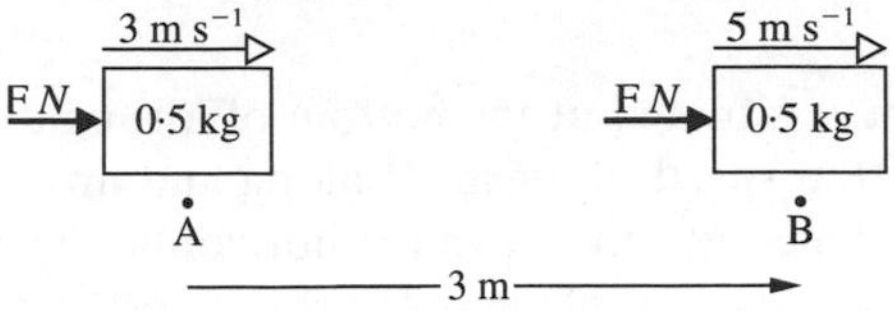

At A: $\quad \text{KE} = \frac{1}{2}mv^2$

$$= \tfrac{1}{2}(0{\cdot}5)(3)^2$$

$$= 2{\cdot}25\ \text{J}$$

At B: $\quad \text{KE} = \frac{1}{2}mv^2$

$$= \tfrac{1}{2}(0{\cdot}5)(5)^2$$

$$= 6{\cdot}25\ \text{J}$$

Gain in energy in moving from A to B $= (6{\cdot}25 \quad 2{\cdot}25)$ J
$= 4$ J

This energy gain has come from the applied force doing work on the body.

Thus: $\quad$ work done by force $= 4$ J

$\therefore \quad$ force × distance moved $= 4$ J

i.e. $\quad F \times 3 = 4$

giving $\quad F = 1\frac{1}{3}$

The magnitude of the force is $1\frac{1}{3}$ N.

Example 12

A body of mass 2 kg falls vertically, passing through two points A and B. The speeds of the body as it passes A and B are $1\ \text{m s}^{-1}$ and $4\ \text{m s}^{-1}$, respectively. The resistance against which the body falls is 9·6 N. Use energy considerations to determine the distance AB.

Let the unknown distance be h m (see diagram).
We will measure potential energies from the level of B.

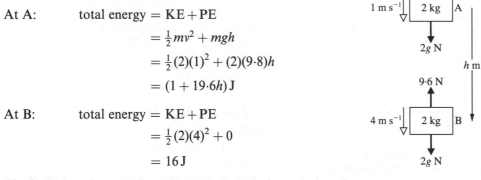

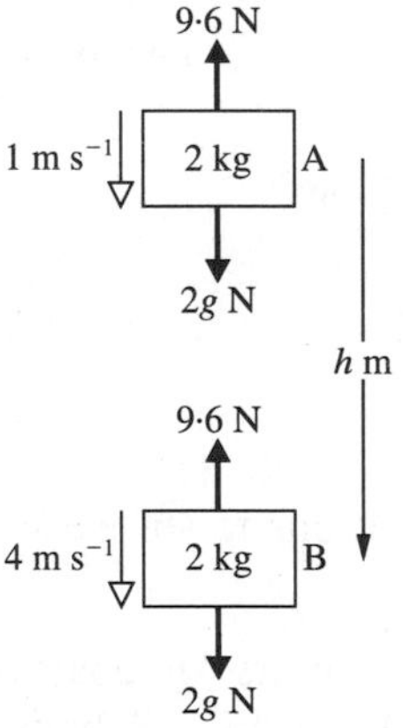

At A: $\quad$ total energy $= \text{KE} + \text{PE}$

$$= \tfrac{1}{2}mv^2 + mgh$$

$$= \tfrac{1}{2}(2)(1)^2 + (2)(9{\cdot}8)h$$

$$= (1 + 19{\cdot}6h)\ \text{J}$$

At B: $\quad$ total energy $= \text{KE} + \text{PE}$

$$= \tfrac{1}{2}(2)(4)^2 + 0$$

$$= 16\ \text{J}$$

The body has done work against a resistance force and so, in falling from A to B, it will have lost energy.
Thus: work done against resistance = loss in energy

$\therefore \quad 9{\cdot}6h = (1 + 19{\cdot}6h) - 16$

giving $\quad h = 1{\cdot}5$

The distance AB is 1·5 m.

Example 13

From the point A situated at the bottom of a rough inclined plane, a body is projected with a speed of $5{\cdot}6\,\mathrm{m\,s^{-1}}$ along and up a line of greatest slope. The plane is inclined at $\tan^{-1}\frac{4}{3}$ to the horizontal. The coefficient of friction between the body and the plane is $\frac{4}{7}$ and the body first comes to rest at point B. By energy considerations, find the distance AB.

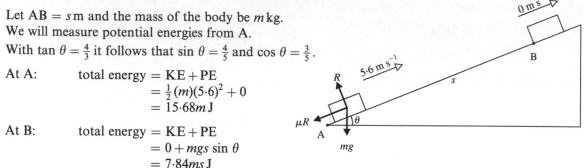

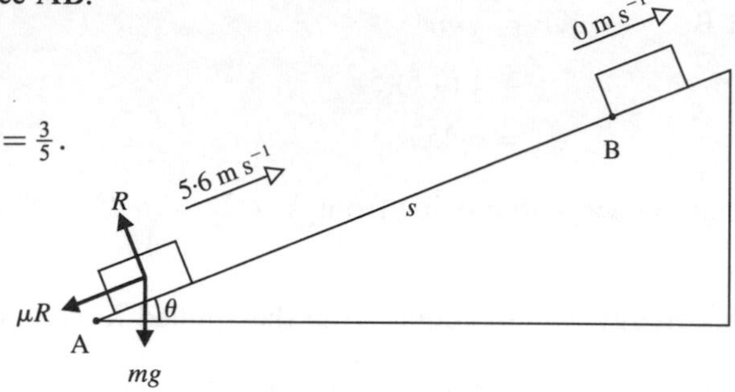

Let AB $= s$ m and the mass of the body be m kg.
We will measure potential energies from A.
With $\tan\theta = \frac{4}{3}$ it follows that $\sin\theta = \frac{4}{5}$ and $\cos\theta = \frac{3}{5}$.

At A: total energy $= \mathrm{KE} + \mathrm{PE}$
$$= \tfrac{1}{2}(m)(5{\cdot}6)^2 + 0$$
$$= 15{\cdot}68m\ \mathrm{J}$$

At B: total energy $= \mathrm{KE} + \mathrm{PE}$
$$= 0 + mgs\sin\theta$$
$$= 7{\cdot}84ms\ \mathrm{J}$$

The body has done work against friction and so, in travelling from A to B, it will have lost energy. Thus:

work done against resistance = loss in energy

$$\therefore \quad \mu Rs = 15{\cdot}68m - 7{\cdot}84ms \quad \ldots[1]$$

Resolving perpendicular to the plane gives:

$$R = mg\cos\theta$$

Thus
$$\mu R = \mu mg\cos\theta$$
$$= \tfrac{4}{7}(m)(9{\cdot}8)\tfrac{3}{5}$$
$$= 3{\cdot}36m\ \mathrm{N}$$

Substituting this into [1] gives:

$$3{\cdot}36ms = 15{\cdot}68m - 7{\cdot}84ms$$

Hence
$$s = 1{\cdot}4$$

The distance AB is $1{\cdot}4$ m.

Exercise 11D

Use energy considerations to solve each of the following.

1. A and B are two points 4 m apart on a smooth horizontal surface. A body of mass 2 kg is initially at rest at A and is pushed by a force of constant magnitude acting in the direction from A to B. The body reaches B with a speed of $4\,\mathrm{m\,s^{-1}}$.
Find the magnitude of the force.
2. A constant force pushes a mass of 4 kg in a straight line across a smooth horizontal surface. The body passes through a point A with a speed of $5\,\mathrm{m\,s^{-1}}$ and then through a point B with a speed of $8\,\mathrm{m\,s^{-1}}$. B is 6 m from A.
Find the magnitude of the force acting on the mass.
3. A and B are two points 3 m apart on a smooth horizontal surface. A body of mass 6 kg is initially at rest at A and is pushed towards B with a constant force of 9 N.
Find the speed of the body when it reaches B.

4. A constant force of magnitude 8 N pushes a body of mass 4 kg in a straight line across a smooth horizontal surface. The body passes through a point A with a speed of $4\,\text{m}\,\text{s}^{-1}$, and then through a point B, 5 m from A. Find the speed of the body at B.

5. A particle of mass 100 g moves in a straight line across a horizontal surface against resistances of constant magnitude. The particle passes through a point A with a speed of $7\,\text{m}\,\text{s}^{-1}$, and then through a point B with a speed of $3\,\text{m}\,\text{s}^{-1}$, B being 2 m from A. Find the magnitude of the resistances.

6. A body of mass 5 kg moves in a straight line across a horizontal surface against a constant resistance of magnitude 10 N. The body passes through a point A and then comes to rest at a point B, 9 m from A. Find the speed of the body when it is at A.

7. A body of mass 5 kg slides over a rough horizontal surface. In sliding 5 m the speed of the body decreases from $8\,\text{m}\,\text{s}^{-1}$ to $6\,\text{m}\,\text{s}^{-1}$. Find:
(a) the work done against friction
(b) the coefficient of friction.

8. A and B are two points 15 m apart in the same vertical line, with A above B. A body of mass 5 kg is released from rest at A and falls vertically against a constant resistance of 25 N.
Find the speed of the body when passing through B.

9. A body of mass 5 kg falls vertically against a constant resistance. The body passes through two points A and B, 2·5 m apart, with A above B. Its speed is $2\,\text{m}\,\text{s}^{-1}$ when passing through A and $6\,\text{m}\,\text{s}^{-1}$ when passing through B. Find the magnitude of the resistance.

10. A body of mass of 2 kg falls vertically against constant resistance of 14 N. The body passes through two points A and B when travelling with speeds $3\,\text{m}\,\text{s}^{-1}$ and $10\,\text{m}\,\text{s}^{-1}$ respectively.
Find the distance AB.

11. A bullet of mass 15 grams is fired towards a fixed wooden block, and enters the block when travelling horizontally at $400\,\text{m}\,\text{s}^{-1}$. It comes to rest after penetrating a distance of 25 cm.
Find the work done against the resistance of the wood, and find the magnitude of the resistance (assumed constant throughout).

12. A bullet of mass 8 g is fired towards a fixed wooden block and enters the block when travelling horizontally at $300\,\text{m}\,\text{s}^{-1}$. Find how far the bullet penetrates if the wood provides a constant resistance of 1800 N.

13. A rough slope of length 5 m is inclined at an angle of 30° to the horizontal. A body of mass 2 kg is released from the top of the slope and travels down the slope against a constant resistance. The body reaches the bottom of the slope with speed $2\,\text{m}\,\text{s}^{-1}$.
Find the work done against the resistance, and the magnitude of the resistance.

14. A rough slope of length 10 m is inclined at an angle of $\tan^{-1}\frac{3}{4}$ to the horizontal. A block of mass 50 kg is released from rest at the top of the slope, and travels down the slope reaching the bottom with a speed of $8\,\text{m}\,\text{s}^{-1}$.
Find:
(a) the work done against friction
(b) the magnitude of the frictional force
(c) the coefficient of friction between the block and the surface.

15. Point A is situated at the bottom of a rough plane which is inclined at an angle $\tan^{-1}\frac{3}{4}$ to the horizontal. A body is projected from A with a speed of $14\,\text{m}\,\text{s}^{-1}$ along and up a line of greatest slope. The coefficient of friction between the body and the plane is 0·25. The body first comes to rest at a point B.
Find the distance AB.

16. Point A is situated at the bottom of a rough plane which is inclined at 45° to the horizontal. A body of mass 0·5 kg is projected from A along and up the line of greatest slope. The coefficient of friction between the body and the plane is $\frac{3}{7}$. The body first comes to rest at a point B, at a distance $4\sqrt{2}$ m from A, before returning to A.
Find:
(a) the work done against friction when the body moves from A to B
(b) the initial speed of the body
(c) the work done against friction when the body moves from A to B and back to A
(d) the speed of the body on return to A.

17. Point A is at the bottom of a rough plane which is inclined at an angle θ to the horizontal. A body of mass m is projected from A, along and up a line of greatest slope. The coefficient of friction between the body and the plane is μ. The body first comes to rest at a point B, a distance x from A. Show that:

(a) the work done against friction when the body moves from A to B and back to A is given by $2\mu mgx\cos\theta$

(b) the initial speed of the body is: $\sqrt{2gx(\sin\theta + \mu\cos\theta)}$

(c) the speed of the body on its return to A is: $\sqrt{2gx(\sin\theta - \mu\cos\theta)}$

(d) Are there any circumstances under which the body will not return to A?

Work done by a force acting in a direction different to that of the motion of the body

Up to this point the work done by a force has been calculated for situations in which the direction of the force has been the same as, or directly opposite to, the direction of motion of the body. This will not always be the case. For example if we drag a load up a hill by a rope slung over the shoulder the direction of motion is up the slope but the force doing the work points in the direction of the rope.

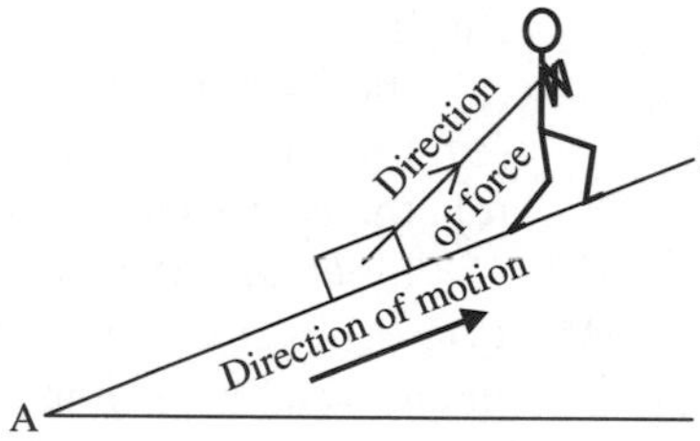

Figure 1 simplifies this picture. The force is represented by a vector **F** at an angle θ to the plane. The length from A to B is AB.

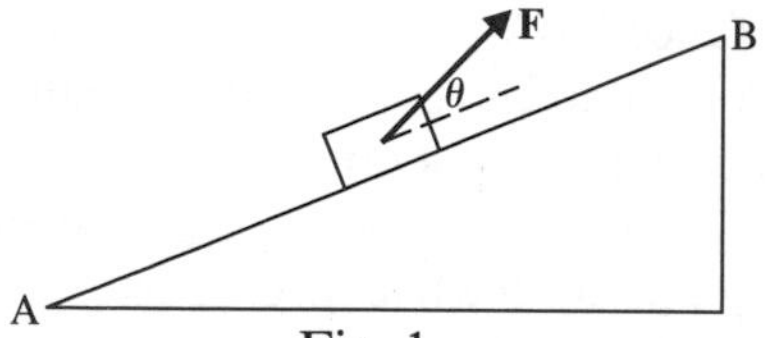

Fig. 1

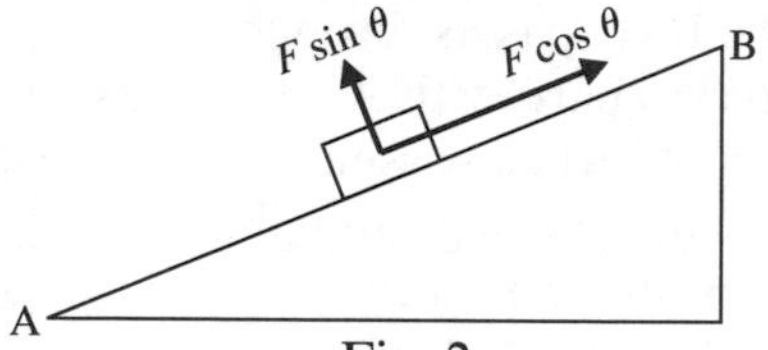

Fig. 2

Figure 2 shows the components of **F** along and perpendicular to the plane. The magnitude of **F** is written F and is assumed constant. The component ($F\sin\theta$) does no work because it is perpendicular to the plane and its point of application has no motion in this direction. Only the component ($F\cos\theta$) contributes to the work done which is given by:

$$\text{work done by } \mathbf{F} = (F\cos\theta)\,AB$$

i.e. the work done is the product of the component of the force in the direction of motion, and the distance moved.

If $\overrightarrow{AB}$ is the vector from A to B, then:

$$\begin{aligned}\text{work done by } \mathbf{F} &= (F\cos\theta)\,\text{AB}\\ &= \mathbf{F}.\overrightarrow{\mathbf{AB}}\end{aligned}$$

where the dot signifies the scalar product of the two vectors, introduced in Chapter 1.
This result applies in one, two and three dimensions.

Remember from Chapter 1 that if $\mathbf{a} = a_1\mathbf{i} + a_2\mathbf{j} + a_3\mathbf{k}$ and $\mathbf{b} = b_1\mathbf{i} + b_2\mathbf{j} + b_3\mathbf{k}$, then

$$\mathbf{a}.\mathbf{b} = a_1b_1 + a_2b_2 + a_3b_3$$

Example 14

The diagrams below show a constant force acting on a particle P. The force continues to act as the particle is made to move along a straight line path from A to B, a distance of 3 metres. Find the work done by the force in each case.

(a)

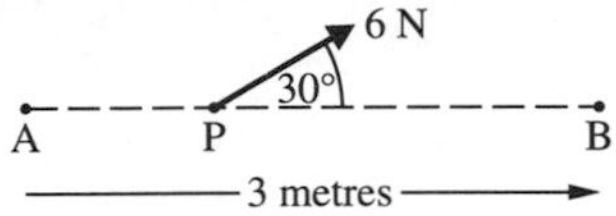

(b)

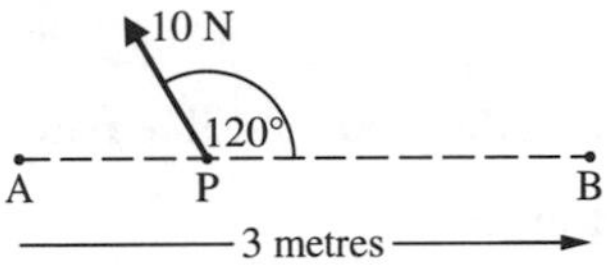

(a) component of the force in the direction of motion

$$\begin{aligned} &= 6\cos 30^\circ\\ &= 3\sqrt{3}\text{ N}\\ \therefore\quad \text{work done by force} &= 3\sqrt{3}\times 3\\ &= 9\sqrt{3}\text{ J}\end{aligned}$$

The force does $9\sqrt{3}$ J of work.

(b) component of the force in the direction of motion

$$\begin{aligned} &= -10\cos 60^\circ\\ &= -5\text{ N}\\ \therefore\quad \text{work done by force} &= -5\times 3\\ &= -15\text{ J}\end{aligned}$$

The force does -15 J of work.

Note. The negative answer for part (b) arises because work has to be done *against* the force.

Example 15

A constant force **F** acts on a particle as it moves along a straight wire from point A to point B. Point A has position vector $(\mathbf{i} - \mathbf{j})$ m, B has position vector $(2\mathbf{i} + 2\mathbf{j})$ m and $\mathbf{F} = (2\mathbf{i} + \mathbf{j})$ N.
Find the work done by **F**.

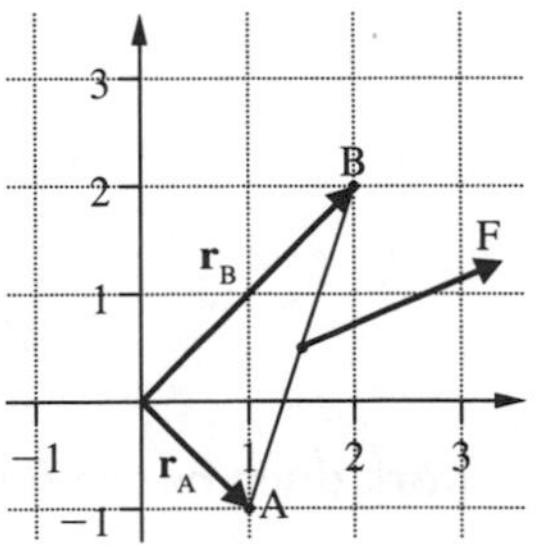

$$\begin{aligned}\overrightarrow{\text{AB}} &= -\mathbf{r}_A + \mathbf{r}_B\\ &= -(\mathbf{i} - \mathbf{j}) + (2\mathbf{i} + 2\mathbf{j})\\ &= \mathbf{i} + 3\mathbf{j}\end{aligned}$$

$$\begin{aligned}\text{Work done by } \mathbf{F} &= \mathbf{F}.\overrightarrow{\text{AB}}\\ &= (2\mathbf{i} + \mathbf{j}).(\mathbf{i} + 3\mathbf{j})\\ &= (2)(1) + (1)(3)\\ &= 5\end{aligned}$$

The work done is 5 J.

Example 16

When $t = 0$ a body of mass 2 kg is at rest at a point A, position vector $(2\mathbf{i} + 3\mathbf{j})$ m. The body is then subject to a constant force $\mathbf{F} = (6\mathbf{i} + 2\mathbf{j})$ N causing it to accelerate and, 3 seconds later, the body passes through the point B.
Find:
(a) the acceleration of the body in vector form
(b) the velocity of the body as it passes through B
(c) the kinetic energy of the body when $t = 3$ seconds
(d) $\overrightarrow{AB}$
(e) the position vector of point B
(f) $\mathbf{F}.(\overrightarrow{AB})$, the work done by the force in these first 3 seconds and verify that this equals the change in the kinetic energy of the body in travelling from A to B.

(a) Using $\mathbf{F} = m\mathbf{a}$, gives the equation of motion:

$$6\mathbf{i} + 2\mathbf{j} = 2\mathbf{a}$$

$$\therefore \quad \mathbf{a} = 3\mathbf{i} + \mathbf{j}$$

The acceleration of the body is $(3\mathbf{i} + \mathbf{j})\,\text{m s}^{-2}$.

(b) Using $\mathbf{v} = \mathbf{u} + \mathbf{a}t$ gives:

$$\begin{aligned}\mathbf{v} &= (0\mathbf{i} + 0\mathbf{j}) + (3\mathbf{i} + \mathbf{j})(3)\\ &= 9\mathbf{i} + 3\mathbf{j}\end{aligned}$$

The velocity of the body as it passes through B is $(9\mathbf{i} + 3\mathbf{j})\,\text{m s}^{-1}$.

(c) When $t = 3$, the speed of the body is

$$|(9\mathbf{i} + 3\mathbf{j})| = \sqrt{90}\ \text{m s}^{-1}$$

$$\begin{aligned}\therefore \quad \text{KE} &= \tfrac{1}{2}mv^2\\ &= \tfrac{1}{2}(2)(90)\\ &= 90\text{ J}\end{aligned}$$

The kinetic energy of the body when $t = 3$ seconds is 90 J.

(d) Using $\mathbf{s} = \mathbf{u}t + \frac{1}{2}\mathbf{a}t^2$ gives:

$$\begin{aligned}\mathbf{s} &= (0\mathbf{i} + 0\mathbf{j})(t) + \tfrac{1}{2}(3\mathbf{i} + \mathbf{j})(3)^2\\ &= 13{\cdot}5\mathbf{i} + 4{\cdot}5\mathbf{j}\end{aligned}$$

Thus $\qquad \overrightarrow{AB} = 13{\cdot}5\mathbf{i} + 4{\cdot}5\mathbf{j}$

(e)

$$\begin{aligned}\overrightarrow{OB} &= \overrightarrow{OA} + \overrightarrow{AB}\\ &= (2\mathbf{i} + 3\mathbf{j}) + (13{\cdot}5\mathbf{i} + 4{\cdot}5\mathbf{j})\\ &= (15{\cdot}5\mathbf{i} + 7{\cdot}5\mathbf{j})\text{ m}\end{aligned}$$

(f) work done by the force in first 3 seconds $= (6\mathbf{i} + 2\mathbf{j}).(\overrightarrow{AB})$

$$\begin{aligned}&= (6\mathbf{i} + 2\mathbf{j}).(13{\cdot}5\mathbf{i} + 4{\cdot}5\mathbf{j})\\ &= 90\text{ J}\end{aligned}$$

The work done by the force in these first 3 seconds is 90 J, which is the same as the increase in the kinetic energy of the body as determined in part (c).

Exercise 11E

Each diagram for questions **1** to **9** shows a constant force acting on a particle P. The force continues to act as the particle moves along a straight line path from A to B, a distance of 3 metres. Find the work done by the force in each case.

1.
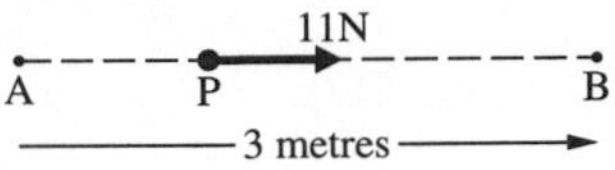

2.
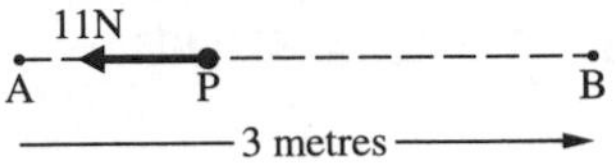

3.
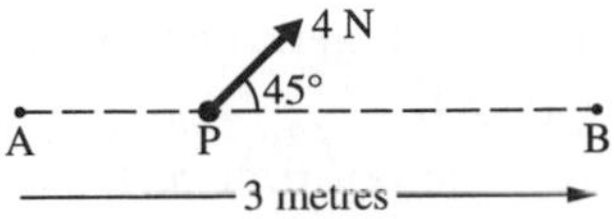

4.
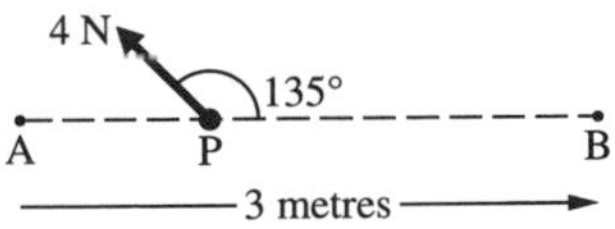

5.
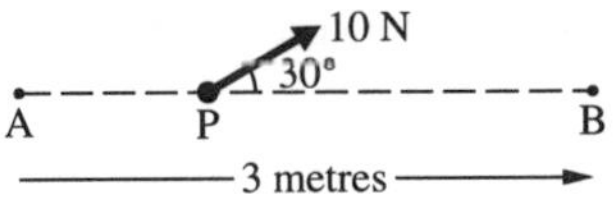

6.
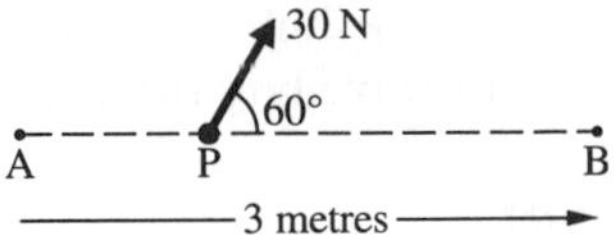

7.
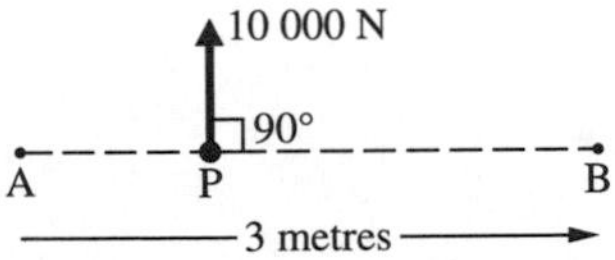

8.
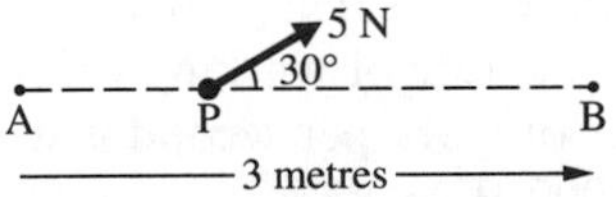

9.
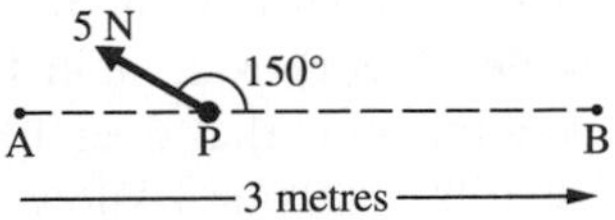

Each of questions **10** to **18** involves a constant force **F** acting on a particle as it moves along a straight wire from point A to point B. The position vectors of A and B are $\mathbf{r}_A$ and $\mathbf{r}_B$ respectively and force **F** is as stated. You may assume that all the forces are measured in newtons and the position vectors in metres.
For each question sketch the appropriate diagram and find the work done by **F**:

10. $\mathbf{r}_A = \mathbf{i}$, $\mathbf{r}_B = 3\mathbf{i}$, $\mathbf{F} = 5\mathbf{i}$
11. $\mathbf{r}_A = -\mathbf{i}$, $\mathbf{r}_B = \mathbf{i}$, $\mathbf{F} = 2\mathbf{i}$
12. $\mathbf{r}_A = \mathbf{i}+\mathbf{j}$, $\mathbf{r}_B = 3\mathbf{i}+\mathbf{j}$, $\mathbf{F} = 5\mathbf{i}$
13. $\mathbf{r}_A = -2\mathbf{j}$, $\mathbf{r}_B = -3\mathbf{j}$, $\mathbf{F} = 3\mathbf{i}$
14. $\mathbf{r}_A = \mathbf{i}+\mathbf{j}$, $\mathbf{r}_B = 2\mathbf{i}+2\mathbf{j}$, $\mathbf{F} = -3\mathbf{i}-3\mathbf{j}$
15. A is at the origin, $\mathbf{r}_B = 3\mathbf{i}$, $\mathbf{F} = 4\mathbf{i}+4\mathbf{j}$
16. $\mathbf{r}_A = \mathbf{i}+\mathbf{j}$, $\mathbf{r}_B = 3\mathbf{i}-\mathbf{j}$, $\mathbf{F} = \mathbf{i}+\mathbf{j}$. (Explain the result.)
17. $\mathbf{r}_A = 2\mathbf{i}-3\mathbf{j}$, $\mathbf{r}_B = -\mathbf{i}$, $\mathbf{F} = -2\mathbf{i}$
18. $\mathbf{r}_A = \mathbf{i}+\mathbf{j}$, $\mathbf{r}_B = -3\mathbf{i}-3\mathbf{j}$, $\mathbf{F} = -5\mathbf{i}+5\mathbf{j}$.

In each of questions **19** to **23** a constant force **F** acts on a particle P which moves in a straight line from A to B. You may assume that all the forces are measured in newtons and the position vectors in metres.
For each question find the work done by **F**:

19. A has position vector $3\mathbf{i}-4\mathbf{j}$, B has position vector $6\mathbf{i}-2\mathbf{j}$, and $\mathbf{F} = \mathbf{i}+4\mathbf{j}$.

20. A has position vector $\begin{pmatrix} 0 \\ 2 \end{pmatrix}$, **B** has position vector $\begin{pmatrix} 5 \\ -2 \end{pmatrix}$, and $\mathbf{F} = \begin{pmatrix} 2 \\ 1 \end{pmatrix}$.

21. A has position vector $2\mathbf{i}+\mathbf{j}+\mathbf{k}$, B has position vector $3\mathbf{i}+3\mathbf{j}+2\mathbf{k}$, and $\mathbf{F} = -30\mathbf{k}$.

22. A has position vector $\begin{pmatrix} 0 \\ 0 \\ 0 \end{pmatrix}$, **B** has position vector $\begin{pmatrix} -1 \\ -1 \\ -2 \end{pmatrix}$, and $\mathbf{F} = \begin{pmatrix} 1 \\ 1 \\ 2 \end{pmatrix}$.

23. A has position vector $-\mathbf{i}-\mathbf{j}-\mathbf{k}$, B has position vector $2\mathbf{i}+2\mathbf{j}+2\mathbf{k}$, and $\mathbf{F} = \mathbf{i}+\mathbf{j}+\mathbf{k}$.

24. When $t = 0$ a body of mass 5 kg is at rest at a point A, position vector $(3\mathbf{i} + 7\mathbf{j} + 2\mathbf{k})$ m. The body is then subject to a constant force $\mathbf{F} = (10\mathbf{i} + 15\mathbf{j} - 5\mathbf{k})$ N causing it to accelerate and, 2 seconds later, the body passes through the point B. Find:
(a) the acceleration of the body in vector form
(b) the velocity of the body as it passes through B
(c) the kinetic energy of the body when $t = 2$ seconds
(d) $\overrightarrow{AB}$
(e) the position vector of point B
(f) $\mathbf{F}.(\overrightarrow{AB})$, the work done by the force in these first 2 seconds, and verify that this equals the change in the kinetic energy of the body in travelling from A to B.

Power

Power is a measure of the rate at which work is being done.
If 1 joule of work is done in 1 second, the rate of working is 1 watt (W).
Hence a machine that does 1000 joules of work per second is working at a rate of 1000 watts or 1 kilowatt (i.e. 1000 W = 1 kW).

Example 17

Each of the diagrams below show a particle moving along a straight wire under the action of a constant force. In each case the particle travels the length of the wire in 5 seconds.
Find the work done by the force and the average rate at which the force is working in each case.

(a)

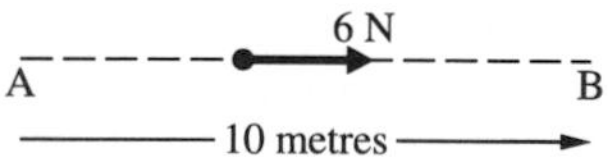

(b)

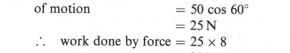

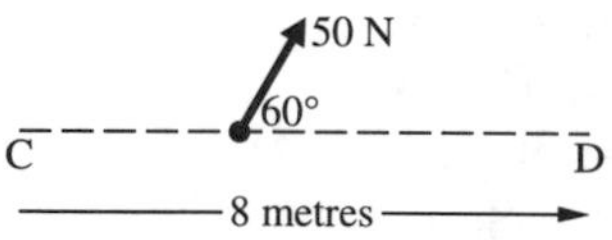

(a) The force of 6 N is in the direction of motion.

$\therefore$ work done by force $= 6 \times 10 = 60$ J

rate of working $= 60 \div 5 = 12$ W

The force does 60 J of work and its average rate of working is 12 watts.

(b) The component of the force in the direction of motion $= 50 \cos 60° = 25$ N

$\therefore$ work done by force $= 25 \times 8 = 200$ J

rate of working $= 200 \div 5 = 40$ W

The force does 200 J of work and its average rate of working is 40 watts.

Example 18

Find the rate at which work is being done when a mass of 20 kg is lifted vertically at a constant speed of 5 m s^{-1}.

work done = increase in PE
= $20gh$, where h is height lifted vertically

$\therefore$ work done in one second $= 20g(5)$ J = 980 J

$\therefore$ rate of working = 980 J per second or 980 W

The rate of working is 980 W.

Example 19

A pump ejects 12 000 kg of water at a speed of $4\,\text{m}\,\text{s}^{-1}$ in 40 seconds. Find the average rate at which the pump is working.

$$\begin{aligned}\text{KE given to water} &= \tfrac{1}{2}mv^2\\ &= \tfrac{1}{2}(12\,000)(4)^2\ \text{J in 40 s}\\ &= \tfrac{1}{2}\left(\tfrac{12\,000}{40}\right)(4)^2\ \text{J in 1 s} = 2400\ \text{J in 1 s}\end{aligned}$$

The average rate of working is 2400 W or 2·4 kW.

Example 20

A constant force $\mathbf{F} = (-2\mathbf{i} + \mathbf{j})$ N acts on a particle as it moves along a straight wire from point A, position vector $(2\mathbf{i} - \mathbf{j})$ m, to point B, position vector $(-\mathbf{i} + 3\mathbf{j})$ m in 4 seconds. Find the average rate at which **F** is working.

$$\begin{aligned}\overrightarrow{AB} &= -\mathbf{r}_A + \mathbf{r}_B\\ &= -(2\mathbf{i} - \mathbf{j}) + (-\mathbf{i} + 3\mathbf{j})\\ &= -3\mathbf{i} + 4\mathbf{j}\end{aligned}$$

$$\begin{aligned}\text{work done by } \mathbf{F} &= \mathbf{F}.\overrightarrow{AB}\\ &= (-2\mathbf{i} + \mathbf{j}).(-3\mathbf{i} + 4\mathbf{j})\\ &= 10\ \text{J}\\ \text{rate of working} &= 10\ \text{J in 2·5 seconds}\\ &= 4\ \text{watts}\end{aligned}$$

The force is working at an average rate of 4 W.

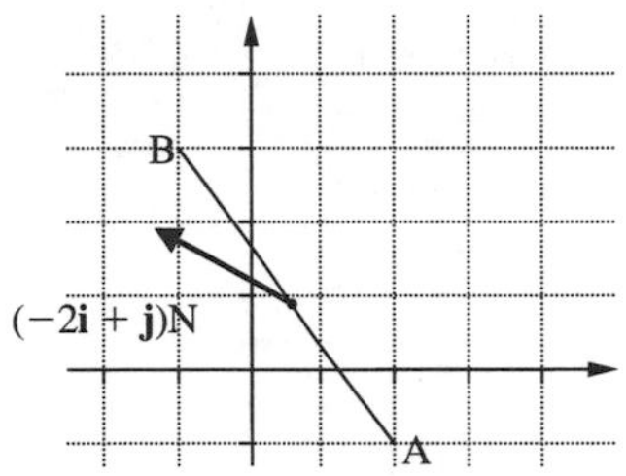

Pump raising and ejecting water

If a pump is used not only to raise water, but also to eject it at a given speed, then the total work being done is the sum of the potential energy and the kinetic energy given to the water each second.

Example 21

A pump draws water from a tank and issues it from the end of a hose which is 2·5 m vertically above the level from which the water is drawn. The cross-sectional area of the hose is $10\,\text{cm}^2$, and the water leaves the end of the hose at a speed of $5\,\text{m}\,\text{s}^{-1}$. Find the rate at which the pump is working. (Take the density of water to be $1000\,\text{kg}\,\text{m}^{-3}$.)

$$\text{volume of water raised and issued per second} = 5 \times \frac{10}{(100)^2}\ \text{m}^3$$

$$\text{mass of water raised and issued per second} = \frac{50}{(100)^2} \times 1000\ \text{kg} = 5\ \text{kg}$$

$$\begin{aligned}\text{PE given to water per second} &= mgh\\ &= 5(9{\cdot}8)(2{\cdot}5)\ \text{J}\\ \text{KE given to water per second} &= \tfrac{1}{2}mv^2\\ &= \tfrac{1}{2}(5)5^2\ \text{J}\\ \text{work done per second, by the pump} &= 5(9{\cdot}8)(2{\cdot}5) + \tfrac{1}{2}(5)5^2\ \text{J} = 185\ \text{J}\end{aligned}$$

The rate at which the pump is working is 185 W.

Exercise 11F

Take the density of water as $1000\,kg\,m^{-3}$.

1. Each of the diagrams below shows a particle moving along a straight wire under the action of a constant force. In each case the particle travels the length of the wire in 4 seconds. Find the work done by the force and the average rate at which the force is working in each case.

(a)

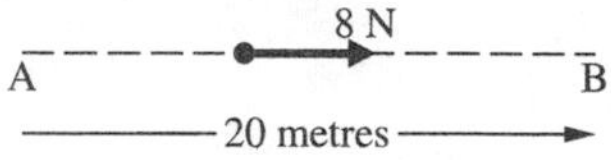

(b)

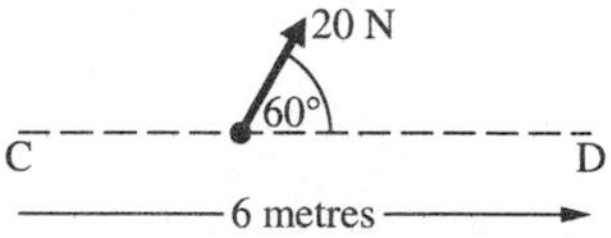

(c)

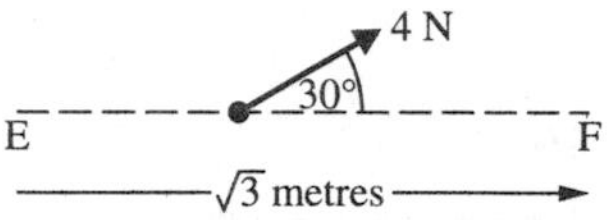

(d)

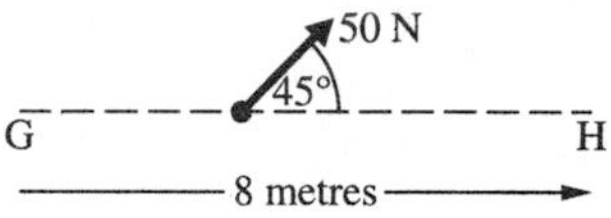

2. What is the average rate at which work must be done in lifting a mass of 100 kg a vertical distance of 5 m in 7 seconds?

3. What is the rate at which work must be done in lifting a mass of 500 kg vertically at a constant speed of $3\,m\,s^{-1}$?

4. In every minute a machine pumps 300 kg of water along a horizontal hose from rest at one end to eject it at a speed of $4\,m\,s^{-1}$ at the other.
Find the average rate at which the machine is working.

5. A bowling machine bowls 6 balls every minute. Each ball has a mass of 150 g and leaves the machine with a speed of $20\,m\,s^{-1}$.
Find the average rate at which the machine is working.

6. Find the average rate at which a climber of mass 80 kg must work when climbing a vertical distance of 30 m in 2 minutes.

7. A constant force $\mathbf{F} = (2\mathbf{i} + 3\mathbf{j})$ N acts on a particle and causes it to move along a straight wire from point A, position vector $(2\mathbf{i} + 2\mathbf{j})$ m, to point B, position vector $(6\mathbf{i} + 8\mathbf{j})$ m in 5 seconds. Find the average rate at which **F** is working.

8. A constant force $\mathbf{F} = (2\mathbf{i} - \mathbf{j})$ N acts on a particle and causes it to move along a straight wire from point A, position vector $5\mathbf{j}$ m, to point B, position vector $6\mathbf{i}$ m in 10 seconds. Find the average rate at which **F** is working.

9. In building a section of wall, a man has to lift 500 bricks a vertical distance of 150 cm. Each brick has a mass of 2 kg and the man completes the section in 5 minutes.
Find the man's average rate of working.

10. A pump raises 75 kg of water a vertical distance of 20 m in 14 seconds.
Find the average rate at which the pump is working.

11. In every minute a pump draws $6\,m^3$ of water from a well and issues it at a speed of $5\,m\,s^{-1}$ from a nozzle situated 4 m above the level from which the water was drawn.
Find the average rate at which the pump is working.

12. In every minute a pump draws $5\,m^3$ of water from a well and issues it at a speed of $6\,m\,s^{-1}$ from a nozzle situated 6 m above the level from which the water was drawn.
Find the average rate at which the pump is working.

13. A pump draws water from a tank and issues it at a speed of $8\,\text{m s}^{-1}$ from the end of a pipe of cross-sectional area $0{\cdot}01\,\text{m}^2$, situated 10 m above the level from which the water is drawn.
Find:
(a) the mass of water issued from the pipe in each second
(b) the rate at which the pump is working.

14. A pump draws water from a tank and issues it at a speed of $10\,\text{m s}^{-1}$ from the end of a hose of cross-sectional area $5\,\text{cm}^2$, situated 4 m above the level from which the water is drawn. Find the rate at which the pump is working.

15. In each minute a pump draws $2{\cdot}4\,\text{m}^3$ of water from a well 5 m below ground, and issues it at ground level through a pipe of cross-sectional area $50\,\text{cm}^2$. Find:
(a) the speed with which the water leaves the pipe
(b) the rate at which the pump is working.
If in fact the pump is only 75% efficient (i.e. 25% of the power is lost in the running of the pump), find the rate at which it must work.

16. In each minute a pump, working at 3·48 kW, raises $1{\cdot}5\,\text{m}^3$ of water from an underground tank and issues it from the end of a pipe situated at ground level. The water leaves the pipe with speed $10\,\text{m s}^{-1}$ and the pump is 50% efficient (i.e. 50% of the power is lost in the running of the pump). Find:
(a) the area of cross-section of the pipe
(b) the depth below ground level from which the water is drawn.

17. In each minute a pump, working at a constant rate of 0·825 kW, draws $0{\cdot}3\,\text{m}^3$ of water from a well and issues it through a nozzle situated 5 m above the level from which the water was drawn. If the pump is 60% efficient (i.e. 40% of the power is lost in the running of the pump), find the velocity with which the water is ejected. Find also the area of cross-section of the nozzle.

Vehicles in motion

Consider a car being driven along a road. The forward force F N which propels the car is supplied by the engine. Suppose the engine is working at a constant rate of P watts. Then:

$$\begin{aligned} P &= \text{work done per second} \\ &= \text{force} \times \text{distance moved per second} \\ &= F \times \text{speed} = Fv \end{aligned}$$

Thus $F = \dfrac{P}{v}$ gives the forward force exerted by the engine working at P watts, when the car is travelling at $v\,\text{m s}^{-1}$.

The expression $\dfrac{P}{v}$ is frequently referred to as the *tractive force* being exerted by the engine at the instant when the vehicle is travelling at $v\,\text{m s}^{-1}$.
For problems involving vehicles in motion, if the power or rate of working is involved, draw a diagram showing the forces acting, including $\dfrac{P}{v}$, the tractive force. Then apply $F = ma$ in the direction of motion; if there is no acceleration, use the fact that the resultant force acting on the vehicle must be zero.

Example 22

The engine of a car is working at a steady rate of 5 kW. The car of mass 1200 kg is being driven along a level road against a constant resistance to motion of 325 N. Find:
(a) the acceleration of the car when its speed is $8\,\text{m s}^{-1}$
(b) the maximum speed of the car.

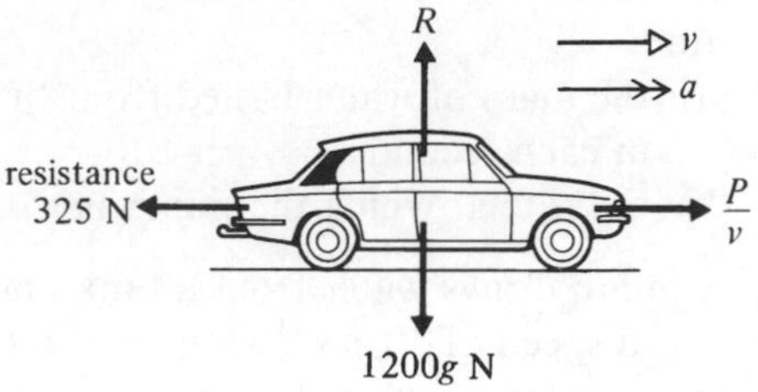

(a) Power $P = 5\,\text{kW} = 5000\,\text{W}$

$$v = 8\,\text{m s}^{-1}$$

Using $F = ma$ gives: $\frac{P}{v} - 325 = 1200a$

$\therefore$ $\frac{5000}{8} - 325 = 1200a$

$\therefore$ $a = 0{\cdot}25\,\text{m s}^{-2}$

The acceleration is $0{\cdot}25\,\text{m s}^{-2}$

(b) To find the maximum speed of the car, consider the situation when the acceleration is zero. Then:

$$\frac{P}{v} = 325$$

$\therefore$ $\frac{5000}{v} = 325$ or $v = 15{\cdot}38\,\text{m s}^{-1}$

The maximum speed of the car is $15{\cdot}4\,\text{m s}^{-1}$.

Example 23

The greatest speed at which a cyclist can travel is $9\,\text{m s}^{-1}$ when on a level road and cycling against a constant resistance of 40 N. Find the maximum rate at which the cyclist can work.

Let the rate of working be P.
At maximum speed there is no acceleration.

$\therefore$ $\frac{P}{v} = 40$

But $v = 9\,\text{m s}^{-1}$

$\therefore$ $P = 360\,\text{W}$

The cyclist can work at 360 W.

Example 24

A car travels along a level road against a constant resistance to motion of 500 N. The mass of the car is 1500 kg and its maximum speed is $40\,\text{m s}^{-1}$. Find the rate at which the engine of the car is working. If the engine of the car works at the same rate and the resistances are unchanged, find the maximum speed of the car when ascending an incline of $\sin^{-1}\frac{1}{49}$.

On the level, let the rate of working be P.
At maximum speed there is no acceleration.

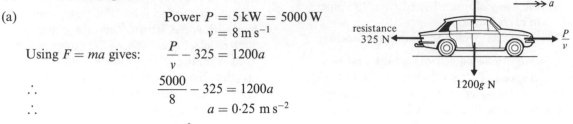

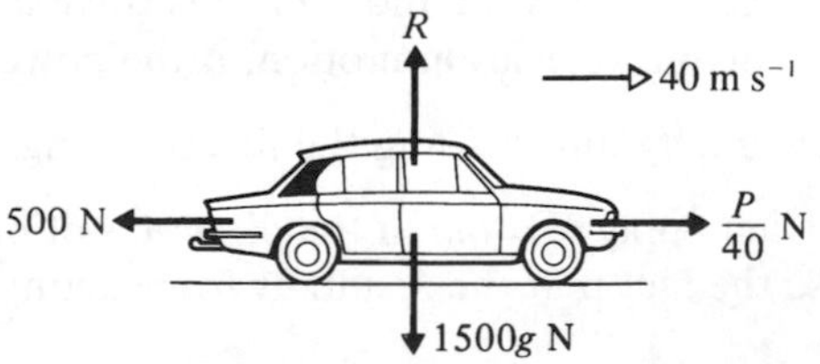

$\therefore$ $\frac{P}{40} = 500$

$\therefore$ $P = 20\,000\,\text{W}$

The car is working at 20 kW.

On the incline, let the maximum speed be v.
At maximum speed there is no acceleration.

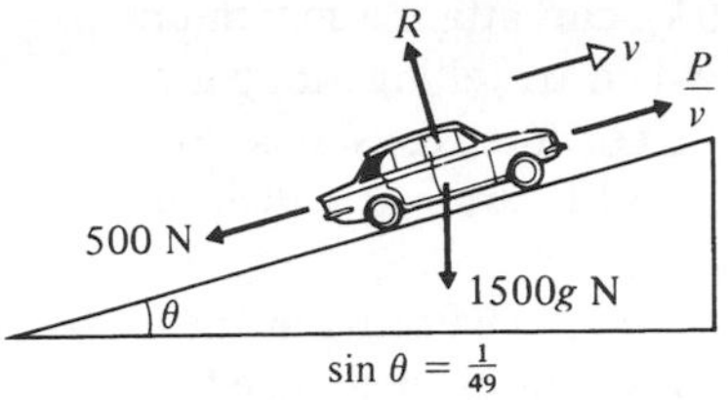

$$\therefore \quad \frac{P}{v} = 500 + 1500g \sin \theta$$

$$\therefore \quad \frac{20\,000}{v} = 500 + 1500(9{\cdot}8)\tfrac{1}{49}$$

$$\therefore \quad v = 25\,\text{m s}^{-1}$$

The maximum speed of the car on the incline is $25\,\text{m s}^{-1}$.

Exercise 11G

1. A car is driven along a level road against a constant resistance to motion of 400 N. Find the maximum speed at which the car can move when its engine works at a steady rate of:
   (a) 4 kW (b) 6 kW (c) 8·8 kW
2. A car of mass 1 tonne is driven along a level road against a constant resistance to motion of 200 N. With the engine of the car working at a steady rate of 8 kW, find:
   (a) the acceleration of the car when its speed is $5\,\text{m s}^{-1}$
   (b) the acceleration of the car when its speed is $10\,\text{m s}^{-1}$
   (c) the maximum speed of the car.
3. A car of mass 900 kg is driven along a level road against a constant resistance to motion of 300 N. With the engine of the car working at a steady rate of 12 kW, find:
   (a) the acceleration of the car when its speed is $4\,\text{m s}^{-1}$
   (b) the acceleration of the car when its speed is $10\,\text{m s}^{-1}$
   (c) the maximum speed of the car.
4. With its engines working at a steady 14 kW, the maximum speed that a car can attain along a level road is $35\,\text{m s}^{-1}$.
   Find the magnitude of the resistances experienced by the car.
5. A cyclist travels along a level road at a constant speed of $8\,\text{m s}^{-1}$.
   If the resistances to motion total 50 N, find the rate at which the cyclist is working.
6. A cyclist finds that when cycling along a level road against a resistance to motion of 20 N, the greatest speed that he can attain is $10\,\text{m s}^{-1}$.
   Find the maximum rate at which the cyclist can work.
7. A train of mass 100 tonnes travels along level track with its engines developing a constant 60 kW of power.
   If the greatest speed that the train can attain along the track is $108\,\text{km h}^{-1}$, find the magnitude of the resistance to motion. Assuming this resistance remains unchanged, find the acceleration of the train when it travels along the track at $54\,\text{km h}^{-1}$ with its engines working at the same rate as before.
8. A cyclist and his bike have a combined mass of 75 kg.
   Find the maximum speed that the cyclist can attain on level ground when working at a constant rate of 210 W against resistances totalling 21 N.
   With the resistances and the rate of working unchanged, the cyclist ascends a slope of inclination $\sin^{-1} \frac{1}{15}$.
   Find the maximum speed of the cyclist up the slope.
9. With its engine working at a constant rate of 15 kW, a car of mass 800 kg ascends a hill of 1 in 98 against a constant resistance to motion of 420 N.
   Find:
   (a) the acceleration of the car up the hill when travelling with a speed of $10\,\text{m s}^{-1}$
   (b) the maximum speed of the car up the hill.
10. With its engines working at a constant rate of 369 kW, a train of mass 100 tonnes ascends a hill of 1 in 50 at a constant speed of $54\,\text{km h}^{-1}$. Find the magnitude of the resistance to motion experienced by the train.

11. A car of mass 900 kg can attain a maximum speed of 48 m s^{-1} when travelling along a level road against a constant resistance to motion of 350 N. Find the rate at which the car engine is working.
With the engine of the car working at the same rate and the resistances unchanged, the car ascends a hill of inclination $\sin^{-1} \frac{1}{18}$.
Find the maximum speed of the car up the hill.

12. A cyclist and his bike have a combined mass of 75 kg and the maximum rate at which the cyclist can work is 0·392 kW.
If the greatest speed with which the cyclist can ride along a level road is 8 m s^{-1}, find the magnitude of the constant resistance to motion. With this resistance unchanged, find the greatest speed at which the cyclist can ascend a hill of inclination $\sin^{-1} \frac{1}{15}$.

13. (a) With its engine working at a steady rate of 32 kW, a car of mass 1 tonne travels at a constant speed of 40 m s^{-1} along a level road.
Find the magnitude of the resistance to motion experienced by the car.
(b) Given that the resistance to motion is directly proportional to the speed at which the car is travelling, find the magnitude of the resistance experienced by the car when travelling at 30 m s^{-1}.
(c) Find the rate at which the engine must work for the car to ascend a slope of 1 in 98 at a constant speed of 20 m s^{-1}, the resistance to motion still obeying the same rule of proportionality.

14. When a car of mass 900 kg has its engine working at a constant rate of 7·35 kW, the car can ascend a hill of 1 in 63 at a constant speed of 15 m s^{-1}. Find:
(a) the resistance to motion experienced by the car
(b) the maximum speed of the car when travelling down the same slope with its engine working at the same rate as before and the resistance to motion unchanged.

15. With its engine working at a constant rate of 9·8 kW, a car of mass 800 kg can descend a slope of 1 in 56 at twice the steady speed that it can ascend the same slope, the resistances to motion remaining the same throughout.
Find the magnitude of the resistance and the speed of ascent.

Exercise 11H Examination questions

(Unless otherwise indicated take $g = 9.8 \text{ m s}^{-2}$ in this exercise.)

1. (Take $g = 10 \text{ m s}^{-2}$ in this question.)
A boy throws a stone straight up into the air. It leaves the boy's hand travelling at a speed of 8 m s^{-1}.
(a) Draw a diagram to show the forces acting on the stone after it has left the boy's hand.
(b) Find the maximum height that the stone would reach if all resistance forces are assumed to be zero.
(c) In fact the stone, which has a mass of 0·1 kg, only reaches a maximum height of 2 m. Find the work done by the stone against air resistance. (AEB, Spec)

2. A rough channel is in the form of a quarter circle PQ. The radii OP and OQ are horizontal and vertical respectively and $OP = OQ = 2$ m. A particle of mass 1·5 kg is released from rest at P. Given that the channel offers a frictional resistance whose magnitude has an average value of 0·8 N, calculate the speed of the particle at Q.

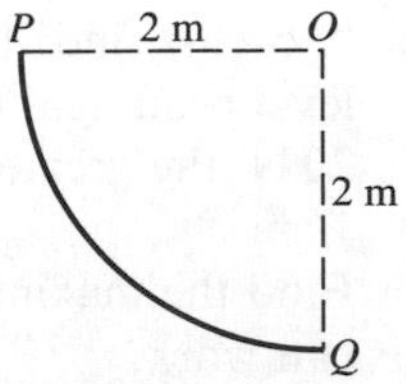

(UCLES)

3. A builder is demolishing a chimney stack. As he takes off the bricks he gets them to his truck by letting them slide 6 m down a plastic trough inclined at 30° to the horizontal.

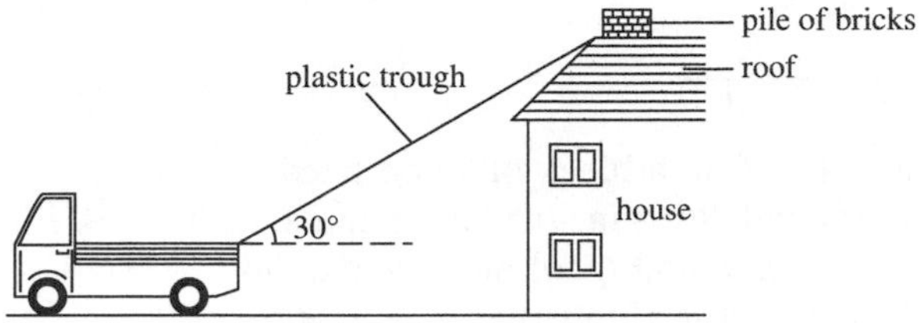

The builder releases a brick of mass 3·5 kg from rest at the top of the trough. When it reaches the bottom, having slid 6 m, its speed is $5\,\text{m}\,\text{s}^{-1}$. Calculate

(i) the potential energy lost by the brick,
(ii) the frictional force (assumed to be constant) which acts on the brick whilst it is sliding,
(iii) the coefficient of friction between the brick and the trough,
(iv) the builder pushes a similar brick down the trough, giving it an initial speed of $2\,\text{m}\,\text{s}^{-1}$. What is the speed of the brick when it has slid 6 m? (OCSEB)

4. The diagram shows a rough circular wire, centre O, radius 1·2 m, fixed in a vertical plane with the top point C vertically above the bottom point A. A particle of mass 0·6 kg is threaded on the wire and projected from A with speed $8\,\text{m}\,\text{s}^{-1}$. The point B is at the same level as O and the particle passes through B with a speed of $6\,\text{m}\,\text{s}^{-1}$. Calculate

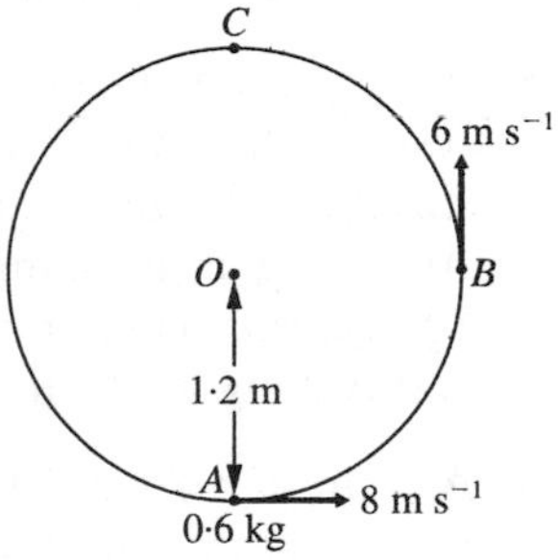

(i) the loss of kinetic energy as the particle travels from A to B,
(ii) the gain in potential energy as the particle travels from A to B,
(iii) the work done in overcoming the frictional resistance of the wire from A to B.

Given that this frictional resistance is constant for the whole wire, find its value and the speed of the particle as it passes through C. (UCLES)

5. A ring of mass $3m$ can slide on a fixed smooth vertical wire. The ring is attached to a light inextensible string which passes over a small smooth pulley fixed at a distance d from the wire. The other end of the string is attached to a particle of mass $5m$ which hangs freely. Given that the system is in equilibrium, find

(a) the length of the string between the ring and the pulley;
(b) the magnitude of the reaction between the ring and the wire;
(c) the magnitude and direction of the reaction on the pulley.

The ring is then lifted and released from rest on the same horizontal level as the pulley.

(d) Use the principle of conservation of energy to find how far the ring will fall before it first comes momentarily to rest. (AEB 1994)

6.

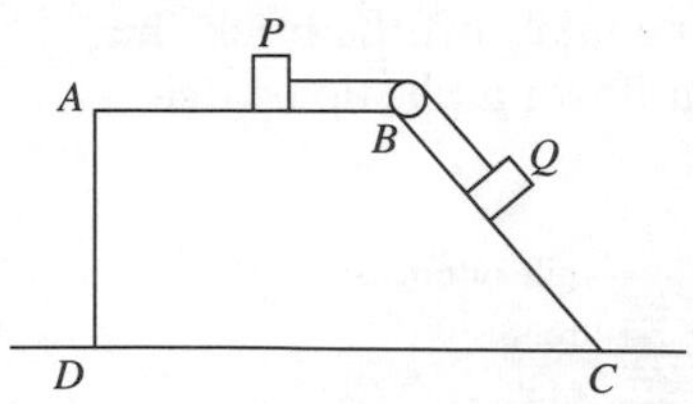

The diagram shows a vertical section $ABCD$ of a block of wood fixed on a horizontal plane. AB is horizontal and BC is inclined at an angle α to the horizontal where $\sin \alpha = \frac{4}{5}$. Particles P and Q, of mass m and $5m$ respectively, are placed on AB and BC and joined together by a light inextensible string passing over a smooth pulley at B. The particles are then released from rest. Find, assuming that P does not reach B and Q does not reach C, the acceleration of the particles and the tension in the string when

(a) AB and BC are smooth,

(b) AB and BC are both equally rough, the coefficient of friction being $\frac{1}{3}$.

Find, for the second case, the loss of energy due to friction after both particles have moved a distance d. (WJEC)

7. A water pump raises 40 kg of water a second through a height of 20 m and ejects it with a speed of $45\,\text{m s}^{-1}$. Find the kinetic energy and potential energy per second given to the water and the effective rate at which the pump is working. (WJEC)

8. A car has an engine of maximum power 15 kW. Calculate the force resisting the motion of the car when it is travelling at its maximum speed of $120\,\text{km h}^{-1}$ on a level road.
Assuming an unchanged resistance, and taking the mass of the car to be 800 kg, calculate, in m s^{-2}, the maximum acceleration of the car when it is travelling at $60\,\text{km h}^{-1}$ on a level road. (UCLES)

9. A train of mass 5×10^5 kg travels along a straight horizontal track. The resistance to motion is constant and has magnitude $2{\cdot}5 \times 10^4$ N.

(a) Find the force produced by the engine when the acceleration is $0{\cdot}1\,\text{m s}^{-2}$ and the power which is developed at the instant when the speed of the train is $10\,\text{m s}^{-1}$.

(b) If the engine then works at a constant rate of 500 kW, determine the greatest possible speed of the train along the same track. (UODLE)

10. A car of mass 800 kg tows a trailer of mass 200 kg. The constant resistances acting on the car and trailer are R and 200 Newtons respectively.

(i) If the car is travelling at a constant speed of $54\,\text{km h}^{-1}$ and the engine is working at a rate of $9{\cdot}75$ kW, find the tension in the tow bar and the value of R.

Due to a fault in the tow bar the maximum tension which the tow bar can take is 350 N. When the car is travelling at $72\,\text{km h}^{-1}$, find

(ii) the maximum acceleration if the resistances remained unchanged, and

(iii) the maximum power at which the engine can work, if the resistances remain unchanged. (NICCEA)

11. A car of mass 800 kg tows a caravan of mass 480 kg along a straight level road. The tow-bar connecting the car and the caravan is horizontal and of negligible mass. With the car's engine working at a rate of 24 kW, the car and caravan are travelling at a constant speed of $25\,\text{m}\,\text{s}^{-1}$.

(a) Calculate the magnitude of the total resistance, in N, to the motion of the car and the caravan.

The resistance to the motion of the car has magnitude 800λ newtons and the resistance to the motion of the caravan has magnitude 480λ newtons, where λ is a constant.

(b) Find the value of λ.

(c) Find the tension, in N, in the tow-bar. (ULEAC)

12. A tractor of mass 700 kg pulls a trailer of mass 300 kg along a straight level road. The total resistance to motion is 1500 N and the tractor is using its full power of 30 kW. Show that when the tractor's speed is $36\,\text{km}\,\text{h}^{-1}$, its acceleration is $1{\cdot}5\,\text{m}\,\text{s}^{-2}$.

Assuming that the resistive force is divided between the tractor and trailer in the ratio of their masses, find the tension in the coupling between the tractor and the trailer when the speed is $36\,\text{km}\,\text{h}^{-1}$.

(AEB 1990)

13. A car of mass 1·5 tonnes moves with constant speed $6\,\text{m}\,\text{s}^{-1}$ up a slope of inclination $\sin^{-1}\left(\frac{1}{7}\right)$. Given that the engine of the car is working at a rate of 18 kW, find, in Newtons, the resistance to the motion.

(AEB 1994)

14. A car of mass 1000 kg, whose engine is working at a rate of P Watts, moves at a constant speed of $20\,\text{m}\,\text{s}^{-1}$ on a horizontal road. Find, in terms of P, the total frictional resistance exerted on the car.

The car then freewheels (i.e. without the engine exerting any force) down a slope of inclination $\sin^{-1}\left(\frac{1}{14}\right)$ to the horizontal at a constant speed. Assuming that the total frictional resistance is unchanged, find P.

Assuming the same total frictional resistance and that the engine is working at the rate P Watts, find the numerical value of the acceleration of the car at the instant when it is moving up the above slope with speed $7\,\text{m}\,\text{s}^{-1}$. (WJEC)

15. A car of mass 800 kg is pulling a trailer of mass 200 kg up a hill inclined at an angle $\sin^{-1}\left(\frac{1}{14}\right)$ to the horizontal. When the total force exerted by the engine is 1000 N the car and trailer move up the hill at a steady speed. Find the total frictional resistance to the motion of the car and trailer during this motion.

When the car and trailer are travelling at a steady speed of $10\,\text{m}\,\text{s}^{-1}$ up the hill, the power exerted by the engine is instantaneously increased by 2 kW. Find

(i) the instantaneous acceleration,

(ii) the instantaneous tension in the coupling between the car and the trailer, given that the total frictional resistance on the trailer is 75 N. (WJEC)

16. (Take $g = 10\,\text{m s}^{-2}$ in this question.)

A car has a mass of 600 kg and a maximum power rating of S watts. The car has a top speed of $40\,\text{m s}^{-1}$ on a flat road and a top speed of $24\,\text{m s}^{-1}$ up a hill inclined at an angle θ to the horizontal, where $\sin\theta = 0{\cdot}2$. Given that the resistance to the motion of the car on both roads is F N where F is constant, write down two equations connecting S and F. Hence find the value of S and of F. (UCLES)

17. When a car is moving on any road with speed $v\,\text{m s}^{-1}$ the resistance to its motion is $(a + bv^2)$ N, where a and b are positive constants. When the car moves on a level road, with the engine working at a steady rate of 53 kW, it moves at a steady speed of $40\,\text{m s}^{-1}$. When the engine is working at a steady rate of 24 kW the car can travel on a level road at a steady speed of $30\,\text{m s}^{-1}$. Find a and b and hence deduce that, when the car is moving with speed $34\,\text{m s}^{-1}$, the resistance to its motion is 992 N. Given that the car has mass 1200 kg find, in m s^{-2} correct to 2 decimal places, its acceleration on a level road at the instant when the engine is working at a rate of 51 kW and the car is moving with speed $34\,\text{m s}^{-1}$ so that the resistance to the motion is 992 N.
The car can ascend a hill at a steady speed of $34\,\text{m s}^{-1}$ with the engine working at a steady rate of 68 kW. Find, in degrees correct to one decimal place, the inclination of the hill to the horizontal.

(AEB 1992)

12 Projectiles

Horizontal projection

In order to investigate the motion of a projectile, we should consider the horizontal and vertical motions separately.
The horizontal velocity of a projectile is constant since there is no force acting in this direction.
The vertical velocity, on the other hand, is subject to the force of gravity, and the usual equations of motion $v = u + at$, $v^2 = u^2 + 2as$, $s = ut + \frac{1}{2}at^2$ and $s = \frac{(u+v)t}{2}$ are used. In some cases **i** and **j**, the unit vectors in a horizontal and vertical direction, are used in stating the position and velocity vectors.

Example 1

A particle is projected horizontally at $36\,\text{m}\,\text{s}^{-1}$ from a point 122·5 m above a horizontal surface. Find the time taken by the particle to reach the surface and the horizontal distance travelled in that time.

horizontal motion	vertical motion
$u = 36\,\text{m}\,\text{s}^{-1}$	$u = 0$
$s = x$	$s = 122{\cdot}5\,\text{m}\downarrow$
$t = t$	$t = t$
no acceleration	$a = 9{\cdot}8\,\text{m}\,\text{s}^{-2}\downarrow$

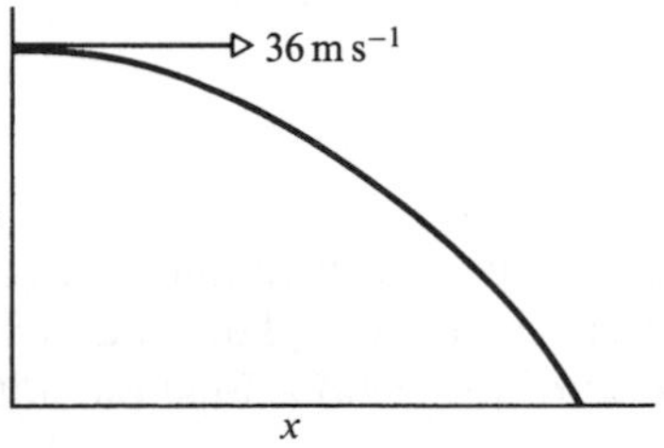

Using $s = ut + \frac{1}{2}at^2$, for a vertical motion gives:

$$122{\cdot}5 = 0 + \tfrac{1}{2}(9{\cdot}8)t^2$$

$\therefore \quad t^2 = \frac{2(122{\cdot}5)}{9{\cdot}8} = 25$

$\therefore \quad t = 5s$

Using distance = velocity × time for horizontal motion gives:

$$x = 36 \times 5$$

$\therefore \quad x = 180\,\text{m}$

The time taken is 5 s and the horizontal distance travelled is 180 m.

Example 2

A particle is projected horizontally with a speed of $14{\cdot}7\,\text{m}\,\text{s}^{-1}$. Find the horizontal and vertical displacements of the particle from the point of projection, after 2 seconds. Find also how far the particle then is from the point of projection.

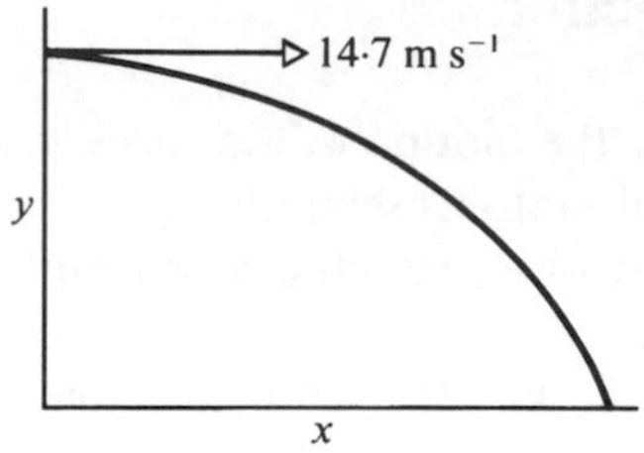

horizontal motion	vertical motion
$u = 14{\cdot}7\,\text{m}\,\text{s}^{-1}$	$u = 0$
$s = x$	$s = y \downarrow$
$t = 2\,\text{s}$	$t = 2\,\text{s}$
no acceleration	$a = 9{\cdot}8\,\text{m}\,\text{s}^{-2} \downarrow$

Using $s = ut + \frac{1}{2}at^2$ for vertical motion gives:

$$y = 0 + \tfrac{1}{2}(9{\cdot}8)2^2$$
$$\therefore \quad y = 19{\cdot}6\,\text{m}$$

Using distance = velocity × time for horizontal motion gives:

$$x = 14{\cdot}7 \times 2$$
$$= 29{\cdot}4\,\text{m}$$

$$\text{distance from point of projection} = \sqrt{(x^2 + y^2)}$$
$$= \sqrt{\{(29{\cdot}4)^2 + (19{\cdot}6)^2\}}$$
$$= 35{\cdot}33\,\text{m}$$

After 2 seconds the horizontal and vertical displacements are 29·4 m and 19·6 m respectively, and the displacement from the point of projection is 35·3 m.

Example 3

A particle is projected horizontally from a point 44·1 m above a horizontal plane. The particle hits the plane at a point which is, horizontally, 39 m from the point of projection. Find the initial speed of the particle.

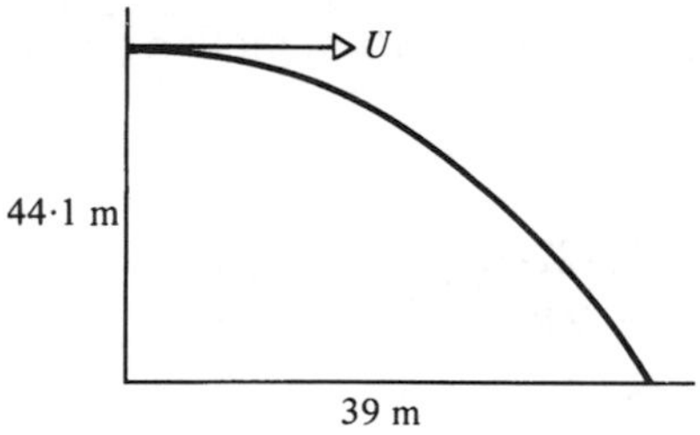

horizontal motion	vertical motion
$u = U$	$u = 0$
$s = 39\,\text{m}$	$s = 44{\cdot}1\,\text{m} \downarrow$
$t = t$	$t = t$
no acceleration	$a = 9{\cdot}8\,\text{m}\,\text{s}^{-2} \downarrow$

Using $s = ut + \frac{1}{2}at^2$, for a vertical motion gives:

$$44{\cdot}1 = 0 + \tfrac{1}{2}(9{\cdot}8)t^2$$
$$\therefore \quad t^2 = 9 \quad \text{i.e.} \quad t = 3\,\text{s}$$

Using distance = velocity × time for horizontal motion gives:

$$39 = U \times 3$$
$$\therefore \quad U = 13\,\text{m}\,\text{s}^{-1}$$

The speed of projection is $13\,\text{m}\,\text{s}^{-1}$.

Example 4

A particle is projected horizontally with a velocity of $39{\cdot}2\,\text{m s}^{-1}$. Find the horizontal and vertical components of the velocity of the particle 3 seconds after projection. Find also the speed and direction of motion of the particle at this time.

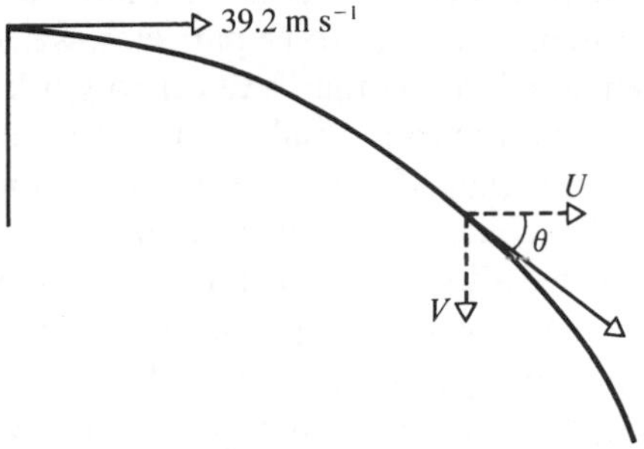

horizontal motion	vertical motion
$u = 39{\cdot}2\,\text{m s}^{-1}$	$u = 0$
horizontal velocity	$v = V \downarrow$
is constant	$t = 3\,\text{s}$
$\therefore\ U = 39{\cdot}2\,\text{m s}^{-1}$	$a = 9{\cdot}8\,\text{m s}^{-2} \downarrow$

Using $v = u + at$ for a vertical motion gives:

$$V = 0 + (9{\cdot}8) \times 3$$

$$\therefore \quad V = 29{\cdot}4\,\text{m s}^{-1}$$

Speed after 3 s is given by: $\sqrt{(U^2 + V^2)} = \sqrt{\{(39{\cdot}2)^2 + (29{\cdot}4)^2\}} = 49\,\text{m s}^{-1}$

Direction of motion is given by: $\tan\theta = \dfrac{V}{U} = \dfrac{29{\cdot}4}{39{\cdot}2}$

$$\therefore \quad \theta = 36{\cdot}87^\circ$$

The horizontal and vertical components are $39{\cdot}2\,\text{m s}^{-1}$ and $29{\cdot}4\,\text{m s}^{-1}$ respectively after 3 s; the speed of the particle is then $49\,\text{m s}^{-1}$ and it is travelling at an angle of $36{\cdot}87^\circ$ below the horizontal.

Example 5

At time $t = 0$ a particle is projected with a velocity of $2\mathbf{i}\,\text{m s}^{-1}$ from a point with position vector $(10\mathbf{i} + 90\mathbf{j})$ m. Find the position vector of the particle when $t = 4$ s.

Initial velocity is $2\mathbf{i}\,\text{m s}^{-1}$

$\therefore$ initial vertical velocity is zero and initial horizontal velocity is $2\,\text{m s}^{-1}$.

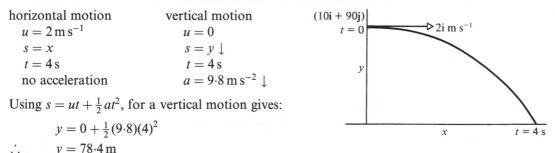

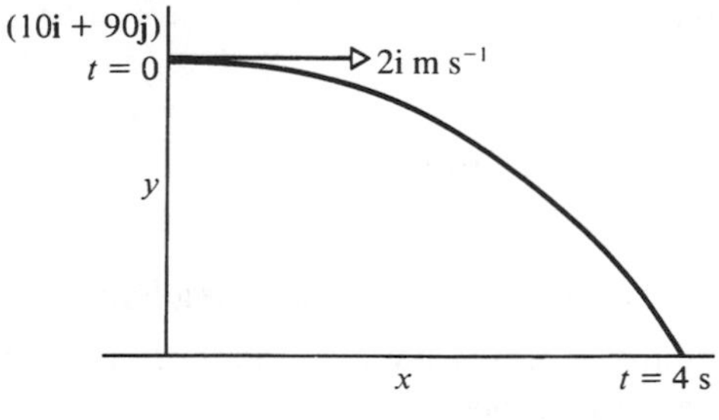

horizontal motion	vertical motion
$u = 2\,\text{m s}^{-1}$	$u = 0$
$s = x$	$s = y \downarrow$
$t = 4\,\text{s}$	$t = 4\,\text{s}$
no acceleration	$a = 9{\cdot}8\,\text{m s}^{-2} \downarrow$

Using $s = ut + \frac{1}{2}at^2$, for a vertical motion gives:

$$y = 0 + \tfrac{1}{2}(9{\cdot}8)(4)^2$$

$$\therefore \quad y = 78{\cdot}4\,\text{m}$$

Using distance = velocity × time for horizontal motion gives:

$$x = 2 \times 4 = 8\,\text{m}$$

Position vector after 4 seconds $= 10\mathbf{i} + 90\mathbf{j} + 8\mathbf{i} - 78{\cdot}4\mathbf{j} = 18\mathbf{i} + 11{\cdot}6\mathbf{j}$

The position vector after 4 seconds is $(18\mathbf{i} + 11{\cdot}6\mathbf{j})$ m.

Modelling real objects as "particles"

Examples **1** to **5** all involved "particles". Earlier in the book we defined a particle as something so small in size that the distance between its extremities is negligible. In practice of course, projected objects such as cricket balls, tennis balls, bullets etc., do have size and are not particles. However, by assuming them to be particles, we simplify the real situation to give a mathematical model that is easier to work with. Particles have no size; they are "point masses" and so cannot rotate. By modelling the flight of a cricket ball as that of a particle we can ignore the rotation of the ball and our model becomes more manageable.
In such models we also tend to assume that there is no air resistance, that the only force acting is the weight and that the motion is in one vertical plane. The appropriateness of the model can then be checked by comparing the outcomes predicted by the model with the real-life situation.

Setting up the mathematical model

The questions of previous chapters have required us to make various assumptions in order to create a simplified, manageable, model of the real situation. You do not need to list all of the assumptions you have made in order to create the model unless the question specifically asks you to do this. Remember in this text a question may do this by asking you to "set up the model". In such cases you should clearly state any assumptions you need to make in order to create the model and to solve the problem, as the next example demonstrates.

Example 6

A ball is dropped from a height of 1 metre in the back of a van which is moving with a constant speed of $20\,\mathrm{m\,s^{-1}}$. Set up a mathematical model of this situation and use it to find the horizontal distance moved by the ball from the point from which it is dropped to the point at which it hits the floor of the van.

Step 1. Set up the model.

- Consider the ball as a particle.
- Assume that any air resistance is negligible.
- Assume that nothing impedes the motion of the ball and that its motion is in one vertical plane.
- Take g, the acceleration due to gravity, as $9{\cdot}8\,\mathrm{m\,s^{-2}}$.
- The initial velocity of the ball will be taken as $20\,\mathrm{m\,s^{-1}}$ horizontally, i.e. the same as that of the van. (Although the ball hits the floor of the van at a point vertically below the point, relative to the van, from which it was dropped. (The ball and the floor of the van have both moved forward in this time.)

Step 2. Apply the mathematics.

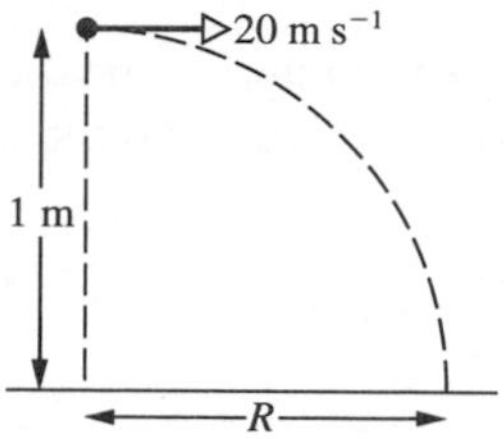

horizontal motion	vertical motion
$u = 20\,\text{m s}^{-1}$	$u = 0$
$t = t\,\text{s}$	$t = t\,\text{s}$
$s = R\,\text{m}$	$s = 1\,\text{m}$
no acceleration	$a = g\,(=9{\cdot}8\,\text{m s}^{-2})$
Using $s = vt$	Using $s = ut + \frac{1}{2}at^2$
gives: $R = 20t$ ...[1]	gives: $1 = \frac{1}{2}gt^2$ ...[2]

Substituting for t from [2] into [1] gives:

$$R = 20\sqrt{\frac{2}{g}}$$

$$= 9{\cdot}04$$

The horizontal distance moved is 9·04 m.

Exercise 12A

1. A particle is projected horizontally, at $20\,\text{m s}^{-1}$, from a point 78·4 m above a horizontal surface.
   Find the time taken for the particle to reach the surface and the horizontal distance travelled in that time.
2. A particle is projected horizontally with a speed of $21\,\text{m s}^{-1}$.
   Find the horizontal and vertical displacements of the particle, from the point of projection, $2\frac{6}{7}$ seconds after projection. Find how far the particle is then from the point of projection.
3. A particle is projected horizontally from a point 2·5 m above a horizontal surface. The particle hits the surface at a point which is, horizontally, 10 m from the point of projection. Find the initial speed of the particle.
4. At time $t = 0$ a particle is projected with velocity $5\mathbf{i}\,\text{m s}^{-1}$ from a point with position vector $20\mathbf{j}\,\text{m}$. Find the position vector of the particle when:
   (a) $t = 1\,\text{s}$ (b) $t = 2\,\text{s}$.
5. At time $t = 0$ a particle is projected with a velocity of $3\mathbf{i}\,\text{m s}^{-1}$ from a point with position vector $(5\mathbf{i} + 25\mathbf{j})\,\text{m}$. Find the position vector of the particle when $t = 2\,\text{s}$.
6. A stone is thrown horizontally at $21\,\text{m s}^{-1}$ from the edge of a vertical cliff and falls to the sea 40 m below.
   Find the horizontal distance from the foot of the cliff to the point where the stone enters the sea.
7. A batsman strikes a ball horizontally when it is 1 m above the ground. The ball is caught 10 cm above the ground by a fielder standing 12 m from the batsman.
   Find the speed with which the batsman hits the ball.
8. A darts player throws a dart horizontally with a speed of $14\,\text{m s}^{-1}$. The dart hits the board at a point 10 cm below the level at which it was released.
   Find the horizontal distance travelled by the dart.
9. A tennis ball is served horizontally with an initial speed of $21\,\text{m s}^{-1}$ from a height of 2·8 m. By what distance does the ball clear a net 1 m high situated 12 m horizontally from the server?
10. A fielder retrieves a cricket ball and throws it horizontally with a speed of $28\,\text{m s}^{-1}$ to the wicket-keeper standing 12 m away.
    If the fielder releases the ball at a height of 2 m above level ground, find the height of the ball when it reaches the wicket-keeper.

11. A particle is projected horizontally at $20\,\text{m s}^{-1}$.
Find the horizontal and vertical components of the particle's velocity two seconds after projection.

12. A particle is projected horizontally at $168\,\text{m s}^{-1}$.
Find the magnitude and direction of the velocity of the particle five seconds after projection.

13. Initially a particle is at an origin O and is projected with a velocity of $a\mathbf{i}\,\text{m s}^{-1}$. After t seconds, the particle is at the point with position vector $(30\mathbf{i} - 10\mathbf{j})$ m.
Find the values of t and a.

14. Two vertical towers stand on horizontal ground and are of heights 40 m and 30 m. A ball is thrown horizontally from the top of the higher tower with a speed of $24{\cdot}5\,\text{m s}^{-1}$ and just clears the smaller tower.
Find the distance:
(a) between the two towers
(b) between the smaller tower and the point on the ground where the ball first lands.

15. The top of a vertical tower is 20 m above ground level. When a ball is thrown horizontally from the top of this tower, it first hits the ground 24 m from the base of the tower.
By how much does the ball clear a vertical wall of height 13 m situated 12 m from the tower?

16. A stone is thrown horizontally with speed u from the edge of a vertical cliff of height h. The stone hits the ground at a point which is a distance d horizontally from the base of the cliff.
Show that $2hu^2 = gd^2$.

17. A vertical tower stands with its base on horizontal ground. Two particles A and B are both projected horizontally and in the same direction from the top of the tower with initial velocities of $14\,\text{m s}^{-1}$ and $17{\cdot}5\,\text{m s}^{-1}$ respectively.
If A and B hit the ground at two points 10 m apart, find the height of the tower.

18. O, A and B are three points with O on level ground and A and B respectively 3·6 metres and 40 metres vertically above O. A particle is projected horizontally from B with a speed of $21\,\text{m s}^{-1}$ and, 2 seconds later, a particle is projected horizontally from A with a speed of $70\,\text{m s}^{-1}$. Show that the two particles reach the ground at the same time and at the same distance from O, and find this distance.

19. A and B are two points on level ground. A vertical tower of height $4h$ has its base at A and a vertical tower of height h has its base at B. When a stone is thrown horizontally with speed v from the top of the taller tower towards the smaller tower, it lands at a point X where $\text{AX} = \frac{3}{4}\text{AB}$. When a stone is thrown horizontally with speed u from the top of the smaller tower towards the taller tower, it also lands at the point X.
Show that $3u = 2v$.

20. A gun is situated at the edge of a cliff of height 90 m and has its barrel horizontal and pointing directly out to sea. The gun fires a shell at $147\,\text{m s}^{-1}$. Set up a mathematical model of this situation and use it to find how far from the base of the cliff the shell strikes the sea.

21. When an aircraft is flying horizontally at a speed of $420\,\text{km h}^{-1}$, it releases a bomb. The moment of release occurs when the aircraft is a distance 2 km horizontally and h km vertically from a target. Set up a mathematical model of this situation and, given that the bomb hits the target, find the value of h.

22. A window in a house is situated 4·9 m above the ground. When a boy throws a ball horizontally from this window with a speed of $14\,\text{m s}^{-1}$, the ball just clears a vertical wall situated 10 m from the house. Set up a mathematical model of this situation and determine the height of the wall and how far beyond the wall the ball first hits the ground.

Particle projected at an angle to the horizontal

Suppose a particle is projected at an angle θ above the horizontal and with a velocity u.

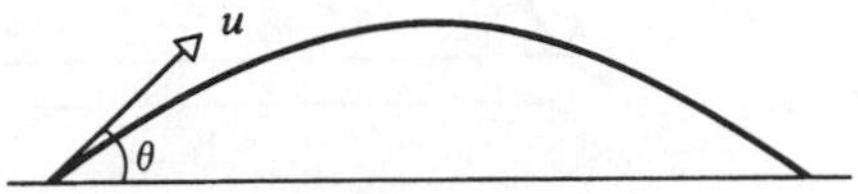

Again it is necessary to consider the motion in a horizontal and in a vertical direction separately.
The horizontal velocity remains constant and is $u \cos \theta$.
The vertical velocity, initially $u \sin \theta$, is reduced by the gravitational constant g.

Example 7

A particle is projected from a point on a horizontal plane and has an initial velocity of $28\sqrt{3}\,\text{m s}^{-1}$ at an angle of elevation of 60°. Find the greatest height reached by the particle and the time taken to reach this point.

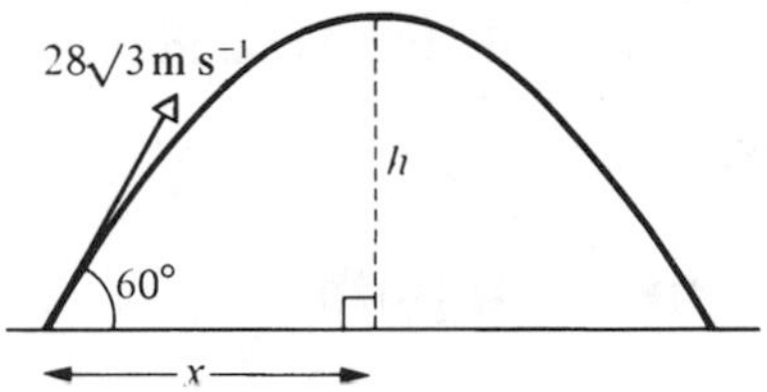

Consider the motion of the particle to the highest point of the path.

horizontal motion
$u = 28\sqrt{3} \cos 60°\,\text{m s}^{-1}$
$s = x$
$t = t$
no acceleration

vertical motion
$u = 28\sqrt{3} \sin 60°\,\text{m s}^{-1} \uparrow$
$s = h \uparrow$
$v = 0$ (at highest point)
$a = 9{\cdot}8\,\text{m s}^{-2} \downarrow = -9{\cdot}8\,\text{m s}^{-2} \uparrow$
$t = t$

Using $v^2 = u^2 + 2as$ for vertical motion gives:

$$0 = (28\sqrt{3} \sin 60°)^2 + 2(-9{\cdot}8)h$$
$$\therefore \quad h = 90\,\text{m}$$

Using $v = u + at$ for vertical motion gives:

$$0 = 28\sqrt{3} \sin 60° + (-9{\cdot}8)t$$
$$\therefore \quad t = 4\tfrac{2}{7}\,\text{s}$$

The greatest height reached is 90 m and the time taken is 4·29 s.

Example 8

A particle is projected from a point on a horizontal plane and has an initial velocity of $45\,\text{m s}^{-1}$ at an angle of elevation of $\tan^{-1} \frac{3}{4}$.
Find the time of flight and the range of the particle on the horizontal plane.

Consider the motion of the particle from A to B.

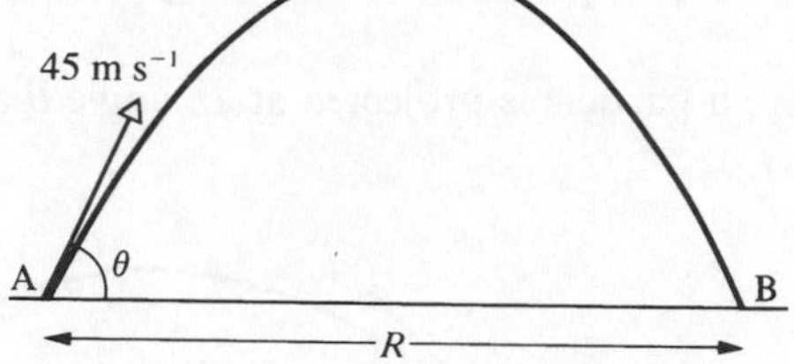

horizontal motion
$u = 45 \cos\theta \text{ m s}^{-1}$
$s = R$
$t = t$
no acceleration

vertical motion
$u = 45 \sin\theta \text{ m s}^{-1} \uparrow$
$s = 0$
$t = t$
$a = 9{\cdot}8 \text{ m s}^{-2} \downarrow$
$= -9{\cdot}8 \text{ m s}^{-2} \uparrow$

Since $\tan\theta = \frac{3}{4}$, $\sin\theta = \frac{3}{5}$ and $\cos\theta = \frac{4}{5}$.
Using $s = ut + \frac{1}{2}at^2$ for vertical motion gives:

$$0 = 27(t) - \tfrac{1}{2}(9{\cdot}8)t^2$$

$$\therefore \qquad t = 0, \text{ at A} \quad \text{or} \quad t = \frac{27 \times 2}{9{\cdot}8} = 5{\cdot}51 \text{ s, at B}$$

Using distance = velocity × time, for horizontal motion gives:

$$R = 36 \times 5{\cdot}51 = 198{\cdot}4 \text{ m}$$

The time of flight is 5·51 s and the range is 198 m.

Symmetrical path

It should be noted that the trajectory, i.e. the path of a particle projected from a point A on a horizontal plane and striking the plane at some point B, is symmetrical.
The time of flight from A to B is twice the time taken by the particle to reach the highest point T. Hence the time of flight may be calculated by doubling the time taken to reach the highest point of the path.
At the highest point T, the vertical velocity is zero and the particle is instantaneously travelling horizontally.

The direction of motion at any other time is found by considering the horizontal and vertical components of the velocity. During the first half of the motion, the particle is travelling at an angle above the horizontal and in the second half, below the horizontal.

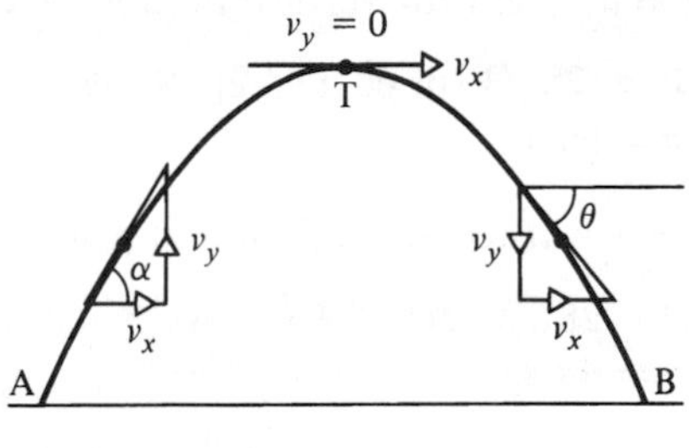

Accuracy and value of *g*

Some of the examples and questions in this chapter instruct the reader to take the value of g to be 10 m s^{-2} so that the arithmetic involved is simplified. The treatment of projectiles in this chapter already ignores air resistances, so the 2% error incurred by taking g as 10 m s^{-2} rather than $9{\cdot}8 \text{ m s}^{-2}$ will not greatly affect the accuracy of the answers obtained from the mathematical model. However, the reader should use the more accurate value of $9{\cdot}8 \text{ m s}^{-2}$ unless instructed otherwise.

Example 9

A particle is projected from an origin O with a velocity of $(30\mathbf{i} + 40\mathbf{j})\ \mathrm{m\,s^{-1}}$.
Find the position and velocity vectors of the particle 5 seconds later.
Hence find the distance of the particle from O and the speed and direction of its motion at this time. Take $g = 10\ \mathrm{m\,s^{-2}}$.

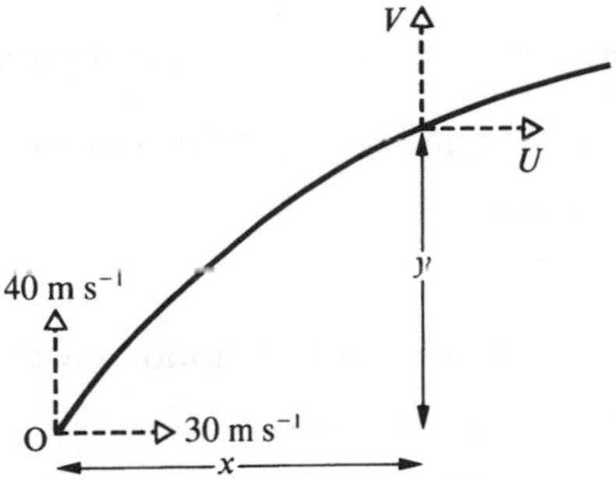

horizontal motion	vertical motion
$u = 30\ \mathrm{m\,s^{-1}}$	$u = 40\ \mathrm{m\,s^{-1}} \uparrow$
$s = x$	$s = y \uparrow$
$t = 5\ \mathrm{s}$	$t = 5\ \mathrm{s}$
no acceleration	$a = 10\ \mathrm{m\,s^{-2}} \downarrow$
$\therefore\ U = 30\ \mathrm{m\,s^{-1}}$	$= -10\ \mathrm{m\,s^{-2}} \uparrow$
	$v = V \uparrow$

Using $s = ut + \frac{1}{2}at^2$, for vertical motion gives:

$$y = 40(5) - \tfrac{1}{2}(10)5^2$$
$$\therefore \quad y = 75\ \mathrm{m}$$

Using $v = u + at$, for vertical motion gives:

$$V = 40 - (10)5$$
$$\therefore \quad V = -10, \text{ i.e. } 10\ \mathrm{m\,s^{-1}} \text{ downwards}$$

Using distance = velocity × time for horizontal motion gives:

$$x = 30 \times 5$$
$$\therefore \quad x = 150\ \mathrm{m}$$

Position vector is: $(x\mathbf{i} + y\mathbf{j}) = (150\mathbf{i} + 75\mathbf{j})\ \mathrm{m}$

Velocity vector is: $(U\mathbf{i} + V\mathbf{j}) = (30\mathbf{i} - 10\mathbf{j})\ \mathrm{m\,s^{-1}}$

Distance of particle from O is: $\sqrt{(x^2 + y^2)} = \sqrt{(150^2 + 75^2)}$
$= 75\sqrt{5}\ \mathrm{m}$

Speed of particle is: $\sqrt{(U^2 + V^2)} = \sqrt{\{(30)^2 + (-10)^2\}}$
$= 10\sqrt{10}\ \mathrm{m\,s^{-1}}$

The direction of motion is at an angle θ below the horizontal, given by:

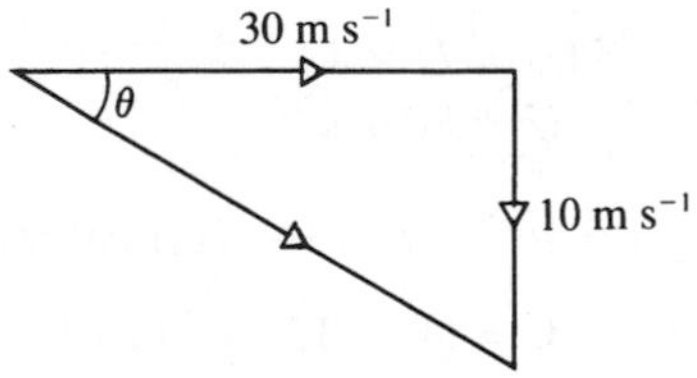

$$\tan\theta = \tfrac{10}{30}$$
$$\therefore \quad \theta = 18{\cdot}43°$$

The position vector is $(150\mathbf{i} + 75\mathbf{j})\ \mathrm{m}$ and the velocity vector is $(30\mathbf{i} - 10\mathbf{j})\ \mathrm{m\,s^{-1}}$; the distance from O at this time is $75\sqrt{5}\ \mathrm{m}$ and the speed of the particle is $10\sqrt{10}\ \mathrm{m\,s^{-1}}$ at an angle of 18·43° below the horizontal.

Example 10

A stone is thrown from the edge of a vertical cliff with a velocity of $50\ \mathrm{m\,s^{-1}}$ at an angle $\tan^{-1}\frac{7}{24}$ above the horizontal. The stone strikes the sea at a point 240 m from the foot of the cliff.
Find the time for which the stone is in the air and the height of the cliff.

Let θ be the angle of projection, then $\tan\theta = \frac{7}{24}$, $\sin\theta = \frac{7}{25}$ and $\cos\theta = \frac{24}{25}$.

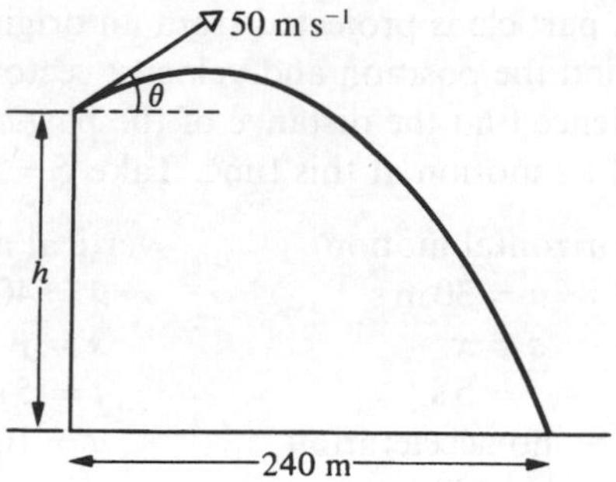

horizontal motion	vertical motion
$u = 50\cos\theta$	$u = 50\sin\theta$
$= 48\,\text{m s}^{-1}$	$= 14\,\text{m s}^{-1}\uparrow = -14\,\text{m s}^{-1}\downarrow$
$s = 240\,\text{m}$	$s = h\downarrow$
$t = t$	$t = t$
no acceleration	$a = 9{\cdot}8\,\text{m s}^{-2}\downarrow$

Using distance = velocity × time for horizontal motion gives:

$$240 = 48t$$
$$\therefore \quad t = 5\,\text{s}$$

Using $s = ut + \frac{1}{2}at^2$ for vertical motion gives:

$$h = -14(5) + \tfrac{1}{2}(9{\cdot}8)5^2$$
$$= 52{\cdot}5\,\text{m}$$

The time of flight of the stone is 5 s and the height of the cliff is 52·5 m.

Example 11

A shell fired from a gun has a horizontal range of 960 m and a time of flight of 12 s. Find the magnitude and direction of the velocity of projection. Take $g = 10\,\text{m s}^{-2}$.

Consider the motion of the shell from A to B.

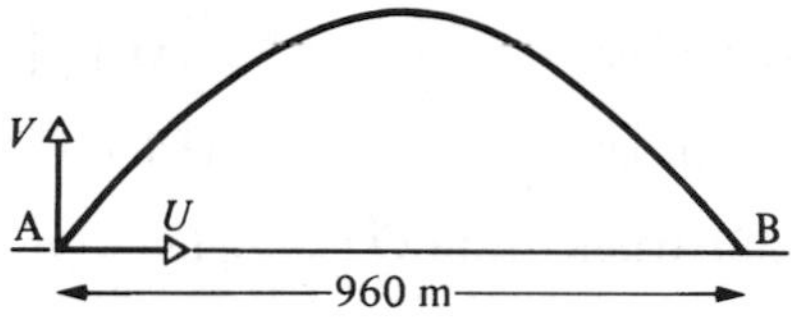

horizontal motion	vertical motion
$u = U$	$u = V\uparrow$
$t = 12\,\text{s}$	$t = 12\,\text{s}$
$s = 960\,\text{m}$	$s = 0$
no acceleration	$a = 10\,\text{m s}^{-2}\downarrow = -10\,\text{m s}^{-2}\uparrow$

Using distance = velocity × time for horizontal motion gives:

$$960 = U \times 12$$
$$\therefore \quad U = 80\,\text{m s}^{-1}$$

Using $s = ut + \frac{1}{2}at^2$, for vertical motion gives:

$$0 = (V \times 12) - \tfrac{1}{2}(10)12^2$$
$$\therefore \quad V = 60\,\text{m s}^{-1}$$

speed of projection $= \sqrt{(U^2 + V^2)} = \sqrt{(80^2 + 60^2)} = 100\,\text{m s}^{-1}$

The angle of projection is θ above the horizontal, given by:

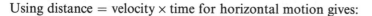

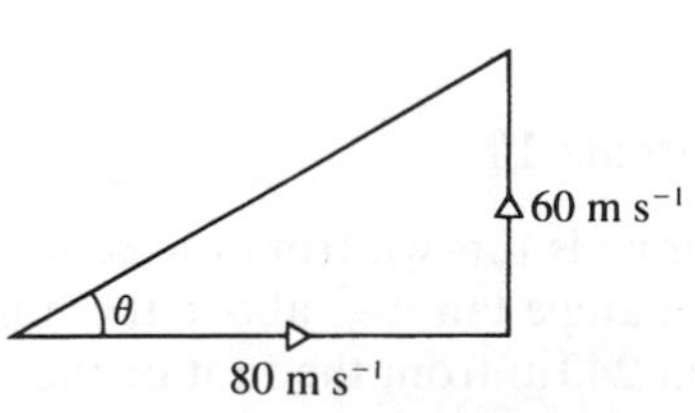

$$\tan\theta = \tfrac{60}{80}$$
or $\quad \theta = 36{\cdot}87°$

The shell is projected at $100\,\text{m s}^{-1}$ at an angle of 36·87° above the horizontal.

Example 12

A ball is projected with a velocity of $28\,\text{m s}^{-1}$. If the range of the ball on the horizontal plane is 64 m, find the two possible angles of projection.

Let the angle of projection be θ.

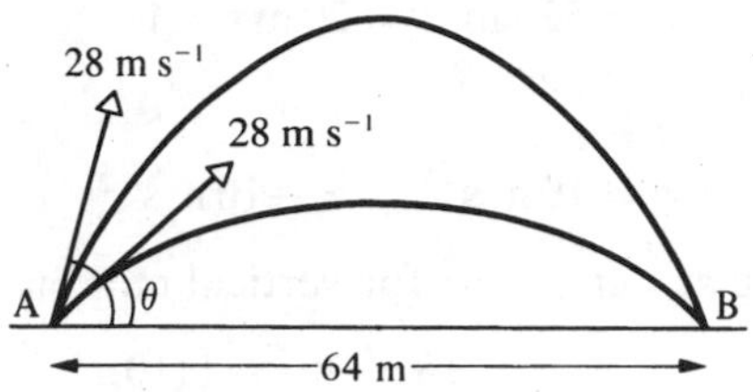

horizontal motion	vertical motion
$u = 28\cos\theta\,\text{m s}^{-1}$	$u = 28\sin\theta\,\text{m s}^{-1}\uparrow$
$s = 64\,\text{m}$	$s = 0$
$t = t$	$t = t$
no acceleration	$a = 9{\cdot}8\,\text{m s}^{-2}\downarrow = -9{\cdot}8\,\text{m s}^{-2}\uparrow$

Using $s = ut + \frac{1}{2}at^2$ for vertical motion gives:

$$0 = (28\sin\theta)t - \tfrac{1}{2}(9{\cdot}8)t^2$$

$$\therefore \quad t = 0 \text{ i.e. at A} \quad \text{or} \quad t = \frac{2(28\sin\theta)}{9{\cdot}8} \text{ at B}$$

Using distance = velocity × time for horizontal motion gives:

$$64 = 28\cos\theta \times \frac{56\sin\theta}{9{\cdot}8}$$

Using $2\sin\theta\cos\theta = \sin 2\theta$ gives:

$$\sin 2\theta = \frac{64 \times 9{\cdot}8}{28 \times 28} = 0{\cdot}8$$

$$\therefore \quad 2\theta = 53{\cdot}13^\circ \quad \text{or} \quad 126{\cdot}87^\circ$$

$$\therefore \quad \theta = 26{\cdot}57^\circ \quad \text{or} \quad 63{\cdot}44^\circ$$

The two possible angles of projection are $26{\cdot}57^\circ$ and $63{\cdot}44^\circ$.

Example 13

A golfer hits a ball with a velocity of $52\,\text{m s}^{-1}$ at an angle α above the horizontal where $\tan\alpha = \frac{5}{12}$. Set up a mathematical model of this situation and determine the time for which the ball is at least 15 m above the ground. (Take $g = 10\,\text{m s}^{-2}$.)

Step 1. Set up the model.

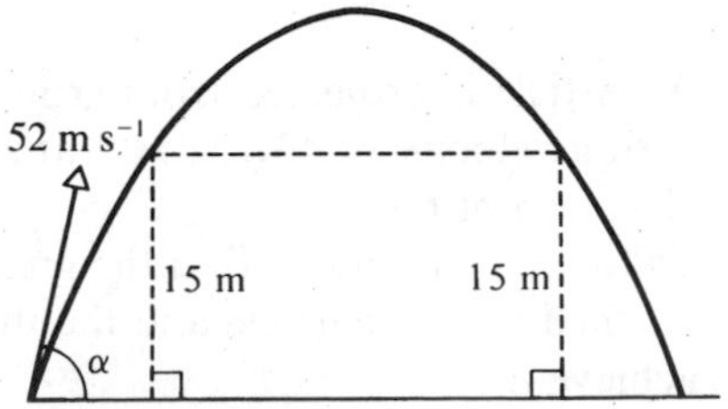

- Consider the golf ball as a particle.
- Assume that any air resistance is negligible.
- Assume the ground to be level.
- Assume that nothing impedes the motion of the ball and that motion occurs in one vertical plane.
- Take g, the acceleration due to gravity, as $10\,\text{m s}^{-2}$.

Step 2. Apply the mathematics.

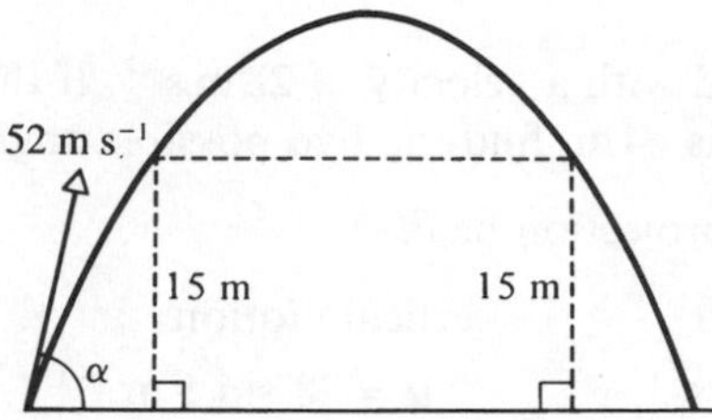

Since $\tan\alpha = \frac{5}{12}$, $\sin\alpha = \frac{5}{13}$ and $\cos\alpha = \frac{12}{13}$

for vertical motion

$$u = 52\sin\alpha = 20\,\text{m s}^{-1}\uparrow$$
$$s = 15\,\text{m}\uparrow$$
$$t = t$$
$$a = 10\,\text{m s}^{-2}\downarrow = -10\,\text{m s}^{-2}\uparrow$$

Using $s = ut + \frac{1}{2}at^2$ for vertical motion gives:

$$15 = 20t - \tfrac{1}{2}(10)t^2$$
$$\therefore \quad t^2 - 4t + 3 = 0$$
$$\therefore \quad (t-3)(t-1) = 0 \text{ i.e. } t = 3 \text{ or } 1\text{ s}$$

Hence the ball is at a height of 15 m 1 second after projection and 3 seconds after projection. The ball is at least 15 m above the ground for 2 seconds.

Exercise 12B

Take g as $9{\cdot}8\,\text{m s}^{-2}$ unless instructed otherwise.

1. A particle is projected from a point on level ground such that its initial velocity is $56\,\text{m s}^{-1}$ at an angle of 30° above the horizontal. Find:
(a) the time taken for the particle to reach its maximum height
(b) the maximum height
(c) the time of flight
(d) the horizontal range of the particle.

2. A particle is projected from a point on level ground such that its initial velocity is $60\,\text{m s}^{-1}$ at an angle of elevation 30°. Taking g as $10\,\text{m s}^{-2}$, find:
(a) the time taken for the particle to reach its maximum height
(b) the maximum height
(c) the time of flight
(d) the horizontal range of the particle.

3. Find the horizontal range of a particle projected with a velocity of $35\,\text{m s}^{-1}$ at an angle of $\sin^{-1}\frac{4}{5}$ above the horizontal.

4. A particle is projected from ground level with a velocity of $40\sqrt{3}\,\text{m s}^{-1}$ at an angle of elevation of 60°.
Taking $g = 10\,\text{m s}^{-2}$, find the greatest height reached by the particle and the time taken to achieve it.

5. A golfer hits a golf ball with a velocity of $44{\cdot}1\,\text{m s}^{-1}$ at an angle of $\sin^{-1}\frac{3}{5}$ above the horizontal. The ball lands on the green at a point which is level with the point of projection.
Find the time for which the golf ball was in the air.

6. A gun has its barrel set at an angle of elevation of 15°.
The gun fires a shell with an initial speed of $210\,\text{m s}^{-1}$.
Find the horizontal range of the shell.

7. In a game of indoor football, a free kick is awarded whenever the ball is kicked above shoulder height.
The ball is initially at rest on the floor when a player kicks it with a velocity of $14\,\text{m s}^{-1}$ at an angle of elevation θ, with $\sin\theta = \frac{2}{5}$.
Taking shoulder height as 1·5 m and assuming nothing prevents the ball from reaching its highest point, find whether a free kick is awarded or not.

8. A ball is projected from horizontal ground and has an initial velocity of $20\,\text{m s}^{-1}$ at an angle of elevation $\tan^{-1}\frac{7}{24}$. When the ball is travelling horizontally, it strikes a vertical wall.
How high above ground level does the impact occur?

9. A particle is projected from an origin O at a velocity of $(4\mathbf{i} + 13\mathbf{j})\,\text{m s}^{-1}$.
Find the position vector of the particle 2 seconds after projection and the distance the particle is then from O. (Take $g = 10\,\text{m s}^{-2}$.)

10. A particle is projected from the origin at a velocity of $(10\mathbf{i} + 20\mathbf{j})\,\text{m s}^{-1}$.
Find the position and velocity vectors of the particle 3 seconds after projection.
(Take $g = 10\,\text{m s}^{-2}$.)

11. A particle is projected at a velocity of $25\,\text{m s}^{-1}$ at an angle of 30° above the horizontal.
Find the horizontal and vertical components of the particle's velocity $2\frac{1}{2}$ seconds after projection. Hence find the speed and direction of motion of the particle at that time.

12. A particle is projected from an origin O and has an initial velocity of $30\sqrt{2}\,\text{m s}^{-1}$ at an angle of 45° above the horizontal.
Find the horizontal and vertical displacements from O of the particle 2 seconds after projection. Hence find its distance from O at that time.

13. A particle is projected from an origin O at a velocity of $(4\mathbf{i} + 11\mathbf{j})\,\text{m s}^{-1}$ and passes through a point P which has a position vector $(8\mathbf{i} + x\mathbf{j})\,\text{m}$. Taking $g = 10\,\text{m s}^{-2}$, find the time taken for the particle to reach P from O and the value of x.

14. A stone is thrown from the edge of a vertical cliff and has an initial velocity of $26\,\text{m s}^{-1}$ at an angle $\tan^{-1}\frac{5}{12}$ below the horizontal. The stone hits the sea at a point level with the base of the cliff and 72 m from it.
Find the height of the cliff and the time for which the stone is in the air.
(Take $g = 10\,\text{m s}^{-2}$.)

15. A batsman hits a ball at a velocity of $17{\cdot}5\,\text{m s}^{-1}$ angled at $\tan^{-1}\frac{3}{4}$ above the horizontal, the ball initially being 60 cm above level ground. The ball is caught by a fielder standing 28 m from the batsman.
Find the time taken for the ball to reach the fielder and the height above ground at which he takes the catch.

16. A vertical tower stands on level ground. A stone is thrown from the top of the tower and has an initial velocity of $24{\cdot}5\,\text{m s}^{-1}$ angled at $\tan^{-1}\frac{4}{3}$ above the horizontal. The stone strikes the ground at a point 73·5 m from the foot of the tower.
Find the time taken for the stone to reach the ground and the height of the tower.

17. A particle projected from a point on level ground has a horizontal range of 240 m and a time of flight of 6 s.
Find the magnitude and direction of the velocity of projection. (Take $g = 10\,\text{m s}^{-2}$.)

18. A particle is projected from ground level with an initial speed of $28\,\text{m s}^{-1}$ and during the course of its motion it must not go higher than 10 m above ground level.
Find the angle of projection that would allow the particle to go as high as possible.

19. Ten seconds after its projection from the origin a particle has a position vector of $(150\mathbf{i} - 200\mathbf{j})\,\text{m}$.
Find, in vector form, the velocity of projection.

20. A football is kicked from a point on level ground, 15 m from a vertical wall. Three seconds later the football hits the wall at a point 6 m above ground.
Find the horizontal and vertical components of the initial velocity of the ball.
(Take $g = 10\,\text{m s}^{-2}$.)

21. A football is kicked from a point O on level ground and, 2 seconds later, just clears a vertical wall of height 2·4 m.
If O is 22 m from the wall, find the velocity with which the ball is kicked.

22. A particle is projected from a horizontal plane and has an initial velocity of $49\,\text{m s}^{-1}$ at an angle of 30° above the horizontal.
For how long is the particle at least 19·6 m above the level of the plane?

23. A particle is projected from a horizontal plane and has an initial velocity of $50\,\text{m s}^{-1}$ at an angle of $\tan^{-1}\frac{4}{3}$ above the horizontal.
For how long is the particle at least 60 m above the level of the plane?
(Take $g = 10\,\text{m s}^{-2}$.)

24. A particle is projected from the origin and has an initial velocity of $(7\mathbf{i} + 5\mathbf{j})\,\text{m s}^{-1}$.
Given that the particle passes through the point P, position vector $(x\mathbf{i} - 30\mathbf{j})\,\text{m}$, find the time taken for this to occur and the value of x.
(Take $g = 10\,\text{m s}^{-2}$.)

25. A stone is thrown from the top of a vertical cliff, 100 m above sea level. The initial velocity of the stone is $13\,\text{m s}^{-1}$ at an angle of elevation of $\tan^{-1}\frac{5}{12}$. Find the time taken for the stone to reach the sea and its horizontal distance from the cliff at that time. ($g = 10\,\text{m s}^{-2}$.)

26. A golfer hits a golf ball with a velocity of $30\,\text{m s}^{-1}$ at an angle of $\tan^{-1}\frac{4}{3}$ above the horizontal. The ball lands on a green 5 m below the level from which it was struck. Find the horizontal distance travelled by the ball. (Take $g = 10\,\text{m s}^{-2}$.)

27. A ball is thrown from the top of a vertical tower, 40 m above level ground. The initial velocity of the ball is $10\sqrt{2}\,\text{m s}^{-1}$ at an angle of 45° below the horizontal. Find the distance from the foot of the tower to the point where the ball first lands. (Take $g = 10\,\text{m s}^{-2}$.)

28. A particle is projected at $84\,\text{m s}^{-1}$ to hit a point 360 m away and on the same horizontal level as the point of projection. Find the two possible angles of projection.

29. A golfer hits a golf ball at $30\,\text{m s}^{-1}$ and wishes it to land at a point 45 m away, on the same horizontal level as the starting point.
Find the two possible angles of projection.

30. A ball is projected from horizontal ground and has an initial speed of $35\,\text{m s}^{-1}$. When the ball is travelling horizontally, it strikes a vertical wall.
If the wall is 25 m from the point of projection, find the two possible angles of projection of the ball.

31. A particle is projected from a point O, and passes through a point A when travelling horizontally.
If A is 10 m horizontally and 8 m vertically from O, find the magnitude and direction of the velocity of projection.

32. Two particles A and B are projected simultaneously, A from the top of a vertical cliff and B from the base. Particle A is projected horizontally with speed $3u\,\text{m s}^{-1}$ and B is projected at angle θ above the horizontal with speed $5u\,\text{m s}^{-1}$. The height of the cliff is 56 m and the particles collide after 2 seconds.
Find the horizontal and vertical distances from the point of collision to the base of the cliff and the values of u and θ.

33. Two stones are thrown from the top of a vertical cliff 50 m above level ground. The stones are thrown at the same time and in the same vertical plane, one at $25\,\text{m s}^{-1}$ and angle of elevation $\tan^{-1}\frac{3}{4}$ and the other at $25\,\text{m s}^{-1}$ and angle of depression $\tan^{-1}\frac{3}{4}$. Find the time interval between the stones hitting the ground and the horizontal distance between their points of impact. (Take $g = 10\,\text{m s}^{-2}$.)

34. Two particles A and B are projected simultaneously from the same point on horizontal ground and both travel in the same vertical plane; A is projected at an angle of elevation of 45° and a speed of $28\,\text{m s}^{-1}$ and B is projected with a speed of $35\,\text{m s}^{-1}$.
If the two particles land at points 15 metres apart, find the four possible angles of projection of particle B.

35. A golfer wishes to hit a ball from a tee to a green which is at the same level as the tee. The front edge of the green is 160 m from the tee and the back edge is 185 m from the tee.
If the golfer hits the ball with a speed of $49\,\text{m s}^{-1}$, find the possible angles of projection.

36. A batsman hits a ball from ground level and gives it an initial velocity of $28\,\text{m s}^{-1}$ at an angle of inclination of 15°. The ball lands 10 m short of the boundary. Set up a mathematical model for this situation and determine, for this angle of projection, the least speed of projection necessary for the ball to clear the boundary.

37. Sarah throws a message, wrapped around a stone, with a speed of $12\,\text{m s}^{-1}$ from the top of a cliff 50 m high, to Callum who is in a boat 40 m from the bottom of the cliff.
Set up a model to determine the possible angles of projection with which Sarah should throw the stone if it is to reach Callum.

38. A fireman is attempting to direct water through the open window of a top room in a building. The window is 25 m above ground level. The water leaves the hose at an angle of 40° to the horizontal at a height 1·5 m above the ground. The fireman finds that if he stands 35 m away from the base of the wall then the jet of water, travelling in an upward direction, goes through the open window.
Set up a model and determine the speed of the water as it leaves the hose.

General results

Certain standard results can be established regarding the motion of a particle which is projected from a point O on a horizontal plane, with an initial velocity of U at an angle α above the horizontal.
The following examples illustrate how this may be done.

Example 14

Find the time of flight T and range R on the horizontal plane.

Consider the motion of the particle from A to B.

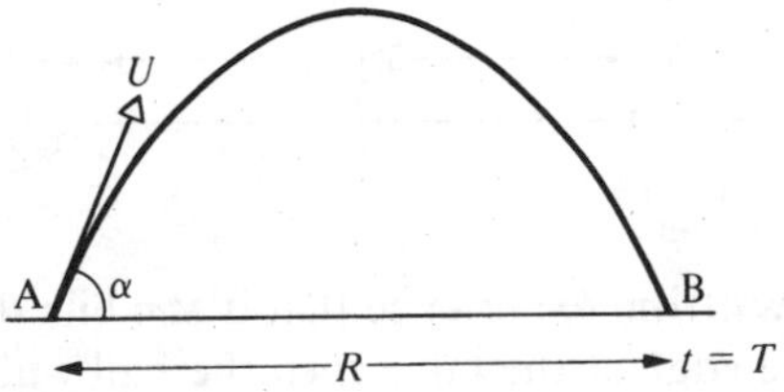

horizontal motion	vertical motion
$u = U\cos\alpha$	$u = U\sin\alpha \uparrow$
$s = R$	$s = 0$
$t = T$	$t = T$
no acceleration	$a = g\downarrow = -g\uparrow$

Using $s = ut + \frac{1}{2}at^2$, for vertical motion gives:

$$0 = (U\sin\alpha)T - \tfrac{1}{2}gT^2$$

$\therefore \quad T = 0$ at A

or $\quad T = \dfrac{2U\sin\alpha}{g}$ at B

Using distance = velocity × time for horizontal motion gives:

$$R = (U\cos\alpha)T$$
$$= \frac{2U^2\sin\alpha\cos\alpha}{g}$$

and using $2\sin\alpha\cos\alpha = \sin 2\alpha$ gives:

$$R = \frac{U^2\sin 2\alpha}{g}$$

The time of flight is $\dfrac{2U\sin\alpha}{g}$ and the range is $\dfrac{U^2\sin 2\alpha}{g}$.

Example 15

Find the maximum range of the particle on the horizontal plane and the angle of projection which gives this range.

From the previous example it is known that the range R is given by:

$$R = \frac{U^2 \sin 2\alpha}{g}$$

For any particular value of U, the maximum value of R occurs when $\sin 2\alpha$ is a maximum. The maximum value of $\sin 2\alpha = 1$ when:

$$2\alpha = 90° \quad \text{or} \quad \alpha = 45°$$

Hence $\quad R_{\text{max}} = \dfrac{U^2 \sin 90°}{g} = \dfrac{U^2}{g}$

The maximum range is $\dfrac{U^2}{g}$ and the angle of projection to give this range is 45°.

Example 16

The diagram below shows the cross-section of a particular fairway on a golf course.

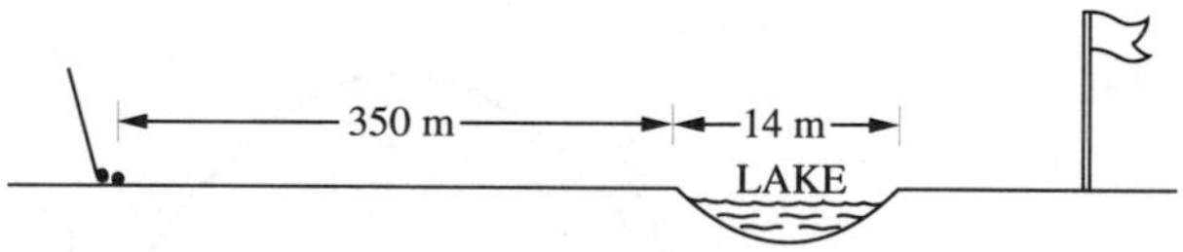

Diana knows from experience that if she hits the ball and it lands within one metre of the edge of the lake, then the ball will roll into the lake.
Set up a model and use it to solve the following problems.

(a) Given that Diana intends to hit the ball so that it travels as far as possible before landing, find the range of values for the initial speed of the ball which Diana should avoid in order not to lose the ball in the lake.

(b) When Diana actually hit the ball the angle of projection was 30° and the ball landed directly in the lake. Find the possible range of values for the initial speed of the ball.

Step 1. Set up the model.

- Assume that the golf ball is a particle.
- Assume that the motion is all in the same vertical plane.
- Assume that the angle of projection to attain maximum range is 45°.
- Assume that there is no air resistance.

Step 2. Apply the mathematics.

(a) From Example 15 the maximum range is given by $\frac{U^2}{g}$, where U is the initial speed of projection.
The diagram below shows that part of the fairway which Diana should avoid.

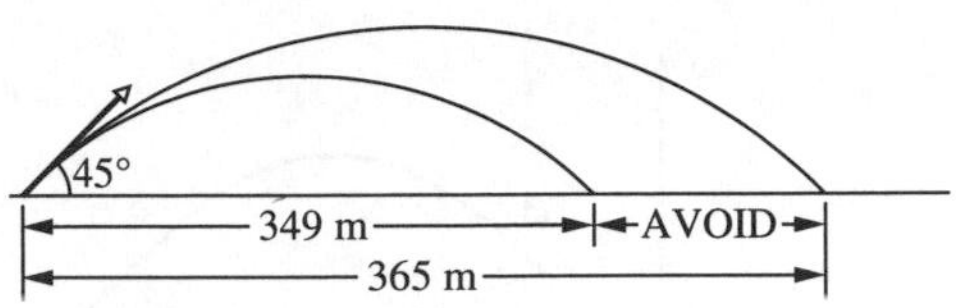

Diana will lose the ball in the lake if:

$$349 < R_{max} < 365$$

$$349 < \frac{U^2}{g} < 365$$

i.e. $58{\cdot}48\,\text{m s}^{-1} < U < 59{\cdot}81\,\text{m s}^{-1}$

Therefore, Diana should avoid giving the ball an initial speed U in the range $58{\cdot}5\,\text{m s}^{-1} < U < 59{\cdot}8\,\text{m s}^{-1}$.

(b) The diagram below shows the golf ball being projected at an angle of 30° to the horizontal with an initial speed of $V\,\text{m s}^{-1}$.

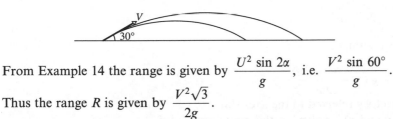

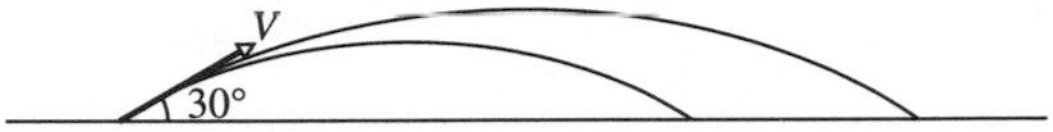

From Example 14 the range is given by $\frac{U^2 \sin 2\alpha}{g}$, i.e. $\frac{V^2 \sin 60°}{g}$.

Thus the range R is given by $\frac{V^2\sqrt{3}}{2g}$.

Since the ball landed directly in the lake, R must satisfy the inequality:

$$350 < R < 364$$

$$\therefore \quad 350 < \frac{V^2\sqrt{3}}{2g} < 364$$

$$\frac{6860}{\sqrt{3}} < V^2 < \frac{7134{\cdot}4}{\sqrt{3}}$$

i.e. $62{\cdot}93\,\text{m s}^{-1} < V < 64{\cdot}18\,\text{m s}^{-1}$

Therefore, Diana must have given the ball an initial speed V in the range $62{\cdot}9\,\text{m s}^{-1} < V < 64{\cdot}2\,\text{m s}^{-1}$.

In order to determine whether this is a realistic model for the situation the figures should be compared with data derived from a real-life situation. Some of the questions to be answered are:

- Is an initial speed of $60\,\text{m s}^{-1}$ realistic?
- Is the range of 350 m realistically attainable with an initial speed of $60\,\text{m s}^{-1}$?

Equation of trajectory

When a particle is projected from a point O on a horizontal plane, the equation of the trajectory may be obtained by taking x and y axes through the point of projection O, as shown in the diagram.

Consider the motion of the particle from O to the point P (x, y).

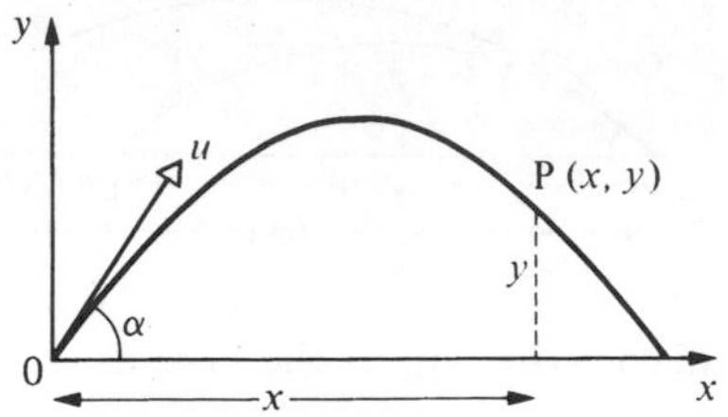

horizontal motion	vertical motion
$u = u \cos \alpha$	$u = u \sin \alpha \uparrow$
$s = x$	$s = y \uparrow$
$t = t$	$t = t$
no acceleration	$a = g \downarrow = -g \uparrow$

Using $s = ut + \frac{1}{2}at^2$ for vertical motion gives:

$$y = (u \sin \alpha)t - \tfrac{1}{2}gt^2 \qquad \ldots[1]$$

Using distance = velocity × time for horizontal motion gives:

$$x = (u \cos \alpha)t$$

$$\therefore \quad t = \frac{x}{u \cos \alpha} \qquad \ldots[2]$$

Substituting for t from equation [2] in equation [1] gives:

$$y = \frac{(u \sin \alpha)x}{u \cos \alpha} - \tfrac{1}{2}g\left(\frac{x}{u \cos \alpha}\right)^2$$

$$\therefore \quad y = x \tan \alpha - \frac{gx^2}{2u^2 \cos^2 \alpha}$$

and since $\dfrac{1}{\cos^2 \alpha} = \sec^2 \alpha = 1 + \tan^2 \alpha$

then $$y = x \tan \alpha - gx^2 \frac{(1 + \tan^2 \alpha)}{2u^2}$$

This is the equation of the trajectory referred to the axes shown.

There are four variables (x, y, u and α) involved in this equation, and if three are known, the fourth can be determined in any particular example.

Exercise 12C

Take g as $9{\cdot}8\,\text{m s}^{-2}$ unless instructed otherwise.

Greatest height, time of flight, range, maximum range

1. A particle is projected from a point on a horizontal plane and has an initial velocity u at an angle of θ above the plane. Show that:
   (a) the greatest height h reached by the particle above the plane is given by $h = \dfrac{u^2 \sin^2 \theta}{2g}$
   (b) the time of flight T is given by $T = \dfrac{2u \sin \theta}{g}$
   (c) the horizontal range R is given by $R = \dfrac{u^2 \sin 2\theta}{g}$
   (d) the maximum horizontal range R_{max} is given by $R_{max} = \dfrac{u^2}{g}$ and find the value of θ for which this occurs.

The results of question **1** may be used without proof to answer questions **2** to **7**.

2. A particle is projected from a point on a horizontal plane and has an initial velocity of $140\,\mathrm{m\,s^{-1}}$ at an angle of elevation of $30°$. Find the greatest height reached by the particle above the plane.

3. A particle is projected from a point on a horizontal plane and has an initial velocity of $70\,\mathrm{m\,s^{-1}}$ at an angle of $10°$ above the plane. Find the range of the particle on the horizontal plane.
At this speed of projection, what is the maximum range the particle could have on the horizontal plane and for what angle of projection would it occur?

4. A particle is projected from a point on a horizontal plane and has an initial velocity of $49\sqrt{2}\,\mathrm{m\,s^{-1}}$ at an angle of $45°$ above the plane. If the particle returns to the plane, find
(a) the time of flight
(b) the time taken to reach the highest point.

5. A bullet fired from a gun has a maximum horizontal range of 2000 m.
Find the muzzle velocity of the gun (i.e. the speed with which the bullet leaves the gun).

6. A gun is on the same horizontal level as a target 484 metres away. A bullet is fired from the gun with a muzzle velocity of $154\,\mathrm{m\,s^{-1}}$. At what angle of elevation should the muzzle of the gun be set in order to ensure the bullet hits the target?

7. A particle projected from a point on a horizontal plane reaches a greatest height h above the plane and has a horizontal range R. If $R = 2h$, find the angle of projection.

Equation of trajectory

8. A particle is projected from a point O and has an initial velocity u at an angle of θ above the horizontal. In the vertical plane of projection take x and y axes as the horizontal and vertical axes respectively and O as the origin. Show that the equation of the trajectory is

$$y = x\tan\theta - \frac{gx^2}{2u^2}\sec^2\theta.$$

If the particle passes through the point with coordinates (a, b), show that

$$ga^2\tan^2\theta - 2u^2 a\tan\theta + 2u^2 b + ga^2 = 0.$$

The equation of the trajectory obtained in question **8** may be used without proof to answer questions **9** to **12**.

9. A particle is projected from a point on a horizontal plane and initially travels in a direction which makes an angle of $\tan^{-1}\frac{3}{4}$ with the horizontal. In the subsequent motion the particle passes through a point above the plane which is 20 m horizontally and 10 m vertically from the point of projection.
Find the speed of projection. (Take $g = 10\,\mathrm{m\,s^{-2}}$.)

10. A particle is projected from a point on a horizontal plane and has an initial speed of $28\,\mathrm{m\,s^{-1}}$.
If the particle passes through a point above the plane, 40 m horizontally and 20 m vertically from the point of projection, find the possible angles of projection.

11. A particle is projected from a point on a horizontal plane and has an initial speed of $42\,\mathrm{m\,s^{-1}}$.
If the particle passes through a point above the plane, 70 m vertically and 60 m horizontally from the point of projection, find the possible angles of projection.

12. An origin O lies on a horizontal plane and a point P lies above the plane, 50 m horizontally and 60 m vertically from O. Show that a particle projected from O with speed $35\,\mathrm{m\,s^{-1}}$ cannot pass through P.

Any of the standard projectile results obtained in question **1** and question **8** of this exercise may be used without proof to answer questions **13** and **14**.

13. In a game of volleyball Jane is 2 m away from the net, which is 3 m high. Jane hits the ball at $10\,\mathrm{m\,s^{-1}}$ at an angle of $40°$ to the horizontal and at a height of 0·75 m above the ground.
Set up a model and determine whether the ball passes over the net. Using your model determine the angle to the horizontal with which the ball should be projected if Jane is to just clear the net still using an initial speed of projection of $10\,\mathrm{m\,s^{-1}}$ and height of projection of 0·75 m.

14. The diagram below shows the cross-section of a particular fairway on a golf course.

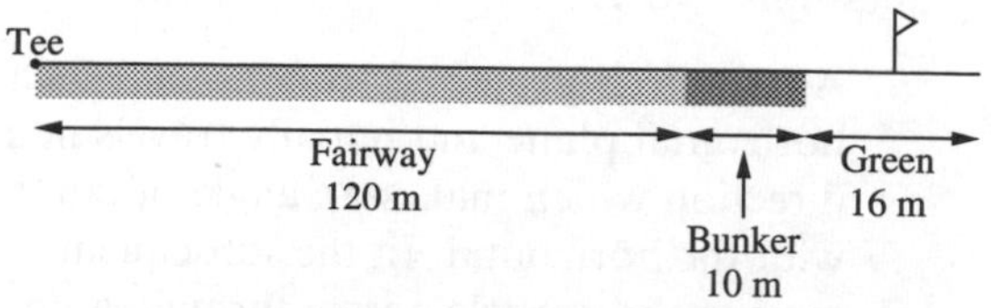

Sean intends to use a club that he knows will give the ball an angle of projection of 55°. At this angle Sean feels that when the ball first lands it will probably roll a further 4 metres before coming to rest (unless it lands in the bunker in which case it will stop immediately). Set up a model and use it to solve the following.

(a) Find the possible range of values for the initial speed of the ball that would see Sean's shot land, and remain, on the green (answer in $m\,s^{-1}$, correct to 1 decimal place).
(b) Find the range of values for the initial speed of the ball that Sean should avoid if his ball is not to end up in the bunker (answer in $m\,s^{-1}$, correct to 1 decimal place).
(c) Sean actually hit the ball at $37{\cdot}3\,m\,s^{-1}$ and his anticipated roll of 4 metres did not occur. Instead the backspin on the ball caused it to roll backwards 3 metres, from where it first landed, towards the tee. Did the shot end up in the bunker?

Projectile on an inclined plane

Suppose that a particle is projected from a point O up a plane inclined at an angle α to the horizontal, and that the motion takes place in a vertical plane containing a line of greatest slope of the plane. Let the particle be projected at an angle β above the plane.
The motion of the particle is best investigated by considering the component of the initial velocity at right angles to the plane i.e. $u \sin \beta$. In this direction, the acceleration due to gravity is $-g \cos \alpha$.
When the particle strikes the plane at the point P, the displacement of the particle at right angles to the plane is zero. The distance OP is called the range of the particle on the inclined plane.

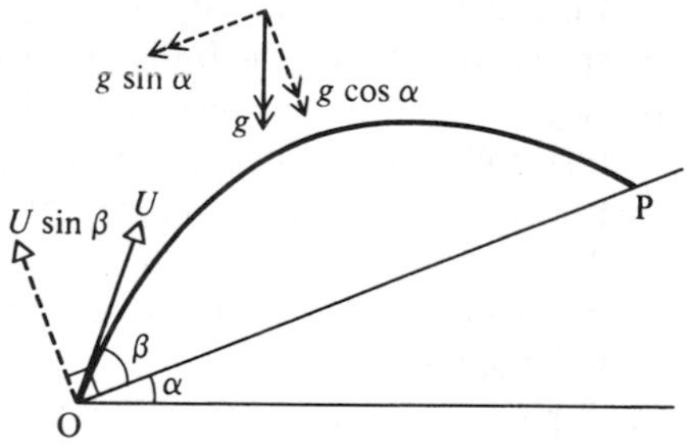

Example 17

A particle is projected up a plane which is inclined at 25° to the horizontal. The particle is projected from a point A on the plane with a velocity of $49\,m\,s^{-1}$ at an angle of 50° to the plane, the vertical plane of the motion containing a line of greatest slope of the plane.
Find the time taken by the particle to strike the plane and the range on the plane.

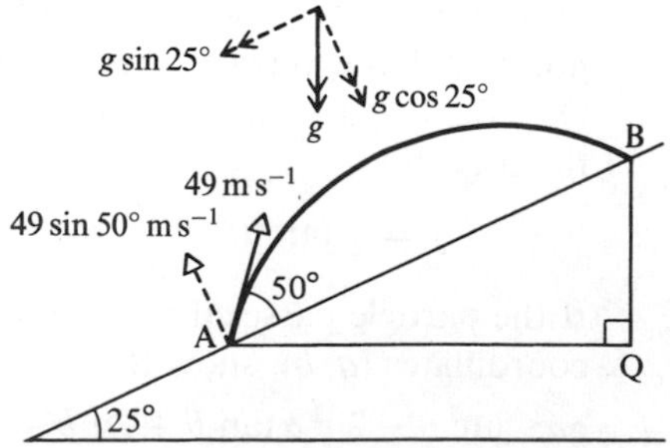

Suppose the particle strikes the plane at B.

horizontal motion	motion at right angles to AB
$u = 49 \cos 75°\,m\,s^{-1}$	$u = 49 \sin 50°\,m\,s^{-1}$
$s = AQ$	$s = 0$
$t = t$	$t = t$
no acceleration	$a = -g \cos 25°\,m\,s^{-2}$

Using $s = ut + \frac{1}{2}at^2$ for motion at right angles to AB gives:

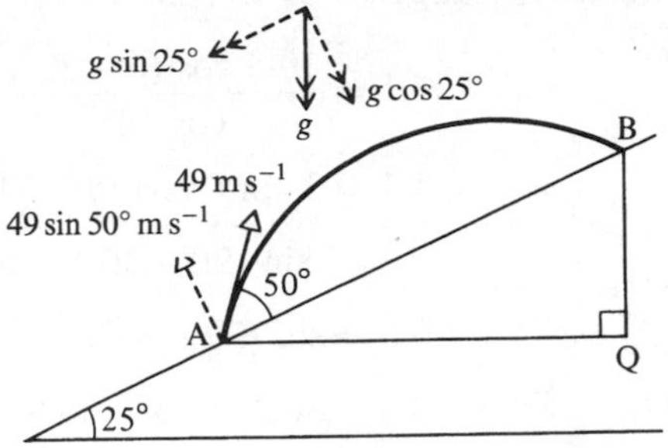

$$0 = (49 \sin 50°)t - \tfrac{1}{2}(g \cos 25°)t^2$$

$$\therefore \quad t = 0 \text{ at A} \quad \text{or} \quad t = \frac{98 \sin 50°}{g \cos 25°} = 8{\cdot}452 \text{ s at B}$$

From the triangle ABQ, the range AB is given by:

$$AB = \frac{AQ}{\cos 25°}$$

But AQ is the horizontal distance travelled by the particle.

$$\therefore \quad AQ = \text{horizontal velocity} \times \text{time}$$
$$= 49 \cos 75° \times 8{\cdot}452$$

$$\text{and } \therefore \quad AB = \frac{49 \cos 75° \times 8{\cdot}452}{\cos 25°}$$
$$= 118{\cdot}3 \text{ m}$$

The time of flight is 8·45 s and the range is 118 m.

Example 18

A particle is projected down a slope which is inclined at 30° to the horizontal. The particle is projected from a point A on the slope and has an initial velocity of 10·5 m s^{-1} at an angle θ to the slope, the vertical plane of the motion containing a line of greatest slope. If the range of the particle down the slope is 16·5 m, find the two possible values of θ.

Suppose the particle strikes the plane at P.

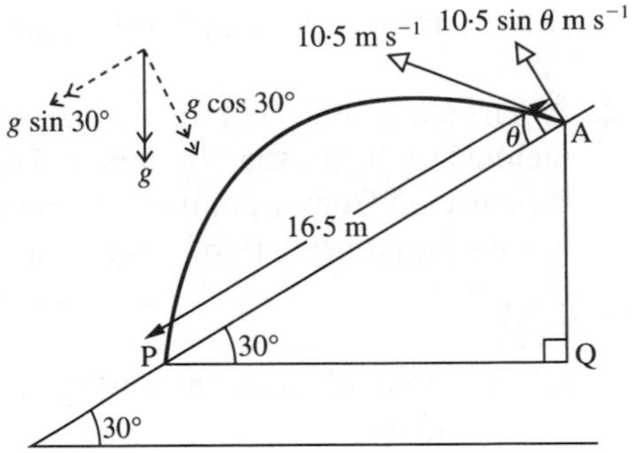

horizontal motion	motion at right angles to AP
$u = 10{\cdot}5 \cos(\theta - 30°)$ m s^{-1}	$u = 10{\cdot}5 \sin\theta$ m s^{-1}
$s = PQ = 16{\cdot}5 \cos 30°$ m	$s = 0$
$t = t$	$t = t$
no acceleration	$a = -g \cos 30°$ m s^{-2}

Using $s = ut + \frac{1}{2}at^2$, for motion at right angles to AP gives:

$$0 = (10{\cdot}5 \sin\theta)t - \tfrac{1}{2}(g \cos 30°)t^2$$

$$\therefore \quad t = 0 \text{ at A} \quad \text{or} \quad t = \frac{21 \sin\theta}{g \cos 30°} \text{ s at P}$$

From the triangle APQ, the range AP is given by:

$$AP = \frac{PQ}{\cos 30°}$$

But PQ is the horizontal distance travelled by the particle.

$$\therefore \quad PQ = \text{horizontal velocity} \times \text{time}$$
$$= 10{\cdot}5 \cos(\theta - 30°) \times \frac{21 \sin\theta}{g \cos 30°}$$

Since the range AP is 16·5 m,

$$16{\cdot}5 = \frac{10{\cdot}5\cos(\theta - 30°)}{\cos 30°} \times \frac{21\sin\theta}{g\cos 30°}$$

$$\therefore \quad 1{\cdot}1 = 2\sin\theta\cos(\theta - 30°)$$

$$= \sin(2\theta - 30°) + \sin 30°$$

$$\therefore \quad 0{\cdot}6 = \sin(2\theta - 30°)$$

$$\therefore \quad (2\theta - 30°) = 36{\cdot}87° \quad \text{or} \quad 143{\cdot}13°$$

$$\therefore \quad \theta = 33{\cdot}44° \quad \text{or} \quad 86{\cdot}57°$$

The possible angles of projection are at 33·44° and 86·57° to the slope of the plane.

Exercise 12D

In questions **1** to **9** it should be assumed that the vertical plane of the motion contains a line of greatest slope, i.e. the motion is either *directly down* or *directly up* the slope.

1. A particle is projected up a plane which is inclined at 30° to the horizontal. The particle is projected from a point on the plane and has an initial velocity of $7\sqrt{3}\,\text{m s}^{-1}$ at 30° to the plane.
Find:
(a) the time taken by the particle to strike the plane
(b) the range up the inclined plane.

2. A particle is projected up a plane which is inclined at 30° to the horizontal. The particle is projected from a point on the plane and has an initial velocity of $30\,\text{m s}^{-1}$ at 20° to the plane.
Find:
(a) the time taken by the particle to strike the plane
(b) the range up the inclined plane.

3. A particle is projected down a plane which is inclined at 30° to the horizontal. The particle is projected from a point on the plane and has an initial velocity of $14{\cdot}7\,\text{m s}^{-1}$ at 60° to the plane.
Find:
(a) the time taken by the particle to strike the plane
(b) the range down the inclined plane.

4. A particle is projected down a plane which is inclined at 45° to the horizontal. The particle is projected from a point on the plane and has an initial velocity of $7\sqrt{2}\,\text{m s}^{-1}$ at 30° to the plane.
Find:
(a) the time taken by the particle to first strike the plane
(b) the range down the inclined plane.

5. A particle is projected from the foot of a slope which is inclined at 15° to the horizontal, the initial direction of motion being at an angle of 30° to the slope.
If the particle first strikes the slope 5 seconds after projection, find the initial speed of projection and the range up the inclined plane.

6. A particle is projected downwards from the top of a slope which is inclined at 30° to the horizontal, the initial direction of motion being at an angle of 45° to the slope.
If the particle first strikes the slope 2 seconds after projection, find the initial speed of projection and the range down the incline.

7. A particle is projected down a slope which is inclined at 45° to the horizontal. The particle is projected from a point on the slope and has an initial velocity of $14\sqrt{2}\,\text{m s}^{-1}$ at an angle θ to the slope. If the particle first hits the slope 4 seconds after projection, find the value of θ.

8. A particle is projected down a slope which is inclined at 30° to the horizontal. The particle is projected from a point on the slope and has an initial velocity of 21 m s^{-1} at an angle θ to the slope.
If the range of the particle down the slope is 60 m, find the two possible values of θ.

9. A particle is projected up a plane of inclination ϕ. It is projected from a point on the plane and has an initial velocity u at an angle θ to the plane.
(a) Show that the particle will hit the plane after a time t where $t = \dfrac{2u \sin \theta}{g \cos \phi}$.
(b) Show that the range R up the plane is given by:

$$R = \frac{2u^2}{g \cos^2 \phi} \sin \theta \cos (\theta + \phi).$$

(c) If ϕ is fixed but θ can vary, show that the maximum range up the plane is given by:

$$R_{max} = \frac{u^2}{g(1 + \sin \phi)}$$

and find the relationship between θ and ϕ for which R_{max} occurs.
Hint. Use the identity $2 \sin A \cos B = \sin (A + B) + \sin (A - B)$.

Exercise 12E Examination questions

(Unless otherwise indicated take $g = 9{\cdot}8$ m s^{-2} in this exercise.)

1. A particle is projected from a point O with a velocity which has a horizontal component U and a vertically upward component V. Show that R, the range on a horizontal plane through O, and H, the maximum height of the particle above the plane, are given by

$$R = \frac{2UV}{g} \quad \text{and} \quad H = \frac{V^2}{2g}.$$

A tennis ball is served horizontally from a point which is 2·5 m vertically above a point A. The ball first strikes the horizontal ground through A at a distance 20 m from A. Show that the ball is served with speed 28 m s^{-1}.
During its flight the ball passes over a net which is a horizontal distance 12 m from A. Find the vertical distance of the ball above the horizontal ground at the instant when it passes over the net. (AEB 1994)

2. (Take $g = 10$ m s^{-2} in this question.)
An aircraft, at a height of 180 m above horizontal ground and flying horizontally with a speed of 70 m s^{-1}, releases emergency supplies. If these supplies are to land at a specified point, at what horizontal distance from this point must the aircraft release them? (UCLES)

3. At time $t = 0$ a particle is projected from a point O on a horizontal plane with speed 14 m s^{-1} in a direction inclined at an angle $\tan^{-1}\left(\frac{3}{4}\right)$ above the horizontal. The particle just clears the top of a vertical wall, the base of which is 8 m from O. Find
(a) the time at which the particle passes over the wall,
(b) the height of the wall. (AEB 1993)

4. A particle is projected from a point O with speed $60\,\text{m s}^{-1}$ at an angle $\cos^{-1}\frac{4}{5}$ above the horizontal. Find

(i) the time the particle takes to reach the point P whose horizontal displacement from O is 96 metres,

(ii) the height of P above O,

(iii) the speed of the particle 2 s after projection. (WJEC)

5. A cannonball fired at an angle of elevation of 10° has a range, on a horizontal plane, of 1·25 km. Ignoring air resistance, find the speed of projection. (If a general formula for the range of a projectile is used then it should be proved.)
If air resistance is taken into account, what can be said about the speed of projection? (UCLES)

6. A particle P is projected from a point O on level ground with speed $50\,\text{m s}^{-1}$ at an angle $\sin^{-1}\left(\frac{7}{25}\right)$ above the horizontal. Find

(a) the height of P at the point where its horizontal displacement from O is 120 m,

(b) the speed of P two seconds after projection,

(c) the times after projection at which P is moving at an angle of $\tan^{-1}\left(\frac{1}{4}\right)$ to the ground. (AEB 1992)

7. (Take $g = 10\,\text{m s}^{-2}$ in this question.)

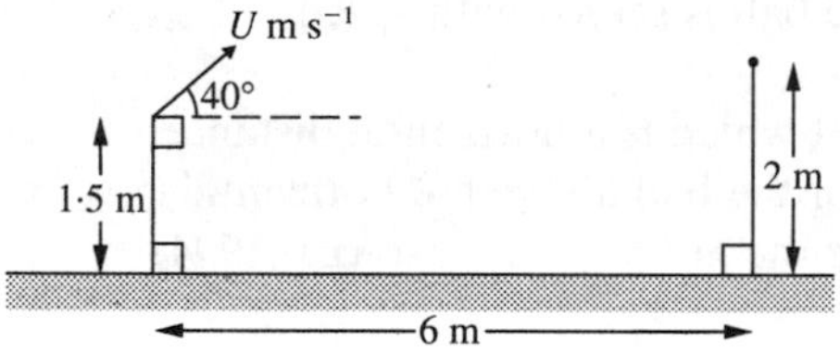

Fig. 1

In a fairground game, a child throws a small ball from a height of 1·5 m above level ground, aiming at a small target. The target is on top of a vertical pole of height 2 m, and the horizontal displacement of the child from the pole is 6 m, as shown in Fig. 1. The initial velocity of the ball has magnitude $U\,\text{m s}^{-1}$ and angle of elevation 40°. The ball moves freely under gravity.

(a) For $U = 10$, find the greatest height above the ground reached by the ball, giving your answer in metres to 1 decimal place.

(b) Calculate, to 1 decimal place, the value of U for which the ball hits the target. (ULEAC)

8. (Take $g = 10\,\text{m}\,\text{s}^{-2}$ in this question.)
A girl throws a stone from a height of 1·5 m above level ground with a speed of $10\,\text{m}\,\text{s}^{-1}$ and hits a bottle standing on a wall 4 m high and 5 m from her.

(i) Show that if α is the angle of projection of the stone as it leaves her hand then

$$1{\cdot}25\tan^2\alpha - 5\tan\alpha + 3{\cdot}75 = 0.$$

(ii) The horizontal component of the stone's velocity has to be at least $6\,\text{m}\,\text{s}^{-1}$ for the bottle to be knocked off. By solving the above equation, or otherwise, show that α has to be 45° for the bottle to be knocked off.

(iii) If α is 45° find the direction in which the stone is moving when it hits the bottle.

(iv) If the bottle has a velocity of $3\,\text{m}\,\text{s}^{-1}$ after being struck find where it hits the ground. (NICCEA)

9. (i) If a particle is projected with speed u at an angle θ to the horizontal show that the horizontal range, R is given by

$$R = \frac{u^2 \sin 2\theta}{g}$$

(ii) A golfer is about to strike a golf ball from a point A. In order to hit the green the ball must cross a lake 30 m wide. The nearest point, N, of the lake is 180 m from the golfer. The golf ball is struck at $50\,\text{m}\,\text{s}^{-1}$ at an angle θ to the horizontal.

(a) Find the two possible values of θ so that the ball lands just before it reaches the lake.

(b) If it is just possible for the ball when hit from A, to reach the nearest point, G, of the green find the distance AG.

(c) In order to reach this point G, the ball, when hit from A, must pass over a tree, T, on the opposite side of the lake from N (see Fig. 1 below). Find the maximum height of this tree if the ball is to reach the green.

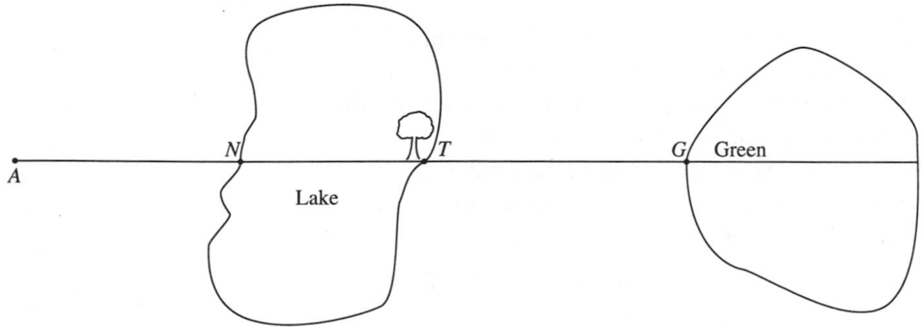

Fig. 1

(NICCEA)

10.

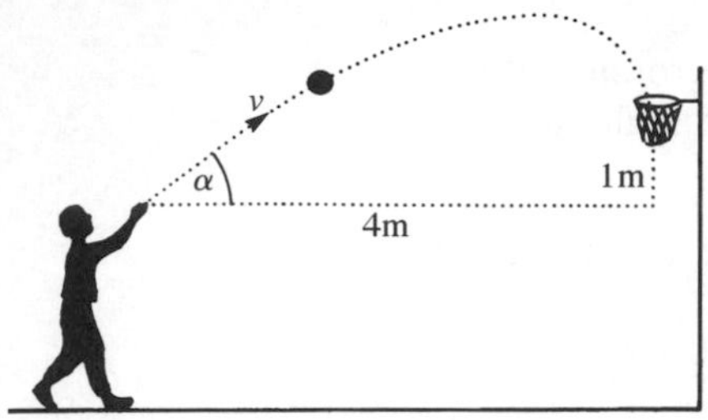

The motion of the ball in a successful free shot in basketball is illustrated opposite.

The model assumes that the ball is a point particle, acted on by constant gravity, g.

The ball is projected from a position, distance 4 m horizontally and 1 m vertically from the basket, with speed $v\,\text{m s}^{-1}$ at an angle $\alpha°$ to the horizontal. The ball falls into the basket.

(a) Show that, taking $g = 9{\cdot}8\,\text{m s}^{-2}$, v and α must satisfy

$$1 = 4\tan\alpha - \frac{78{\cdot}4}{v^2}\sec^2\alpha.$$

(b) Use this equation to find the required speed of projection when angle α equals 45°.

(c) Also, use this equation to find the two possible trajectories when $v = 8{\cdot}0\,\text{m s}^{-1}$.

For the ball to fall through the basket, the angle made with the vertical at the basket should be as small as possible. Which of your two solutions above would be preferred? (AEB Spec)

11. (a) An object is projected from an origin O with initial speed V and at an angle of inclination α above the horizontal. Given that the object lands at the same horizontal level as the point of projection, and that resistance to motion is negligible, show that the range attained by the object is R, where $R = \dfrac{V^2 \sin 2\alpha}{g}$

State, with justification, the value of α for which R is a maximum.

(b) Use the result of (a) to find the least speed with which an object needs to be projected so as to travel a horizontal distance of 4 m before first hitting the ground. Give your answer correct to three significant figures.

(c) A basketball is released from a player's hands with a speed of $8\,\text{m s}^{-1}$ at an inclination of α degrees above the horizontal so as to land in the centre of the basket, which is 4 m horizontally from the point of release and a vertical height of 0·5 m above it.
Taking the origin, O, to be the point of release, and taking $g = 10\,\text{m s}^{-2}$, show that α satisfies the quadratic equation

$$5\tan^2\alpha - 16\tan\alpha + 7 = 0.$$

Given that the player throws the ball at the larger angle of projection, find:

(i) α correct to the nearest degree

(ii) the time taken from the moment of release of the ball from the player's hands until the ball lands in the basket. (UODLE)

12. (Take $g = 10\,\text{m}\,\text{s}^{-2}$ in this question.)
A shot-putter (illustrated in the diagram) finds that he can project a shot a maximum distance of 20 m. He wishes to determine the shot's projection speed.

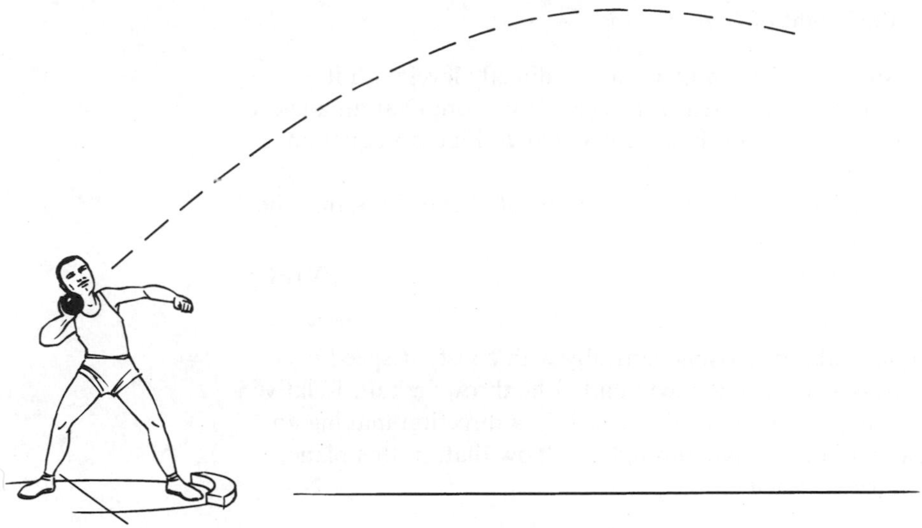

(The shot will be modelled as a particle travelling in a vacuum.)

(a) For his first attempt he decides to consider the shot to be projected from the level ground. Using this model determine the minimum possible speed of projection V_1.

(b) Not satisfied with this value for the projection speed, he decides to take account of the height from which the shot is projected. He measures this to be 1·5 m.

(i) Show that the range R m of the shot along the level ground, when projected from a height of 1·5 m, satisfies the equation

$$10R^2 \tan^2\theta - 2v^2 R \tan\theta + 10R^2 - 3v^2 = 0,$$

where $v\,\text{m}\,\text{s}^{-1}$ is the speed of projection of the shot and θ is its angle of projection above the horizontal.

(ii) If the equation, regarded as a quadratic in $\tan\theta$, gives real values of $\tan\theta$, show that

$$R \leqslant \frac{v}{10}\sqrt{(30 + v^2)}.$$

(iii) Use the result in (ii) to determine the minimum speed of projection V_2 for the shot in this model.

(iv) If the shot is projected with speed V_2, would you expect that more than one angle of projection would be possible for the range of 20 m? (UODLE)

13. A point O is vertically above a fixed point A of a horizontal plane. A particle P is projected from O with speed $5V$ at an angle $\cos^{-1}\frac{3}{5}$ above the horizontal and hits the plane at a point B at a distance $\dfrac{48V^2}{g}$ from A.

(i) Show that the height of O above A is $\dfrac{64V^2}{g}$.

(ii) Find the distance of P from O when it is directly level with it.

A second particle is now projected with speed $24W$ from O at an angle α above the horizontal and it also hits the plane at B. Find an equation involving V, W, and α.

Given that one value of α is 45° find W in terms of V and show that the other value of α is such that

$$7\tan^2\alpha - 6\tan\alpha - 1 = 0.$$

(WJEC)

14. A man in a hot air balloon is rising vertically with constant speed u. When the man passes through a fixed point O he throws a ball. Relative to the man, the ball is projected with speed V in a direction making an angle α with the horizontal plane through O. Show that, in this plane, the ball has a range R, where

$$gR = V(2u\cos\alpha + V\sin 2\alpha).$$

Show that, when u and V are both equal to a constant W and α varies, the maximum value of R occurs when $\alpha = \dfrac{\pi}{6}$.

Given that $u = V = W$ and $\alpha = \dfrac{\pi}{6}$, find, in terms of W and g, the distance between the man and the ball at the instant when the ball reaches the horizontal plane through O. (AEB 1993)

13 Circular motion

Radians

Angles are usually measured in degrees, minutes and seconds.
A radian is a larger unit which is sometimes used. It is defined as the angle subtended at the centre of a circle by an arc equal to the radius of the circle.

In the diagram, suppose that the radius OP is of length 1 unit and that the arc PQ is also of length 1 unit. Then the angle POQ is 1 radian. Since the circumference of this circle is 2π units, the angle at O subtended by the whole circumference is 2π radians.

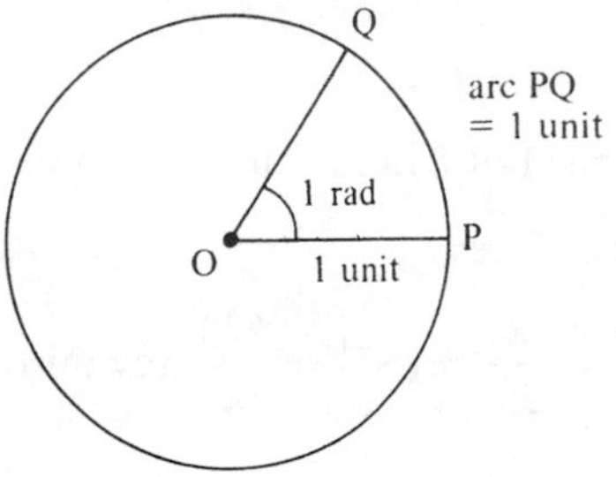

Hence 1 revolution $= 360° = 2\pi$ radians

or $\frac{1}{4}$ revolution $= 90° = \frac{\pi}{2}$ radians.

As 2π radians $= 360°$

$$1 \text{ radian} = \frac{360°}{2\pi} \text{ which is approximately } 57°.$$

It is very seldom necessary, or desirable, to use this approximate result.

It is better to use one of the exact relationships such as $\frac{\pi}{2}$ radians $= 90°$.

Linear and angular speed

The linear speed of a body, i.e. its speed in a straight line, is usually measured in $\mathrm{m\,s^{-1}}$, $\mathrm{km\,h^{-1}}$ or some similar unit.
When a body is moving on a circular path it is often useful to measure its speed as the rate of change of the angle at the centre of the circle.

Suppose a particle P moves along the circumference of a circle, centre O, at a constant speed. If the body moves along the arc from A to B in one second, then the angle AOB gives the change in angle per second.
Thus, if angle AOB is in radians, the angular speed of the body is in radians per second ($\mathrm{rad\,s^{-1}}$).
If the particle makes 5 revolutions in one minute, its angular speed is $5\,\mathrm{rev\,min^{-1}}$.

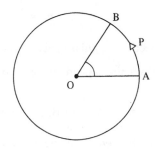

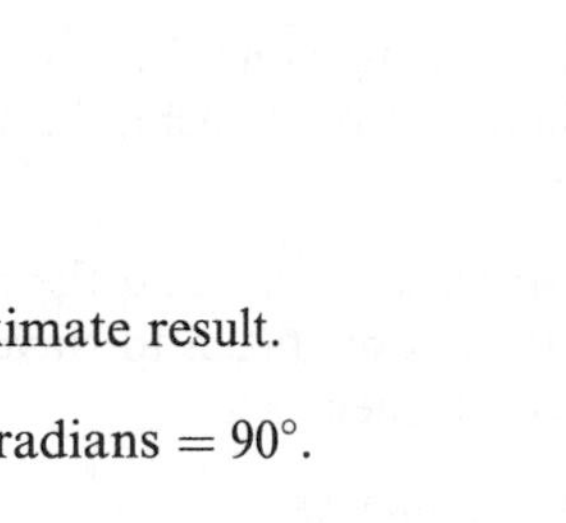

Example 1

Express an angular speed of (a) $15\,\text{rev}\,\text{min}^{-1}$ in $\text{rad}\,\text{min}^{-1}$,
(b) $3\,\text{rev}\,\text{s}^{-1}$ in $\text{rad}\,\text{min}^{-1}$.

(a)
$$1 \text{ rev} = 2\pi \text{ rad}$$
$$\therefore \quad 15 \text{ rev min}^{-1} = 15 \times 2\pi \text{ rad min}^{-1} = 30\pi \text{ rad min}^{-1}$$

(b)
$$1 \text{ rev} = 2\pi \text{ rad}$$
$$\therefore \quad 3 \text{ rev s}^{-1} = 3 \times 2\pi \times 60 \text{ rad min}^{-1} = 360\pi \text{ rad min}^{-1}$$

Example 2

Express an angular speed of $8\,\text{rad}\,\text{s}^{-1}$ in $\text{rev}\,\text{min}^{-1}$.

$$2\pi \text{ rad} = 1 \text{ rev}$$

$$\therefore \quad 8 \text{ rad s}^{-1} = \frac{8}{2\pi} \text{ rev s}^{-1} = \frac{240}{\pi} \text{ rev min}^{-1}$$

Relationship between linear and angular speed

When a particle P is moving along the circumference of a circle, centre O, at a constant speed, there is a relationship between the linear and angular speeds of the particle.

Suppose the linear speed is $v\,\text{m}\,\text{s}^{-1}$ and the radius of the circle is r m. If the time taken to travel form A to B is 1 second, then:

$$\text{arc AB} = v \times 1 \text{ metres}$$

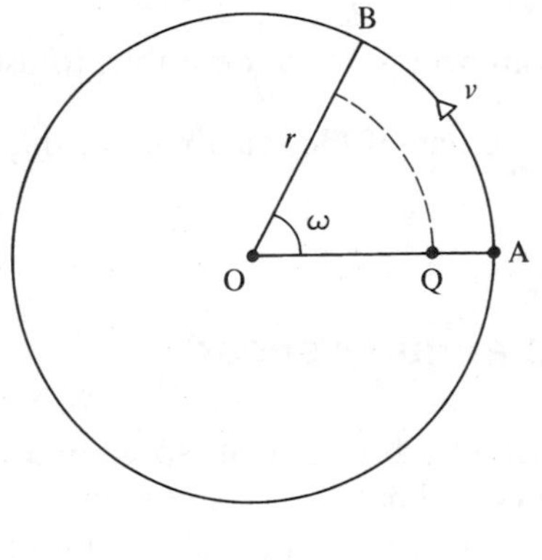

Suppose the angular speed is $\omega\,\text{rad}\,\text{s}^{-1}$, then as the time to travel from A to B is one second, angle AOB is ω radians.

Then:
$$\frac{\text{arc AB}}{2\pi r} = \frac{\omega}{2\pi}$$

Substituting for arc AB gives:
$$\frac{v}{2\pi r} = \frac{\omega}{2\pi}$$

$$\therefore \quad \omega = \frac{v}{r} \quad \text{or} \quad v = r\omega$$

Hence the linear speed = angular speed × radius of circle.

Note that this relationship depends upon ω being measured in *radians* per unit of time.

As the particle P moves from A to B, all points on the radius OA will move to new positions on the radius OB.
The angular speed of P is equal to the angular speed of the point Q shown in the diagram, but the linear speed of P will be greater than that of Q.

Example 3

Find the speed in $m\,s^{-1}$ of a particle which is moving on a circular path of radius 30 cm with an angular speed of (a) $4\,rad\,s^{-1}$, (b) $5\,rev\,min^{-1}$.

(a) $\omega = 4\,rad\,s^{-1}$ $\quad r = 30\text{ cm} = \frac{30}{100}\,m$

Using $v = r\omega$ gives:

$$v = \tfrac{30}{100} \times 4 = 1{\cdot}2\,m\,s^{-1}$$

The speed is $1{\cdot}2\,m\,s^{-1}$.

(b) $\omega = 5\,rev\,min^{-1}$ $\quad r = 30\text{ cm} = \frac{30}{100}\,m$

$$= 5 \times 2\pi\text{ rad min}^{-1}$$

$$= \frac{5 \times 2\pi}{60}\text{ rad s}^{-1}$$

Using $v = r\omega$ gives:

$$v = \frac{30}{100} \times \frac{5 \times 2\pi}{60} = \frac{\pi}{20}\,m\,s^{-1}$$

The speed is $\frac{\pi}{20}\,m\,s^{-1}$.

Example 4

A particle moving on a circular path of radius 4 m has a speed of $36\,m\,s^{-1}$. Find the angular speed of the particle in (a) $rad\,s^{-1}$, (b) $rev\,min^{-1}$.

(a) $v = 36\,m\,s^{-1}$ $\quad r = 4\,m$

Using $v = r\omega$ gives:

$$36 = 4 \times \omega \quad \text{or} \quad \omega = 9\,rad\,s^{-1}$$

(b) $2\pi\text{ rad} = 1\text{ rev}$

$$\therefore \quad 9\text{ rad s}^{-1} = \frac{9}{2\pi}\text{ rev s}^{-1} = \frac{9}{2\pi} \times 60\text{ rev min}^{-1} = \frac{270}{\pi}\text{ rev min}^{-1}$$

The speed is $9\,rad\,s^{-1}$ or $\frac{270}{\pi}\,rev\,min^{-1}$.

Exercise 13A

(Leave π in the answers.)

1. Express an angular speed of $100\,rev\,min^{-1}$ in $rad\,min^{-1}$.
2. Express an angular speed of $90\,rev\,min^{-1}$ in $rad\,s^{-1}$.
3. Express an angular speed of $10\,rad\,s^{-1}$ in $rev\,min^{-1}$.
4. A uniformly rotating disc completes one revolution every ten seconds. Find the angular speed of the disc:
   (a) in $rev\,min^{-1}$ (b) in $rad\,min^{-1}$ (c) in $rad\,s^{-1}$.
5. Find in $rad\,s^{-1}$ the angular speed of a record revolving at:
   (a) $45\,rev\,min^{-1}$ (b) $33\frac{1}{3}\,rev\,min^{-1}$.
6. Find the speed in $m\,s^{-1}$ of a particle which is moving on a circular path of radius 5 m with an angular speed of:
   (a) $2\,rad\,s^{-1}$ (b) $300\,rev\,min^{-1}$.

7. Find the speed in $m\,s^{-1}$ of a particle which is moving on a circular path of radius 50 cm with an angular speed of:
   (a) $10\,rad\,s^{-1}$
   (b) $20\,rev\,s^{-1}$
8. A particle moving in a circular path of radius 50 cm has a speed of $12\,m\,s^{-1}$. Find the angular speed of the particle:
   (a) in $rad\,s^{-1}$ (b) in $rev\,min^{-1}$.
9. A particle moving in a circular path of radius 20 cm has a speed of $20\,cm\,s^{-1}$. Find the angular speed of the particle:
   (a) in $rad\,s^{-1}$ (b) in $rev\,min^{-1}$.
10. Sue and Pam stand on a playground roundabout at distances of 1 m and 1·5 m respectively from the centre of rotation. If the roundabout revolves at $12\,rev\,min^{-1}$, find the speeds with which Sue and Pam are moving.
11. Tom and John stand on a playground roundabout respectively 50 cm and 175 cm from the centre. The roundabout moves with constant angular velocity and Tom's speed is found to be $1\,m\,s^{-1}$. Find:
    (a) the angular speed of the roundabout in $rad\,s^{-1}$
    (b) the time taken for the roundabout to complete ten revolutions
    (c) John's speed.

Motion on a circular path

When the speed of a car travelling in a straight line changes from $12\,m\,s^{-1}$ to $15\,m\,s^{-1}$, it is said to have an acceleration because the velocity has changed. If a body moves on a circular path, radius r, at a constant speed v, the *direction* of the motion is changing. The *velocity* of the body is therefore changing and again the body is said to have an acceleration, even though the magnitude of the velocity (i.e. its speed) is constant.

The acceleration of such a body at any instant is $\dfrac{v^2}{r}$ and is directed towards the centre of the circular path.

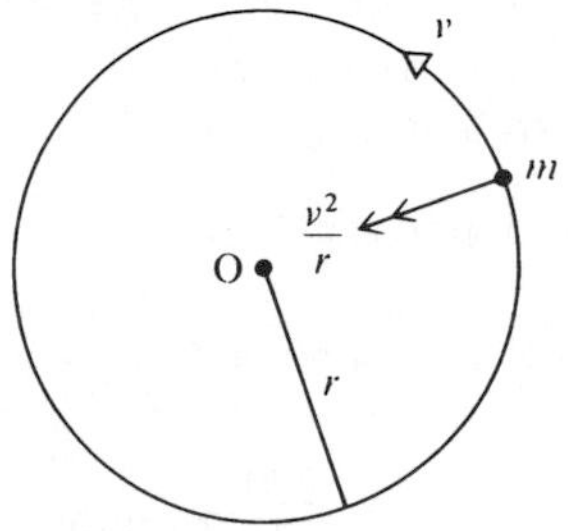

A proof of the above statement follows.

Suppose a particle is moving on a circular path of radius r with a constant speed v.

Let the time taken for the particle to travel from P to Q be δt, the angle POQ be $\delta\theta$ and the arc PQ be δs.

The velocity at P is v along the tangent PA and the velocity at Q is v along the tangent AQ.

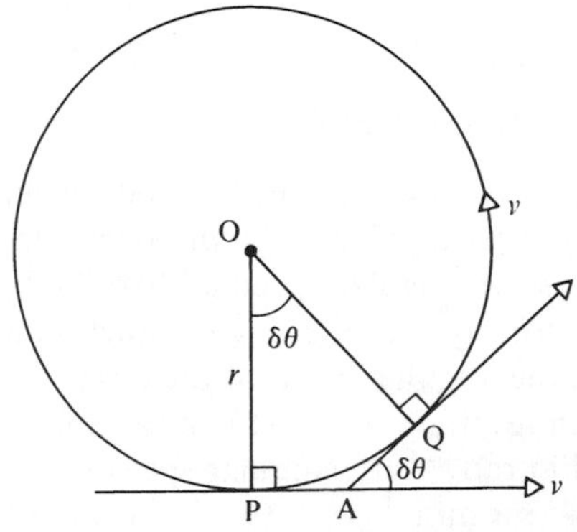

The velocity at Q can be resolved into components of:

$v\cos\delta\theta$ parallel to PA
and $v\sin\delta\theta$ perpendicular to PA.

$$\text{acceleration} = \frac{\text{change in velocity}}{\text{time}}$$

For the acceleration at any instant, we must let

$\delta t \to 0$ and hence $\delta\theta \to 0$

$$\text{acceleration along PA} = \lim_{\delta t \to 0}\left(\frac{v\cos\delta\theta - v}{\delta t}\right) = \lim_{\delta t \to 0}\frac{v(\cos\delta\theta - 1)}{\delta t}$$

but as $\delta\theta \to 0$, $(\cos\delta\theta - 1) \to 0$

$\therefore$ acceleration along PA = 0

$$\text{acceleration along PO} = \lim_{\delta t \to 0}\left(\frac{v\sin\delta\theta - 0}{\delta t}\right) = \lim_{\delta t \to 0}\frac{v\sin\delta\theta}{\delta t}$$

but as $\delta\theta \to 0$, $\sin\delta\theta \to \delta\theta$

$$\therefore \quad \text{acceleration along PO} = \lim_{\delta t \to 0}\frac{v\delta\theta}{\delta t} = v\omega$$

but $v = r\omega$

$\therefore$ acceleration $= \dfrac{v^2}{r}$ and is towards the centre of the circular path.

Example 5

A particle describes a horizontal circle of radius 2 metres at a speed of $3\,\text{m}\,\text{s}^{-1}$. Find the acceleration of the particle.

$$\text{acceleration towards centre of circle} = \frac{v^2}{r}$$

Given $v = 3\,\text{m}\,\text{s}^{-1}$ and $r = 2\,\text{m}$:

$$\text{acceleration} = \frac{(3)^2}{2} = 4{\cdot}5\,\text{m}\,\text{s}^{-2}$$

The acceleration of the particle is $4{\cdot}5\,\text{m}\,\text{s}^{-2}$ towards the centre of the circle.

Central force

For a body to have an acceleration, there has to be a force acting on the body. In the case of a body following a circular path, since the acceleration is directed towards the centre of the circle, the force must also be in this direction.

The magnitude of the force will be:

mass $\times$ acceleration $= m\dfrac{v^2}{r}$, where m is the mass of the body.

Example 6

A body of mass 250 g moves with constant angular speed of $4\,\text{rad}\,\text{s}^{-1}$ in a horizontal circle of radius 3·5 m. Find the force that must act on the body towards the centre of the circle.

force = mass $\times$ acceleration

$$= m\frac{v^2}{r} = mrw^2 \qquad \text{since } v = r\omega$$

$$\therefore \quad \text{force} = \frac{250}{1000} \times 3{\cdot}5 \times 4^2 = 14\,\text{N}$$

The force acting on the body towards the centre of the circle is 14 N.

The force which acts upon a body so that it follows a circular path may be provided in various ways.

(i) Particle on a string

A particle is attached to one end of a string, the other end of the string being attached at a point O on a smooth horizontal surface. If the particle describes circles on the surface, the necessary force towards the centre of the circle is provided by the tension in the string.

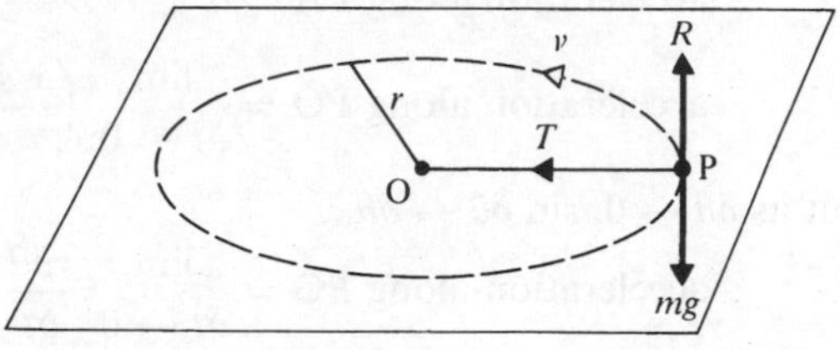

$$T = m\frac{v^2}{r}$$

(ii) Bead on a circular wire

If a bead is threaded on a smooth horizontal circular wire and moves at a speed v, the necessary force towards the centre of the circular wire is provided by the force between the bead and the wire. In addition, the wire supports the weight of the bead, and the vertical reaction R equals mg.

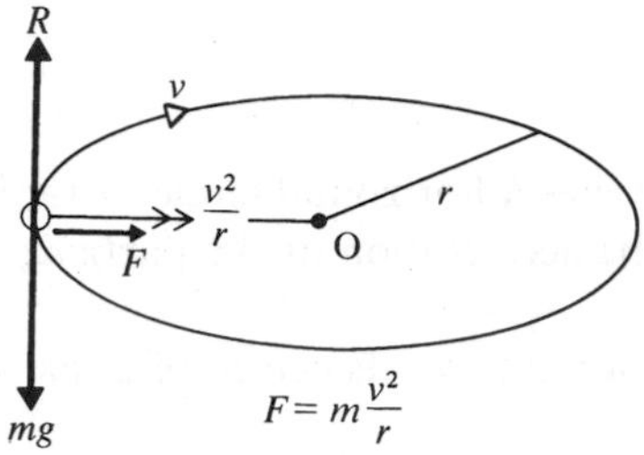

$$F = m\frac{v^2}{r}$$

(iii) Particle on a rotating disc

If a particle rests on the surface of a rotating horizontal disc, the only horizontal force acting on the particle is the frictional force between the particle and the surface of the disc. This frictional force provides the necessary force towards the centre of rotation.

For any particular surface, there will be a maximum value of F, i.e. $F_{max} = \mu R$, and then the particle will be on the point of slipping.

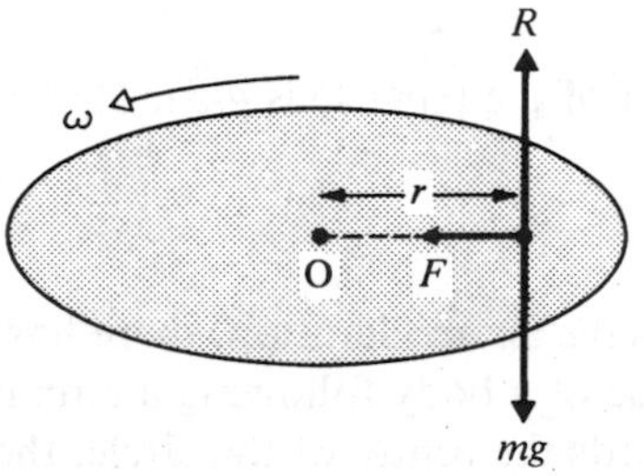

(iv) Car on circular path

Again, the necessary force towards the centre of the circular path is provided by the frictional force between the tyres of the car and the surface of the road.

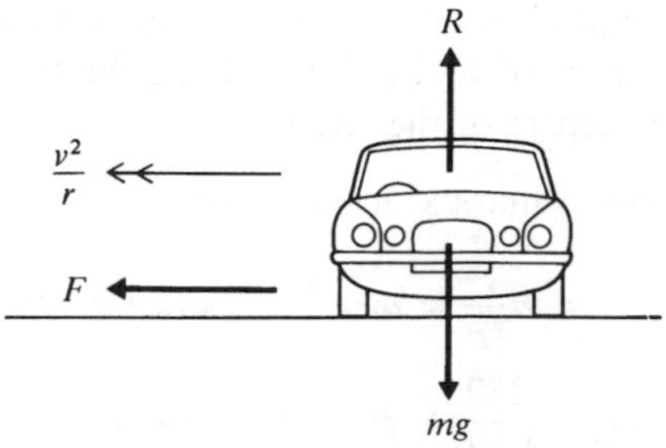

$$F = m\frac{v^2}{r}$$

Example 7

A particle of mass 300 g is attached to one end of a light inextensible string of length 40 cm, the other end of the string being fixed at O on a smooth horizontal surface. If the particle describes circles, centre O, find the tension in the string when

(a) the speed of the particle is $2\sqrt{2}\,\text{m s}^{-1}$

(b) the angular speed of the particle is $5\,\text{rad s}^{-1}$

(a) acceleration towards centre O $= \dfrac{v^2}{r}$

$\therefore$ force towards O, i.e. the tension in the string,

$$= \frac{mv^2}{r}$$

$$\therefore \quad \text{tension} = \frac{300}{1000} \times \frac{(2\sqrt{2})^2}{0{\cdot}4}$$

$$= 6\,\text{N}$$

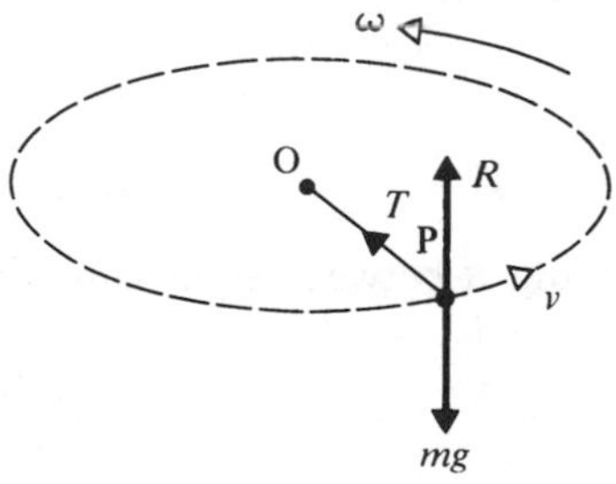

The tension in the string is 6 N.

(b) Again, $\quad \text{tension} = m\dfrac{v^2}{r} = mr\omega^2$

$$\therefore \quad \text{tension} = \frac{300}{1000} \times 0{\cdot}4 \times (5)^2 = 3\,\text{N}$$

The tension in the string is 3 N.

Example 8

A car travels along a horizontal road which is an arc of a circle of radius 125 m. The greatest speed at which the car can travel without slipping is $42\,\text{km h}^{-1}$. Find the coefficient of friction between the tyres of the car and the surface of the road.

Let the mass of the car be m kg

Resolving vertically gives: $\quad R = mg \quad \ldots[1]$

Friction provides the force towards the centre of the circle and since the car is on the point of slipping,

$$F_{\text{max}} = \mu R = m\frac{v^2}{r} \quad \ldots[2]$$

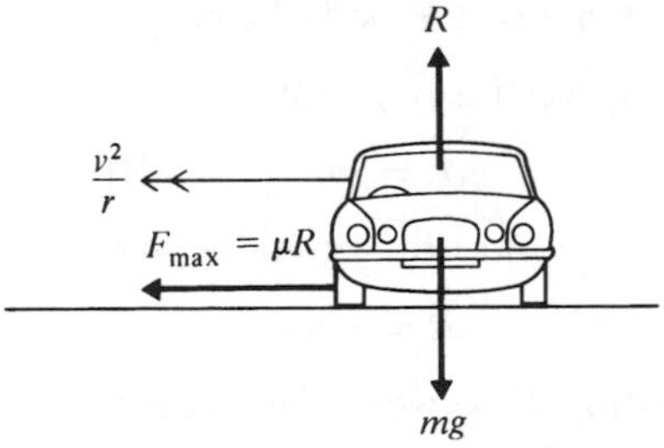

Eliminating R between equations [1] and [2] gives:

$$\mu mg = m\frac{v^2}{r} \quad \text{or} \quad \mu = \frac{v^2}{rg}$$

But $\quad r = 125\,\text{m} \quad$ and $\quad v = 42\,\text{km h}^{-1} = \dfrac{42 \times 1000}{60 \times 60}\,\text{m s}^{-1}$

$$\therefore \quad \mu = \left(\frac{42 \times 1000}{60 \times 60}\right)^2 \times \frac{1}{125(9{\cdot}8)} = \frac{1}{9}$$

The coefficient of friction between the tyres and the road is $\frac{1}{9}$.

Example 9

A small box is placed on the surface of a horizontal disc at a point 5 cm from the centre of the disc. The box is on the point of slipping when the disc rotates at $1\frac{2}{5}$ rev s^{-1}.
Find the coefficient of friction between the box and the surface of the disc.

Let the mass of the box be m kg.

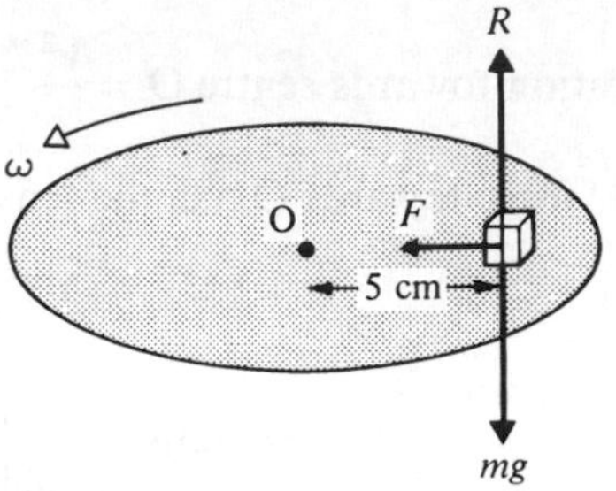

Resolving vertically gives: $R = mg$ ... [1]

Friction provides the force towards O, and since the box is on the point of slipping,

$$F_{max} = \mu R = m\frac{v^2}{r}$$

or $\mu R = mr\omega^2$ ... [2]

Eliminating R between equations [1] and [2] gives:

$$\mu = \frac{r\omega^2}{g}$$

but $r = 5\,\text{cm} = \frac{5}{100}$ m and $\omega = 1\frac{2}{5}\,\text{rev}\,s^{-1} = \frac{7}{5} \times 2\pi\,\text{rad}\,s^{-1}$

$$\therefore \qquad \mu = \left(\frac{5}{100}\right) \times \left(\frac{14\pi}{5}\right)^2 \times \left(\frac{1}{9{\cdot}8}\right) = 0{\cdot}395$$

The coefficient of friction between the box and the disc is 0·395.

Example 10

A hollow circular cylinder of radius 2 m rotates at 7 rad s^{-1} about its axis of symmetry which is vertical. A body rotates, without slipping, on the inner surface of the cylinder. Given that the body is on the point of sliding down the cylinder find the coefficient of friction between the body and the cylinder.

When the body is on the point of slipping the maximum frictional force μR acts to prevent the body from slipping down the inner surface of the cylinder.

Let the mass of the body be m kg.

Resolving vertically gives:

$$\mu R = mg \qquad \ldots [1]$$

Using $F = ma$ horizontally gives:

$$R = mr\omega^2 \qquad \ldots [2]$$

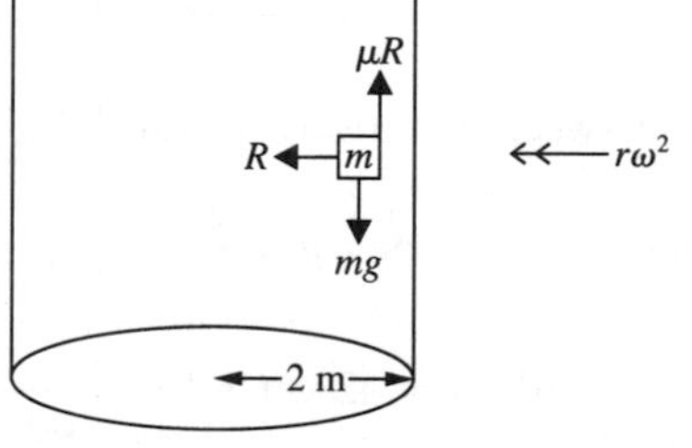

Eliminating R between these equations gives:

$$\mu mr\omega^2 = mg$$

$$\therefore \qquad \mu r\omega^2 = g$$

Substituting for r, ω and g gives:

$$\mu \times 2 \times 7^2 = 9{\cdot}8$$

$$\therefore \qquad \mu = 0{\cdot}1$$

The coefficient of friction between the body and the cylinder is 0·1.

Exercise 13B

1. A particle moves with a constant angular speed of $2\,\text{rad s}^{-1}$ around a circular path of radius 1·5 m.
Find the acceleration of the particle.
2. A particle moves with a constant speed of $3\,\text{m s}^{-1}$ around a circular path of radius 50 cm.
Find the acceleration of the particle.
3. A record is revolving at $45\,\text{rev min}^{-1}$. Points A, B and C lie on the record at 5 cm, 10 cm and 20 cm from the centre respectively.
Find the accelerations of A, B and C.
4. A body of mass 2 kg moves with a constant angular speed of $5\,\text{rad s}^{-1}$ around a horizontal circle of radius 10 cm.
Find the magnitude of the horizontal force that must be acting on the body towards the centre of the circle.
5. A body of mass 0·5 kg moves with a constant speed of $4\,\text{m s}^{-1}$ around a horizontal circle of radius 1 m.
Find the magnitude of the horizontal force that must be acting on the body towards the centre of the circle.
6. A body of mass 1·5 kg moves with a constant angular speed of $600\,\text{rev min}^{-1}$ around a horizontal circle of radius 10 cm.
Find the magnitude of the horizontal force that must be acting on the body towards the centre of the circle.
7. A body of mass 100 g moves with a constant angular speed around a horizontal circle of radius 50 cm.
If the body completes 50 revolutions in 3 minutes, find the magnitude of the horizontal force that must be acting on the body towards the centre of the circle.
8. A particle of mass 250 g lies on a smooth horizontal surface and is connected to a point O on the surface by a light inextensible string of length 20 cm. With the string taut, the particle describes a horizontal circle, centre O, with constant angular speed ω.
Find the tension in the string when ω is
(a) $10\,\text{rad s}^{-1}$
(b) $20\,\text{rad s}^{-1}$
(c) $100\,\text{rev min}^{-1}$.
9. A particle of mass 125 g lies on a smooth horizontal surface and is connected to a point O on the surface by a light inextensible string of length 50 cm. With the string taut, the particle describes a horizontal circle, centre O, with constant speed v.
Find the tension in the string when v is
(a) $2\,\text{m s}^{-1}$ (b) $10\,\text{m s}^{-1}$ (c) $20\,\text{m s}^{-1}$
10. A body of mass 2 kg lies on a smooth horizontal surface and is connected to a point O on the surface by a light inextensible string of length 25 cm. With the string taut, the body describes a horizontal circle with centre O. If the tension in the string is 18 N, find the angular speed of the body.
11. A body of mass 3 kg lies on a smooth horizontal surface and is connected to a point O on the surface by a light inextensible string of length 1·5 m.
With the string taut, the body describes a horizontal circle with centre O.
If the tension in the string is 32 N, find the speed of the body. Find the greatest possible speed of the body around the circle given that the string will break if the tension exceeds 98N.
12. A smooth wire is in the form of a circle of radius 20 cm and centre O. The wire has a bead of mass 100 g threaded on it. The wire is held horizontally and the bead moves along the wire with a constant speed of $3\,\text{m s}^{-1}$. Find:
(a) the vertical force experienced by the wire due to the bead (state up or down)
(b) the vertical force experienced by the bead due to the wire (state up or down)
(c) the horizontal force experienced by the wire due the bead (state whether towards or away from O)
(d) the horizontal force experienced by the bead due to the wire (state whether towards or away from O).
13. A bend in a level road forms a circular arc of radius 50 m. A car travels around the bend at a speed of $14\,\text{m s}^{-1}$ without slipping occurring. What information does this give about the value of the coefficient of friction between the tyres of the car and the road surface?

14. A bend in a level road forms a circular arc of radius 75 m. The greatest speed at which a car can travel around the bend without slipping occurring is 63 km h^{-1}.
Find the coefficient of friction between the tyres of the car and the road surface.

15. A bend in a level road forms a circular arc of radius 54 m.
Find the greatest speed at which a car can travel around the bend without slipping occurring if the coefficient of friction between the tyres of the car and the road surface is 0·3.

16. A body of mass 200 g rests without slipping on a horizontal disc which is rotating at 5·6 rad s^{-1}. The coefficient of friction between the body and the disc is 0·4.
Find the greatest possible distance between the body and the centre of rotation.
Repeat the question for a body of mass 400 g.

17. A body is placed on a horizontal disc at a point which is 15 cm from the centre of the disc. When the disc rotates at 30 rev min^{-1}, the body is just on the point of slipping.
Find the coefficient of friction between the body and the surface of the disc.

18. A level road is to be constructed; one design requirement is that any vehicle travelling around any bend in the road, can travel at any speed up to 21 m s^{-1} without slipping occurring. It is calculated that under the worst conditions, the coefficient of friction between a vehicle's tyres and the surface of the road could be as low as 0·125.
Find the minimum radius that any bend in the road may have.

19. A level railway track is in the form of a circular arc of radius 200 m.
Find the horizontal force acting on the rails when a train of mass 15 000 kg travels around the bend with a speed of 20 m s^{-1}.

20. A hollow circular cylinder of radius 4 m rotates at 3·5 rad s^{-1} about its axis of symmetry which is vertical. A body rotates, without slipping, on the inner surface of the cylinder.
Given that the body is on the point of sliding down the cylinder find the coefficient of friction between the body and the cylinder.

21. A hollow circular cylinder of radius $\left(\frac{50}{\pi^2}\right)$ m rotates at 14 revolutions per minute about its axis of symmetry which is vertical. A body rotates, without slipping, on the inner surface of the cylinder.
Given that the body is on the point of sliding down the cylinder find the coefficient of friction between the body and the cylinder.

22. A ride at a fair consists of a hollow circular cylinder of radius 3 m which rotates about its axis of symmetry which is vertical. People remain, without slipping, in contact with the inner surface of the cylinder.
Given that the coefficient of friction between a person and the inside of the cylinder is 0·6, find the least angular velocity with which the cylinder can rotate.

23. Two particles A and B, of masses 200 g and 50 g respectively, are connected by a light inextensible string. Particle A, at one end of the string, lies on a smooth horizontal table. The string passes smoothly through a small hole O in the table, and particle B hangs freely at the other end of the string. When particle A describes a horizontal circle about O with angular speed 3·5 rad s^{-1}, particle B hangs 30 cm below the level of the surface of the table.
Find the length of the string.

24. Two particles A and B, of masses 100 g and 250 g respectively, are connected by a light inextensible string of length 20 cm. Particle A at one end of the string lies on a smooth horizontal table. The string passes smoothly through a small hole O in the table and particle B hangs freely at the other end of the string.
Find the speed (in m s^{-1}) with which A must be made to perform horizontal circles about O in order to keep B in equilibrium at a position 7·5 cm below the level of the surface of the table.

25. A light inextensible string of length l has one end fixed at a point on a smooth horizontal surface, and the other end attached to a body of mass m lying on the surface. With the string taut, the body is given an initial speed of $2\sqrt{(gl)}$ in a direction parallel to the plane of the surface and perpendicular to the string. Show that the tension in the string during the ensuing circular motion will be four times the weight of the particle.

26. A car is just on the point of slipping when travelling on level ground at a speed v around a bend of radius r. Under the same road surface conditions, the car is just on the point of slipping when travelling on level ground at a speed $2v$ around a bend of radius R. Show that $R = 4r$.

27. Two particles A and B of masses m_1 and m_2 respectively are connected by a light inextensible string of length l. Particle A, at one end of the string, lies on a smooth horizontal table. The string passes smoothly through a small hole O in the table and particle B hangs freely at the other end of the string.

If A follows a horizontal circular path with centre O and angular speed ω, show that particle B will rest in equilibrium at a point which is a distance x below the level of the surface of the table, where

$$x = \frac{lm_1\omega^2 - m_2 g}{m_1\omega^2}$$

Conical pendulum

A particle is attached to the lower end of a light inextensible string, the upper end of which is fixed. When the particle describes a horizontal circle, the string describes the curved surface of a cone. This arrangement is known as a *conical pendulum*.

Example 11

A conical pendulum consists of a light inextensible string AB, fixed at A and carrying a particle of mass 50 g at B. The particle moves in a horizontal circle of radius $\sqrt{3}$ m and centre vertically below A. If the angle between the string and the vertical is 30°, find the tension in the string and the angular speed of the particle.

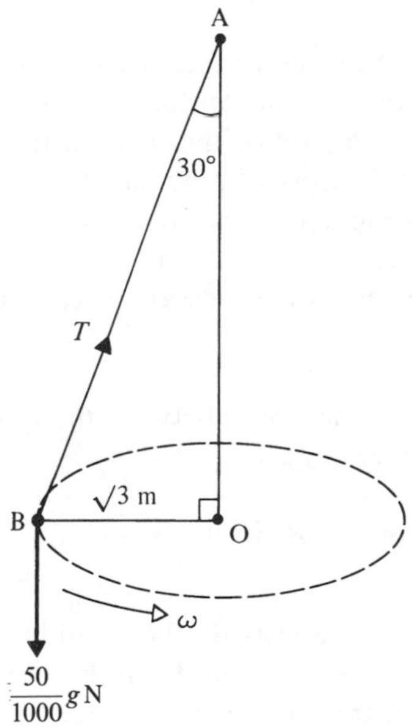

Resolving vertically gives:

$$T\cos 30° = \frac{50}{1000}g \qquad \ldots [1]$$

$$\therefore \qquad T = 0{\cdot}5658\text{ N}$$

The horizontal component of the tension provides the force towards the centre O of the circle.

Using $F = ma$ gives:

$$T\cos 60° = mr\omega^2$$

$$= \frac{50}{1000} \times \sqrt{3} \times \omega^2$$

and using $T = \frac{50}{1000} g \times \frac{1}{\cos 30°}$ from equation [1] gives:

$$\frac{50}{1000} g \times \frac{\cos 60°}{\cos 30°} = \frac{50}{1000} \times \sqrt{3} \times \omega^2$$

$$\therefore \qquad \omega = 1{\cdot}807 \text{ rad s}^{-1}$$

The tension in the string is 0·566 N and the angular speed of the particle is 1·81 rad s^{-1}.

Motion of a car rounding a banked curve

When a car travels along a circular arc on a horizontal road, the frictional force between the tyres and the road provides the necessary horizontal force F towards the centre of the circular arc.
If the speed of the car exceeds a particular value then the force required $\left(\text{i.e. } m\frac{v^2}{r}\right)$ may be greater than can be provided by the maximum frictional force, and the car will slip.

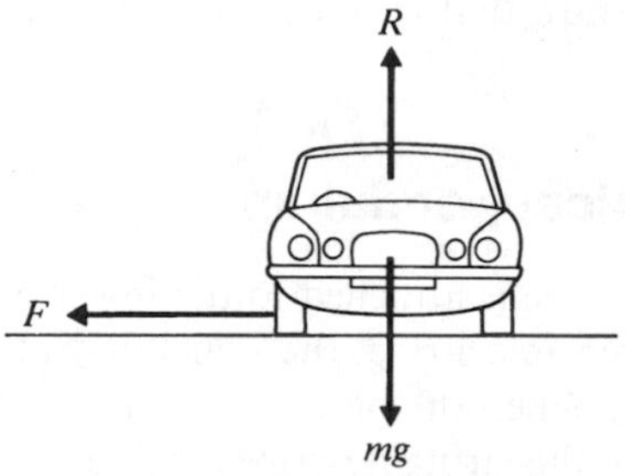

In practice, a bend in a road or race track may be banked, i.e. the level of the road on the inside of the bend is lower than the level of the road on the outside of the bend. The normal reaction R between the car and the road is then no longer vertical. The horizontal component of R can therefore provide or help to provide the necessary force, directed towards the centre.

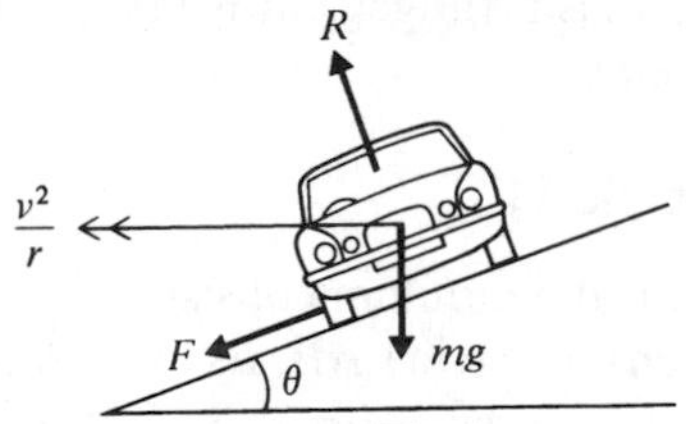

Fig. 1

If F is the frictional force between the tyres and the road surface (Fig. 1):

$$F \cos \theta + R \sin \theta = m\frac{v^2}{r}$$

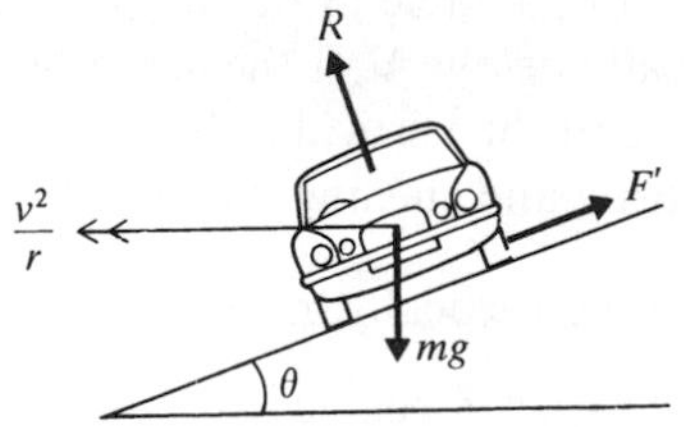

Fig. 2

If the car travels too slowly there will be a tendency for the car to slip down the slope. To prevent this, a frictional force F' will act (Fig. 2):

$$R \sin \theta - F' \cos \theta = m\frac{v^2}{r}$$

If the car has no tendency to slip, either up or down the slope, then there will be no frictional force acting between the tyres and the road and the horizontal component of R provides the necessary central force.

In this case: $\qquad R \sin \theta = m\frac{v^2}{r}$

Example 12

A car travels around a bend of radius 400 m on a road which is banked at an angle θ to the horizontal. If the car has no tendency to slip when travelling at $35\,\text{m}\,\text{s}^{-1}$, find θ.

Since there is no tendency to slip, there is no frictional force between the tyres and the road.

Resolving vertically gives:

$$R\cos\theta = mg \qquad \ldots[1]$$

Using $F = ma$, the force towards the centre of the bend is provided by a component of R.

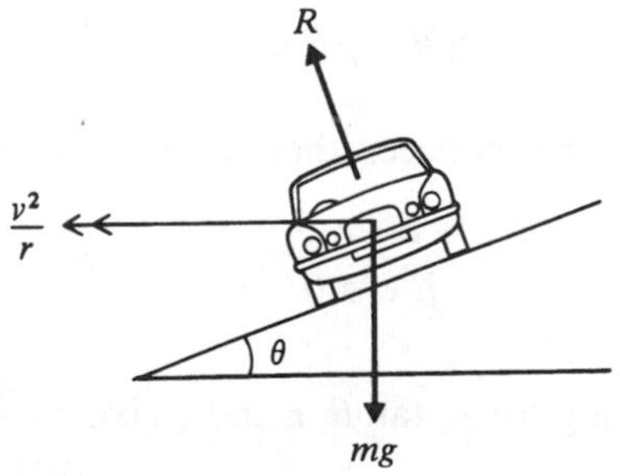

$$\therefore \qquad R\sin\theta = m\frac{v^2}{r} \qquad \ldots[2]$$

Dividing equation [2] by equation [1] gives:

$$\tan\theta = \frac{v^2}{rg}$$

Substituting for r and v gives: $\tan\theta = \dfrac{(35)^2}{400 \times 9{\cdot}8} = \dfrac{5}{16}$

$$\therefore \qquad \theta = 17{\cdot}35^\circ$$

The angle θ is $17{\cdot}35^\circ$.

Example 13

A car travels around a bend in a road which is a circular arc of radius 62·5 m. The road is banked at an angle $\tan^{-1}\frac{5}{12}$ to the horizontal. If the coefficient of friction between the tyres of the car and the road surface is 0·4, find:

(a) the greatest speed at which the car can be driven around the bend without slipping occuring

(b) the least speed at which this can happen.

(a) When the car is travelling as fast as possible, the maximum frictional force μR acts so as to prevent the car slipping up the slope.

Resolving vertically gives:

$$R\cos\theta = mg + \mu R\sin\theta \qquad \ldots[1]$$

Using $F = ma$ horizontally gives:

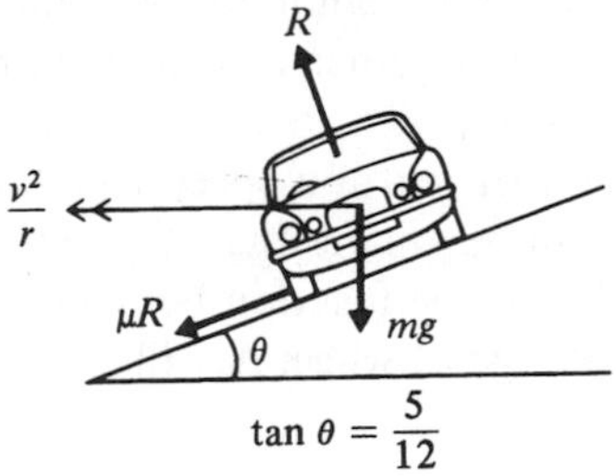

$$R\sin\theta + \mu R\cos\theta = m\frac{v^2}{r} \qquad \ldots[2]$$

Eliminating R between these equations gives:

$$\frac{\cos\theta - \mu\sin\theta}{\sin\theta + \mu\cos\theta} = \frac{rg}{v^2} \quad \text{or} \quad \frac{1 - \mu\tan\theta}{\tan\theta + \mu} = \frac{rg}{v^2}$$

Substituting for μ, $\tan\theta$, r and g gives: $\dfrac{1 - (0{\cdot}4) \times \frac{5}{12}}{\frac{5}{12} + 0{\cdot}4} = \dfrac{62{\cdot}5 \times 9{\cdot}8}{v^2}$

$$\therefore \qquad v = 24{\cdot}5\,\text{m}\,\text{s}^{-1}$$

(b) When the car is travelling as slowly as possible, the maximum frictional force μR acts so as to prevent the car slipping down the slope.

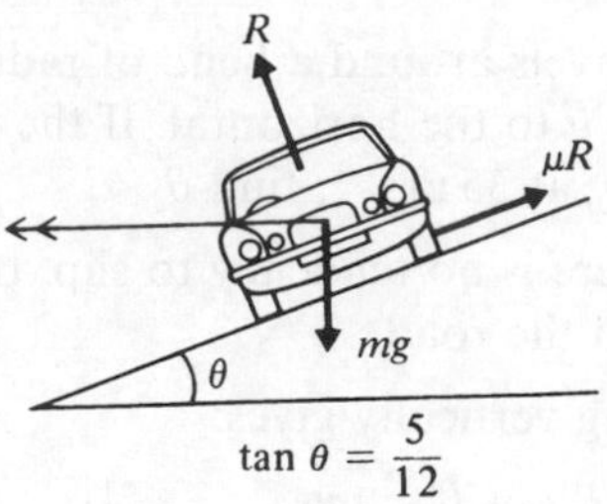

Resolving vertically gives:

$$R\cos\theta + \mu R\sin\theta = mg \qquad \ldots[1]$$

Using $F = ma$ horizontally gives:

$$R\sin\theta - \mu R\cos\theta = m\frac{v^2}{r} \qquad \ldots[2]$$

Eliminating R between these equations gives:

$$\frac{\cos\theta + \mu\sin\theta}{\sin\theta - \mu\cos\theta} = \frac{rg}{v^2} \quad \text{or} \quad \frac{1 + \mu\tan\theta}{\tan\theta - \mu} = \frac{rg}{v^2}$$

Substituting for μ, $\tan\theta$, r and g gives: $\dfrac{1 + (0{\cdot}4) \times \frac{5}{12}}{\frac{5}{12} - 0{\cdot}4} = \dfrac{62{\cdot}5 \times 9{\cdot}8}{v^2}$

$\therefore$ $\qquad v = 2{\cdot}958\,\text{m s}^{-1}$

The greatest and least speeds at which the car can travel without slipping are $24{\cdot}5\,\text{m s}^{-1}$ and $2{\cdot}96\,\text{m s}^{-1}$ respectively.

Motion of a train rounding a banked curve

If the outer rail is raised above the level of the inner rail, this will have a similar effect to the banking on a road. The normal reaction again has a horizontal component $R\sin\theta$ acting towards the centre of the circular arc. If the train travels at the speed v for which the banking was designed, then the force $R\sin\theta$ will provide the necessary central force towards the centre of the curve. If the train travels at a speed greater than v, then the force $R\sin\theta$ will be insufficient to provide the necessary force. In such a situation there will be a reaction between the outer rail and the flange of wheel B, and the horizontal component of this reaction will, together with $R\sin\theta$, provide the necessary force.

If the train travels at a speed less than v, then $R\sin\theta$ will be greater than the central force required, and the reaction between the inner rail and the flange of wheel A will compensate.

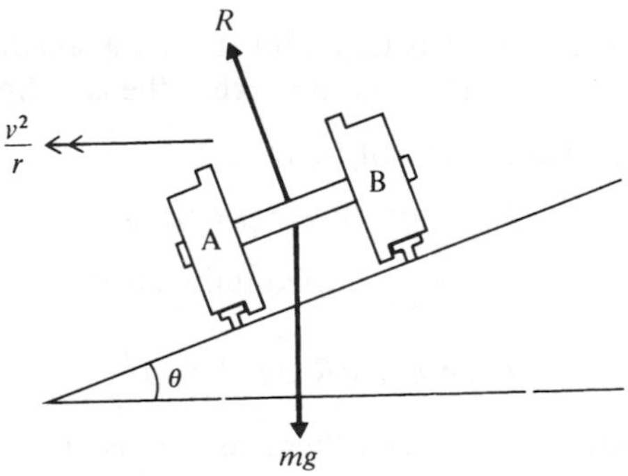

Example 14

A train of mass 50 tonnes travels around a bend of radius 900 m. The outer rail of the track is raised 7 cm above the inner rail and the distance between the rails is 1·4 m. Calculate the speed at which the train should travel so that there is no force acting between the flanges on the wheels and the rails.

Let the angle of the banking be θ. Then, since θ is small:

$$\sin\theta = \tfrac{7}{140} = \tfrac{1}{20} \approx \tan\theta$$

At $v\,\text{m s}^{-1}$, there is no reaction between the wheel flanges and the rails.

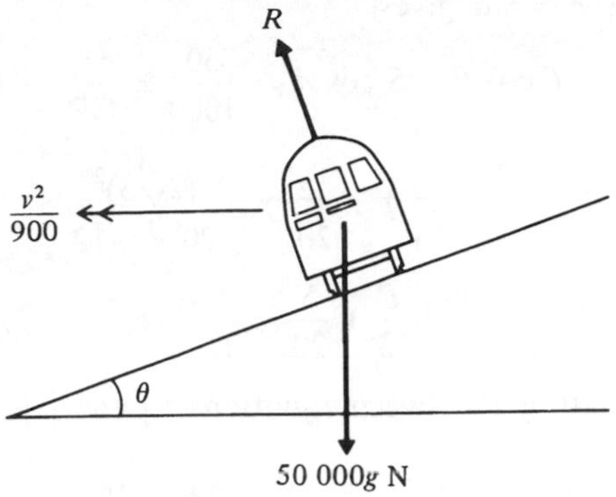

Resolving vertically gives:

$$R\cos\theta = 50\,000g \qquad \ldots[1]$$

Using $F = ma$, gives the force towards the centre of curve provided by component of normal reaction R as:

$$R\sin\theta = 50\,000 \times \frac{v^2}{900} \qquad \ldots[2]$$

Dividing equation [2] by equation [1] gives:

$$\tan\theta = \frac{v^2}{900g}$$

$$\therefore \quad \frac{900}{20}g = v^2 \quad \text{substituting for } \tan\theta$$

$$\therefore \quad v = 21\,\text{m s}^{-1}$$

The train should travel at $21\,\text{m s}^{-1}$ for there to be no force between the flanges and the rails.

Example 15

A light inextensible string AB of length 33 cm has a particle of mass 50 g attached to it at a point P, 13 cm form the end A. The ends of the string are attached at two fixed points in the same vertical line with A 21 cm above B. The particle moves in a horizontal circle, 5 cm vertically below A with both parts of the string taut, at a constant speed of $2\sqrt{3}\,\text{m s}^{-1}$.
Find the tensions in the two parts of the string.

Let O be the centre of the horizontal circle. Then:

$$\text{OP}^2 = 13^2 - 5^2 = 144$$

$$\therefore \quad \text{radius of circle} = 12\,\text{cm} = 0{\cdot}12\,\text{m}$$

Resolving vertically gives:

$$T\sin\theta = \frac{50}{1000}g + S\sin\phi$$

or

$$\frac{5}{13}T = \frac{g}{20} + \frac{16}{20}S \qquad \ldots[1]$$

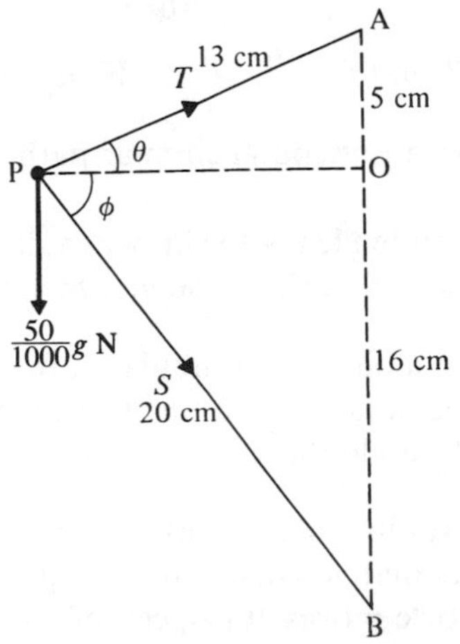

The horizontal components of the tension in the two parts of the string provide the necessary central force, towards O.

Using $F = ma$ gives:

$$T\cos\theta + S\cos\phi = \frac{50}{1000} \times \frac{v^2}{\text{OP}}$$

or
$$\frac{12}{13}T + \frac{12}{20}S = \frac{(2\sqrt{3})^2}{20 \times 0{\cdot}12}$$

$\therefore$
$$\frac{T}{13} + \frac{S}{20} = \frac{5}{12} \qquad \ldots [2]$$

Eliminating T between equations [1] and [2] gives:

$$5\left(\frac{5}{12} - \frac{S}{20}\right) = \frac{9{\cdot}8}{20} + \frac{16}{20}S \qquad \therefore \qquad S = 1{\cdot}517\,\text{N}$$

Substituting into equation [2] for S gives: $T = 4{\cdot}431\,\text{N}$

The tensions in the upper and lower parts of the string are 4·43 N and 1·52 N respectively.

Exercise 13C

1. Each of the following diagrams shows a conical pendulum consisting of a light inextensible string AB fixed at A and carrying a bob of mass 10 kg at B. The bob moves in a horizontal circle, centre vertically below A, with a constant angular speed ω. The tension in the string is T.

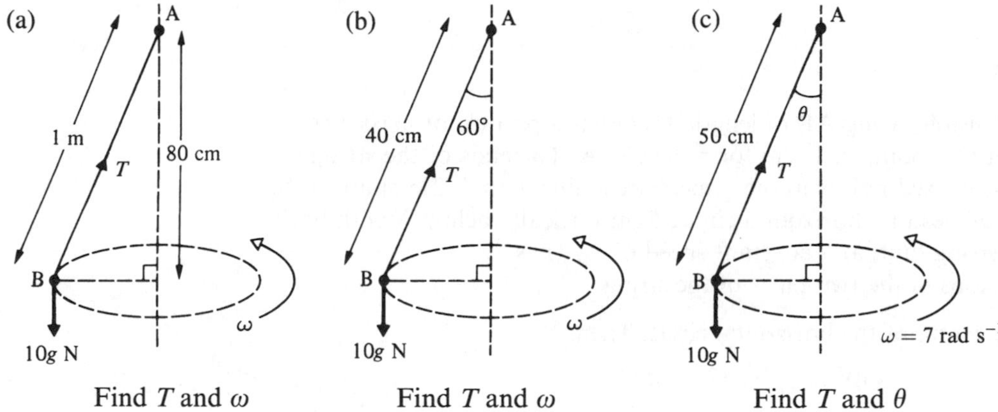

Find T and ω Find T and ω Find T and θ

2. A car moves in a horizontal circular path of radius 50 m around a banked track.
If the car has no tendency to slip when driven around the track at a speed of $14\,\text{m}\,\text{s}^{-1}$, find the angle at which the track is banked.

3. A car moves on a horizontal circular path of radius 75 m around a bend which is banked at $\tan^{-1}\frac{5}{12}$ to the horizontal.
At what speed should the car be driven if it is to have no tendency to slip?

4. A particle moves in a horizontal circular path of radius 20 cm on a smooth surface that is banked at an angel θ to the horizontal.
When the particle moves at a speed of $70\,\text{cm}\,\text{s}^{-1}$ it does not slip; find θ.

5. A car moves in a horizontal circle of radius 75 m around a bend which is banked at an angle of 20° to the horizontal.
At what speed should the car be driven if it is to have no tendency to slip?

6. The bend in a road is to be constructed of radius 150 m and such that, under any road conditions, a car can be driven in a horizontal circular path around the bend at 63 km h^{-1} without slipping.
At what angle to the horizontal should the road be banked?

7. A conical pendulum consists of a light inextensible string AB of length 50 cm fixed at A and carrying a bob of mass 2 kg at B. The bob describes a horizontal circle about the vertical through A with a constant angular speed of 5 rad s^{-1}.
Find the tension in the string.

8. A particle is attached to one end of a light inextensible string which has its other end attached to a fixed point A. With the string taut, the particle describes a horizontal circle with constant angular speed 2·8 rad s^{-1}, the centre of the circle being at a point O vertically below A.
Find the distance OA.

9. A bend in a railway track forms a horizontal circular arc of radius 1·25 km. A train of mass 40 tonnes travels round the bend at a constant speed of 63 km h^{-1}. The distance between the rails is 1·5 m.

(a) With the rails set at the same horizontal level (see Fig. 1), find the force exerted on the side of the rails.

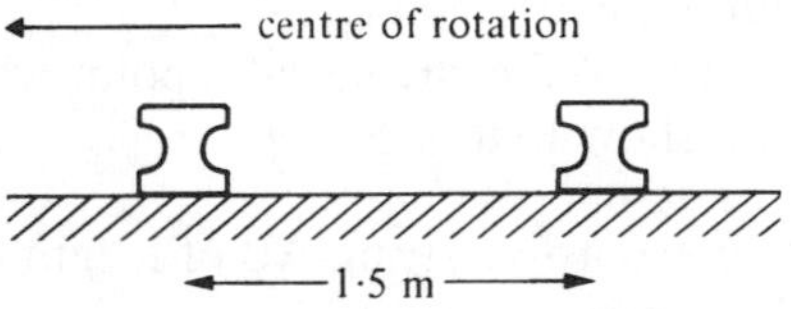

Fig. 1

(b) With the outer rail raised a distance x above the inner rail (see Fig. 2), the force on the side of the rails is reduced to zero.
Find x.

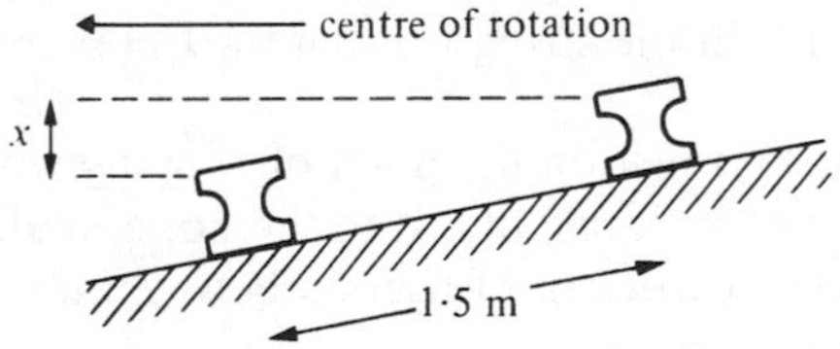

Fig. 2

10. Repeat question **9** for a train of mass 50 tonnes travelling at 50·4 km h^{-1} around a bend of radius 2 km.

11. A car moves on a horizontal circular path of radius 60 m around a bend that is banked at an angle $\tan^{-1} \frac{7}{24}$ to the horizontal.
If the coefficient of friction between the tyres of the car and the road surface is $\frac{3}{7}$, find the greatest speed with which the car can be driven around the bend without slipping occurring.

12. A car moves in a horizontal circular path of radius 60 m around a bend that is banked at an angle $\tan^{-1} 0{\cdot}5$ to the horizontal. Without slipping occurring, the maximum speed with which the car can travel around the bend is $28\ \text{m s}^{-1}$.
Find the coefficient of friction between the tyres of the car and the road surface.

13. A film stuntman has to drive a car in a horizontal circular path of radius 105 m around a bend that is banked at 45° to the horizontal. The stuntman finds that he must drive with a speed of at least $21\ \text{m s}^{-1}$ if he is to avoid slipping sideways down the slope.
Find the coefficient of friction between the tyres of the car and the road surface.

14. A car moves in a horizontal circular path of radius 140 m around a banked corner of a race track. The greatest speed with which the car can be driven around the corner without slipping occurring is $42\ \text{m s}^{-1}$.
If the coefficient of friction between the tyres of the car and the surface of the track is $\frac{1}{3}$, find the angle of banking.

15. A bend in a road is in the form of a horizontal circular arc of radius r, with the road surface banked at an angle θ to the horizontal.
Show that a car will have no tendency to slip when driven around the bend with speed $\sqrt{(rg \tan \theta)}$.

16. A light inextensible string AB has a particle attached at end B and A is fixed. With the string taut, the particle describes a horizontal circle with constant angular speed ω.
If the centre of the circle is at a point which is a distance x vertically below A, show that $\omega^2 x = g$.

17. A light inextensible string AB of length l has end A fixed and carries a particle of mass m at B.
With the string taut, the particle describes a horizontal circle about the vertical axis through A, with constant angular speed ω. Show that the tension T in the string is given by $T = ml\omega^2$.

18. A vehicle is just on the point of slipping when parked on a bend that is banked at an angle of 20° to the horizontal.
Find the coefficient of friction between the vehicle's tyres and the surface of the road.
If the vehicle were driven around this bend in a horizontal circular path of radius 60 m, find the greatest speed it could attain without slipping occurring.

19. One corner of a race track is banked at 45° to the horizontal. A car is to be driven around the corner on a horizontal circular path of radius 60 m. The coefficient of friction between the tyres of the caravan and the surface of the track is $0{\cdot}5$.
Find the greatest and least speeds with which the car can travel without slipping.

20. A point A lies 25 cm above a smooth horizontal surface. A light inextensible string of length 50 cm has one end fixed at A and carries a body B of mass 0·5 kg at its other end. With the string taut, B moves in a circular path on the surface, the centre of the circle being vertically below A.
If B moves with constant angular speed of 4 rad s^{-1}, find the magnitude of:

(a) the tension in the string

(b) the reaction force between B and the surface.

21. A smooth bead of mass 100 g is threaded on a light inextensible string of length 70 cm. The string has one end attached to a fixed point A and the other to a fixed point B 50 cm vertically below A. The bead moves in a horizontal circle about the line AB with a constant angular speed of ω rad s^{-1}, and the string taut. If the bead is at a point C on the string with AC = 40 cm, find the value of ω and the tension in the string.

22. A smooth hemispherical shell of radius r is fixed with its rim horizontal. A small ball bearing of mass m lies inside the shell and performs horizontal circles on the inner surface of the shell, the plane of these circles lying 5 cm below the level of the rim.
Find the angular speed of the ball bearing in rad s^{-1}.

23. A light inextensible string AB of length 90 cm has a particle of mass 600 g fastened to it at a point C. The ends A and B are attached to two fixed points in the same vertical line as each other, with A 60 cm above B. The particle moves on a horizontal circle at a constant angular speed of 5 rad s^{-1} with both parts of the string taut and CB horizontal.
Find the tensions in the two parts of the string.

24. A light inextensible string AB of length $2l$ has a particle attached to its mid-point C. The ends A and B of the string are fastened to two fixed points with A distance l vertically above B. With both parts of the string taut, the particle describes a horizontal circle about the line AB with constant angular speed ω.
If the tension in CA is three times that in CB, show that $\omega = 2\sqrt{\frac{g}{l}}$.

25. At any speed greater than v, a car will slip when driven on a horizontal circular path of radius r around a track that is banked at an angle θ to the horizontal.
If μ is the coefficient of friction between the tyres of the car and the surface of the track, show that $\mu = \dfrac{v^2 - rg \tan \theta}{rg + v^2 \tan \theta}$.

Motion in a vertical circle

The circular motion questions considered so far have all involved particles moving around *horizontal* circles with *constant* speed. Consider now a particle moving in a circular path at a *varying* speed. A common example of this would be a particle moving in a *vertical* circle with the speed of the particle decreasing as the height of the particle increases.

The particle is moving in a circle so, as before, the acceleration towards the centre of the circle at any instant must be $\frac{v^2}{r}$, where v is the speed of the particle at that instant.

However, the particle will also have acceleration due to the fact that v itself is changing.
We say that the acceleration has two components:

(1) towards the centre of the circular path and of magnitude $\frac{v^2}{r}$

(2) in the direction of motion (i.e. along the tangent) and equal to the rate of change of v.

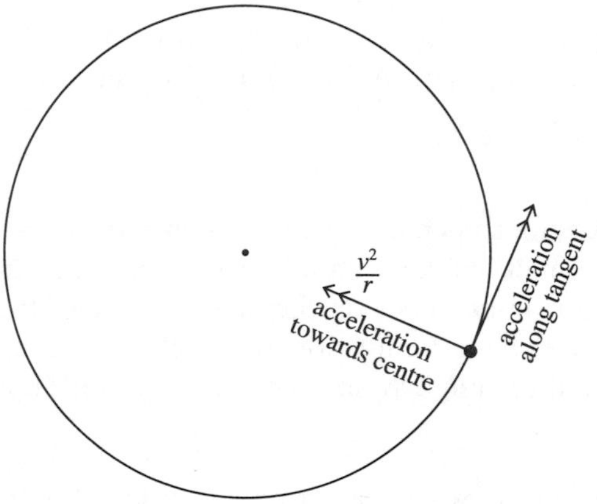

The next example considers a particle suspended from a fixed point by a light inextensible string. The particle is projected from its lowest point with a horizontal speed of $4\,\text{m s}^{-1}$ (see diagram on right). We assume that in the subsequent circular motion the only force that will resist the motion of the particle is its weight. As the particle gains height and increases its potential energy, its kinetic energy will decrease, as will its speed. The total energy of the system will remain unchanged. As the reader will see, the problem is solved using the Conservation of Energy Principle and by applying $F = ma$ towards the centre of the circle.

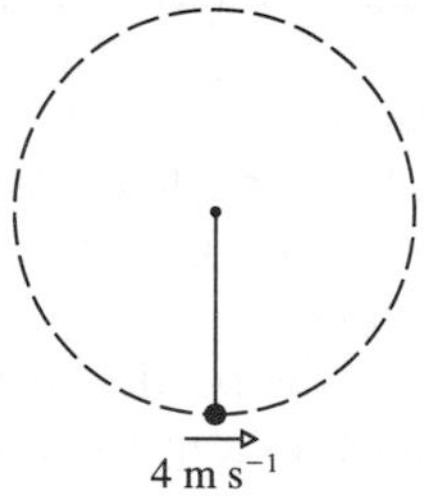

Once again we are modelling the real situation. Strings are not light and inextensible and other forces, e.g. air resistance, will be acting, reducing the speed of the particle. However, by considering these to be insignificant compared to the concepts we have allowed for, such as the mass of the particle itself and the gravitational force acting on it, we have a mathematical model that is simple enough for us to consider. If the outcomes predicted by our model were not sufficiently close to the real-life outcomes, we would need to examine any assumptions we have made and perhaps produce a more accurate model that takes into account considerations such as the mass of the string and air resistance.

Example 16

A particle P of mass 5 kg is suspended from a fixed point O by a light inextensible string of length 1 m. The particle is projected from its lowest position at the point A, with a horizontal speed of $4\,\text{m s}^{-1}$. When angle AOP = 60° find:

(a) the speed of P

(b) the tension in the string.

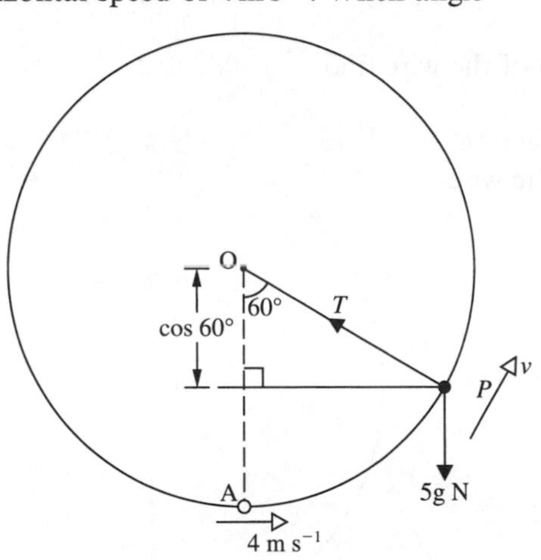

(a) We choose to measure the potential energy from the level of A.

At A: $\text{KE} = \frac{1}{2}(5)(4)^2$

$= 40\,\text{J}$

$\text{PE} = 0$

When P is in the position shown:

$\text{KE} = \frac{1}{2}(5)v^2\,\text{J}$

$\text{PE} = 5g(1 - \cos 60°)$

$= \dfrac{5g}{2}\,\text{J}$

By the Principle of Conservation of Energy:

$$40 = \frac{5v^2}{2} + \frac{5g}{2}$$

$5v^2 + 5g = 80$

$\therefore \quad v^2 = 6{\cdot}2$

$\therefore \quad v = 2{\cdot}49\,\text{m s}^{-1}$

When the angle AOP = 60° the speed of P is $2{\cdot}49\,\text{m s}^{-1}$.

(b) Applying $F = ma$ along PO gives:

$$T - 5g\cos(60°) = \frac{5v^2}{(1)}$$

$$T - 5g \times \tfrac{1}{2} = 5(6{\cdot}2)$$

$\therefore \quad T = 55{\cdot}5\,\text{N}$

When angle AOP = 60° the tension in the string is $55{\cdot}5$ N.

Example 17

A small bead of mass 3 kg is threaded on to a smooth wire in the shape of a circle, radius 2 m and centre O. The circular wire is fixed in a vertical plane and the bead passes through the lowest point A on the wire with speed $10\,\text{m s}^{-1}$.

When the bead is at the top of the wire find:

(a) the velocity of the bead

(b) the magnitude of the reaction between the bead and the wire.

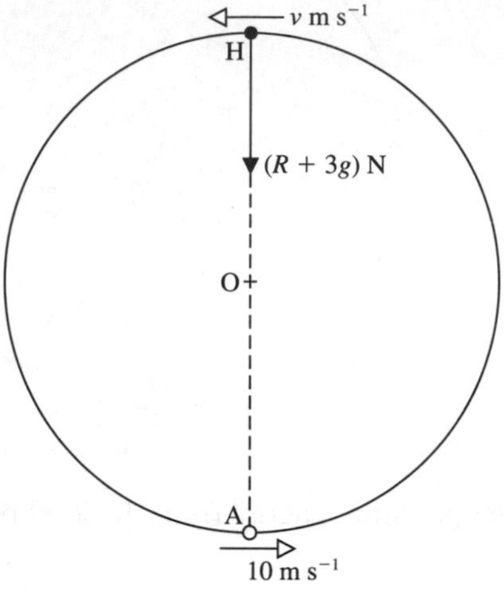

(a) Let the speed of the bead at the top of the wire be $v\,\text{m s}^{-1}$ and measure potential energy from the level of A.

At A: $\text{KE} = \frac{1}{2}(3)(10)^2$

$= 150\,\text{J}$

$\text{PE} = 0$

At H: $\text{KE} = \frac{1}{2}(3)v^2\,\text{J}$

$\text{PE} = 3g(4)$

$= 12g\,\text{J}$

By the Principle of Conservation of Energy:

$$150 = \frac{3v^2}{2} + 12g$$

$\therefore \quad v^2 = 21{\cdot}6$

$\therefore \quad v = 4{\cdot}65$

At the top of the wire the velocity of the bead is $4{\cdot}65\,\text{m s}^{-1}$ horizontally.

(b) Applying $F = ma$ along HO gives:

$$R + 3g = \frac{3v^2}{2}$$

$\therefore \quad R = 3\text{ N}$

When the bead is at the top of the wire the reaction between the bead and the wire is 3 N.

Conditions for a particle to perform complete circles

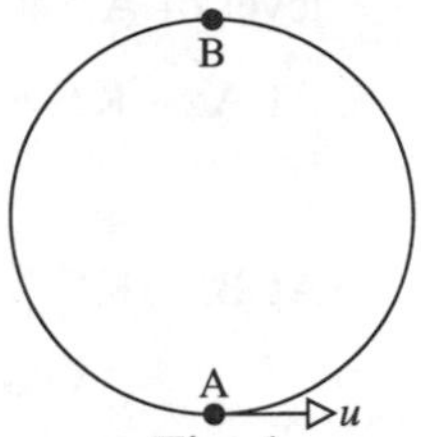
Fig. 1

Suppose a bead is threaded on a smooth circular *ring*. The ring is held in a vertical plane with the bead at the lowest point A (see Fig. 1). The bead is then projected horizontally with speed u. If u is sufficiently large, the bead will still possess some forward speed when it reaches the highest point B and will continue on to A and begin another revolution. If u is not large enough for this to happen, the bead will either just reach B with zero speed, and remain there, or it will not reach B at all and will instead fall back and oscillate about the lowest point A.

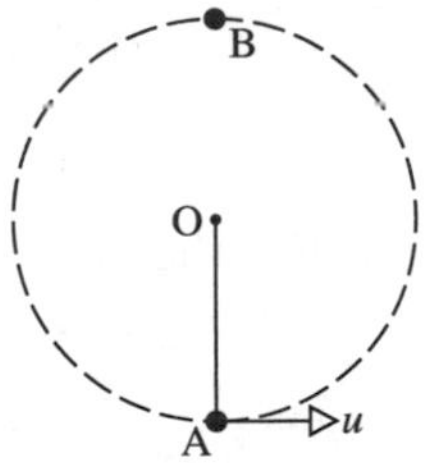
Fig. 2

A similar set of possibilities exists for a particle attached to one end of a light *rod* OA (see Fig. 2). The rod is held vertically with A below O and is able to rotate in a vertical circle, centre O. The particle is given an initial horizontal speed u. If u is sufficiently large, the particle will still possess some forward speed when it reaches the highest point, B, and will continue on to A and begin another revolution. If u is not large enough for this to happen, the particle will either reach B with zero speed, and remain there, or it will not reach B at all and will instead fall back and oscillate about the lowest point.

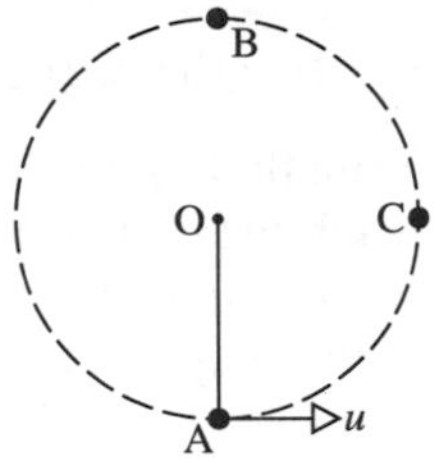
Fig. 3

Consider now a particle suspended from a fixed point O by a light inextensible *string*. This situation is a little different. If the initial speed u is sufficiently large, the particle will perform complete circles. However, it is now not sufficient for the particle to simply reach the highest point, B, with *some* forward speed. The speed the particle has when it reaches B must be sufficient to keep the string taut and thus maintain the circular motion. If it is not sufficient, the string will go slack and the particle will behave as a projectile in free flight. For the particle to perform complete circles we now require it to reach the highest point with the string still taut.

If u is sufficient to enable the particle to go beyond C (see Fig. 3), but not to reach B, the string will go slack at some stage when the particle is between C and B and it will then act as a projectile in free flight. If u is such that the particle is not able to reach C, there will be a point somewhere between A and C where the particle will momentarily stop and then return through A and perform oscillations about A.

Example 18

A light rod of length 2 m is pivoted at one end, O, and has a particle of mass 8 kg attached at the other end. The rod is held vertically with the particle at point A, directly below O, and the particle is given an initial horizontal speed $u\,\text{m s}^{-1}$.

Find:

(a) an expression in terms of u and θ for the speed of the particle when at point B where $\angle AOB = \theta$ (see diagram)

(b) the restriction on u^2 if the particle is to perform complete circles.

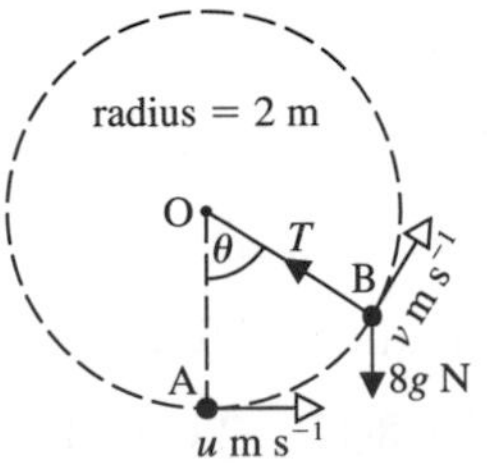

(a) Let the speed of the particle at B be v m s^{-1} and measure PE from the level of A.

At A: $\text{KE} = \frac{1}{2}(8)(u)^2 = 4u^2$ J $\quad$ PE = 0

At B: $\text{KE} = \frac{1}{2}(8)(v)^2 = 4v^2$ J $\quad$ $\text{PE} = 8g(2 - 2\cos\theta)$ J

By the Principle of Conservation of Energy:

$$4u^2 = 4v^2 + 8g(2 - 2\cos\theta)$$

$\therefore$ $\quad v = \sqrt{u^2 - 4g(1 - \cos\theta)}$ is the required expression.

(b) For the particle to perform complete circles we require $v > 0$ when $\theta = 180°$,

i.e. $\quad u^2 - 4g(1 - \cos 180°) > 0$

$\therefore$ $\quad u^2 > 8g$

For the particle to perform complete circles $u^2 > 8g$.

Note that, in the last example, if $u = \sqrt{8g}$, the particle would in theory *just* reach the highest point. The force in the rod would then have to be a thrust, so as to support the weight of the stationary particle. It is possible for a rod to act in this way, i.e. as a strut, but this would not be possible if the rod were replaced by a string. In such a case an initial speed of projection of $\sqrt{8g}$ m s^{-1} would not be sufficient to maintain a positive tension in the string and the particle would not complete the vertical circle, as the next example confirms.

Example 19

A particle P of mass 8 kg is suspended from a fixed point O by an inextensible string of length 2 m. The particle is projected from its lowest position A with a horizontal speed of u m s^{-1}.
Find:

(a) an expression in terms of u and θ for the tension in the string when the particle is at a point B where $\angle AOB = \theta$ (see diagram)

(b) the restriction on u^2 if the particle is to perform complete circles.

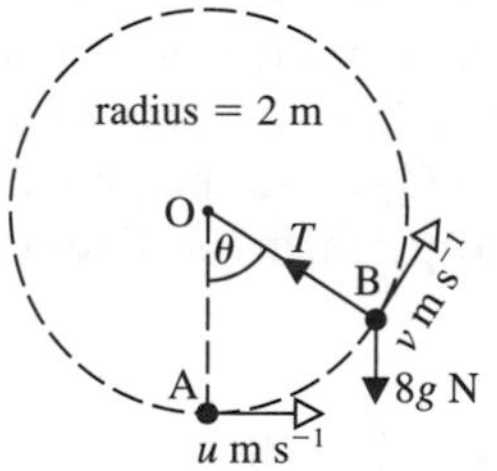

(a) Let the speed of the particle at B be v m s^{-1} and measure PE from the level of A.

At A: $\text{KE} = \frac{1}{2}(8)(u)^2 = 4u^2$ J $\quad$ PE = 0

At B: $\text{KE} = \frac{1}{2}(8)(v)^2 = 4v^2$ J $\quad$ $\text{PE} = 8g(2 - 2\cos\theta)$ J

By the Principle of Conservation of Energy:

$$4u^2 = 4v^2 + 8g(2 - 2\cos\theta) \quad \ldots [1]$$

Applying $F = ma$ along BO gives:

$$T - 8g\cos\theta = 8\frac{v^2}{2}$$

$$\therefore \quad T - 8g\cos\theta = 4v^2 \qquad \ldots [2]$$

From equations [1] and [2] it follows that:

$$4u^2 = T - 8g\cos\theta + 8g(2 - 2\cos\theta)$$

$$\therefore \quad T = 4u^2 + 24g\cos\theta - 16g \quad \text{is the required expression.}$$

(b) For the particle to perform complete circles we require $T > 0$ when $\theta = 180°$.

We require: $4u^2 + 24g\cos 180° - 16g > 0$

i.e. $u^2 > 10g$

For the particle to perform complete circles we require $u^2 > 10g$.

Example 20

A particle of mass 5 kg is slightly disturbed from rest on the top of a smooth hemisphere, radius 4 m and centre O, resting with its plane face on horizontal ground.

(a) Show that the particle leaves the surface of the hemisphere at the point P, where the angle between the radius PO and the upward vertical is $\cos^{-1}\left(\frac{2}{3}\right)$.

(b) After leaving the surface of the hemisphere the particle hits a vertical wall situated perpendicular to the horizontal diameter through O and at a distance of $\frac{5\sqrt{5}}{3}$ m from the centre of the hemisphere. Find the height of the particle above the ground when it hits the wall.

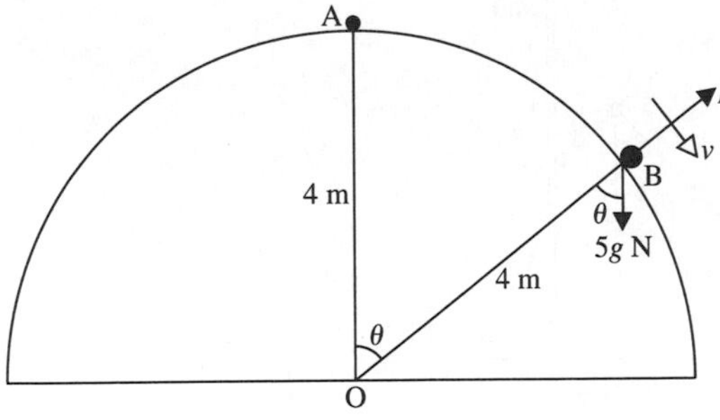

(a) Consider the particle at a general point B (see diagram) and choose to measure PE from the level of B.

At A: KE $= 0$

PE $= 5g(4 - 4\cos\theta)$ J

At B: KE $= \frac{1}{2}(5)v^2$ J

PE $= 0$

By the Principle of Conservation of Energy

$$\frac{5v^2}{2} = 5g(4 - 4\cos\theta)$$

$\therefore \qquad v^2 = 2g(4 - 4\cos\theta) \qquad \ldots[1]$

Applying $F = ma$ along BO gives: $5g\cos\theta - R = \dfrac{5v^2}{4}$

When the particle leaves the surface of the hemisphere $R = 0$, giving

$$5g\cos\theta = \frac{5v^2}{4}$$

i.e. $\qquad v^2 = 4g\cos\theta \qquad \ldots[2]$

Substituting from [2] into [1] gives: $4g\cos\theta = 2g(4 - 4\cos\theta)$

$\therefore \qquad \cos\theta = \frac{2}{3}$ and $\theta = \cos^{-1}\left(\frac{2}{3}\right)$

Therefore when the particle leaves the surface of the hemisphere the angle between the radius and the upward vertical is $\cos^{-1}\left(\frac{2}{3}\right)$.

(b) Since from [2] $v^2 = 4g\cos\theta$ at the point at which the particle leaves the surface of the hemisphere:

$$v^2 = 4g\left(\frac{2}{3}\right)$$

$\therefore \qquad v^2 = \dfrac{8g}{3}$

When the particle leaves the surface of the hemisphere it becomes a projectile

with initial velocity $\sqrt{\dfrac{8g}{3}}$ at an angle θ below the horizontal.

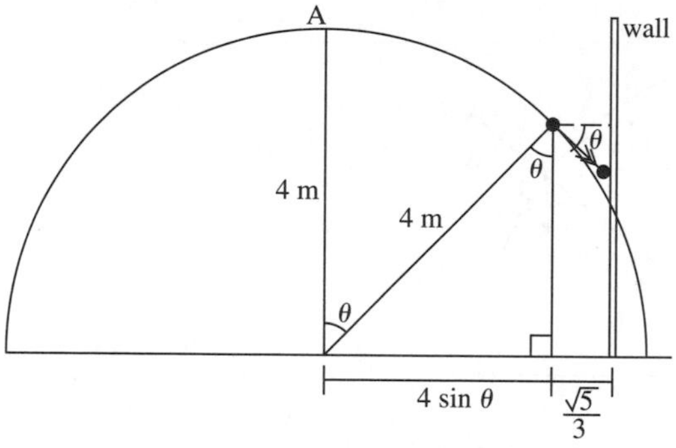

Note: • The horizontal distance travelled by the particle whilst in contact with the surface of the hemisphere is $4\sin\theta = \dfrac{4\sqrt{5}}{3}$ m.

• The horizontal distance travelled by the particle after it leaves the surface of the hemisphere is $\dfrac{5\sqrt{5}}{3} - \dfrac{4\sqrt{5}}{3} = \dfrac{\sqrt{5}}{3}$ m.

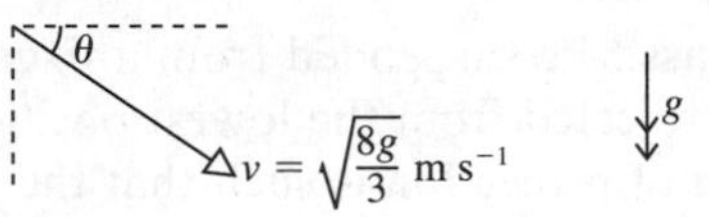

horizontal motion

$u = \sqrt{\frac{8g}{3}} \cos\theta \text{ m s}^{-1}$

$t = t\text{ s}$

$s = \frac{\sqrt{5}}{3}\text{ m}$

no acceleration

vertical motion

$v = \sqrt{\frac{8g}{3}} \sin\theta \text{ m s}^{-1} \downarrow$

$t = t\text{ s}$

$s = s\text{ m}$

$a = g\text{ m s}^{-2} \downarrow$

Using $s = ut + \frac{1}{2}at^2$ horizontally gives:

$$\frac{\sqrt{5}}{3} = \left(\sqrt{\frac{8g}{3}} \cos\theta\right)t$$

$$\therefore \quad \frac{\sqrt{5}}{3} = \left(\sqrt{\frac{8g}{3}} \times \frac{2}{3}\right)t$$

$$\therefore \quad t = \frac{1}{2}\sqrt{\frac{15}{8g}}$$

Using $s = ut + \frac{1}{2}at^2$ vertically gives:

$$s = \sqrt{\frac{8g}{3}} \sin\theta \cdot \frac{1}{2}\sqrt{\frac{15}{8g}} + \frac{1}{2}g\left(\frac{1}{4} \cdot \frac{15}{8g}\right)$$

$$\therefore \quad s = \frac{1}{2} \cdot \frac{5}{3} + \frac{15}{64}$$

$$\therefore \quad s = \frac{205}{192}\text{ m}$$

$(s \approx 1{\cdot}1\text{ m})$

The particle falls $\frac{205}{192}$ m from leaving the hemisphere to hitting the wall.

Therefore the particle is $\frac{307}{192}$ m above the ground when it hits the wall.

Exercise 13D

Questions **1**, **2** and **3** each involve a particle of mass 8 kg suspended from a fixed point O by a light inextensible string of length 25 cm. The particle is projected from the lowest point, A, with the horizontal speed indicated. In each case the speed of projection is sufficient for the particle to execute complete circles about O.

For each case find the speed of the particle as it passes through its highest position, B.

1.

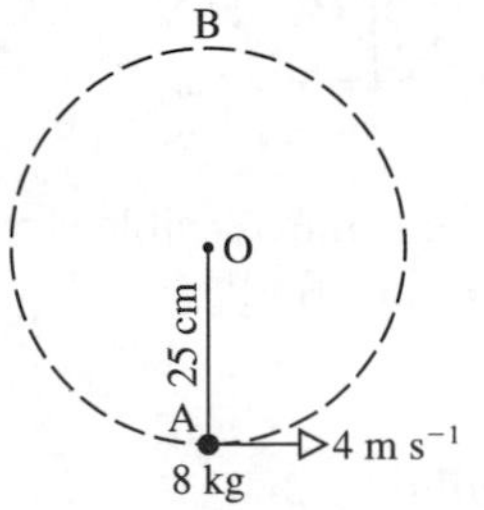

2.

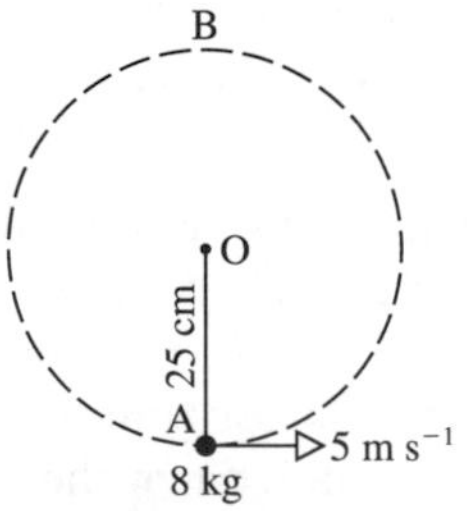

3.

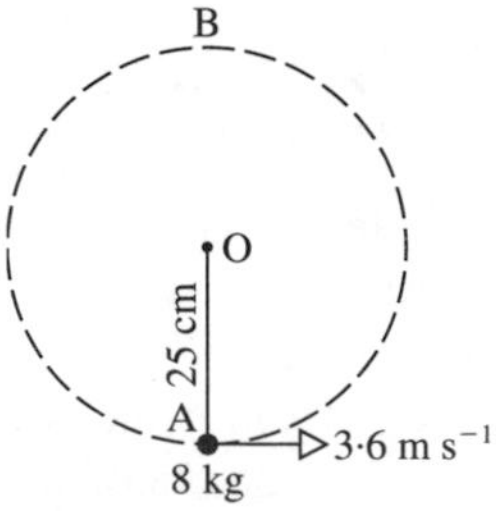

Questions **4**, **5** and **6** each involve a particle of mass 5 kg suspended from a fixed point O by a light inextensible string of length 2 m. The particle is projected from the lowest position, A, with the horizontal speed indicated. In each case the speed of projection is such that the particle just reaches point B and then returns through A and oscillates about the lowest position.
For each case determine the size of angle AOB, to the nearest degree.

4.

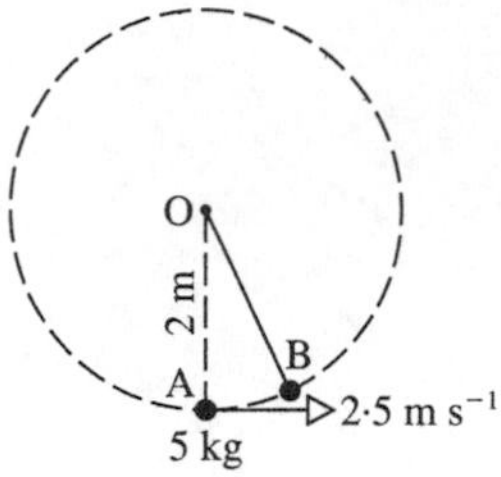

5.

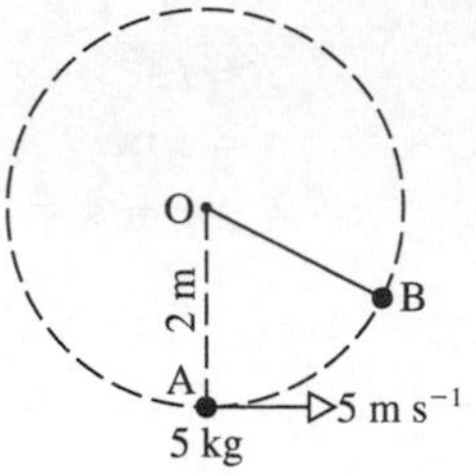

6.

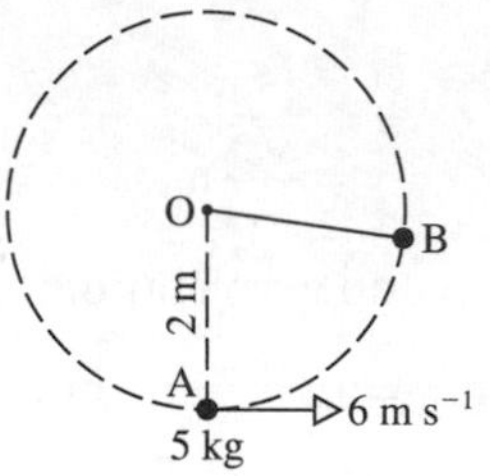

Questions **7**, **8** and **9** each involve a particle of mass 2 kg suspended from a fixed point O by a light inextensible string of length 20 cm. The particle is projected from the lowest position, A, with a horizontal speed of $5\,\mathrm{m\,s^{-1}}$.
Determine the speed of the particle and the tension in the string for each of the positions shown below.

7.

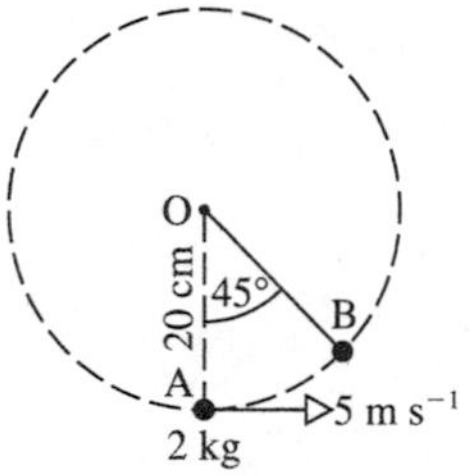

8.

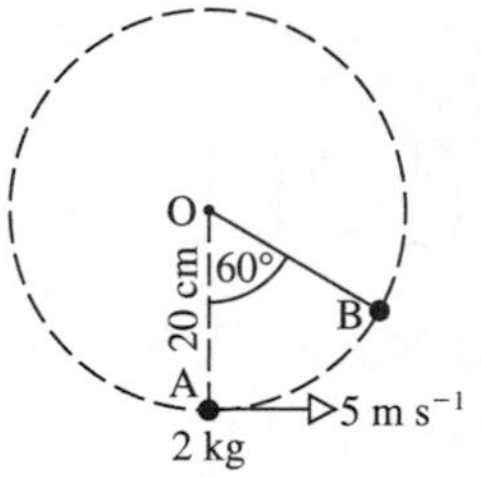

9.

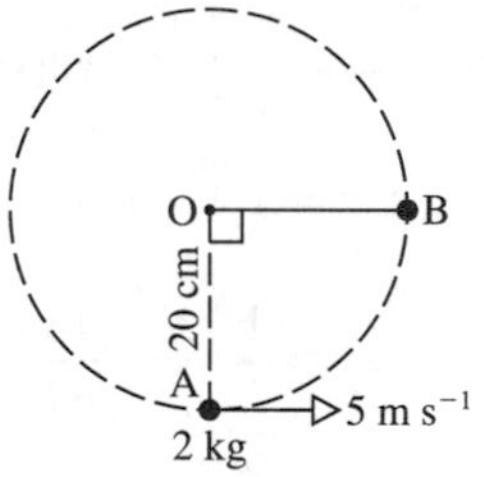

Questions **10**, **11** and **12** each involve a particle of mass 5 kg suspended from a fixed point O by a light rod of length 50 cm. The particle is projected from the lowest position A with the horizontal speed indicated. In each case the speed of projection is such that the particle executes vertical circles, centre O.
For each situation, determine the speed of the particle as it passes through the highest point, B, and determine the force in the rod at that instant, stating whether the rod is in tension or in thrust.

10.

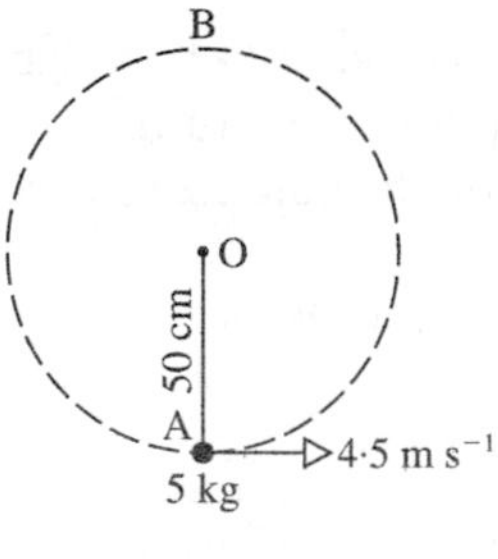

11.

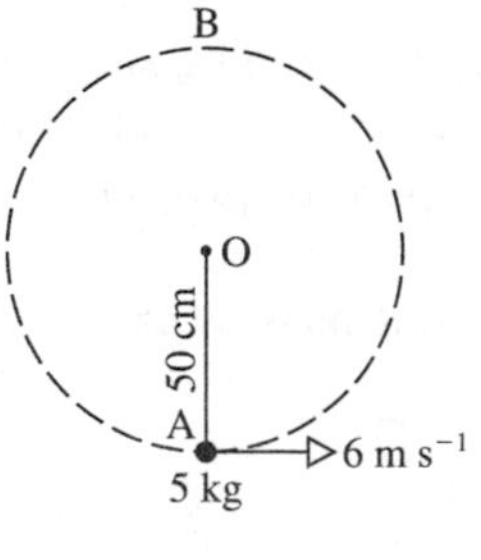

12.

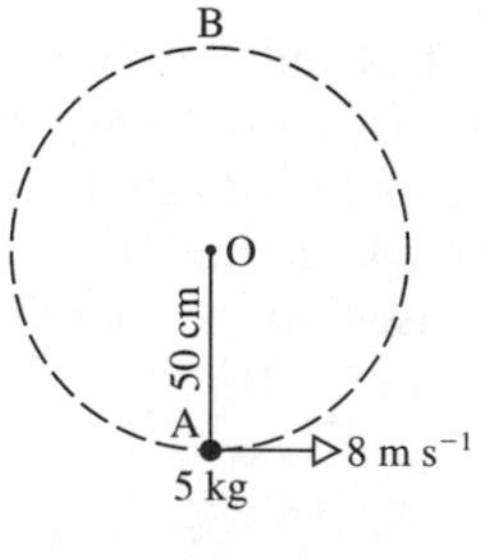

13. A particle of mass 5 kg describes complete vertical circles on the end of a light inextensible string of length 2 m. Given that the speed of the particle is $5\,\mathrm{m\,s^{-1}}$ at its highest point, find:
(a) its speed at its lowest point
(b) the tension in the string when the string is horizontal
(c) the magnitude of the tangential acceleration when the string is horizontal.

14. A particle of mass m describes complete vertical circles on the end of a light inextensible string of length 0·5 metres.
Given that when the particle is at the lowest point of the circle the tension in the string is twice that when the particle is at the highest point on the circle, find the speed of the particle at the lowest and highest points of the circle.

15. A particle of mass m describes complete vertical circles on the end of a light inextensible string of length r.
Given that the speed of the particle at the lowest point is twice the speed at the highest point, find:
(a) the speed of the particle at the lowest point
(b) the tension in the string when the particle is at its highest point.

16. A light rod of length 1·5 m is pivoted at one end O and has a particle P of mass 5 kg attached to the other end. Given that when the rod is vertical with P below O the speed of the particle is 8 m s^{-1}, find the location of P when the tension in the rod is zero and determine the speed of the particle at that point.

17. A particle of mass 5 kg is suspended from a fixed point O by a light inextensible string of length a metres. The particle is projected from the lowest point, A, with horizontal speed u m s^{-1}. At the instant shown in the diagram the particle is at point B, the particle's speed is x m s^{-1}, the tension in the string is T N and OB makes an angle θ with the downward vertical.

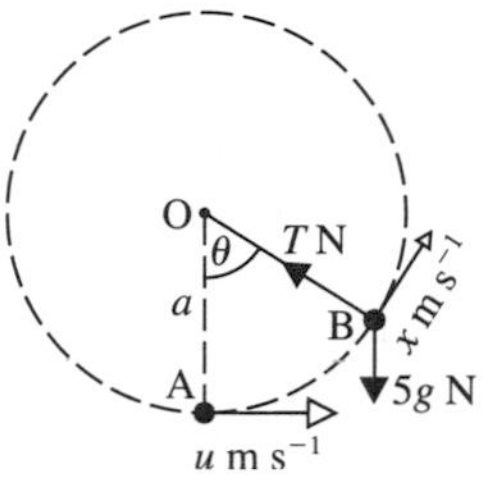

Show that $$x^2 = u^2 - 2ga + 2ga \cos \theta$$
and $$T = 15g \cos \theta + \frac{5u^2}{a} - 10g.$$

Hence determine the value of θ when the string goes slack given that $u = 2{\cdot}8$ and $a = 0{\cdot}2$, giving your answer to the nearest degree.

18. A particle P of mass m is attached to one end of a light inextensible string of length r, the other end being attached to a fixed point O. When the particle is hanging in equilibrium, with P at A, it is given a horizontal velocity of magnitude u. Obtain an expression for the speed of the particle and for the magnitude of the tension in the string at the instant that OP makes an angle θ with the downward vertical. Hence show that if:

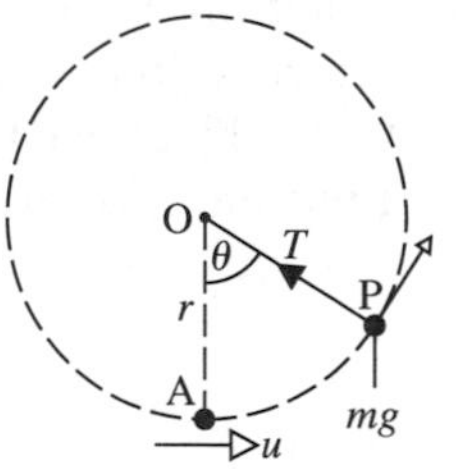

$u^2 < 2gr$ the particle's speed will be zero at some point for which $0 \leqslant \theta < 90°$.

$u^2 = 2gr$ P will just reach a point level with O, at which instant the string will momentarily be slack (i.e. zero tension).

$u^2 > 5gr$ the tension in the string is always positive (i.e. the string is always taut and P will complete vertical circles).

If the string is replaced by a rod, complete circles will be possible provided P still has some forward speed when it reaches the top of the circle.
Show that this will occur provided $u^2 > 4gr$.

19. The diagram shows a particle of mass 8 kg at rest inside a smooth vertical circular rim of radius 1·5 m, centre O. The particle is projected horizontally with speed $10\,\text{m s}^{-1}$ from the lowest point, A. Find the speed of the particle and the magnitude of the resultant acceleration when the particle is at the point P where angle POA is:

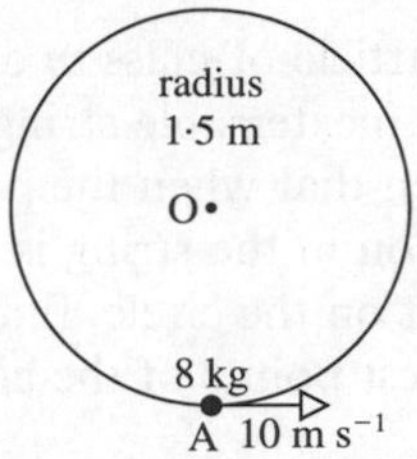

(a) 30°

(b) 45°

(c) 120°.

20. A small bead P of mass 3 kg is threaded on a smooth circular wire of radius 2 m and centre O, fixed in the vertical plane. The bead is gently disturbed from rest at the highest point H of the wire.
Find the angular speed of the bead and the magnitude of the reaction between the bead and the wire when:

(a) angle HOP = 30°

(b) angle HOP = 60°

(c) angle HOP = 90°.

21. A small bead P of mass 2 kg is threaded on a smooth circular wire radius 1 m and centre O, fixed in the vertical plane. The bead is projected from its lowest position A with a horizontal speed of $10\,\text{m s}^{-1}$. Find:

(a) an expression, in terms of θ, for the magnitude of the reaction between the bead and the wire when the bead is at a point such that angle AOP $= \theta$

(b) θ when the reaction between the bead and the wire is 200 N.

22. A light rod is pivoted at one end O and has a particle of mass 2 kg attached to the other end. The system is held at rest with the particle vertically above O and released gently.
Find the tension in the rod when the rod makes an angle of 60° with the upward vertical.

23. A particle of mass m kg is slightly disturbed form rest at the top of a smooth sphere radius 3 metres and centre O. Find, to one decimal place, the horizontal and vertical distances travelled by the particle from the time it was disturbed until 0·5 seconds after it left the surface.

24. A particle of mass m is projected from the top of a smooth sphere of radius r. In the subsequent motion the particle slides down the outside surface of the sphere and leaves the surface of the sphere with a speed of $2\sqrt{\frac{gr}{5}}$. Find:

(a) the vertical distance travelled by the particle while it is in contact with the sphere
(b) the initial speed of projection
(c) the speed of the particle when it is level with the horizontal diameter of the sphere.

25. A small bead P of mass m is threaded on a smooth circular wire of radius a and centre O, fixed in the vertical plane. Initially the bead is held at rest at a point B, where OB is inclined at an angle of 30° to the downward vertical through O. The bead is projected from B, perpendicular to OB with speed u so that P starts describing a vertical circle about O. A point C is on the wire and OC makes an angle θ with the upward vertical through O. Show that:

(a) the reaction between the wire and the bead at C has magnitude

$$\left|\frac{mu^2}{a} - mg\left(\sqrt{3} + 3\cos\theta\right)\right|$$

(b) the difference between the reaction of the wire on the bead at B and the reaction of the wire on the bead at C is

$$\left|3mg\left(\cos\theta + \frac{\sqrt{3}}{2}\right)\right|.$$

26. A particle P of mass m is released from rest at a point B on the surface of a smooth sphere, centre O and radius a. The line OB is inclined at an angle of 30° to the upward vertical through O. Show that when OP makes an angle θ with the upward vertical:

(a) the reaction between the particle and the sphere is given by $mg\,(3\cos\theta - \sqrt{3})$
(b) the square of the angular speed of P is $\frac{g}{a}\left(\sqrt{3} - 2\cos\theta\right)$.

Find, also, the angular speed of the particle when it leaves the surface of the sphere.

Exercise 13E Examination questions

(Take $g = 9{\cdot}8\,\text{m s}^{-2}$ throughout this exercise.)

1. On a level race track a car can just go round a bend of radius 125 m at a speed of $25\,\text{m s}^{-1}$.
Find the coefficient of friction between the car and the track.
(NICCEA)

2. A particle moves with constant speed u in a horizontal circle of radius a on the inside of a fixed smooth spherical shell of internal radius $2a$.
Show that $u^2\sqrt{3} = ag$. (AEB 1994)

3. A circular track of radius r is banked at an angle α to the horizontal. A motorcyclist travels around the track at speed V without slipping.
The coefficient of friction between the tyres and the track is μ.
(i) Show that the least value of V is given by

$$V^2 = rg\frac{(\sin\alpha - \mu\cos\alpha)}{(\cos\alpha + \mu\sin\alpha)}$$

A fairground show called "The Wall of Death" consist of a large cylindrical cage mounted with the axis vertical. A motorcyclist starts off on the floor of the cage and, as his speed increases, steers the motorcycle up onto the walls of the cage until he is riding around on the vertical inside surface of the cage.

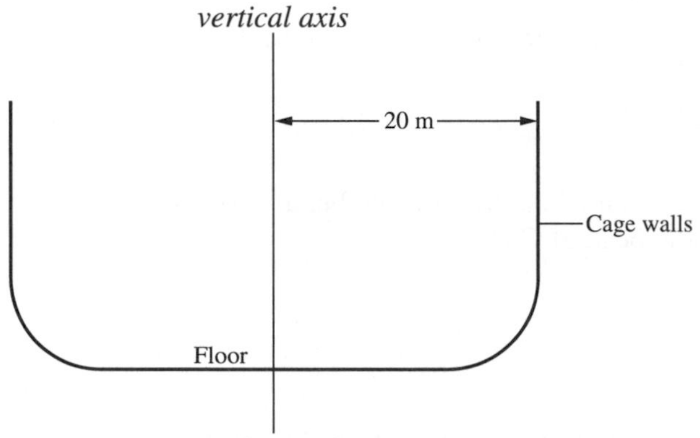

The radius of the cage is 20 metres and the coefficient of friction between the tyres and the walls of the cage is 0·8.
(ii) Using the result of (i), or otherwise, find the least speed he must maintain if he is to stay on the walls of the cage. (NICCEA)

4. A device in a fun-fair consists of a hollow circular cylinder of radius 3 m, with a horizontal floor and vertical sides. A small child stands inside the cylinder and against the vertical side. The cylinder is rotated about its vertical axis of symmetry.
When the cylinder is rotating at a steady angular speed of 30 revs/min the floor of the cylinder is lowered, so that the child is in contact only with the vertical side.
Given that the child does not slip, find, correct to two decimal places, the coefficient of friction between the child and the side. (AEB 1994)

5. A child of mass 30 kg keeps herself amused by swinging on a 5 m rope attached to an overhanging tree. She is holding on to the lower end of the rope and "swinging" in a horizontal circle of radius 3 m.

(a) Draw a diagram to show the forces acting on the girl.

(b) Find the tension in the rope.

(c) Show that the time she takes to complete a circle is approximately 4 seconds.

(d) State any assumptions that you have made about the rope.

(e) The girl's older brother then swings, on his own, on the rope in a horizontal circle of the same radius. Show that the tension in the rope is now $\frac{5mg}{4}$ where m is his mass.
Find the time that it takes for him to complete one circle.
(AEB Spec)

6. A particle is attached to the end A of a light inextensible string OA of length 1 m. The end O is fixed at a distance of 1 m above the horizontal ground. The particle describes a horizontal circle at a height of 0·5 m above the ground.

(a) Calculate the steady speed of the particle.

(b) The string is then released at O.
  (i) Describe in words the initial direction of the subsequent motion of the particle.
  (ii) Calculate the horizontal displacement of the particle from its position when the string is released to where it first strikes the ground. (UODLE)

7. A particle P of mass m is attached to one end of a light inextensible string of length d. The other end of the string is fixed at a height h, where $h < d$, vertically above a point O on a smooth horizontal table. The particle describes a circle, centre O, on the surface of the table, with constant angular speed ω.
Find expressions, in terms of m, d, h, ω and g as appropriate, for

(a) the tension in the string

(b) the magnitude of the force exerted on P by the plane.

(c) Hence show that

$$\omega \leqslant \sqrt{\left(\frac{g}{h}\right)}.$$

The angular speed of P is now increased to $\sqrt{\left(\frac{3g}{h}\right)}$, and P now describes a horizontal circle, centre Q, *above* the surface of the table, with this constant angular speed.
(d) Determine, in terms of h, the distance OQ. (ULEAC)

8. A particle P, of mass M, moves on the smooth inner surface of a fixed hollow spherical bowl, centre O and inner radius r, describing a horizontal circle at constant speed. The centre C of this circle is at a depth $\frac{1}{3}r$ vertically below O. Determine

(a) the magnitude of the force exerted by the surface of the sphere on P

(b) the speed of P. (ULEAC)

9. One section of a "Loop the Loop" ride at an Adventure Park takes passengers round a vertical circle.

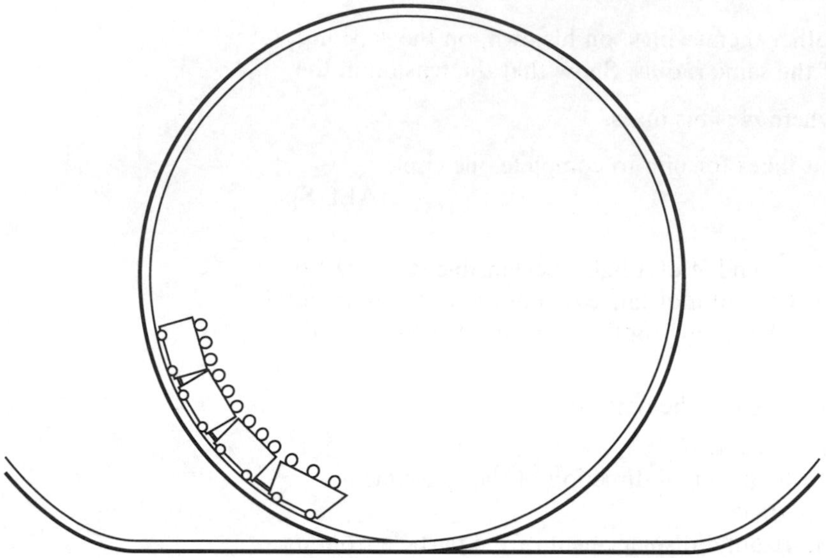

The situation is modelled by considering a small bead P of mass m threaded on a fixed smooth circular wire. The circular wire has centre O and radius a, and its plane is vertical. The bead is projected from the lowest point of the wire with speed $\sqrt{ag}$. When OP makes an angle θ with the downward vertical, find

(a) the speed of the bead,

(b) the reaction of the wire on the bead. (AEB Spec)

10. A smooth loop of wire in the form of a circle, centre O and of radius 0·3 m, is fixed in a vertical plane. A bead of mass 0·5 kg is threaded on the wire and projected with speed $u\,\text{m}\,\text{s}^{-1}$ from the lowest point of the wire so that it comes to instantaneous rest at a height of 0·1 m above the level of O. Find

(i) the value of u,

(ii) the reaction of the wire on the particle when the particle is level with O. (WJEC)

11. A particle is free to slide on the smooth outer surface of a fixed sphere of centre O. The particle is released from rest at a point A on the sphere, where OA is inclined at α to the upward vertical, $0 < \alpha < \dfrac{\pi}{2}$.

Given that the particle leaves the sphere at a point B, where OB is inclined at β to the upward vertical, show that $\dfrac{\cos\beta}{\cos\alpha} = \dfrac{2}{3}$.

(AEB 1994)

12.

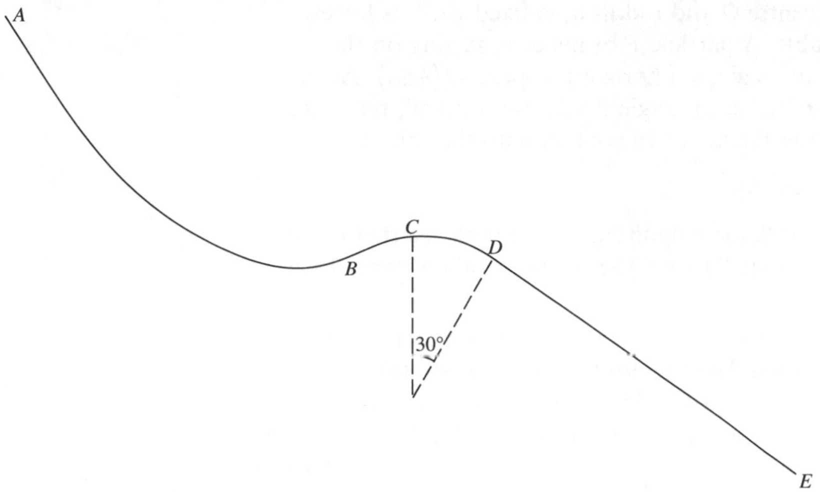

The diagram shows the shape of a "slide" for a children's playground. The section DE is straight and BCD is a circular arc of radius 5 m. C is the highest point of the arc and CD subtends an angle of 30° at the centre. For safety reasons, children should not be sliding so fast that they lose contact with the slide at any point. Neglecting any resistances to motion, find:

(i) the child's speed as it passes through C, given that it is on the point of losing contact at C,

(ii) the child's speed as it passes through D, given that it is on the point of losing contact as it reaches D.

Find the greatest possible height of the starting point A of the slide above the level of D, if a child starting from rest at A is not to lose contact with the slide at any point.

Explain briefly whether taking resistances into account would lead to a larger or smaller value for the greatest "safe height" above D. (UCLES)

13. A particle A is free to move on the smooth inner surface of a fixed spherical shell of internal radius a and centre O. Given that A passes through the lowest point of the spherical surface with speed u and that A leaves the surface when OA is inclined at an angle α to the horizontal, and A is above the horizontal through O, show that

$$u^2 = ga(2 + 3 \sin \alpha).$$

Given also that A passes through O before next meeting the surface, show that $\sin^2\alpha = \frac{1}{3}$. (AEB 1993)

14. A light inextensible string of length a has one end attached to a fixed point O and the other end is attached to a particle of mass m.

When the particle is hanging in equilibrium it is given a horizontal velocity of magnitude $2\sqrt{(ga)}$. Obtain an expression for the magnitude of the tension in the string when the particle is still on its circular path and the string has rotated through an angle θ.

Deduce that the tension in the string vanishes when the particle is at a point at a height $\frac{2}{3}a$ above the level of O.

Show that in the subsequent motion the particle passes through a point distant $\frac{9}{16}a$ directly above O. (AEB 1994)

15. A smooth sphere, with centre O and radius a, is fixed with its lowest point on a horizontal table. A particle P of mass m, resting on the sphere at its highest point, is given a horizontal speed $\sqrt{(\frac{1}{4}ga)}$. At the instant when the line OP makes an angle θ with the vertical, the speed of P is v. Show that, while P remains in contact with the sphere,

$$v^2 = \tfrac{1}{4}ga(9 - 8\cos\theta).$$

Find, in terms of m, g and θ, the magnitude of the force exerted by the sphere on P. Hence show that P leaves the surface of the sphere when $\theta = \cos^{-1}\left(\frac{3}{4}\right)$.

After leaving the surface of the sphere, P moves freely under gravity until it strikes the table. Find the magnitude of the horizontal component of the velocity of P just before it strikes the table and show that the magnitude of the vertical component of the velocity of P at this instant is $\frac{7}{8}\sqrt{(5ga)}$. (UCLES)

16. The axis of a smooth circular cylinder of radius a is horizontal. Two particles P of mass m and Q of mass $3m$ are joined together by a light string of length $\frac{1}{2}\pi a$ and placed on the surface of the cylinder in a vertical plane perpendicular to the axis. The point on the axis in the vertical plane through both P and Q is denoted by O. Initially the position of the particles is symmetrical about the vertical plane containing the axis of the cylinder with both OP and OQ inclined at an angle of 45° to the upward vertical through O. The particles are released from rest.

(a) Show, by using the equation of energy or otherwise, that the speed v of the particles when the angle between OQ and the upwards vertical is θ is given by

$$2v^2 = ga\left(2\sqrt{2} - \sin\theta - 3\cos\theta\right).$$

(b) (i) Show that in this position the reaction of the cylinder on Q is

$$\tfrac{3}{2}mg\left(\sin\theta + 5\cos\theta - 2\sqrt{2}\right).$$

(ii) Find, in terms of m, g, a and θ, the reaction of the cylinder on P and the tension in the string.

(c) Show, assuming that Q leaves the cylinder before P, that it leaves the cylinder before $\theta = 70°$. (UODLE)

14 Momentum and impulse

Momentum

The momentum of a body of mass m, having a velocity v, is mv. If the units of mass and velocity are kg and $\mathrm{m\,s^{-1}}$ respectively, then the units of momentum are newton-seconds (N s). There is no named unit for momentum in the way that there is for force (newton) and energy (joulc). Since the momentum of a body depends upon the *velocity* which which the body is moving, momentum is a vector quantity.

Example 1

Find the magnitude of the momentum of
(a) a cricket ball of mass 420 g thrown at $20\,\mathrm{m\,s^{-1}}$
(b) a steam-roller of mass 6 tonnes moving at $0{\cdot}4\,\mathrm{m\,s^{-1}}$.

(a)
$$\text{momentum} = \text{mass} \times \text{velocity}$$
$$\text{magnitude of momentum} = \tfrac{420}{1000} \times 20 = 8{\cdot}4\,\mathrm{N\,s}$$

The magnitude of the momentum is 8·4 N s.

(b)
$$\text{momentum} = \text{mass} \times \text{velocity}$$
$$\text{magnitude of momentum} = (6 \times 1000) \times \tfrac{4}{10} = 2400\,\mathrm{N\,s}$$

The magnitude of the momentum is 2400 N s.

Changes in momentum

If the velocity of a body changes from u to v, then its momentum also changes. The change in momentum can be found by considering the initial momentum mu and the final momentum mv.

Example 2

Find the change in the momentum of a body of mass 2 kg when its speed changes from:
(a) $6\,\mathrm{m\,s^{-1}}$ to $15\,\mathrm{m\,s^{-1}}$ in the same direction
(b) $5\,\mathrm{m\,s^{-1}}$ to $3\,\mathrm{m\,s^{-1}}$ in the opposite direction.

(a) Draw two diagrams.

Taking velocities to the right as positive, we find:

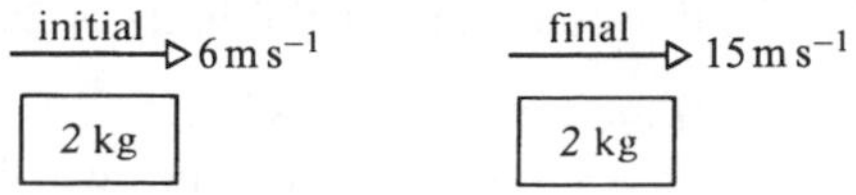

$$\text{initial momentum} = 2 \times 6 = 12\,\mathrm{N\,s}$$
$$\text{final momentum} = 2 \times 15 = 30\,\mathrm{N\,s}$$

Thus the change in the momentum is 18 N s.

(b) Draw two diagrams.

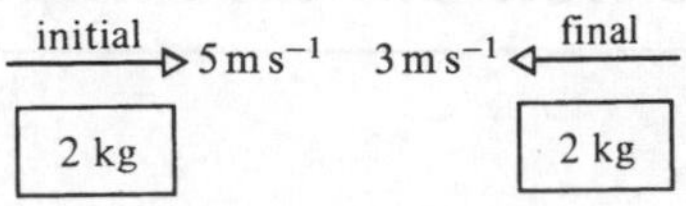

Taking velocities to the right as positive, we find:

$$\text{initial moment} = 2 \times 5 = 10\,\text{N s}$$
$$\text{final momentum} = 2 \times (-3) = -6\,\text{N s}$$

Thus the change in momentum is 16 N s.

Impulse

The impulse of a constant force F is defined as $F \times t$, where t is the time for which the force is acting.

$$\text{impulse} = F \times t$$

But $\quad F = m \times a$

$$\therefore \quad \text{impulse} = ma \times t$$
$$= mv - mu \qquad (\text{from} \quad v = u + at)$$
$$= \text{change in momentum}$$

Thus: impulse of a force = change in momentum produced.

It should be noted that in the case of a hammer hitting a nail, or a batsman hitting a ball, the force involved is large but it only acts for a short time. In other cases the force may continue to act for a much greater length of time.

Example 3

A body of mass 4 kg is initially at rest on a smooth horizontal surface. A horizontal force of 3·5 N acts on the body for 8 seconds. Find:
(a) the magnitude of the impulse given to the body
(b) the magnitude of the final momentum of the body
(c) the final speed of the body.

(a) Draw two diagrams.

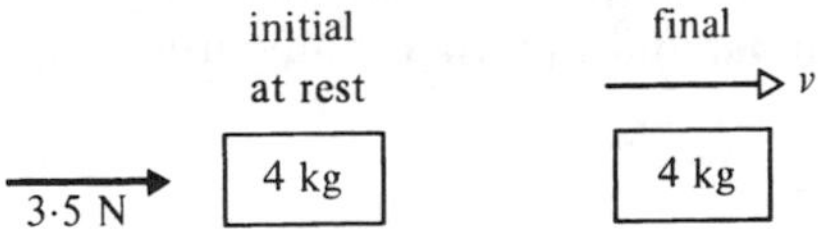

$$\text{impulse} = \text{force} \times \text{time}$$
$$= 3{\cdot}5 \times 8 = 28\,\text{N s}$$

The impulse given to the body is 28 N s.

(b) Body initially at rest, so initial momentum is zero.

$$\text{impulse} = \text{change in momentum}$$
$$= \text{final momentum} - \text{initial momentum}$$
$$\therefore \quad 28 = \text{final momentum} - 0$$

The final momentum of the body is 28 N s.

(c) $\quad \text{final momentum} = 28\,\text{N s}$

$$\therefore \quad 4 \times v = 28 \quad \text{or} \quad v = 7\,\text{m s}^{-1}$$

The final speed of the body is $7\,\text{m s}^{-1}$

Example 4

A body of mass 3 kg is initially moving with a constant velocity of $15\mathbf{i}\,\text{m s}^{-1}$. It experiences a force of $8\mathbf{i}\,\text{N}$ for 6 seconds. Find:

(a) the impulse given to the body by the force

(b) the velocity of the body when the force ceases to act.

(a)

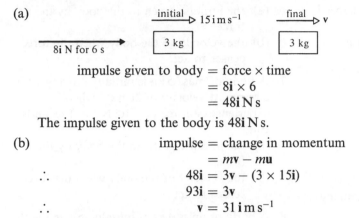

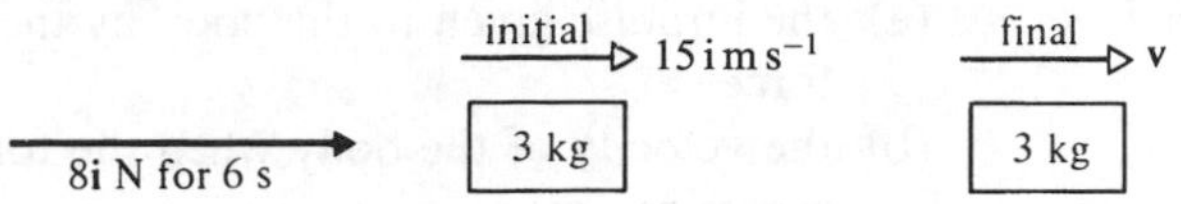

$$\begin{aligned}\text{impulse given to body} &= \text{force} \times \text{time}\\ &= 8\mathbf{i} \times 6\\ &= 48\mathbf{i}\,\text{N s}\end{aligned}$$

The impulse given to the body is $48\mathbf{i}\,\text{N s}$.

(b)

$$\begin{aligned}\text{impulse} &= \text{change in momentum}\\ &= m\mathbf{v} - m\mathbf{u}\\ \therefore \quad 48\mathbf{i} &= 3\mathbf{v} - (3 \times 15\mathbf{i})\\ 93\mathbf{i} &= 3\mathbf{v}\\ \therefore \quad \mathbf{v} &= 31\mathbf{i}\,\text{m s}^{-1}\end{aligned}$$

The final velocity of the body is $31\mathbf{i}\,\text{m s}^{-1}$.

Example 5

A force of $(2\mathbf{i} + 7\mathbf{j})\,\text{N}$ acts on a body of mass 5 kg for 10 seconds. The body was initially moving with constant velocity of $(\mathbf{i} - 2\mathbf{j})\,\text{m s}^{-1}$. Find the final velocity of the body in vector form, and hence obtain its final speed.

Let $\mathbf{v}$ be the final velocity.

$$\begin{aligned}\text{impulse} &= \text{change in momentum}\\ &= m\mathbf{v} - m\mathbf{u}\\ \therefore \quad (2\mathbf{i} + 7\mathbf{j}) \times 10 &= 5\mathbf{v} - 5(\mathbf{i} - 2\mathbf{j})\\ \therefore \quad 25\mathbf{i} + 60\mathbf{j} &= 5\mathbf{v}\\ \text{so} \quad \mathbf{v} &= 5\mathbf{i} + 12\mathbf{j}\end{aligned}$$

The final velocity is $(5\mathbf{i} + 12\mathbf{j})\,\text{m s}^{-1}$.

$$\begin{aligned}\text{speed of body} &= \sqrt{5^2 + 12^2}\\ &= 13\,\text{m s}^{-1}\end{aligned}$$

The final speed is $13\,\text{m s}^{-1}$.

Example 6

A body of mass 3 kg is moving with velocity $(5\mathbf{i} - 2\mathbf{j})\,\text{m s}^{-1}$ when it is given an impulse of $(6\mathbf{i} + 9\mathbf{j})\,\text{N s}$. Find the velocity of the body after the impulse has been applied.

Let $\mathbf{v}$ be the final velocity.

$$\begin{aligned}\text{impulse} &= \text{change in momentum}\\ &= m\mathbf{v} - m\mathbf{u}\\ \therefore \quad 6\mathbf{i} + 9\mathbf{j} &= 3\mathbf{v} - 3(5\mathbf{i} - 2\mathbf{j})\\ \therefore \quad 21\mathbf{i} + 3\mathbf{j} &= 3\mathbf{v}\\ \text{so} \quad \mathbf{v} &= 7\mathbf{i} + \mathbf{j}\end{aligned}$$

The final velocity is $(7\mathbf{i} + \mathbf{j})\,\text{m s}^{-1}$.

Exercise 14A

1. Find the magnitude of the momentum possessed by:
   (a) a car of mass 700 kg moving with a speed of 20 m s^{-1}
   (b) a ball of mass 200 g moving with a speed of 15 m s^{-1}
   (c) a man of mass 75 kg running with a speed of 4 m s^{-1}
   (d) a train of mass 200 tonnes moving with a speed of 18 m s^{-1}
   (e) a bullet of mass 15 g moving with a speed of 400 m s^{-1}.

2. Find the change in the momentum of a body of mass 5 kg when
   (a) its speed changes from 4 m s^{-1} to 7 m s^{-1} without its direction of motion changing
   (b) its speed changes from 2 m s^{-1} to 12 m s^{-1} without its direction of motion changing
   (c) its speed changes from 4 m s^{-1} to 7 m s^{-1} and its direction of motion is reversed
   (d) its speed changes from 2 m s^{-1} to 12 m s^{-1} and its direction of motion is reversed.

3. A body of mass 2 kg is initially at rest on a smooth horizontal surface. A horizontal force of 6 N acts on the body for 3 seconds. Find:
   (a) the magnitude of the impulse given to the body
   (b) the magnitude of the final momentum of the body
   (c) the final speed of the body.

4. A body is initially at rest on a smooth horizontal surface when a horizontal force of 7 N acts on it for 5 seconds. Find the magnitude of the final momentum of the body.

5. A body of mass 12 kg is initially moving with a constant velocity of 2**i** m s^{-1} when it is acted upon by a force of 6**i** N for 10 seconds. Find:
   (a) the impulse given to the body by the force
   (b) the velocity of the body when the force ceases to act.

6. A body of mass 5 kg is initially moving with a constant velocity of 8**i** m s^{-1} when it experiences a force of −10**i** N for 2 seconds. Find:
   (a) the impulse given to the body by the force
   (b) the velocity of the body when the force ceases to act.

7. A body of mass 5 kg is initially moving with a constant velocity of 2**i** m s^{-1} when it experiences a force of −10**i** N for 2 seconds. Find:
   (a) the impulse given to the body by the force
   (b) the velocity of the body when the force ceases to act.

8. A body of mass 4 kg is initially moving with a constant velocity of 2**i** m s^{-1}. After a force of a**i** N has acted on the body for 5 seconds, the body has velocity 12**i** m s^{-1}. Find:
   (a) the initial momentum of the body
   (b) the momentum of the body at the end of the 5 seconds
   (c) the value of a.

9. A body of mass 6 kg is initially moving with a constant velocity of 9**i** m s^{-1}. After a force of b**i** N has acted on the body for 9 seconds, the body has velocity −3**i** m s^{-1}. Find the value of b.

10. A body of mass 5 kg is initially moving with a constant velocity of 7**i** m s^{-1}. After a force of 10**i** N has acted on the body for t seconds, the body has a velocity of 13**i** m s^{-1}. Find the value of t.

11. A body of mass 500 g is initially moving with a velocity of 8**i** m s^{-1}. After a force of −2**i** N has acted on the body for t seconds, the body has the same speed as before but its direction of motion has been reversed. Find the value of t.

12. A body of mass 4 kg, initially at rest on a smooth horizontal surface, experiences a force of (5**i** − 12**j**) N for 2 seconds. Find the final velocity of the body in vector form and hence obtain its final speed.

13. A body of mass 10 kg is initially moving with constant velocity $(2\mathbf{i} - 7\mathbf{j})\,\text{m s}^{-1}$. The body experiences a force of $(5\mathbf{i} + 10\mathbf{j})$ N for 4 seconds.
Find the final velocity of the body in vector form and hence obtain its final speed.

14. A body of mass 750 g is initially moving with constant velocity $(4\mathbf{i} + 6\mathbf{j})\,\text{m s}^{-1}$. After a force of $(a\mathbf{i} + b\mathbf{j})$ N has acted on the body for 6 seconds, the body has a velocity of $(12\mathbf{i} - 10\mathbf{j})\,\text{m s}^{-1}$.
Find the values of a and b.

15. A body of mass 2 kg is moving with velocity $(2\mathbf{i} - 3\mathbf{j})\,\text{m s}^{-1}$ when it is given an impulse of $(6\mathbf{i} + 2\mathbf{j})$ N s.
Find the velocity of the body after the impulse has been applied.

16. A body of mass 6 kg is moving with velocity $(-3\mathbf{i} + 5\mathbf{j})\,\text{m s}^{-1}$ when it is given an impulse of $(12\mathbf{i} - 18\mathbf{j})$ N s.
Find the velocity of the body after the impulse has been applied.

17. A body of mass 3 kg is moving with velocity $(\mathbf{i} + 4\mathbf{j})\,\text{m s}^{-1}$ when an impulse is applied. The impulse causes its velocity to change to $(3\mathbf{i} + 2\mathbf{j})\,\text{m s}^{-1}$.
Find the impulse.

18. A body of mass 200 g is moving with velocity $(40\mathbf{i} - 30\mathbf{j})\,\text{m s}^{-1}$ when an impulse is applied. The impulse causes its velocity to change to $(10\mathbf{i} + 15\mathbf{j})\,\text{m s}^{-1}$.
Find the impulse.

19. Each part of this questions involves a smooth sphere colliding normally with a fixed vertical wall. The diagrams show the situations before and after collision. For each part take $\mathbf{i}$ as a unit vector in the direction → and find in vector form:
(i) the impulse given to the ball by the wall
(ii) the impulse given to the wall by the ball.

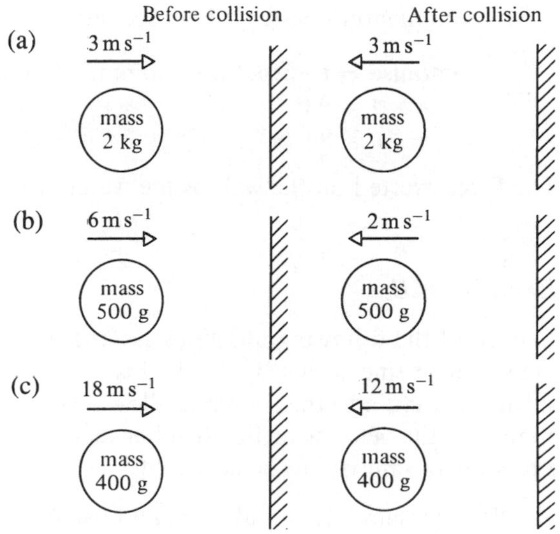

Force on a surface

When a jet of water strikes a surface and does not rebound, the surface destroys the momentum of the jet of water. The loss in momentum is equal to the impulse of the force exerted on the water by the surface. By Newton's Third Law, this force is equal in magnitude but opposite in direction to the force exerted on the surface by the water.
If the amount of water hitting the surface per second is known, then the momentum destroyed per second can be calculated.

Since impulse = momentum destroyed
and also impulse = force × time
force × 1 = momentum destroyed per second
or force = momentum destroyed per second

If the water does rebound from the surface, the direction of motion of the jet is reversed and therefore the change in momentum is increased.

Example 7

Water issues horizontally from a pipe and strikes a nearby wall without rebounding. The cross-sectional area of the pipe is $6\,\text{cm}^2$ and the water travels at $4\,\text{m}\,\text{s}^{-1}$. Find the magnitude of the force acting on the wall. (Take density of water as $1000\,\text{kg}\,\text{m}^{-3}$.)

$$\text{volume of water hitting wall in 1 second} = \frac{6}{(100)^2} \times 4\,\text{m}^3$$
$$\text{mass of water hitting wall in 1 second} = \text{volume of water} \times \text{density}$$

Since the wall is nearby, it may be assumed that there is no significant difference between the velocity of the water as it leaves the pipe and when it hits the wall.

$$\text{momentum destroyed per second} = \frac{6}{(100)^2} \times 4 \times 1000 \times 4 = 9{\cdot}6\,\text{N}\,\text{s}$$
$$\text{impulse} = \text{force} \times \text{time} = \text{momentum destroyed}$$
$$\therefore \quad F \times 1 = 9{\cdot}6$$
$$\therefore \quad F = 9{\cdot}6\,\text{N}$$

The force exerted on the wall by the water is 9·6 N.

Exercise 14B

In each of the following questions assume that the surface being struck by the fluid is sufficiently close to the source to allow any change in the velocity of the fluid between emission and impact to be neglected.

1. Water issues horizontally from a hose of cross-sectional area $5\,\text{cm}^2$ at a speed of $10\,\text{m}\,\text{s}^{-1}$. The water hits a vertical wall without rebounding.
   Find the magnitude of the force acting on the wall. (Take the density of water as $1000\,\text{kg}\,\text{m}^{-3}$.)
2. Water issues horizontally from a pipe of cross-sectional area $0{\cdot}1\,\text{m}^2$ at a speed of $8\,\text{m}\,\text{s}^{-1}$. The water hits a vertical wall without rebounding.
   Find the magnitude of the force acting on the wall.
   (Take the density of water as $1000\,\text{kg}\,\text{m}^{-3}$.)
3. A water cannon emits water at a speed of $16\,\text{m}\,\text{s}^{-1}$ through a pipe of cross-sectional area $0{\cdot}015\,\text{m}^2$.
   If the water hits a nearby vertical surface, find the force on the surface, assuming that the water does not rebound after impact. (Take the density of water as $1000\,\text{kg}\,\text{m}^{-3}$.)
4. During a water pistol fight, Mike squirts water into the face of Jack. The water is emitted through a nozzle of cross-sectional area $1\,\text{mm}^2$ and travels horizontally with a speed of $8\,\text{m}\,\text{s}^{-1}$ to hit Jack's face at right angles without rebounding.
   Find the magnitude of the force acting on Jack's face. (Take the density of water as $1000\,\text{kg}\,\text{m}^{-3}$.)
5. A factory waste pipe emits fluid of density $1200\,\text{kg}\,\text{m}^{-3}$ through a pipe of cross-sectional area $25\,\text{cm}^2$. The fluid leaves the pipe horizontally with a speed of $5\,\text{m}\,\text{s}^{-1}$ and strikes a vertical wall without rebounding. Find the magnitude of the force acting on the wall.
6. In order to wash a wound in a person's arm a nurse squirts saline solution, density $1{\cdot}03\,\text{g}\,\text{cm}^{-3}$, onto the wound from a syringe. The needle of the syringe emits the solution horizontally with a speed of $4\,\text{m}\,\text{s}^{-1}$ through a hole of cross-sectional area $2{\cdot}5\,\text{mm}^2$. The horizontal jet of solution hits the vertical surface of the wound and does not rebound. Find the magnitude of the force acting on the surface of the wound.

Impact

When a collision occurs between two bodies A and B, the force exerted on B by A is equal and opposite to the force exerted on A by B. This is another example of Newton's Third Law. In the absence of any other forces acting on the two bodies, the changes in the momenta of A and B will be equal in magnitude, but opposite in direction. The gain in the momentum of one body will equal the loss in momentum of the other body; hence, the sum of the momenta of A and B before the impact will be equal to the sum of their momenta after the impact. This is referred to as the *Principle of Conservation of Linear Momentum.*

In dealing with examples on collision it is usually advisable to draw two diagrams; one showing the situation before the collision and the other showing the situation after the collision.

Since momentum is a vector quantity, the direction of motion of each body must be carefully and clearly indicated so that the correct sign may be attached to the momentum. In some instances after colliding, two bodies are said to coalesce. In such instances the bodies do not rebound from each other, but they have a common velocity after the collision and move as a single body.

Example 8

A body of mass 2 kg moving on a smooth horizontal surface at $3\,\mathrm{m\,s^{-1}}$, collides with a second body of mass 1 kg which is at rest. After the collision the bodies coalesce. Find the common speed of the bodies after impact.

Let v be the common speed of the bodies after the collision.

Draw two diagrams.

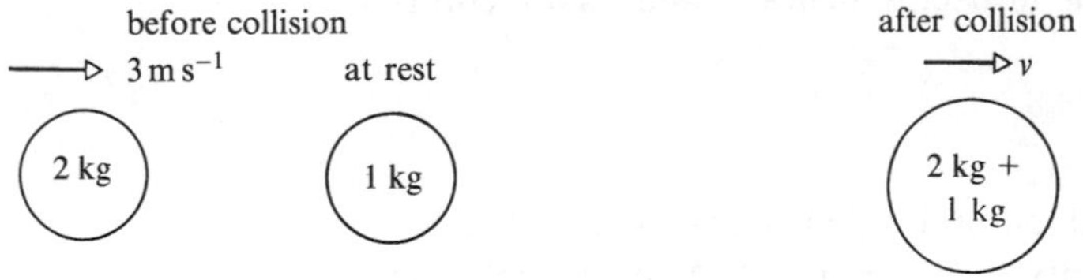

Taking velocities to the right as positive, we find:

$$\text{momentum before collision} = (2 \times 3) + (1 \times 0)$$

$$\text{momentum after collision} = (2 + 1) \times v$$

By the Principle of Conservation of Momentum:

$$6 + 0 = 3v \quad \text{or} \quad v = 2\,\mathrm{m\,s^{-1}}.$$

The common speed of the two bodies after the impact is $2\,\mathrm{m\,s^{-1}}$.

Example 9

The two bodies shown collide on a smooth horizontal surface. Find the value of v, the speed of the lighter body after impact.

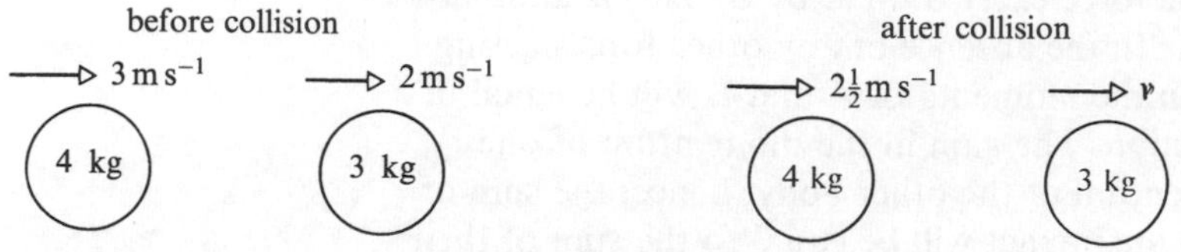

Taking velocities to the right as positive, we find:

$$\text{momentum before collision} = (4 \times 3) + (3 \times 2)$$
$$\text{momentum after collision} = (4 \times 2{\cdot}5) + (3 \times v)$$

By Conservation of Momentum: $12 + 6 = 10 + 3v$ or $v = 2\frac{2}{3}\ \text{m s}^{-1}$

After the impact the lighter body has a speed of $2\frac{2}{3}\ \text{m s}^{-1}$.

Example 10

The two bodies shown collide on a horizontal surface. Find the speed v of the lighter body after impact.

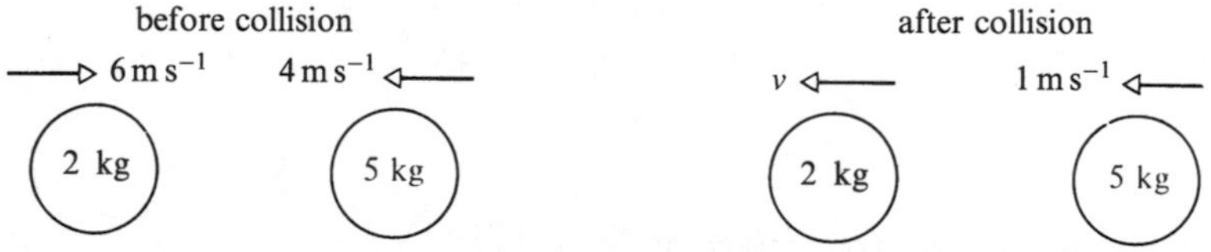

Taking velocities to the right as positive, we find:

$$\text{momentum before collision} = (2 \times 6) + (5 \times (-4))$$
$$\text{momentum after collision} = (2 \times (-v)) + (5 \times (-1))$$

By Conservation of Momentum: $12 - 20 = -2v - 5$ or $2v - 3$

so $v = 1\frac{1}{2}\ \text{m s}^{-1}$

The speed of the lighter body after the impact is $1\frac{1}{2}\ \text{m s}^{-1}$, and its direction of motion is reversed.

Example 11

A body of mass 3 kg, moving with velocity $(2\mathbf{i} - \mathbf{j})\ \text{m s}^{-1}$, collides with a body of mass 4 kg, moving with velocity $(-5\mathbf{i} + 6\mathbf{j})\ \text{m s}^{-1}$. After the collision the two bodies coalesce. Find the common velocity of the bodies after the impact.

Let the common velocity of the bodies after the impact be $\mathbf{v}\ \text{m s}^{-1}$.

$$\text{momentum before collision} = 3(2\mathbf{i} - \mathbf{j}) + 4(-5\mathbf{i} + 6\mathbf{j})$$
$$= -14\mathbf{i} + 21\mathbf{j}$$
$$\text{momentum after collision} = 7\mathbf{v}$$

By Conservation of Momentum: $-14\mathbf{i} + 21\mathbf{j} = 7\mathbf{v}$

so $\mathbf{v} = -2\mathbf{i} + 3\mathbf{j}$

The common velocity of the two bodies after impact is $(-2\mathbf{i} + 3\mathbf{j})\ \text{m s}^{-1}$.

Loss of energy

When bodies collide, there is no loss of momentum but there is a loss of kinetic energy. Some of the kinetic energy possessed by the bodies is transformed into other forms of energy at the impact, e.g. heat and sound energy.

Example 12

Two smooth spheres A and B, of masses 150 g and 350 g, are travelling towards each other along the same horizontal line with speeds of $4\,\text{m s}^{-1}$ and $2\,\text{m s}^{-1}$ respectively. After the collision, the direction of motion of B has been reversed and it is travelling at a speed of $1\,\text{m s}^{-1}$. Find the speed of A after the collision and the loss of kinetic energy due to the collision.

Draw two diagrams showing the situation before and after the collision.

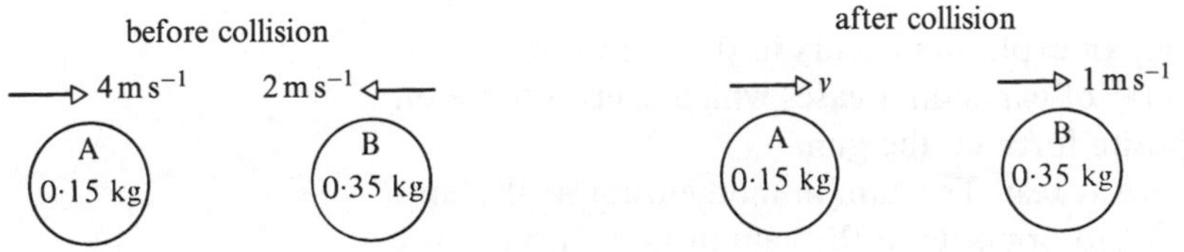

Taking velocities to the right as positive, we find:

$$\text{momentum before collision} = 0{\cdot}15(4) + 0{\cdot}35(-2)$$
$$\text{momentum after collision} = 0{\cdot}15(v) + 0{\cdot}35(1)$$

By Conservation of Momentum: $0{\cdot}6 - 0{\cdot}7 = 0{\cdot}15v + 0{\cdot}35$ or $v = -3\,\text{m s}^{-1}$

Thus the speed of A after the impact is $3\,\text{m s}^{-1}$. The negative sign indicates that the 'after collision' diagram shows the incorrect direction of motion of sphere A.

$$\text{loss of kinetic energy} = \tfrac{1}{2}(0{\cdot}15)4^2 + \tfrac{1}{2}(0{\cdot}35)2^2 - [\tfrac{1}{2}(0{\cdot}15)3^2 + \tfrac{1}{2}(0{\cdot}35)1^2]$$
$$= 1{\cdot}05\,\text{J}$$

The speed of A after impact is $3\,\text{m s}^{-1}$ and the loss of kinetic energy is $1{\cdot}05\,\text{J}$.

Example 13

A body A, of mass 5 kg, is moving with velocity $(-4\mathbf{i} + 3\mathbf{j})\,\text{m s}^{-1}$ when it collides with a body B, of mass 2 kg, moving with velocity $(3\mathbf{i} - \mathbf{j})\,\text{m s}^{-1}$. Immediately after the collision the velocity of A is $(-2\mathbf{i} + \mathbf{j})\,\text{m s}^{-1}$. Find:

(a) the velocity of B after the collision
(b) the loss of kinetic energy of the system due to the collision.

(a) Let the velocity of B after the impact be $\mathbf{v}\,\text{m s}^{-1}$.

$$\text{momentum before collision} = 5(-4\mathbf{i} + 3\mathbf{j}) + 2(3\mathbf{i} - \mathbf{j})$$
$$= -14\mathbf{i} + 13\mathbf{j}$$
$$\text{momentum after collision} = 5(-2\mathbf{i} + \mathbf{j}) + 2\mathbf{v}$$

By Conservation of Momentum: $-14\mathbf{i} + 13\mathbf{j} = -10\mathbf{i} + 5\mathbf{j} + 2\mathbf{v}$

$\therefore$ $-4\mathbf{i} + 8\mathbf{j} = 2\mathbf{v}$

so $\mathbf{v} = -2\mathbf{i} + 4\mathbf{j}$

The velocity of B after the impact is $(-2\mathbf{i} + 4\mathbf{j})\,\text{m s}^{-1}$.

(b)
$$\text{initial KE} = \tfrac{1}{2} \times 5 \times (4^2 + 3^2) + \tfrac{1}{2} \times 2 \times (3^2 + 1^2)$$
$$= 72{\cdot}5\text{ J}$$
$$\text{final KE} = \tfrac{1}{2} \times 5 \times (2^2 + 1^2) + \tfrac{1}{2} \times 2 \times (2^2 + 4^2)$$
$$= 32{\cdot}5\text{ J}$$
$$\text{loss of KE} = 72{\cdot}5 - 32{\cdot}5$$
$$= 40\text{ J}$$

The loss of KE is 40 J.

Recoil of a gun

When a shot is fired from a gun, an explosion occurs in the barrel of the gun. The explosion takes the form of expanding gases which exert a force on the shot and an equal and opposite force on the gun.
Initially, the shot and the gun are at rest. The gain in momentum of the shot after the explosion will be equal and opposite to the gain in momentum of the gun.
In a horizontal direction therefore, the forward momentum of the shot is equal to the backward momentum of the gun.

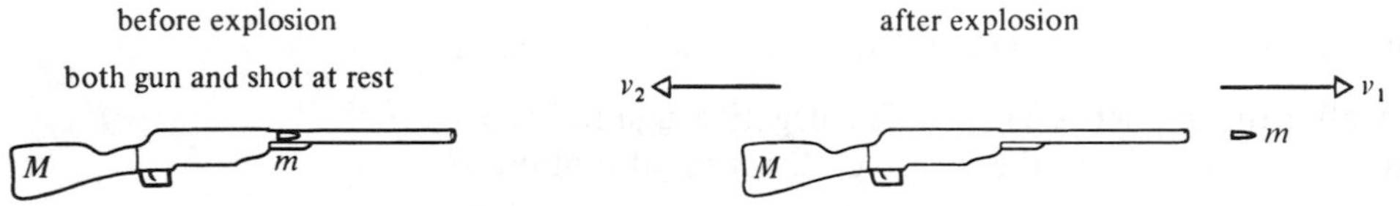

$$\text{momentum before explosion} = M(0) + m(0)$$
$$\text{momentum after explosion} = m(v_1) + M(-v_2)$$

By Conservation of Momentum: $0 = mv_1 - Mv_2$ or $Mv_2 = mv_1$

Example 14

A bullet is fired from a gun with a horizontal velocity of 400 m s^{-1}. The mass of the gun is 3 kg and the mass of the bullet is 60 g. Find the initial speed of recoil of the gun and the gain in the kinetic energy of the system.

Draw two diagrams.

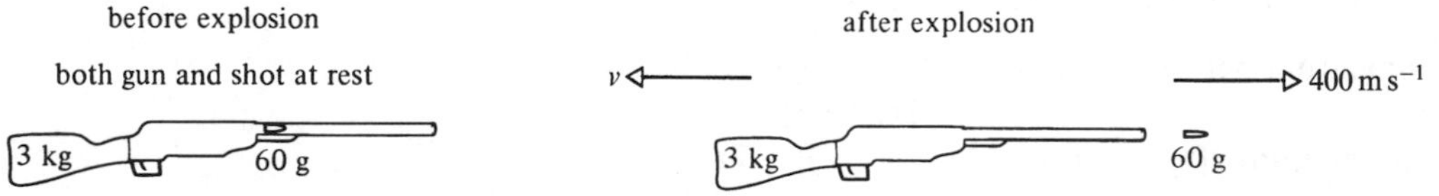

Taking velocities to the right as positive

$$\text{momentum before explosion} = 3(0) + (0{\cdot}06)(0)$$
$$\text{momentum after explosion} = 3(-v) + (0{\cdot}06)(400)$$

By Conservation of Momentum: $0 = -3v + 24$

$\therefore \quad v = 8\,\mathrm{m\,s^{-1}}$

$$\text{initial KE} = \tfrac{1}{2}(3)(0)^2 + \tfrac{1}{2}(0{\cdot}06)(0)^2$$
$$\text{final KE} = \tfrac{1}{2}(3)(8)^2 + \tfrac{1}{2}(0{\cdot}06)(400)^2$$
$$\text{gain in KE} = \tfrac{1}{2}(192) + \tfrac{1}{2}(9600) - 0 = 4896\,\mathrm{J}$$

The initial speed of recoil of the gun is $8\,\mathrm{m\,s^{-1}}$ and the gain in kinetic energy due to the explosion is 4896 J.

Exercise 14C

In questions **1** to **7**, the two diagrams show the situation before and after a collision between two bodies A and B moving along the same straight line on a smooth horizontal surface. Find the speed v in each case.

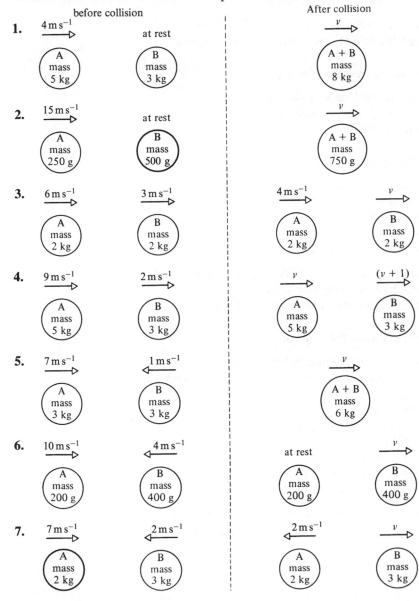

1.

2.
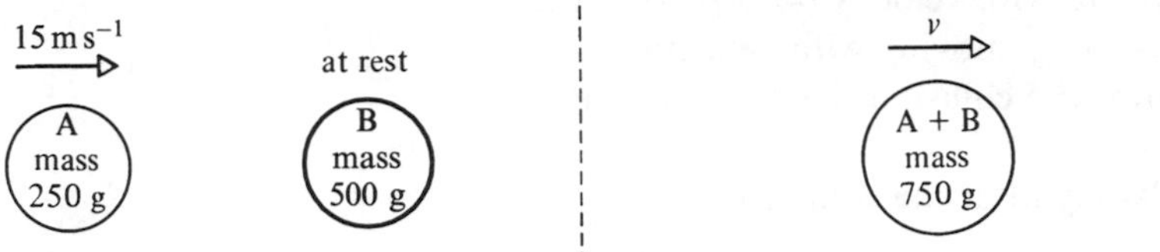

3.

4.

5.

6.
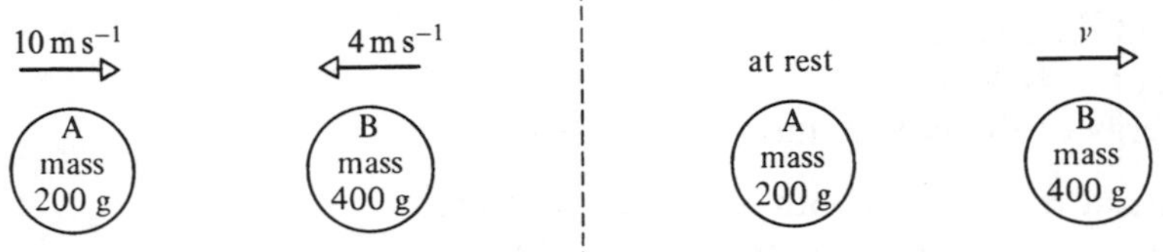

7.
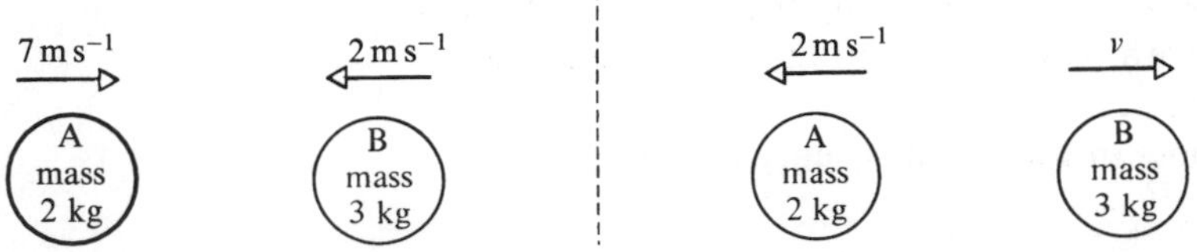

8. Two identical railway trucks are travelling in the same direction along the same straight piece of track with constant speeds of $6\,\mathrm{m\,s^{-1}}$ and $2\,\mathrm{m\,s^{-1}}$. The faster truck catches up with the other one and on collision, the two trucks automatically couple together.
Find the common speed of the trucks after collision.

9. A body of mass 2 kg, moving with velocity $(8\mathbf{i} - \mathbf{j})\,\mathrm{m\,s^{-1}}$, collides with a body of mass 3 kg, moving with velocity $(-2\mathbf{i} + 4\mathbf{j})\,\mathrm{m\,s^{-1}}$. On collision the two bodies coalesce.
Find the common velocity of the bodies after the impact.

10. A body of mass 3 kg, moving with velocity $(2\mathbf{i} - \mathbf{j})\,\mathrm{m\,s^{-1}}$, collides with a body of mass 1 kg, moving with velocity $(6\mathbf{i} + 3\mathbf{j})\,\mathrm{m\,s^{-1}}$. On collision the two bodies coalesce.
Find the common velocity of the bodies after the impact.

11. A body of mass 750 g, moving with velocity $(-\mathbf{i} - 2\mathbf{j})\,\mathrm{m\,s^{-1}}$, collides with a body of mass 500 g, moving with velocity $(9\mathbf{i} + 8\mathbf{j})\,\mathrm{m\,s^{-1}}$. On collision the two bodies coalesce.
Find the common velocity of the bodies after the impact.

12. A body P, of mass 2 kg, is moving with velocity $(4\mathbf{i} + \mathbf{j})\,\mathrm{m\,s^{-1}}$ when it collides with a body Q, of mass 3 kg, moving with velocity $(5\mathbf{i} - 2\mathbf{j})\,\mathrm{m\,s^{-1}}$. Immediately after the collision the velocity of P is $(\mathbf{i} - 2\mathbf{j})\,\mathrm{m\,s^{-1}}$.
Find the velocity of Q immediately after the collision.

13. A body X, of mass 5 kg, is moving with velocity $(-12\mathbf{i} + 6\mathbf{j})\,\mathrm{m\,s^{-1}}$ when it collides with a body Y, of mass 4 kg, moving with velocity $(10\mathbf{i} - 5\mathbf{j})\,\mathrm{m\,s^{-1}}$. Immediately after the collision the velocity of X is $(-8\mathbf{i} + 4\mathbf{j})\,\mathrm{m\,s^{-1}}$.
Find the velocity of Y immediately after the collision.

14. A body S, of mass 3 kg, is moving with velocity $(-2\mathbf{i} + 5\mathbf{j})\,\mathrm{m\,s^{-1}}$ when it collides with a body T, of mass 5 kg, moving with velocity $(3\mathbf{i} - 6\mathbf{j})\,\mathrm{m\,s^{-1}}$. Immediately after the collision the velocity of S is $(3\mathbf{i} - 5\mathbf{j})\,\mathrm{m\,s^{-1}}$.
Find the velocity of T immediately after the collision.

15. A particle A of mass 300 g lies at rest on a smooth horizontal surface. A second particle B of mass 200 g is projected along the surface with speed $6\,\mathrm{m\,s^{-1}}$ and collides directly with A.
If the collision reduces B to rest, find:
(a) the speed with which A moves after the collision
(b) the initial kinetic energy of the system
(c) the final kinetic energy of the system
(d) the loss in the kinetic energy of the system during the collision.

16. A particle A of mass 150 g lies at rest on a smooth horizontal surface. A second particle B of mass 100 g is projected along the surface with speed $u\,\mathrm{m\,s^{-1}}$ and collides directly with A. On collision the masses coalesce and move on with speed $4\,\mathrm{m\,s^{-1}}$.
Find the value of u and the loss in the kinetic energy of the system during impact.

17. Two smooth spheres A and B, of equal radii and masses 3 kg and $1\frac{1}{2}$ kg respectively, are travelling along the same horizontal line. The velocities of A and B are $6\mathbf{i}\,\text{m s}^{-1}$ and $-2\mathbf{i}\,\text{m s}^{-1}$ respectively. The spheres collide and after collision B has a velocity of $4\mathbf{i}\,\text{m s}^{-1}$. Find the velocity of A after the collision.

18. Two smooth spheres A and B, of equal radii and masses 180 g and 100 g respectively, are travelling directly towards each other along a horizontal path. The initial speeds of A and B are $2\,\text{m s}^{-1}$ and $6\,\text{m s}^{-1}$ respectively. After collision both spheres have reversed their original directions of motion and B now has a speed of $3\,\text{m s}^{-1}$. Find the speed of A after impact and the loss in the kinetic energy of the system.

19. A body of mass 6 kg, moving with velocity $(8\mathbf{i} - 4\mathbf{j})\,\text{m s}^{-1}$ collides with a body of mass 2 kg which is at rest. On collision the two bodies coalesce. Find the common velocity of the bodies after the impact, and the loss of kinetic energy of the system due to the collision.

20. A body of mass 500 g, moving with velocity $(2\mathbf{i} - 4\mathbf{j})\,\text{m s}^{-1}$ collides with a body of mass 1500 g, moving with velocity $(6\mathbf{i} + 8\mathbf{j})\,\text{m s}^{-1}$. On collision the two bodies coalesce. Find the common velocity of the bodies after the impact, and the loss of kinetic energy of the system due to the collision.

21. A body P, of mass 4 kg, is moving with velocity $(2\mathbf{i} + 3\mathbf{j})\,\text{m s}^{-1}$ when it collides with a body Q, of mass 3 kg, moving with velocity $(5\mathbf{i} - 6\mathbf{j})\,\text{m s}^{-1}$. Immediately after the collision the velocity of P is $5\mathbf{i}\,\text{m s}^{-1}$. Find:
(a) the velocity of Q after the collision
(b) the loss of kinetic energy of the system due to the collision
(c) the impulse of P on Q due to the collision.

22. A body A, of mass 2 kg, is moving with velocity $(-2\mathbf{i} + 3\mathbf{j})\,\text{m s}^{-1}$ when it collides with a body B, of mass 5 kg, moving with velocity $(6\mathbf{i} - 10\mathbf{j})\,\text{m s}^{-1}$. Immediately after the collision the velocity of A is $(3\mathbf{i} - 2\mathbf{j})\,\text{m s}^{-1}$. Find:
(a) the velocity of B after the collision
(b) the loss of kinetic energy of the system due to the collision
(c) the impulse of A on B due to the collision.

23. A body X, of mass 250 g, is moving with velocity $(-2\mathbf{i} + 3\mathbf{j})\,\text{m s}^{-1}$ when it collides with a body Y, of mass 750 g, moving with velocity $(5\mathbf{i} + 8\mathbf{j})\,\text{m s}^{-1}$. Immediately after the collision the velocity of X is $(\mathbf{i} + 9\mathbf{j})\,\text{m s}^{-1}$. Find:
(a) the velocity of Y after the collision
(b) the loss of kinetic energy of the system due to the collision
(c) the impulse of X on Y due to the collision.

24. A bullet of mass 20 g is fired from a gun of mass 2·5 kg. The bullet leaves the gun with a speed of $500\,\text{m s}^{-1}$.
Find the initial speed of recoil of the gun and the gain in the kinetic energy of the system.

25. A shell of mass 5 kg is fired from a gun of mass 2000 kg. The shell leaves the gun with a speed of $400\,\text{m}\,\text{s}^{-1}$.
Find the initial speed of recoil of the gun and the gain in the kinetic energy of the system.

26. A wooden stake of mass 4 kg is to be driven vertically downwards into the ground using a mallet of mass 6 kg. The speed of the mallet just prior to impact is $10\,\text{m}\,\text{s}^{-1}$. After impact the mallet remains in contact with the stake (i.e. the weight of both mallet and stake aid penetration). Find the speed with which the stake begins to enter the ground.
If the ground offers a constant resistance to motion of 1000 N, how far will the stake penetrate on each blow? (Take $g = 10\,\text{m}\,\text{s}^{-2}$.)

27. Each part of the this question involves three smooth spheres A, B and C, of equal radii, moving along the same straight line. A collides with B and then B collides with C. The diagrams show the situations before any collision, after A has collided with B, and after B has collided with C. Find the unknown speeds u and v in each case.

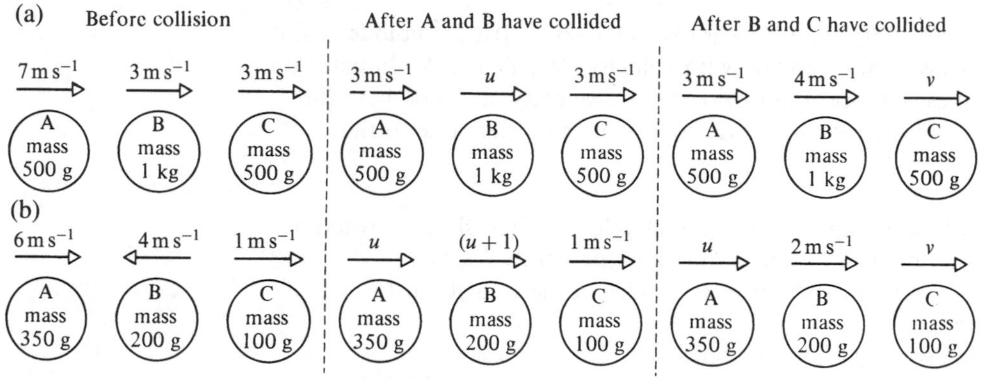

28. Two smooth spheres A and B, of equal radii and masses of 750 g and 1 kg respectively, are initially at rest on a smooth horizontal surface. A is projected directly towards B with speed $5\,\text{m}\,\text{s}^{-1}$ and after collision A has not changed its direction of motion but has a speed of $1\,\text{m}\,\text{s}^{-1}$. The collision sets B into motion and it goes on to strike a fixed wall at right angles, the impact reversing B's direction of motion and halving its speed. B then collides again with A and this collision reduces B to rest. Find the final speed of A and the total loss in kinetic energy due to the collisions.

29. Two smooth spheres A and B, of equal radii and masses m_1 and m_2 respectively, are moving towards each other along the same horizontal line each with speed u. After collision both spheres have reversed their original directions of motion and A now travels with speed $\dfrac{u}{2}$.
Show that $3m_1 > 2m_2$.
Sphere B then strikes a fixed wall at right angles, the impact reversing the direction of motion of B and halving its speed. Show that B will again collide with A provided $3m_1 > 4m_2$.

30. Three smooth spheres A, B and C, of equal radii and masses m_1, m_2 and m_3 respectively, lie at rest in a straight line on a horizontal surface with B between A and C. Sphere B is projected towards C with speed u and C is projected towards B with speed $3u$. The collision between B and C reverses the direction of motion of sphere C which then travels with speed $2u$. Show that B will collide with A provided $m_3 > \frac{m_2}{5}$.
What would be the necessary relationship between m_2 and m_3 for this second collision to occur had A been given an initial speed of u away from the other spheres?

Exercise 14D Examination questions

(Unless otherwise indicated take $g = 9{\cdot}8\,\mathrm{m\,s^{-2}}$ throughout this exercise.)

1. A block of wood, of mass 2 kg, is at rest on a smooth horizontal table. A bullet, of mass 0·1 kg, moving horizontally at a speed of $420\,\mathrm{m\,s^{-1}}$, strikes the block and becomes embedded in it. Find the speed of the block after the impact. (UCLES)

2. A particle of mass 0·05 kg is moving with velocity $1{\cdot}5\mathbf{i}\,\mathrm{m\,s^{-1}}$ when it is given an impulse $\frac{3\sqrt{3}}{40}\mathbf{j}\,\mathrm{N\,s}$, where $\mathbf{i}$ and $\mathbf{j}$ are unit perpendicular vectors. Find the speed of the particle after the impulse has been applied.
After the impulse has been applied, find the angle between the direction of motion of the particle and the vector $\mathbf{i}$. (AEB 1993)

3. A particle of mass 0·2 kg moving with velocity $(4\mathbf{i} + 6\mathbf{j})\,\mathrm{m\,s^{-1}}$ receives an impulse which changes its velocity to $(7\mathbf{i} + 2\mathbf{j})\,\mathrm{m\,s^{-1}}$. Find the impulse and the change in the kinetic energy of the particle. (WJEC)

4. A particle of mass 0·2 kg is moving with velocity $(5\mathbf{i} + 7\mathbf{j})\,\mathrm{m\,s^{-1}}$ when an impulse $\mathbf{J}$ is applied to it so that its velocity becomes $(8\mathbf{i} - 3\mathbf{j})\,\mathrm{m\,s^{-1}}$. Find $\mathbf{J}$ and the kinetic energy of the particle immediately after the impulse has been applied. (AEB 1990)

5. A body A, of mass 2 kg, is moving with velocity $(-2\mathbf{i} + 4\mathbf{j})\,\mathrm{m\,s^{-1}}$ when it collides with a body B, of mass 3 kg, moving with a velocity $(3\mathbf{i} + 4\mathbf{j})\,\mathrm{m\,s^{-1}}$. During the collision the two bodies coalesce.
(a) Find the velocity of the combined body immediately after the collision, in terms of $\mathbf{i}$ and $\mathbf{j}$.
(b) Calculate the loss of kinetic energy as a result of the collision.
(c) Find the impulse exerted on the body B by the body A in the collision. (UODLE)

6. A toy train consists of an engine of mass 0·2 kg and a carriage of mass 0·1 kg. While travelling along a straight section of track at a constant speed of $0{\cdot}6\,\mathrm{m\,s^{-1}}$ the engine collides with the stationary carriage. After the collision the engine and the carriage continue along the track together.
(i) Find the speed with which the combined engine and carriage move after the collision.
(ii) Find the impulse exerted by the engine on the carriage. (NICCEA)

7. A van of mass 1200 kg and a lorry of mass 3200 kg collide. Just before the crash they are moving directly towards each other, and each has speed $12\,\text{m s}^{-1}$. Immediately after the crash they move with the same velocity. Making a suitable assumption, which should be stated, find the loss in kinetic energy. (UCLES)

8. (a) A child projects a toy railway truck P along a smooth horizontal straight track so that it collides with and couples to an identical truck Q which is stationary and free to move. Show that the speed of P is halved by the impact.
The pair of trucks subsequently collides with a third identical truck, which is stationary before the impact, and all three trucks then move together. Show that the total kinetic energy of the three trucks after the second impact is one third of the original kinetic energy of P.

(b) A stationary ship of mass $1{\cdot}5 \times 10^7$ kg is jerked into motion when the cable which connects it to a tug becomes taut. Given that the impulse imparted by the cable is $7{\cdot}5 \times 10^5$ N s, calculate the speed with which the ship starts to move.
Given that the cable is taut for 1·25 seconds, calculate the tension in the cable whilst it is taut, assuming this tension to be constant.
(UCLES)

9. Particles A and B have masses kM and M respectively, where k is a positive constant. The particles are moving in the same direction, along the same straight line, with speeds $2u$ and u respectively. There is a collision between A and B, which halves the speed of A. After the collision, both A and B continue to move in the same direction as before.
(a) Find, in terms of k and u, the speed of B after the collision.
In the collision, one-twelfth of the total kinetic energy is lost.
(b) Find the possible values of k. (ULEAC)

10. (Take $g = 10\,\text{m s}^{-2}$ in this question.)
The diagram shows two particles, A and B, on a rough horizontal surface, where the coefficient of friction between each particle and the surface is 0·1. The masses of A and B are 2 kg and 0·5 kg respectively. Particle A is 4·5 m from B and is given an initial speed of $5\,\text{m s}^{-1}$ towards B.

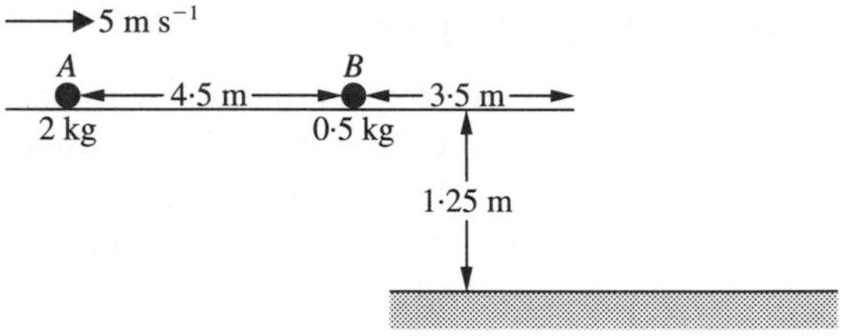

(i) Find the speed of A immediately before impact.
On impact the two particles coalesce.
(ii) Show that the speed of the combined particles immediately after impact is $3{\cdot}2\,\text{m s}^{-1}$.
The combined particles then travel 3·5 m to the edge of the surface, which is 1·25 m above the horizontal ground. Find:
(iii) the speed with which the combined particles reach the edge of the surface
(iv) the horizontal distance from the edge of the surface at which the combined particles strike the ground. (UCLES)

11. (Take $g = 10\,\text{m}\,\text{s}^{-2}$ in this question.)
Particles P, of mass 0·3 kg and Q, of mass 0·2 kg, are attached to the ends of a light inextensible string. The string passes over a small fixed smooth peg which is at a height of 1 m above the horizontal floor. Initially the system is held at rest with the string taut, so that Q is on the floor and P is at a height of 0·5 m above the floor (see diagram). The system is released. Find the acceleration of P and the tension in the string while the particles are in motion.
When P hits the floor it does not rebound. Find the magnitude of the impulse exerted by the floor on P in this impact.
At the instant when P hits the floor, the string becomes detached from P. Find the greatest height above the floor reached by Q, and find also the speed of Q when it returns to the floor.
When Q hits the floor it does not rebound. Sketch the velocity–time graph for the whole of the motion of Q. (UCLES)

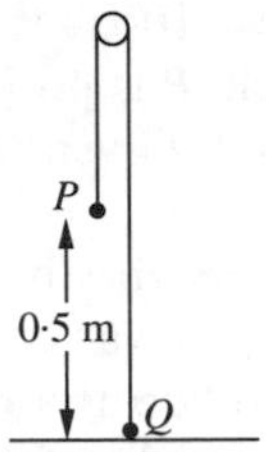

12. A small smooth pulley O is fixed at a height $4a$ above a horizontal table. Masses m and M ($m < M$) are fastened to the ends of a light inextensible string of length $7a$, and the string passes over the pulley. At time $t = 0$, when the mass M is resting on the table vertically below O, the mass m is released from rest at O. Assuming that the length of the string in contact with the pulley is negligible, find, in terms of m, M, a and g, the speed of the mass M when it is jerked into motion.
Determine the ratio m/M so that the mass m may just reach the table.
Given that $m/M = 1/2$, find, in terms of a and g, the time at which the mass M first returns to the table. (AEB 1993)

13. The figure shows a particle A of mass $10\,m$ on a smooth plane inclined at an angle θ to the horizontal, where $\sin\theta = 0{\cdot}6$. The particle is attached to one end of a light inelastic string which passes over a smooth pulley P at the highest point of the plane. A particle B of mass $10\,m$ is attached to the other end of the string. Initially the system is held at rest with the string parallel to a line of greatest slope of the plane but not in contact with the plane and BP vertical, with B being at a distance d above an inelastic horizontal floor.

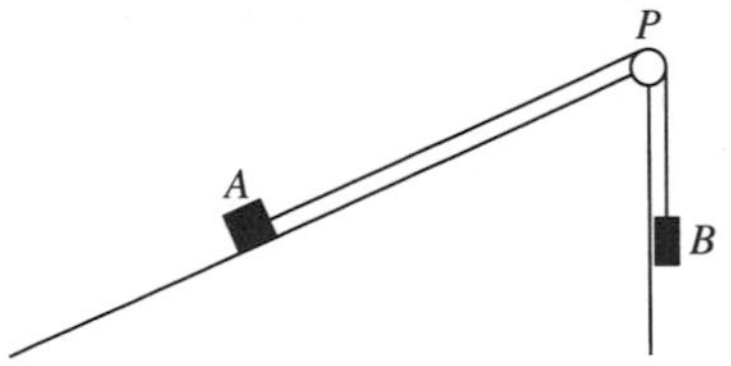

The system is released from rest so that B starts to fall. Find the acceleration of B and the tension in the string.

Determine:
(a) the further distance that A moves up the plane after B has struck the floor (you may assume that A does not reach the pulley)
(b) the time that B remains in contact with the floor
(c) the speed with which B is jerked off the floor. (AEB 1991)

14. (Take $g = 10\,\text{m}\,\text{s}^{-2}$ in this question.)
A particle P is projected vertically upwards from a point X with a speed of $V\,\text{m}\,\text{s}^{-1}$. Given that it reaches a maximum height of 45 m, find the value of V.
When P reaches its maximum height, a second particle Q is projected vertically upwards, also from X. Given that, 2 s after Q is projected, P and Q collide at a point Y, find
(i) the height of Y above the ground,
(ii) the speed of projection of Q.
Given that P is of mass 2 kg and Q of mass 4 kg and that they coalesce on impact, find the speed of the combined particle
(iii) immediately after impact,
(iv) on hitting the ground. (UCLES)

15 Elasticity

Elastic strings

In the situations in preceding chapters the strings connecting bodies, passing over pulleys or maintaining equilibrium etc. have been said to be *inextensible*. This assumption allowed us to model the situation and to use our mathematics to determine solutions. However, in some cases it would be quite inappropriate for our model to assume a string to be inextensible. It could be that the "stretchiness" of a string is the very property that makes it suitable for a particular situation.

In this chapter we are going to consider strings that are said to be *elastic*. This implies that they can be stretched and will regain their natural length once the stretching force has been removed. In this way a more "refined" model of a situation can be considered.

The reader should note that our model will still involve many assumptions. The fact that the string will regain its natural length once the stretching force has been removed is an assumption. We continue to neglect air resistance and, when applying energy principles, we will make no allowance for energy expended as noise and heat. However, our aim is not necessarily to produce a model that is correct in every detail but rather to find one that takes account of all the significant features and that allows us to use our mathematics to determine solutions. The appropriateness of the model can then be checked by comparing the outcomes predicted by the model with the real-life situation.

Hooke's Law

This law states that the tension in a stretched string is proportional to the extension of the string from its natural (or unstretched) length:

tension $\propto$ extension

This is usually written in the form:

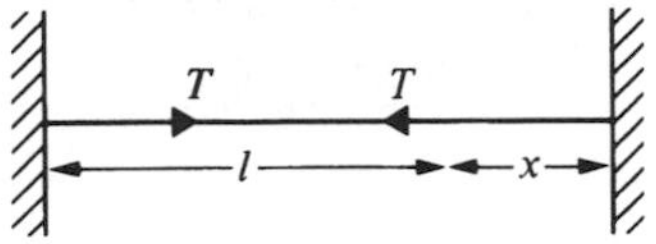

$$T = \lambda \frac{x}{l}$$

where T is the tension in the string, x is the extension and l is the natural length of the string. The constant λ is called the modulus of the string; by considering the units in the above equation, it is seen that the units of λ are those of force, i.e. newtons.

Example 1

An elastic string is of natural length 3 m and modulus 15 N. Find:
(a) the tension in the string when the extension is 40 cm
(b) the extension of the string when the tension is 3 N.

(a) extension = 40 cm = 0·4 m, natural length = 3 m, modulus = 15 N, tension = T

Using Hooke's Law gives:

$$T = \lambda \frac{x}{l}$$

$$\therefore \quad T = 15 \times \frac{(0{\cdot}4)}{3} = 2\,\text{N}$$

The tension in the string is 2 N.

(b) extension = x, natural length = 3 m, modulus = 15 N, tension = 3 N

Using Hooke's Law gives:

$$T = \lambda \frac{x}{l}$$

$$\therefore \quad 3 = 15 \times \frac{x}{3}$$

$$\therefore \quad x = 0{\cdot}6\,\text{m}$$

The extension of the string is 60 cm.

Elastic springs

Hooke's Law also applies to an elastic spring which is either stretched or compressed.
When a spring is compressed Hooke's Law gives the *thrust* in the spring due to its compression to a length which is less than its natural length.

Example 2

A spring is of natural length 1·5 m and modulus 25 N. Find the thrust in the spring when it is compressed to a length of 1·2 m.

natural length = 1·5 m, modulus = 25 N

compression = 1·5 − 1·2 = 0·3 m, thrust = T

Using Hooke's Law gives:

$$T = \lambda \frac{x}{l}$$

$$\therefore \quad T = 25 \times \frac{0{\cdot}3}{1{\cdot}5}$$

$$\therefore \quad T = 5\,\text{N}$$

The thrust in the spring is 5 N.

Equilibrium of a suspended body

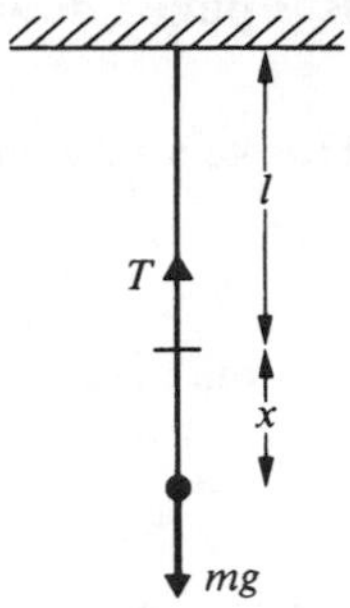

When an elastic string has one end fixed and a mass attached to its other end so that the mass is suspended in equilibrium, then the string is stretched by the force due to the mass.

Resolving vertically gives: $T = mg$

Using Hooke's Law gives: $T = \lambda \frac{x}{l}$

Hence $mg = \lambda \frac{x}{l}$

Example 3

A light elastic string of natural length 75 cm has one end fixed and a mass of 800 g freely suspended from the other end. Find the modulus of the string if the total length of the string in the equilibrium position is 95 cm.

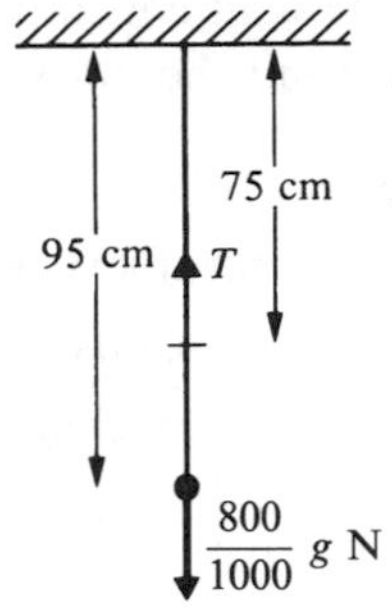

In the equilibrium position

Resolving vertically gives: $T = 0{\cdot}8g$

Using Hooke's Law gives: $T = \lambda \frac{x}{l}$

$\therefore$ $0{\cdot}8g = \lambda \frac{(0{\cdot}95 - 0{\cdot}75)}{0{\cdot}75}$

$\therefore$ $\lambda = 3g$ N

The modulus of the string is $3g$ N.

Example 4

A light elastic spring has its upper end A fixed and a body of mass 0·6 kg attached to its other end B. If the modulus of the spring is $4{\cdot}5g$ N and its natural length 1·5 m, find the extension of the spring when the body hangs in equilibrium.
The end B of the spring is pulled vertically downwards to C, where BC = 10 cm. Find the initial acceleration of the body when it is released from this position.

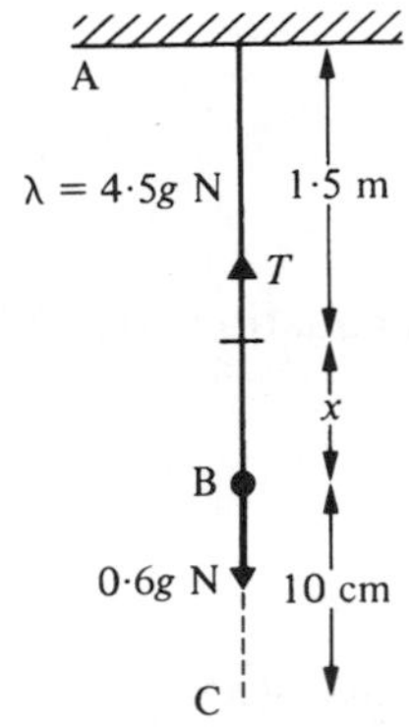

In the equlibrium position

Resolving vertically gives: $T = 0{\cdot}6g$

Using Hooke's Law gives: $T = \lambda \frac{x}{l}$

$\therefore$ $0{\cdot}6g = 4{\cdot}5g \times \frac{x}{1{\cdot}5}$

$\therefore$ $x = 0{\cdot}2$ m

When B is pulled down 10 cm, the total extension is then 30 cm.

Using Hooke's Law gives: $T = 4{\cdot}5g \times \dfrac{0{\cdot}3}{1{\cdot}5}$

$= 0{\cdot}9g$

Using $F = ma$ gives: $0{\cdot}9g - 0{\cdot}6g = 0{\cdot}6a$

$\therefore \quad a = \dfrac{g}{2} \quad \text{or} \quad 4{\cdot}9\,\text{m s}^{-2}$

The extension in the equilibrium position is 20 cm and the initial acceleration when the body is pulled down and released is $4{\cdot}9\,\text{m s}^{-2}$.

Example 5

A body of mass M kg lies on a smooth horizontal surface and is connected to a point O on the surface by a light elastic string of natural length 50 cm and modulus 70 N. When the body moves in a horizontal circular path about O with constant speed of $3{\cdot}5\,\text{m s}^{-1}$, the extension in the string is 20 cm. Find the mass of the body.

String

natural length = 50 cm = 0·5 m
extension = 20 cm = 0·2 m
modulus = 70 N
tension = T

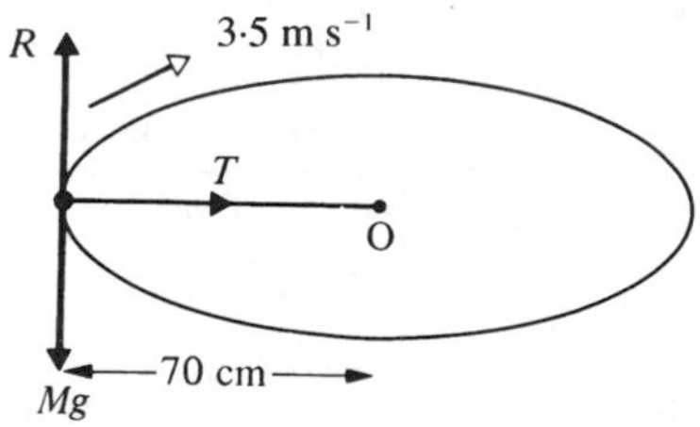

Using Hooke's Law gives:

$$T = \lambda \frac{x}{l}$$

$$\therefore \quad T = 70 \times \frac{0{\cdot}2}{0{\cdot}5} = 28\,\text{N} \quad \ldots[1]$$

But the tension T provides the force towards O necessary for circular motion

$$\therefore \quad T = \frac{mv^2}{r}$$

$$\therefore \quad T = M \times \frac{(3{\cdot}5)^2}{0{\cdot}7} \quad \ldots[2]$$

From equations [1] and [2]

$$28 = M \times \frac{(3{\cdot}5)^2}{0{\cdot}7}$$

$$\therefore \quad M = 1{\cdot}6$$

The mass of the body is 1·6 kg.

Exercise 15A

1. An elastic string is of natural length 1 m and modulus 20 N.
Find the tension in the string when the extension is 20 cm.
2. A spring is of natural length 50 cm and modulus 10 N.
Find the thrust in the spring when it is compressed to a length of 40 cm.
3. When the tension in a spring is 5 N, the length of the spring is 25 cm longer than its natural length of 75 cm.
Find the modulus of the spring.
4. When the length of a spring is 60% of its unstretched length, the thrust in the spring is 10 N.
Find the modulus of the spring.
5. Find the extension of an elastic string of natural length 60 cm and modulus 18 N when the tension in the string is 6 N.
6. The tension in an elastic string is 8 N when the extension in the string is 25 cm.
If the modulus of the string is 8 N, find the unstretched length of the string.
7. When an elastic string is stretched to a length of 18 cm, the tension in the string is 14 N.
If the modulus of the string is 28 N, find the natural length of the string.
8. A light elastic string of natural length 20 cm and modulus $2g$ N has one end fixed and a mass of 500 g freely suspended from the other end.
Find the total length of the string.
9. When a mass of 5 kg is freely suspended from one end of a light elastic string, the other end of which is fixed, the string extends to twice its natural length.
Find the modulus of the string.
10. A light spring of natural length 60 cm and modulus $3g$ N has one end fixed and a body of mass 2 kg freely suspended from its other end.
Find the extension of the spring.
What mass would cause the same length of extension when suspended from a spring of natural length 50 cm and modulus $2g$ N?
11. An elastic string is of natural length 24 cm and modulus 18 N. One end of the string is attached to a fixed point O on a smooth horizontal surface and to the other end is attached a body of mass 2 kg. The body is held at rest on the surface at a point 32 cm away from O and then released.
Find the initial acceleration of the body.
12. An elastic string is of natural length 45 cm and modulus 27 N. One end of the string is attached to a fixed point O on a rough horizontal surface and to the other end is attached a body of mass 5 kg. The body is held at rest on the surface at a point 60 cm away from O and then released.
If the coefficient of friction between the body and the surface is $\frac{1}{7}$, find the initial acceleration of the body.
13. The diagram shows a body of mass 10 kg freely suspended from a light elastic spring of natural length 50 cm and modulus $25g$ N.

Find the extension in the spring when
(a) the body is at rest
(b) the top of the spring is moved upwards with constant velocity
(c) the top of the spring is moved upwards with a constant acceleration of $4{\cdot}9\ \text{m s}^{-2}$
(d) the top of the spring is moved downwards with a constant acceleration of $4{\cdot}9\ \text{m s}^{-2}$.
14. A smooth surface is inclined at an angle of 30° to the horizontal. A body A of mass 2 kg is held at rest on the surface by a light elastic string which has one end attached to A and the other to a point on the surface 1·5 m away from A up a line of greatest slope. If the modulus of the string is $2g$ N, find its natural length.

15. A particle of mass 5 kg is attached to one end of a light elastic string of natural length 1 m and modulus $4g$ N. The other end of the string is fastened to a fixed point O at the top of a smooth slope that is inclined at $\tan^{-1}\frac{3}{4}$ to the horizontal. The particle is held on the slope at a point that is 2·5 m from O down a line of greatest slope.
If the particle is released from rest, find its initial acceleration towards O.
What would the acceleration have been had the slope been rough, coefficient of friction 0·25?

16. A light elastic string has one end fixed and a body of mass $\sqrt{3}$ kg freely suspended from its other end. With a horizontal force of X N acting on the body, the system is in equilibrium with the string extended to twice its natural length and making an angle of 30° with the downward vertical. Find the modulus of the string and the value of X.

17. A body of mass 4 kg lies on a smooth horizontal surface and is connected to a point O on the surface by a light elastic string of natural length 64 cm and modulus 25 N. When the body moves with constant speed v m s^{-1} in a horizontal circle with centre O, the extension in the string is 36 cm. Find v.

18. A body of mass 5 kg lies on a smooth horizontal surface and is connected to a point O on the surface by a light elastic string of natural length 2 m and modulus 30 N.
Find the extension in the string when the body moves in a horizontal circle with centre O at a constant speed of 3 m s^{-1}.

19. A and B are two fixed points on the same horizontal level and distance 48 cm apart. A light elastic string of natural length 40 cm has one end attached to A and the other to B. A body of mass 200 g is attached to the mid-point of the string and hangs in equilibrium at a point 7 cm below the level of A and B.
Find the modulus of the string.

20. A body of mass m lies on a smooth horizontal surface and is connected to a point O on the surface by a light elastic string of natural length l and modulus λ. When the body moves with constant speed v around a horizontal circular path, centre at O, the extension in the string is $\frac{1}{4}l$.

Show that $\lambda = \dfrac{16mv^2}{5l}$.

Potential energy stored in an elastic string

The work done in stretching an elastic string is stored in the string. This stored energy can later be recovered by allowing the string to do some work and, by so doing, return to its natural length. This is the way a catapult works. It is first stretched, storing energy. Then when the elastic is allowed to return to its natural length the energy reappears as the kinetic energy of the stone fired from the catapult. The stored energy is another form of potential energy because the stretched spring has the potential to do work.

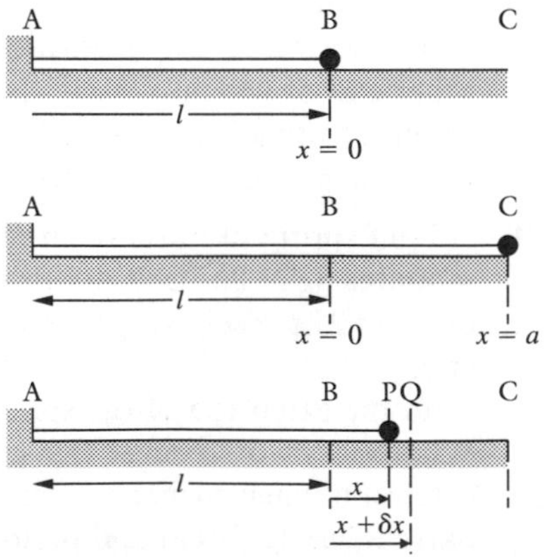

Consider an elastic string of modulus λ and natural length l, with one end attached to a point A and the other to a particle resting on a smooth surface (see diagram).

If x is the extension of the string then consider the particle initially at a point B, where $x = 0$, and pulled to a point C, where $x = a$.

To determine the energy stored in the string we need to find the work done by the force in stretching the string.

Consider BC to be divided into a large number of very short steps, each of length δx. A typical step is shown (enlarged) as PQ in the diagram on the right.

Over the very short step PQ the applied force needed to stretch the string is almost constant, its value being nearly equal to its value at P which is equal to $\frac{\lambda x}{l}$, by Hooke's Law.

Thus: work done in moving the particle from P to Q = force × distance

$$\approx \frac{\lambda x}{l}\delta x$$

The total work done is obtained by adding the contributions of all the steps and seeing what this summation approaches as δx gets smaller and smaller (i.e. as $\delta x \to 0$).

$$\text{total work done in extending the string from B to C} = \lim_{\delta x \to 0} \sum_{x=0}^{x=a} \frac{\lambda x}{l}\delta x$$

Using calculus this can be written as:

$$\text{total work done in extending the string} = \int_0^a \frac{\lambda x}{l}\,\mathrm{d}x$$

$$= \frac{\lambda a^2}{2l}$$

General result

The elastic potential energy stored in a string stretched a distance a from its natural length l is equal to $\frac{\lambda a^2}{2l}$.

Readers who are not familiar with the ideas of integral calculus should accept and use the above result for now and should return to read the above ideas again when they have met them in their Pure Mathematics studies.

Example 6

An elastic string is of natural length 3·6 m and modulus 45 N. Find the work done in stretching it from a length of 4 m to a length of 4·2 m.

When string is of length 4 m, the extension is 0·4 m.

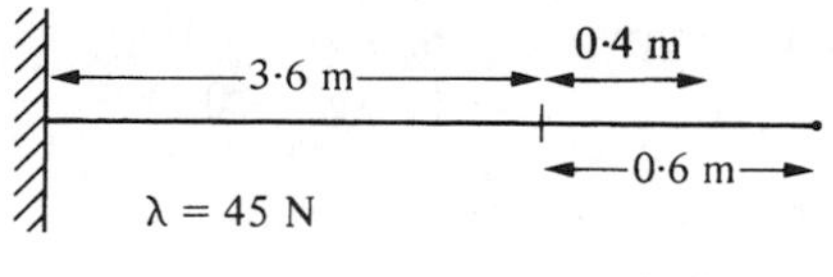

$\therefore$ energy stored $= \frac{45(0{\cdot}4)^2}{2(3{\cdot}6)} = 1\,\text{J}$

When string is of length 4·2 m, the extension is 0·6 m

$\therefore$ energy stored $= \frac{45(0{\cdot}6)^2}{2(3{\cdot}6)} = 2{\cdot}25\,\text{J}$

$\therefore$ work done in stretching the string from 4 m to 4·2 m is 1·25 J.

Conservation of energy for an elastic string

We can introduce the potential energy of an elastic string into the Principle of Conservation of Energy. If there is no work used in overcoming friction and the only external force which does work is gravity, then

total energy = [potential energy due to gravity] + [potential energy stored in spring] + [kinetic energy] = constant

For solving certain types of problem it is useful to write the Principle of Conservation of Energy in the form:

initial total energy = final total energy

Example 7

A light inelastic string has natural length 2 m and modulus $15g$ N. One end of the string is attached to a fixed point, and a body of mass 3 kg hangs from the other end.

(a) Find the extension of the string when the body is in equilibrium.

(b) The body is pulled down a further 10 cm and then released. Use the Principle of Conservation of Energy to determine the speed of the body as it passes through the equilibrium position.

(a) Consider the body at the equilibrium position, E, and let the unknown extension be x metres.

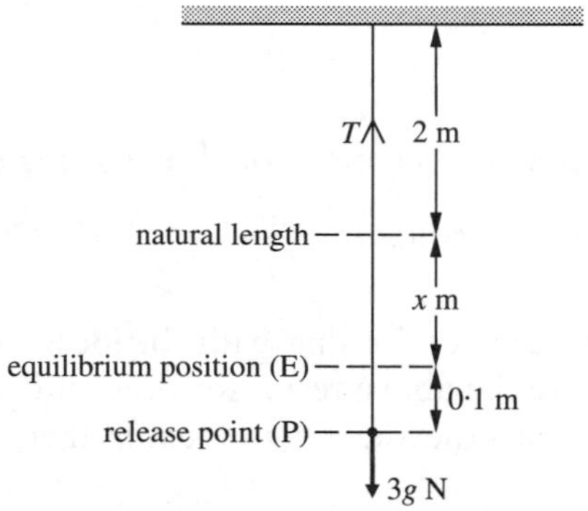

Resolving vertically gives: $T = 3g$

Using Hooke's Law gives: $T = \dfrac{(15g)(x)}{2}$

$\therefore \quad \dfrac{(15g)(x)}{2} = 3g$

$\therefore \quad x = 0{\cdot}4$

The extension in the equilibrium position is 40 cm.

(b) Let the required speed be $v\,\mathrm{m\,s^{-1}}$ and measure the gravitational potential energy from E.

<u>At P</u>

$$\begin{aligned} \text{KE} &= \tfrac{1}{2}mv^2 \\ &= \tfrac{1}{2}(3)(0)^2 \\ &= 0\text{ J} \\ \text{PE due to gravity} &= mgh \\ &= (3)\,(g)\,(-0{\cdot}1) \\ &= -\ 0{\cdot}3g\text{ J} \\ \text{PE in the string} &= \frac{\lambda a^2}{2l} \\ &= \frac{(15g)(0{\cdot}4 + 0{\cdot}1)^2}{2(2)} \\ &= 0{\cdot}9375g\text{ J} \\ \text{total energy at P} &= 0{\cdot}6375g\text{ J} \end{aligned}$$

<u>At E</u>

$$\begin{aligned} \text{KE} &= \tfrac{1}{2}mv^2 \\ &= \frac{3v^2}{2}\text{ J} \\ \text{PE due to gravity} &= mgh \\ &= (3)\,(g)\,(0) \\ &= 0 \\ \text{PE in the string} &= \frac{\lambda a^2}{2l} \\ &= \frac{(15g)(0{\cdot}4)^2}{2(2)} \\ &= 0{\cdot}6\,g \\ \text{total energy at E} &= \left(0{\cdot}6g + \frac{3v^2}{2}\right)\text{ J} \end{aligned}$$

By the Principle of Conservation of Energy:

total energy at P = total energy at E

i.e. $$0{\cdot}6375g = 0{\cdot}6g + \frac{3v^2}{2}$$

$$\therefore \quad v^2 = 0{\cdot}245$$
$$v = 0{\cdot}495$$

The speed of the body as it passes through the equilibrium position is $0{\cdot}495\,\text{m s}^{-1}$.

Potential energy of an elastic spring

The work done in compressing a spring by a distance a from its natural length is the same as the work done in extending it a distance a. The same expression for stored energy is therefore obtained whether the spring is stretched or compressed. If the extension or compression is a, then the energy is $\frac{\lambda a^2}{2l}$. If x is the *coordinate* of the end-point measured from the unstrained position, then x may be positive (for stretching) or negative (for compressing), but the energy, $\frac{\lambda x^2}{2l}$, is always positive.

Example 8

A body of mass 3 kg slides on a smooth horizontal surface, and is attached to a fixed point A by a horizontal spring of natural length 0·5 m and modulus 400 N. The body is pulled out so that the length of the spring becomes 0·6 m, and is given a velocity of $4\,\text{m s}^{-1}$ towards A.
Find:
(a) the maximum compression of the spring in the subsequent motion
(b) the maximum extension of the spring in the subsequent motion.

(a) The diagram below shows the position of the body located by a coordinate x measured from the equilibrium position.

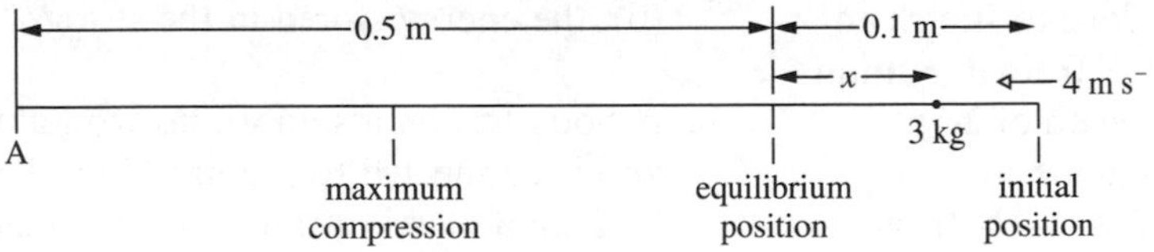

During the motion the total energy, which in this case consists only of the kinetic energy and the energy in the spring, remains constant.

At initial position:

$$\begin{aligned}\text{total energy} &= \text{KE} + \text{energy stored in spring}\\ &= \frac{mv^2}{2} + \frac{\lambda x^2}{2l}\\ &= \frac{(3)(4)^2}{2} + \frac{(400)(0{\cdot}1)^2}{2(0{\cdot}5)}\\ &= 28\,\text{J}\end{aligned}$$

At maximum compression the velocity of the particle is zero:

$$\begin{aligned}\text{total energy} &= \text{KE} + \text{energy stored in spring}\\ &= \frac{mv^2}{2} + \frac{\lambda x^2}{2l}\\ &= \frac{(3)(0)^2}{2} + \frac{(400)\,x^2}{2(0{\cdot}5)}\\ &= 400\,x^2\end{aligned}$$

The total energy does not change,

so $400\,x^2 = 28$ ... [1]

giving $x = \pm 0{\cdot}26$

Thus for maximum compression:

$x = -0{\cdot}26$ (negative due to compression)

The maximum compression is 0·26 m.

(b) At maximum extension the speed, and hence the kinetic energy, is zero, just as it was at maximum compression. Thus equating the initial energy with that at maximum extension will again given equation [1] above.

Thus, for maximum extension: $x = +0{\cdot}26$ (positive due to extension)

The maximum extension is 0·26 m.

Exercise 15B

1. An elastic string is of natural length 2 m and modulus 10 N.
   Find the energy stored in the string when it is extended to a length of 3 m.
2. An elastic string is of natural length 1 m and modulus 20 N.
   Find the energy stored in the string when it is extended to a length of 1·3 m.
3. Find the work that must be done to stretch an elastic string of modulus 200 N from its natural length of 2 m to a stretched length of 2·5 m.
4. Find the work that must be done to compress a spring of modulus 500 N from its natural length of 10 cm to a shortened length of 8 cm.
5. A spring is of natural length 50 cm and modulus 60 N.
   How much energy is released when the length of the spring is reduced from 1·5 m to 1 m?
6. An elastic string is of natural length 4 m and modulus 24 N.
   Find the work that must be done to stretch the string from a length of 5 m to a length of 6 m.
7. A light elastic string is of natural length 50 cm and modulus 147 N. One end of the string is attached to a fixed point and a body of mass 3 kg is freely suspended from the other end.
   Find:
   (a) the extension of the string in the equilibrium position
   (b) the energy stored in the string.
8. A body lies on a smooth horizontal table and is connected to a point O on the table by a light elastic string of natural length 1·5 m and modulus 24 N. Initially the body lies at a point P 1·5 m from O. The body is pulled directly away from O and held at a point Q, 2 m from O, and then released.
   Find:
   (a) the initial energy stored in the string when the body is at P
   (b) the energy stored in the string when the body is held at Q
   (c) the kinetic energy of the body as it passes through P after release from Q.

9.

The diagram shows a body of mass 2 kg freely suspended from a spring of natural length 75 cm and modulus $6g$ N, the other end of which is fixed to a point A. The body initially hangs freely in equilibrium at a point B. It is then pulled down a further distance of 25 cm to a point C and released from rest.

Find:

(a) the distance AB
(b) the energy stored in the spring when the body rests at B
(c) the energy stored in the spring when the body is held at C
(d) the kinetic energy of the body when it passes through B after release from C.

10.

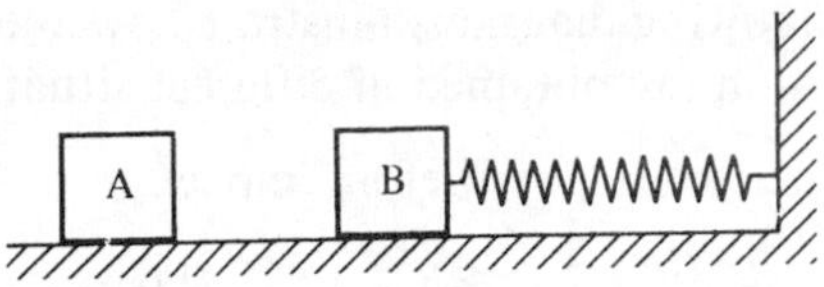

The diagram shows a body A of mass 480 g projected along a smooth horizontal surface with speed $2\,\text{m}\,\text{s}^{-1}$ to collide directly with a body B of mass 320 g, initially at rest. B is attached to a fixed wall 20 cm away by a spring of natural length 20 cm and modulus 36 N. After the collision, A and B move on together and the thrust in the spring brings them momentarily to rest before accelerating them away from the wall.

Find:

(a) the common velocity of A and B immediately after collision
(b) the length of the spring when A and B are momentarily at rest.

Elastic impact

In the last chapter the collision of two elastic bodies in motion was considered, using the principle of conservation of linear momentum. There is another law which involves the coefficient of restitution of the bodies and their relative speeds before and after impact. This is known as *Newton's Experimental Law.*

The coefficient of restitution e for any two bodies is a measure of their elasticity.

It depends upon both of the bodies and the material of which they are made.

Newton's Experimental Law states that:

$$e = \frac{\text{speed of separation of the bodies}}{\text{speed of approach of the bodies}}$$

With the law stated in this form, both the numerator and the denominator of the fraction are positive quantities.

The following diagrams illustrate how these speeds of approach and separation are obtained in different situations:

before impact | after impact

(a)

speed of approach $= u_1 - 0 = u_1$ | speed of separation $= v_2 - 0 = v_2$

(b)

speed of approach $u_1 - u_2$
($u_1 > u_2$, otherwise no collision)

speed of separation $= v_2 - v_1$
($v_2 > v_1$ or bodies do not separate)

(c)

speed of approach $= u_1 + u_2$ | speed of separation $= v_1 + v_2$

Value of coefficient of restitution

For two perfectly elastic bodies in collision, $e = 1$. For two inelastic bodies, i.e. ones which coalesce on impact, $e = 0$. In all other cases e will be between 0 and 1.

Example 9

Two spheres collide when moving along the same straight line on a horizontal surface. The speeds of the spheres before and after the collision are as shown in the diagram. Find the value of e in each case.

(a) before impact

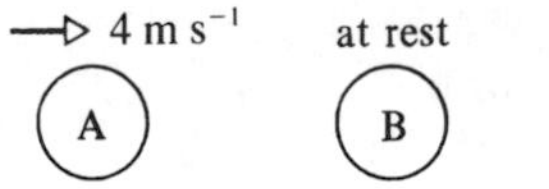

$$e = \frac{\text{speed of separation}}{\text{speed of approach}} = \frac{6-3}{4-0} = \frac{3}{4}$$

after impact

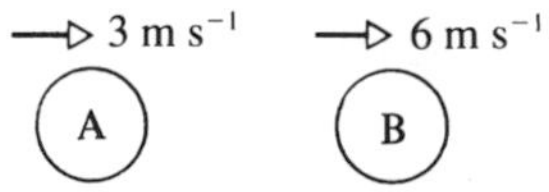

The coefficient of restitution is $\frac{3}{4}$.

(b) before impact

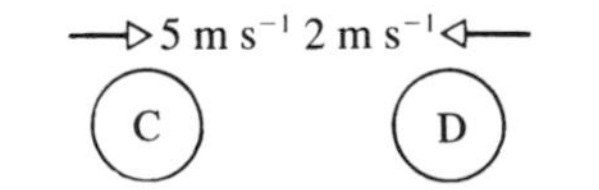

$$e = \frac{\text{speed of separation}}{\text{speed of approach}} = \frac{4+1}{5+2} = \frac{5}{7}$$

after impact

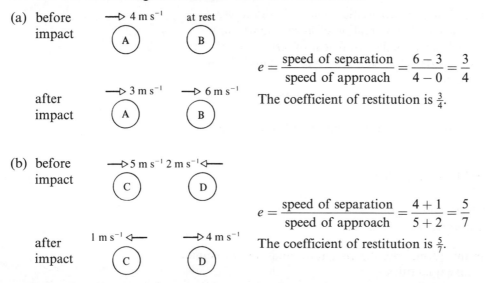

The coefficient of restitution is $\frac{5}{7}$.

Example 10

Find the unknown speed for the given spheres if their speeds before and after impact and their coefficient of restitution e are as shown.

(a)

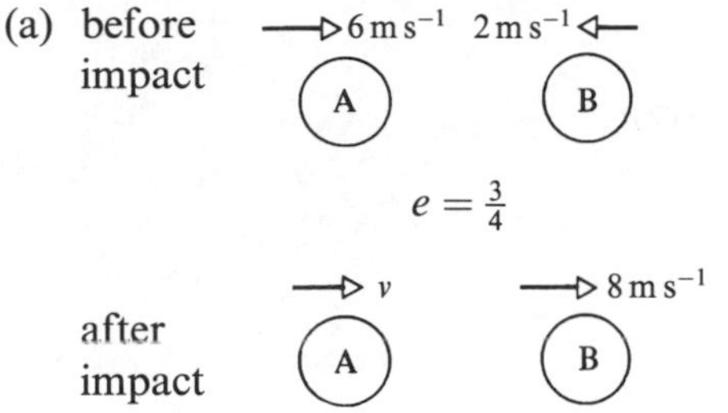

$$e = \frac{\text{speed of separation}}{\text{speed of approach}} \qquad \therefore \quad \frac{3}{4} = \frac{8 - v}{6 + 2}$$

$$3(8) = 4(8 - v)$$

$$\therefore \quad v = 2\,\text{m s}^{-1}$$

The unknown speed after impact is $2\,\text{m s}^{-1}$ in the direction shown.

(b)

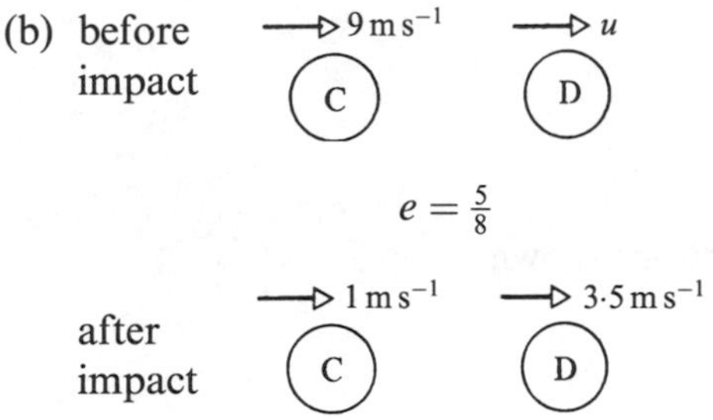

$$e = \frac{\text{speed of separation}}{\text{speed of approach}} \qquad \therefore \quad \frac{5}{8} = \frac{3{\cdot}5 - 1}{9 - u}$$

$$5(9 - u) = 8(3{\cdot}5 - 1)$$

$$\therefore \quad u = 5\,\text{m s}^{-1}$$

The unknown speed after impact is $5\,\text{m s}^{-1}$ in the direction shown.

It should be noted that when the diagrams giving the "before" and "after" impact situations are not given, care must be taken to translate accurately the given data on to the appropriate diagram.
In some cases where there is more than one unknown quantity, both the law of conservation of linear momentum and Newton's Experimental Law need to be used.

Example 11

Two spheres, A and B, are of equal radii and masses 2 kg and $1\frac{1}{2}$ kg respectively. The spheres A and B move towards each other along the same straight line on a smooth horizontal surface with speeds of $4\,\text{m s}^{-1}$ and $6\,\text{m s}^{-1}$ respectively.
If the coefficient of restitution between the spheres is $\frac{2}{5}$, find their speeds and directions after impact.

Draw two diagrams.

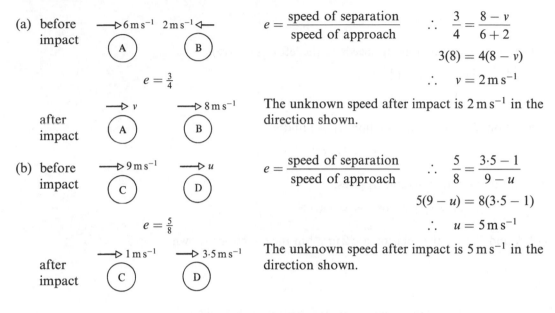

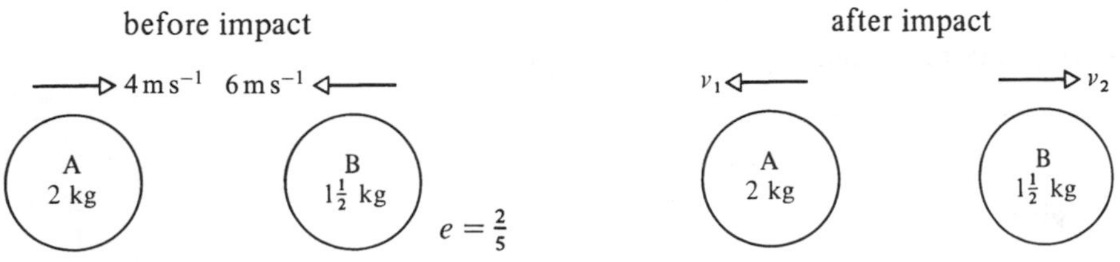

Let v_1 and v_2 (in the directions shown) be the speeds of the spheres after impact.

Newton's Law gives: $\frac{2}{5} = \frac{\text{speed of separation}}{\text{speed of approach}}$

$$= \frac{v_1 + v_2}{4+6}$$

$$\therefore \quad 4 = v_1 + v_2 \qquad \ldots [1]$$

By Conservation of Momentum (taking speeds to the left as positive):

$$2(-4) + 1\tfrac{1}{2}(6) = 2v_1 + 1\tfrac{1}{2}(-v_2)$$

$$\therefore \quad 1 = 2v_1 - \tfrac{3}{2}v_2 \qquad \ldots [2]$$

Subtracting equation [2] from twice equation [1], we find:

$$7 = \tfrac{7}{2}v_2 \quad \text{or} \quad v_2 = 2\,\text{m s}^{-1} \qquad \ldots [3]$$

Substituting equation [3] into equation [1] gives:

$$4 = v_1 + 2 \quad \text{or} \quad v_1 = 2\,\text{m s}^{-1}$$

The speeds of A and B after the collision are $2\,\text{m s}^{-1}$ in the directions shown.

Example 12

Three particles A, B and C, of equal mass m are (in that order) travelling along the same straight line on a smooth horizontal surface. Initially, A and B have speeds of $3\,\text{m s}^{-1}$ and $2\,\text{m s}^{-1}$ respectively and are moving towards each other. The initial speed of C is $1{\cdot}5\,\text{m s}^{-1}$ and it is moving away from A and B. The coefficient of restitution between any two particles is $\frac{4}{5}$.
Find the final speeds of A, B and C.

Draw two diagrams for the first impact between A and B.

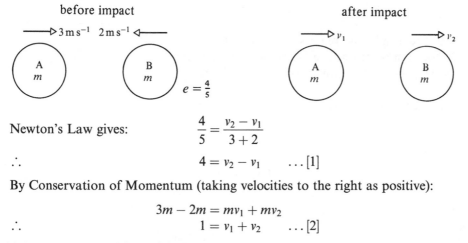

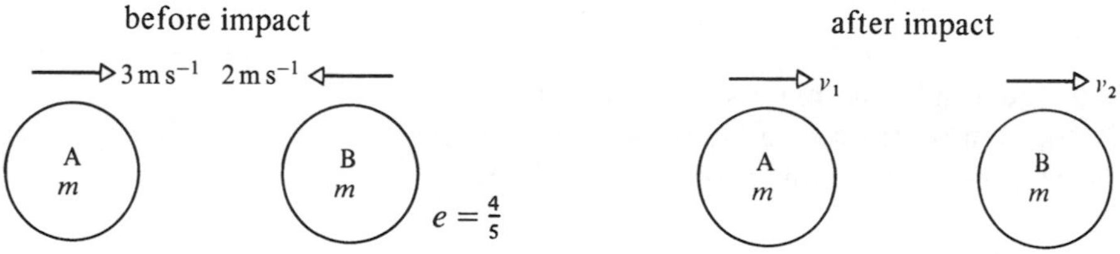

Newton's Law gives: $\frac{4}{5} = \frac{v_2 - v_1}{3+2}$

$$\therefore \quad 4 = v_2 - v_1 \qquad \ldots [1]$$

By Conservation of Momentum (taking velocities to the right as positive):

$$3m - 2m = mv_1 + mv_2$$

$$\therefore \quad 1 = v_1 + v_2 \qquad \ldots [2]$$

Adding equations [1] and [2] gives:

$$v_2 = 2\tfrac{1}{2}\,\text{m s}^{-1}$$

and by substitution: $v_1 = -1\tfrac{1}{2}\,\text{m s}^{-1}$

Hence after this impact, A is moving to the left in the diagram at $1\frac{1}{2}\,\mathrm{m\,s^{-1}}$ and B is moving towards C at $2\frac{1}{2}\,\mathrm{m\,s^{-1}}$. A second impact will occur between B and C since the velocity of B is greater than that of C.

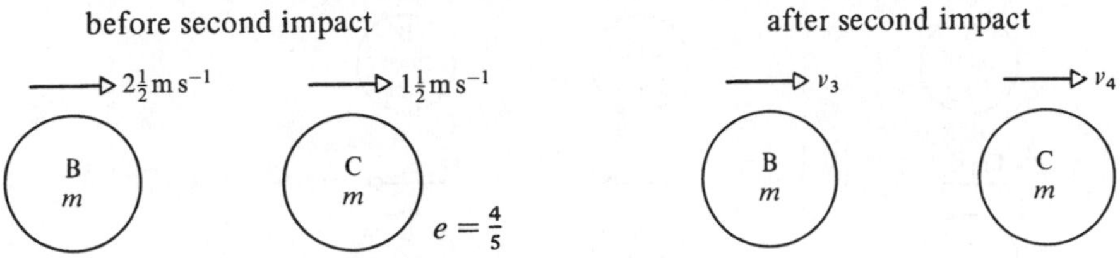

Newton's Law gives: $\quad \dfrac{4}{5} = \dfrac{v_4 - v_3}{2\frac{1}{2} - 1\frac{1}{2}}$

$\therefore \qquad 4 = 5v_4 - 5v_3 \qquad \ldots[3]$

By Conservation of Momentum:

$$2\tfrac{1}{2}m + 1\tfrac{1}{2}m = mv_3 + mv_4$$

$\therefore \qquad 4 = v_3 + v_4 \qquad \ldots[4]$

Solving equations [3] and [4] simultaneously, we find that:

$$v_3 = 1{\cdot}6\,\mathrm{m\,s^{-1}} \quad \text{and} \quad v_4 = 2{\cdot}4\,\mathrm{m\,s^{-1}}$$

From the values of v_1, v_3 and v_4, it can be seen that no further collisions take place.

After the two impacts which occur the speeds of A, B and C are 1·5, 1·6 and $2{\cdot}4\,\mathrm{m\,s^{-1}}$ respectively.

Exercise 15C

1. In each part of this question the two diagrams show the situations before and after a collision between two spheres A and B of equal radii moving along the same straight line on a smooth horizontal surface. For (a), (b) and (c), find the value of the coefficient of restitution e.

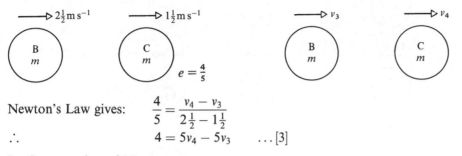

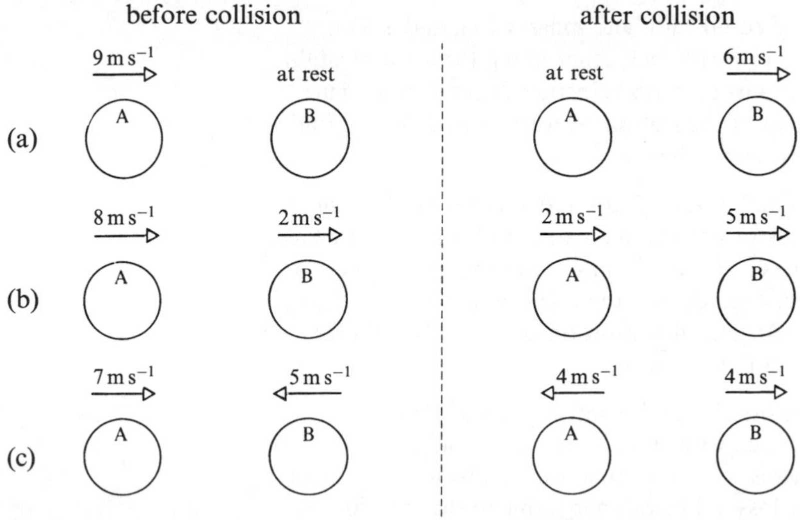

For (d) to (h), find the speeds u and v (e is given for each part).

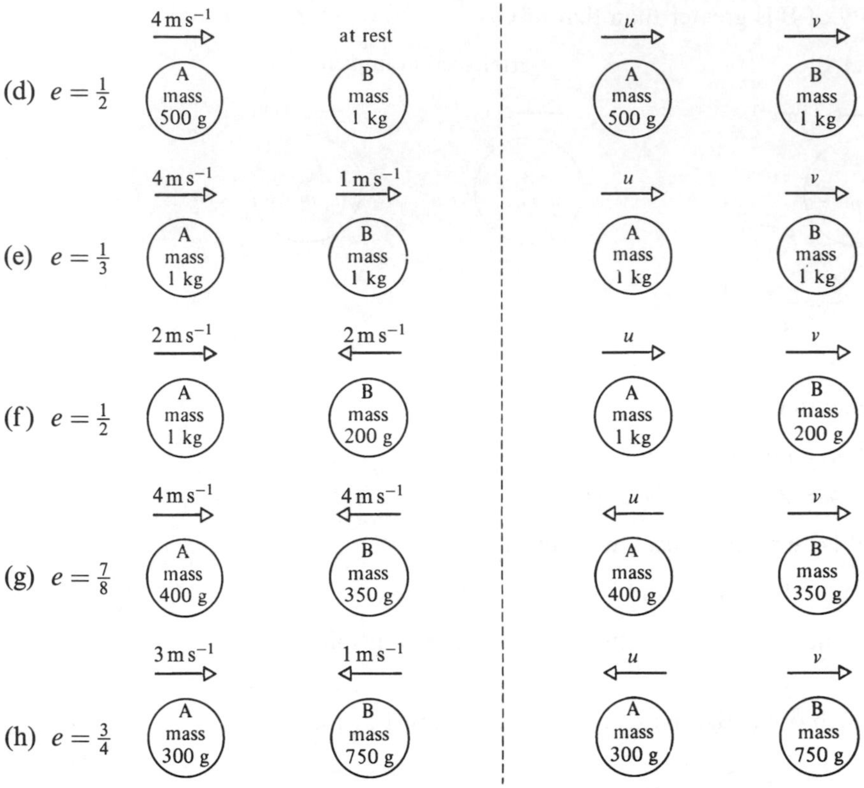

2. Two particles A and B of masses 200 g and 500 g respectively, are travelling along the same straight line on a smooth horizontal surface. Particle A, initially travelling at $6\,\text{m s}^{-1}$, catches up and collides with B which was initially travelling at $2\,\text{m s}^{-1}$. After the collision, A has not changed its direction of motion but now has a speed of $1\,\text{m s}^{-1}$. Find the speed of B after collision and the coefficient of restitution for the particles.

3. Two spheres A and B are of equal radii and masses 1 kg and 1·5 kg respectively. A and B move towards each other along the same straight line on a smooth horizontal surface with velocities $2\mathbf{i}\,\text{m s}^{-1}$ and $-\mathbf{i}\,\text{m s}^{-1}$ respectively. If the coefficient of restitution between A and B is $\frac{2}{3}$, find the velocities of the spheres after collision.

4. Two spheres of equal radii and masses 250 g and 150 g are travelling towards each other along a straight line on a smooth horizontal surface. Initially, the 250 g sphere has a speed of $3\,\text{m s}^{-1}$ and the 150 g sphere a speed of $2\,\text{m s}^{-1}$. The spheres collide and the collision reduces the 250 g sphere to rest. Find the coefficient of restitution between the spheres and the kinetic energy lost in the collision.

5. Two particles A and B of masses $2m$ and m respectively are travelling directly towards each other each with speed u. If the coefficient of restitution between the spheres is $\frac{1}{2}$, show that after collision, A is at rest and B has speed u. Find the loss in kinetic energy due to the collision.

6. Each part of this question involves three smooth spheres A, B and C, of equal radii, moving along the same straight line. A collides with B, and then B collides with C. The diagrams show the situations before any collision, after A has collided with B, and after B has collided with C. Find the speeds u, v, w and x.
(a) $e = \frac{3}{4}$ between any two spheres in collision.

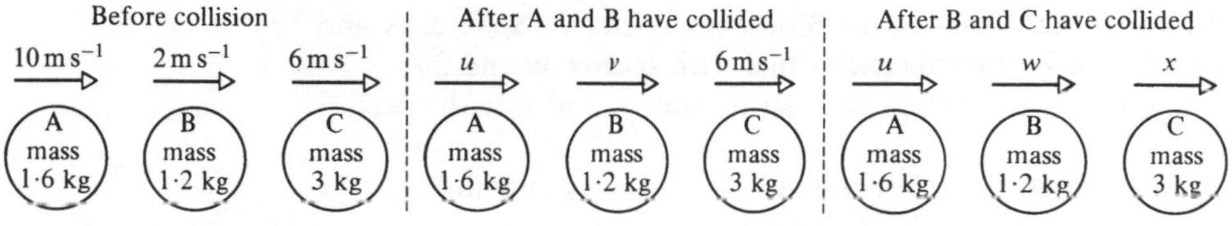

(b) Between A and B $e = \frac{2}{3}$
Between B and C $e = \frac{1}{2}$

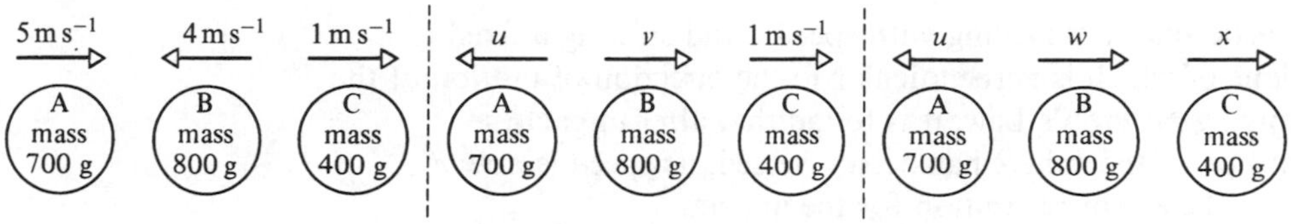

7. Three particles A, B and C of masses 80 g, 200 g and 500 g respectively, are all travelling in the same direction along the same straight line on a smooth horizontal surface. Initially the speeds of A, B and C are $6\,\text{m s}^{-1}$, $2\,\text{m s}^{-1}$ and $2\,\text{m s}^{-1}$ respectively, and B lies between A and C; A collides with B and after this collision, B collides with C.
If the coefficient of restitution between any two particles colliding is $\frac{3}{4}$, find the final speeds of A, B and C.

8. A, B and C are three spheres of equal radii and of masses of 750 g, 500 g and 1 kg respectively. The spheres are travelling along the same straight line on a smooth horizontal surface with B between A and C. Initially the velocities of A, B and C are $5\mathbf{i}\,\text{m s}^{-1}$, $-3\mathbf{i}\,\text{m s}^{-1}$ and $4\mathbf{i}\,\text{m s}^{-1}$ respectively, where $\mathbf{i}$ is a unit vector in the direction ABC. The coefficient of restitution between A and B is $\frac{7}{8}$ and between B and C is $\frac{1}{2}$.
Find the velocities of A, B and C after all collisions have taken place.

9. Two particles A and B are travelling in the same direction along the same straight line on a smooth horizontal surface with speeds of $2u$ and u respectively. Particle A catches up and collides with B, coefficient of restitution e.
If the mass of A is twice that of B, find expressions for the speeds of A and B after collision.

10. Two identical spheres each of mass m are projected directly towards each other on a smooth horizontal surface. Each sphere is given an initial speed of u and the coefficient of restitution between the spheres is e.
Show that the collision between the two spheres causes a loss in kinetic energy of $mu^2(1 - e^2)$.

11. Two identical particles each of mass m are projected towards each other with speeds of $3u$ and u. Show that after collision, the directions of motion of both particles will be reversed, provided $e > \frac{1}{2}$. Find an expression for the loss in the kinetic energy of the particles due to this impact.

12. Three identical spheres A, B and C, each of mass m, lie in a straight line on a smooth horizontal surface with B between A and C. Spheres A and B are projected directly towards each other with speeds $3u$ and $2u$ respectively and C is projected directly away from A and B with speed $2u$.
If the coefficient of restitution between any two spheres in collision is e, show that B will only collide with C provided $e > \frac{3}{5}$.

Impact with a surface

Consider an elastic sphere travelling with speed v and striking a fixed surface, the plane of which is perpendicular to the direction of motion of the sphere. By applying Newton's Law, it is found that after impact the direction of motion of the sphere has been reversed; its speed is now ev, where e is the coefficient of restitution for the impact.

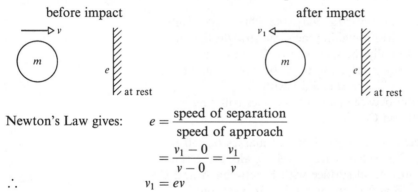

Newton's Law gives: $$e = \frac{\text{speed of separation}}{\text{speed of approach}} = \frac{v_1 - 0}{v - 0} = \frac{v_1}{v}$$

$\therefore \quad v_1 = ev$

A body is said to collide *normally* with a fixed surface if its direction of motion prior to the impact is at right angles to the surface.

Example 13

(a) A smooth sphere collides normally with a fixed vertical wall. From the information in the diagram find the coefficient of restitution e.

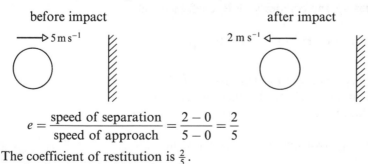

$$e = \frac{\text{speed of separation}}{\text{speed of approach}} = \frac{2-0}{5-0} = \frac{2}{5}$$

The coefficient of restitution is $\frac{2}{5}$.

(b) A smooth sphere collides normally with a fixed vertical wall. From the information in the diagram find the speed v of the sphere after impact.

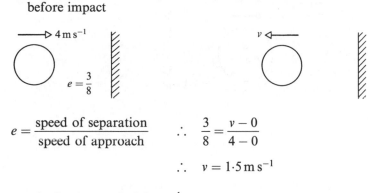

before impact

$$e = \frac{\text{speed of separation}}{\text{speed of approach}} \qquad \therefore \quad \frac{3}{8} = \frac{v - 0}{4 - 0}$$

$$\therefore \quad v = 1{\cdot}5\,\text{m s}^{-1}$$

The speed after impact is $1{\cdot}5\,\text{m s}^{-1}$.

Example 14

Two spheres A and B of equal radii are initially at rest on a smooth horizontal surface; they have masses of 0·5 kg and 0·4 kg respectively. They are projected towards each other with speeds of $3\,\text{m s}^{-1}$ and $2\,\text{m s}^{-1}$ respectively, the coefficient of restitution being $\frac{4}{5}$. After collision, B collides normally with a fixed vertical wall, the coefficient of restitution between B and the wall being $\frac{2}{3}$. Find the velocities of A and B after the first impact between them and after the second impact between them.

before impact — after impact

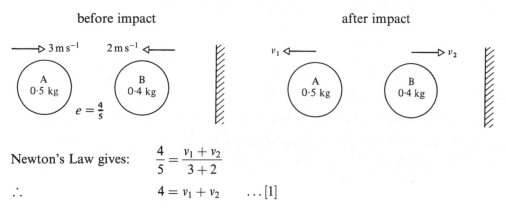

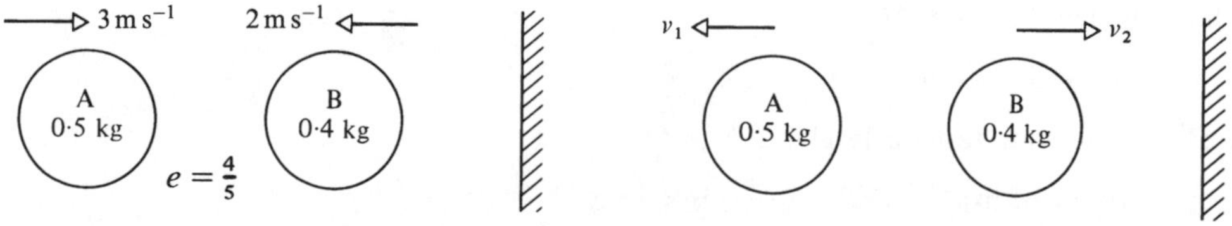

Newton's Law gives: $$\frac{4}{5} = \frac{v_1 + v_2}{3 + 2}$$

$$\therefore \qquad 4 = v_1 + v_2 \qquad \ldots [1]$$

By the Principle of Conservation of Momentum (taking velocities to the right as positive):

$$0{\cdot}5(3) + 0{\cdot}4(-2) = 0{\cdot}5(-v_1) + 0{\cdot}4(v_2)$$

$$\therefore \qquad 7 = -5v_1 + 4v_2 \qquad \ldots [2]$$

Solving equations [1] and [2] simultaneously gives:

$$v_2 = 3\,\text{m s}^{-1} \quad \text{and} \quad v_1 = 1\,\text{m s}^{-1}$$

Hence B now approaches the vertical wall with a speed of $3\,\text{m}\,\text{s}^{-1}$.

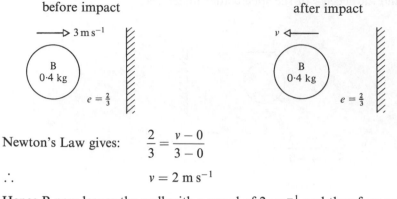

Newton's Law gives: $\dfrac{2}{3} = \dfrac{v-0}{3-0}$

$\therefore$ $v = 2\,\text{m}\,\text{s}^{-1}$

Hence B now leaves the wall with a speed of $2\,\text{ms}^{-1}$ and therefore collides again with A which has a speed of $1\,\text{m}\,\text{s}^{-1}$

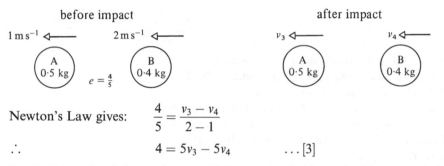

Newton's Law gives: $\dfrac{4}{5} = \dfrac{v_3 - v_4}{2-1}$

$\therefore$ $4 = 5v_3 - 5v_4$ ... [3]

By the Principle of Conservation of Momentum (taking velocities to the left as positive):

$$0{\cdot}5(1) + 0{\cdot}4(2) = 0{\cdot}5v_3 + 0{\cdot}4v_4$$

$\therefore$ $13 = 5v_3 + 4v_4$ ... [4]

Solving equations [3] and [4] simultaneously gives:

$$v_3 = 1{\cdot}8\,\text{m}\,\text{s}^{-1} \quad \text{and} \quad v_4 = 1\,\text{m}\,\text{s}^{-1}$$

The velocities of A and B after the first impact between them are $1\,\text{m}\,\text{s}^{-1}$ and $3\,\text{m}\,\text{s}^{-1}$, and after the second impact between them are $1{\cdot}8\,\text{m}\,\text{s}^{-1}$ and $1\,\text{m}\,\text{s}^{-1}$ respectively, in the directions indicated in the diagrams.

Setting up the model

Remember that if a question requires you to "set up the model" you should clearly state any assumptions you are making.

Example 15

A ball is dropped from a height of 1 m. Set up a mathematical model to determine an expression for the time for which the ball will bounce in terms of *e*, the coefficient of restitution between the ball and the ground. Clearly state any assumptions you make.

Step 1. Set up the model.

- Assume that there is no air resistance.
- Assume that the motion is vertical.
- Consider the ball as a particle so that rotational considerations can be neglected.
- Assume that whilst the ball is in flight (i.e. excluding collisions with ground) energy is conserved, and that such motion only involves gravitational potential energy and kinetic energy.
- Assume the ball bounces on a "hard" horizontal surface.

Step 2. Apply the mathematics.

Suppose the ball is released from some point A, see diagram. Point B is on the ground and vertically below A. Suppose the ball reaches the ground with speed v_1.

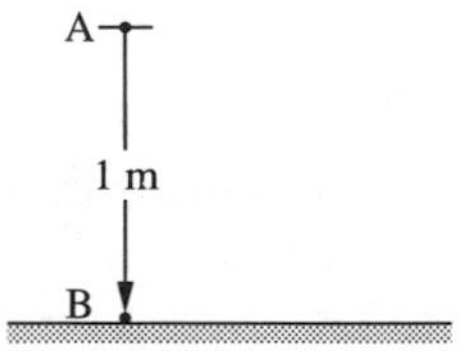

vertical motion, from A to B.

$$u = 0$$
$$v = v_1 \text{ m s}^{-1}\downarrow$$
$$s = 1 \text{ m}\downarrow$$
$$a = g \text{ m s}^{-2}\downarrow$$
$$t = t_1 \text{ s}$$

Use $v^2 = u^2 + 2as$

$$v_1{}^2 = 0^2 + 2g(1)$$
$$v_1 = \sqrt{2g}$$

Use $s = ut + \frac{1}{2}at^2$

$$1 = (0)t_1 + \tfrac{1}{2}g{t_1}^2$$
$$t_1 = \sqrt{\frac{2}{g}}$$

Let v_2 be the speed immediately after collision (see diagram):

Newton's Law gives: $v_2 = ev_1$

$\therefore$ $v_2 = e\sqrt{2g}$

before impact after impact

Suppose that the ball just reaches point C on its first rebound.

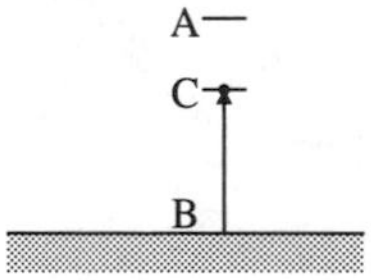

vertical motion, from B to C.

$$u = e\sqrt{2g} \text{ m s}^{-1}\uparrow$$
$$v = 0$$
$$a = g \text{ m s}^{-2}\downarrow = -g \text{ m s}^{-2}\uparrow$$
$$t = t_2 \text{ s}$$

Use $v = u + at$

$$0 = e\sqrt{2g} - gt_2$$

$$\therefore \quad t_2 = e\sqrt{\frac{2}{g}}$$

The time for the ball to travel from C back to B will also be t_2. Therefore the total time for the ball to travel from B to C and back to B is $2e\sqrt{\frac{2}{g}}$.

Repeating this process for subsequent bounces gives $t_3 = e^2\sqrt{\frac{2}{g}}$, $t_4 = e^3\sqrt{\frac{2}{g}}$ etc.

The total time T will be: $t_1 + 2(t_2 + t_3 + t_4 + \ldots)$.

$$\therefore \quad T = \sqrt{\frac{2}{g}} + 2\left(e\sqrt{\frac{2}{g}} + e^2\sqrt{\frac{2}{g}} + e^3\sqrt{\frac{2}{g}} + \ldots\right)$$

$$= \sqrt{\frac{2}{g}}\,[1 + 2e(1 + e + e^2 \ldots)]$$

$(1 + e + e^2 + \ldots)$ is an infinite geometric progression with first term 1 and common ratio e, so

$$T = \sqrt{\frac{2}{g}}\left[1 + 2e\left(\frac{1}{1-e}\right)\right]$$

$$= \sqrt{\frac{2}{g}}\left(\frac{1+e}{1-e}\right)$$

In this particular model an assumption is being made that the ball will make an infinite number of bounces in a finite time.

A graph of T against e is shown below:

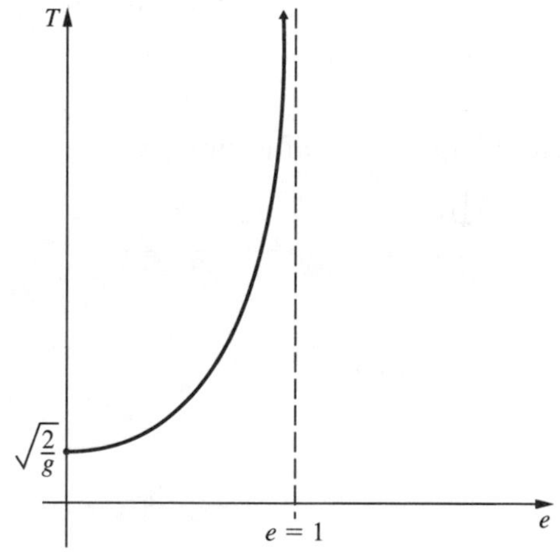

The graph shows how T will increase as $e \to 1$.

Practically, this means that if a rubber 'bouncy ball' is used with $e \approx 0{\cdot}9$, then the total time of bouncing will be greater than if a snooker ball is used with $e \approx 0{\cdot}1$ and on the same surface.

In order to determine whether this is a realistic model the experiment could be carried out practically. However, in a practical situation, it is difficult to determine when the motion of the ball has changed from bouncing to rolling.

Example 16

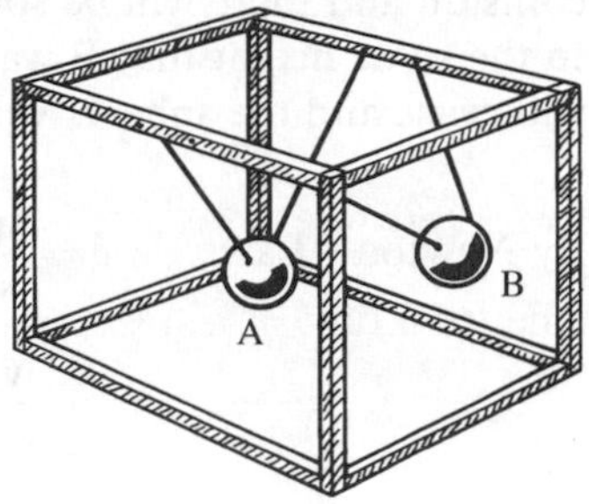

Simon is intrigued by the mathematics of his Newton's Cradle, a device involving two suspended spheres A and B (see diagram). He explores various mathematical models for the situation to see what each model predicts for the behaviour of the two spheres after collision. In one such model he assumes that both linear momentum and kinetic energy are conserved in the collision.Set up such a model, determine the behaviour of the masses after collision, as predicted by the model, and investigate what these predictions mean for the value of e the coefficient of restitution for the spheres.

Step 1. Set up the model.

- Assume that there is no air resistance.
- Consider each sphere as a particle.
- Assume the two spheres are identical, each with mass m.
- Assume that the collision involves no loss of kinetic energy.
- Assume that the collision involves no loss of linear momentum.

Step 2. Apply the mathematics.

Suppose the before and after situations are as follows:

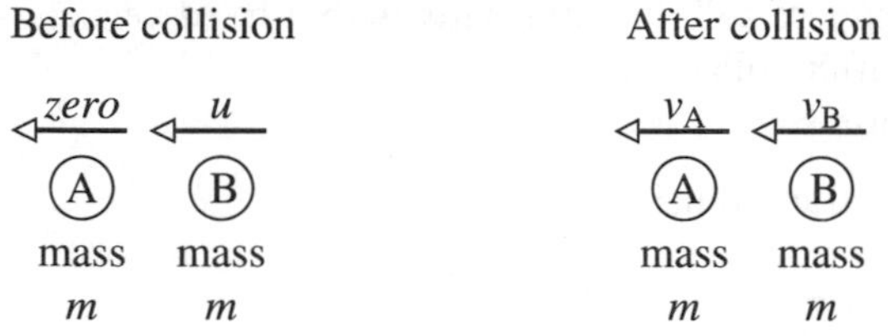

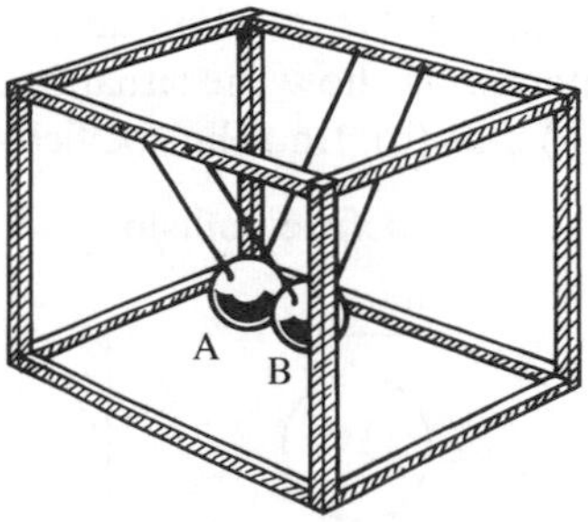

By the Principle of Conservation of Linear Momentum (taking velocities to the left as positive):

$$mu = mv_A + mv_B$$

$$\therefore \quad u = v_A + v_B \qquad \ldots[1]$$

By the Principle of Conservation of Kinetic Energy:

$$\tfrac{1}{2}mu^2 = \tfrac{1}{2}mv_A{}^2 + \tfrac{1}{2}mv_B{}^2$$

$$\therefore \quad u^2 = v_A{}^2 + v_B{}^2 \qquad \ldots[2]$$

Substituting u from [1] into [2] gives:

$$(v_A + v_B)^2 = v_A{}^2 + v_B{}^2$$

$$\text{i.e.} \quad v_A{}^2 + 2v_Av_B + v_B{}^2 = v_A{}^2 + v_B{}^2$$

$$v_Av_B = 0$$

Thus either $v_A = 0$ and, from [1], $v_B = u$ (this refers to the situation just before collision)
or $v_B = 0$ and, from [1], $v_A = u$

Thus, according to this model, after the collision A moves with speed u and B is stationary. The collision reduces B to rest and it does not swing past its vertical position. Those familiar with this cradle idea will know that this is a good approximation to what happens. However, some energy will be lost in collision and there will be some resistance to motion so A will not quite rise to the same height that B was released from and eventually the collisions will cease and the spheres will come to rest.

By Newton's Law:

$$e = \frac{\text{speed of separation}}{\text{speed of approach}}$$

$$= \frac{v_A - v_B}{u}$$

$$= \frac{u - 0}{u}$$

$$= 1$$

The assumptions made by our model require the coefficient of restitution between the spheres to be 1.

Exercise 15D

1. Each part of this question involves a smooth sphere colliding normally with a fixed vertical wall. The diagrams show the situations before and after collision.
For (a) and (b), find the coefficient of restitution e.

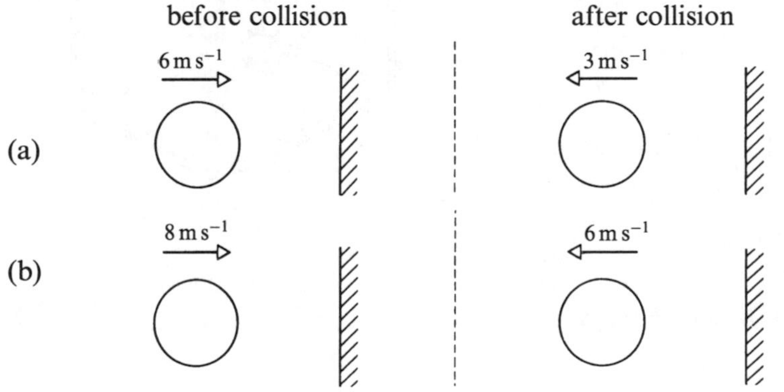

For (c) and (d), find the speed v after impact (e is given for each part).

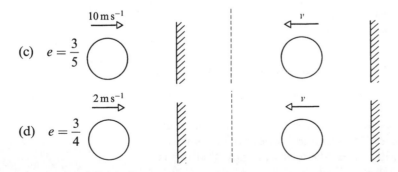

For (e) and (f), find the speed u before impact (e is given for each part).

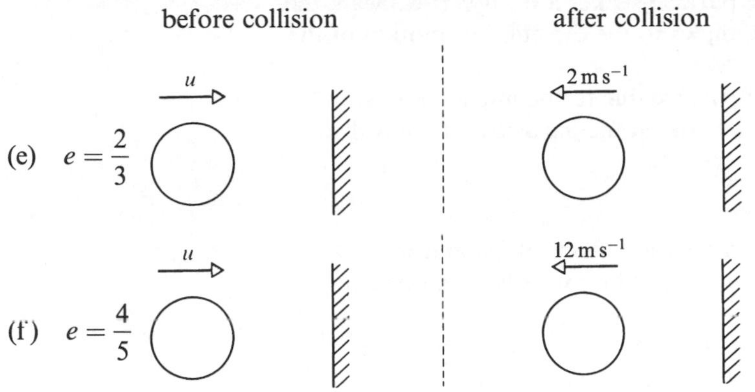

2. A body moves horizontally with velocity $6\mathbf{i}\ \mathrm{m\,s^{-1}}$ and strikes a vertical wall normally.
If the coefficient of restitution between the body and the wall is 0·75, find the velocity of the body after impact.

3. A body is moving horizontally with velocity $5\mathbf{i}\ \mathrm{m\,s^{-1}}$ when it strikes a vertical wall normally and rebounds with velocity $-4\mathbf{i}\ \mathrm{m\,s^{-1}}$.
Find the coefficient of restitution between the body and the wall.

4. A particle is initially at a point A on a smooth horizontal surface midway between two vertical walls which are parallel to each other and 2 metres apart. The particle is projected from A with a speed of $2\ \mathrm{m\,s^{-1}}$ in a direction perpendicular to the walls.
If the coefficient of restitution between the particle and each wall is $\frac{1}{2}$, find the time taken for the body to return to A having touched each wall once and once only.

5. Two spheres A and B of equal radii are initially at rest on a smooth horizontal surface. A is projected directly towards B with a speed of $10\ \mathrm{m\,s^{-1}}$ and the collision between the two spheres reduces A to rest. B continues after collision to strike a fixed vertical surface at right angles and rebounds to hit A again. The coefficient of restitution between A and B is $\frac{4}{5}$ and between B and the wall is $\frac{5}{8}$.
If A has a mass of 200 g, find the mass of B and the speeds of the spheres after the second collision between them.

6. A and B are two spheres of equal radii and masses 750 g and 600 g respectively. The spheres are moving directly towards each other along a smooth horizontal surface with A travelling with a speed of $6\ \mathrm{m\,s^{-1}}$ and B with a speed of $4\ \mathrm{m\,s^{-1}}$. The spheres collide (coefficient of restitution $\frac{4}{5}$) and the collision reverses the direction of motion of sphere B which then strikes a fixed vertical wall at right angles.
If the coefficient of restitution between B and the wall is $\frac{2}{3}$, show that B will collide again with A and find the speeds of A and B after this second collision.

7. A particle of mass m is travelling in a straight line with speed u along a smooth horizontal surface. The particle strikes a fixed vertical wall, the plane of the wall being at right angles to the direction of motion of the particle.
If the kinetic energy lost by the particle due to the impact is E, show that the coefficient of restitution between the particle and the wall is given by $\sqrt{\left(\frac{mu^2 - 2E}{mu^2}\right)}$.

8. Two spheres A and B, of equal radii and masses of $2m$ and m respectively, are each travelling with speed u towards each other on a smooth horizontal surface.
If the coefficient of restitution between A and B is e_1, show that:
(a) if $e_1 = \frac{1}{2}$, then A is reduced to rest by the collision
(b) if $e_1 > \frac{1}{2}$, both spheres will have their original directions of motion reversed by the collision.
With $e_1 > \frac{1}{2}$ and B going on after collision to hit a fixed vertical wall at right angles, show that B will collide again with A provided $e_2 > \dfrac{2e_1 - 1}{1 + 4e_1}$, where e_2 is the coefficient of restitution between B and the wall.

9. Two spheres A and B, of equal radii and masses 500 g and 200 g respectively, are initially travelling along the same straight line towards each other on a smooth horizontal surface. Sphere A has an initial speed of $5\,\text{m}\,\text{s}^{-1}$ and sphere B an initial speed of $7\,\text{m}\,\text{s}^{-1}$. The spheres collide and the coefficient of restitution between them is 0·75. The collision reverses the direction of motion of B which then strikes a wall normally, coefficient of restitution 0·5. The first collision between A and B takes place 3 m from the wall.
(a) Show that there will be a second collision between A and B.
(b) Find the time interval between the first and second collisions between the spheres.
(c) Find how far from the wall the second collision between A and B occurs.
(d) Find the speeds of A and B after the second collision between them.

10. Sandra releases a ball from a height h above the floor, and John measures the height r to which the ball rises. The results of their experiment are given below for two different types of balls:

type A	type B
$h = 1\,\text{m}$	$h = 1\,\text{m}$
$r = 0{\cdot}58\,\text{m}$	$r = 0{\cdot}47\,\text{m}$

Set up a model to determine the coefficient of restitution between the two different types of ball and the floor.

11. A ball is dropped from a height of 1 m. Set up a mathematical model to determine an expression for the total distance the ball travels in terms of e, the coefficient of restitution between the ball and the ground. Clearly state any assumptions you make.

12. Simon continues to be intrigued by his Newton's Cradle, a device involving two suspended spheres A and B, see diagram. He explores various mathematical models for the situation to see what each model predicts for the behaviour of the two spheres after collision. In one such model he assumes that the coefficient of restitution between the spheres is 0·75 and that linear momentum is conserved in the collision.
Set up such a model, determine the speeds of the masses after collision, as predicted by the model, and, based on these predictions, determine the percentage loss in the kinetic energy of the system due to the collision.

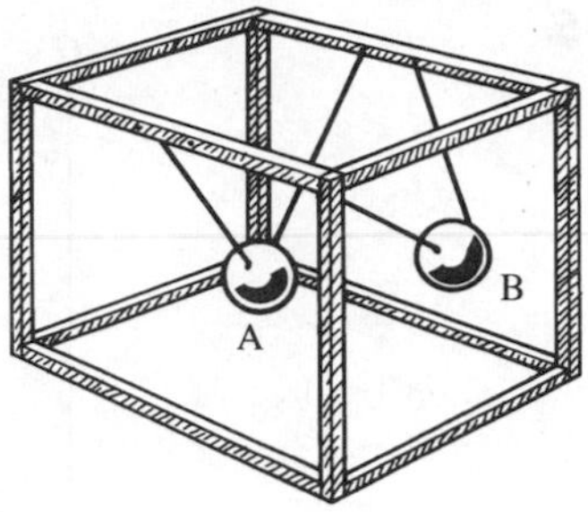

Exercise 15E Examination questions

(Unless otherwise indicated take $g = 9{\cdot}8\,\text{m}\,\text{s}^{-2}$ in this exercise.)

1. A light elastic string of natural length 0·3 m has one end fixed to a point on a ceiling. To the other end of the string is attached a particle of mass M. When the particle is hanging in equilibrium, the length of the string is 0·4 m.
   (a) Determine, in terms of M and g, the modulus of elasticity of the string.
   A horizontal force is applied to the particle so that it is held in equilibrium with the string making an angle α with the downward vertical. The length of the string is now 0·45 m.
   (b) Find α, to the nearest degree. (ULEAC)

2. A light elastic string, of natural length $3a$ and modulus $2mg$, has one end attached to a fixed point O. The other end is attached to a small light ring R which is threaded on a smooth straight wire fixed in a vertical position at a distance $4a$ from O. Give a reason why the equilibrium position of R is at the same horizontal level as O. Find the tension in the string in this position.
   The ring R is replaced by a small bead B, of mass qm, which is threaded on the wire and attached to the end of the string. In the new equilibrium position the string is inclined at 30° to the vertical. Find the value of q and find also, in terms of m and g, the magnitude of the force exerted on the bead by the wire.
   Give a sketch of the triangle of forces for the forces acting on the bead.
   A vertical force of magnitude P is now applied to the bead and in the new equilibrium position the string is again horizontal. Find P in terms of m and g.
   (UCLES)

3. A particle of weight 60 N is attached to two inextensible strings each of length 13 cm. The other ends of the strings are attached to two points A and B on the same horizontal level at a distance of 24 cm apart. Find the tension in the strings when the particle hangs in equilibrium.
   The inextensible strings are then replaced by elastic ones, each of natural length 13 cm and of the same modulus of elasticity. The particle then hangs in equilibrium 9 cm below the line AB. Find the elastic modulus of the strings. (WJEC)

4. A light spring has natural length a and modulus $4mg$. One end of the spring is fixed at a point O on a smooth horizontal table and a particle of mass m is attached to the other end. The particle moves on the table in a circular path of radius $\dfrac{5a}{4}$ and centre O, with constant speed u. Find u in terms of a and g. (AEB 1991)

5.

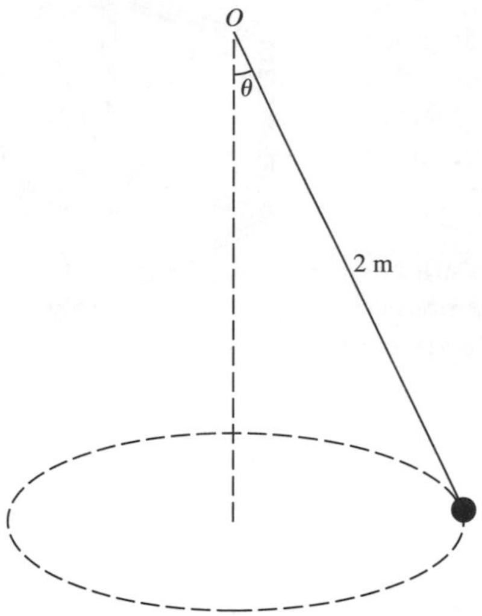

In a simple model of a 'rotating swing', a particle of mass 30 kg is attached to one end of a light inextensible rope of length 2 m. The other end is attached to a fixed point O. The particle moves in a horizontal circle at a constant angular speed of 3 rad s^{-1}. The rope is inclined at a constant angle θ to the vertical. Find θ.
In a more refined model, the rope is assumed to be elastic, with natural length 2 m and modulus of elasticity 10 000 N. The angular speed is the same as before. Find the angle θ in this case. (UCLES)

6.

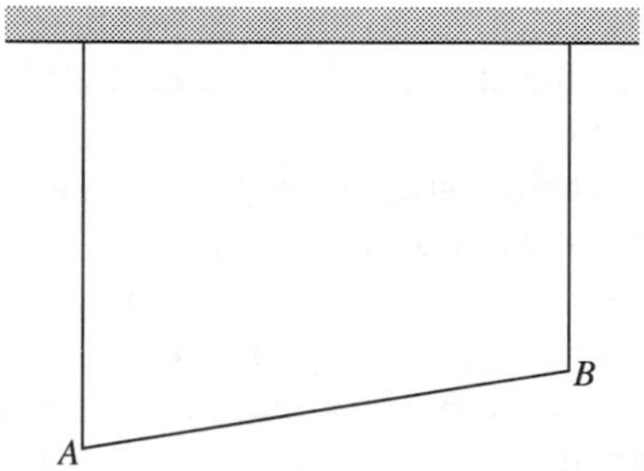

The diagram shows a non-uniform rod AB of length l hanging in equilibrium, suspended by two vertical elastic strings. The strings have the same unstretched length and the same modulus of elasticity. In the equilibrium position shown, the extension of the string attached to A is twice the extension of the string attached to B. Find the distance of the centre of gravity of the rod AB from A. (UCLES)

7. A ball B of mass m is attached to one end of a light elastic string of natural length a and modulus $2mg$. The other end of the string is attached to a fixed point O. The ball is projected vertically upwards from O with speed $\sqrt{(8ga)}$. Find the speed of the ball when $OB = 2a$.
Given that the string breaks when $OB = 2a$, find the speed of the ball when it returns to O. (AEB 1992)

8. In the dangerous sport of bungee diving an individual attaches one end of an elastic rope to a fixed point on a river bridge. He/she is attached to the other end and jumps over the bridge so as to fall vertically downwards towards the water. The rope should be such that the diver comes to rest just above the surface of the water. In order to find out which particular ropes are suitable experiments are carried out with weights attached to the rope rather than people. In one experiment it was found that when a weight of mass m was attached to a particular rope of natural length a and dropped from a bridge at a height of $3a$ above the water level then the weight just reached the level of the water. Show that the modulus of elasticity of the rope is $\dfrac{3mg}{2}$.
The weight of mass m is then removed and a weight of mass $\dfrac{5m}{2}$ is then attached to this rope and dropped from the same height. Find the speed of the weight just as it reaches the water.
When the weight emerges from the water its speed has been reduced to zero by the resistance of the water. Show by using conservation of energy or otherwise and assuming that the rope does not slacken, that the subsequent speed v of the weight at height h above the water level is given by

$$v^2 = \frac{gh}{5a}(2a - 3h).$$

Describe the subsequent motion of the weight. (WJEC)

9. A railway truck has two identical buffers. Each buffer can be modelled as a light spring which is held horizontally with one end attached to a fixed point on the truck. Each spring exerts a total force of magnitude 37 500 N for every centimetre it is compressed. A railway truck of mass 3000 kg runs with a speed of $2{\cdot}5\,\mathrm{m\,s^{-1}}$ against a fixed wall and the buffers are equally compressed. Find, in centimetres, the maximum compression of the springs when the truck is brought to rest. (AEB 1993)

10.

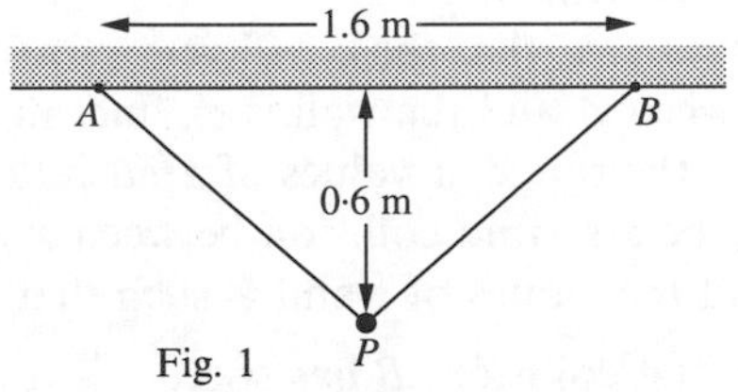

Fig. 1

Two light elastic strings each have natural length 0·8 m and modulus λ N. One end of each string is fastened to a particle P, of mass 0·2 kg, and the other ends are attached to fixed points A and B on a horizontal ceiling, where $AB = 1{\cdot}6$ m. Initially, P is held on the ceiling at the mid-point of AB. After being released from rest in this position, P falls vertically and comes instantaneously to rest at a point 0·6 m below the level of the ceiling, as shown in Fig. 1.
Using the principle of conservation of energy,
(a) show that $\lambda = 23{\cdot}52$
(b) find the speed of P at the instant when the length of each string is 0·9 m, giving your answer to 2 significant figures.
(ULEAC)

11. An ice-hockey puck, of mass 250 grams, moving across the ice at a speed of $24\,\mathrm{m\,s^{-1}}$, strikes a fixed vertical barrier at right angles to its path. The coefficient of restitution is $\frac{1}{3}$. Find the impulse exerted on the puck in the impact. (UCLES)

12. A large box of mass 2 kg rests on a smooth horizontal surface. A boy decides to propel the box along the surface by throwing a ball against the box. The ball, when travelling horizontally with speed $10\,\mathrm{m\,s^{-1}}$, strikes the box normally at the centre of a plane face, and then returns to the boy.
(a) The ball is of mass 0·4 kg and the coefficient of restitution between the ball and the box is 0·5.
 (i) Show that after the ball has been thrown once against the box, the box's speed will be $2{\cdot}5\,\mathrm{m\,s^{-1}}$.
 (ii) The ball is thrown a second time against the box, so that it strikes it as before. Show that the resulting speed of the box will then be greater than $4\,\mathrm{m\,s^{-1}}$.
(b) Give a reason why, if the boy uses a ball of mass 2 kg, throwing it as before, he will not be able to use the same ball more than once if he remains in the same position. (UODLE)

13. Two small uniform smooth spheres A and B of equal size and of mass m and $4m$ respectively are moving directly towards each other with speeds $2u$ and $6u$ respectively. The coefficient of restitution between the spheres is $\frac{1}{2}$. Find the speed of B immediately after the spheres collide. (AEB 1991)

14. A small smooth sphere P of mass $6m$ moving on a smooth plane in a straight line with constant speed $8u$ collides directly with a small smooth sphere Q of the same radius but of mass $4m$ and moving in the same direction with speed $6u$. The direction of motion of sphere Q is unchanged by the collision and immediately after collision it moves with speed $8u$. Find:
(i) the speed of P immediately after collision,
(ii) the coefficient of restitution. (WJEC)

15. Three identical spheres A, B and C lie on a smooth horizontal table with their centres in a straight line and with B between A and C. Given that A is projected towards B with speed u, show that, after impact, B moves with speed $\frac{1}{2}u(1+e)$, where e is the coefficient of restitution between each of the spheres.
When C first moves, it is found to have speed $\dfrac{9u}{16}$. Find e.
In this case show that when C begins to move, 75/128 of the initial kinetic energy has been lost. (AEB 1989)

16. The diagram shows two small, smooth spheres, P and Q, moving towards each other in the same straight line on a smooth horizontal surface. The mass of P is $2m$ kg, and the mass of Q is $5m$ kg. Their speeds, prior to the ensuing collision, are $u\,\text{m s}^{-1}$ and $3u\,\text{m s}^{-1}$ respectively. The coefficient of restitution between the spheres is e.

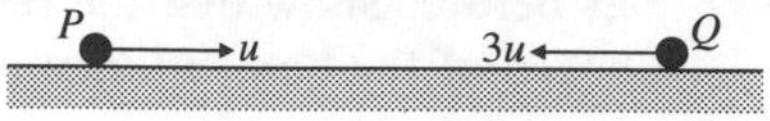

In the collision the direction of the motion of P is reversed. Thereafter P moves at $v\,\text{m s}^{-1}$ and Q at $w\,\text{m s}^{-1}$, both in the direction QP.
Show that $v = \frac{1}{7}u(20e + 13)$, and find w in terms of e and u.

(a) (i) Determine the value of e for which $v = 3u$.
(ii) Show that there is no value of e for which Q is brought to rest by the collision, and determine in terms of u the least possible value of w.

(b) Find, in terms of m and u only, an expression for the total kinetic energy of both particles before the collision.
Hence show that, in the case when $e = 0{\cdot}75$, the kinetic energy lost during the collision is $5mu^2$ joules.
State one way in which energy can be lost during a collision of this kind.

(c) In the case when $e = 0{\cdot}4$, determine the magnitude of the impulse which acts on Q.
(UODLE)

17. A sphere P, of mass m, is moving in a straight line with speed u on the surface of a smooth horizontal table. Another sphere Q, of mass $5m$ and having the same radius as P, is initially at rest on the table. The sphere P strikes the sphere Q directly, and the direction of motion of P is reversed by the impact. The coefficient of restitution between P and Q is e.

(a) Find an expression, in terms of u and e, for the speed of P after the impact.

(b) Find the set of possible values of e.
(ULEAC)

18. Two uniform smooth spheres A and B have equal radii but are of masses m and $4m$ respectively. They are at rest on a smooth horizontal floor with their line of centres perpendicular to a smooth vertical wall and with A lying between B and the wall.
A is projected away from the wall along the lines of centres towards B with speed u. The coefficient of restitution between the spheres is e. Find the speed of B after the impact and find the condition that e has to satisfy in order that A moves towards the wall after impact with B.
Given also, that the coefficient of restitution between A and the wall is e', find, in terms of e, the range of values of e' such that there will be a second collision between A and B.
Find the values of e and e' such that, after the first collision, B has speed $\frac{7u}{20}$ and, after the collision with the wall, A has speed $\frac{2u}{5}$.
Determine, in this case, the kinetic energy of the system immediately before the second impact of the spheres. (AEB 1990)

19. Three small bodies, A of mass m, B of mass $2m$, C of mass $3m$, are such that A is connected to B and B to C by two equal light inextensible strings. The bodies lie together at rest with A, B, and C in that order in a straight line on a smooth horizontal surface. Body C is given a speed u along the surface away from B and A. The strings do not impede the motion of the particles when slack.

(a) Find the speed with which all three bodies begin to move together.

(b) The motion of A, B, and C is in a line perpendicular to a fixed plane barrier and the body C is next in a perfectly elastic collision with the barrier. The coefficient of restitution between the bodies is $\frac{1}{2}$.

(i) Show that after C's collision with B its speed is $\frac{1}{10}u$ and determine the speeds of A and B.

(ii) Show that the total momentum of the bodies after C's collision with B is zero and say where the momentum of the bodies has been lost.

(c) What happens next? Give reasons. (UODLE)

20. Three small smooth spheres A, B and C, of equal radii and masses m, $2m$ and $3m$ respectively, are placed at rest with their centres in a straight line, l, on a smooth horizontal table with B between A and C. The sphere A is now projected along l towards B with speed $5u$. Given that, after the collision between A and B, B moves towards C with speed $3u$, find

(a) the magnitude and direction of the velocity of A after impact,

(b) the coefficient of restitution between A and B,

(c) the loss in kinetic energy due to the collision between A and B.

The sphere B now moves to collide with C and, as a result, C receives an impulse of magnitude $4mu$. Find the velocities of B and C after their collision and the coefficient of restitution between them. (AEB 1991)

21. A circular groove, of radius 0·5 m, has been cut in a horizontal surface; the points A and B are at the opposite ends of a diameter of the groove. Two small smooth spheres P, Q, with equal radii and masses 0·01 kg and 0·02 kg respectively, are constrained to move round the groove. Initially P and Q are at rest at A and B respectively. The sphere P is projected from A, along the groove, so that its speed immediately before collision with Q is $u\,\mathrm{m\,s^{-1}}$. The coefficient of restitution between the spheres is $\frac{1}{8}$. Find, in terms of u, the speeds of P and Q immediately after collision.

(a) Assuming that the groove is smooth and that $u = 12$, find the time that elapses before the spheres next collide.

(b) Assume now that the groove is rough and that P and Q are at rest at A and B respectively. The sphere P is projected from A with speed $12\,\mathrm{m\,s^{-1}}$ and the speed of Q immediately after collision is $3\,\mathrm{m\,s^{-1}}$. Find

(i) the impulse acting on P during its collision with Q

(ii) the work done by friction as P travelled from A to B. (WJEC)

22. A small rubber ball is held at height h above a smooth level floor and released from rest at time $t = 0$. If the coefficient of restitution between the ball and the floor is e, show that after the first bounce the ball rises to a height h_1, where $h_1 = e^2h$.

The ball continues to bounce until it comes to rest. Show that the total distance travelled by the ball from initial release to rest is

$$\frac{1+e^2}{1-e^2}\,h.$$

Find

(i) the time when the ball first hits the floor,

(ii) the time between the first and second impacts of the ball on the floor,

Show that the ball comes to rest when

$$t = \frac{1+e}{1-e}\sqrt{\left(\frac{2h}{g}\right)}. \qquad \text{(OCSEB)}$$

Hint. $1 + x + x^2 + \ldots = \dfrac{1}{1-x}$ for $|x| < 1$.

16 Use of calculus

In Chapter 2 the motion of a body in a straight line with uniform or constant acceleration was considered. The four uniform acceleration formulae could be used:

$$v = u + at \qquad s = ut + \tfrac{1}{2}at^2$$

$$v^2 = u^2 + 2as \qquad s = \frac{(u+v)t}{2}$$

If the acceleration is non-uniform, differential equations and calculus must be used in place of these equations:

velocity = rate of change of distance

or $$v = \frac{ds}{dt}$$

The dot notation is used to show differentiation with respect to time, for example:

$$\text{velocity} = \frac{ds}{dt} = \dot{s}$$

In a similar way, $\ddot{s}$ stands for the second differential of s with respect to time:

acceleration = rate of change of velocity

or $$a = \frac{dv}{dt} = \dot{v}$$

and $$a = \frac{d}{dt}\left(\frac{ds}{dt}\right) = \frac{d^2s}{dt^2} = \ddot{s}$$

In all the worked examples in this chapter, the letters a, v, s and t have their usual meaning and SI units are used throughout.

Example 1

A body moves in a straight line such that $s = 4t^2 - t^3$.
Find:
(a) s when $t = 3$
(b) v when $t = 2$
(c) a when $t = 1$.

(a) Given $s = 4t^2 - t^3$
when $t = 3$, $s = 4(3)^2 - (3)^3 = 36 - 27 = 9\,\text{m}$

The displacement s is 9 m when $t = 3$ s.

(b) Given $s = 4t^2 - t^3$

$$v = \frac{ds}{dt} = 8t - 3t^2$$

when $t = 2$, $v = 8(2) - 3(2)^2 = 4\,\text{m}\,\text{s}^{-1}$

The velocity v is $4\,\text{m}\,\text{s}^{-1}$ when $t = 2$ s.

(c) Given $s = 4t^2 - t^3$

$$v = \frac{ds}{dt} = 8t - 3t^2$$

and $$a = \frac{d^2s}{dt^2} = 8 - 6t$$

When $t = 1,\ a = 8 - 6(1) = 2\,\text{m s}^{-2}$

The acceleration a is $2\,\text{m s}^{-2}$ when $t = 1$ s.

Acceleration as a function of time

In order to obtain an expression for the velocity when the acceleration is given as a function of time, the process of integration has to be employed:

$$a = \frac{dv}{dt} = f(t)$$

$$\therefore \quad \int dv = \int f(t)\,dt$$

or $$v = \int f(t)\,dt + \text{constant}$$

The value of the constant can only be determined if more information is given in the question.

Example 2

A body moves in a straight line such that $a = 4t$, and initially, i.e. when $t = 0$, the velocity of the body is $3\,\text{m s}^{-1}$. Find:
(a) a when $t = 5$ s
(b) v when $t = 2$ s.

(a) Given $a = 4t$
when $t = 5,\ a = 4(5) = 20\,\text{m s}^{-2}$

The acceleration a is $20\,\text{m s}^{-2}$ when $t = 5$ s.

(b) Given $$a = \frac{dv}{dt} = 4t$$

$$\int dv = \int 4t\,dt$$

or $$v = 2t^2 + C$$

but when $t = 0,\ v = 3$

Thus $$3 = 2(0)^2 + C$$
$$\therefore \quad C = 3$$
$$\therefore \quad v = 2t^2 + 3 \quad \text{on substituting for } C$$
When $t = 2,\ v = 2(2)^2 + 3 = 11\,\text{m s}^{-1}$

The velocity v is $11\,\text{m s}^{-1}$ when $t = 2$ s.

The relationships between displacement s, velocity v and acceleration a may be shown as follows.

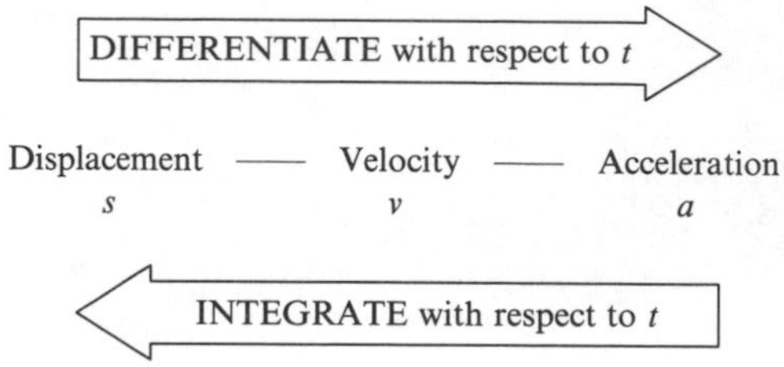

Thus if s is given as a function of time, differentiation will give an expression for velocity, and a second differentiation will give an expression for acceleration.
On the other hand, if a is given as a function of time, integration will give an expression for velocity, and a second integration will give an expression for displacement.
Again, it should be remembered that a constant must be introduced at each integration.

Example 3

A body moves in a straight line such that $v = 2t^2 - 11t + 14$. Initially, i.e. when $t = 0$, the displacement of the body from some fixed point O on the line is 50 m. Find:
(a) the initial velocity of the body
(b) the values of t when the body is at rest
(c) the acceleration of the body when $t = 5$ s
(d) the displacement of the body from O when $t = 6$ s.

(a) Given $\quad v = 2t^2 - 11t + 14$

when $t = 0$, $\quad v = 2(0) - 11(0) + 14 = 14$

$\therefore \quad v = 14\,\text{m s}^{-1}$

The initial velocity of the body is $14\,\text{m s}^{-1}$.

(b) Given $\quad v = 2t^2 - 11t + 14$

the body is at rest when $v = 0$

$\therefore \quad 0 = 2t^2 - 11t + 14$

giving $\quad t = 3{\cdot}5$ or 2 s

The body is at rest when $t = 3{\cdot}5$ s and when $t = 2$ s.

(c) Given $\quad v = 2t^2 - 11t + 14$

acceleration $\quad a = \dfrac{dv}{dt} = \dot{v}$

$\quad = 4t - 11$

when $t = 5$, $\quad a = 4(5) - 11$

$\therefore \quad a = 9\,\text{m s}^{-2}$

When $t = 5$ s the acceleration of the body is $9\,\text{m s}^{-2}$.

(d) Given $v = 2t^2 - 11t + 14$

by integration, $s = \dfrac{2t^3}{3} - \dfrac{11t^2}{2} + 14t + C$

But $s = 50$ when $t = 0$

$\therefore \quad 50 = \frac{2}{3}(0) - \frac{11}{2}(0) + 14(0) + C \quad \text{or} \quad C = 50$

Hence $s = \frac{2}{3}t^3 - \frac{11}{2}t^2 + 14t + 50$

When $t = 6$, $s = \frac{2}{3}(6)^3 - \frac{11}{2}(6)^2 + 14(6) + 50$

$= 144 - 198 + 84 + 50$

$\therefore \quad s = 80\,\text{m}$

When $t = 6\,\text{s}$, the displacement of the body is 80 m.

Motion in the i–j plane

If motion with non-uniform acceleration takes place in the **i–j** plane and vector notation is employed, the same methods as in the previous examples may be used and each component can be differentiated or integrated separately.

Suppose: $\mathbf{s} = x\mathbf{i} + y\mathbf{j} = \begin{pmatrix} x \\ y \end{pmatrix}$ where x and y are each functions of time.

It follows that: $\mathbf{v} = \dot{x}\mathbf{i} + \dot{y}\mathbf{j} = \begin{pmatrix} \dot{x} \\ \dot{y} \end{pmatrix}$

and $\mathbf{a} = \ddot{x}\mathbf{i} + \ddot{y}\mathbf{j} = \begin{pmatrix} \ddot{x} \\ \ddot{y} \end{pmatrix}$

Example 4

If $\mathbf{v} = t^2\mathbf{i} + 3t\mathbf{j}$ and, when $t = 0$, $\mathbf{s} = 18\mathbf{i} - 24\mathbf{j}$, find:

(a) **a** when $t = 2\,\text{s}$

(b) **s** when $t = 6\,\text{s}$.

(a) Given $\mathbf{v} = t^2\mathbf{i} + 3t\mathbf{j}$

$$\mathbf{a} = \frac{d\mathbf{v}}{dt} = 2t\mathbf{i} + 3\mathbf{j}$$

when $t = 2$, $\mathbf{a} = 2(2)\mathbf{i} + 3\mathbf{j} = 4\mathbf{i} + 3\mathrm{i}$

When $t = 2\,\text{s}$, the acceleration is $(4\mathbf{i} + 3\mathbf{j})\,\text{m s}^{-2}$.

(b) Given $\mathbf{v} = t^2\mathbf{i} + 3t\mathbf{j}$

$$\frac{d\mathbf{s}}{dt} = t^2\mathbf{i} + 3t\mathbf{j}$$

$$\therefore \quad \int d\mathbf{s} = \int (t^2\mathbf{i} + 3t\mathbf{j})\,dt$$

$$\therefore \quad \mathbf{s} = \frac{t^3}{3}\mathbf{i} + \frac{3t^2}{2}\mathbf{j} + A\mathbf{i} + B\mathbf{j}$$

But $\mathbf{s} = 18\mathbf{i} - 24\mathbf{j}$ when $t = 0$,

$\therefore \quad 18\mathbf{i} - 24\mathbf{j} = 0\mathbf{i} + 0\mathbf{j} + A\mathbf{i} + B\mathbf{j}$

or $\quad 18\mathbf{i} - 24\mathbf{j} = A\mathbf{i} + B\mathbf{j}$

$\therefore \quad A = 18 \quad \text{and} \quad B = -24$

$$\mathbf{s} = \left(\frac{t^3}{3} + 18\right)\mathbf{i} + \left(\frac{3t^2}{2} - 24\right)\mathbf{j}$$

When $t = 6\,\text{s}$ $\quad \mathbf{s} = \left(\frac{6^3}{3} + 18\right)\mathbf{i} + \left(\frac{3(6)^2}{2} - 24\right)\mathbf{j}$

or $\quad \mathbf{s} = 90\mathbf{i} + 30\mathbf{j}$

When $t = 6\,\text{s}$, the displacement $\mathbf{s}$ is $(90\mathbf{i} + 30\mathbf{j})\,\text{m}$.

Example 5

A particle P of mass 3 kg moves such that its displacement $\mathbf{s}$ at time t seconds is given by:

$$\mathbf{s} = \begin{pmatrix} 3 + \sin 2t \\ \cos 2t \end{pmatrix} \text{ metres.}$$

(a) Calculate the value of t when P crosses the x-axis for the first time.
(b) By first finding an expression for $\mathbf{v}$ in terms of t, show that the speed of P is constant.
(c) By first finding an expression for $\mathbf{a}$ in terms of t, calculate the force acting on P when $t = \frac{\pi}{2}$.

(a) P crosses the x-axis when the $\mathbf{j}$ component of its displacement is zero,

i.e. $\quad \cos 2t = 0$

$\therefore \quad 2t = \frac{\pi}{2}$

giving $\quad t = \frac{\pi}{4}$

P first crosses the x-axis when $t = \frac{\pi}{4}$ seconds.

(b) Given that $\quad \mathbf{s} = \begin{pmatrix} 3 + \sin 2t \\ \cos 2t \end{pmatrix}$

it follows that $\quad \mathbf{v} = \frac{d\mathbf{s}}{dt} = \begin{pmatrix} 2\cos 2t \\ -2\sin 2t \end{pmatrix}$

$\therefore$
$$|\mathbf{v}| = \sqrt{(2\cos 2t)^2 + (-2\sin 2t)^2}$$
$$= \sqrt{4}$$
$$= 2$$

Hence the speed of P is always $2\,\text{m}\,\text{s}^{-1}$.

(c) Given that $\mathbf{v} = \begin{pmatrix} 2\cos 2t \\ -2\sin 2t \end{pmatrix}$

it follows that $\mathbf{a} = \dfrac{d\mathbf{v}}{dt} = \begin{pmatrix} -4\sin 2t \\ -4\cos 2t \end{pmatrix}$

When $t = \dfrac{\pi}{2}$ $\mathbf{a} = \begin{pmatrix} -4\sin \pi \\ -4\cos \pi \end{pmatrix}$

$= \begin{pmatrix} 0 \\ 4 \end{pmatrix}$

Using $\mathbf{F} = m\mathbf{a}$ gives: $\mathbf{F} = 3\begin{pmatrix} 0 \\ 4 \end{pmatrix} = \begin{pmatrix} 0 \\ 12 \end{pmatrix}$

The force acting on P when $t = \dfrac{\pi}{2}$ is $\begin{pmatrix} 0 \\ 12 \end{pmatrix}$ N.

Motion in i–j–k space

The technique for the **i–j** plane can be extended to **i–j–k** space.

If $\mathbf{s} = x\mathbf{i} + y\mathbf{j} + z\mathbf{k} = \begin{pmatrix} x \\ y \\ z \end{pmatrix}$ where x, y and z are each functions of time,

then $\mathbf{v} = \dot{x}\mathbf{i} + \dot{y}\mathbf{j} + \dot{z}\mathbf{k} = \begin{pmatrix} \dot{x} \\ \dot{y} \\ \dot{z} \end{pmatrix}$

and $\mathbf{a} = \ddot{x}\mathbf{i} + \ddot{y}\mathbf{j} + \ddot{z}\mathbf{k} = \begin{pmatrix} \ddot{x} \\ \ddot{y} \\ \ddot{z} \end{pmatrix}$

Example 6

If $\mathbf{a} = 4\mathbf{i} - 6t\mathbf{j} + 12t^2\mathbf{k}$ and, when $t = 1$, $\mathbf{v} = 2\mathbf{i} + 3\mathbf{j} + 4\mathbf{k}$ and $\mathbf{s} = 6\mathbf{i} - 5\mathbf{j} + 10\mathbf{k}$, find:

(a) **v** when $t = 3$

(b) **s** when $t = 2$.

(a) Given $\mathbf{a} = 4\mathbf{i} - 6t\mathbf{j} + 12t^2\mathbf{k}$

i.e. $\dfrac{d\mathbf{v}}{dt} = 4\mathbf{i} - 6t\mathbf{j} + 12t^2\mathbf{k}$

it follows that $\mathbf{v} = \int (4\mathbf{i} - 6t\mathbf{j} + 12t^2\mathbf{k})\,dt$

$= 4t\mathbf{i} - 3t^2\mathbf{j} + 4t^3\mathbf{k} + \mathbf{c}$ for some constant vector **c**.

But when $t = 1$ $\mathbf{v} = 2\mathbf{i} + 3\mathbf{j} + 4\mathbf{k}$

$\therefore$ $2\mathbf{i} + 3\mathbf{j} + 4\mathbf{k} = 4(1)\mathbf{i} - 3(1)^2\mathbf{j} + 4(1)^3\mathbf{k} + \mathbf{c}$

from which $\mathbf{c} = -2\mathbf{i} + 6\mathbf{j}$

$\therefore$ $\mathbf{v} = (4t - 2)\mathbf{i} + (6 - 3t^2)\mathbf{j} + 4t^3\mathbf{k}$ ... [1]

When $t = 3$, $\mathbf{v} = (4(3) - 2)\mathbf{i} + (6 - 3(3)^2)\mathbf{j} + 4(3)^3\mathbf{k}$

$= 10\mathbf{i} - 21\mathbf{j} + 108\mathbf{k}$

When $t = 3$ seconds the velocity **v** is $(10\mathbf{i} - 21\mathbf{j} + 108\mathbf{k})\,\text{m}\,\text{s}^{-1}$.

(b) Given [1] $\mathbf{v} = (4t - 2)\mathbf{i} + (6 - 3t^2)\mathbf{j} + 4t^3\mathbf{k}$

i.e. $\dfrac{d\mathbf{s}}{dt} = (4t - 2)\mathbf{i} + (6 - 3t^2)\mathbf{j} + 4t^3\mathbf{k}$

it follows that $\mathbf{s} = \displaystyle\int ((4t - 2)\mathbf{i} + (6 - 3t^2)\mathbf{j} + 4t^3\mathbf{k})\,dt$

$= (2t^2 - 2t)\mathbf{i} + (6t - t^3)\mathbf{j} + t^4\mathbf{k} + \mathbf{e}$

for some constant vector **e**.

But when $t = 1$ $\mathbf{s} = 6\mathbf{i} - 5\mathbf{j} + 10\mathbf{k}$

$\therefore$ $6\mathbf{i} - 5\mathbf{j} + 10\mathbf{k} = 0\mathbf{i} + 5\mathbf{j} + \mathbf{k} + \mathbf{e}$

from which $\mathbf{e} = 6\mathbf{i} - 10\mathbf{j} + 9\mathbf{k}$

$\therefore$ $\mathbf{s} = (2t^2 - 2t + 6)\mathbf{i} + (6t - t^3 - 10)\mathbf{j} + (t^4 + 9)\mathbf{k}$

When $t = 2$, $\mathbf{s} = 10\mathbf{i} - 6\mathbf{j} + 25\mathbf{k}$

When $t = 2$ seconds the displacement **s** is $(10\mathbf{i} - 6\mathbf{j} + 25\mathbf{k})$ m.

Example 7

A particle P moves such that at time t seconds its displacement, **s** metres, is given by:

$$\mathbf{s} = \begin{pmatrix} t^3 - 4 \\ t^2 - t \\ t^2 - 8t + 5 \end{pmatrix}$$

(a) By first finding an expression for **v**, the velocity of the particle, in terms of t calculate the speed of P when $t = 2$.
(b) Find the value of t for which the z-coordinate of **s** is a minimum and find the displacement of P at this time.
(c) Find the average velocity of P from $t = 1$ to $t=3$.

(a) Given $\mathbf{s} = \begin{pmatrix} t^3 - 4 \\ t^2 - t \\ t^2 - 8t + 5 \end{pmatrix}$

it follows that $\mathbf{v} = \dfrac{d\mathbf{s}}{dt} = \begin{pmatrix} 3t^2 \\ 2t - 1 \\ 2t - 8 \end{pmatrix}$

When $t = 2$ $\mathbf{v} = \begin{pmatrix} 12 \\ 3 \\ -4 \end{pmatrix}$

$\therefore$ $|\mathbf{v}| = \sqrt{(12)^2 + (3)^2 + (-4)^2}$

$= \sqrt{169}$

$= 13$

When $t = 2$ the speed of P is $13\,\text{m s}^{-1}$.

(b) By differentiation, $t^2 - 8t + 5$ is a minimum when $2t - 8 = 0$
i.e. when $t = 4$

(**Note.** It is a minimum because the second derivative is positive.)

When $t = 4$ $\quad \mathbf{s} = \begin{pmatrix} 60 \\ 12 \\ -11 \end{pmatrix}$

The z-coordinate of $\mathbf{s}$ is a minimum when $t = 4$ and, at that time,

$$\mathbf{s} = \begin{pmatrix} 60 \\ 12 \\ -11 \end{pmatrix}$$

(c) When $t = 1$, $\mathbf{s} = \begin{pmatrix} -3 \\ 0 \\ -2 \end{pmatrix}$ and when $t = 3$, $\mathbf{s} = \begin{pmatrix} 23 \\ 6 \\ -10 \end{pmatrix}$

The change in the displacement of the particle, from $t = 1$ to $t = 3$ is

$$\begin{pmatrix} 26 \\ 6 \\ -8 \end{pmatrix}$$

The average velocity in this time is $\begin{pmatrix} 13 \\ 3 \\ -4 \end{pmatrix}$ m s^{-1}.

Exercise 16A

All units in this exercise are in SI units. Questions **1** to **14** involve motion along a straight line. The letter s represents the displacement of the body at time t from a fixed point O on the line. The letters v and a represent the velocity and the acceleration of the body at time t ($t \geqslant 0$).

1. If $s = 5t^3$, find s when $t = 2$ s.
2. If $s = t^3 + t$, find v when $t = 3$ s.
3. If $s = 5t^2 - t^3$, find a when $t = 1$ s.
4. If $v = 6t^2$, find v when $t = 2$ s.
5. If $v = t^3$, find a when $t = 2$ s.
6. If $v = 4t + 5$ and $s = 10$ m when $t = 1$ s, find s when $t = 2$ s.
7. If $a = 6t$, find a when $t = 5$ s.
8. If $a = 6t$ and the body is initially at rest, find v when $t = 4$ s.
9. If $a = \frac{2}{3}t$, find s when $t = 6$ s given that $v = 4$ m s^{-1}, and $s = 10$ m when $t = 3$ s.
10. If $s = t^2 - 3$ find:
(a) s when $t = 2$ s
(b) an expression for the velocity of the body at time t
(c) the velocity when $t = 2$ s
(d) the value of t when velocity is 8 m s^{-1}
(e) the displacement of the body from O when $v = 8$ m s^{-1}.

11. If $s = 2t^3 - 21t^2 + 60t$, find:
(a) s when $t = 3$ s
(b) the values of t when the body is at rest
(c) the initial velocity of the body
(d) an expression for the acceleration of the body at time t
(e) the initial acceleration of the body.

12. If $v = 8t - 3t^2$ and the body is initially at O, find:
(a) v when $t = 2$ s
(b) an expression for the acceleration of the body at time t
(c) the acceleration when $t = 3$ s
(d) an expression for the displacement of the body from O at time t
(e) how far the body is from O when $t = 3$ s.

13. If $v = (3t - 2)(t - 4)$ and $s = 8$ m when $t = 1$ s, find:
(a) the initial velocity of the body
(b) the values of t when the body is at rest
(c) the acceleration of the body when $t = 3$ s
(d) the distance the body is from O when $t = 2$ s.

14. If $a = 2t$ and initially the body is at rest at O, find the velocity of the body when $t = 3$ s and the distance the body is then from O.

Questions **15** to **32** involve motion in the **i**–**j** plane with **s**, **v** and **a** representing displacement, velocity and acceleration vectors at time t ($t \geqslant 0$).

15. If $\mathbf{s} = t^3\mathbf{i} + 2t^2\mathbf{j}$, find **a** when $t = 1$ s.

16. If $\mathbf{s} = t^3\mathbf{i} + 9t\mathbf{j}$, find the speed of the body when $t = 2$ s.

17. If $v = 2t^2\mathbf{i} + 6\mathbf{j}$, find **a** when $t = 3$ s.

18. If $v = 3t^2\mathbf{i} + 10t\mathbf{j}$, find the distance the body is from the origin when $t = 2$ s given that $\mathbf{s} = (4\mathbf{i} - 4\mathbf{j})$ m when $t = 0$.

19. If $\mathbf{a} = (4t + 2)\mathbf{i} - 3\mathbf{j}$ and the body is initially at rest, find **v** when $t = 3$ s.

20. If $\mathbf{a} = 6t\mathbf{i} + 2\mathbf{j}$, find **s** when $t = 3$ s given that $\mathbf{v} = (4\mathbf{i} - \mathbf{j})\,\text{m s}^{-1}$ and $\mathbf{s} = (2\mathbf{i} + 3\mathbf{j})$ m when $t = 1$ s.

21. If $\mathbf{s} = 2\sin t\,\mathbf{i} + 3\cos t\,\mathbf{j}$, find **a** when $t = \dfrac{\pi}{2}$ s.

22. If $\mathbf{s} = 2\sqrt{3}\sin t\,\mathbf{i} + 8\cos t\,\mathbf{j}$, find the speed of the body when $t = \dfrac{\pi}{6}$ s.

23. If $\mathbf{v} = \sin 2t\,\mathbf{i} - \cos t\,\mathbf{j}$, find **a** when $t = \dfrac{\pi}{2}$ s.

24. If $\mathbf{v} = 4\cos 2t\,\mathbf{i} + 2\sin 2t\,\mathbf{j}$, find the distance of the body from the origin when $t = \pi$ s, given that $\mathbf{s} = 6\mathbf{i} - 2\mathbf{j}$ when $t = \dfrac{\pi}{4}$ s.

25. If $\mathbf{a} = 9\sin 3t\,\mathbf{i} + 2\cos t\,\mathbf{j}$, and the body is initially at rest, find **v** when $t = \dfrac{\pi}{6}$ s.

26. If $\mathbf{a} = 6\sin 6t\,\mathbf{i} + 9\cos 3t\,\mathbf{j}$, find **s** when $t = \dfrac{\pi}{3}$ s, given that $\mathbf{v} = (\mathbf{i} + 3\mathbf{j})\,\text{m s}^{-1}$ and $\mathbf{s} = (5\mathbf{i} + 2\mathbf{j})$ m when $t = \dfrac{\pi}{6}$ s.

27. A particle of mass 6 kg moves such that

$$\mathbf{s} = \begin{pmatrix} t^2 - 5 \\ t^2 - 3t + 2 \end{pmatrix}$$

(a) Calculate the times when the particle crosses the x-axis.
(b) Find an expression for **v** in terms of t, and hence calculate the speed of the particle at $t = 6$ s.
(c) Find **a** and hence calculate the force acting on the particle.

28. A particle of mass 2 kg moves such that

$$\mathbf{s} = \begin{pmatrix} t^2 - 4t - 5 \\ t^2 - 4t + 3 \end{pmatrix}$$

(a) Calculate the time when the particle crosses the y-axis.
(b) Find an expression for **v** in terms of t, and hence calculate the speed of the particle at $t = 2$ s.
(c) Find **a**, and hence calculate the force acting on the particle.

29. A particle of mass 4 kg moves such that

$$\mathbf{s} = \begin{pmatrix} t^3 - t^2 - 4t + 3 \\ t^3 - 2t^2 + 3t - 7 \end{pmatrix}$$

(a) Calculate the times when the particle crosses the line $y = x$.
(b) Find the velocity of the particle at $t = 4$ s.
(c) Find an expression for **a** in terms of t, and hence calculate the force acting on the particle at $t = \frac{2}{3}$ s.

30. A particle of mass 0·5 kg moves such that

$$\mathbf{s} = \begin{pmatrix} 4\sin 2t \\ 2\cos t - 1 \end{pmatrix}$$

(a) Calculate the time when the particle first crosses the x-axis.
(b) Find the velocity of the particle at $t = \dfrac{\pi}{6}$ s.
(c) Find an expression for the force acting on the particle at a time t.

31. A particle of mass 2 kg moves such that

$$\mathbf{s} = \begin{pmatrix} 2 - \cos 3t \\ 6\sin 2t \end{pmatrix}$$

(a) Show that the particle never crosses the y-axis.
(b) Find the velocity of the particle at $t = \dfrac{\pi}{6}$ s.
(c) Find the force acting on the particle at $t = \pi$ s.

32. A particle moves such that
$$\mathbf{s} = \begin{pmatrix} 2\sin t + \sin 2t \\ 4\cos t + \cos 2t \end{pmatrix}$$
(a) Find the velocity of the particle at $t = \dfrac{\pi}{3}$ s.
(b) Find the acceleration of the particle at $t = \dfrac{\pi}{2}$ s.
(c) Show that the force acting on the particle at $t = \dfrac{\pi}{3}$ s is parallel to the x-axis.

Questions **33** to **43** involve motion in **i**, **j**, **k** space with **s**, **v** and **a** representing displacement, velocity and acceleration vectors at time t, ($t \geqslant 0$).

33. If $\mathbf{a} = 6t\mathbf{i} + 6\mathbf{j} - 2\mathbf{k}$ and, when $t = 1$, $\mathbf{v} = (3\mathbf{i} + 6\mathbf{j} - 3\mathbf{k})\,\text{m s}^{-1}$ and $\mathbf{s} = (2\mathbf{i} + 5\mathbf{j} - 2\mathbf{k})$ m, find
(a) $\mathbf{v}$ when $t = 2$ s
(b) $\mathbf{s}$ when $t = 3$ s.

34. If $\mathbf{a} = 2\mathbf{i} + 6t\mathbf{j} + 12t^2\mathbf{k}$ and, when $t = 0$, $\mathbf{v} = (-3\mathbf{i} + \mathbf{k})\,\text{m s}^{-1}$ and $\mathbf{s} = (-\mathbf{i} + \mathbf{k})$ m, find
(a) $\mathbf{v}$ when $t = 1$ s
(b) $\mathbf{s}$ when $t = 2$ s.

35. If $\mathbf{a} = 6t\mathbf{i} - 2\mathbf{k}$ and, when $t = 2$, $\mathbf{v} = (\mathbf{i} + 12\mathbf{j} - 4\mathbf{k})\,\text{m s}^{-1}$ and $\mathbf{s} = (3\mathbf{i} + 6\mathbf{j})$ m, find
(a) $\mathbf{v}$ when $t = 4$ s
(b) $\mathbf{s}$ when $t = 3$ s.

36. If $\mathbf{s} = 4t^3\mathbf{i} + (6 - t^2)\mathbf{j} + 2t\,\mathbf{k}$, find
(a) $\mathbf{v}$ when $t = 1$ s
(b) $\mathbf{a}$ when $t = 2$ s.

37. If $\mathbf{v} = 4t^3\mathbf{i} + 6t\,\mathbf{j} - 3t^2\mathbf{k}$ and, when $t = 1$, $\mathbf{s} = (14\mathbf{i} + 6\mathbf{j} - 3\mathbf{k})$ m, find
(a) $\mathbf{a}$ when $t = 3$ s
(b) $\mathbf{s}$ when $t = 0$ s.

38. If $\mathbf{v} = (3t^2 - 10t)\mathbf{i} + 2\mathbf{j} - 6t\mathbf{k}$ and, when $t = 2$, $\mathbf{s} = (-9\mathbf{i} + 3\mathbf{j} - 13\mathbf{k})$ m, find
(a) $\mathbf{a}$ when $t = 5$ s
(b) $\mathbf{s}$ when $t = 3$ s.

39. A particle moves such that $\mathbf{s} = \begin{pmatrix} t^3 \\ 4t - t^2 \\ t^4 - 1 \end{pmatrix}$
Find:
(a) the velocity when $t = 3$ s
(b) the acceleration when $t = 2$ s
(c) the maximum value of the y-coordinate of **s**
(d) the average velocity between $t = 1$ s and $t = 3$ s.

40. A particle moves such that
$$\mathbf{s} = \begin{pmatrix} t - t^2 \\ 4t \\ t^3 - 2t^2 + t + 3 \end{pmatrix}$$
Find:
(a) the velocity when $t = 5$ s
(b) the acceleration when $t = 3$ s
(c) the minimum value of the z-coordinate of **s**
(d) the average velocity between $t = 0$ s and $t = 5$ s.

41. A particle moves such that
$$\mathbf{s} = \begin{pmatrix} \sin 2t \\ t + 1 \\ \cos t + \sin t \end{pmatrix}$$
Find:
(a) the velocity when $t = \dfrac{\pi}{2}$ s
(b) the acceleration when $t = \pi$ s
(c) the maximum value of the z coordinate of **s**
(d) the average velocity between $t = 0$ s and $t = \pi$ s.

42. A particle moves such that
$$\mathbf{v} = \begin{pmatrix} 2\cos 2t + 11 \\ 3\sin 3t \\ 4 \end{pmatrix},$$ and the particle passes through the origin at $t = 0$ s.
Find:
(a) the speed when $t = \dfrac{\pi}{6}$ s
(b) the displacement when $t = \dfrac{\pi}{2}$ s
(c) the acceleration when $t = \pi$ s.

43. A particle starts from rest at (2, 0, 0) and moves such that $\mathbf{a} = \begin{pmatrix} 16\cos 4t \\ 8\sin 2t \\ \sin t - 2\sin 2t \end{pmatrix}$

Find:
(a) the acceleration when $t = \pi$ s
(b) the velocity when $t = \dfrac{\pi}{2}$ s
(c) the displacement when $t = \dfrac{\pi}{4}$ s.

44. The acceleration of a particle moving in a straight line is governed by the relationship $a = \frac{3t^2}{2}$ where t is the time in seconds. When $t = 2$ s the particle passes through the point O with velocity $1\,\text{m s}^{-1}$.
Find the velocity of the particle when $t = 4$ s and the distance the particle is then from O.

45. A particle moves in a straight line with its acceleration a obeying the rule $a = 2t$, where t is the time in seconds. When $t = 0$ both s and v are equal to zero, where s is the distance of the particle from a point O and v is the velocity of the particle. For what value of t are s and v again numerically equal?

46. A particle moves along a straight line and t seconds after passing through an origin O the velocity of the body is given by $v = kt^2 - ct$, where k and c are constants. When $t = 2$ the body is again at O and has an acceleration of $6\,\text{m s}^{-2}$. Find the values of k and c.

47. A body moves along a straight line with acceleration given by $a = 2t$, where t is the time in seconds. When $t = 0$ the body is at rest at an origin O. The acceleration continues until $t = 3$, whereupon it ceases and the body is uniformly retarded to rest. During this retardation the acceleration is given by $a = -2\,\text{m s}^{-2}$. Find the value of t when the body comes to rest and the displacement of the body from O at that time.

48. A body moves along a straight line with acceleration given by $a = \frac{7t}{36}$, where t is the time in seconds. When $t = 0$ the body is at rest at an origin O. The acceleration continues until $t = 6$, whereupon it ceases and the body is retarded to rest. During this retardation the acceleration is given by $a = -\frac{t}{4}$. Find the value of t when the body comes to rest and the displacement of the body from O at that time.

Velocity as a function of displacement

In the examples so far considered s, v and a have always been given as functions of time t. This may not always be the case; dependent upon the information given and what is required, one of the following should be used:

$$a = \frac{v\,\mathrm{d}v}{\mathrm{d}s} \quad \text{or} \quad v = \frac{\mathrm{d}s}{\mathrm{d}t}$$

The first of these relationships may be obtained as follows:

$$a = \frac{\mathrm{d}v}{\mathrm{d}t} = \frac{\mathrm{d}v}{\mathrm{d}s} \times \frac{\mathrm{d}s}{\mathrm{d}t} = \frac{\mathrm{d}v}{\mathrm{d}s} \times v = v\frac{\mathrm{d}v}{\mathrm{d}s}$$

Example 8

If $v = 3s^2 - 4s$, find: (a) v when $s = 2$ m (b) a when $s = 2$ m.

(a) Given $v = 3s^2 - 4s$
when $s = 2$ m, $v = 3(2)^2 - 4(2) = 4\,\text{m s}^{-1}$

(b) We are given $v = 3s^2 - 4s$. To find a we use $a = v\frac{\mathrm{d}v}{\mathrm{d}s}$.

Now $\frac{\mathrm{d}v}{\mathrm{d}s} = 6s - 4$

$\therefore$ $a = v\frac{\mathrm{d}v}{\mathrm{d}s} = (3s^2 - 4s)(6s - 4)$

When $s = 2$ m, $a = (12 - 8)(12 - 4) = 32\,\text{m s}^{-2}$

Example 9

If $v = \dfrac{20}{3s - 2}$, find:

(a) v when $s = 4$ m
(b) s when $v = 5\ \text{m s}^{-1}$
(c) t when $s = 20$ m
given that $s = 0$ when $t = 0$

(a) Given $$v = \frac{20}{3s - 2}$$

when $s = 4\,\text{m}$, $$v = \frac{20}{3(4) - 2} = \frac{20}{10}$$

$$\therefore \quad v = 2\,\text{m}\,\text{s}^{-1}$$

When $s = 4\,\text{m}$, the velocity v is $2\,\text{m}\,\text{s}^{-1}$.

(b) Given $$v = \frac{20}{3s - 2}$$

when $v = 5\,\text{m}\,\text{s}^{-1}$, $$5 = \frac{20}{3s - 2}$$

$$\therefore \quad 3s - 2 = 4 \quad \text{or} \quad s = 2\,\text{m}$$

When $v = 5\,\text{m}\,\text{s}^{-1}$, the displacement s is $2\,\text{m}$.

(c) To find t, given v as a function of s, we must use $v = \dfrac{\text{d}s}{\text{d}t}$ as this expression involves t.

Hence $$\frac{\text{d}s}{\text{d}t} = \frac{20}{3s - 2}$$

$$\therefore \quad \int (3s - 2)\,\text{d}s = \int 20\,\text{d}t$$

or $$\tfrac{3}{2}s^2 - 2s = 20t + C$$

But $s = 0$ when $t = 0$ $\quad \therefore \quad 0 = 0 + C \quad$ or $\quad C = 0$

$$\therefore \quad \tfrac{3}{2}s^2 - 2s = 20t$$

When $s = 20$ m, $\tfrac{3}{2}(20)^2 - 2(20) = 20t$
$$t = 28\ \text{s}$$

When the displacement s is 20 m, then $t = 28s$.

Acceleration as a function of displacement

In this case, an expression for the velocity v is found by again using $a = v\dfrac{dv}{ds}$, as shown below.

acceleration $\qquad a = f(s)$

$$\therefore \qquad v\frac{dv}{ds} = f(s)$$

$$\therefore \qquad \int v\,dv = \int f(s)\,ds$$

$$\therefore \qquad \frac{v^2}{2} = \int f(s)\,ds + \text{constant}$$

Example 10

If $a = 3s + 5$ and initially $v = 1\,\text{m s}^{-1}$ when $s = 1\,\text{m}$, find v when $s = 2\,\text{m}$.

Given $\qquad a = 3s + 5$

it follows that: $\qquad v\dfrac{dv}{ds} = 3s + 5$

or $\qquad \displaystyle\int v\,dv = \int (3s + 5)\,ds$

$$\therefore \qquad \frac{v^2}{2} = \frac{3s^2}{2} + 5s + C$$

But $v = 1\,\text{m s}^{-1}$ when $s = 1\,\text{m}$

$$\therefore \qquad \tfrac{1}{2} = \tfrac{3}{2} + 5 + C \quad \text{or} \quad C = -6$$

$$\therefore \qquad \frac{v^2}{2} = \frac{3s^2}{2} + 5s - 6$$

and when $s = 2\,\text{m}$, $\qquad \dfrac{v^2}{2} = \dfrac{3(4)}{2} + 5(2) - 6$

$$\therefore \qquad v^2 = 20 \quad \text{or} \quad v = 2\sqrt{5}\,\text{m s}^{-1}$$

When the displacement s is $2\,\text{m}$, the velocity is $2\sqrt{5}\,\text{m s}^{-1}$.

Having found v in terms of s, it is then possible to find s in terms of t by substituting $\dfrac{ds}{dt}$ for v and integrating.

Example 11

If $a = s + 5$ and initially $s = 0$ and $v = 5\,\mathrm{m\,s^{-1}}$, find
(a) an expression for v as a function of s
(b) an expression for t as a function of s.

(a) $$a = s + 5$$

$$\therefore \quad v\frac{dv}{ds} = s + 5$$

$$\int v\,dv = \int (s + 5)\,ds$$

$$\therefore \quad \frac{v^2}{2} = \frac{s^2}{2} + 5s + C$$

$v = 5$ when $s = 0$ gives $C = \frac{25}{2}$

$$\therefore \quad v^2 = s^2 + 10s + 25 = (s + 5)^2$$

$$\therefore \quad v = s + 5$$ (The negative root is not applicable.)

The expression for v as a function of s is $v = s + 5$.

(b) Since $$v = s + 5$$

it follows that: $$\frac{ds}{dt} = s + 5$$

$$\int \frac{ds}{s + 5} = \int dt$$

$$\therefore \quad \ln(s + 5) = t + C'$$

$s = 0$ when $t = 0$ gives $C' = \ln 5$

Thus $$\ln\left(\frac{s + 5}{5}\right) = t$$

The expression for t as a function of s is $t = \ln\left(\frac{s + 5}{5}\right)$

Acceleration as a function of velocity

When the acceleration is given as a function of the velocity there are two possibilities:

(i) use $a = v\dfrac{dv}{ds}$ which will lead to a relationship between v and s

or

(ii) use $a = \dfrac{dv}{dt}$ which will lead to a relationship between v and t.

The choice will depend on the further information given and also on what relationship is required.

Example 12

If $a = \dfrac{6}{v^2}$, find s when $v = 4\,\text{m s}^{-1}$ given that $s = 0{\cdot}5\,\text{m}$ when $v = 2\,\text{m s}^{-1}$.

We are given $\quad a = \dfrac{6}{v^2}$

Since the data given and the answer required involve s, use $a = v\dfrac{dv}{ds}$

thus $\quad v\dfrac{dv}{ds} = \dfrac{6}{v^2}$

$\therefore \quad \int v^3\,dv = \int 6\,ds \quad \text{or} \quad \dfrac{v^4}{4} = 6s + C$

But $v = 2\,\text{m s}^{-1}$ when $s = 0{\cdot}5\,\text{m}$

$\therefore \quad C = 1$

Thus $\quad v^4 = 4(6s + 1)$

When $v = 4$, $\quad 4^4 = 4(6s + 1)$

$\therefore \quad 64 = 6s + 1 \quad \text{or} \quad s = 10{\cdot}5\,\text{m}$

When $v = 4\,\text{m s}^{-1}$ the displacement s is $10{\cdot}5\,\text{m}$.

Example 13

If $a = \dfrac{4}{v^3}$, find t when $v = 2\,\text{m s}^{-1}$ given that when $t = 0$, $v = 0$.

We are given. $\quad a = \dfrac{4}{v^3}$

Since the data given and the answer required involve t, use $a = \dfrac{dv}{dt}$.

Thus $\quad \dfrac{dv}{dt} = \dfrac{4}{v^3}$

$$\int v^3\,dv = \int 4\,dt$$

$\therefore \quad \dfrac{v^4}{4} = 4t + C$

But $v = 0$ when $t = 0$

$\therefore \quad C = 0$

$\therefore \quad \dfrac{v^4}{4} = 4t$

and when $\quad v = 2\,\text{m s}^{-1}, \quad \dfrac{2^4}{4} = 4t \quad \text{or} \quad t = 1\,\text{s}$

When $v = 2\,\text{m s}^{-1}$, $t = 1\,\text{s}$.

Example 14

If $a = 4 + 3v$, find s when $v = 2\,\text{m s}^{-1}$ given that $s = 0$ when $v = 0$.

Since the data given and the answer required involve s, use $a = v\dfrac{dv}{ds}$.

Thus $$v\frac{dv}{ds} = 4 + 3v$$

$$\int ds = \int \frac{v\,dv}{4+3v}$$

$$= \frac{1}{3}\int \frac{4+3v-4}{4+3v}\,dv$$ (This rearrangement is necessary to produce an expression which can be integrated.)

$$= \frac{1}{3}\int\left(1 - \frac{4}{4+3v}\right)dv$$

$$s = \frac{v}{3} - \frac{4}{9}\ln(4+3v) + C$$

$s = 0$ when $v = 0$ gives: $C = \dfrac{4}{9}\ln 4$

Hence: $$s = \frac{v}{3} - \frac{4}{9}\ln(4+3v) + \frac{4}{9}\ln 4$$

or $$s = \frac{v}{3} + \frac{4}{9}\ln\left(\frac{4}{4+3v}\right)$$

When $v = 2\,\text{m s}^{-1}$, $$s = \frac{2}{3} + \frac{4}{9}\ln\left(\frac{4}{4+6}\right) = 0{\cdot}2594\,\text{m}$$

When $v = 2\,\text{m s}^{-1}$, the displacement s is $0{\cdot}259\,\text{m}$.

Exercise 16B

Questions **1** to **18** involve bodies moving along a straight line. The letter s represents the displacement of the body at time t from a fixed point O on the line. The letters v and a represent the velocity and the acceleration of the body at time t. All units are in SI units.

v given as a function of s.

1. If $v = \dfrac{4}{1+s}$, find v when $s = 1\,\text{m}$.

2. If $v = 2s - 3$, find a when $s = 4\,\text{m}$.

3. If $v = \dfrac{s^2}{15}$, find t when $s = 10\,\text{m}$ given that $s = 3\,\text{m}$ when $t = 0$.

4. If $v = 4s - 2$, find s when $v = 8\,\text{m s}^{-1}$.

5. If $v = s - 2$, find s when $a = 3\,\text{m s}^{-2}$.

6. If $v = 2s + 3$, find t when $s = 3\,\text{m}$ given that $s = 0$ when $t = 0$.

a given as a function of s.

7. If $a = \dfrac{1}{2s-1}$, find a when $s = 3\,\text{m}$.

8. If $a = \dfrac{40}{s^2}$, find v when $s = 20\,\text{m}$, given that initially $v = 0$ and $s = 10\,\text{m}$.

9. If $a = 2s - 3$, find s when $a = 1\,\text{m s}^{-2}$.

10. If $a = \dfrac{1}{s+2}$, find s when $v = 1{\cdot}5\,\text{m s}^{-1}$, given that initially $v = 0$ and $s = 0$.

11. If $a = s + 2$ and initially $s = 0$ and $v = 2\,\text{m s}^{-1}$, find
(a) an expression for v as a function of s
(b) an expression for t as a function of s

12. If $a = 4s + 2$ and initially $s = 0$ and $v = 1\,\text{m s}^{-1}$, find
(a) the value of v when $s = 3\,\text{m}$
(b) the value of t when $s = 3\,\text{m}$

a given as a function of v.

13. If $a = 10 - v^2$, find a when $v = 3\,\text{m s}^{-1}$.
14. If $a = \dfrac{7}{v}$, find s when $v = 6\,\text{m s}^{-1}$, given that $v = 3\,\text{m s}^{-1}$ when $s = 0$.
15. If $a = \dfrac{2}{3v^2}$, find t when $v = 4\,\text{m s}^{-1}$ given that $v = 0$ when $t = 0$.
16. If $a = 2 + v$, find t when $v = 12\,\text{m s}^{-1}$ given that $v = 0$ when $t = 0$.
17. If $a = \dfrac{v^2}{5}$, find s when $v = 6\,\text{m s}^{-1}$ given that initially $s = 0$ and $v = 1\,\text{m s}^{-1}$.
18. If $a = 5 + 2v$, find s when $v = 5\,\text{m s}^{-1}$ given that initially $s = 0$ and $v = 0$.

Miscellaneous.

19. Four points O, A, B and C lie in that order on a straight line with OA = 2 m, OB = 4 m and OC = 8 m. Initially a body is at A and moves in the direction OC such that its velocity v at any instant is given by $v = \dfrac{3}{s}\,\text{m s}^{-1}$, where s is the distance in metres that the body is from O at that instant. Find the time taken for the body to go from B to C.
20. OAB is a straight line with A between O and B and OA = 2 m. A body is initially at rest at A and moves towards B with its acceleration at any instant given by $a = \dfrac{k}{s}\,\text{m s}^{-2}$, where s metres is the distance that the body is from O at that instant and k is a constant. Show that if $v\,\text{m s}^{-1}$ is the velocity of the body, then $v^2 = 2k \ln \dfrac{s}{2}$.
21. A body is initially at rest at an origin O. It subsequently moves in a straight line away from O with acceleration at any instant given by $a = k(1 + v)$, where v is the velocity of the particle at that instant and k is a constant. If s is the displacement of the particle from O at any instant, show that $s = \dfrac{1}{k}\,[v - \ln(1 + v)]$.
22. Three points O, A and B lie in that order on a straight line with OA = d m and AB = 4 m. A particle travels along the line from O towards B with its velocity v at any instant given by $v = \dfrac{8}{3 + s}\,\text{m s}^{-1}$, where s metres is the distance that the particle is from O at that instant. If the particle takes three seconds to travel from A to B, find the value of d.
23. A particle moves along a straight line with its acceleration at any instant given by $a = \alpha + \beta v^2$, where v is the velocity of the particle at that instant and α and β are constants. Initially the particle is at rest at a point O on the line. If the displacement of the particle from O during the motion is s, show that $v^2 = \dfrac{\alpha}{\beta}\,(e^{2\beta s} - 1)$.
24. A body moves along a straight line with its acceleration at any time t given by $a = 5 - 2v$, where v is the velocity of the body at time t. (All units are SI units.) When $t = 0$, the body is at rest. Show that $t = \tfrac{1}{2} \ln \left(\dfrac{5}{5 - 2v}\right)$ s, and hence obtain an expression for v as a function of t. Show that as the motion continues, the velocity of the body approaches a maximum value and find this maximum value.

Variable force

If a force, which is constant in direction but variable in magnitude, acts upon a body, then the acceleration produced by the force will be variable. If Newton's Second Law, $F = ma$ is used, one of the differential forms $v\dfrac{dv}{ds}$ or $\dfrac{dv}{dt}$ must be used for the acceleration. Which one is used will depend upon the data given and the type of relationship required.

The worked examples in this section will, unless stated otherwise, involve a horizontal force F of constant direction and variable magnitude acting upon a body of constant mass m resting on a smooth horizontal surface.
The displacement s of the body from a fixed point O and the velocity v of the body after time t will be in SI units.

Example 15

If $F = 3t + 1$, $m = 4\,\text{kg}$ and the body is initially at rest at a point O, find:
(a) v when $t = 2\,\text{s}$ (b) s when $t = 2\,\text{s}$.

(a) Given $F = 3t + 1$, we need to find v when $t = 2\,\text{s}$
$m = 4\,\text{kg}$
$s = 0$ and $v = 0$ when $t = 0$

Since v and t are involved, use $\dfrac{dv}{dt}$ for a in Newton's Second Law equation:

$$F = ma$$

$$3t + 1 = 4\frac{dv}{dt}$$

$$\int (3t + 1)\,dt = \int 4\,dt$$

$$\frac{3t^2}{2} + t = 4v + C$$

$v = 0$ when $t = 0$, hence $C = 0$

$$\therefore \quad 4v = \frac{3t^2}{2} + t$$

When $t = 2\,\text{s}$, $$4v = \frac{3(4)}{2} + 2$$

$$\therefore \quad v = 2\,\text{m}\,\text{s}^{-1}$$

When $t = 2\text{s}$, the velocity v is $2\,\text{m}\,\text{s}^{-1}$.

(b) We need to find s when $t = 2\,\text{s}$

From part (a) $$4v = \frac{3t^2}{2} + t$$

Thus $$4\frac{ds}{dt} = \frac{3t^2}{2} + t$$

$$\therefore \quad \int 4\,ds = \int \left(\frac{3t^2}{2} + t\right)dt$$

$$\therefore \quad 4s = \frac{t^3}{2} + \frac{t^2}{2} + C'$$

$s = 0$ when $t = 0$, hence $C' = 0$

$$\therefore \quad 4s = \frac{t^3}{2} + \frac{t^2}{2}$$

When $t = 2\,\text{s}$, $$4s = \frac{8}{2} + \frac{4}{2}$$

$$\therefore \quad s = 1{\cdot}5\,\text{m}$$

When $t = 2\,\text{s}$, the displacement s is $1{\cdot}5\,\text{m}$.

Example 16

If $F = 5s + 6$, $m = 1$ kg and the body is initially at rest at a point O, find
(a) v when $s = 4$ m
(b) s when $v = 9\,\text{m s}^{-1}$.

(a) Given $F = 5s + 6$, we need to find v when $s = 4$ m
$m = 1$ kg
$s = 0$ and $v = 0$ when $t = 0$

Since v and s are involved, use $v\dfrac{dv}{ds}$ for a in Newton's Second Law equation:

$$F = ma$$

$$5s + 6 = (1)\,v\frac{dv}{ds}$$

$$\int (5s+6)\,ds = \int v\,dv$$

$$\therefore \quad \frac{5s^2}{2} + 6s = \frac{v^2}{2} + C$$

$v = 0$ when $s = 0$, hence $C = 0$

$$\therefore \quad \frac{v^2}{2} = \frac{5s^2}{2} + 6s$$

When $s = 4$ m, $$\frac{v^2}{2} = \frac{5(16)}{2} + 6(4)$$

$$\therefore \quad v = 8\sqrt{2}\,\text{m s}^{-1}$$

When $s = 4$ m, the velocity is $8\sqrt{2}\,\text{m s}^{-1}$.

(b) We need to find s when $v = 9\,\text{m s}^{-1}$
From part (a) $v^2 = 5s^2 + 12s$
Hence, when $v = 9\,\text{m s}^{-1}$, $81 = 5s^2 + 12s$
$\therefore \quad 5s^2 + 12s - 81 = 0$
or $\quad (5s + 27)(s - 3) = 0$
$\therefore \quad s = -5\tfrac{2}{5}$ or $+3$

Taking the positive answer, when the velocity is $9\,\text{m s}^{-1}$, then $s = 3$ m.

Example 17

If $F = \dfrac{3}{2v+1}$, $m = 2$ kg and the body is initially moving with a velocity of $2\,\text{m s}^{-1}$, find t when $v = 6\,\text{m s}^{-1}$.

Given $F = \dfrac{3}{2v+1}$, we need to find t when $v = 6\,\text{m s}^{-1}$

$s = 0$ and $v = 2$ when $t = 0$

Since t is required use $\dfrac{dv}{dt}$ for a in $F = ma$:

$$\frac{3}{2v+1} = 2\frac{dv}{dt}$$

$\therefore$ $$\int 3\,dt = \int 2(2v+1)\,dv \qquad \ldots[1]$$

or $$3t = 2(v^2+v) + C$$

$v = 2$ when $t = 0$, hence $C = -12$

$\therefore$ $$3t = 2(v^2+v) - 12$$

when $v = 6\,\text{m s}^{-1}$ $$3t = 2(36+6) - 12$$
$$= 72$$

$\therefore$ $$t = 24$$

When the velocity is $6\,\text{m s}^{-1}$, $t = 24$ s.

Alternative method

In this method, instead of finding a constant of integration, we introduce the initial condition that $v = 2\,\text{m s}^{-1}$ when $t = 0$ s as limits in the integral.

So, from Equation [1]:

$$\int_0^T 3\,dt = \int_2^6 2(2v+1)\,dv$$

The lower limits of 0 and 2 refer to the fact that when $t = 0$, $v = 2$, and the upper limits of T and 6 refer to the fact that we are looking for the value, T, of t when $v = 6$.

$\therefore$ $$[3t]_0^T = [2(v^2+v)]_2^6$$

$\therefore$ $$3T - 0 = 2(36+6) - 2(4+2)$$

$\therefore$ $$3T = 72$$

$\therefore$ $$T = 24$$

When the velocity is $6\,\text{m s}^{-1}$, $t = 24$ s.

Example 18

A force **F** acts on a body of mass 250 g which is initially at rest at a fixed point O. If $\mathbf{F} = [(5t-2)\mathbf{i} + 4t\mathbf{j}]$ N, where t is the time for which the force has been acting on the body, find expressions for:
(a) the velocity vector of the body at time t
(b) the position vector of the body at time t.

(a) Given $\mathbf{F} = [(5t-2)\mathbf{i} + 4t\mathbf{j}]$ N, we need to find **v**
$m = 250$ g
$\mathbf{s} = 0$ and $\mathbf{v} = 0$ when $t = 0$

Since **v** is required, and t is involved, use $\frac{d\mathbf{v}}{dt}$ for **a** in $F = m\mathbf{a}$:

$$(5t-2)\mathbf{i} + 4t\mathbf{j} = \frac{1}{4}\frac{d\mathbf{v}}{dt}$$

$\therefore$ $$\int [(5t-2)\mathbf{i} + 4t\mathbf{j}]\,dt = \int \frac{1}{4}d\mathbf{v}$$

$\therefore$ $$(\tfrac{5}{2}t^2 - 2t)\mathbf{i} + 2t^2\mathbf{j} = \frac{\mathbf{v}}{4} + A\mathbf{i} + B\mathbf{j}$$

Since $\mathbf{v} = 0$ when $t = 0$, $0\mathbf{i} + 0\mathbf{j} = \mathbf{0} + A\mathbf{i} + B\mathbf{j}$

Equating coefficients of **i** and **j**, $A = 0$ and $B = 0$ gives:

$$\left(\frac{5t^2}{2} - 2t\right)\mathbf{i} + 2t^2\mathbf{j} = \frac{\mathbf{v}}{4}$$

The velocity vector $\mathbf{v} = [(10t^2 - 8t)\mathbf{i} + 8t^2\mathbf{j}]\ \text{m s}^{-1}$.

(b) Using $\frac{d\mathbf{s}}{dt}$ for **v** from part (a) gives:

$$\frac{d\mathbf{s}}{dt} = (10t^2 - 8t)\mathbf{i} + 8t^2\mathbf{j}$$

$$\therefore \qquad \mathbf{s} = \left(\frac{10t^3}{3} - 4t^2\right)\mathbf{i} + \frac{8t^3}{3}\mathbf{j} + C\mathbf{i} + D\mathbf{j}$$

$\mathbf{s} = 0$ when $t = 0$, hence $C = 0$ and $D = 0$

The position vector $\mathbf{s} = \left[\left(\frac{10t^3}{3} - 4t^2\right)\mathbf{i} + \frac{8t^3}{3}\mathbf{j}\right]$ m.

Exercise 16C

Questions **1** to **10** involve a horizontal force F of constant direction and variable magnitude, acting on a body of mass m initially at rest at a point O on a smooth horizontal surface. After a time t, the displacement of the body from O is s, and the body has a velocity v (SI units used throughout).

1. If $F = 2t$ and $m = 5$ kg, find v when $t = 4$ s.
2. If $F = 6t - 4$ and $m = 2$ kg, find s when $t = 4$ s.
3. If $F = 2s + 1$ and $m = 1{\cdot}5$ kg, find v when $s = 3$ m.
4. If $F = \frac{1}{4s + 5}$ and $m = 250$ g, find v when $s = 8$ m.
5. If $F = \frac{2}{s + 5}$ and $m = 470$ g, find s when $v = 2\ \text{m s}^{-1}$.
6. If $F = \frac{1}{v + 2}$ and $m = 250$ g, find t when $v = 4\ \text{m s}^{-1}$.
7. If $F = 8 - 3v^2$ and $m = 6$ kg, find s when $v = 1\ \text{m s}^{-1}$.
8. If $F = 5 - 3v$ and $m = 6$ kg, find t when $v = 1\ \text{m s}^{-1}$.
9. If $F = 4 + v^2$ and $m = 4$ kg, find v when $t = \frac{\pi}{2}$ s.

   $\left(\textbf{Hint.} \quad \int \frac{dx}{\alpha^2 + x^2} = \frac{1}{\alpha} \tan^{-1} \frac{x}{\alpha} + c\right)$

10. If $F = 40 - v$ and $m = 5$ kg, find s when $v = 15\ \text{m s}^{-1}$.

11. A resultant force **F** acts on a body of mass 500 g initially at rest at an origin. $\mathbf{F} = (3t\mathbf{i} + \mathbf{j})$ N, where t seconds is the time for which the force has been acting on the body.
    Find expressions for:
    (a) the velocity vector of the body at time t
    (b) the position vector of the body at time t.

12. A resultant force **F** acts on a body of mass 500 g initially at rest at an origin O. $\mathbf{F} = [(4t - 1)\mathbf{i} + 4\mathbf{j}]$ N, where t seconds is the time for which the force has been acting on the body.
    Find the body's speed and distance from O when $t = 2$.

13. A body of mass 750 g is initially at rest at a point O on a smooth horizontal surface. A horizontal force F acts on the body and causes it to move in a straight line across the surface. The magnitude of F is given by $F = (3t + 1)$ N where t seconds is the time for which the force has been acting on the body.
    Find the speed of the body when $t = 3$ and its distance from O at that time.

Resisting medium

In earlier chapters, in order to construct a manageable mathematical model of a real situation in which a resistance force may exist, we have tended to simplify the real situation by assuming the force of resistance to be constant, or even to assume that it does not exist at all. In some situations this assumption may not be appropriate. The resistance to the motion of a body may be significant and may indeed vary. For example, it could depend upon the speed of the body. The calculus techniques we have considered so far in this chapter enable us to cope with the mathematics that will arise from a model that assumes a variable resistance force is acting.

Example 19

A body of mass 5 kg is projected along a horizontal track with an initial velocity of 21 m s^{-1}. Whilst it is moving the body experiences a resisting force of $(0{\cdot}15 + 0{\cdot}05v)$ N. Find the time taken for the body to come to rest. Give the possible physical interpretation of a resisting force of the form $(a + bv)$ N.

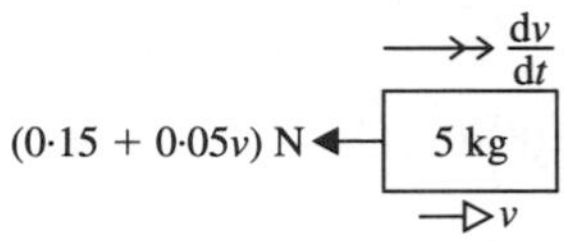

Since the data given and the answer required involve v and t, use $a = \dfrac{dv}{dt}$.

Apply $F = ma$: $$-(0{\cdot}15 + 0{\cdot}05v) = 5\,\frac{dv}{dt}$$

$\therefore$ $$-(3 + v) = 100\,\frac{dv}{dt}$$

$\therefore$ $$\int_0^T dt = \int_{21}^{0} \frac{-100}{3+v}\,dv$$

where T is the time at which the body comes to rest.

$\therefore$ $$\Big[t\Big]_0^T = \Big[-100\ln(3+v)\Big]_{21}^{0}$$

$\therefore$ $$\begin{aligned} T &= -100\ln 3 + 100\ln 24 \\ &= 100\ln 8 \\ &\simeq 208 \end{aligned}$$

The body comes to rest after 208 seconds.

A resistance force of the form $(a + bv)$ N consists of two components: one constant and one variable. The constant force of a N is independent of the speed of the body and could perhaps be due to friction between the body and the track. The variable force, bv N, increases with increasing speed and could be caused by air resistance.

Example 20

A person on a bicycle, combined mass 80 kg, cycles along a horizontal road and produces a constant forward thrust of 200 N against a variable force of resistance given by $(4v^2)$ N, where v is the person's speed in m s^{-1}, and that the person starts from rest, find how fast they are travelling after 20 metres.

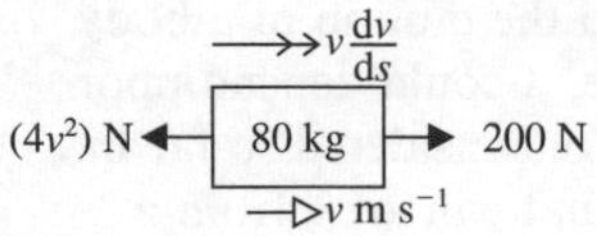

Since the data given and the answer required involve v and s, use $a = v\dfrac{dv}{ds}$.

Apply $F = ma$ $$200 - 4v^2 = 80v\frac{dv}{ds}$$

$\therefore$ $$50 - v^2 = 20v\frac{dv}{ds}$$

$\therefore$ $$\int_0^{20} ds = \int_0^V \frac{20v}{50 - v^2}\,dv$$

where V m s^{-1} is the person's velocity when they have gone 20 m.

Integrate: $$[s]_0^{20} = [-10\ln(50 - v^2)]_0^V$$

$\therefore$ $$20 = -10\ln(50 - V^2) + 10\ln(50)$$

$$= 10\ln\left(\frac{50}{50 - V^2}\right)$$

$\therefore$ $$2 = \ln\left(\frac{50}{50 - V^2}\right)$$

giving $$e^2 = \frac{50}{50 - V^2}$$

i.e. $$50 - V^2 = 50e^{-2}$$

which leads to $$V = \sqrt{50(1 - e^{-2})}$$

Thus $$V = 6{\cdot}58$$

When they have travelled 20 m the person is moving at 6·58 m s^{-1}.

Example 21

A toy train of mass 2 kg produces a forward thrust of 0·5 N against a variable force of resistance of $(6v)$ N, where v is the speed of the train in m s^{-1}. Given and that the train starts from rest at time $t = 0$ s, find:

(a) an expression for v as a function of t and sketch the velocity–time graph
(b) an expression for s, the displacement of the train at time t.

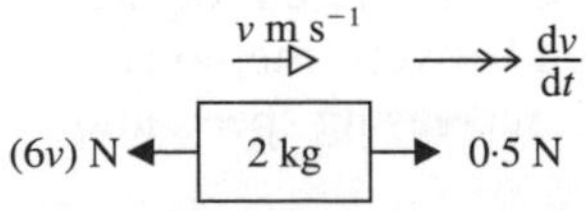

(a) Since the data given and the answer required involve v and t, use $a = \dfrac{dv}{dt}$.

Apply $F = ma$: $\quad 0{\cdot}5 - 6v = 2\dfrac{dv}{dt}$

$\therefore \quad 1 - 12v = 4\dfrac{dv}{dt}$

$\therefore \quad \int dt = \int \dfrac{4}{1 - 12v}\, dv$

$\therefore \quad t = -\frac{1}{3}\ln(1 - 12v) + c$

At $t = 0$, $v = 0$, $\quad \therefore \quad c = 0$,

so $\quad e^{-3t} = 1 - 12v$

$\therefore \quad v = \dfrac{(1 - e^{-3t})}{12}\ \text{m s}^{-1}$

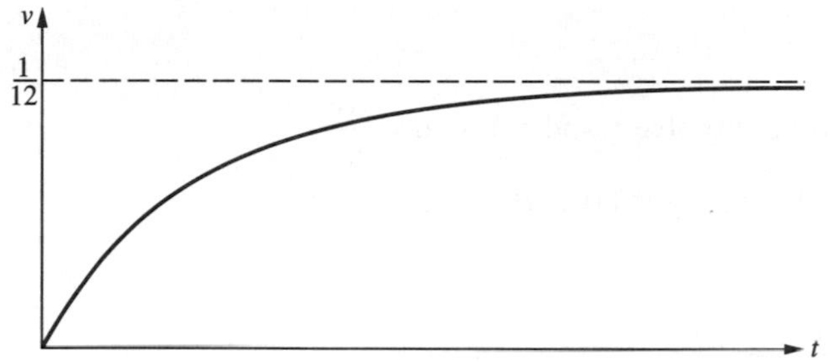

(b) Use $\quad \dfrac{ds}{dt} = \dfrac{1 - e^{-3t}}{12}$

$\therefore \quad \int ds = \int \dfrac{1 - e^{-3t}}{12}\, dt$

$\therefore \quad s = \frac{1}{12}t + \frac{1}{36}e^{-3t} + c$

At $t = 0$, $s = 0$, $\quad \therefore \quad c = -\frac{1}{36}$,

so $\quad s = \dfrac{(3t + e^{-3t} - 1)}{36}\ \text{m}$

Motion under gravity

When modelling motion under gravity in earlier chapters we have assumed the gravitational force to be constant and any resistance to motion to be constant (and often zero). The effect of any resistance, for example air resistance, is to limit the acceleration of the body. When the gravititational force downwards is "balanced" by the upward resistance force the body has reached its *terminal* velocity. When modelling such motion our model can now include the idea of variable forces acting, the skills we have developed in this chapter allowing us to cope with the mathematics that will arise from such a model.

Example 22

A body of mass 5 kg falls under gravity and reaches a terminal velocity $V\,\mathrm{m\,s^{-1}}$ downwards. Given the body experiences a resisting force of $(0{\cdot}04v^2)$ N, where v is the speed of the body in $\mathrm{m\,s^{-1}}$, determine V and find the time taken for the body to reach a speed of $\frac{4V}{5}\,\mathrm{m\,s^{-1}}$.

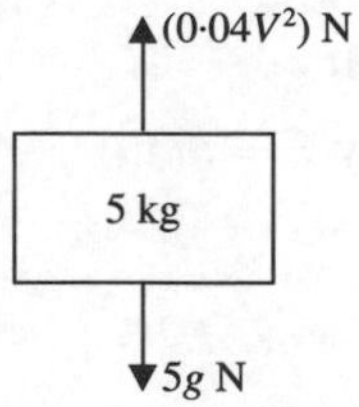

At the terminal velocity $\quad 0{\cdot}04\,V^2 = 5g$

$\therefore \quad V = 35.$

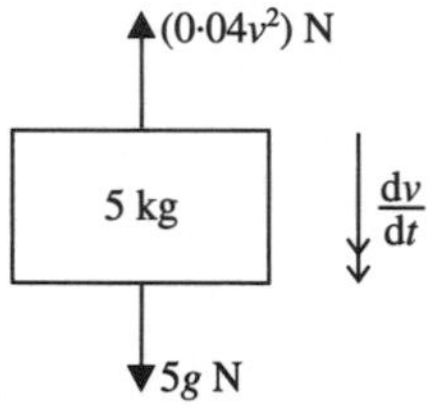

Since the data given and the required answer involve v and t, use $a = \frac{\mathrm{d}v}{\mathrm{d}t}$ and since $\frac{4V}{5} = 28$ we need to work out how long it takes the body to reach a velocity of $28\,\mathrm{m\,s^{-1}}$.

Thus, applying $F = ma$ gives: $\quad 5g - 0{\cdot}04\,v^2 = 5\,\frac{\mathrm{d}v}{\mathrm{d}t}$

$$\therefore \quad \int_0^T \mathrm{d}t = \int_0^{28} \frac{5}{49 - 0{\cdot}04\,v^2}\,\mathrm{d}v$$

where T is the time taken to reach a speed of $28\,\mathrm{m\,s^{-1}}$.

$$\therefore \quad \int_0^T \mathrm{d}t = \int_0^{28} \frac{125}{1225 - v^2}\,\mathrm{d}v$$

$$= \frac{125}{2 \times 35} \int_0^{28} \left(\frac{1}{35 + v} + \frac{1}{35 - v}\right) \mathrm{d}v$$

$$\therefore \quad \left[t\right]_0^T = \frac{125}{70}\left[\ln(35 + v) - \ln(35 - v)\right]_0^{28}$$

$$= \frac{25}{14}\left[\ln\left(\frac{35 + v}{35 - v}\right)\right]_0^{28}$$

$$\therefore \quad T = \frac{25}{14} \ln 9$$

$$\simeq 3{\cdot}92$$

It reaches a velocity which is four-fifths of its terminal velocity after 3·92 s.

Example 23

A body of mass 10 kg is projected vertically upwards through a viscous liquid with an initial velocity of $18\,\text{m s}^{-1}$. It experiences a resisting force of $(3v)$ N.

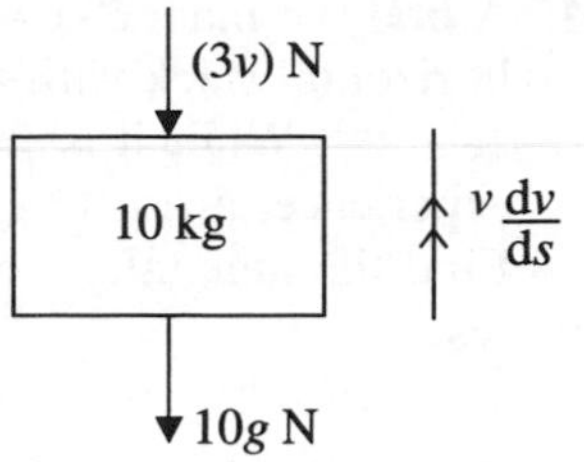

Find the distance above its point of projection at which it comes instantaneously to rest.

Since the data given and the answer required involve v and s, use $a = v\dfrac{dv}{ds}$.

Thus, applying $F = ma$ gives: $\quad -10g - 3v = 10v\dfrac{dv}{ds}$

$$\therefore \qquad \int_0^h -ds = \int_{18}^0 \frac{10v}{98+3v}\,dv$$

where h is the height at which it comes instantaneously to rest.

$$\therefore \qquad \Big[-s\Big]_0^h = \left[\frac{10}{3}v - \frac{980}{9}\ln(98+3v)\right]_{18}^0$$

$$\therefore \qquad -h = -\frac{980}{9}\ln(98) - \frac{10}{3}(18) + \frac{980}{9}\ln[98 + 3(18)]$$

$$\therefore \qquad h \simeq 12{\cdot}2$$

The body comes instantaneously to rest 12·2 m above the point of projection.

Appendix of useful standard integrals

$$\int \frac{1}{a+bx}\,dx = \frac{1}{b}\ln(a+bx) + c$$

$$\int \frac{x}{a+bx}\,dx = \frac{x}{b} - \frac{a}{b^2}\ln(a+bx) + c$$

$$\int \frac{1}{a^2-b^2x^2}\,dx = \frac{1}{2ab}\ln\left(\frac{a+bx}{a-bx}\right) + c$$

$$\int \frac{x}{a^2 \pm b^2x^2}\,dx = \pm\frac{1}{2b^2}\ln(a^2 \pm b^2x^2) + c$$

$$\int \frac{1}{a^2+b^2x^2}\,dx = \frac{1}{ab}\tan^{-1}\left(\frac{bx}{a}\right) + c$$

Exercise 16D

1. A body of mass 6 kg is projected along a horizontal track with an initial velocity of $28\,\text{m s}^{-1}$. While it is moving the body experiences a resisting force $(0{\cdot}12 + 0{\cdot}06v)$ N. Find the time taken for the body to come to rest.

2. A box of mass 0·2 kg is projected across a horizontal floor with an initial velocity of $2\,\text{m s}^{-1}$. While it is moving the box experiences a resisting force $(0{\cdot}4 + 0{\cdot}6v)$ N, where v is the speed of the body. Find the time taken for the box to come to rest.

3. A sledge and rider, combined mass 120 kg, slide across a horizontal surface against a variable force of resistance given by $(3 + 0{\cdot}2v)$ N, where v is their speed in m s^{-1}. When they pass a particular marker their speed is $12\,\text{m s}^{-1}$.
   How many seconds later do they come to rest?

4. A man in a go-cart, combined mass 90 kg, is travelling along a horizontal surface at $24\,\text{m s}^{-1}$ when he sees a red light directly ahead. He immediately disengages the engine and applies the brakes which, together with other resistances, produce a total resisting force of $(40 + 5v)$ N, where $v\,\text{m s}^{-1}$ is the speed of the cart.
   Find the further distance he travels before coming to rest.

5. A toy boat of mass 0·2 kg starts on one side of a pond and is given an initial velocity of $0{\cdot}5\,\text{m s}^{-1}$ towards the other side. The boat experiences a resisting force of $(0{\cdot}001 + 0{\cdot}002v)$ N, where $v\,\text{m s}^{-1}$ is the speed of the boat.
   Given that the boat just reaches the other side, find the width of the pond.

6. A cyclist, mass (including bicycle) 120 kg, is travelling along a horizontal cycle path at $5\,\text{m s}^{-1}$ when she sees an obstacle ahead. The cyclist immediately applies the brakes which, together with other resistances, produce a total resisting force of $(20 + 30v)$ N, where $v\,\text{m s}^{-1}$ is the cyclist's speed.
   How far will the cyclist travel before coming to rest?
   Realising that she is not going to stop before hitting the obstacle the cyclist drags her feet on the ground. How is this likely to effect the resisting force?

7. A motor-cycle and rider of combined mass 1000 kg travel along a horizontal road with the engine producing a forward thrust of 1500 N against a variable force of resistance of $(5v^2)$ N, where $v\,\text{m s}^{-1}$ is the speed of the motor-cycle. Given that the motor-cycle starts from rest, find how fast it is going when it has travelled 200 m.

8. A body of mass 6 kg is propelled along a horizontal surface by a constant horizontal force of 30 N acting against a variable resistance force of $(3v^2)$ N, where $v\,\text{m s}^{-1}$ is the speed of the body. Calculate the distance the body travels in accelerating from a velocity of $1\,\text{m s}^{-1}$ to a velocity of $3\,\text{m s}^{-1}$.
   Without doing any further calculations state whether the distance is greater or less than the distance travelled in reaching a velocity of $2\,\text{m s}^{-1}$ from rest, and give a reason for your answer.

9. A speed-boat and driver of combined mass 1000 kg produce a forward thrust of 900 N against a variable force of resistance given by $(2v^2)$ N, where $v\,\text{m s}^{-1}$ is the speed of the boat. Find the distance travelled in accelerating from a velocity of $5\,\text{m s}^{-1}$ to a velocity of $20\,\text{m s}^{-1}$.
   If the driver wishes to increase the maximum speed of his boat, suggest two ways in which he might do so.

10. A toy car of mass 2 kg is propelled along a horizontal surface by a constant horizontal force of 0·1 N acting against a variable resistance force of $(0{\cdot}1v^2)$ N, where $v\,\text{m s}^{-1}$ is the speed of the car. Find the time it takes for the car to attain a velocity of $0{\cdot}5\,\text{m s}^{-1}$ from rest.

11. A body of mass 5 kg is propelled along a horizontal surface by a constant horizontal force of 90 N acting against a variable resistance force of $(10v^2)$ N, where $v\,\mathrm{m\,s^{-1}}$ is the speed of the body.
Find the time it takes for the body to accelerate from a velocity of $1\,\mathrm{m\,s^{-1}}$ to a velocity of $2\,\mathrm{m\,s^{-1}}$.

12. A truck of mass 3000 kg travels along a horizontal road and produces a forward thrust of 6000 N against a variable force of resistance given by $(15v^2)$ N, where $v\,\mathrm{m\,s^{-1}}$ is the speed of the truck. Find the time it takes the truck to accelerate from a velocity of $10\,\mathrm{m\,s^{-1}}$ to a velocity of $15\,\mathrm{m\,s^{-1}}$.

13. A toy train of mass 2 kg travels along a horizontal track and produces a forward thrust of 1 N against a variable force of resistance of $(10v)$ N, where $v\,\mathrm{m\,s^{-1}}$ is the speed of the train. The train starts from rest at a point O at time $t = 0$ seconds. Find:
(a) an expression for v as a function of t and sketch the velocity–time graph
(b) an expression for s, the displacement of the train from O (in metres) at time t.

14. A body of mass 5 kg is propelled along a horizontal surface by a constant horizontal force of 100 N against a variable force of resistance of $(10v)$ N, where $v\,\mathrm{m\,s^{-1}}$ is the speed of the body. When $t = 0$ seconds the body is at rest at a point O. Find:
(a) an expression for v as a function of t and sketch the velocity–time graph
(b) an expression for s, the displacement of the body from O (in metres) at time t.

15. A man on a bicycle, combined mass 100 kg, cycles along a horizontal road and produces a forward thrust of 100 N against a variable resistance force of $(1v^2)$ N, where $v\,\mathrm{m\,s^{-1}}$ is the speed of the man and bike.
If they start from rest at an origin, O, find an expression for v in terms of s, the displacement from O (in metres), and sketch the velocity–displacement graph.

16. A body of mass 12 kg is propelled along a horizontal surface by a constant horizontal force of 480 N against a variable force of resistance of $(30v^2)$ N, where $v\,\mathrm{m\,s^{-1}}$ is the speed of the body. The body starts from rest at an origin, O.
Find an expression for v in terms of s, the displacement of the body from O (in metres), and sketch the velocity–displacement graph.

17. A body of mass 8 kg falls from rest under gravity. It experiences a resisting force of $(0{\cdot}1v^2)$ N, where $v\,\mathrm{m\,s^{-1}}$ is the speed of the body.
Find the time taken to reach a speed of $21\,\mathrm{m\,s^{-1}}$.

18. A body of mass 15 kg falls under gravity. It experiences a resisting force of $(3v)$ N, where $v\,\mathrm{m\,s^{-1}}$ is the speed of the body.
Find the time taken for its speed to increase from $29\,\mathrm{m\,s^{-1}}$ to $39\,\mathrm{m\,s^{-1}}$.

19. A body of mass 20 kg falls under gravity. It experiences a resisting force of $(0{\cdot}16v^2)$ N, where $v\,\mathrm{m\,s^{-1}}$ is the speed of the body.
Find V, the terminal speed of the body, and determine the time taken to reach a speed of $0{\cdot}6V$ from rest.

20. A body of mass 10 kg falls under gravity. It experiences a resisting force of $(2v)$ N, where $v\,\mathrm{m\,s^{-1}}$ is the speed of the body.
Find V, the terminal speed of the body, and determine the time taken to reach a speed of $\dfrac{6V}{7}$ from rest.

21. A body of mass 50 kg falls under gravity. It experiences a resisting force of $(10v)$ N, where $v\,\mathrm{m\,s^{-1}}$ is the speed of the body.
Calculate the distance the body travels in accelerating from a speed of $10\,\mathrm{m\,s^{-1}}$ to one of $40\,\mathrm{m\,s^{-1}}$.

22. A body of mass 4 kg falls under gravity. It experiences a resisting force of $(0{\cdot}01v^2)$ N, where $v\,\mathrm{m\,s^{-1}}$ is the speed of the body.
Calculate the distance the body travels in accelerating from a speed of $30\,\mathrm{m\,s^{-1}}$ to one of $60\,\mathrm{m\,s^{-1}}$.

23. A body of mass 8 kg is projected vertically upwards into the air, with an initial speed of $10\,\mathrm{m\,s^{-1}}$. The body experiences a resisting force of $(0{\cdot}2v)$ N, where $v\,\mathrm{m\,s^{-1}}$ is the speed of the body.
Find the height above its point of projection at which the body instantaneously comes to rest.

24. A body of mass 3 kg is projected vertically upwards into the air, with an initial speed of $100\,\text{m}\,\text{s}^{-1}$. The body experiences a resisting force of $(0{\cdot}1v)$ N, where $v\,\text{m}\,\text{s}^{-1}$ is the speed of the body.
Find the height above its point of projection at which the body instantaneously comes to rest.

25. A body of mass 10 kg is projected vertically upwards through a viscous liquid, with an initial speed of $12\,\text{m}\,\text{s}^{-1}$. The body experiences a resisting force of $(0{\cdot}5v^2)$ N, where $v\,\text{m}\,\text{s}^{-1}$ is the speed of the body.
Find the distance above its point of projection at which the body instantaneously comes to rest.

26. A body of mass 3 kg is projected vertically upwards through a viscous liquid, with an initial speed of $15\,\text{m}\,\text{s}^{-1}$. The body experiences a resisting force of $(0{\cdot}1v^2)$ N, where $v\,\text{m}\,\text{s}^{-1}$ is the speed of the body.
Find the distance above its point of projection at which the body instantaneously comes to rest.

27. A body of mass 6 kg is projected vertically upwards into the air, with an initial speed of $200\,\text{m}\,\text{s}^{-1}$. The body experiences a resisting force of $(0{\cdot}2v)$ N, where $v\,\text{m}\,\text{s}^{-1}$ is the speed of the body.
Determine the time that elapses from the moment of projection until the body instantaneously comes to rest.

28. A body of mass 40 kg is projected vertically upwards into the air, with an initial speed of $60\,\text{m}\,\text{s}^{-1}$. The body experiences a resisting force of $(2v)$ N, where $v\,\text{m}\,\text{s}^{-1}$ is the speed of the body.
Determine the time that elapses from the moment of projection until the body instantaneously comes to rest.

29. A body of mass 20 kg is projected vertically upwards through a viscous liquid, with an initial speed of $7\,\text{m}\,\text{s}^{-1}$. The body experiences a resisting force of $(4v^2)$ N, where $v\,\text{m}\,\text{s}^{-1}$ is the speed of the body.
Determine the time that elapses from the moment of projection until the body instantaneously comes to rest.

30. A body of mass 5 kg is projected vertically upwards into the air, with an initial speed of $24\,\text{m}\,\text{s}^{-1}$. The body experiences a resistance force of $(0{\cdot}49v^2)$ N, where $v\,\text{m}\,\text{s}^{-1}$ is the speed of the body.
Determine the time that elapses from the moment of projection until the body instantaneously comes to rest.

31. A body of mass 1 kg is released from rest and falls under gravity against a resistance of $\frac{7}{s+1}$ N, where s is the distance (in metres) that the body has fallen since release.
Find the speed of the particle when it has fallen a distance of 6·4 m.
(Take $g = 10\,\text{m}\,\text{s}^{-2}$.)

Further examples

Example 24

A vehicle of mass 500 kg, travelling on a horizontal surface, has its engine working at a constant rate of 10 kW against a resisting force of $(25v)$ N, where v is the speed in $\text{m}\,\text{s}^{-1}$. Find:

(a) the maximum speed of the car

(b) the time taken for the car to increase its speed from $5\,\text{m}\,\text{s}^{-1}$ to $15\,\text{m}\,\text{s}^{-1}$.

(a) Since power = force × velocity, the force produced by the engine is $\frac{10\,000}{v}$ N.

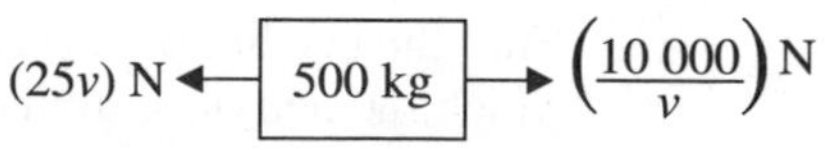

At maximum speed, $V\,\mathrm{m\,s^{-1}}$, the force produced by the engine balances the resisting force:

$$25V = \frac{10\,000}{V}$$

$$\therefore \quad V = 20$$

The maximum speed of the car is $20\,\mathrm{m\,s^{-1}}$.

(b)

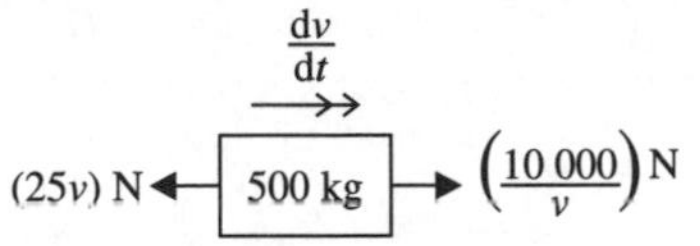

Since the data and the answer involve v and t, use $a = \dfrac{dv}{dt}$.

Apply $F = ma$ to give $\dfrac{10\,000}{v} - 25v = 500\dfrac{dv}{dt}$

$$\therefore \quad 400 - v^2 = 20v\frac{dv}{dt}$$

$$\therefore \quad \int_0^T dt = \int_5^{15} \frac{20v}{400 - v^2}\,dv$$

where T is the time taken for the speed to increase from $5\,\mathrm{m\,s^{-1}}$ to $15\,\mathrm{m\,s^{-1}}$.

This gives: $T \simeq 7{\cdot}62$

It takes $7{\cdot}62$ s to accelerate from $5\,\mathrm{m\,s^{-1}}$ to $15\,\mathrm{m\,s^{-1}}$.

Example 25

The force acting on a body of mass 10 kg is inversely proportional to the square of the distance of the body from the origin and is directed away from the origin. When the body is moving away from the origin, through a point at a distance 2 m from the origin, with speed $3\,\mathrm{m\,s^{-1}}$, the force acting on the body is 20 N.
Calculate the speed of the body when it is 8 m from the origin.

Under the inverse square rule the force, F N, and the displacement, s m, are related by:

$$F = \frac{k}{s^2} \quad \text{for some constant } k.$$

Given $F = 20$ when $s = 2$, $\quad 20 = \dfrac{k}{2^2}$

$$\therefore \quad k = 80$$

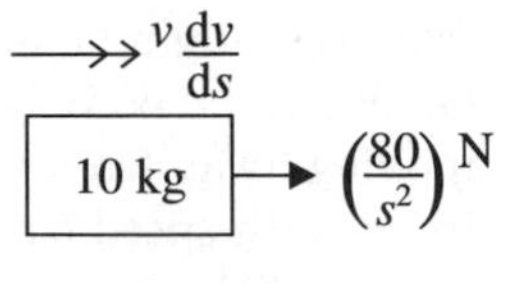

Since the data given and the answer involve v and s, use $a = v\dfrac{dv}{ds}$.

Applying $F = ma$ gives $$\frac{80}{s^2} = 10v\frac{dv}{ds}$$

$$\therefore \qquad \int_2^8 \frac{8}{s^2}\,ds = \int_3^V v\,dv$$

where V is the speed of the body when $s = 8$ m.

This gives $$V \simeq 3{\cdot}87$$

When $s = 8$ m the speed of the body is $3{\cdot}87\,\text{m s}^{-1}$.

Exercise 16E

1. A man on a bicycle, combined mass 120 kg, works at a constant rate of 200 W whilst travelling along a straight horizontal surface, against a resisting force of $(8v)$ N, where $v\,\text{m s}^{-1}$ is the cyclist's speed. Find:
   (a) the maximum speed
   (b) the time taken for the speed to increase from $2\,\text{m s}^{-1}$ to $4\,\text{m s}^{-1}$.

2. A train of mass 2·4 tonnes travels along a straight horizontal surface with its engine working at a constant rate of 400 kW, against a resisting force of $(160v)$ N, where $v\,\text{m s}^{-1}$ is the speed of the train. Find:
   (a) the maximum speed
   (b) the time taken for the speed to increase from $10\,\text{m s}^{-1}$ to $40\,\text{m s}^{-1}$.

3. A go-cart and driver, combined mass 140 kg, travel along a straight horizontal surface with the engine working at a constant rate of 800 W.
   Given that the resisting force is proportional to the speed of the go-cart, and that the maximum speed of the go-cart is $20\,\text{m s}^{-1}$, find the go-cart's speed 7 seconds after starting from rest.

4. A motor-boat and driver of combined mass 400 kg has its engine working at a constant rate of 1·8 kW.
   Given that the resisting force is proportional to the speed of the boat, and that the maximum speed of the boat is $15\,\text{m s}^{-1}$, find the boat's speed 4 seconds after starting from rest.

5. A vehicle of mass 600 kg travels along a straight horizontal road with its engine working at a constant rate of 24 kW.
   Given that the resisting force is proportional to the square of the speed, and that the maximum speed is $20\,\text{m s}^{-1}$, find the distance travelled by the car in accelerating from rest to $16\,\text{m s}^{-1}$.

6. A cruiser of mass 9 tonnes has its engine working at a constant rate of 17·28 kW.
   Given that the resisting force is proportional to the square of the speed of the cruiser, and that the maximum speed is $12\,\text{m s}^{-1}$, find the distance travelled by the cruiser in accelerating from $5\,\text{m s}^{-1}$ to $10\,\text{m s}^{-1}$.

7. The force acting on a body of mass 8 kg is inversely proportional to the square of the distance from the body to the origin, and is directed away from the origin. When the body is 4 m from the origin and moving away from the origin, in the direction of the force, with speed $2\,\text{m s}^{-1}$, the force on the body is 50 N.
   Calculate the speed of the body when it is 10 m from the origin.

8. The force acting on a body of mass 1 kg is proportional to the distance of the body from the origin O, and is directed away from O. When the body is 2 m from O and moving away from O, in the direction of the force, with speed $6\,\text{m s}^{-1}$, the force on the body is 20 N. Calculate the speed of the body when it is 8 m from the origin.

9. The force acting on a body of mass 5 kg is inversely proportional to the square of the distance the body is from an origin, O, and is directed away from O. When the body is 6 m from O and moving away from O, in the direction of the force, with speed $10\,\text{m s}^{-1}$, the force on the body is 30 N.

(a) Show that $v^2 = 172 - \dfrac{432}{s}$

where $v\,\text{m s}^{-1}$ is the speed of the body when it is s m from O.

(b) Determine the speed the body approaches, but does not exceed, as it moves further and further from O.

10. The force acting on a body of mass 4 kg is proportional to the distance of the body from the origin O, and is directed away from O. When the body is 5 m from O and moving away from O, in the direction of the force, with speed $10\,\text{m s}^{-1}$, the force on the body is 80 N.
Show that $v = 2s$, where $v\,\text{m s}^{-1}$ is the speed of the body when it is s m from O.

11. A horizontal force F is applied to a body of mass 5 kg, initially at rest at a point O on a rough horizontal surface, coefficient of friction $\frac{2}{7}$. The force causes the body to move in a straight line across the surface. The magnitude of the force is given by $F = (5s + 25)$ N, where s is the distance in metres that the body is from O.
Find the speed of the body when $s = 10$ m.
If the applied force were $(5s + 12)$ N, would the body move?

12. With its engines developing a constant 100 kW of power, a train of mass 90 tonnes accelerates up an incline of 1 in 98. If the air and frictional resistances are a constant 1000 N, find:

(a) $v_{\max}$, the maximum speed of the train up the incline

(b) the time taken for the train to accelerate up the incline from rest to a speed equal to $\dfrac{v_{\max}}{2}$.

13. A horizontal force F of variable magnitude and constant direction acts on a body of mass m which is initially at rest at a point O on a smooth horizontal surface. The magnitude of F is given by $F = \beta + \alpha t$, where t is the time for which the force has been acting on the body, and α and β are positive constants. If s is the distance the body is from O at time t,

show that $s = \dfrac{t^2}{6m}(3\beta + \alpha t)$.

14. A body of mass m is initially at rest at a point O on a smooth horizontal surface. A horizontal force F is applied to the body and causes it to move in a straight line across the surface. The magnitude of F is given by $F = \dfrac{1}{s + \alpha}$ where s is the distance of the body from O and α is a positive constant. If v is the speed of the body at any moment, show that $s = \alpha(e^{\frac{1}{2}mv^2} - 1)$.

15. A body of mass m is released from rest and falls under gravity against a resistance to motion of mkv, where v is the speed of the body at a time t after release and k is a positive constant. Show that:

(a) $kt = \ln\left(\dfrac{g}{g - kv}\right)$

(b) as the motion continues, v approaches a maximum value of $\dfrac{g}{k}$.

Work done by a variable force

In Chapter 11 we saw that if the point of application of a constant force F N moves through a distance s metres in the direction of the force then the work done is defined to be:

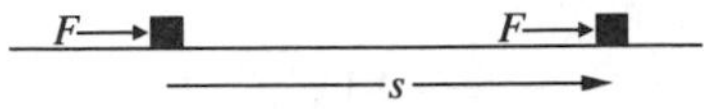

$$\text{work done} = Fs \text{ joules.}$$

If the force is variable we can still calculate the work done but must now use calculus (as we saw in Chapter 15 when determining the work done when stretching an elastic string).

Consider a variable force F and suppose that P, the point of application of the force, moves in the positive direction of the x-axis from $x = 0$ (the point O) to $x = s$ (the point S), and that F acts in the positive direction of the x-axis.

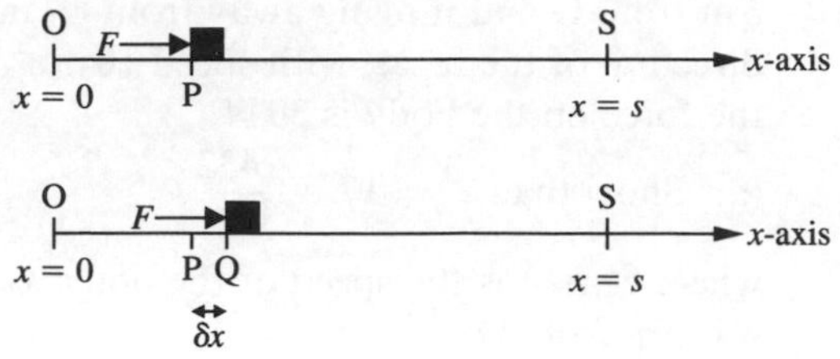

To find the work done we divide OS into a large number of very short steps. Each of these are of length δx and a typical step, PQ, is shown enlarged in the diagram.
Over the very small step PQ the force F is almost constant, its value being nearly equal to its value at P. Thus:

$$\text{work done by } F \text{ over PQ} \approx F\delta x$$

The total work done over OS is obtained by adding the contributions of all the steps and seeing what this summation approaches as δx gets smaller and smaller (i.e. as $\delta x \to 0$).

$$\text{work done by } F \text{ over OS} = \lim_{\delta x \to 0} \sum_{x=0}^{x=s} F\delta x$$

Using calculus this can be written:

$$\text{work done by } F \text{ over OS} = \int_{x=0}^{x=s} F\,\mathrm{d}x$$

Example 26

A particle moves in the direction of the positive x-axis under the influence of a varying force $F = 0{\cdot}2x$ N, where x metres is the distance of the particle from the origin. Find the work done on the particle by F when the particle moves from:

(a) $x = 0$ to $x = 10$ (b) $x = 1$ to $x = 5$.

(a) 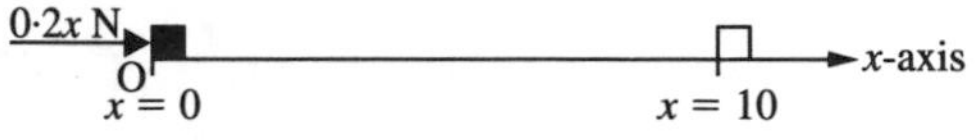

$$\text{work done} = \int_0^{10} 0{\cdot}2x\,\mathrm{d}x$$

$$= \left[0{\cdot}1x^2\right]_0^{10}$$

$$= 10\text{ J}$$

The work done is 10 J.

(b) 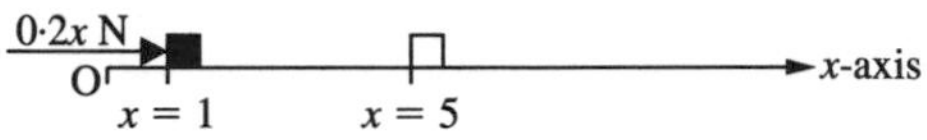

$$\text{work done} = \int_1^{5} 0{\cdot}2x\,\mathrm{d}x$$

$$= \left[0{\cdot}1x^2\right]_1^{5}$$

$$= 2{\cdot}4\text{ J}$$

The work done is 2·4 J.

Example 27

After release from rest at point O, a body of mass 1 kg falls under gravity against a resistance of $\frac{24}{25}s$ N, where s metres is the distance the body is below O at any instant.
Find the amount of work done by the body against the resistance, from release until it passes through a point P, 10 m below O, and find the speed of the body at that instant.

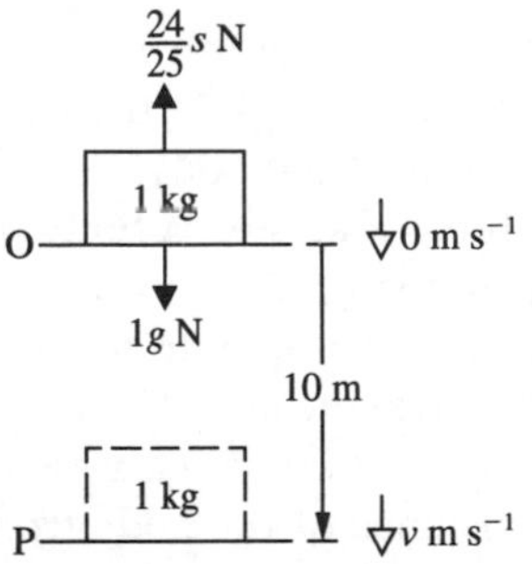

We will measure potential energies from the level of P.

At O: total energy $= \text{KE} + \text{PE}$

$$= \tfrac{1}{2}(1)0^2 + 1(g)10$$
$$= 98\text{ J}$$

At P: total energy $= \text{KE} + \text{PE}$

$$= \tfrac{1}{2}(1)v^2 + 0$$
$$= \tfrac{1}{2}v^2$$

$$\text{work done against resistance} = \int_0^{10} \tfrac{24}{25}s\,ds$$
$$= \left[\tfrac{12}{25}s^2\right]_0^{10}$$
$$= 48\text{ J}$$

The body has done work against the resistance and so, in travelling from O to P, the body will have lost energy:

work done against resistance = loss in energy

$\therefore$ $48 = 98 - \tfrac{1}{2}v^2$

giving $v = 10$

In travelling from O to P the body does 48 J of work against the resistance and passes through the point P with speed 10 m s^{-1}.

Exercise 16F

1. A particle moves in the direction of the positive x-axis under the influence of a force $F = 2x$ N.
Find the work done on the particle by F when the particle moves from $x = 0$ to $x = 2$.

2. A particle moves in the direction of the positive x-axis under the influence of a force $F = 0{\cdot}5x$ N.
Find the work done on the particle by F when the particle moves from $x = 2$ to $x = 4$.

3. A particle moves in the direction of the positive x-axis under the influence of a force $F = x(1 - x)$ N.
Find the work done on the particle by F when the particle moves from $x = 0$ to $x = 1$.

4. A particle moves in the direction of the positive x-axis under the influence of a force $F = \cos x$ N.
Find the work done on the particle by F when the particle moves from $x = 0$ to $x = \dfrac{\pi}{2}$.

5. A particle moves in the direction of the positive x-axis under the influence of a force $F = \sin 2x$ N.
Find the work done on the particle by F when the particle moves from $x = \dfrac{\pi}{4}$ to $x = \dfrac{\pi}{2}$.

6. A horizontal force F acts on a body of mass 6 kg initially at rest on a smooth horizontal surface and causes the body to move across the surface in a straight line. The magnitude of F is given by $F = (2s + 1)$ N, where s metres is the distance of the body from its original position.
If the force ceases when $s = 3$, find:
(a) the work done by the force
(b) the final kinetic energy of the body
(c) the final speed of the body.

7. A horizontal force F acts on a body of mass 250 g initially at rest on a smooth horizontal surface and causes the body to move across the surface in a straight line. The magnitude of F is given by $F = (4s + 5)$ N, where s metres is the distance of the body from its original position.
If the force ceases when $s = 2$, find:
(a) the work done by the force
(b) the final kinetic energy of the body
(c) the final speed of the body.

8. A body of mass 500 g lies at a point O on a horizontal surface. The body is projected from O with a speed of $3\,\text{m}\,\text{s}^{-1}$ and moves against a resistance of $2s$ N, where s metres is the distance of the body from O.
If the body comes to rest at a point B, find:
(a) the loss in the kinetic energy possessed by the body
(b) the distance OB.

9. A body of mass 2 kg is released from rest at a point O and falls under gravity against a resistance of $\frac{16}{5}s$ N, where s metres is the distance of the body from O at any instant. When the body has fallen 3·5 m, find:
(a) the loss in potential energy
(b) the work done against the resistance
(c) the gain in the kinetic energy
(d) the speed of the body.

10. A body of mass 5 mg is projected vertically upwards from a point O and comes to instantaneous rest at a point B, 6 m from O. In addition to the gravitational pull slowing the body it experiences a resistance of $\frac{11}{3}s$ N, where s metres is the distance the body is from O. For the motion from O to B, find:
(a) the gain in potential energy
(b) the work done against the resistance
(c) the speed of projection.

11. A body of mass 4 kg is released from rest at the top of a slope of length 10 m which is inclined at an angle θ above the horizontal where $\sin\theta = \frac{1}{14}$. The body moves down the slope against a resistance of $\frac{2}{5}s$ N, where s metres is the distance of the body from the top. For the motion of the body from the top to the bottom of the slope, find:
(a) the loss in potential energy
(b) the work done against the resistance
(c) the final speed of the body.

Variable force acting along a curved path

Consider a variable force **F** acting on a particle P which moves along a curved path from a point A to a point B. Divide the path AB into a large number of very short steps and consider one such small step PQ, (see diagram).
Let us suppose that the particle moves this short distance PQ in a short interval of time δt. For this short interval the force **F** and the velocity **v** will hardly change from their values at P. Thus essentially we have a constant force vector acting on a particle moving in a straight line.

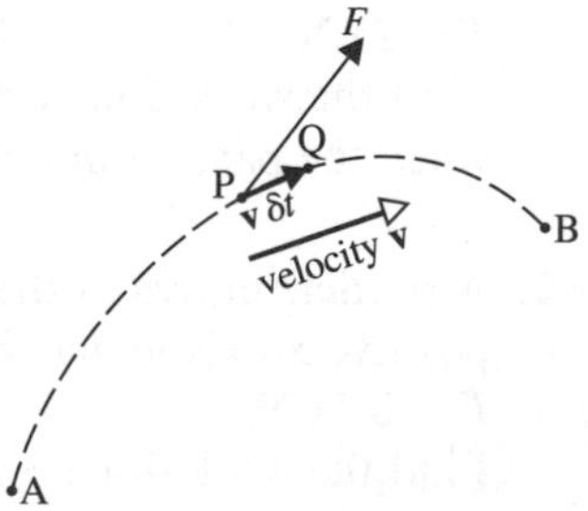

If the velocity at P is **v** then $\quad (\overrightarrow{PQ}) \approx \mathbf{v}\,\delta t$

Work done by **F** as particle moves from P to Q $\approx \mathbf{F}.(\overrightarrow{PQ})$

$$\approx \mathbf{F}.(\mathbf{v}\,\delta t) = (\mathbf{F}.\mathbf{v})\,\delta t \quad \ldots[1]$$

this approximation being more accurate as $\delta t \to 0$.

Now power is the rate at which work is done. Thus:

rate at which the force is working when the particle is at P

$$= \lim_{\delta t \to 0} \frac{(F.\mathbf{v})\,\delta t}{\delta t}$$

$$= \mathbf{F}.\mathbf{v}$$

> For a variable force $\mathbf{F}(t)$ acting on a particle and causing it to move with velocity $\mathbf{v}(t)$ the power of the force at time t is given by:
>
> $$\text{power} = \mathbf{F}.\mathbf{v}$$

Equation [1] gave us the work done by **F** as the particle moves from P to Q. From this we can find the total work done over AB by adding the contributions of all the steps and seeing what this summation approaches as δt gets smaller and smaller.

$$\text{work done by } \mathbf{F} \text{ over AB} = \lim_{\delta t \to 0} \sum_{t_A}^{t_B} (\mathbf{F}.\mathbf{v})\,\delta t$$

where t_A and t_B are the times when the particle is at A and B respectively.

Using calculus this can be written as $\int_{t_A}^{t_B} (\mathbf{F}.\mathbf{v})\,dt$

> For a variable force $\mathbf{F}(t)$ acting on a particle and causing it to move with velocity $\mathbf{v}(t)$ the work done by the force in the time interval from t_A to t_B is given by:
>
> $$\text{work done} = \int_{t_A}^{t_B} (\mathbf{F}.\mathbf{v})\,dt$$

Example 28

A body of mass 3 kg moves along a curve under the action of a resultant force **F** N. At time t seconds the position vector **r** m of the body is $\mathbf{r} = 2t\mathbf{i} + t^3\mathbf{j} + t^2\mathbf{k}$.

(a) Find an expression for **F** in terms of t.
(b) Find an expression for $P(t)$, the power of the force at time t seconds.
(c) Calculate the work done by **F** between $t = 0$ and $t = 2$.
(d) Verify that the work done calculated in (c) is equal to the change in kinetic energy of the body over the same interval.

(a) $\mathbf{r} = 2t\mathbf{i} + t^3\mathbf{j} + t^2\mathbf{k}$

Thus $\mathbf{v} = \frac{d\mathbf{r}}{dt} = 2\mathbf{i} + 3t^2\mathbf{j} + 2t\mathbf{k}$

and $\mathbf{a} = \frac{d\mathbf{v}}{dt} = 6t\mathbf{j} + 2\mathbf{k}$

Using $\mathbf{F} = m\mathbf{a}$ gives: $\mathbf{F} = 3(6t\mathbf{j} + 2\mathbf{k})$
$= 18t\mathbf{j} + 6\mathbf{k}$

Thus $\mathbf{F} = 18t\mathbf{j} + 6\mathbf{k}$.

(b)
$$\begin{aligned} P(t) &= \mathbf{F}\,.\,\mathbf{v} \\ &= (18t\mathbf{j} + 6\mathbf{k})\,.\,(2\mathbf{i} + 3t^2\mathbf{j} + 2t\mathbf{k}) \\ &= 54t^3 + 12t \end{aligned}$$

The power of the force at time t seconds is $(54t^3 + 12t)$ W.

(c) Work done
$$\begin{aligned} \int_0^2 (\mathbf{F}\,.\,\mathbf{v})\,\mathrm{d}t &= \int_0^2 (54t^3 + 12t)\,\mathrm{d}t \\ &= \left[13{\cdot}5t^4 + 6t^2\right]_0^2 \\ &= 240\ \text{J} \end{aligned}$$

The work done is 240 J.

(d) At time $t = 0$, $\mathbf{v} = 2\mathbf{i}$ $\quad$ At time $t = 2$, $\mathbf{v} = 2\mathbf{i} + 12\mathbf{j} + 4\mathbf{k}$

$\therefore$ $|\mathbf{v}| = 2$ $\quad$ $|\mathbf{v}| = \sqrt{164}$

KE = 6 J $\quad$ KE = 246 J

Hence the change in kinetic energy is 240 J (= 246 J – 6 J) which agrees with the work done calculated in part (c).

Exercise 16G

1. A body of mass 4 kg moves along a curve under the action of a resultant force **F** N. At time t seconds the velocity vector, $\mathbf{v}$ m s^{-1}, of the body is given by $\mathbf{v} = 2t\mathbf{i} + t^3\mathbf{j} + 3\mathbf{k}$.
   (a) Find the speed and the kinetic energy of the body when $t = 0$.
   (b) Find the speed and the kinetic energy of the body when $t = 3$.
   (c) From your previous answers determine how much work **F** has done on the body in the time interval from $t = 0$ to $t = 3$.
   (d) Find an expression for **F** in terms of t.
   (e) Find an expression for $P(t)$, the power of the force at time t.
   (f) Use calculus and your answer to part (e) to determine the work done by **F** in the time interval from $t = 0$ to $t = 3$.

2. A body of mass 5 kg moves along a curve under the action of a resultant force **F** N. At time t seconds the position vector, **r** m, of the body is $\mathbf{r} = 4t\mathbf{i} + t^3\mathbf{j} + \frac{1}{2}t^2\mathbf{k}$.
   (a) Find an expression for **F** in terms of t.
   (b) Find an expression for $P(t)$, the power of the force at time t.
   (c) Use calculus to determine the work done by **F** between $t = 1$ and $t = 3$.
   (d) Verify that the work done calculated in (c) is equal to the change in kinetic energy of the body over the same interval.

3. A body of mass 5 kg moves along a curve under the action of a resultant force **F** N. At time t seconds the position vector **r** m of the body is $\mathbf{r} = (t\mathbf{i} + t^2\mathbf{j})$ m.
   (a) Use integration to determine the work done by **F** between $t = 1$ and $t = 2$.
   (b) Verify that the work done calculated in (a) is equal to the change in kinetic energy of the body over the same interval.

4. A body of mass 3 kg moves along a curve under the influence of a resultant force **F** N. At time t seconds the position vector **r** m of the body is $\mathbf{r} = (t^2\mathbf{i} + t\mathbf{j} + 2t\mathbf{k})$ m. Use integration to determine the work done by **F** in the interval from $t = 0$ to $t = 3$.

5. A body of mass 5 kg moves along a curve under the influence of a resultant force **F** N and such that the position vector of the body at time t seconds is $\mathbf{r} = (t^2\mathbf{i} + t^3\mathbf{j} + t^2\mathbf{k})$ m. Use integration to determine the work done by **F** in the interval from $t = 1$ to $t = 2$.

6. A body of mass 2 kg moves along a curve under the influence of a resultant force **F** N and such that the position vector of the body at time t seconds is $\mathbf{r} = (\sin 2t\mathbf{i} - 2\cos 2t\mathbf{j})$ m. Use integration to determine the work done by **F** in the interval from $t = 0$ to $t = \frac{1}{8}\pi$.

7. A particle P of mass 2 kg is acted upon by two forces and the velocity $\mathbf{v}\,\mathrm{m\,s^{-1}}$ of P at time t seconds is given by $\mathbf{v} = 2t\mathbf{i} + 3t^2\mathbf{j}$.

(a) Find $\mathbf{R}(t)$, the resultant force acting on the particle (in the form $(a\mathbf{i} + b\mathbf{j})$ N).

(b) Find an expression in terms of t for the rate at which $\mathbf{R}$ is doing work on the particle at time t seconds.

(c) Calculate the work done by $\mathbf{R}$ on the particle from $t = 1$ to $t = 3$.

(d) The two forces making up $\mathbf{R}$ are $\mathbf{F} = (3t\mathbf{i} + 2t\mathbf{j})$ N and $\mathbf{T}$. Find $\mathbf{T}$ and determine the work each of $\mathbf{F}$ and $\mathbf{T}$ do on the particle from $t = 1$ to $t = 3$.

8. A force $\mathbf{F} = (-3\cos t\mathbf{i} - 3\sin t\mathbf{j})$ N is acting at a point P and the position vector $\mathbf{r}$ m of P at time t seconds is given by: $\mathbf{r} = 2\cos t\mathbf{i} + 2\sin t\mathbf{j}$.
Find an expression for the power of the force at time t. Explain your result.

Exercise 16H Examination questions

(Unless otherwise indicated take $g = 9{\cdot}8\,\mathrm{m\,s^{-2}}$ in this exercise.)

1. The velocity, $v\,\mathrm{m\,s^{-1}}$, of a particle moving in a straight line, t seconds after passing through a fixed point O, is given by

$$v = 6t^2 - 5t + 3.$$

Find:

(i) the acceleration of the particle when $t = 3$

(ii) the distance of the particle from O when $t = 4$. (UCLES)

2. A particle moves in a straight line so that, t seconds after passing through a fixed point O, its velocity, $v\,\mathrm{m\,s^{-1}}$, is given by

$$v = 3t^2 - 30t + 72.$$

Find:

(i) the initial acceleration of the particle

(ii) the values of t when the particle is at instantaneous rest

(iii) the distance moved by the particle during the interval between these two values

(iv) the total distance moved by the particle between $t = 0$ and $t = 7$.

(UCLES)

3. Two boys, Mark and John, run a 100 m race. Their velocity–time (v–t) graphs for the race are shown in the diagram.

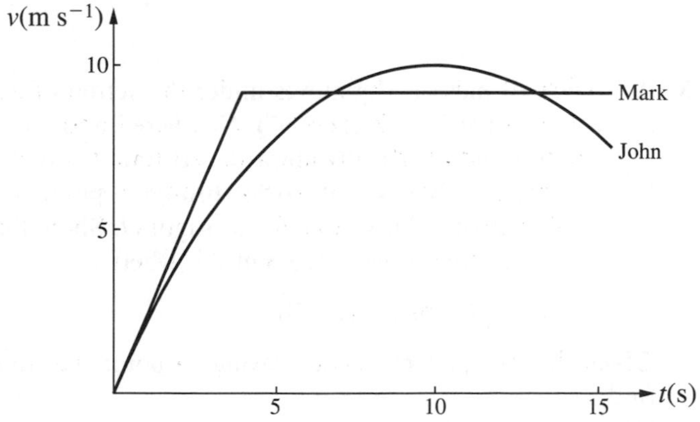

Mark's race consists of two periods of uniform motion. In the first he accelerates from rest to a velocity of $9\,\mathrm{m\,s^{-1}}$ in 4 s; in the second he runs at constant velocity.
John's race is non-uniform. The equation of his velocity–time curve is

$$v = 0{\cdot}1(20 - t)t.$$

(a) Show, by calculation, that Mark completes the race in a time of $13\frac{1}{9}$ s.

(b) By further calculation, determine who wins the race.

(c) Without any further calculation, give a brief commentary of the race.

(UODLE)

4. A dot moves on the screen of an oscilloscope so that its position relative to a fixed origin is given by

$$r = 2t\mathbf{i} + \sin\left(\frac{\pi t}{2}\right)\mathbf{j}$$

(a) Sketch the path of the dot for $0 \leqslant t \leqslant 4$.
(b) Find the velocity and acceleration of the dot when $t = 3$. Draw vectors on your diagram to show these two quantities. (AEB Spec)

5. A particle P moves such that, at time t seconds, its position vector $\mathbf{r}$ metres relative to a fixed origin O is given by

$$\mathbf{r} = ct^2\mathbf{i} + (t^3 - 4t)\mathbf{j},$$

where c is a positive constant. When $t = 2$, the speed of P is $10\,\mathrm{m\,s^{-1}}$.
(a) Find the value of c.
(b) Find the acceleration vector of P when $t = 2$. (ULEAC)

6. A particle P, of mass 2 kg is moving under the influence of a variable force F. At time t seconds, the velocity $\mathbf{v}\,\mathrm{m\,s^{-1}}$ of P is given by

$$\mathbf{v} = 2t\mathbf{i} + \mathrm{e}^{-t}\mathbf{j}.$$

(a) Find the acceleration, $\mathbf{a}\,\mathrm{m\,s^{-2}}$, of P at time t seconds.
(b) Calculate, in N to 2 decimal places, the magnitude of $\mathbf{F}$ when $t = 0{\cdot}2$. (ULEAC)

7. The position vector $\mathbf{r}$, relative to a fixed origin O, of a particle P is given at time t seconds by

$$\mathbf{r} = \mathrm{e}^{-t}(\cos t\,\mathbf{i} + \sin t\,\mathbf{j})\,\mathrm{m},$$

where $\mathbf{i}$ and $\mathbf{j}$ are fixed perpendicular unit vectors. Find similar expressions for the velocity and acceleration of P at any time t. Show that the acceleration is always perpendicular to the position vector. (AEB 1989)

8. A particle of mass $1{\cdot}5$ kg moves under the action of its own weight and a constant force $\mathbf{F} = (2{\cdot}25\mathbf{i} + 5{\cdot}7\mathbf{j})$ N, where $\mathbf{i}$ and $\mathbf{j}$ are unit vectors with $\mathbf{i}$ horizontal and $\mathbf{j}$ vertically upwards. At time $t = 0$, when the particle has velocity $(\mathbf{i} + 3\mathbf{j})\,\mathrm{m\,s^{-1}}$, it passes through a point A with position vector $(4\mathbf{i} + 2\mathbf{j})$ m relative to a fixed origin O. Show that the velocity of the particle at time t seconds is $\mathbf{v}\,\mathrm{m\,s^{-1}}$, where

$$\mathbf{v} = \left(\tfrac{3}{2}\mathbf{i} - 6\mathbf{j}\right)t + (\mathbf{i} + 3\mathbf{j}).$$

Given that the particle passes through a point B at time $t = 4$ seconds, find:
(a) the speed of the particle at B,
(b) the position vector of B relative to O,
(c) the work done by $\mathbf{F}$ as the particle moves from A to B,
(d) the time at which the particle is instantaneously moving parallel to AB. (AEB 1993)

9. (Take $g = 10\,\text{m}\,\text{s}^{-2}$ in this question.)
A bead B, of mass 0·2 kg, is released from rest in a barrel of oil. The bead moves under gravity and the action of a resistance, of magnitude $0{\cdot}1v^2$ N, where $v\,\text{m}\,\text{s}^{-1}$ is the speed of the bead. Find the distance through which the bead has fallen when its speed is $4\,\text{m}\,\text{s}^{-1}$.
What can be said about the speed of the bead when it has fallen a large distance? (UCLES)

10. A charged particle P, of mass m kg, is repelled from a fixed point O by a force of magnitude $\dfrac{25m}{x^2}$ N, where x m is the distance of P from O.
No other forces act on P. The particle is projected, directly towards O, from a point A, where $OA = 0{\cdot}5$ m, with initial speed $10\,\text{m}\,\text{s}^{-1}$. Obtain a differential equation relating the velocity, $v\,\text{m}\,\text{s}^{-1}$, of P, to x.
Show that P comes to rest at a distance 0·25 m from O.
Briefly describe the ensuing motion. (UCLES)

11. A particle of mass 0·2 kg moves in a horizontal straight line under the action of a resistive force directly proportional to its speed. The force is 2 N when the particle is moving with speed $10\,\text{m}\,\text{s}^{-1}$. Given that the speed is $v\,\text{m}\,\text{s}^{-1}$ at time t seconds, show that

$$\frac{\mathrm{d}v}{\mathrm{d}t} = -v.$$

Find:
(i) the time taken for the speed to decrease from $4\,\text{m}\,\text{s}^{-1}$ to $2\,\text{m}\,\text{s}^{-1}$,
(ii) the distance travelled during this period. (WJEC)

12. A particle of mass 0·2 kg is moving in a straight line under the action of a resistive force which, when the particle has speed $v\,\text{m}\,\text{s}^{-1}$, is of magnitude kv^3 N, where k is a constant. Given that the force is of magnitude 80 N when $v = 2$, find k. Show that, at time t seconds, v satisfies

$$\frac{\mathrm{d}v}{\mathrm{d}t} = -50v^3.$$

Find the time taken for the speed to decrease from $2\,\text{m}\,\text{s}^{-1}$ to $1\,\text{m}\,\text{s}^{-1}$.
(AEB 1991)

13. A vehicle of mass 1000 kg accelerates from $10\,\text{m}\,\text{s}^{-1}$ to $20\,\text{m}\,\text{s}^{-1}$. During this period the force produced by the engine is $\dfrac{15\,000}{v}$ newton where $v\,\text{m}\,\text{s}^{-1}$ is the speed of the vehicle at time t seconds, and resistance can be ignored.
Use Newton's Second Law to write down a differential equation connecting $\dfrac{\mathrm{d}v}{\mathrm{d}t}$ and v. Solve this equation by separating the variables and show that the vehicle takes 10 seconds to achieve this change of speed. (OCSEB)

14. A particle P moves in a straight line and experiences a retardation of $0{\cdot}005\,v^4\,\text{m s}^{-2}$, where $v\,\text{m s}^{-1}$ is the speed of P. Given that P passes through a point O with speed $10\,\text{m s}^{-1}$, show that, when it is a distance 3 m from O, its speed is $5\,\text{m s}^{-1}$.
Find the time taken for the speed of P to be reduced from $10\,\text{m s}^{-1}$ to $5\,\text{m s}^{-1}$. (AEB 1994)

15. A particle of mass 0·2 kg moving on the positive x-axis has displacement x metres and velocity $v\,\text{m s}^{-1}$ at time t seconds. At time $t = 0$, $v = 0$ and $x = 1$. The particle moves under the action of a force in the direction of x increasing and of magnitude $\dfrac{4}{x}$ N.

(i) Assuming that no other forces act on the particle, show that $v = \sqrt{(40 \ln x)}$.

(ii) Assuming that a constant resisting force of magnitude 2 N acts on the particle whenever the particle is in motion, show that the greatest speed reached by the particle is $\sqrt{\{20(2 \ln 2 - 1)\}}\,\text{m s}^{-1}$. (UCLES)

16. An animal A runs in a straight line and has an acceleration of $0{\cdot}05(20v - v^2)\,\text{m s}^{-2}$, where $v\,\text{m s}^{-1}$ is the speed of A. Show that, at time t seconds,

$$\frac{dv}{dt}\left(\frac{1}{v} + \frac{1}{20 - v}\right) = 1.$$

Given that at time $t = 0$, A passes through a point O with speed $4\,\text{m s}^{-1}$, show that at time t seconds the speed of A is given by

$$v = \frac{20\,e^t}{4 + e^t}.$$

Hence, or otherwise, find the distance of the animal from O when $t = 9$. (AEB 1994)

17. (Take $g = 10\,\text{m s}^{-2}$ in this question.)
In the Highway Code the stopping distance is defined as the total distance travelled by a vehicle between the driver seeing a hazard and the vehicle coming to rest. It is expressed as the sum of two distances: the thinking distance and the braking distance.
The thinking distance is the distance travelled in the time interval T between the hazard being seen and the brakes being applied. It is assumed that during this interval the speed of the vehicle is unchanged.
The braking distance is the distance travelled between the brakes being applied and the vehicle coming to rest.

(a) It is assumed that the brakes always produce a retardation equal to that when the vehicle slides along the road, the coefficient of friction between the tyres and the road being denoted by μ.
Find, correct to the nearest metre, the stopping distances for $T = 2$ s when the initial speed is $25\,\text{m s}^{-1}$ and $\mu = 0{\cdot}3$.

(b) A more realistic model of the braking effect is that the retardation depends on the speed of the vehicle. Assuming that the retardation at speed $v\,\mathrm{m\,s^{-1}}$ is $\dfrac{600}{v+175}\,\mathrm{m\,s^{-2}}$ and that the brakes are applied when the speed is $25\,\mathrm{m\,s^{-1}}$, find:

(i) the time to come to rest

(ii) the braking distance in this case. (WJEC)

18. A particle of mass m is projected upwards from a point P on the earth's surface. It experiences a force, of magnitude $\dfrac{(mg\,R^2)}{x^2}$, directed towards the centre O of the earth, where R is the radius of the earth and x is the distance of the particle from O. Given that the particle is projected with speed $\sqrt{gR}$ in the direction OP, find how far it will have travelled when its speed is $\frac{1}{2}\sqrt{gR}$. (AEB 1989)

19. A parachutist of mass m falls vertically under the action of the constant gravitational force and a resisting force of magnitude $\dfrac{mv^2}{k}$, where v is the speed of the parachutist and k is a positive constant. Given that the initial vertical speed of the parachutist is zero, show that after falling a distance x his speed is given by

$$v^2 = gk(1 - e^{-2x/k}).$$

Explain why it is advisable to make k as small as possible, and suggest a way in which this may be achieved. (AEB 1994)

20. At time $t = 0$, a parcel of mass m is released from a helicopter, which is hovering at rest relative to the ground. The parcel falls vertically through the air. At time t, its motion is resisted by a force mkv, where v is its speed at that time and k is a positive constant.

(a) Show that

$$v = \frac{g}{k}(1 - e^{-kt}).$$

(b) Hence show that v approaches a limiting value, and write down that value. (ULEAC)

21. At time $t = 0$ a particle of mass m falls from rest at the point A which is at a height h above a horizontal plane. The particle is subject to a resistance of magnitude mkv^2, where v is the speed of the particle at time t and k is a positive constant. The particle strikes the plane with speed V. Show that

$$kV^2 = g(1 - e^{-2kh}).$$ (ULEAC)

22. A particle P, of mass m, is projected vertically upwards from horizontal ground level with speed U.
The motion takes place in a medium in which the resistance is of magnitude $\dfrac{mgv^2}{k^2}$, where v is the speed of P and k is a positive constant.
Show that P reaches its maximum height above the ground after a time T given by

$$T = \frac{k}{g}\arctan\left(\frac{U}{k}\right).$$ (ULEAC)

23. Assume that the gravitational attraction of the earth on an object of mass m at a distance r from the centre of the earth is $\frac{km}{r^2}$, where k is a positive constant. A rocket is launched from the earth's surface, and it travels vertically upwards. When the fuel is exhausted, the distance of the rocket from the centre of the earth is a and the speed of the rocket is u. Some time later, the distance of the rocket from the centre of the earth is x and the speed of the rocket is v. Neglecting any forces other than the gravitational attraction of the earth, find an expression for v.
Deduce that, if $u^2 \geqslant \frac{2k}{a}$, the rocket will never fall back to earth. (UCLES)

24. A particle P of mass 3 kg moves under the action of a force $\mathbf{F}$. At time t seconds, the velocity $\mathbf{v}\,\mathrm{m\,s^{-1}}$ of the particle is given by

$$\mathbf{v} = (1 - \sin 2t)\mathbf{i} + (-1 + \cos 2t)\mathbf{j}.$$

(a) Find the force $\mathbf{F}$ at time t seconds.
(b) Find the kinetic energy of P at time t seconds.
(c) Verify that the rate of change of the kinetic energy is equal to the power of the force $\mathbf{F}$. (AEB Spec 1993)

25. The velocity $\mathbf{v}\,\mathrm{m\,s^{-1}}$ of a particle of mass 0·4 kg at time t seconds is given by

$$\mathbf{v} = 3 \sin 2t\,\mathbf{i} + 5e^{4t}\mathbf{j}.$$

Find the force $\mathbf{F}$ acting on the particle at time t seconds. By taking the scalar product of $\mathbf{F}$ with an appropriate vector, find the rate of working of this force at time t seconds. (WJEC)

26. (Take $g = 10\,\mathrm{m\,s^{-2}}$ in this question.)
A railway engine of mass 20 tonnes is working at a rate of 300 kW against constant resistances to motion of R newtons per tonne.
(i) If it can maintain a speed of $15\,\mathrm{m\,s^{-1}}$ on level ground find the value of R.
(ii) Find the acceleration of the engine when it is travelling at $12\,\mathrm{m\,s^{-1}}$, if the power and resistances remain unchanged.
(iii) Travelling on the level track, the train reaches the bottom of a slope of inclination α where $\sin\alpha = \frac{1}{20}$. It then begins to climb the slope. Assuming that the power and resistances remain unchanged, show that, when the train is travelling at speed $v\,\mathrm{m\,s^{-1}}$, the acceleration, $\frac{dv}{dt}$, is given by $\frac{dv}{dt} = \frac{3(10 - v)}{2v}$.
(iv) Hence, or otherwise, find the maximum speed which the engine can maintain on this incline.
(v) If the train is travelling at $5\,\mathrm{m\,s^{-1}}$ when it reaches the bottom of the slope, find the time it takes to reach a speed of $8\,\mathrm{m\,s^{-1}}$ up this slope. (NICCEA)

27. (Take $g = 10\,\text{m}\,\text{s}^{-2}$ in this question.)

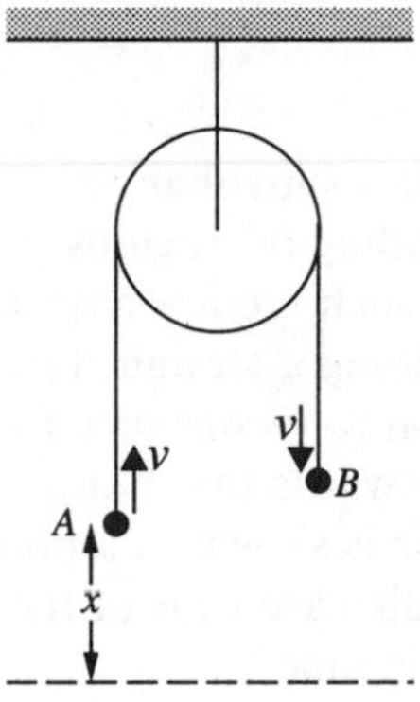

Two particles A and B, of masses 2 kg and 8 kg respectively, are connected by a light inextensible string which passes over a smooth fixed pulley. The resistance to motion is $4v$ N on each particle, where $v\,\text{m}\,\text{s}^{-1}$ is the speed of each particle. Initially A is held at rest, with B hanging freely. The parts of the string between the particles and the pulley are vertical. A is now released and when its speed is $v\,\text{m}\,\text{s}^{-1}$ it has ascended a distance x m (see diagram). Given that A has not reached the pulley, show that

$$v\frac{dv}{dx} = \frac{30 - 4v}{5}.$$

(i) Find the tension in the string at the instant when $v = 5$.
(ii) Find x when $v = 5$, giving your answer correct to two decimal places. (UCLES)

17 Simple harmonic motion

In Chapter 16 we set up mathematical models of situations involving variable acceleration and variable force. Our understanding of calculus enabled us to cope with the differential equations that such models gave rise to. Let us now consider the case of a particle moving along a straight line with the acceleration of the particle always proportional to its distance from some fixed point, O, on the line, and always directed towards that point. The diagram below shows such a situation. The particle is shown at a point between O and a point P and moving towards P. The displacement of the particle from O at time t is x, and v is its velocity at this time

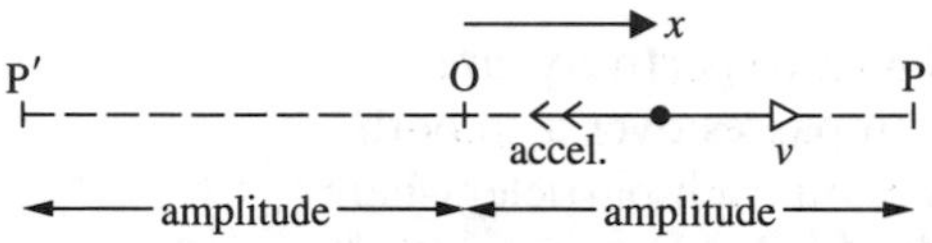

Since the acceleration is directed *towards* O the particle's speed will decrease as it approaches P. If the particle just reaches P the acceleration will then cause it to travel back through O to just reach P′ and then return through O again to just reach P. This process continues, the particle oscillating about O. At points P and P′ the particle will have zero velocity. The distance OP′ equals that of OP and is called the *amplitude* of the motion. The particle completes one *cycle* in travelling from P to P′ and back to P.

Motion of this type is called Simple Harmonic Motion (SHM).

With the accleration proportional to x we can write: $\ddot{x} \propto x$

Introducing the constant of proportionality, n^2, gives: $\ddot{x} = -n^2 x \quad \ldots [1]$

Equation [1] is the differential equation that is characteristic of SHM. A squared constant is used to ease later integration, and the negative sign appears because the acceleration is towards O, i.e. in the direction of decreasing x.

Equation [1] gives the acceleration as a function of displacement and so, using the techniques developed in Chapter 16, we use $v\dfrac{dv}{dx}$ for the acceleration. Thus:

$$v\frac{dv}{dx} = -n^2 x$$

$$\therefore \quad \int v \, dv = \int -n^2 x \, dx$$

$$\therefore \quad \frac{v^2}{2} = \frac{-n^2 x^2}{2} + C$$

But $v = 0$ when the particle is at P or P′, i.e. when $x = \pm a$, the amplitude. Thus:

$$0 = \frac{-n^2 a^2}{2} + C$$

$$\therefore \quad v^2 = n^2(a^2 - x^2) \quad \ldots [2]$$

From equation [2] it can be seen that the maximum speed of the particle occurs when $x = 0$. Thus when the particle is at O:

$$v_{max} = na \qquad \ldots [3]$$

Again, from equation [2]: $v = \pm\sqrt{n^2(a^2 - x^2)}$

Taking the positive root gives: $\dfrac{dx}{dt} = \sqrt{n^2(a^2 - x^2)}$

$\therefore$ $$\int \frac{dx}{\sqrt{a^2 - x^2}} = \int n \, dt$$

giving $$\sin^{-1}\frac{x}{a} = nt + \varepsilon$$

where ε is the constant of integration.

$\therefore$ $$x = a \sin(nt + \varepsilon) \qquad \ldots [4]$$

Differentiating with respect to t gives $v = an\cos(nt + \varepsilon) \qquad \ldots [5]$

(It can also be shown that taking the negative root above would lead to the same results.)

From our understanding of the sine function we know that if we graph $y = b\sin(cx + d)$ the graph is periodic, performing c cycles in 2π radians and the amplitude of the graph is b.
Thus $x = a\sin(nt + \varepsilon)$ is periodic, performing n cycles in 2π radians with amplitude a. The period T (or periodic time) of the motion is the time for one complete cycle.
Since n cycles are performed in 2π rad it follows that: $T = \dfrac{2\pi}{n}$. [6]

The constant ε is called the *phase angle* and depends on the position of the particle when timing commences. It is often wise to choose to commence timing when the particle is at a position which gives ε a convenient value.
For example:

Commence timing when particle is at O
i.e. when $t = 0, x = 0$.
From equation [4] $0 = a\sin\varepsilon$
allowing ε to be taken as zero.
Plotting x against t with t on the horizontal axis:

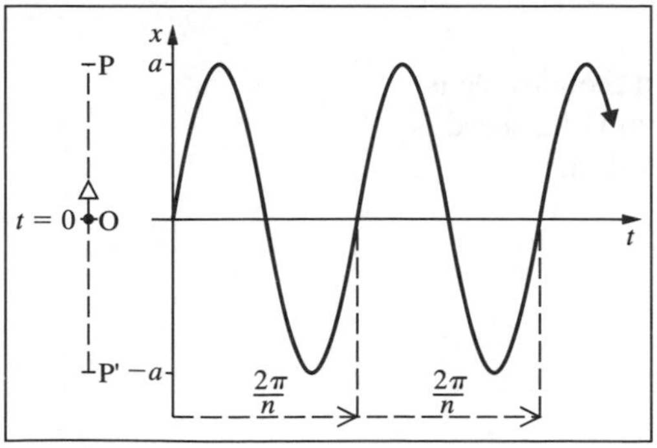

In this case:
equation [4] becomes $x = a\sin nt$
equation [5] becomes $v = an\cos nt$

Commence timing when particle is at P
i.e. when $t = 0, x = a$.
From equation [4] $a = a\sin\varepsilon$
allowing ε to be taken as $\dfrac{\pi}{2}$ radians.

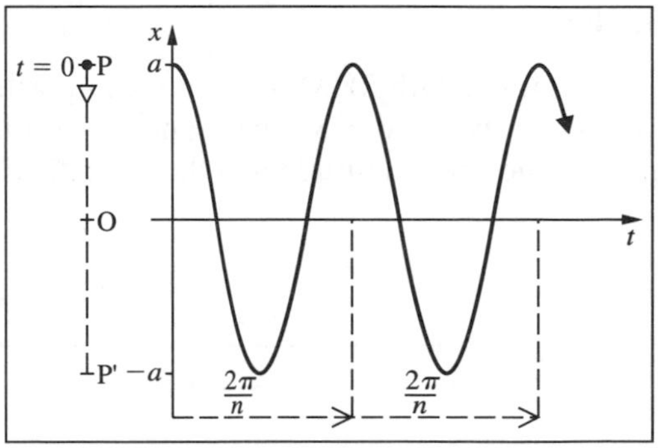

In this case:
equation [4] becomes $x = a\cos nt$
equation [5] becomes $v = -an\sin nt$

Summary

For a particle moving with SHM the following equations apply:

1. $\ddot{x} = -n^2 x$
2. $v^2 = n^2(a^2 - x^2)$
3. $v_{max} = na$
4. $x = a \sin (nt + \varepsilon)$
5. $v = an \cos (nt + \varepsilon)$
6. $T = \dfrac{2\pi}{n}$

These are important equations and they should be memorized.

Example 1

Find the period of the SHM defined by the equation $\ddot{x} = -25x$.

Comparing $\ddot{x} = -25x$ with $\ddot{x} = -n^2x$ gives $n = 5$

$\therefore$ Using $T = \dfrac{2\pi}{n}$ gives $T = \dfrac{2\pi}{5}$

The period is $\dfrac{2\pi}{5}$ s.

Example 2

Find the maximum speed and the maximum acceleration of a particle moving with SHM of period $\dfrac{\pi}{4}$ s and amplitude 25 cm.

We are given the time period $\quad \dfrac{\pi}{4} = \dfrac{2\pi}{n}$

$\therefore \quad n = 8$

Also $\quad a = \frac{1}{4}$ m

Since $\quad v_{max} = na$

$v_{max} = 8\left(\frac{1}{4}\right) = 2 \text{ m s}^{-1}$

From $\quad \ddot{x} = -n^2x$

$\ddot{x}$ is a maximum when $\quad x = -a$

$\therefore \quad \ddot{x}_{max} = -(8)^2\left(-\frac{1}{4}\right) = 16 \text{ m s}^{-2}$

The maximum speed is 2 m s^{-1} and the maximum acceleration is 16 m s^{-2}.

Example 3

A particle moves with SHM about a mean position O. When the particle is 50 cm from O its speed is $3{\cdot}6 \text{ m s}^{-1}$ and when it is 120 cm from O its speed is $1{\cdot}5 \text{ m s}^{-1}$. Find the amplitude and the periodic time of the motion.

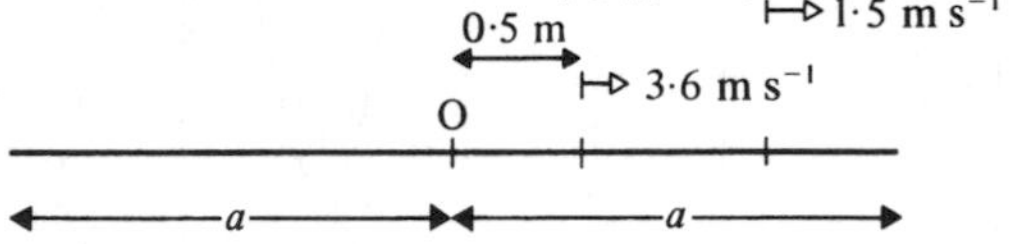

$v = 3{\cdot}6 \text{ m s}^{-1}$ when $x = 0{\cdot}5$ m,
so using $v^2 = n^2(a^2 - x^2)$ gives:

$$3{\cdot}6^2 = n^2(a^2 - 0{\cdot}5^2) \qquad \ldots [1]$$

$v = 1{\cdot}5\,\text{m s}^{-1}$ when $x = 1{\cdot}2\,\text{m}$, so using $v^2 = n^2(a^2 - x^2)$ gives:

$$1{\cdot}5^2 = n^2(a^2 - 1{\cdot}2^2) \qquad \ldots [2]$$

Dividing equation [1] by equation [2] gives: $\dfrac{3{\cdot}6^2}{1{\cdot}5^2} = \dfrac{a^2 - 0{\cdot}5^2}{a^2 - 1{\cdot}2^2}$

with the solution $a = 1{\cdot}3\,\text{m}$

Substituting this value in equation [2] gives: $1{\cdot}5^2 = n^2(1{\cdot}3^2 - 1{\cdot}2^2)$

$\therefore \quad n = 3$

$$\text{periodic time} = \frac{2\pi}{n} = \frac{2\pi}{3}\,\text{s}$$

The amplitude of the motion is $1{\cdot}3\,\text{m}$ and the periodic time is $\dfrac{2\pi}{3}$ s.

Example 4

A particle is projected from a point O at time $t = 0$ and performs SHM with O as the centre of oscillation. The motion is of amplitude 20 cm and time period 4 s. Find:
(a) the speed of projection
(b) the speed of the particle when $t = 1{\cdot}5$ s
(c) the value of t when the particle is first at a point 10 cm from O.

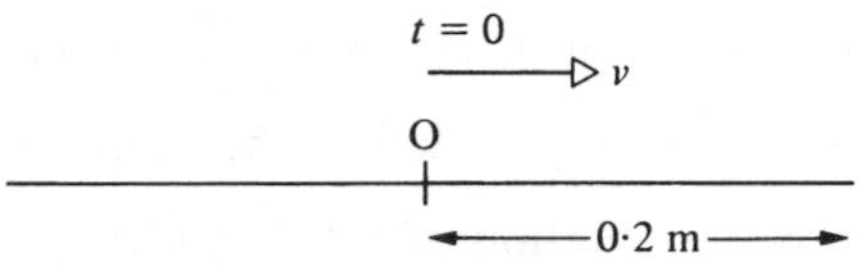

(a) amplitude $= 0{\cdot}2\,\text{m}$ $\quad$ period $T = \dfrac{2\pi}{n} = 4\,\text{s}$

$\therefore \quad n = \dfrac{\pi}{2}$

At O, $\quad$ velocity $v = v_{max}$

$$= na$$

$\therefore \quad v_{max} = \dfrac{\pi}{2}(0{\cdot}2) = \dfrac{\pi}{10}$

The speed of projection is $\dfrac{\pi}{10}\,\text{m s}^{-1}$.

(b) To find v when $t = 1{\cdot}5$ s, use $v = an\cos(nt + \varepsilon)$

since timing starts, i.e. $t = 0$, when the particle is at O, $\varepsilon = 0$:

Thus $\quad v = an\cos nt$

$$= (0{\cdot}2)\left(\frac{\pi}{2}\right)\cos\left(\frac{\pi}{2}\times\frac{3}{2}\right) = \frac{\pi}{10}\cos\frac{3\pi}{4}$$

$$= \frac{\pi}{10}\left(\frac{-1}{\sqrt{2}}\right)$$

When $t = 1{\cdot}5$ s, the speed of the particle is $\dfrac{\pi\sqrt{2}}{20}\,\text{m s}^{-1}$.

(c) To find t when $x = 0{\cdot}1$ m, use $x = a \sin(nt + \varepsilon)$

since timing starts when the particle is at O, $\varepsilon = 0$

Thus $$x = a \sin nt$$

$\therefore$ $$0{\cdot}1 = (0{\cdot}2) \sin \frac{\pi t}{2}$$

or $$\sin \frac{\pi t}{2} = \frac{1}{2}$$

$\therefore$ $$\frac{\pi t}{2} = \frac{\pi}{6}, \frac{5\pi}{6}, \ldots$$

or $$t = \tfrac{1}{3}, \tfrac{5}{3}, \ldots$$

The particle is first at a point 10 cm from O when $t = \frac{1}{3}$ s.

Circular representation of SHM

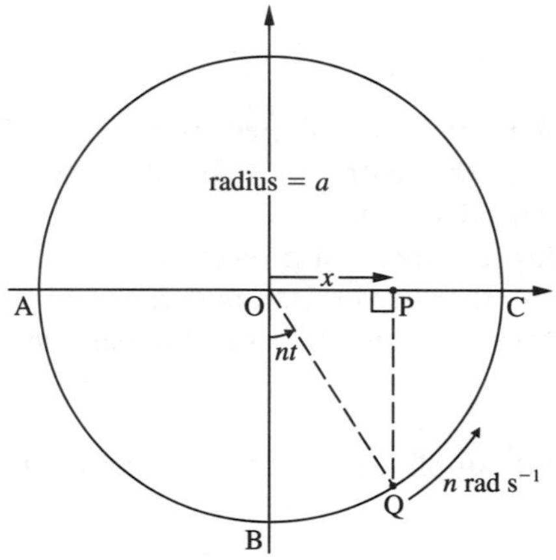

Consider a particle, Q, which starts at B and moves around a circle, centre O and radius a m, with constant angular speed n rad s^{-1}.
At time t the angle $B\hat{O}Q = nt$, and hence $x = a \sin nt$.
However, as we saw earlier in the chapter, the equation $x = a \sin nt$ satisfies: $\ddot{x} = -n^2 x$
Therefore as the particle Q moves around the circle the point P, the projection of Q on to the diameter AC, performs SHM along AC. The time period of this SHM is $\frac{2\pi}{n}$, where n is the angular speed of Q, and the amplitude is a, the radius of the circle.

This circular representation of SHM can be used to solve some questions, as the following example, Method 1, shows.

Example 5

A particle performs SHM of period 6 s and amplitude 4 cm about a centre O. Find the time it takes the particle to travel to a point P, a distance of 2 cm from O.

Method 1

Represent the SHM as a projection of circular motion:

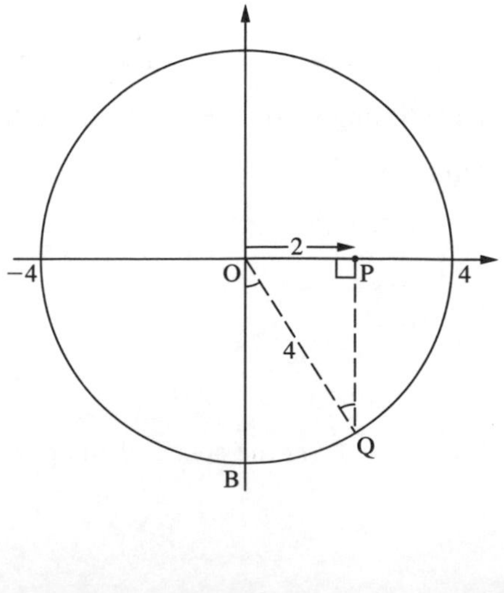

It then follows that $\sin B\hat{O}Q = \sin P\hat{Q}O$
$= 0{\cdot}5$

Therefore $B\hat{O}Q = \frac{\pi}{6}$

So the time taken to travel from O to P, as a fraction of the time period, is $\frac{\pi/6}{2\pi}$, i.e. $\frac{1}{12}$.

Hence the time taken is $\frac{1}{12} \times 6 = 0{\cdot}5$

The particle takes 0·5 s to travel from O to P.

Method 2

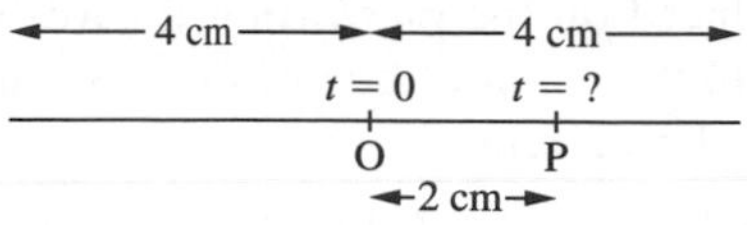

Period $= \dfrac{2\pi}{n} = 6$ s, therefore $n = \dfrac{\pi}{3}$

To find t when $x = 2$ cm, use $x = a \sin(nt + \varepsilon)$

Since timing starts when the particle is at O, $\varepsilon = 0$,

and so $x = a \sin nt$

Substituting values gives: $0{\cdot}02 = 0{\cdot}04 \sin \dfrac{\pi t}{3}$

i.e. $0{\cdot}5 = \sin \dfrac{\pi t}{3}$

$\therefore$ $\dfrac{\pi t}{3} = \dfrac{\pi}{6}$

from which $t = 0{\cdot}5$ as in Method 1.

Exercise 17A

1. Find the periodic time of the simple harmonic motion governed by each of the following equations. (All the equations use SI units.)
   (a) $\ddot{x} = -x$ (b) $\ddot{x} = -4x$ (c) $\ddot{x} = -9x$.
2. A particle moves with SHM about a mean position O with a periodic time of $\dfrac{\pi}{2}$ s.
   Find the magnitude of the acceleration of the particle when 1 metre from O.
3. A particle moves with SHM about a mean position O. When the particle is 25 cm from O, it is accelerating at 1 m s^{-2} towards O. Find the periodic time of the motion and the magnitude of the acceleration of the particle when 20 cm from O.
4. A particle moves with SHM of time period $\dfrac{\pi}{2}$ s and has a maximum speed of 3 m s^{-1}.
   Find the maximum acceleration experienced by the particle.
5. A particle moves with SHM of time period $\dfrac{\pi}{2}$ s and amplitude 2 m.
   Find the maximum speed of the particle.
6. A particle moves with SHM about a mean position O. The amplitude of the motion is 5 m and the period is 8π s.
   Find the maximum speed of the particle and its speed when 3 m from O.
7. A body of mass 500 g moves horizontally with SHM of amplitude 1 m and time period $\dfrac{\pi}{2}$ s.
   Find the magnitude of the greatest horizontal force experienced by the body during the motion.
8. A body of mass 100 g moves horizontally with SHM about a mean position O. When the body is 50 cm from O the horizontal force on the body is of magnitude 5 N. Find the time period of the motion.
9. A body of mass 5 kg is placed on a rough horizontal surface, coefficient of friction $\frac{11}{49}$. State whether or not the body will slide across the surface when the surface is moved horizontally with simple harmonic motion of amplitude 55 cm and:
   (a) a time period of $\dfrac{\pi}{2}$ s
   (b) a time period of $\dfrac{3\pi}{2}$ s
   (c) a frequency of 15 oscillations per minute
   (d) a frequency of 20 oscillations per minute.
10. A horizontal platform is made to move vertically up and down with simple harmonic motion according to the relationship $\dfrac{d^2x}{dt^2} = -49x$, where x m is the vertical displacement of the platform from

its mean position at time t s. Show that any mass placed on the platform will leave the platform if the amplitude of the motion is greater than 20 cm.

11. A particle moves with SHM about a mean position O. The amplitude of the motion is 65 cm and the time period is $\frac{\pi}{4}$ s.
Find how far the particle is from O when its speed is $2\,\text{m}\,\text{s}^{-1}$.

12. A particle moves with SHM about a mean position O. The particle has zero velocity at a point which is 50 cm from O and a speed of $3\,\text{m}\,\text{s}^{-1}$ at O.
Find:
(a) the maximum speed of the particle
(b) the amplitude of the motion
(c) the periodic time of the motion.

13. A particle moves with SHM about a mean position O. When the particle is 60 cm from O, its speed is $1{\cdot}6\,\text{m}\,\text{s}^{-1}$, and when it is 80 cm from O, its speed is $1{\cdot}2\,\text{m}\,\text{s}^{-1}$.
Find the amplitude and period of the motion.

14. The effect of the waves on an empty oil drum floating in the sea is to make it bob up and down with SHM.
If the drum encounters 20 waves every minute and for each wave the vertical distance from peak to trough is 80 cm, find the amplitude and period of the motion and the maximum speed of the drum.

15. A particle moves with SHM about a mean position O. When passing through two points which are 2 m and 2·4 m from O the particle has speeds of $3\,\text{m}\,\text{s}^{-1}$ and $1{\cdot}4\,\text{m}\,\text{s}^{-1}$ respectively.
Find the amplitude of the motion and the greatest speed attained by the particle.

16. A particle moves with SHM about a mean position O. The particle is initially projected from O with speed $\frac{\pi}{6}\,\text{m}\,\text{s}^{-1}$ and just reaches a point A, 2 m from O.
Find how far the particle is from O three seconds after projection.
How many seconds after projection is the particle a distance of 1 m from O:
(a) for the first time
(b) for the second time
(c) for the third time?

17. A particle is released from rest at a point A, 1 m from a second point O. The particle accelerates towards O and moves with simple harmonic motion of time period 12 s and O as the centre of oscillation.
Find how far the particle is from O one second after release.
How many seconds after release is the particle at the mid-point of OA:
(a) for the first time
(b) for the second time?

18. A particle is projected from a point A at time $t = 0$ and performs SHM with A as the centre of oscillation. The amplitude of the motion is 50 cm and the periodic time is 3 s.
Find:
(a) the speed of projection
(b) the speed of the particle when $t = 1$ s
(c) the speed of the particle when $t = 2$ s
(d) the distance of the particle from A when $t = 2$ s.

19. A particle is released from rest at a point A at time $t = 0$ and performs SHM about a mean position B. The particle just returns to A during each oscillation and $AB = 2\sqrt{2}$ m.
If the particle passes through B with speed $\pi\sqrt{2}\,\text{m}\,\text{s}^{-1}$, find the value of t when the particle is first travelling with a speed of $\pi\,\text{m}\,\text{s}^{-1}$. How far from B is the particle then?

20. The head of a piston moves with SHM of amplitude $\frac{\sqrt{3}}{10}$ m about a mean position O.
How far from O is the head of the piston when it is travelling with a speed equal to half of its maximum speed?

21. A particle is fastened to the mid-point of a stretched spring lying on a smooth horizontal surface. The particle is set in motion so that it moves with SHM about a mean position O. If one metre is the greatest distance the particle is from O during the motion, find how far from O the particle is when it is travelling with a speed equal to four-fifths of its greatest speed.

22. A particle performs SHM of period 3 s and amplitude 6 cm about a centre O.
Find the time it takes the particle to travel from O to a point P, a distance of 3 cm from O.

23. A particle performs SHM of period 4 s and amplitude 2 cm about a centre O.
Find the time it takes the particle to travel from O to a point P, a distance of $\sqrt{2}$ cm from O.

24. A particle performs SHM of period 10 s and amplitude 8 cm about a centre O. After passing through O the particle moves through a point A which is 2 cm from O to a point B which is 6 cm from O.
Find the time taken to travel from A to B.

25. A particle performs SHM of period 3 s and amplitude 3 cm about a centre O. After passing through O the particle moves through a point A which is 1 cm from O to a point B which is 2 cm from O.
Find the time taken to travel from A to B.

26. A particle performs SHM of period 4·5 s and amplitude 6 cm about a centre O. The particle passes through a point P which is 3 cm from O, moving away from O.
Find the time which elapses before the particle next passes through P.

27. A particle performs SHM of period 2 s and amplitude 4 cm about a centre O. The particle passes through a point P which is 1 cm from O, moving away from O.
Find the time which elapses before the particle next passes through P.

28. The points A, O, B, C lie in that order on a straight line with AO = OC = 4 cm, and OB = 2 cm. A particle performs SHM of period 6 s and amplitude 4 cm between A and C.
Find the time taken for the particle to travel from A to B.

29. The points A, O, B, C lie in that order on a straight line with AO = OC = 6 cm, and OB = 5 cm. A particle performs SHM of period 3 s and amplitude 6 cm between A and C.
Find the time taken for the particle to travel from A to B.

Springs and SHM

Consider a light spring with one end attached to a fixed point A on a smooth horizontal surface with a body of mass m attached to the other end P. Point O is the position at which the body would rest in equilibrium, with the string neither extended nor compressed.

Spring in extension

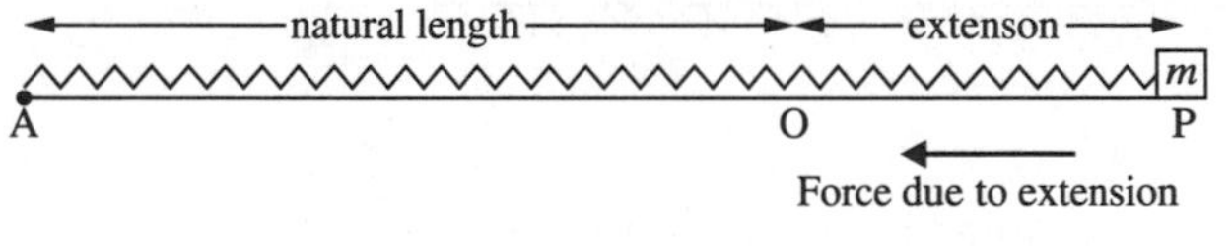

Spring in compression

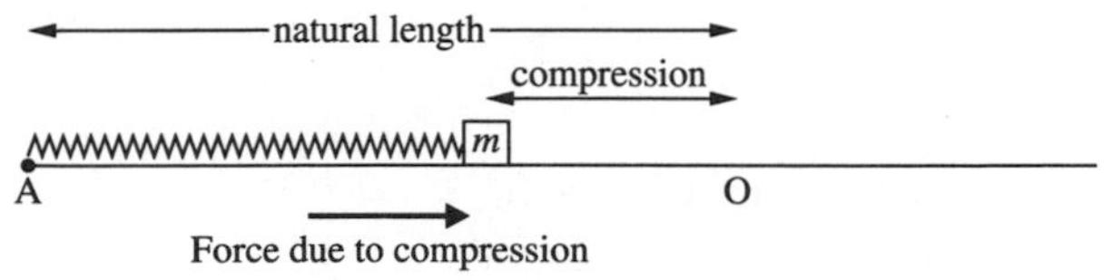

If the body is pulled to the right of O and then released, the horizontal force acting on it will be the tension in the spring, which draws the body towards O. When the body arrives at O the string ceases to be extended, and as the

body passes through O the spring begins to be compressed, so the direction of the horizontal force on the body is reversed. It can be seen that, throughout the motion, the direction of the force acting on the body is always towards O, the centre of motion.
Suppose now that the spring has natural length l and modulus λ. Consider the situation when the spring is extended a distance x.

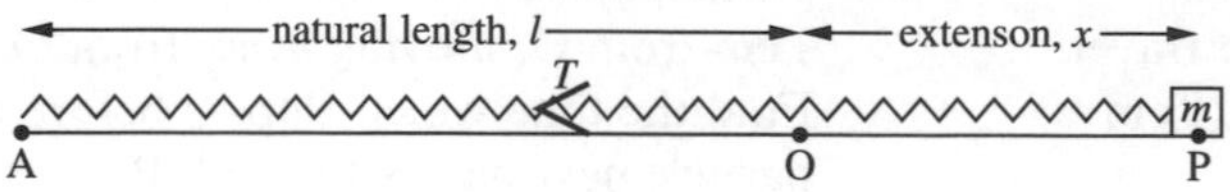

By Hooke's Law $\quad T = \lambda\frac{x}{l}$

Using $F = ma$ gives: $\quad -\lambda\frac{x}{l} = m\ddot{x}$

$\therefore \quad \ddot{x} = -\frac{\lambda}{ml}x$

This is of the form $\ddot{x} = -n^2x$, and so the motion is simple harmonic about $x = 0$, i.e. about the point O, and $n = \sqrt{\frac{\lambda}{ml}}$.

Example 6

One end of a light elastic spring of natural length 2 m and modulus 10 N is fixed to a point A on a smooth horizontal surface. A body of mass 200 g is attached to the other end of the spring and is held at rest at a point B on the surface, causing the spring to be extended by 30 cm. Show that, when released, the body will move with SHM and find the amplitude of the motion.

Let O be the end of the spring when it is unstretched.

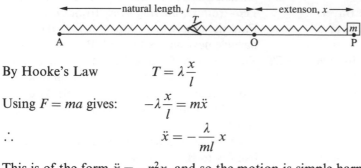

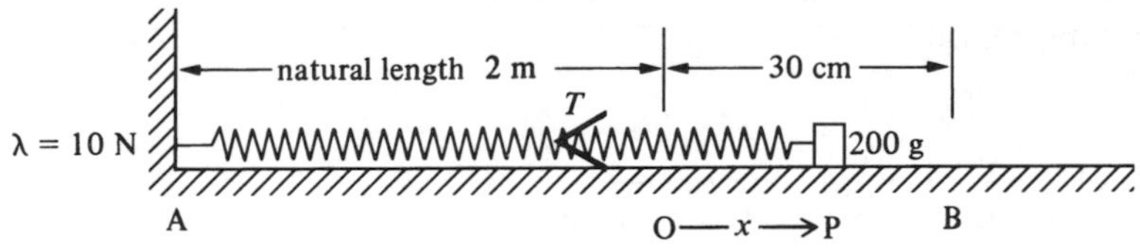

Consider the situation when the body is at a point P and the spring is extended a distance x.

By Hooke's Law $\quad T = \lambda\frac{x}{l} = 10\frac{x}{2}$

$\therefore \quad T = 5x$

Using $F = ma$ gives: $\quad T = -\frac{200}{1000}\ddot{x} \quad$ (*negative* since force is in direction of decreasing x)

$\therefore \quad 5x = -\frac{200}{1000}\ddot{x}$

or $\quad \ddot{x} = -25x$

This is of the form $\ddot{x} = -n^2x$ and thus the motion is simple harmonic about $x = 0$, i.e. about the point O, and $n = 5$.

During SHM the body is furthest from the mean position when its speed is zero, i.e. at B. Thus OB is the amplitude of the motion.

The amplitude of the motion is 30 cm.

Example 7

A light elastic string of natural length 2·4 m and modulus 15 N is stretched between two points A and B, 3 m apart on a smooth horizontal surface. A body of mass 4 kg attached to the mid-point of the string is pulled 10 cm towards B and then released. Show that the subsequent motion is simple harmonic and find the speed of the body when it is 158 cm from A.

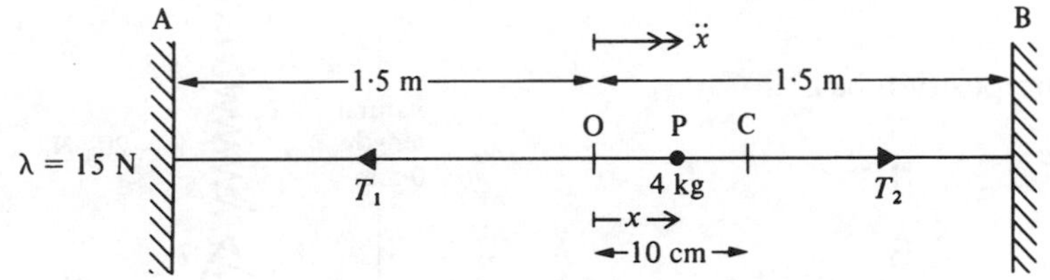

Let O be the centre of AB; C is the point from which the body is released and P is a point such that OP = x metres.
When the body is at P, there are two horizontal forces acting on it: the tensions in the two parts of the string.

$$\text{force tending to increase } x = T_2 - T_1$$

Applying $F = ma$ gives: $\quad T_2 - T_1 = m\ddot{x} \quad \ldots[1]$

If AP and PB are considered as separate strings of modulus 15 N and natural length 1·2 m, and we use Hooke's Law, equation [1] becomes:

$$\lambda\frac{(0\cdot3 - x)}{1\cdot2} - \lambda\frac{(0\cdot3 + x)}{1\cdot2} = 4\ddot{x}$$

$$\therefore \quad 0\cdot3\lambda - \lambda x - 0\cdot3\lambda - \lambda x = 4\cdot8\ddot{x}$$

$$\therefore \quad -2\lambda x = 4\cdot8\ddot{x}$$

and using $\lambda = 15$ N gives: $\quad \ddot{x} = -\frac{25}{4}x$

This is of the form $\ddot{x} = -n^2x$, and thus the motion is simple harmonic about $x = 0$, i.e. about the point O, and $n = \frac{5}{2}$.

Note that, as the body is released from rest at C and it oscillates about a mean position O, then CO is the amplitude of the motion. Thus neither string goes slack at any stage of the motion.

When the body is 158 cm from A, $x = \frac{8}{100}$ m.

amplitude = CO = 10 cm

Using $v^2 = n^2(a^2 - x^2)$ gives: $\quad v^2 = \frac{25}{4}\left[\left(\frac{1}{10}\right)^2 - \left(\frac{8}{100}\right)^2\right]$

and $\quad v = 0\cdot15\,\text{m s}^{-1}$

The motion is simple harmonic, and when the body is 158 cm from A its speed is $0\cdot15\,\text{m s}^{-1}$.

Example 8

A light elastic spring, of natural length 50 cm and modulus $20g$ N, hangs vertically with its upper end fixed and a body of mass 6 kg attached to its lower end. The body initially rests in equilibrium and is then pulled down a distance of 25 cm and released.
Show that the ensuing motion will be simple harmonic, and find the period of the motion and the maximum speed of the body.
Would the answers have been the same had an elastic string been used in place of the spring?

Method 1

In this method, first find the equilibrium position, which is then expected to be the centre of the oscillation.

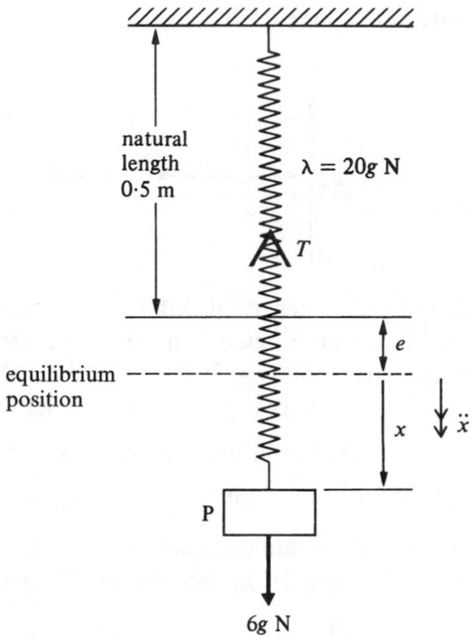

Let the extension in the equilibrium position be e

By Hooke's Law $\quad T = \dfrac{\lambda e}{l} \quad \ldots[1]$

In equilibrium $\quad T = 6g \quad \ldots[2]$

From these equations $\quad \dfrac{\lambda e}{l} = 6g$

Substituting $\lambda = 20g$ and $l = 0{\cdot}5$ gives:

$$e = 15\,\text{cm}$$

Consider the body at a point P, distance x below the equilibrium position.

Applying $F = ma$ gives: $\quad 6g - T = 6\ddot{x}$

Using Hooke's Law gives: $\quad 6g - \dfrac{\lambda(x+e)}{l} = 6\ddot{x}$

$\therefore \quad 6g - 20g\dfrac{(x + 0{\cdot}15)}{0{\cdot}5} = 6\ddot{x}$

$\therefore \quad 6g - 40gx - 6g = 6\ddot{x}$

or $\quad \ddot{x} = -\frac{20}{3}gx$

Hence the motion is simple harmonic about $x = 0$, the equilibrium position, and:

$$n = \sqrt{\left(\tfrac{20}{3}g\right)} = \frac{14}{\sqrt{3}}$$

The body is released from a point 25 cm below the equilibrium position, i.e. the body is at rest when 25 cm from the centre of the oscillation; hence the amplitude of the motion is 25 cm.

Using $T = \dfrac{2\pi}{n}$ and $v_{max} = na$ gives:

$$T = \frac{\pi\sqrt{3}}{7}\text{ s} \quad \text{and} \quad v_{max} = \frac{7}{6}\sqrt{3}\,\text{m s}^{-1}$$

Method 2

In this method, take the general position of the body to be when the spring is extended a distance x from its unstretched length.

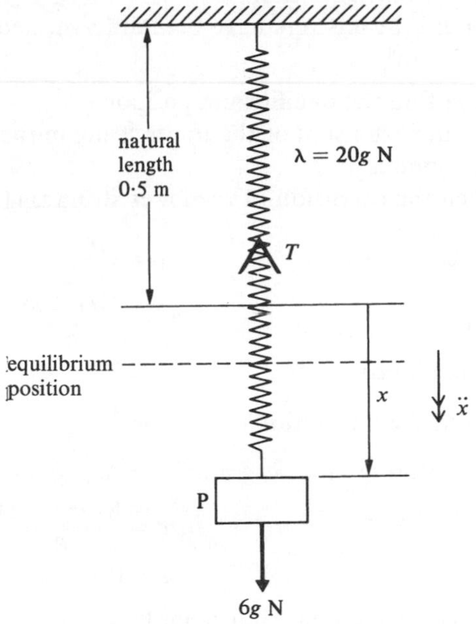

Applying $F = ma$ gives: $6g - T = 6\ddot{x}$

Using Hooke's Law gives: $6g - \lambda\dfrac{x}{l} = 6\ddot{x}$

Substituting $\lambda = 20g$ and $l = 0{\cdot}5$ gives:

$$\ddot{x} = g - \tfrac{20}{3}gx$$

which can be written as: $\ddot{x} = -\tfrac{20}{3}g\left(x - \tfrac{3}{20}\right)$

Substituting $y = x - \tfrac{3}{20}$ gives: $\ddot{y} = -\tfrac{20}{3}gy$

Hence the motion is simple harmonic about $y = 0$, i.e. about $x = 15\,\text{cm}$ as in Method 1.

Again $n = \sqrt{\left(\tfrac{20}{3}g\right)}$ and the amplitude and time period can be obtained as in Method 1.

As will be seen from the diagram on the right, during part of the motion the body will be above the point at which the spring has its natural length, i.e. the spring is being compressed. Thus, were a string to replace the spring, the body would be in free flight for some period of time. The motion would still be periodic but the period of the motion would be different.
The maximum speed occurs when the body passes through the equilibrium position and this would be unchanged if a string replaced the spring.

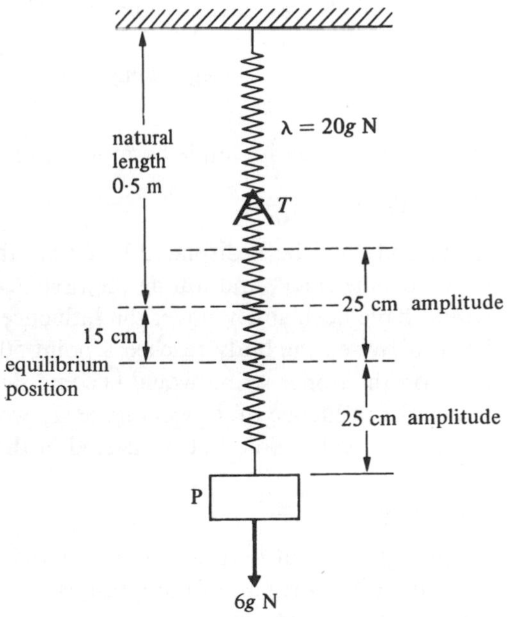

Example 9

A and C are two fixed points 4 m apart with A vertically above C. A body of mass m kg rests between A and C, held in position by two vertical strings. The upper string has modulus $6mg$ N and natural length 1 m. The lower string has modulus $8mg$ N and natural length 2 m. The body is displaced 20 cm below its equilibrium position and released from rest.

(a) Show that the ensuing motion will be simple harmonic, and find the period of the motion.
(b) Would the answers have been different had the body been displaced by 60 cm?

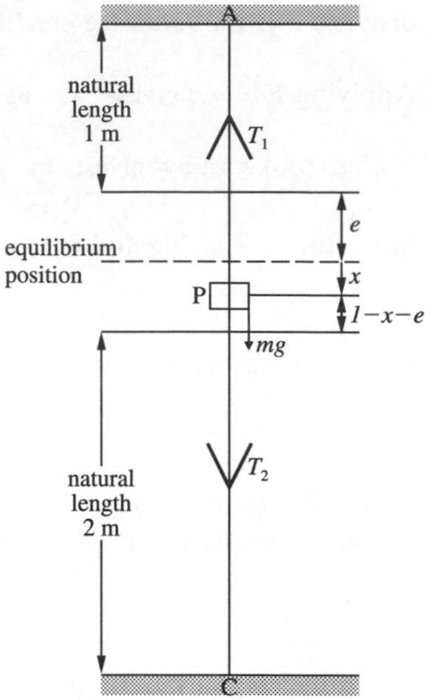

(a) First find the equilibrium position.
Let the extension of the upper string in the equilibrium position be e metres.
Then the extension of the lower string is $(1-e)$.

By Hooke's Law $\quad T_1 = \dfrac{\lambda_1 e}{1} \quad \ldots[1]$

and $\quad T_2 = \dfrac{\lambda_2(1-e)}{2} \quad \ldots[2]$

In equilibrium $\quad T_1 = T_2 + mg \quad \ldots[3]$

From these equations $\quad \lambda_1 e = \dfrac{\lambda_2(1-e)}{2} + mg$

Substituting $\lambda_1 = 6mg$ and $\lambda_2 = 8mg$ gives:

$$6mge = \frac{8mg(1-e)}{2} + mg$$

$$\therefore \quad e = 0{\cdot}5$$

Consider the body at a point P, distance x below the equilibrium position.

Applying $F = ma$ gives: $\quad mg + T_2 - T_1 = m\ddot{x}$

Using Hooke's Law gives: $\quad mg + \lambda_2\dfrac{(1-x-e)}{2} - \lambda_1\dfrac{(x+e)}{1} = m\ddot{x}$

$$\therefore \quad mg + 8mg\frac{(1-x-0{\cdot}5)}{2} - 6mg(x+0{\cdot}5) = m\ddot{x}$$

$$\therefore \quad -10gx = \ddot{x}$$

Hence the motion is simple harmonic about the equilibrium position and of period $\dfrac{2\pi}{\sqrt{10g}}$.

(b) If the body had been displaced by 60 cm, then the lower string would have become slack, and during the first 10 cm of the motion the body would have been solely under the influence of T_1 and mg, as T_2 would be zero. When the body reached a point 50 cm above the equilibrium position the upper string would become slack, leaving the body solely under the influence of T_2 and mg, as T_1 would then be zero. The motion would still be periodic but the period of the motion would be different.

Exercise 17B

1. A light spring is of natural length 1 m and modulus 2 N. One end of the spring is attached to a fixed point A on a smooth horizontal surface, and to the other end is attached a body of mass 500 g. The body is held at rest on the surface at a distance of 1·25 m from A. Show that on release the body will move with SHM and find the amplitude and time period of the motion.

2. Two points A and B are 1 m apart on a smooth horizontal table. A light spring, of natural length 75 cm and modulus 54 N, has one end fastened to the table at A and the other end fastened to a body of mass 8 kg which is held at rest at B.
Show that when the body is released it moves with SHM and find the maximum speed of the body during the motion.

3. A body of mass 2 kg is fixed to the mid-point of a light elastic string of natural length 1 m and modulus 18 N. The ends of the string are attached to two points A and B 2 m apart on a smooth horizontal surface. The body is pulled a distance y towards A ($y < 50$ cm) and released.
Show that the subsequent motion is simple harmonic and find the time period of the motion.
If the maximum speed of the body is $1{\cdot}5\,\text{m s}^{-1}$, find the value of y.

4. A light elastic string of natural length 1·5 m and modulus 12 N is stretched between two points A and B 2 m apart on a smooth horizontal surface. A body of mass 2 kg is attached to the mid-point of the string, pulled 20 cm towards A and released.
Show that the subsequent motion is simple harmonic and find the speed of the body when 88 cm from A.

5. A and B are two fixed points on a smooth horizontal surface with AB = 2 m. A body of mass 4 kg lies on the line AB at a point P and is in equilibrium with a light elastic string of natural length 75 cm and modulus 18 N connecting it to A and a light elastic string of natural length 50 cm and modulus 6 N connecting it to B.
Show that P is midway between A and B.
If the body is then pulled 20 cm towards A and released, show that the subsequent motion is simple harmonic and find the maximum speed of the body during the motion.

6. A light elastic string is of natural length 60 cm and modulus $3mg$ N. The string hangs vertically with its top end fixed and a body of mass m kg fastened to the other end.
Find the extension in the string when the body hangs in equilibrium.
If the body is then pulled vertically downwards a distance of 10 cm and released, show that the ensuing motion will be simple harmonic, and find the time period of the motion and the maximum speed of the body.

7. A body of mass 500 g is attached to end B of a light elastic string AB of natural length 50 cm. The system rests in equilibrium with the string vertical and end A fixed. The body is then pulled vertically downwards a small distance and released.
If the ensuing motion is simple harmonic of time period $\frac{1}{5}\pi$ s, find the modulus of the string.

8. A light spring hangs vertically with its top end fixed and a body of mass m kg attached to the other end. The spring is of natural length 1 metre and modulus $5mg$ N and initially the system rests in equilibrium. The body is then pulled vertically downwards a distance of 30 cm and released.
Show that the subsequent motion is simple harmonic and find the greatest acceleration experienced by the body. Would the motion have been simple harmonic if an elastic string had been used in place of the spring?

9. A light elastic string of natural length 20 cm and modulus 40 N has one end attached to a fixed point A on a smooth horizontal surface and a body of mass 2 kg attached to the other end. The body is held on the surface at a point which is 40 cm from A, and released.
Show that the subsequent motion will be periodic and find the time period of the motion and the speed of the body as it passes through A.

10. A light elastic string of natural length 50 cm hangs vertically with its top end fixed and a body of mass 2 kg attached to the other end. With the body hanging in equilibrium, the string has a total length of 70 cm.
Find the modulus of the string.
If the body is then pulled vertically downwards a distance of 10 cm and released, show that the subsequent motion is simple harmonic and find the speed of the particle when it is 2 cm above the point of release.

11. A light spring of natural length 40 cm and modulus $2g$ N hangs vertically with its upper end fixed and a particle attached to its other end. When the particle hangs in equilibrium the extension of the string is 5 cm. The spring is now replaced by a different spring of natural length 50 cm. The system is again allowed to settle in a position of equilibrium and then the particle is pulled vertically downwards a short distance and released. If the subsequent motion is simple harmonic with time period $\frac{1}{10}\pi$ s, find the mass of the particle and the modulus of this second spring.

12. A and B are two points 25 cm apart on a smooth horizontal surface. A particle of mass 500 g lies at A and is connected to B by a light spring of natural length 25 cm and modulus 50 N.
If the particle is projected directly towards B with speed $4\,\text{m s}^{-1}$, show that the ensuing motion will be simple harmonic and find the time period and amplitude.

13. A light spring of natural length 50 cm and modulus 147 N hangs vertically with its upper end fixed and a body of mass 1·5 kg attached to the lower end. With the system resting in equilibrium, the body is projected vertically downwards with a speed of $1{\cdot}4\,\text{m s}^{-1}$.
Show that the resulting motion will be simple harmonic and find the amplitude of the motion.

14. A light elastic string of natural length 1 m and modulus $2mg$ N hangs vertically with its upper end fixed and a body of mass m kg attached to its other end.
Find the total length of the string when the mass hangs in equilibrium.
The body is then pulled vertically downwards a distance d metres and released. Show that the body will move with SHM provided $d \leqslant 0{\cdot}5$.

15. A and C are two fixed points 5 m apart with A vertically above C. A body of mass m kg rests between A and C, held in position by two identical vertical strings, each of modulus $4mg$ N and natural length 2 m, one string linking the body to A and the other linking it to C. The body is displaced 10 cm below its equilibrium position and released from rest. Show that the ensuing motion will be simple harmonic, and find the period of the motion.

16. A and C are two fixed points 6 m apart with A vertically above C. A body of mass m kg rests between A and C, held in position by two identical vertical strings, each of modulus $0{\cdot}5mg$ N and natural length 1 m, one string linking the body to A and the other linking it to C. The body is displaced 50 cm below its equilibrium position and released from rest.
Show that the ensuing motion will be simple harmonic, and find the period of the motion.

17. A body of mass 2 kg is attached to the midpoint of a light elastic string of modulus $20g$ N and natural length 2 m. One end of the string is attached to a fixed point P and the other end of the string is attached to a fixed point Q, where P is 3 m vertically above Q. The body rests in equilibrium at a point O. Find the distance PO.
The body is now pulled down to a point R which is a distance of 10 cm below O, and released from rest.
Find the time taken to travel from R to O.

18. A and C are two fixed points 5 m apart with A vertically above C. A body of mass m kg rests in equilibrium between A and C, held in position by two vertical strings. The upper string has modulus $3mg$ N and natural length 1 m. The lower string has modulus $4mg$ N and natural length 2 m. The body is displaced 30 cm below its equilibrium position and released from rest.
Show that the ensuing motion will be simple harmonic, and find the period of the motion.

19. A and C are two fixed points 10 m apart with A vertically above C. A body of mass m kg rests in equilibrium between A and C, held in position by two vertical strings. The upper string has modulus $6mg$ N and natural length 3 m. The lower string has modulus $12mg$ N and natural length 4 m. The body is pulled down a distance of 40 cm and released from rest.
Find the time that elapses until the body is next at rest.

More examples involving strings

We have already seen in the previous section, and in some of the questions of Exercise 17B, that while a string behaves in exactly the same way as a spring when in extension it behaves quite differently if the situation requires compression. A spring can continue to exert a force when in compression but a string offers no such force. It is still possible to model the motion of a particle attached to a string even if the string is slack during part of the motion as the following examples, and the questions of Exercise 17C, will show.

Example 10

One end of a light elastic string of natural length 9 m and modulus 108 N is attached to a fixed point A on a smooth horizontal surface. A body of mass 3 kg is attached to the other end of the string and is held at rest at a point C on the surface which is at a distance of 12 m from A. The body is released from rest. Find the time that elapses before the body reaches A, and sketch the velocity–time graph for this phase of the motion.

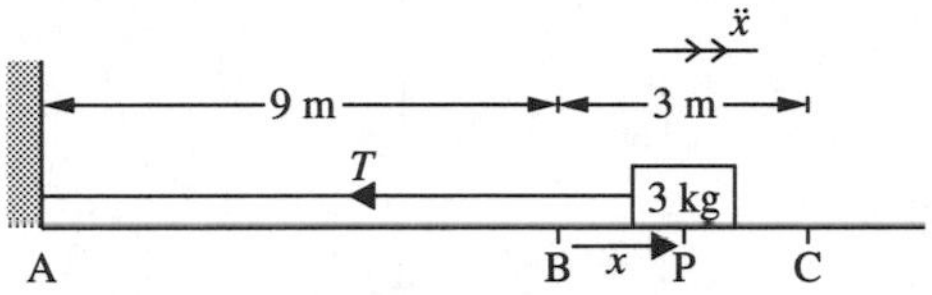

Consider the situation when the body is at a point P and the string is extended a distance x.

By Hooke's Law $\quad T = \lambda\dfrac{x}{l} = 108\dfrac{x}{9}$

$\therefore \quad T = 12x$

Using $F = ma$ gives: $\quad -T = 3\ddot{x}$

$\therefore \quad -12x = 3\ddot{x}$

or $\quad \ddot{x} = -4x \qquad \ldots[1]$

Thus while the body is under the influence of the tension in the string its motion is simple harmonic, with B as the centre of the motion. The time taken for the body to travel from C to B is one quarter of the value of the period to which [1] would give rise were it to apply throughout the motion.

Hence the time taken for the body to travel from C to B is:

$$\frac{1}{4}\times\frac{2\pi}{n} = \frac{1}{4}\times\frac{2\pi}{2} = \frac{\pi}{4}\text{ s} \qquad \ldots[2]$$

In order to find the speed of the body as it passes through B we use the equation $v^2 = n^2(a^2 - x^2)$ with $n = 2$, $a = 3$ and $x = 0$.

This gives: $\quad v^2 = 2^2(3^2 - 0^2)$

i.e. $\quad v = 6$

(We could equally well have used $v_{max} = na$)

After passing through B the body maintains a constant speed of $6\,\mathrm{m\,s^{-1}}$ as it continues towards A, and the time taken to travel from B to A is:

$$\frac{9}{6} = \frac{3}{2}\,\mathrm{s} \qquad \ldots[3]$$

By adding [2] and [3] we find that the total time taken for the body to travel from C to A is $\frac{1}{4}(\pi + 6)$ s.

The velocity–time graph is shown sketched below.

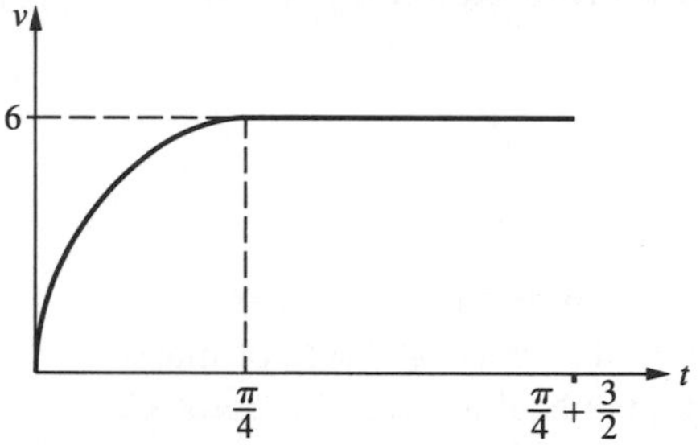

Example 11

A light elastic string of natural length 10 m, hangs vertically with its upper end fixed at a point A and a body of mass 5 kg attached to its lower end. In equilibrium the body rests at a point O where AO = 11 m. The body is then pulled down a further distance of 2 m to a point C and released from rest. Find the period of the ensuing motion.

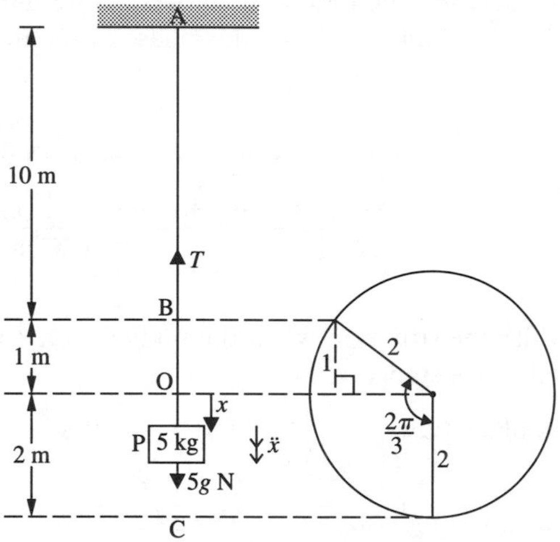

Let the modulus of the string be λ.

Then, applying Hooke's Law in the equilibrium position, we have:

$$5g = \lambda \tfrac{1}{10}$$

$$\therefore \quad \lambda = 50g$$

Consider the body at a point P, distance x below O.

Applying $F = ma$ gives: $\quad 5g - T = 5\ddot{x}$

Using Hooke's Law gives: $\quad 5g - \lambda\dfrac{(x+1)}{10} = 5\ddot{x}$

$$\therefore \quad 5g - 50g\frac{(x+1)}{10} = 5\ddot{x}$$

$$\therefore \quad 5g - 5gx - 5g = 5\ddot{x}$$

$$\therefore \quad \ddot{x} = -gx \qquad \ldots[1]$$

Considering the circular representation of the motion whilst under the influence of the string, we can see that the time taken for the body to travel from C to the unstretched position B is one-third of the value for the period to which [1] would give rise were it to apply throughout the motion.

Hence the time taken for the body to travel from C to B is:

$$\frac{1}{3}\times\frac{2\pi}{n}=\frac{1}{3}\times\frac{2\pi}{\sqrt{g}}$$
$$=\frac{2\pi}{3\sqrt{g}} \qquad \ldots[2]$$

In order to find the speed of the body as it passes through B we use the equation $v^2 = n^2(a^2 - x^2)$ with $n = \sqrt{g}$, $a = 2$ and $x = -1$.

This gives: $v^2 = g(2^2 - (-1)^2)$

i.e. $v = \sqrt{3g}$

After passing through B, the body is a free projectile under the influence of gravity, and the time taken for it to return to B can be found by using the equation $s = ut + \frac{1}{2}at^2$ with $s = 0$, $u = \sqrt{3g}$, and $a = -g$.

This gives:

time taken for the body to travel from B, back to B $= 2\sqrt{\frac{3}{g}}$ $\qquad \ldots[3]$

On returning to B the body continues to C under the action of the string. By symmetry, the time taken for the body to travel from B to C is the same as the time taken for the body to travel from C to B.
Thus, using [2] and [3] gives the period of the motion as:

$$2\left(\frac{2\pi}{3\sqrt{g}}\right)+2\sqrt{\frac{3}{g}}, \quad \text{i.e.} \quad \frac{2(2\pi+3\sqrt{3})}{3\sqrt{g}}$$

Example 12

A man of mass 75 kg is attached to one end of a light elastic rope of natural length 14·4 m and modulus $216g$ N. The other end of the rope is attached to a bridge which is high above a river. The man steps gently from the bridge. Show that the time that elapses before the man is instantaneously at rest is given by:

$$\frac{1}{14}\left[24+5\left(\pi+2\sin^{-1}\left(\frac{5}{13}\right)\right)\right]$$

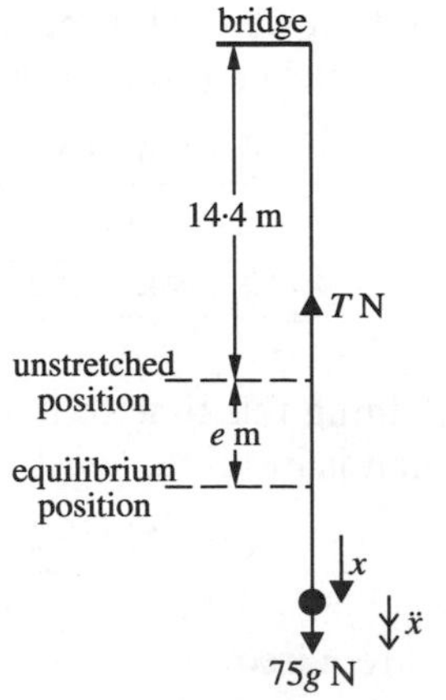

As in Example 11 there are two stages to the motion. For the first 14·4 m the man falls as a free body under gravity. After falling 14·4 m the man begins to experience the resisting force of the string, and it is this force which eventually brings him to instantaneous rest.

<u>For vertical "free fall" motion.</u>

$s = 14{\cdot}4\,\text{m}$
$u = 0\,\text{m s}^{-1} \downarrow$
$a = 9{\cdot}8\,\text{m s}^{-2} \downarrow$
$v = v\,\text{m s}^{-1}$
$t = t\,\text{s}$

Use $s = ut + \frac{1}{2}at^2$

$14{\cdot}4 = \frac{1}{2}(9{\cdot}8)t^2$

$t^2 = \dfrac{28{\cdot}8}{9{\cdot}8} = \dfrac{144}{49}$

$\therefore \quad t = \dfrac{12}{7}$ ... [1]

Use $v^2 = u^2 + 2as$

$v^2 = 2(9{\cdot}8)(14{\cdot}4)$

$\therefore \quad v = 16{\cdot}8$ ... [2]

For motion under the influence of the elastic rope.

Let e m be the extension in the equilibrium situation.

By Hooke's Law: $T = 216g\,\dfrac{e}{14{\cdot}4}$

Thus, in equilibrium: $75g = 216g\,\dfrac{e}{14{\cdot}4}$

$\therefore \quad e = 5$

Consider the body at a point P, distance x m below the equilibrium position.

Applying $F = ma$: $75g - T = 75\ddot{x}$

Using Hooke's Law gives: $75g - 216g\,\dfrac{(x+5)}{14{\cdot}4} = 75\ddot{x}$

which simplifies to: $-147x = 75\ddot{x}$

i.e. $\ddot{x} = -1{\cdot}96x$

Thus this part of the motion is simple harmonic with $n = 1{\cdot}4$.
When this part of the motion commences:

$x = -5$

and $v = 16{\cdot}8$ (from [2]).

Use $v^2 = n^2(a^2 - x^2)$

$(16{\cdot}8)^2 = (1{\cdot}4)^2(a^2 - 5^2)$

$\therefore \quad a = 13$

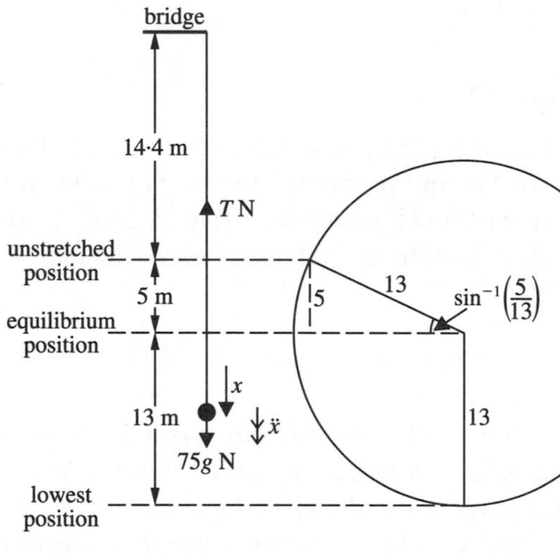

Thus the man first comes to instantaneous rest at a point 13 m below the equilibrium position (see diagram).
Considering the circular representation of the motion we obtain the time taken to travel from the unstretched position to the lowest position as:

$$\frac{\frac{\pi}{2} + \sin^{-1}\left(\frac{5}{13}\right)}{2\pi} \times \frac{2\pi}{n}$$

$$= \frac{\frac{\pi}{2} + \sin^{-1}\left(\frac{5}{13}\right)}{1{\cdot}4} = \frac{5\left(\pi + 2\sin^{-1}\left(\frac{5}{13}\right)\right)}{14}$$

Adding this time to that given by [1] gives the time that elapses before the man is instantaneously at rest as

$$\frac{12}{7} + \frac{5\left(\pi + 2\sin^{-1}\left(\frac{5}{13}\right)\right)}{14} = \frac{1}{14}\left[24 + 5\left(\pi + 2\sin^{-1}\left(\frac{5}{13}\right)\right)\right]$$

as required.

Exercise 17C

1. One end of a light elastic string of natural length 6 m and modulus 48 N is attached to a fixed point A on a smooth horizontal surface. A body of mass 2 kg is attached to the other end of the string and is held at rest at a point B on the surface which is at a distance of 9 m from A. The body is released from rest.
   Find the time that elapses before the body reaches A.
2. One end of a light elastic string of natural length 10 m and modulus 450 N is attached to a fixed point A on a smooth horizontal surface. A body of mass 5 kg is attached to the other end of the string and is held at rest at a point B on the surface which is at a distance of 12 m from A. The body is released from rest.
   Find the time that elapses before the body reaches A.
3. One end of a light elastic string of natural length 2 m and modulus 2 N is attached to a fixed point A on a smooth horizontal surface. A body of mass 250 g is attached to the other end of the string and is held at rest at a point B on the surface which is at a distance of 6 m from A. The body is released from rest.
   (a) Find the time that elapses before the body reaches A.
   (b) Sketch the velocity–time graph for this phase of the motion.
4. A body B, of mass 2 kg, rests on a smooth horizontal surface and is attached to a fixed point P by a light elastic string of natural length 5 m and modulus 10 N. In this position with the string just taut B is given a velocity of 3 m s^{-1} in the direction PB.
   Find the time that elapses before B reaches P.
5. A body B, of mass 2 kg, rests on a smooth horizontal surface and is attached to a fixed point P by a light elastic string of natural length 4 m and modulus 8 N. In this position with the string just taut B is given an impulse of 6 Ns in the direction PB.
   (a) Find the time that elapses before B reaches P.
   (b) Sketch the velocity–time graph for this phase of the motion.
6. One end of a light elastic string of natural length 10 m and modulus 10 N is attached to a fixed point A on a smooth horizontal surface. A body of mass 4 kg is attached to the other end of the string. The body is projected across the surface from A with speed 4 m s^{-1}.
   Find:
   (a) the distance of the body from A when it first comes instantaneously to rest
   (b) the time that elapses until the body returns to A.
7. One end of a light elastic string of natural length 1 m and modulus 10 N is attached to a fixed point A on a smooth horizontal surface. A body of mass 100 grams is attached to the other end of the string. The body is resting at A when it is given a horizontal impulse of 0·4 N s.
   (a) Calculate the distance of the body from A when it first comes instantaneously to rest.
   (b) Calculate the time that elapses before the body returns to A.
   (c) Sketch the velocity–time graph for this phase of the motion.
8. A light elastic string of natural length 5 m, hangs vertically with its upper end fixed at a point A and a body of mass 4 kg attached to its lower end. In equilibrium the body rests at a point O where AO = 6·25 m. The body is then pulled down a further distance of 2·5 m to a point C and released from rest. Show that the period of the ensuing motion is given by:
   $$\tfrac{5}{21}[2\pi + 3\sqrt{3}]\text{ s.}$$
9. A light elastic string of natural length 2 m, hangs vertically with its upper end fixed at a point A and a body of mass 2 kg attached to its lower end. In equilibrium the body rests at a point O where AO = 2·2 m. The body is then pulled down a further distance of 0·25 m to a point C and released from rest. Show that the period of the ensuing motion is given by:
   $$\tfrac{1}{14}[2\pi + 4\sin^{-1}(\tfrac{4}{5}) + 3]\text{ s.}$$

10. A light elastic string hangs vertically with its upper end fixed and a body attached to its lower end. The body initially rests in equilibrium with the string stretched 5 cm beyond its unstretched length. The body is pulled down a further distance 10 cm and released from rest. Show that the period of the ensuing motion is given by:

$$\tfrac{1}{21}[2\pi + 3\sqrt{3}] \text{ s.}$$

11. A light elastic string, of natural length a metres and modulus mg N, has one end attached to a fixed point O, and the other end attached to a particle of mass m kg. The particle is released from rest from a point A which is a vertical distance of $\frac{a}{2}$ metres below O. Show that the particle first returns to A after a time:

$$\left(\frac{3\pi}{2}+2\right)\sqrt{\frac{a}{g}} \text{ s.}$$

12. A particle of mass m kg is suspended from a ceiling by a light elastic string of natural length l metres and modulus $4mg$ N. The particle is pulled vertically downwards from its equilibrium position to a depth of $3l$ metres below the ceiling and released from rest.

(a) Find the speed with which the particle strikes the ceiling.

(b) Show that the time from the moment of release until the instant when the string goes slack is:

$$\sqrt{\frac{l}{4g}}\left[\frac{\pi}{2}+\sin^{-1}\left(\frac{1}{7}\right)\right] \text{ s.}$$

13. A child of mass 30 kg is attached to one end of a light elastic rope of natural length 10 m and modulus $60g$ N. The other end of the rope is attached to a bridge which is high above a river. The child steps gently from the bridge. Show that the time that elapses before the child is instantaneously at rest is given by:

$$\tfrac{5}{14}\left[4+\pi+2\tan^{-1}\left(\tfrac{1}{2}\right)\right] \text{ s.}$$

14. A body of mass 500 grams is attached to one end of a light elastic string of natural length 2·5 m and modulus $\frac{1}{4}g$ N. The other end of the string is attached to a fixed point P. The body is released from rest at P and falls vertically. Show that the time that elapses before the body is instantaneously at rest is given by:

$$\tfrac{5}{28}[3\pi + 4] \text{ s.}$$

The simple pendulum

A particle of mass m kg is suspended from a fixed point A by a light inextensible string of length l metres.

O is the point directly below A.

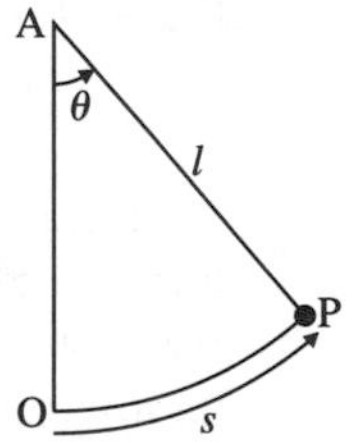

Consider the situation when the particle is at a point P where $\text{O}\hat{\text{A}}\text{P} = \theta$. Let s denote the distance, measured in metres, along the arc from O to P.

Then $\quad s = l\theta,\quad$ where θ is measured in radians

$\therefore \quad \dot{s} = l\dot{\theta},$

and $\quad \ddot{s} = l\ddot{\theta}$

i.e. the component of the acceleration of the particle in the direction of the tangent at P is given by $l\ddot{\theta}$.

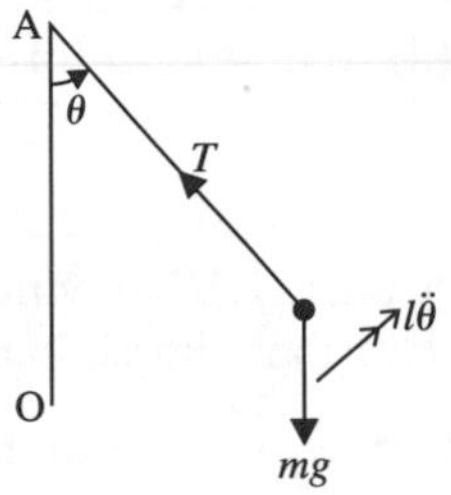

Applying $F = ma$, perpendicular to the string, in the direction of increasing θ, gives:

$$-mg \sin \theta = ml\ddot{\theta}$$

But for small θ $\quad \sin \theta \simeq \theta$

$\therefore \quad -g\theta \simeq l\ddot{\theta}$

i.e. $\quad \ddot{\theta} \simeq -\frac{g}{l}\,\theta$

So, for small oscillations, the motion of a simple pendulum is approximately simple harmonic with $n^2 = \frac{g}{l}$.

That is, the SHM is of period $2\pi\sqrt{\frac{l}{g}}$

Exercise 17D

1. (a) If a simple pendulum of length l makes small oscillations, show that the motion approximates to simple harmonic motion of time period $2\pi\sqrt{\left(\frac{l}{g}\right)}$.

(b) Find the time period if:
   (i) $l = 80$ cm
   (ii) $l = 1{\cdot}25$ m
   (iii) $l = 50$ cm.

(c) Find l if the time period is:
   (i) $\frac{2\pi}{7}$ s
   (ii) $\frac{\pi}{\sqrt{5}}$ s
   (iii) 1 s.

2. A simple pendulum completes 25 complete oscillations each minute. Calculate the length of the string.

3. Find the time period of a simple pendulum of length 2 m performing small oscillations. What is the increase in the time period when the length of the pendulum is increased by 10%?

4. A simple pendulum performs small oscillations with time period $2T$. By what percentage should the string be shortened for the time period to be T?

5. A simple pendulum performs small oscillations with time period 5T. By what percentage should the string be shortened for the time period to be $4T$?

6. A simple pendulum has length 9·8 cm. The bob is pulled aside so that the string is inclined at 0·03 radians to the downward vertical. It is held at rest, then gently released at time $t = 0$.
Show that if θ is the angle the string makes with the downward vertical at time t, then:
(a) $\ddot{\theta} \simeq -100\theta$
(b) $\theta \simeq 0{\cdot}03 \cos 10t$.

Exercise 17E Examination questions

1. A musician wishes to use a simple pendulum as a makeshift metronome. Regarding a complete oscillation as two "beats", find, to the nearest centimetre, the length of pendulum required to produce
(a) 50 beats per minute
(b) 100 beats per minute. (UODLE)

2. A body P is moving in Simple Harmonic Motion, in a straight line, with centre O and amplitude 3 m. Given that the speed of P when $OP = 2$ m is $1\,\text{m s}^{-1}$, find the period of the motion. (UCLES)

3. A particle describes simple harmonic motion about a point O as centre and the amplitude of the motion is a metres. Given that the period of the motion is $\frac{\pi}{4}$ seconds and that the maximum speed of the particle is $16\,\text{m s}^{-1}$ find:
(a) the speed of the particle at a point B, a distance $\frac{1}{2}a$ from O
(b) the time taken to travel directly from O to B. (AEB 1992)

4. A particle moves with simple harmonic motion about a mean position O (i.e. acceleration $= -\omega^2 x$, where x is the displacement from O). When passing through two points 1·5 m and 2·0 m from O the particle has speeds $4\,\text{m s}^{-1}$ and $3\,\text{m s}^{-1}$ respectively.
(i) Find ω.
(ii) Find the amplitude of the motion.
(iii) Find the periodic time of the motion. (NICCEA)

5. The three points O, B, C lie, in that order, on a straight line l on a smooth horizontal plane with $OB = 0{\cdot}3$ m, $OC = 0{\cdot}4$ m.
A particle P describes simple harmonic motion with centre O along the line l. At B the speed of the particle is $12\,\text{m s}^{-1}$ and at C its speed is $9\,\text{m s}^{-1}$. Find:
(a) the amplitude of the motion
(b) the period of the motion
(c) the maximum speed of P
(d) the time to travel from O to C. (AEB 1991)

6. A particle P moving along the x-axis describes simple harmonic motion of period $\frac{2\pi}{\omega}$ and amplitude a with the origin O as centre. Given that P is at $x = a$ at time $t = 0$, write down an expression in terms of a, ω and t for the displacement x of P from O at any subsequent time t.
Find, in terms of ω, the time taken for P to travel
(i) from the point $x = a$ directly to the point $x = \frac{a}{2}$
(ii) from the point $x = \frac{a}{2}$ directly to the point $x = -\frac{a}{\sqrt{2}}$. (WJEC)

7. A particle describes simple harmonic motion about a centre O. When at a distance of 5 cm from O its speed is $24\,\text{cm s}^{-1}$ and when at a distance of 12 cm from O its speed is $10\,\text{cm s}^{-1}$.
Find the period of the motion and the amplitude of the oscillation.
Determine the time in seconds, to two decimal places, for the particle to travel a distance of 3 cm from O. (AEB 1990)

8. A particle P describes simple harmonic motion, making 3 complete oscillations per second. At a certain instant, P is at the point O and is moving at its maximum speed of $5\,\text{m s}^{-1}$.
(a) Find the speed of P 0·05 s after it passes through O, giving your answer to 3 significant figures.
After passing through O, P first comes to instantaneous rest at the point A.
(b) Find the average speed of P as it moves from O to A, giving your answer to 3 significant figures. (ULEAC)

9. A particle P moves horizontally along the x-axis and describes simple harmonic motion with centre O. At a particular instant, $x = 0{\cdot}04\,\text{m}$ and the magnitudes of the velocity and acceleration of P are $0{\cdot}2\,\text{m}\,\text{s}^{-1}$ and $1\,\text{m}\,\text{s}^{-2}$ respectively. Find:
(i) the period of the motion
(ii) the amplitude of the motion
At time $t = 0$ seconds the particle is passing through O in the direction of increasing x. Find:
(iii) x at any subsequent time t
(iv) the least positive value of t (correct to two decimal places) when $x = 0{\cdot}04\,\text{m}$.

The simple harmonic motion is produced by a light elastic spring one end of which is attached to P and the other end to the point $x = -0{\cdot}5\,\text{m}$. Given that the mass of P is $0{\cdot}3\,\text{kg}$, find the elastic modulus of the spring.

(WJEC)

10. A light elastic string of natural length a and modulus mg has one end fastened to a particle P, of mass m, and the other end to a fixed point A. The particle is held, with the string vertical, at a point O which is at a distance a below A. At a given instant, P is projected vertically downwards.

(a) Write down the equation of motion of P when it is a distance x, $(x > 0)$, below O.

(b) Show that, whilst the string is taut, P performs simple harmonic motion centred on a point B, distance a below O.

(c) Given that the initial speed at O is $\sqrt{(8ga)}$, use the principle of conservation of energy to show that when the string has total length $a + x$, where $x > 0$, the speed of P is given by v, where

$$av^2 = 2ag(x + 4a) - gx^2.$$

(d) Hence, or otherwise, find the maximum distance of P below O, and find the magnitude of the greatest tension in the string. (AEB 1994)

(Note for question 11 that the stiffness of a string is $\frac{\lambda}{l}$, i.e. the modulus per unit length.)

11. A particle A of mass m is attached at one end of a light elastic string of natural length l and stiffness $\frac{12mg}{l}$. The other end of the string is attached to a fixed point O.

(a) Initially, A is held at O and then allowed to fall from rest under gravity. Find its speed when the string first becomes taut. Using energy considerations, show that A first comes instantaneously to rest at a distance $\frac{3}{2}l$ below O.

(b) If, instead, A is allowed to hang freely in equilibrium below O, supported only by the string, determine the length of the string in this equilibrium position.
The particle is now pulled down a further distance $\frac{l}{4}$ beyond this equilibrium position, E, and released. The displacement, measured downwards, from E is x at time t after release. Show that the immediately ensuing motion is simple-harmonic with period

$$T = \pi\sqrt{\left(\frac{l}{3g}\right)}$$

Show that when the particle A reaches the point distance l below O, its speed is given by $v^2 = \frac{2}{3}gl$.
Describe briefly, and without further calculation, the subsequent motion of A.

(UODLE)

12. A light elastic spring has one end attached to a fixed point O and the other end to a seat of mass m. When a young girl of mass M sits on the seat and the girl and seat hang vertically below O, the extension of the spring is a. The girl and seat are released from rest in a position where the extension of the spring is c $(c > a)$. Show that at time t, when the girl is still on the seat, the extension, x, of the spring satisfies the differential equation

$$\frac{d^2x}{dt^2} = \frac{g}{a}(a - x).$$

Show also that the speed of the girl is v, where

$$av^2 = g(c^2 - 2ac + 2ax - x^2).$$

Find, in terms of M, g, x and a, an expression for the reaction between the girl and the seat and deduce that if the girl leaves the seat, she will do so when the spring is momentarily at its natural length. (AEB 1993)

13. A particle P of mass m is attached, at the point C, to two light elastic strings AC and BC. The other ends of the strings are attached to two fixed points A and B on a smooth horizontal table, where $AB = 4a$. Both of the strings have the same natural length a and the same modulus. When the particle is in its equilibrium position the tension in each string is mg. Show that when the particle performs oscillations along the line AB in which neither string slackens, the motion is simple harmonic with period $\pi\sqrt{\dfrac{2a}{g}}$.

The breaking tension of each string has magnitude $\dfrac{3mg}{2}$. Show that when the particle is performing complete simple harmonic oscillations the amplitude of the motion must be less than $\frac{1}{2}a$.
Given that the amplitude of the simple harmonic oscillations is $\frac{1}{4}a$, find the maximum speed of the particle. (AEB 1992)

14.

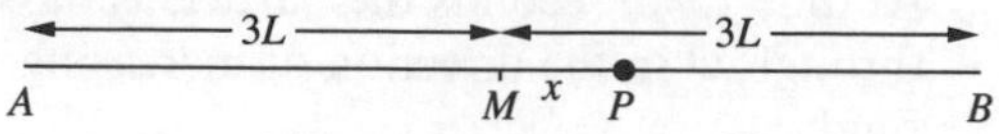

The ends of a light elastic string, of natural length $2L$ and modulus of elasticity $3mg$, are attached to two fixed points A and B, at a distance $6L$ apart, on a smooth horizontal table. A particle P, of mass m, is in equilibrium at the mid-point M of AB. It is projected from M with speed $\sqrt{(\frac{2}{3}gL)}$, directly towards B. At time t the displacement of P from M is x (see diagram). Given that P moves along the line AB, with both strings taut, obtain a differential equation relating x and t. Deduce that the motion is simple harmonic with period $2\pi\sqrt{\left(\dfrac{L}{6g}\right)}$.
Find an expression for x in terms of t. (UCLES)

18 Compound bodies and frameworks

Jointed rods in equilibrium

When two rods are smoothly jointed together, each rod will exert a force on the other at the joint. If the joint is in equilibrium, with no external forces acting, the forces on the rods must be equal and opposite to each other. These forces may, in general, be in any direction. To find these forces it is convenient to "separate" the rods at the joints and suppose that these forces have horizontal and vertical components, X and Y:

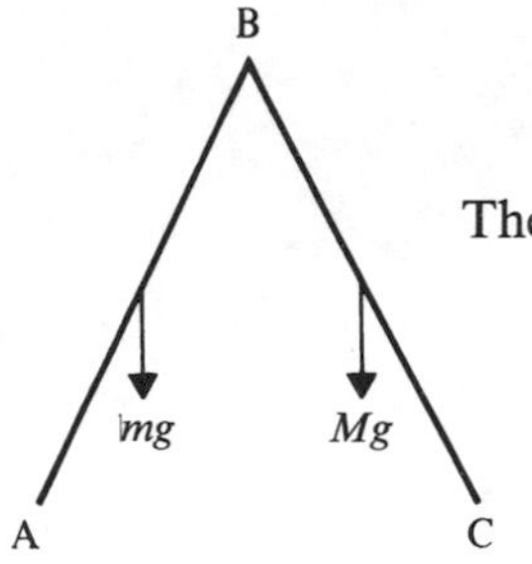

The rods AB and BC can be separated:

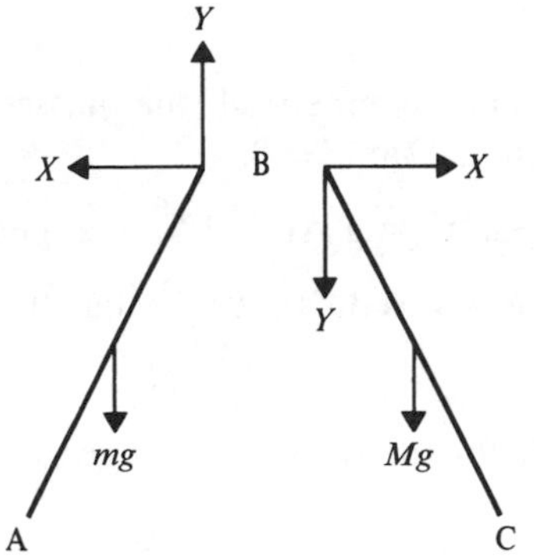

Since the joint B is in equilibrium, the forces exerted on rod AB by BC, must be equal and opposite to those exerted on rod BC by rod AB.

It is sometimes possible to evaluate some of these forces by symmetry considerations. In the above diagram, if the system is symmetrical, i.e. AB = BC in length and mass (i.e. $m = M$), then there must still be this symmetry after "separation". Thus $Y = 0$.

In addition to considering the equilibrium of each rod separately, the equilibrium of the whole system may also be considered.

Example 1

The diagram shows two uniform rods AB and BC, of equal length, smoothly jointed together at B. The mass of AB is 6 kg and the mass of BC is 8 kg. The end C is freely hinged to a rough horizontal surface and end A rests on the same surface, coefficient of friction μ. The points A, B and C lie in the same vertical plane and angle BAC = 30°.

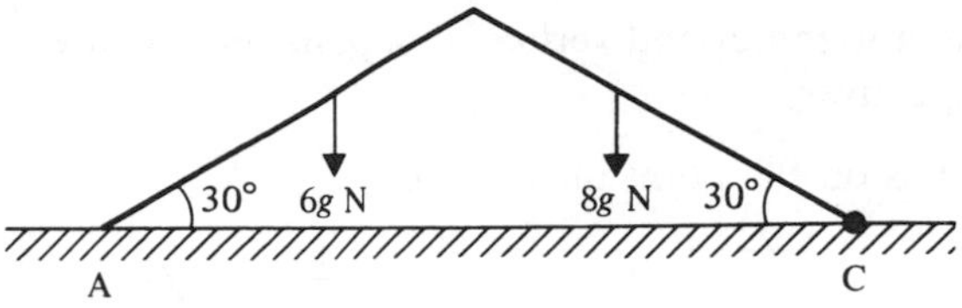

If A is on the point of slipping, find:
(a) the horizontal and vertical components of the reaction at B
(b) the horizontal and vertical components of the reaction at C
(c) the value of μ.

First, draw the "separated" diagram:

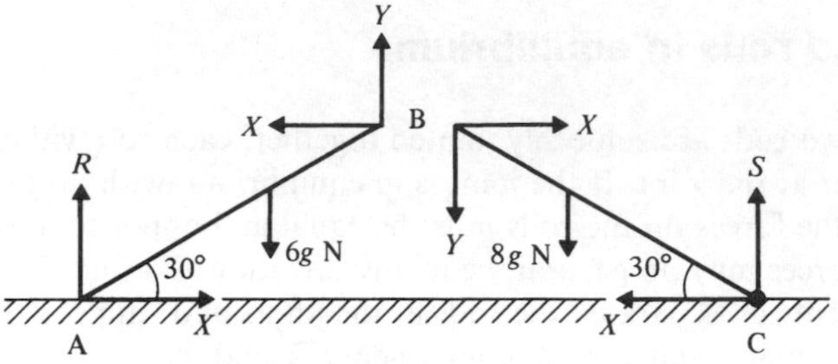

Note that:
(1) the forces on the ends of the rods at B are shown to be equal and opposite, since the joint is in equilibrium,
(2) the horizontal forces at A and C must be X as shown, so as to ensure horizontal equilibrium for each rod considered separately,
(3) as the system is not symmetrical (the masses of the rods are unequal) it cannot be assumed that $Y = 0$.

For the whole system, taking AB = BC = $2l$ gives:

C↻ $$R(4l \cos 30°) = 6g(3l \cos 30°) + 8g(l \cos 30°)$$

$$\therefore \quad R = \frac{13}{2} g \text{ N}$$

Resolving vertically for the whole system gives:

$$R + S = 6g + 8g$$

$$\therefore \quad S = \frac{15}{2} g \text{ N}$$

For rod AB:

B↻ $$R(2l \cos 30°) = 6g(l \cos 30°) + X(2l \sin 30°)$$

Substituting for R gives: $X = \tfrac{7}{2} g\sqrt{3}$ N

Resolving vertically for rod AB then gives:

$$Y + R = 6g$$

$$\therefore \quad Y = -\frac{g}{2} \text{ N}$$ (a negative sign indicates that the force is in the opposite direction to that shown on diagram.)

The horizontal and vertical components at B are $\frac{7}{2} g\sqrt{3}$ N and $\frac{g}{2}$ N respectively.

The horizontal and vertical components at C are $\frac{7}{2} g\sqrt{3}$ N and $\frac{15}{2} g$ N respectively.

If A is on the point of slipping:

$$\mu = \frac{X}{R} = \tfrac{7}{13}\sqrt{3}$$

The value of μ is $\tfrac{7}{13}\sqrt{3}$.

Example 2

The diagram shows two identical uniform rods AB and AC each of mass 4 kg. The third uniform rod BC is of mass 6 kg and angle CAB is 90°. The rods are smoothly jointed to form a framework which hangs in equilibrium in a vertical plane, freely suspended by a string attached at A. Find:

(a) the tension in the string
(b) the horizontal and vertical components of the reaction at C.

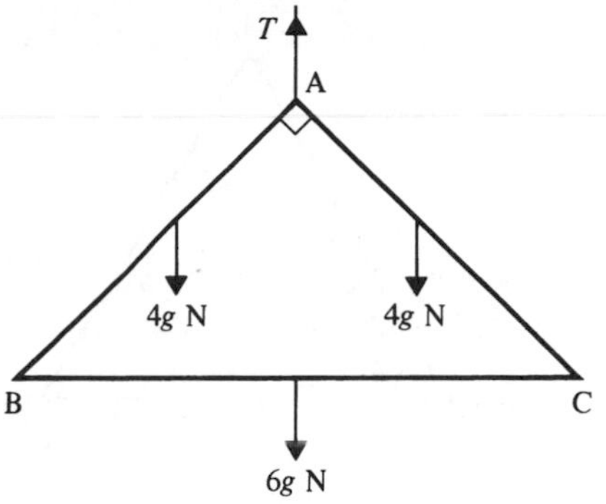

First, draw the "separated" diagram:

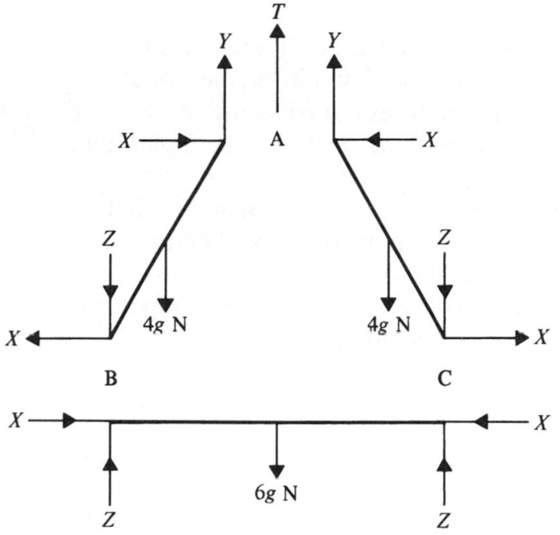

Note that, with the external force T acting at A, the vertical forces of reaction on each rod need not be equal and opposite. Symmetry consideration enables them to be as shown and:

$$T = 2Y \qquad \ldots [1]$$

(a) For the whole framework, resolving vertically gives:

$$T = 4g + 4g + 6g$$

$$\therefore \quad T = 14g$$

The tension in the string is $14g$ N.

(b) From equation [1],

$$Y = \frac{T}{2} = 7g \text{ N}$$

For rod AC, taking AC = AB = $2l$ gives:

$\overset{\frown}{\text{C}}$ $\quad X(2l \sin 45°) + 4g(l \cos 45°) = Y(2l \cos 45°)$

or $\quad X + 2g = Y$

Substituting for Y gives: $\quad X + 2g = 7g$

$\therefore \quad X = 5g$ N

Resolving vertically for rod AC gives: $\quad Y = 4g + Z$

Hence $\quad Z = 3g$ N

The horizontal and vertical components of the reaction at C are $5g$ N and $3g$ N respectively.

Exercise 18A

1.

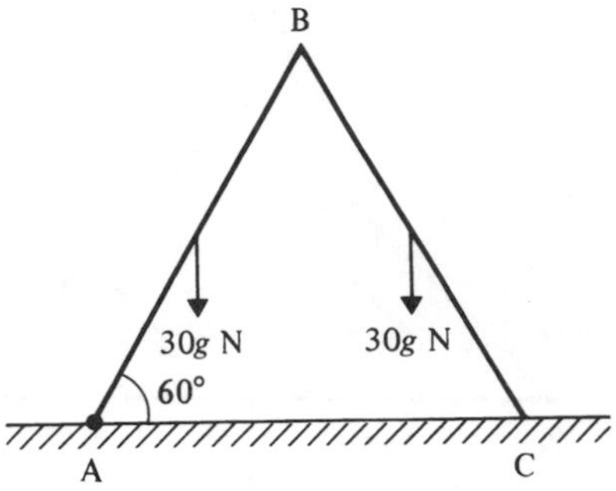

The diagram shows two identical uniform rods AB and BC smoothly jointed together at B.
End A is freely hinged to a rough horizontal surface and end C stands on the same surface, coefficient of friction μ. A, B and C all lie in the same vertical plane. Each rod is of mass 30 kg and angle BAC = 60°.
If C is just on the point of slipping, find:
(a) the horizontal and vertical components of the reaction at B
(b) the horizontal and vertical components of the reaction at A
(c) the value of μ.

2.

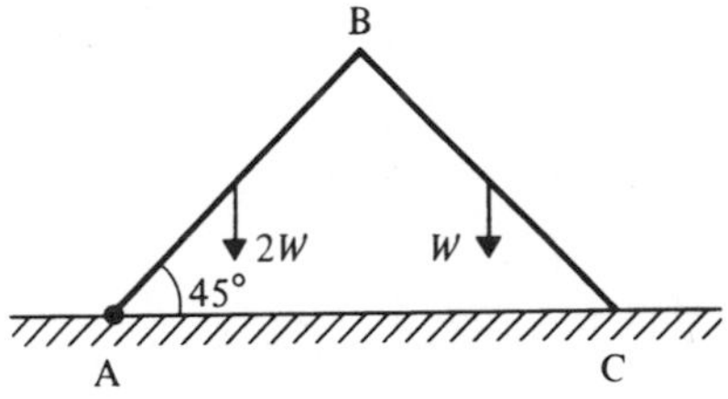

The diagram shows two uniform rods AB and BC of equal length, smoothly jointed together at B.
End A is freely hinged to a rough horizontal surface and C stands on the same surface, coefficient of friction μ; A, B and C all lie in the same vertical plane. AB and BC have weights of $2W$ and W respectively and angle BAC = 45°.
Show that slipping will not occur provided $\mu \geqslant 0{\cdot}6$.
Find the horizontal and vertical components of the reaction at B.

3.

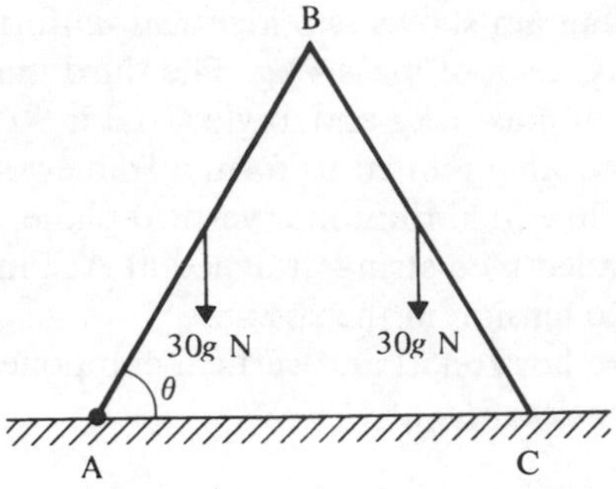

The diagram shows two identical uniform rods AB and BC, each of mass 30 kg and lying in the same vertical plane. The rods are smoothly jointed together at B and ends A and C rest on rough horizontal ground, coefficient of friction μ. If angle BAC = θ, where $\tan\theta = \frac{3}{2}$, and the rods are just on the point of slipping, find the value of μ and the magnitude of the reaction at B.

4. Two identical uniform heavy rods, AB and BC, are smoothly jointed together at B and have ends A and C resting on rough horizontal ground, coefficient of friction μ. The points A, B and C lie in the same vertical plane and angle BAC = α.
Show that slipping will not occur provided $\mu \geqslant \frac{1}{2}\cot\alpha$.

5.

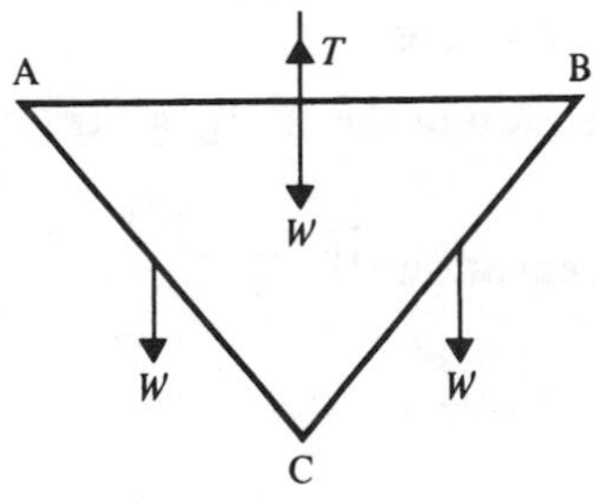

The diagram shows a framework of three rods AB, BC and AC each of weight W, smoothly jointed together at their ends and lying in the same vertical plane.
The framework hangs in equilibrium, suspended by a string attached to the mid-point of AB. The rods AB, BC and AC have lengths $l\sqrt{2}$, l and l respectively.
Find:
(a) the tension in the string
(b) the magnitude of the reaction at C
(c) the magnitude of the reaction at B.

6.

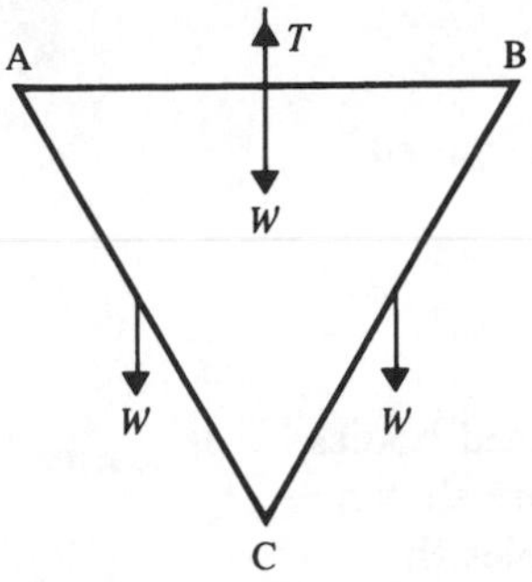

The diagram shows three identical uniform rods AB, BC and CA smoothly jointed together at A, B and C and lying in the same vertical plane. The framework is suspended by a string attached to the mid-point of AB. The weight of each rod is W.
With the system in equilibrium, find:
(a) the tension in the string
(b) the magnitude of the reaction at C
(c) the magnitude of the reaction at B.

7. Two rods AB and BC, of equal length and weights of $5W$ and W respectively, are smoothly jointed together at B. Ends A and C rest on rough horizontal ground and B is vertically above AC. The coefficient of friction between the ground and the rods is $\frac{3}{8}$ and angle BAC $= \alpha$. Show that neither rod will slip provided $\tan \alpha \geqslant 2$.
With $\tan \alpha = 2$, find the magnitude of the reaction at B.

8. Two uniform heavy rods AB and BC, each of length l, lie in the same vertical plane and are smoothly jointed together at B. Ends A and C rest on rough horizontal ground and angle BAC $= \alpha$. The rod AB is twice as heavy as rod BC. The coefficients of friction at A and C are μ_1 and μ_2 respectively. Show that neither rod will slip provided:

$$\mu_1 \geqslant \frac{3}{7 \tan \alpha} \quad \text{and} \quad \mu_2 \geqslant \frac{3}{5 \tan \alpha}$$

Suppose now, with these conditions fulfilled, $\mu_1 = \mu_2$ and α is gradually decreased (both A and C remaining in contact with the ground).
Show that slipping will tend to occur at C before it does at A.

9.

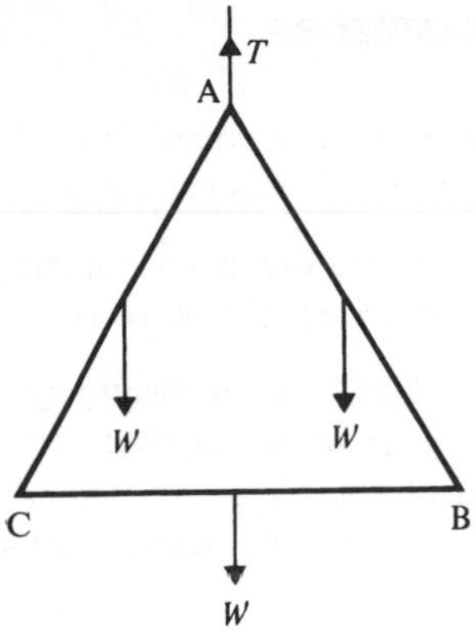

The diagram shows three identical uniform rods, AB, BC and AC, each of weight W. The rods are smoothly jointed at their ends to form a framework which hangs in equilibrium in a vertical plane, freely suspended by a string attached at A.
Find:

(a) the tension in the string

(b) the horizontal and vertical components of the reaction at B.

10.

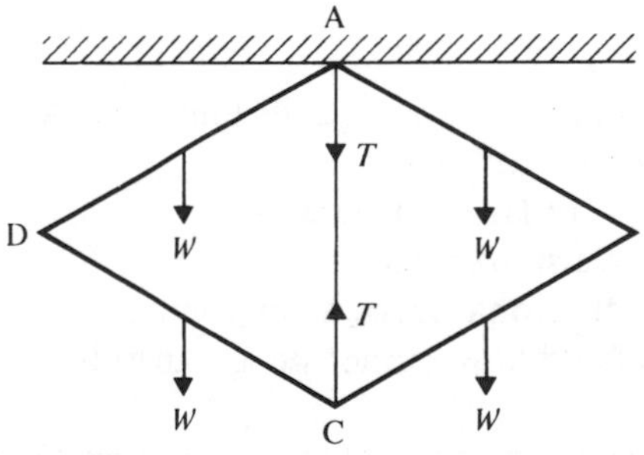

The diagram shows four identical uniform rods each of weight W and length l, freely hinged together at their ends. A light inextensible string of length l connects A to C and the framework hangs in equilibrium in a vertical plane, with A freely hinged to a horizontal ceiling.
Find the tension T in the string, and the horizontal and vertical components of the reaction at B.

Further examples

The examples encountered so far in this chapter have all involved jointed *rods* in equilibrium. However the techniques of:

1) considering any symmetry that may exist

and 2) "separating" the rods

can also be applied to equilibrium systems involving other connected bodies, as the following examples demonstrate. In Example 3 the forces are shown with due regard to the symmetry of the system and in later examples the technique of showing the bodies "separated" is shown.

Example 3

The diagram below shows two rings, A and B, each of mass 1 kg, which are free to move on a rough horizontal rail, coefficient of friction μ. They are connected by two identical, light, inextensible strings to a body C, of mass 2 kg, hanging freely.

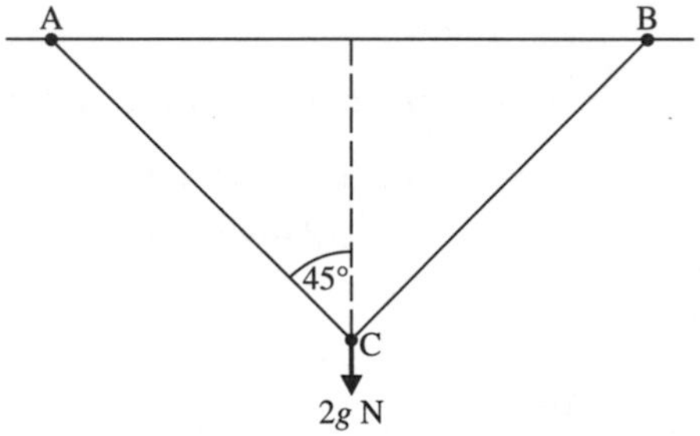

If the system rests in equilibrium, with AC making an angle of 45° with the upward vertical, find:
(a) the normal reaction at A
(b) the tension in each string
(c) the frictional force acting on A.
Show also that μ cannot be less than 0·5.

First draw a diagram showing the forces acting on the rings and the suspended body paying due regard to the symmetry of the system (to keep the number of unknowns needed to a minimum).

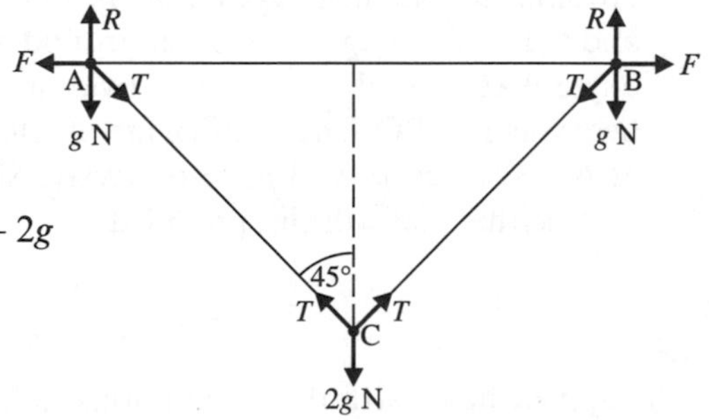

(a) For whole system, resolve vertically: $R + R = g + g + 2g$

$\therefore \quad R = 2g$

The normal reaction at A is $2g$ N.

(b) For C, resolve vertically: $2T \cos 45° = 2g$

$\therefore \quad 2T \times \dfrac{1}{\sqrt{2}} = 2g$

giving $T = g\sqrt{2}$

The tension in each string is $g\sqrt{2}$ N.

(c) For A, resolve horizontally: $F = T \sin 45°$

Substituting for T gives: $F = g\sqrt{2} \times \dfrac{1}{\sqrt{2}}$

$\therefore$ $F = g$

The frictional force acting on A is g N.

For equilibrium: $F \leqslant \mu R$

$\therefore$ $g \leqslant \mu\, 2g$

i.e. $\mu \geqslant 0{\cdot}5$, as required.

Example 4

The diagram shows the cross-section of three smooth cylinders which rest inside a smooth gutter. The centres of the circular ends of the cylinders A, B and C are A, B and C respectively. Cylinders A and B are identical and lie against the sides and on the base of the gutter, each having mass 5 kg and radius 6 cm. Cylinder C, which rests on top of A and B, has a mass of 8 kg and radius 8 cm.

If the distance between the centres of A and B is 14 cm, find:

(a) the normal reaction between A and C

(b) the normal reaction between A and the side of the gutter

(c) the normal reaction between A and the base of the gutter.

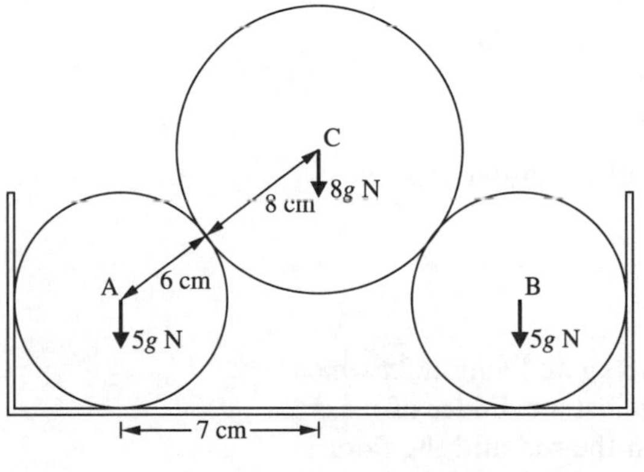

First, draw the "separated" diagram. Let the normal reaction between A and C be R, the normal reaction between A and the side of the gutter be S and the normal reaction between A and the base of the gutter be T.

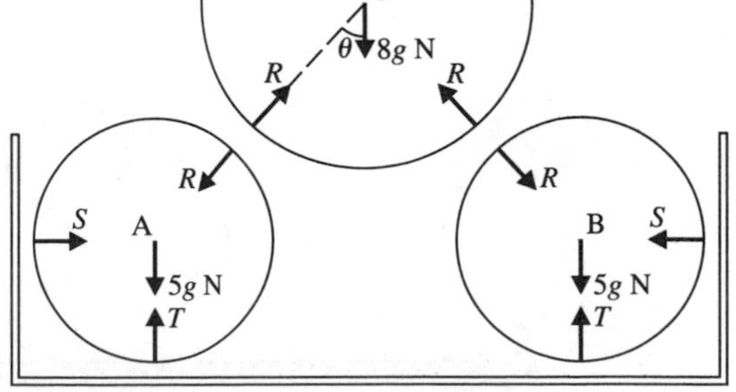

Note that:

(1) by symmetry, the normal reaction between A and C is the same as the normal reaction between B and C

(2) if we let θ be the angle AC makes with the downward vertical, then by trigonometry $\sin\theta = \frac{7}{14}$

$\therefore \qquad \theta = 30°.$

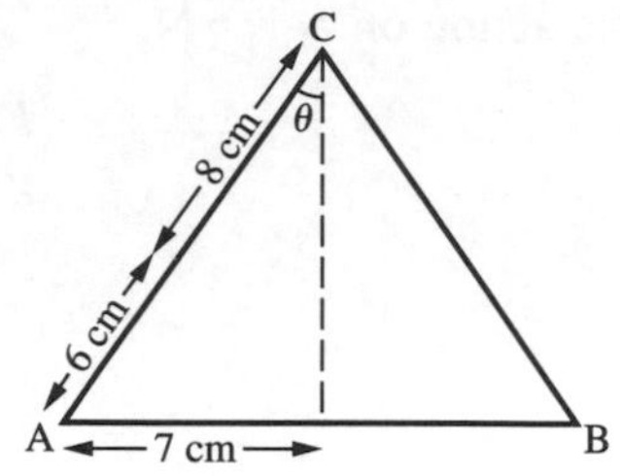

(a) Resolve vertically for C: $\quad 2R\cos 30° = 8g$

$\therefore \qquad R = \dfrac{8g}{\sqrt{3}}$

The normal reaction between cylinders A and C is $\dfrac{8\sqrt{3}g}{3}$ N.

(b) Resolve horizontally for A: $\quad R\sin 30° = S$

$\therefore \qquad S = \dfrac{4g}{\sqrt{3}}$

The normal reaction between cylinder A and the side of the gutter is $\dfrac{4\sqrt{3}g}{3}$ N.

(c) Resolve vertically for A: $\quad T = 5g + R\cos 30°$
Substituting for R gives: $\quad T = 5g + 4g$
$\therefore \qquad T = 9g$

The normal reaction between A and the base of the gutter is $9g$ N.

Example 5

The diagram below shows a uniform rod of mass 8 kg and length $2a$ which is hinged to the floor at point A. The rod rests against a cylinder of mass 10 kg and radius a. The angle of elevation between the rod and the floor is 60° and the contacts between the rod and cylinder and between the cylinder and floor are both rough. The rod lies in a vertical plane at right angles to the axis of the cylinder.

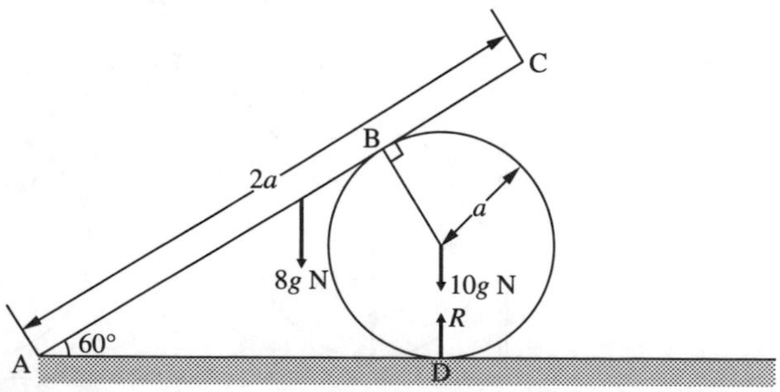

If the system rests in equilibrium and is on the point of slipping at both points B and D, find:
(a) the normal reaction between the cylinder and the floor
(b) the normal reaction between the cylinder and the rod
(c) the value of μ, the coefficient of friction between the cylinder and the ground
(d) the horizontal and vertical components of the reaction at the hinge.

Firstly, draw the separated diagram. Let the horizontal and vertical components of the reaction at the hinge be forces X and Y respectively, the normal reactions at B and D be forces S and R respectively and the frictional forces at D and B be F_1 and F_2 respectively.

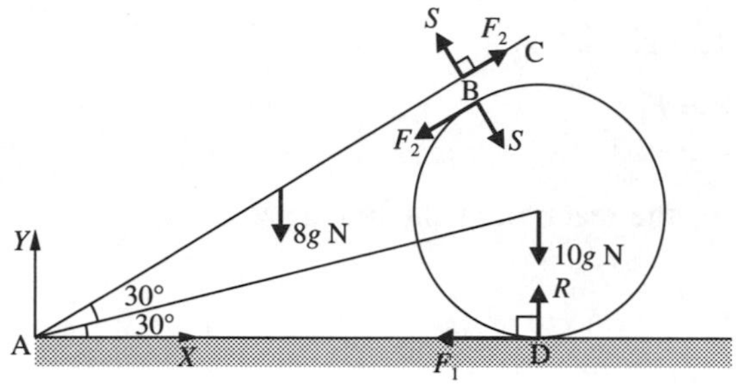

Note that $\tan 30° = \dfrac{a}{\text{AB}}$

$\therefore$ $\text{AB} = \text{AD} = \sqrt{3}a$

(a) For the whole system, taking moments about A gives:

$$8g \times \frac{\text{AC}}{2} \times \cos 60° + 10g \times \text{AD} = R \times \text{AD}$$

$$\therefore \quad 8g \times a \times \frac{1}{2} + 10g \times a\sqrt{3} = R \times a\sqrt{3}$$

$$\therefore \quad R = \frac{4g + 10\sqrt{3}g}{\sqrt{3}}$$

$$\therefore \quad R \simeq 120{\cdot}6\,\text{N}$$

The normal reaction between the cylinder and the floor is 120·6 N.

(b) For rod AC, taking moments about A gives:

$$8g(a \sin 30°) = \sqrt{3}a \times S$$

$$\therefore \quad 4g = \sqrt{3} \times S$$

$$\therefore \quad S \simeq 22{\cdot}6\,\text{N}$$

The normal reaction between the rod and cylinder is 22·6 N.

(c) For the cylinder, taking moments about B gives:

$$a \cos 30° \times 10g + a\sqrt{3} \sin 60° \times F_1 = a \cos 30° \times R$$

$$\therefore \quad F_1 \simeq 13{\cdot}1$$

Since the system is on the point of slipping:

$$F_1 = \mu R$$

$$\therefore \quad \mu \simeq 0{\cdot}108$$

The value of μ is 0·108.

(d) Resolving vertically for the whole system gives:

$$Y + R = 8g + 10g$$

$$\therefore \quad Y = 18g - 120{\cdot}6$$

$$\therefore \quad Y \simeq 55{\cdot}8$$

Resolving horizontally for the whole system gives:

$$X = F_1$$

$$\therefore \quad X \simeq 13{\cdot}1$$

The horizontal and vertical components of the reaction at the hinge are 13·1 N and 55·8 N respectively.

Example 6

The cylinders in Example 4 are now rough and lie unsupported on a rough horizontal surface, as shown in the diagram below.

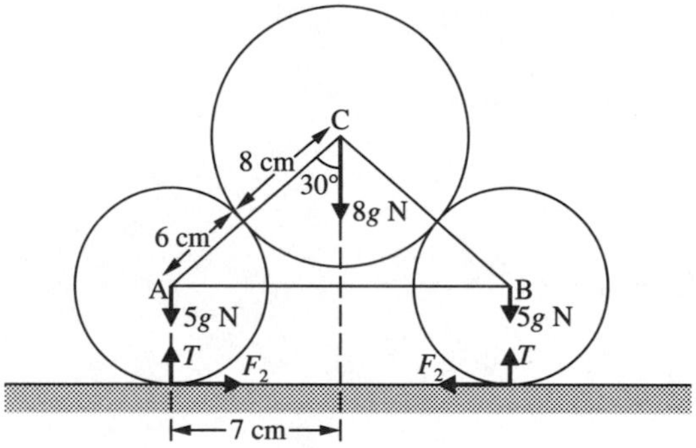

Given that AC makes an angle of 30° with the downward vertical, find:

(a) The normal reaction between cylinder A and the horizontal surface

(b) The normal reaction between cylinder A and cylinder C.

Show also that, if μ_1 is the coefficient of friction between each lower cylinder and the upper one and μ_2 is the coefficient of friction between each lower cylinder and the horizontal surface, then:

$$\mu_1 \geqslant 0{\cdot}268 \quad \text{and} \quad \mu_2 \geqslant 0{\cdot}119$$

Draw the "separated" diagram and include the forces with due regard to symmetry.

Let F_1 be the frictional force between each lower cylinder and the upper cylinder, F_2 the frictional force between each lower cylinder and the horizontal surface and R and T be the normal reactions as shown.

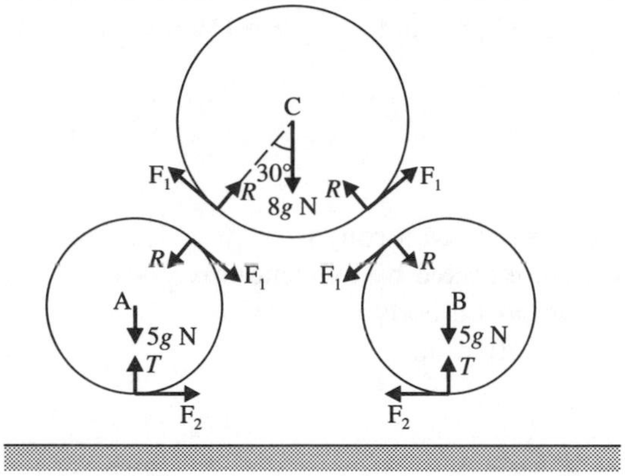

(a) For the whole system, resolving vertically gives:

$$2T = 18g$$
$$\therefore \quad T = 9g$$

The normal reaction between cylinder A and the horizontal surface is 88·2 N (to 3 s.f.).

(b) We now require R. Resolving vertically for cylinder C will give an equation involving R and F_1. From this one equation, involving two unknowns, we cannot find R so more information is needed. If we consider cylinder A and resolve horizontally and take moments, we will then have three equations involving three unknowns, R, F_1 and F_2, and hence R can be determined.

For cylinder C, resolving vertically gives:

$$2R \cos 30° + 2F_1 \cos 60° = 8g$$
$$\therefore \quad \sqrt{3}R + F_1 = 8g \qquad \ldots [1]$$

For cylinder A, resolving horizontally gives:

$$F_1 \cos 30° + F_2 - R \cos 60° = 0$$
$$\therefore \quad \frac{\sqrt{3}}{2} F_1 + F_2 - \frac{R}{2} = 0 \qquad \ldots [2]$$

and taking moments about the axis of cylinder A gives:

$$F_1 \times 0{\cdot}06 = F_2 \times 0{\cdot}06$$
$$\therefore \quad F_1 = F_2 \qquad \ldots [3]$$

Substituting [3] into [2] and then solving simultaneously with [1] gives:

$$R \simeq 39{\cdot}2$$

The normal reaction between cylinders A and C is 39·2 N.

From this value for R it follows that	$F_1 = 10{\cdot}5$	and	$F_2 = 10{\cdot}5$
But for equilibrium	$F_1 \leqslant \mu_1 R$	and	$F_2 \leqslant \mu_2 T$
$\therefore$	$10{\cdot}5 \leqslant \mu_1 39{\cdot}2$	and	$10{\cdot}5 \leqslant \mu_2 88{\cdot}2$
i.e.	$\mu_1 \geqslant 0{\cdot}268$	and	$\mu_2 \geqslant 0{\cdot}119$ (all to 3 s.f.)

as required.

Exercise 18B

In questions **1** to **4**, two rings A and B are free to move on a rough horizontal rail, coefficient of friction μ. They are connected by two smooth light inextensible strings of equal length which support a body C which hangs at the centre. The diagram below shows the situation.

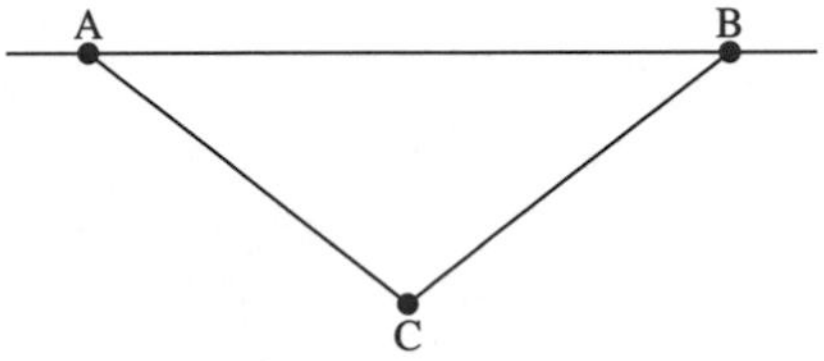

1. The rings A and B are each of mass 2 kg, body C has mass 6 kg and the angle that AC makes with the upward vertical is 30°. If the system is in limiting equilibrium, find:
(a) the normal reaction at A
(b) the tension in each string
(c) the frictional force at A
(d) the value of μ.

2. The rings A and B are each of mass 4 kg, body C has mass 9 kg and the angle that AC makes with the upward vertical is 40°. If the system is in limiting equilibrium, find:
(a) the normal reaction at A
(b) the tension in each string
(c) the value of μ.

3. The rings A and B are each of mass M kg and body C has mass $4M$ kg and angle ACB is 120°. If the system is in equilibrium, find, in terms of M:
(a) the normal reaction at A,
(b) the tension in each string,
(c) the frictional force at A.
(d) Show that $\mu \geqslant \dfrac{2\sqrt{3}}{3}$.

4. The rings A and B are each of mass 4 kg, AC makes an angle of 45° with the upward vertical and $\mu = 0{\cdot}4$. If the system is in limiting equilibrium, find
(a) the normal reaction at A,
(b) the tension in each string,
(c) the mass of body C.

In questions **5** and **6**, three smooth cylinders A, B and C rest inside a smooth gutter. The centres of the circular ends of cylinders A, B and C are A, B and C respectively. Cylinders A and B are identical and lie against the side and base of the gutter. Cylinder C rests on top of cylinders A and B.

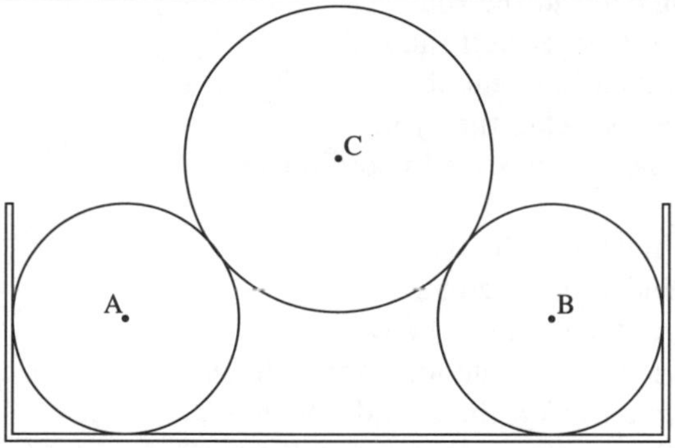

5. Given that cylinders A and B each have mass 6 kg and radius $5\sqrt{2}$ cm, cylinder C has a mass of 9 kg and radius $6\sqrt{2}$ cm, and the distance between the centres of A and B is 22 cm, find:
(a) the normal reaction between A and C
(b) the normal reaction between A and the side of the gutter
(c) the normal reaction between A and the base of the gutter.

6. Given that cylinders A and B each have mass 7 kg and radius 7 cm, cylinder C has a mass of 11 kg and radius 10 cm, and the distance between the centres of A and B is 16 cm, find:
(a) the normal reaction between A and C
(b) the normal reaction between A and the side of the gutter
(c) the normal reaction between A and the base of the gutter.

In questions **7** to **12**, a uniform rod AC is hinged to the floor at point A. The rod rests against a cylinder and θ is the angle of elevation that the rod makes with the floor. Both the contacts between the rod and cylinder and cylinder and floor are rough. The diagram below shows the situation.

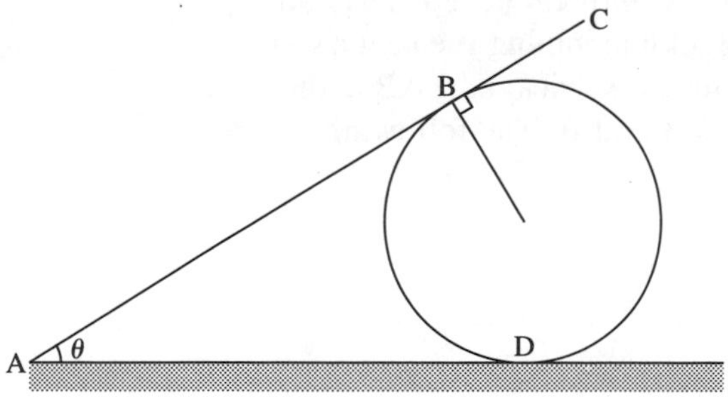

7. Given that the rod is of length $2a$ m and mass 10 kg, the cylinder is of radius a m and mass 12 kg, $\theta = 60°$, and the system rests in equilibrium, with the cylinder on the point of slipping at the ground, find:
(a) the normal reaction between the cylinder and the floor
(b) the normal reaction between the cylinder and the rod
(c) the coefficient of friction between the cylinder and the floor
(d) the horizontal and vertical components of the reaction at the hinge.

8. Given that the rod is of length 3 m and mass 14 kg, the cylinder is of radius 1 m and mass 15 kg, $\theta = 60°$, and the system rests in equilibrium, on the point of slipping at B, find:
(a) the normal reaction between the cylinder and the floor
(b) the normal reaction between the cylinder and the rod
(c) the coefficient of friction between the cylinder and the rod
(d) the horizontal and vertical components of the reaction at the hinge.

9. Given that the rod is of length 4 m and mass 12 kg, the cylinder is of radius 1 m and mass 14 kg, $\theta = 50°$ and the system rests in equilibrium, on the point of slipping at B, find:
(a) the normal reaction between the cylinder and the floor
(b) the normal reaction between the cylinder and the rod
(c) the coefficient of friction between the cylinder and the rod
(d) the horizontal and vertical components of the reaction at the hinge.

10. Given that the rod is of length 4 m and mass 16 kg, the cylinder is of radius 2 m and mass 6 kg, $\theta = 60°$, and the system rests in equilibrium, find:
(a) the normal reaction between the cylinder and the floor
(b) the normal reaction between the cylinder and the rod
(c) the minimum possible value of the coefficient of friction between the cylinder and rod.

11. The rod is of length 4 m and mass of $5M$ kg, the cylinder is of radius 2 m and mass $6M$ kg, $\theta = 60°$ and the system rests in equilibrium.
(a) Find the normal reaction between the cylinder and the floor (in terms of M).
(b) Show that the minimum value of the coefficient of friction between the cylinder and the floor is $\dfrac{5}{5\sqrt{3}+36}$.

12. If the system is on the point of slipping at B show that θ depends only on the value of μ, the coefficient of friction between the cylinder and the rod and prove that $\theta = 2\tan^{-1}\mu$.

In questions **13** to **15**, two identical rough cylinders A and B rest in equilibrium on a rough horizontal floor. A third cylinder C rests on top of A and B. θ is the angle that the straight line joining the centres of cylinders A and C makes with the downward vertical and AB is the distance between the centres of cylinders A and B. The following diagram shows the situation.

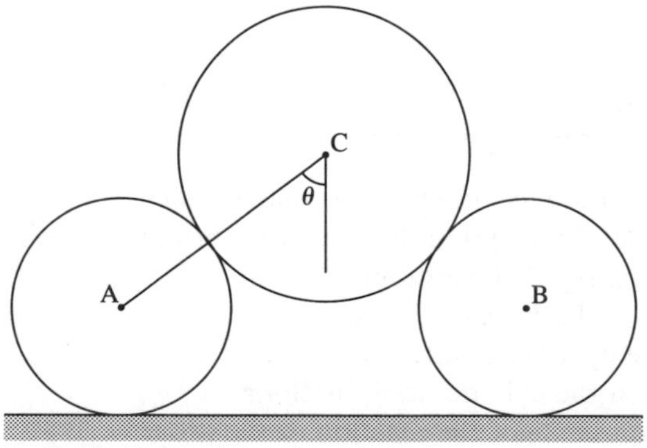

13. Cylinders A and B each have mass 7 kg and radius 7 cm, cylinder C has a mass of 11 kg and radius 10 cm and AB = 16 cm. Find:
(a) the value of θ
(b) the normal reaction between A and C
(c) the normal reaction between A and the floor
(d) the minimum possible value of the coefficient of friction between each lower cylinder and the upper cylinder
(e) the minimum possible value of the coefficient of friction between each lower cylinder and the floor.

14. Given that cylinders A and B each have mass 9 kg and radius 8 cm, cylinder C has a mass of 12 kg and radius 12 cm and AB = 24 cm, find:
(a) the normal reaction between A and C
(b) the normal reaction between A and the floor
(c) the minimum possible value of the coefficient of friction between each lower cylinder and the upper cylinder
(d) the minimum possible value of the coefficient of friction between each lower cylinder and the floor.

15. Cylinders A and B each have mass M kg and radius $2a$ cm, cylinder C has a mass of $2M$ kg and radius $3a$ cm and AB = $6a$ cm. Find:
(a) the normal reaction between A and C (in terms of M)
(b) the normal reaction between A and the floor (in terms of M)
(c) the minimum possible value of the coefficient of friction between each lower cylinder and the upper cylinder
(d) the minimum possible value of the coefficient of friction between each lower cylinder and the floor.

Light rods

A framework may consist of a number of rods that are said to be *light*. This means that when we create our mathematical model of the situation we can ignore the weight of these rods. Of course the rods are not weightless but, compared to the other forces acting and the other assumptions we make, the weights are sufficiently small that ignoring them will not greatly affect the accuracy of our answer. If our model then gives answers that are not accurate enough, or that are too far removed from the real situation, we would need to reconsider any assumptions we made when creating our model.

Consider two such *light* rods AB and BC, freely hinged together at B:

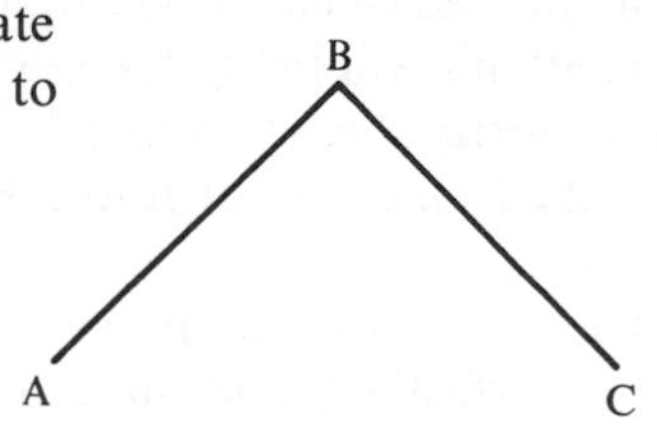

We can "separate" each rod, and show the reaction on each, due to the other, at B as a single force R:

As the rod AB has no weight, it can be seen that for equilibrium R must have no turning effect about A, i.e. R must act along AB. Thus all the reaction forces between light rods will act *along* the rods.

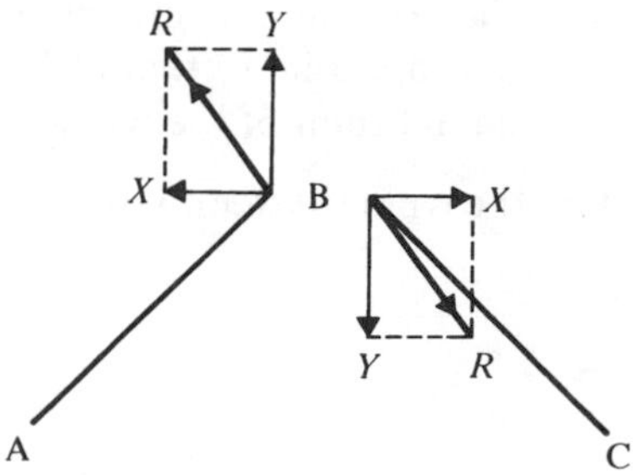

A particular light rod AB may be acting as a *tie* between the points A and B, i.e. preventing A and B from moving apart, and the rod is then in *tension*.

Alternatively, the rod AB may be acting as a *strut*, i.e. preventing points A and B moving closer together, and the rod is then exerting a *thrust*.

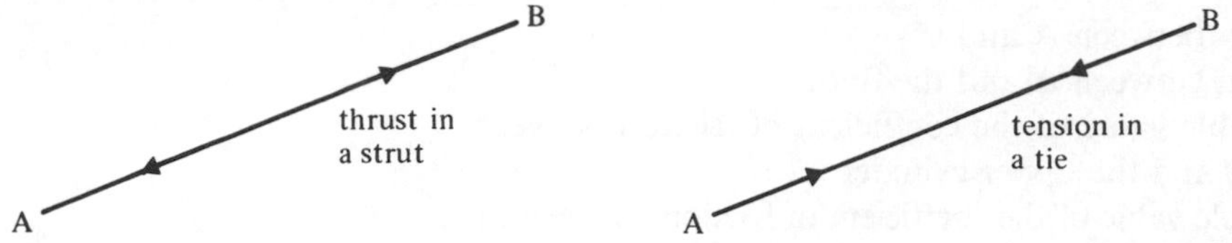

A light rod which is acting as a *tie* could be replaced by a string, but a *strut* could not.

For a framework in equilibrium, equations can be formed by:
(i) taking moments about a convenient point, for the whole system
(ii) resolving horizontally and vertically for the whole system
(iii) resolving horizontally and vertically at the ends of each rod.

In addition the forces in some rods may be determined by considering the symmetry of the framework.

Example 7

The diagram shows a framework of four light rods smoothly jointed together, freely hinged at A and B to a vertical wall, and carrying a load of 100 N at D. BC and AD are horizontal, $C\hat{A}D = C\hat{D}A = 30°$ and $CD = l$. Find:
(a) the force of reaction exerted on the rods at A
(b) the force of reaction exerted on the rods at B
(c) the stresses in each rod stating whether in tension or in thrust.

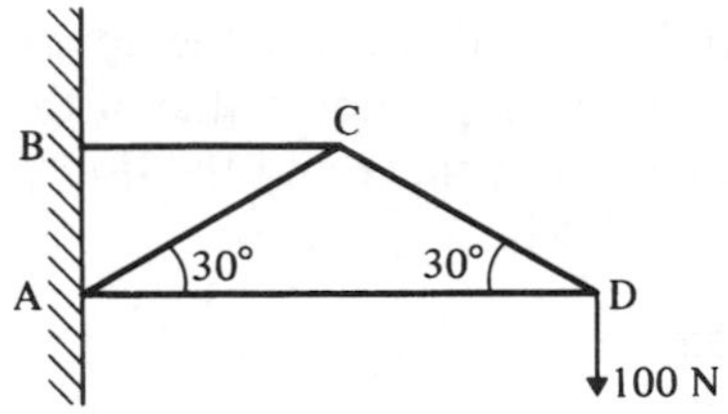

First draw a diagram showing each rod in tension. If any of these tensions are found to have a negative value, this will imply there is a thrust in that rod. The horizontal and vertical components of the reactions on the framework at A and B are also shown. Note that:
(i) the vertical component at B must be zero for vertical equilibrium at B
(ii) the horizontal component at A must be equal and opposite to that at B for horizontal equilibrium of the whole framework.

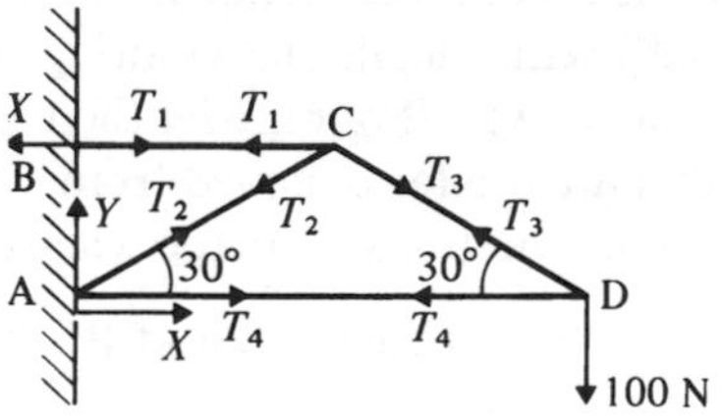

For the whole system, resolve vertically:

$$Y = 100\,\text{N}$$

$\overset{\curvearrowright}{A}$ $$Xl \sin 30° = 100 \times 2 \times l \cos 30°$$

$\therefore$ $$X = 200\sqrt{3}\,\text{N}$$

The reaction at A is $\sqrt{(X^2 + Y^2)}$ at $\tan^{-1} \dfrac{Y}{X}$ to AD,

i.e. $100\sqrt{13}$ N at $\tan^{-1} \dfrac{\sqrt{3}}{6}$ to AD

At B, resolve horizontally: $\quad T_1 = X = 200\sqrt{3}$ N

At A, resolve vertically: $\quad T_2 \sin 30° + Y = 0$

$\therefore \quad T_2 = -200$ N

Resolve horizontally: $\quad T_2 \cos 30° + T_4 + X = 0$

$\therefore \quad T_4 = -100\sqrt{3}$ N

At D, resolve vertically: $\quad T_3 \sin 30° = 100$

$\therefore \quad T_3 = 200$ N

(a) The force of reaction at A is $100\sqrt{3}$ N at $\tan^{-1} \dfrac{\sqrt{3}}{6}$ to AD

(b) The force of reaction at B is $200\sqrt{3}$ N in direction CB

(c) BC is in tension of $200\sqrt{3}$ N $\quad$ CD is in tension of 200 N
CA is in thrust of 200 N $\quad$ AD is in thrust of $100\sqrt{3}$ N

Example 8

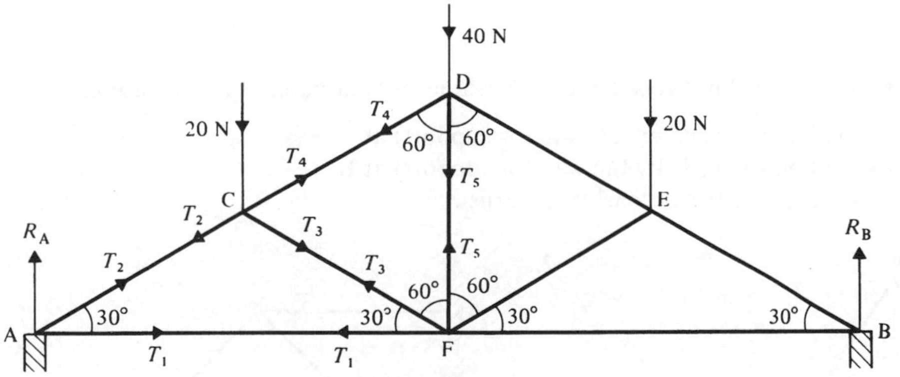

The framework shown is smoothly supported at A and B and carries loads of 20 N, 40 N and 20 N at C, D and E respectively. AB is horizontal and the other angles are as shown. Find the forces of reaction on the rods at A and at B, and the stresses in each rod.

Let the tensions in the rods be as shown (as the framework is symmetrical, the forces in some rods can be determined by symmetry).

For the whole framework, resolve vertically:

$$R_A + R_B = 20 + 40 + 20$$

By symmetry, $\quad R_A = R_B$

$\therefore \quad R_A = R_B = 40$ N

(Alternatively, R_A and R_B may be found by taking moments about A or B for the whole system.)

At A, resolve horizontally: $T_2 \cos 30° + T_1 = 0$ ... [1]

At A, resolve vertically: $T_2 \cos 60° + R_A = 0$ ... [2]

Hence $T_2 = -80\,\text{N}$ and $T_1 = 40\sqrt{3}\,\text{N}$

At C, resolve horizontally: $T_4 \cos 30° + T_3 \cos 30° = T_2 \cos 30°$
or $T_4 + T_3 = T_2$ ... [3]

At C, resolve vertically: $20 + T_3 \cos 60° + T_2 \cos 60° = T_4 \cos 60°$
or $40 + T_3 + T_2 = T_4$ ... [4]

Solving equations [3] and [4], using the values of T_2 and T_1 gives:

$T_3 = -20\,\text{N}$ and $T_4 = -60\,\text{N}$

At D, resolving vertically, and remembering that by symmetry the stress in CD will be the same as that in DE:

$$40 + T_5 + T_4 \cos 60° + T_4 \cos 60° = 0$$

which by substitution gives: $T_5 = 20\,\text{N}$

The reactions at A and B are both 40 N.

The stresses in the rods are as follows.

In AF and FB a tension of $40\sqrt{3}$ N
In DF a tension of 20 N
In AC and BE a thrust of 80 N
In CD and ED a thrust of 60 N
In CF and EF a thrust of 20 N.

Exercise 18C

Each question in this exercise involves light rods lying in the same vertical plane. In each case find:

(a) the force of reaction exerted on the rods by the wall (or support) at A
(b) the force of reaction exerted on the rods by the wall (or support) at B
(c) the stresses in each rod stating whether in tension or thrust.

1.

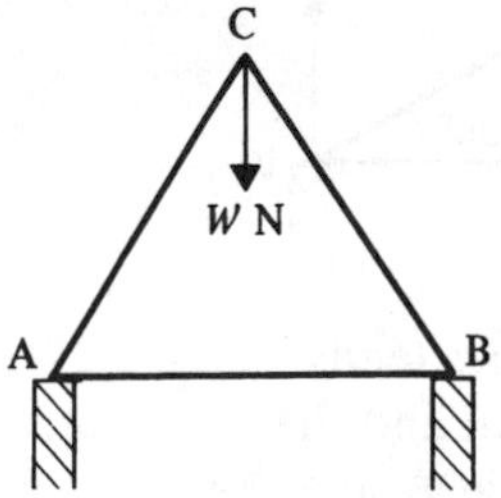

The diagram shows a framework of three identical light rods smoothly jointed together and resting on smooth supports at A and B. AB is horizontal and a load of W N is suspended from C.

2.

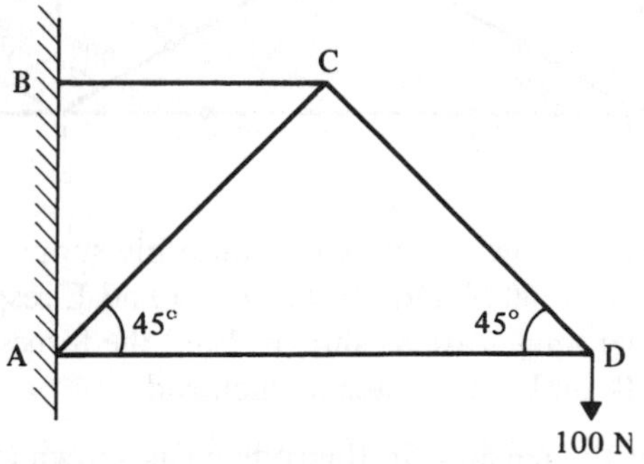

The diagram shows a framework of four light rods smoothly jointed together, freely hinged at A and B to a vertical wall, and carrying a load of 100 N at D.
$C\hat{A}D = C\hat{D}A = 45°$; BC and AD are horizontal.

3.

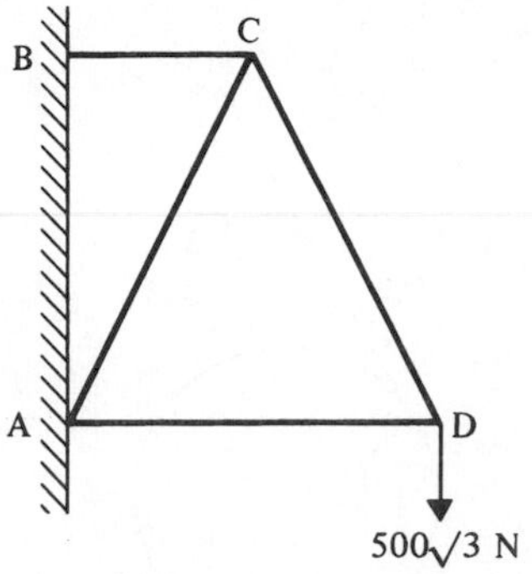

The diagram shows a framework of four light rods smoothly jointed together, freely hinged at A and B to a vertical wall, and carrying a load of $500\sqrt{3}$ N at D. AC = CD = AD = 2BC and AD and BC are horizontal.

4.

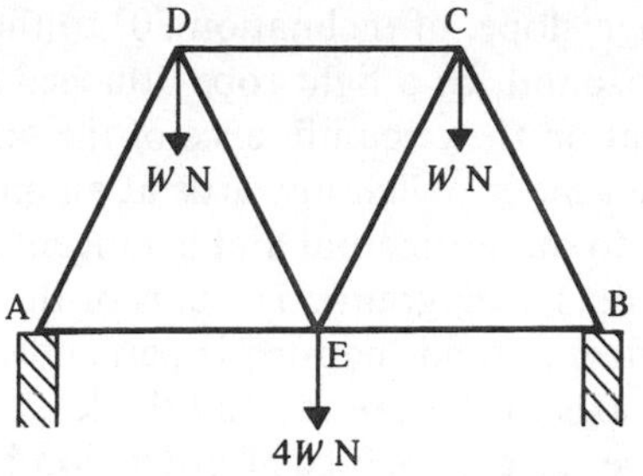

The diagram shows a framework of seven identical light rods smoothly jointed together and resting on smooth supports at A and B; AB is horizontal and loads of W N, W N and $4W$ N are suspended from D, C and E respectively.

5.

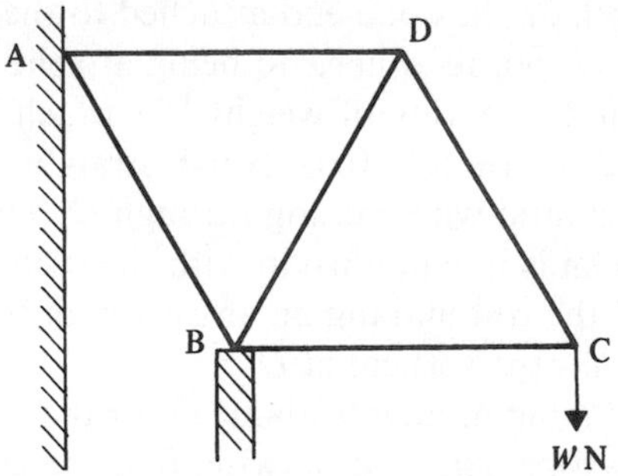

The diagram shows a framework of five identical light rods smoothly jointed together. The framework is freely hinged at A to a vertical wall, rests on a smooth support at B, and carries a load of W N at C; BC is horizontal.

6.

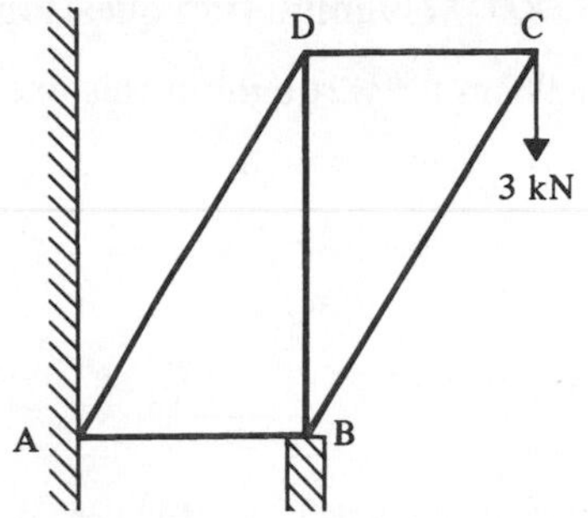

The diagram shows a framework of five light rods smoothly jointed together; ABCD is a parallelogram, AB is horizontal and angle DAB = 60°. The framework is freely hinged to a vertical wall at A, rests on a smooth support at B, and carries a load of 3 kN at C.

7.

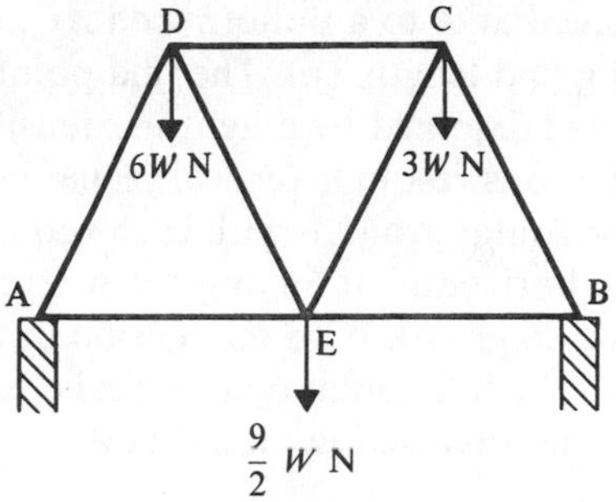

The diagram shows a framework of seven identical rods smoothly jointed together and resting on smooth supports at A and B; AB is horizontal and loads of $6W$ N, $\frac{9}{2}W$ N and $3W$ N are suspended from D, E and C respectively.

8.

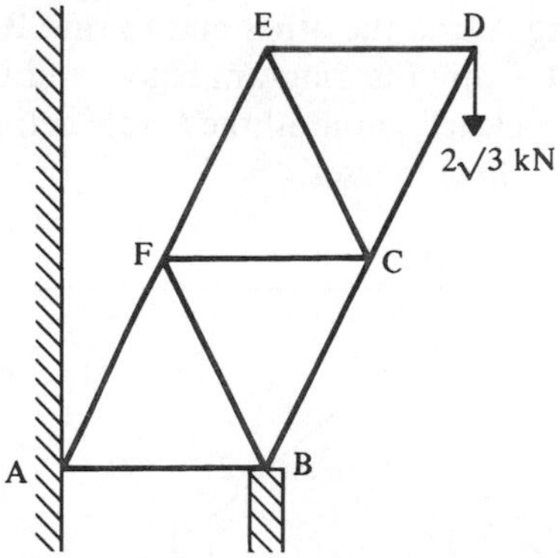

The diagram shows a framework of nine identical light rods smoothly jointed together. The framework is freely hinged at A to a vertical wall, rests on a smooth support at B, and carries a load of $2\sqrt{3}$ kN at D; AB is horizontal.

Exercise 18D Examination questions

(Take $g = 9{\cdot}8\,\mathrm{m\,s^{-2}}$ throughout this exercise.)

1.

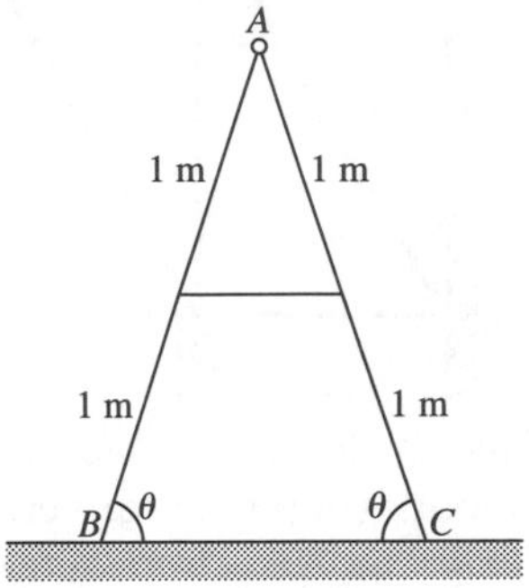

A pair of steps can be modelled as a uniform rod AB, of mass 24 kg and length 2 m, freely hinged at A to a uniform rod AC, of mass 6 kg and length 2 m. The mid-points of the rods are joined by a light inextensible string. The rods rest in a vertical plane on smooth horizontal ground, with each rod inclined to the horizontal at an angle θ, where $\tan\theta = \frac{5}{4}$ (see diagram). Find the tension in the string and the horizontal and vertical components of the force acting on AB at A. (UCLES)

2. Two identical small rings A and B, each of mass M, are permitted to slide on a rough horizontal rail. The coefficient of friction between the rings and the rail is 0·5. A ring of mass $4M$ is threaded onto a string of length $2L$. One end of the string is fixed to ring A and the other end to ring B.

(i) Copy the diagram below and indicate clearly on it all the forces acting on the three masses.

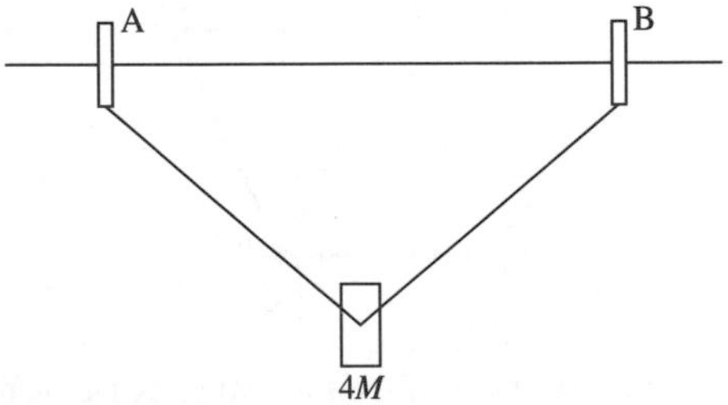

(ii) With the system in limiting equilibrium show that the distance between the rings A and B is $\dfrac{6L}{5}$. (NICCEA)

3.

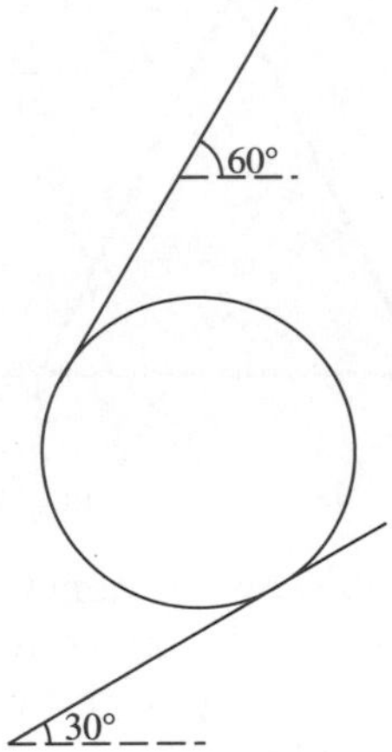

A uniform cylindrical oil drum is held in equilibrium, with its curved surface on a rough slope, of inclination 30° to the horizontal, by a light rope attached to a point on the circumference of the oil drum. The rope is pulled upwards at an angle of 60° to the horizontal and is tangential to the drum (see diagram). The axis of the drum is horizontal and the rope is perpendicular to the axis of the drum. Find the least possible value for the coefficient of friction between the oil drum and the slope. (UCLES)

4. A uniform rod OA of weight W and length $2a$ is pivoted smoothly to a fixed point at the end O. A light inextensible string, also of length $2a$, has one end attached to the end A of the rod, its other end being attached to a small heavy ring of weight $\frac{1}{2}W$ which is threaded on to a fixed rough straight horizontal wire passing through O. The system is in equilibrium with the string taut and the rod making an angle α with the downward vertical at O.
By taking moments about O for the rod and ring together, or otherwise, find the normal component R of the reaction exerted by the wire on the ring. Find also the frictional force F on the ring.
The equilibrium is limiting when $\alpha = 60°$.
Determine the coefficient of friction between the wire and the ring. (OCSEB)

5.

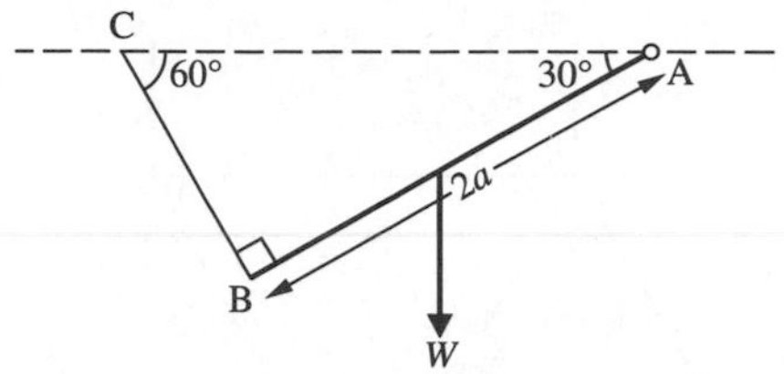

The diagram shows a uniform rod AB, of length $2a$ and weight W, with a small ring attached to the end A which can slide on a rough horizontal wire passing through a point C. From C a string CB is attached to the other end B of the rod. The system is in equilibrium with B vertically below the wire; $C\hat{A}B = 30°$ and the string is perpendicular to the rod. Find T, the tension in the string, in terms of W.
Given that the ring is about to slide on the wire, find μ, the coefficient of friction between the ring and the wire. (OCSEB)

6.

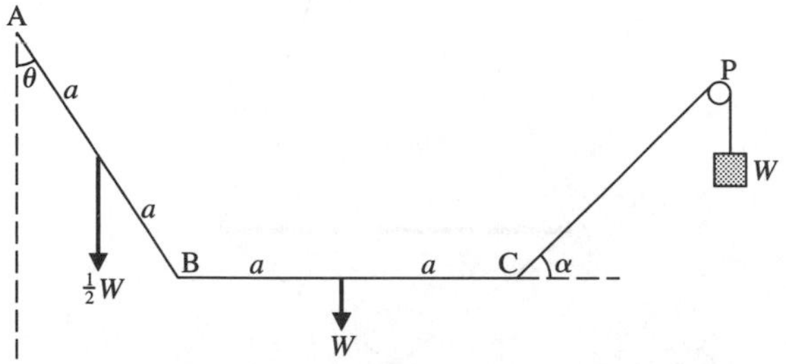

The diagram shows a uniform rod AB of length $2a$ and weight $\frac{1}{2}W$ freely pivoted to a fixed support at A. A uniform rod BC, also of length $2a$ but weight W, is freely pivoted to AB at B. A light string is attached to C, passes over a smooth fixed peg P, and carries a load W hanging freely at its other end. The system is in equilibrium with BC horizontal, AB inclined at an angle θ to the vertical and the string making an angle α with the horizontal. By considering the equilibrium of the rod BC (or otherwise), show that $\alpha = 30°$, and find the reaction exerted by the rod AB on the rod BC at B. Find also the reaction exerted by the support at A on the rod AB and the value of θ.
(OCSEB)

7. A light ring attached to the end A of a uniform rod AB of weight W and length $2a$ can slide along a fixed rough horizontal bar. A string joins B to a fixed point C of the bar, where $BC = 2a$. A particle of weight $\frac{1}{2}W$ is fixed to the rod at a distance ka from A, where $0 \leqslant k \leqslant 2$. The system is in equilibrium with AB making an angle θ with the downward vertical at A.
Find, in terms of W, k and θ
(a) the tension T in the string,
(b) the normal reaction R and the frictional force F exerted by the bar on the ring.
Show that

$$\frac{F}{R} = \frac{(2+k)\tan\theta}{(10-k)}.$$

The coefficient of friction between the ring and the bar is μ. If $\theta = 45°$, by considering the ratio $\frac{F}{R}$ for variable k, show that equilibrium is possible for all positions of the particle on AB, provided that $\mu \geqslant \frac{1}{2}$.
For the same value of θ, find the range of values of k for which equilibrium is possible if $\mu = \frac{1}{3}$. (OCSEB)

8. A uniform sphere of radius a and weight $\frac{W}{\sqrt{3}}$ rests on a rough horizontal table. A uniform rod AB of weight $2W$ and length $2a$ is freely hinged at A to a fixed point on the table and leans against the sphere so that the centre of the sphere and the rod lie in a vertical plane. The rod makes an angle of 60° with the horizontal. Show that the frictional force between the rod and the sphere is $\frac{1}{3}W$.
The coefficient of friction at each point of contact is μ. What is the smallest value of μ which makes equilibrium possible? (OCSEB)

9. A uniform sphere of weight W and radius a rests on a horizontal table touching it at A. A uniform rod BCD of weight $2W$ and length $\frac{4a}{\sqrt{3}}$ rests on the table at B and leans against the sphere at C, the angle ABC being 60°. The points A, B, C lie in a vertical plane and the contacts at these three points are rough with the same coefficient of friction μ. The normal reactions at A, B, C have magnitudes R, P, S respectively, and the

corresponding frictional forces have magnitudes F_1, F_2, F_3. By considering the equilibrium of the sphere and rod together, and also the equilibrium of the sphere itself, prove that $F_1 = F_2 = F_3$.
Calculate the values of R, P, S and F_1 in terms of W.
Find the smallest value of μ for which this equilibrium position is possible. (OCSEB)

10. An equilateral triangular frame ABC is made up of three light rods AB, BC and CA which are smoothly jointed together at their end points. The frame rests on smooth supports at A and B, with AB horizontal and with C vertically above the rod AB. A load of 50 N is suspended from C. Find the reactions at the supports and the tensions or thrusts in the rods. (AEB 1991)

11.

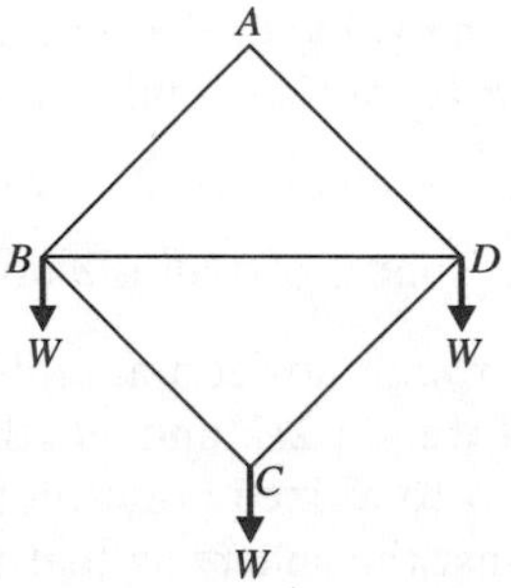

The diagram shows a square framework $ABCD$ of five freely jointed light rods. The framework is freely suspended from the point A and weights of magnitude W are hung from B, C and D respectively. Find the stress in the rod BD, stating whether it is in compression or tension. (AEB 1994)

12. The figure shows a framework consisting of seven equal smoothly jointed light rods AB, BC, DC, DE, AE, EB and EC. The framework is in a vertical plane with AE, ED and BC horizontal and is simply supported at A and D. It carries vertical loads of 50 N and 90 N at B and C respectively.

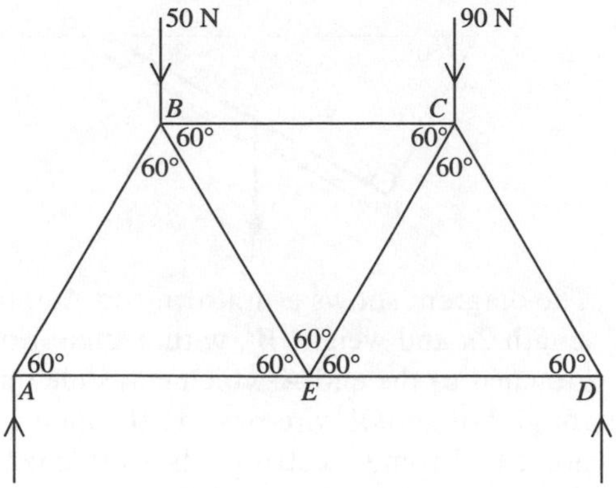

Find
(a) the reactions at A and D,
(b) the magnitudes of the forces in AB, AE and BC. (AEB 1991)

13.

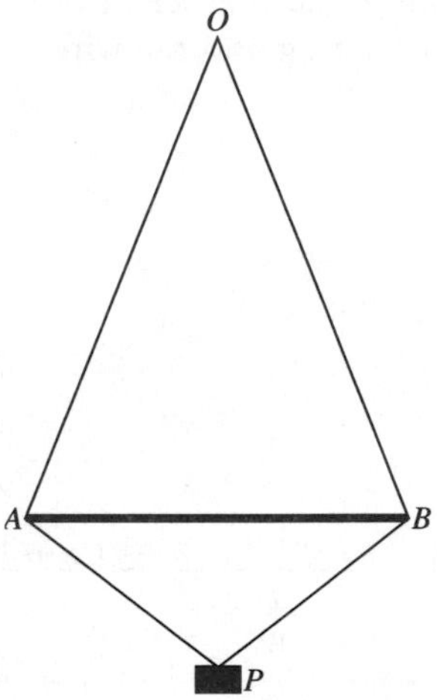

The figure shows a body P, of mass 13 kg, which is attached to a continuous inextensible string of length 15·4 m. The string passes over a small smooth peg O and the hanging portions of the string are separated by a heavy uniform horizontal rigid rod AB, which is 4 m long, the string fitting into small smooth grooves at the ends of the rod. Given that in the equilibrium position $AP = BP = 2{\cdot}5$ m, show that
(a) $\sin P\hat{A}B = \frac{3}{5}$ and $\sin O\hat{A}B = \frac{12}{13}$,
(b) the tension in the string has magnitude $\dfrac{637}{6}$ N
(c) Hence find the mass of the bar AB.
(AEB 1992)

14.

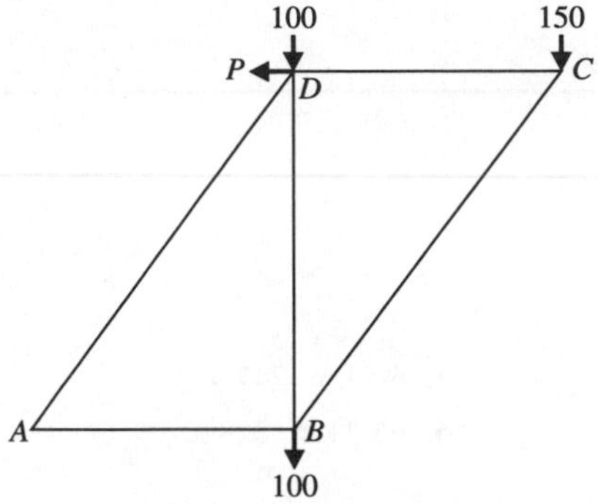

The diagram shows a framework *A*, *B*, *C*, *D* of freely hinged light rods loaded at *B*, *C* and *D* with weights of magnitude 100 N, 150 N and 100 N respectively. The framework is in a vertical plane, smoothly hinged at the fixed point *A* and equilibrium is maintained by a horizontal force of magnitude *P* Newtons at *D*. The rods *AD* and *BC* are inclined at $\frac{\pi}{4}$ to the horizontal, the rods *AB* and *CD* are horizontal and rod *BD* is vertical.
Find
(a) the value of *P*,
(b) the magnitude and direction of the reaction at the hinge *A*,
(c) the magnitude of the forces in each of the rods; stating in each case whether the rod is in tension or compression.
(AEB 1994)

15. The diagram shows a framework of seven freely-jointed light rods, the angles between the rods being as indicated. The framework rests on smooth horizontal ground at *A* and *D* and carries loads 500 N at *B* and *C*.
By drawing force diagrams, or otherwise, find the stresses in the rods, distinguishing between those rods in tension and those in compression.

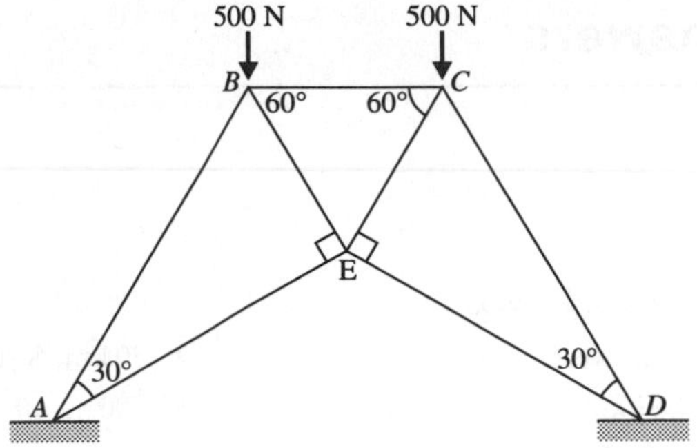

(OCSEB)

16.

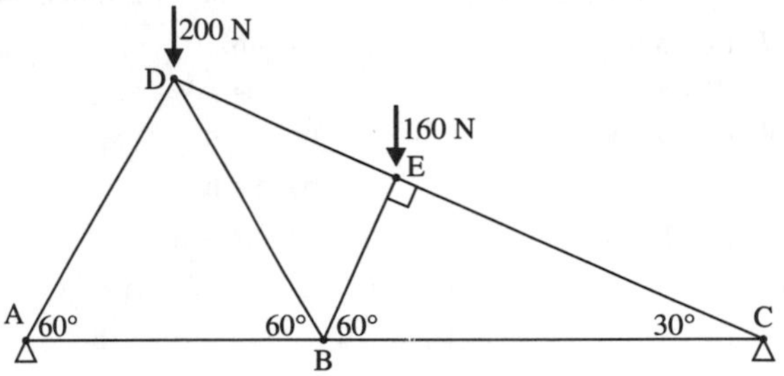

The diagram shows a framework of seven freely jointed light rods resting in a vertical plane on fixed smooth supports at A and C, with AB and BC horizontal. The rods CE and ED are in the same straight line and the angles between the rods are as indicated in the diagram. Loads of 200 N, 160 N are applied vertically at D, E respectively.
Find the reactions at A and C.
By drawing a force diagram, or otherwise, find which rod bears the largest tension and which rod bears the largest thrust, stating the magnitude of these stresses. (OCSEB)

Answers

Exercise 1A   page 2

1. 2 km west
2. 50 km, S 36·9° E
3. 6·5 km, 247·4°
4. 7·81 km, N 56·3° E, yes
5. 700 m, 101·8°
6. 65·7 km, 328·2°
7. N 78·7° W, S 11·3° E
8. 6·8 km, N 77° E

Exercise 1B   page 5

1. 6·77 units, 083·7°
2. 10·8 units, 249·1°
3. 8·06 units, 100·3°
4. 5·74 units, 299·1°
5. 11·2 units, 083°
6. 4·1 units, 112°
7. (a) $\mathbf{b}$ (b) $-\mathbf{b}$ (c) $\mathbf{a}$ (d) $-\mathbf{a}$
(e) $\mathbf{a}+\mathbf{b}$ (f) $-\mathbf{a}-\mathbf{b}$ (g) $\mathbf{b}-\mathbf{a}$ (h) $\mathbf{a}-\mathbf{b}$
8. (a) $\mathbf{b}-\mathbf{a}$ (b) $\mathbf{a}-\mathbf{b}$ (c) $\frac{1}{2}(\mathbf{b}-\mathbf{a})$ (d) $\frac{1}{2}(\mathbf{a}+\mathbf{b})$
9. (a) $\mathbf{b}$ (b) $\mathbf{a}+\mathbf{b}$ (c) $\mathbf{b}-\mathbf{a}$
10. (a) $\mathbf{b}-\mathbf{a}$ (b) $\frac{2}{3}(\mathbf{b}-\mathbf{a})$ (c) $\frac{1}{3}(\mathbf{a}-\mathbf{b})$ (d) $\frac{1}{3}(\mathbf{a}+2\mathbf{b})$
11. (a) $\frac{1}{3}\mathbf{c}$ (b) $\frac{1}{4}\mathbf{a}$ (c) $\mathbf{a}+\frac{1}{3}\mathbf{c}$ (d) $\mathbf{c}+\frac{1}{4}\mathbf{a}$ (e) $\mathbf{c}-\frac{3}{4}\mathbf{a}$ (f) $\frac{2}{3}\mathbf{c}-\frac{3}{4}\mathbf{a}$
12. (a) $\frac{1}{2}\mathbf{a}$ (b) $\mathbf{b}-\mathbf{a}$ (c) $\frac{3}{4}(\mathbf{b}-\mathbf{a})$ (d) $\frac{1}{4}(\mathbf{a}+3\mathbf{b})$ (e) $\mathbf{b}-\frac{1}{2}\mathbf{a}$ (f) $\frac{1}{4}(3\mathbf{b}-\mathbf{a})$

Exercise 1C   page 10

1. (a) (i) $4\mathbf{i}+3\mathbf{j}$ (ii) 5 units (iii) 36·9° (b) (i) $4\mathbf{j}$ (ii) 4 units (iii) 90°
(c) (i) $\mathbf{i}+4\mathbf{j}$ (ii) $\sqrt{17}$ units (iii) 76·0° (d) (i) $5\mathbf{i}+5\mathbf{j}$ (ii) $5\sqrt{2}$ units (iii) 45°
(e) (i) $-3\mathbf{i}+3\mathbf{j}$ (ii) $3\sqrt{2}$ units (iii) 135° (f) (i) $-3\mathbf{i}-\mathbf{j}$ (ii) $\sqrt{10}$ units (iii) 198·4°
(g) (i) $-2\mathbf{i}-3\mathbf{j}$ (ii) $\sqrt{13}$ units (iii) 236·3° (h) (i) $3\mathbf{i}-2\mathbf{j}$ (ii) $\sqrt{13}$ units (iii) 326·3°
(i) (i) $-3\mathbf{j}$ (ii) 3 units (iii) 270°
2. $\mathbf{a}=5\sqrt{3}\mathbf{i}+5\mathbf{j}$ $\mathbf{b}=2\mathbf{i}+2\sqrt{3}\mathbf{j}$ $\mathbf{c}=3\sqrt{2}\mathbf{i}+3\sqrt{2}\mathbf{j}$ $\mathbf{d}=-4\sqrt{3}\mathbf{i}+4\mathbf{j}$ $\mathbf{e}=-5\mathbf{i}+5\sqrt{3}\mathbf{j}$
$\mathbf{f}=2\sqrt{3}\mathbf{i}-2\mathbf{j}$ $\mathbf{g}=-3\mathbf{i}-3\sqrt{3}\mathbf{j}$ $\mathbf{h}=-4\sqrt{2}\mathbf{i}-4\sqrt{2}\mathbf{j}$
3. $\mathbf{a}=4\mathbf{i}$ $\mathbf{b}=-7\mathbf{j}$ $\mathbf{c}=5\mathbf{i}+5\mathbf{j}$ $\mathbf{d}=5\sqrt{3}\mathbf{i}+5\mathbf{j}$ $\mathbf{e}=-3\sqrt{3}\mathbf{i}-3\mathbf{j}$ $\mathbf{f}=-4{\cdot}23\mathbf{i}+9{\cdot}06\mathbf{j}$
4. $\mathbf{p}=2{\cdot}4\mathbf{i}+3{\cdot}6\mathbf{j}+4{\cdot}8\mathbf{k}$
5. $\mathbf{q}=4{\cdot}9\mathbf{i}+3\mathbf{j}+2{\cdot}1\mathbf{k}$
6. (a) (i) $3\mathbf{i}+3\mathbf{j}+\mathbf{k}$ (ii) $\sqrt{19}$ units (iii) 46·5°, 46·5°, 76·7°
(b) (i) $\mathbf{i}+2\mathbf{j}+3\mathbf{k}$ (ii) $\sqrt{14}$ units (iii) 74·5°, 57·7°, 36·7°
(c) (i) $2\mathbf{i}+3\mathbf{k}$ (ii) $\sqrt{13}$ units (iii) 56·3°, 90°, 33·7°
(d) (i) $3\mathbf{i}+\mathbf{j}-2\mathbf{k}$ (ii) $\sqrt{14}$ units (iii) 36·7°, 74·5°, 57·7°
7. (a) $7\mathbf{i}+24\mathbf{j}$ (b) $8\mathbf{i}-15\mathbf{j}$ (c) 5 units (d) $4\sqrt{26}$ units
(e) 25 units (f) $9\mathbf{i}+12\mathbf{j}$ (g) $28\mathbf{i}+96\mathbf{j}$
8. (a) $-5\mathbf{i}+12\mathbf{j}$ (b) $9\mathbf{i}+12\mathbf{j}$ (c) $\sqrt{29}$ units (d) $7\sqrt{2}$ units
(e) 14 units (f) 15 units (g) $10\mathbf{i}+25\mathbf{j}$ (h) $54\mathbf{i}+72\mathbf{j}$
9. (a) $6\mathbf{i}+\mathbf{j}+2\mathbf{k}$ (b) $9\mathbf{i}+2\mathbf{j}+6\mathbf{k}$ (c) $\sqrt{14}$ units (d) $\sqrt{41}$ units
(e) $\sqrt{26}$ units (f) 11 units (g) $2\sqrt{14}(\mathbf{i}-3\mathbf{j}+2\mathbf{k})$ (h) $\frac{5}{11}(9\mathbf{i}+2\mathbf{j}+6\mathbf{k})$
10. (a) $\begin{pmatrix}8\\4\\9\end{pmatrix}$ (b) $\begin{pmatrix}2\\3\\4\end{pmatrix}$ (c) $\sqrt{102}$ units (d) 7 units (e) 5 units (f) 10 units (g) $\begin{pmatrix}40\\0\\30\end{pmatrix}$

Exercise 1D page 15

1. (a) 9·64 units (b) 28 units (c) $-40\sqrt{3}$ units (d) 30·78 units (e) $28\sqrt{2}$ units (f) 0
2. (a) scalar (b) vector (c) vector (d) scalar (e) scalar (f) scalar (g) vector (h) scalar (i) vector (j) scalar (k) vector (l) scalar
3. (a) 0 (b) 0 (c) 0 (d) 1 (e) 1 (f) 1 (g) 6 (h) 0
4. -1 **5.** 16 **6.** -6 **7.** 7 **8.** 39 **9.** 7 **10.** -13 **11.** 13
12. 4 **13.** 6 **14.** 8 **15.** 2 **16.** 121° **17.** 90° **18** 90° **19.** 25°
20. 104° **21.** 48° **22.** 180° **23** 54° **24** 82° **25.** 92° **26.** 120° **27.** 73°
28. **c** and **d** **29.** $\frac{1}{3}$ **30.** 0

Exercise 2A page 20

1. $10\,\text{m s}^{-1}$ **2.** $22{\cdot}5\,\text{m s}^{-1}$ **3.** $126\,\text{km h}^{-1}$ **4.** $79{\cdot}2\,\text{km h}^{-1}$
5. $100\,\text{m s}^{-1}$ **6.** $25\,\text{m s}^{-1}$ **7.** $25\,\text{m s}^{-1}$ ($90\,\text{km h}^{-1}$) **8.** 16 m
9. 200 m **10.** 6 s **11.** 12 m, $5\frac{1}{2}$ s
12. (a) 80 000 km (b) 20 000 km (c) $22\frac{2}{9}$ km
13. 40·8 km **14.** 8 min 20 s **15.** 1·7 km **16.** $7{\cdot}04\,\text{m s}^{-1}$
17. (a) 45 min (b) $12\,\text{km h}^{-1}$ **18.** (a) $2\,\text{m s}^{-1}$ (b) $1\,\text{m s}^{-1}$ north
19. (a) $80\,\text{km h}^{-1}$ (b) zero
20. (a) $5\,\text{m s}^{-1}$ (b) $8\,\text{m s}^{-1}$ (c) 10 s (d) $3\,\text{m s}^{-1}$ in direction $\overrightarrow{AB}$

Exercise 2B page 23

1. 25 m
2. (a) 17 m (b) 13 m (c) $(-3\mathbf{i}+3\mathbf{j})$ m (d) $3\sqrt{2}$
3. (a) $5\sqrt{2}$ m (b) $\sqrt{137}$ m (c) 5 m (d) $(5\mathbf{i}+12\mathbf{j}-8\mathbf{k})$ m (e) $(-4\mathbf{i}-8\mathbf{j}+6\mathbf{k})$ m (f) $(4\mathbf{i}+8\mathbf{j}-6\mathbf{k})$ m (g) $\sqrt{233}$ m (h) $2\sqrt{29}$ m
4. $10\,\text{m s}^{-1}$ **5.** $25\,\text{m s}^{-1}$ **6.** $\sqrt{17}\,\text{m s}^{-1}$ **7.** $3\sqrt{13}\,\text{m s}^{-1}$
8. $\sqrt{59}\,\text{m s}^{-1}$ **9.** B **10.** $\pm 4{\cdot}8$ **11.** -15 or 8
12. (a) $(7\mathbf{i}+7\mathbf{j})$ m (b) $(9\mathbf{i}+11\mathbf{j})$ m **13.** (a) $(11\mathbf{i}+\mathbf{j})$ m (b) $(15\mathbf{i}-\mathbf{j})$ m
14. 1, -2 **15.** $13\,\text{m s}^{-1}$, $(16\mathbf{i}-30\mathbf{j})$ m, 34 m
16. (a) $[(4+3t)\mathbf{i}+(3-2t)\mathbf{j}+(9-5t)\mathbf{k}]$ m (b) $(19\mathbf{i}-7\mathbf{j}-16\mathbf{k})$ m, $3\sqrt{74}$ m
17. $a=1$, $b=-1$, $c=3$. $\sqrt{329}$ m **18.** -5, 2, 3 **19.** $-2\mathbf{i}+3\mathbf{j}+2\mathbf{k}$

Exercise 2C page 27

1. 128 m **2.** $15\,\text{m s}^{-1}$ **3.** 16 m **4.** 40 m **5.** $4\,\text{m s}^{-2}$
6. $11\,\text{m s}^{-1}$ **7.** $-4\,\text{m s}^{-2}$ **8.** $5\,\text{m s}^{-1}$ **9.** 4 s **10.** 28 m
11. $3\,\text{m s}^{-2}$ **12.** $9{\cdot}8\,\text{m s}^{-2}$ **13.** 12 m **14.** 5 s **15.** 2 s
16. $2\,\text{m s}^{-1}$ **17.** 30 m **18.** $15\,\text{m s}^{-1}$ **19.** 16 m **20.** $5\,\text{m s}^{-1}$
21. 300 m, 20 s **22.** $18\,\text{m s}^{-2}$ **23.** $2{\cdot}5\,\text{m s}^{-2}$ **24.** $31\,500\,\text{m s}^{-2}$, $\frac{1}{150}$ s
25. $6\frac{2}{3}\,\text{m s}^{-2}$, 3 s **26.** (a) 8 m (b) 18 m, 10 m **27.** $1{\cdot}6\,\text{m s}^{-2}$, 100 m
28. (a) $7\frac{1}{2}$ m (b) $5\frac{1}{2}$ m **29.** (a) 21 m (b) 25 m, 5 s (c) 29 m
30. 10 s, 29 m **31.** 2·5 m, $15\,\text{m s}^{-1}$ **32.** (a) 1·82 km (b) 1 min 44 s (c) $17{\cdot}5\,\text{m s}^{-1}$
33. $2\,\text{m s}^{-2}$, $3\,\text{m s}^{-1}$ **34.** 4, 6 **35.** 300 m

Exercise 2D page 31

1. $5{\cdot}6\,\text{m s}^{-1}$ **2.** $2\frac{6}{7}$ s **3.** 10 m **4.** (a) 27·6 m (b) $23{\cdot}6\,\text{m s}^{-1}$
5. (a) 16·1 m (b) 22·4 m (c) 18·9 m **6.** (a) $28\,\text{m s}^{-1}$ (b) 41 m
7. $4\frac{2}{7}$ s **8.** 10 m, $2\frac{6}{7}$ s **9.** 22·5 **10.** (a) 1 s (b) 4 s, 3 s
11. (a) 40 cm (b) 10 cm **12.** 6 s **13.** 10 s
14. (a) 9·1 m (b) 8·4 m, 2·5 m **15.** 180 m

Exercise 2E page 36

1. (a) (i) zero (ii) $6\,\text{m s}^{-1}$ (iii) $1{\cdot}5\,\text{m s}^{-2}$ (iv) 12 m
(b) (i) $2\,\text{m s}^{-1}$ (ii) $6\,\text{m s}^{-1}$ (iii) $1\,\text{m s}^{-2}$ (iv) 16 m
(c) (i) $6\,\text{m s}^{-1}$ (ii) $3\,\text{m s}^{-1}$ (iii) $-0{\cdot}75\,\text{m s}^{-2}$ (iv) 18 m

2. (a) (i) $2\,m\,s^{-2}$ (ii) $3\,m\,s^{-2}$ (iii) 39 m
(b) (i) $4\,m\,s^{-2}$ (ii) $1\frac{1}{3}\,m\,s^{-2}$ (iii) 20 m
3. (a) $1\frac{1}{2}\,m\,s^{-2}$ (b) $2\,m\,s^{-2}$ (c) 324 m
4. (a) $6\,m\,s^{-2}$ (b) $20\,m\,s^{-1}$ (c) $2\,m\,s^{-2}$ (d) 240 m
5. 70 s
7. Approx 200
8. No (He was travelling at $12{\cdot}1\,m\,s^{-1}$, or 27·2 miles per hour.)
9. $10{\cdot}5\,m\,s^{-2}$, cannot complete without accelerating, needs $7{\cdot}41\,m\,s^{-2}$

Exercise 2F *page 38*

1. (a) 50 s (b) $24{\cdot}2\,m\,s^{-1}$
2. 7 s
3. (i) $6\,m\,s^{-2}$
(ii) 180 m
(iii) $18\,m\,s^{-1}$
(iv) 12 s

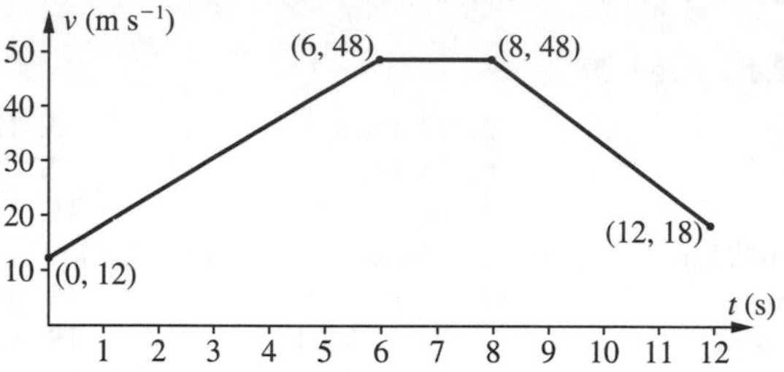

4. $12\,m\,s^{-1}$, 16 s
5. (i) $9\,m\,s^{-1}$, (ii) 10 s
6. $\frac{u}{12}(8+u)$ m
7. $30\,m\,s^{-1}$, 45 m
8. (i) 36·45 m
(ii) $27\,m\,s^{-1}$
(iii) 3 s
(iv) $9\,m\,s^{-1}$

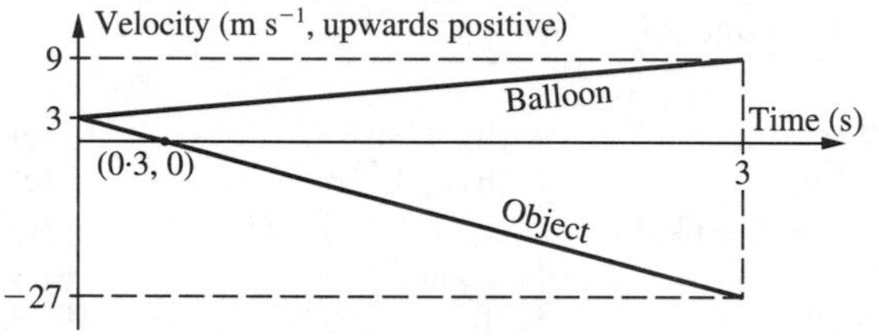

9. (a) (i) 0·3 s (iii) $7\,m\,s^{-1}$
(b) (i) Point A is when the velocity is first equal to zero, i.e. when the ball first reaches the highest point.
(ii) When the ball hits the ground it rebounds. The velocity instantly jumps from a negative value (question defines downwards as negative) to a positive value. Hence the discontinuity.
10. (a)

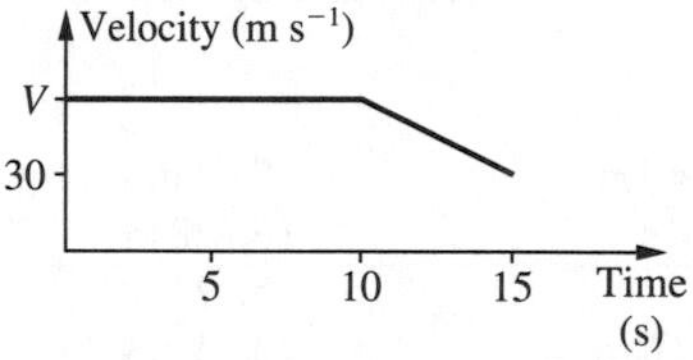

(b) $42\,m\,s^{-1}$ (c) $2{\cdot}4\,m\,s^{-2}$
11. $1{\cdot}2\,m\,s^{-2}$, $1{\cdot}8\,m\,s^{-2}$, 85 s, 1305 m
12. (i) 15, 0·5, 25 (ii)
(iii) 2·5 km, 135 s

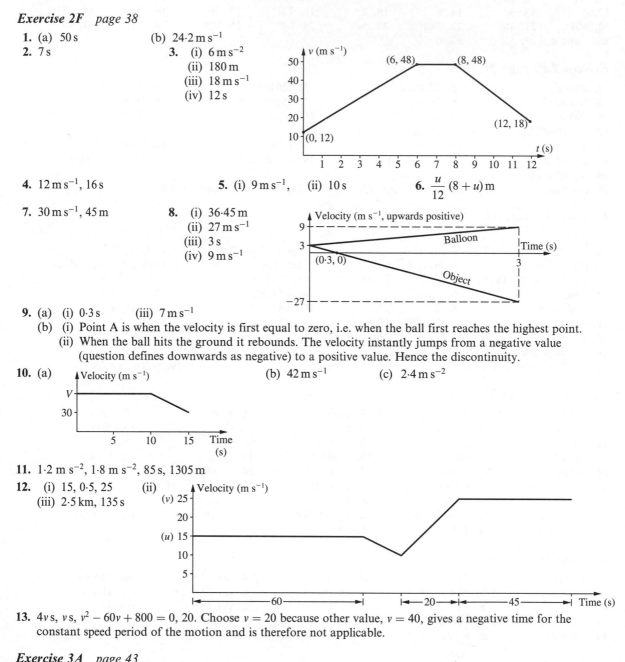

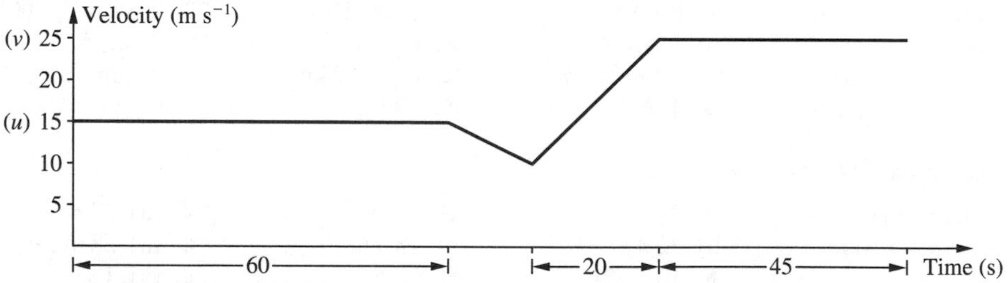

13. $4v$ s, v s, $v^2 - 60v + 800 = 0$, 20. Choose $v = 20$ because other value, $v = 40$, gives a negative time for the constant speed period of the motion and is therefore not applicable.

Exercise 3A *page 43*

1. (a) 50 N (b) 20 N, 98 N (c) 90 N (d) 60 N, 50 N
2. (a) 30 N (b) 10 N, 25 N (c) 40 N, 20 N (d) 60 N, 10 N
3. $2\,m\,s^{-2}$
4. 6 N
5. 8 kg
6. $50\,m\,s^{-2}$

7. $(10\mathbf{i}+4\mathbf{j})$ N **8.** $(6\mathbf{i}+6\mathbf{j})$ m s^{-2} **9.** $(1{\cdot}5\mathbf{i}+\mathbf{j}+3\mathbf{k})$ m s^{-2} **10.** 3 m s^{-2}, 1800 N
11. (a) 6 m s^{-2} (b) 4 m
12. (a) 118 N (b) 78 N (c) 49 N, 25 N (d) 49 N, 30 N (e) 49 N, 10 N (f) 240 N
13. 500 N **14.** 100 N **15.** 0·75 m s^{-2} **16.** $2\sqrt{10}$ N
17. 8**i** **18.** 2, $-\frac{1}{2}$ **19.** $a=6, b=3, c=1$
20. (a) 1250 N (b) 1600 N
21. 770 N **22.** 1200 m **23.** 4050 m

Exercise 3B page 47

1. 39·2 N **2.** 500 kg **3.** 0·98 N
4. (a) 58 N (b) 138 N (c) 49 N, 25 N (d) 5 kg (e) 15 kg (f) 3 kg
(g) 5 m s^{-2} (h) 0·2 m s^{-2} (i) 4·8 m s^{-2}
5. (a) 148 N (b) 48 N (c) 98 N (d) 98 N
6. 1·1 N **7.** 500 N **8.** (a) 0·4 m s^{-2} (b) 10 200 N (c) 9400 N
9. 2575 N **10.** 0·2 N **11.** 3 N **12.** 5810 N
13. (a) 24 N (b) 147 N

Exercise 3C page 52

1. 98 N **2.** 1·96 N **3.** 39·2 N **4.** 49 N, 68·6 N **5.** 19·6 N, 78·4 N
6. (a) $F=(m_1+m_2)a$ (b) $F-T=m_1 a$ (c) $T=m_2 a,\ R=m_1 g,\ R_2=m_2 g$
7. (a) $m_1 g-T=m_1 a$ (b) $T-m_2 g=m_2 a$
8. (a) $T=m_1 a$ (b) $m_2 g-T=m_2 a$
9. (a) $m_1 g-T_1=m_1 a$ (b) $T_1-T_2=m_2 a$ (c) $T_2-m_3 g=m_3 a$
10. (a) $T-mg-Mg=(m+M)a$ (b) $T-R-Mg=Ma$ (c) $R-mg=ma$
11. 1·96 m s^{-2}, 47·04 N **12.** 510 N **13.** 660 N
14. 1·4 m s^{-2}, 13·44 N **15.** 1·5 m s^{-2}, 1050 N
16. (a) (i) 4·2 m s^{-2} (ii) 28 N (iii) 56 N
(b) (i) zero (ii) 49 N (iii) 98 N
(c) (i) 2·8 m s^{-2} (ii) 6·3 N (iii) 12·6 N
17. 4·9 m s^{-2}, 9·8 m
18. (a) 1·4 m s^{-2}, 44·8 N, 58·8 N (b) 1·4 m s^{-2}, 16·8 N, 11·2 N
19. (a) 0·7 m s^{-2} (b) 0·0455 N (c) 1·4 m
20. 1080 N, 980 N, 860 N **21.** 1300 N, $1\frac{2}{3}$ m s^{-2}, $\frac{2}{3}$ m s^{-2}. Thrust of 100 N.

Exercise 3D page 55

3. $\frac{1}{10}g$, $\frac{33}{10}mg$, $\frac{18}{5}mg$ **4.** (a) 2100 N (b) 700 N **6.** 4200 N, 1050 N
7. 1·96 m s^{-2} (a) 1·4 m s^{-1} (b) 30 cm **8.** $\frac{11}{14}$ **9.** 66 cm

Exercise 3E page 59

1. $\frac{4}{11}g$ m s^{-2} ↑, $\frac{15}{11}g$ N **2.** $\frac{1}{9}g$ m s^{-2} ↓, $\frac{8}{3}g$ N
3. (a) 1·4 m s^{-2} ↑ (b) 7 m s^{-2} ↑, 4·2 m s^{-2} ↓, 1·4 m s^{-2} ↓ (c) 33·6 N, 16·8 N
4. $\frac{9}{37}g$ m s^{-2} ↓, $\frac{224}{37}g$ N, $\frac{112}{37}g$ N **5.** 1·4 m s^{-2}, 42 N **6.** $\frac{1}{17}g$ m s^{-2} ↓, $\frac{48}{17}mg$, $\frac{24}{17}mg$
8. $\frac{3}{19}g$ m s^{-2} ↓, $\frac{48}{19}mg$ **9.** 8 kg **10.** 1 kg

Exercise 3F page 61

1. 87·5 kN **2.** $(6{\cdot}5\mathbf{i}+14\mathbf{j})$ m **3.** (a) (i) $(4\mathbf{i}+2\mathbf{j})$ N (ii) $(2\mathbf{i}+\mathbf{j})$ m s^{-2} (b) (ii) 2

4.
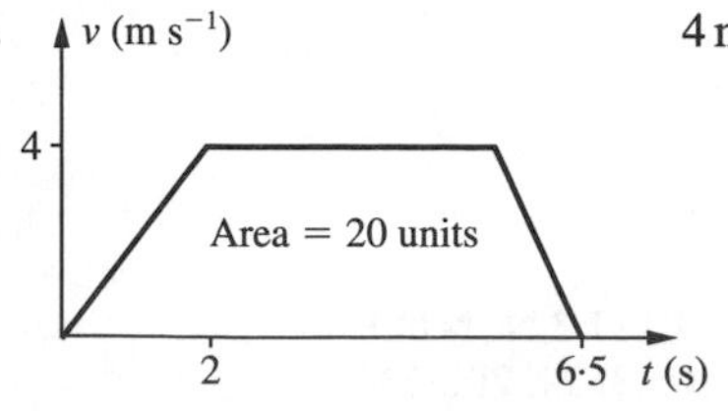

4 m s^{-2}, 5900 N

5. (a) (i) 9800 N
(ii) 490 N upwards
(b) (i) 7800 N
(ii) 390 N upwards

6. $\frac{12}{5}g$ N **7.** $\frac{24}{5}mg$ **8.** (i) 15 (ii) 3 m
9. $a = 2{\cdot}1$, $P = 46{\cdot}2$, $Q = 35{\cdot}7$, 0·756 s **10.** $M = 4000$, $\lambda = 0{\cdot}25$

Exercise 4A page 69

1. (a) 7·2 N, 34° (b) 11·6 N, 12° (c) 4·4 N, 37°
2. (a) 5 N, 36·9° (b) 6·71 N, 26·6° (c) 4·47 N, 26·6°
(d) 7·61 N, 28·4° (e) 17·9 N, 18·9° (f) 13·5 N, 45·7°
3. (a) 9·85 N, 15·3° to 8 N force (b) 10·2 N, 13·0° to 8 N force (c) 5·28 N, 11·2° to 8 N force
4. 1·62 N, 38·3° **5.** 5·29 N, 40·9° **6.** 70·5° **7.** 130·5°
8. 5·29 **9.** $F\sqrt{7}$ N, 40·9° **10.** 8 **11.** 8·24
12. (a) 17·2 N, 32° (b) 3·8 N, 172° (c) 4·1 N, 52° **13.** 7·2 N, N 9° E
14. 5·7 N, 101° **15.** 8·2 N, 76° **16.** 10 N, 60° **17.** 740 N, 025·8°
18. 60 N, 6° **19.** 20°

Exercise 4B page 73

1. (a) (i) $4\sqrt{3}$ N (ii) 4 N (b) (i) 0 (ii) 10 N
(c) (i) −7·66 N (ii) 6·43 N (d) (i) $2\sqrt{2}$ N (ii) $-2\sqrt{2}$ N
(e) (i) $\sqrt{3}$ N (ii) 3 N (f) (i) $P\cos\theta$ (ii) $P\sin\theta$
2. (a) $(8\mathbf{i} + 8\sqrt{3}\mathbf{j})$ N (b) $(3\mathbf{i} + 3\mathbf{j})$ N (c) $(-3\mathbf{i} + 3\mathbf{j})$ N (d) $(-5\mathbf{i} - 5\sqrt{3}\mathbf{j})$ N
(e) $P\cos\alpha\,\mathbf{i} + P\sin\alpha\,\mathbf{j}$ (f) $-Q\cos\phi\,\mathbf{i} - Q\sin\phi\,\mathbf{j}$
3. (a) (i) −5 N (ii) $-5\sqrt{3}$ N (b) (i) $5\sqrt{3}$ N (ii) −5 N
(c) (i) $-5\sqrt{2}$ N (ii) $-5\sqrt{2}$ N (d) (i) 3·42 N (ii) −9·40 N
(e) (i) −6·43 N (ii) −7·66 N (f) (i) 1·74 N (ii) −9·85 N
4. (a) (i) $\sqrt{3}$ N (ii) 7 N (b) (i) 12·3 N (ii) −0·071 N
(c) (i) $P\cos\theta + R - Q\sin\phi$ (ii) $P\sin\theta - Q\cos\phi$
5. (a) (i) 1 N (ii) −3·66 N (b) (i) −8·86 N (ii) −0·66 N
(c) (i) $-11\sqrt{2}$ N (ii) $\sqrt{2}$ N (d) (i) 13·4 N (ii) 0·60 N
(e) (i) −17 N (ii) $7\sqrt{3}$ N (f) (i) $R\sin\phi + Q\cos\theta$ (ii) $P + Q\sin\theta - R\cos\phi$

Exercise 4C page 77

1. (a) $5\mathbf{i}$ N (b) $(4\mathbf{i} - 3\mathbf{j})$ N (c) $4\mathbf{i}$ N (d) $(4\mathbf{i} - 6\mathbf{j})$ N
2. (a) $(4\mathbf{i} + 7\mathbf{j} - 5\mathbf{k})$ N (b) $(13\mathbf{i} - 6\mathbf{j} + 10\mathbf{k})$ N (c) $(4\mathbf{i} + 6\mathbf{j} + 2\mathbf{k})$ N
3. −4, 1 **4.** 4, 2 **5.** $a = 5$, $b = 5$, $c = 1$
6. 5 N, 36·9° **7.** 4·47 N, 116·6° **8.** $5\sqrt{2}$ N, 64·9°
9. (a) 5·66 N, 45° (b) 7·07 N, 98·1° (c) 3·61 N, 326·3° (d) 2 N, 180°
10. 7 N, 31·0°
11. (a) $(13\mathbf{i} + 8{\cdot}66\mathbf{j})$ N, 15·6 N, 33·7° (b) $5\mathbf{j}$ N, 5 N, 90° (c) $(5{\cdot}2\mathbf{i} - 4{\cdot}46\mathbf{j})$ N, 319·3°
12. $\sqrt{101}$ N, 354·3° **13.** $-6{\cdot}59\mathbf{i} + 13{\cdot}49\mathbf{j}$ **14.** 12·1 N, 145·6°
15. 5·66 N, 45° on same side as C **16.** 4 N, 30° on other side from C
17. 2·83 N, 45° with BA on other side from C **18.** 2 N parallel to AB
19. 13·4 N, 26·6° to AB on same side as C **20.** 20 N, 60° with AB on same side as C

Exercise 4D page 78

1. $100\sqrt{3}$ N **2.** $2\sqrt{3}$ N at 30° to BA **3.** $(3\sqrt{3} - 2)$ N, in direction of $\mathbf{j}$
4. $\sqrt{13}$ N, 124° **5.** $\sqrt{61}$ N, 093·7° **6.** (i) $P = 20$ N, $Q = 25$ N (ii) 143·1°
7. 50 **8.** 150°, 2

Exercise 5A page 85

1. (a) 32 N, 151° (b) 6 N, 141° (c) 3 N, 5·2 N
2. (a) 7·4 N, 105° (b) 4·8 N, 35° (c) 6·3 N, 8·9 N
3. (a) 5 N, 126·9° (b) 6·56 N, 67·6° (c) 8·89 N, 77°
4. (a) 5 N, 8·66 N (b) 2·18 N, 4·89 N (c) 2·59 N, 7·07 N
5. 8·89 N, 43° **6.** 36·9°, 36·75 N, 61·25 N **7.** 11·3 N, N 6° E
8. 9·8 N, 17·0 N **9.** 11·3 N, 22·6 N **10.** 35 N, $28\sqrt{2}$ N

Exercise 5B *page 91*

1. (a) $P\cos 30° = 4\sqrt{3}$ (b) $P\sin 30° + Q = 6$ (c) $P = 8\,\text{N}, Q = 2\,\text{N}$
2. (a) $P\cos 45° = Q$ (b) $P\sin 45° = 5\sqrt{2}$ (c) $P = 10\,\text{N}, Q = 5\sqrt{2}\,\text{N}$
3. (a) $4 = P\cos\theta$ (b) $3 = P\sin\theta$ (c) $P = 5\,\text{N}, \theta = 36{\cdot}9°$
4. (a) $5 + P\cos\theta = 10\cos 20° + 10\sin 20°$ (b) $P\sin\theta + 10\sin 20° = 10\cos 20°$
(c) $P = 9{\cdot}84\,\text{N}, \theta = 37{\cdot}4°$
5. (a) $P\cos 30° = Q\cos 60° + 10\cos 60°$ (b) $P\sin 30° + Q\sin 60° = 10\sin 60° + 8$
(c) $P = (5\sqrt{3} + 4)\,\text{N}, Q = (5 + 4\sqrt{3})\,\text{N}$
6. (a) $P = 10\cos 60°$ (b) $Q = 10\sin 60°$ (c) $P = 5\,\text{N}, Q = 5\sqrt{3}\,\text{N}$
7. (a) $P\sin 30° = Q$ (b) $P\cos 30° = 12$ (c) $P = 8\sqrt{3}\,\text{N}, Q = 4\sqrt{3}\,\text{N}$
8. (a) $10\cos 30° + 10\cos 60° = Q$ (b) $P + 10\sin 30° = 10\sin 60°$
(c) $P = (5\sqrt{3} - 5)\,\text{N}, Q = (5\sqrt{3} + 5)\,\text{N}$
9. (a) $P\cos\theta = 3 + 5\sqrt{2}\cos 45°$ (b) $P\sin\theta + 1 = 5\sqrt{2}\sin 45°$
(c) $P = 8{\cdot}94\,\text{N}, \theta = 26{\cdot}6°$
10. (a) $-4, 1$ (b) $-8, -5$ (c) $5, 2$ (d) $1, 2$ (e) $7, -14$
11. (a) $a = -5, b = 2, c = -1$ (b) $a = 2, b = -4, c = 4$ (c) $a = 4, b = -2, c = 0$
13. 58·8 N, 98 N **14.** $\dfrac{5mg}{13}, \dfrac{12mg}{13}$ **15.** 3·16 N, N 71·6° E **16.** 49 N, 84·9 N
17. 56·6 N, 113 N **18.** 51·7° **19.** (a) 19·5° (b) 9·8 N (c) 27·7 N
20. 44·4 N, 49 N, 28·3 N **21.** 30°, 41·8° **22.** 29·0°, 75·5°

Exercise 5C *page 98*

1. (a) $R + 50\sin 30° = 10g$ (b) $50\cos 30° = 10a$ (c) $R = 73\,\text{N}, a = \dfrac{5\sqrt{3}}{2}\,\text{m}\,\text{s}^{-2}$
2. (a) $R + P\sin 30° = 10g$ (b) $P\cos 30° = 50\sqrt{3}$ (c) $P = 100\,\text{N}, R = 48\,\text{N}$
3. (a) $R + 50\sin 60° = 10g$ (b) $50\cos 60° - 10 = 10a$ (c) $R = 54{\cdot}7\,\text{N}, a = 1{\cdot}5\,\text{m}\,\text{s}^{-2}$
4. (a) $R + 50\sin 20° = 10g + 50\sin 20°$ (b) $50\cos 20° + 50\cos 20° = 10a$
(c) $R = 98\,\text{N}, a = 9{\cdot}40\,\text{m}\,\text{s}^{-2}$
5. (a) $88 + P\sin\theta = 10g$ (b) $P\cos\theta = 10\sqrt{3}$ (c) $P = 20\,\text{N}, \theta = 30°$
6. (a) $R = 10g\cos 30°$ (b) $10g\sin 30° = 10a$ (c) $R = 84{\cdot}9\,\text{N}, a = 4{\cdot}9\,\text{m}\,\text{s}^{-2}$
7. (a) $R = 10g\cos 30°$ (b) $P - 10g\sin 30° = 20$ (c) $R = 84{\cdot}9\,\text{N}, P = 69\,\text{N}$
8. (a) $R + 50\sin 40° = 10g\cos 40°$ (b) $50\cos 40° + 10g\sin 40° = 10a$ (c) $R = 42{\cdot}9\,\text{N}, a = 10{\cdot}1\,\text{m}\,\text{s}^{-2}$
9. (a) $49 = 10g\cos\theta$ (b) $10g\sin\theta - 50 = 10a$ (c) $a = 3{\cdot}49\,\text{m}\,\text{s}^{-2}, \theta = 60°$
10. 65·3 N, 1·63 m s^{-2}, 84·9 N **11.** 34·6°, 80·7 N, 25·6 N **12.** 2 m s^{-2}, 9 m
13. 0·8 m s^{-2}, $4\sqrt{2}$ N, 45 N **14.** 69·2 N, 2·12 m s^{-2}, 16·96 m **15.** 14 m s^{-1}, 14 m s^{-1}
16. 38 N **17.** 28 N **18.** 0·74 s **19.** 3·8 m s^{-2}, 7·6 m
20. 31 N **21.** 30° **22.** 0·663 s

Exercise 5D *page 100*

12. 10 m **15.** 2·42 m s^{-1}

Exercise 5E *page 102*

1. $4, \sqrt{3}, \dfrac{\sqrt{3}}{4}, \sqrt{19}$ **2.** (i) 24·5, 22·6 (ii) 45·2 in opposite direction to Q
3. (a) 2, −6 (b) $2\sqrt{10}$ N (c) 18° **4.** (i) 1·2 N (ii) 1·7 N
5. 2·2 **6.** 45·8 N at 109·1° to 40 N
7. (a) $(190{\cdot}78\mathbf{i} + 19{\cdot}35\mathbf{j})$ N (b) $-(190{\cdot}78\mathbf{i} + 19{\cdot}35\mathbf{j})$ N (d)

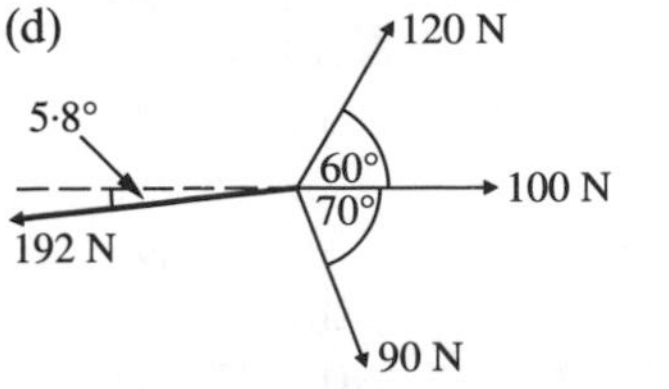

(e) 16·5°
8. (i) 5 N (ii) 0·8 **9.** (a) 8 N, $4\sqrt{3}$ (b) $\sqrt{97}$ N
10. 161 N, 192 N, 360 N 20° left of downward vertical
11. (i) 681 N (ii) 49·7 N (iv) 281 N (v) approx 4°
12. 9 m s^{-1}, 3 s **13.** 2·8 m s^{-1}
14. (i) 1·5 m s^{-2} (ii) 13·5 N (iii) 1·8
15. (i) 5 m s^{-2} (ii) 6 m s^{-1} (iii) 14·4 m

Exercise 6A *page 110*

1. (a) 10 N, rest (b) 14 N, rest (c) 14 N, accelerate (d) 10 N, rest
(e) 14 N, rest (f) 18 N, accelerate (g) 10 N, rest (h) 10 N, accelerate
(i) 10 N, accelerate (j) 12 N, accelerate (k) 12·12 N, rest (l) 22 N, accelerate
2. (a) $\frac{2}{7}$ (b) $\frac{1}{2}$ (c) $\frac{1}{4}$ (d) 0·33 (e) 0·348 (f) 0·677
3. $\frac{4}{7}$ **4.** $\frac{1}{10}$ **5.** 49 N, yes, 0·05 m s^{-2} **6.** 50 N, no
7. 0·49 N, yes, 1·02 m s^{-2} **8.** 19·8 N, 10 N **9.** $\frac{1}{4}$ **10.** 1·02 m
11. (a) 0·7 m s^{-2} (b) 3·15 N (c) 2·45 m s^{-2} (d) 11·25 m
12. (a) 9·8 (b) 8·78 (c) 15·9
13. (a) 3·92 (b) 4·62 (c) 6·93 **14.** 0·58, 0·63 N
15. (a) no sliding (b) no sliding (c) sliding
16. (a) no sliding (b) sliding (c) sliding
17. (a) no sliding (b) no sliding (c) sliding
18. Yes **19.** 9·8 N, no
20. (a) 1·4 m s^{-2} (b) 0·42 N (c) 70 cm
21. (a) 1·09 m s^{-2} (b) $2\frac{1}{3}$ m s^{-2} (c) 1·48 m s^{-1}
23. 2·94 m s^{-2} **24.** $\frac{1}{7}$

Exercise 6B *page 118*

1. (a) 33·5 N, rest (b) 42·4 N accelerate (c) 37·5 N, accelerate
(d) 17·0 N, rest (e) 23·7 N, rest (f) 23·0 N accelerate
2. (a) 1·61 N (b) 10·2 N (c) 18·4 N (d) 19·7 N (e) 18·3 N (f) 23·2 N
3. (a) 19·3 N (b) 23·2 N (c) 27·7 N (d) 33·2 N (e) 27·3 N (f) 22·0 N
4. (a) 0·15 (b) 0·20 (c) 0·24
5. 2·25 N, body will slide **6.** 0·202 **7.** 0·279 **8.** 0·523
9. (a) 2·78 N (b) 7·02 N (c) 8·52 N
10. (a) $2g$ N (b) $18g$ N (c) $6(8 + 3g)$ N
11. $\frac{1}{2}$ **12.** yes, $2\frac{1}{3}$ m s^{-1} **13.** 0·26, 4 s **14.** 0·23 **15.** 3 s, 4·2 m s^{-1}
16. (b) yes (c) no (d) 0·577
17. 0·29 **18.** 5·6 m s^{-1} **19.** 1·96, 0·1 **20.** 0·98 m s^{-2}, 44·1 N

Exercise 6C *page 125*

1. 11·3° **2.** 0·577
3. (a) 51·0 N, 15·9° (b) 52·9 N, 22·2° (c) 44·4 N, 24·1° (d) 5·60 N
(e) 5·60 N (f) 5·25 N (g) 76·6 N (h) 63·4 N
(i) 76·6 N (j) $\frac{19\sqrt{3}}{3}$ N, 30° (k) 44·1 N, 15·8° (l) 28·2 N, 26·2°
4. 20·2 N, 14·0° **5.** 14·9 N, 2·16 m s^{-2}
6. (b) (i) yes (ii) no (c) 25°, 0·466
7. 13·5 N **8.** 10·5 N **9.** 28·0 N **10.** 25·9°, remain at rest
11. (a) 17·0 N (b) 16·8 N (c) 17·8 N
13. (a) 4·12 N (b) 4·09 N (c) 4·17 N
14. (a) 20·8 N (b) 20·7 N (c) 21·4 N
16. (a) 15·2 N (b) 15·0 N (c) 15·2 N
17. (a) 24·5 N (b) 24·9 N (c) 26·1 N

Exercise 6D *page 128*

1. (i) 8·54 N (ii) 2·19 N, 0·26 **2.** (i) 29·6 (ii) 0·47

3. Various possibilities – e.g. give sledge a push on horizontal ground and observe that it does eventually come to rest, place sledge on slight incline and observe that it does not slide down.

(a)

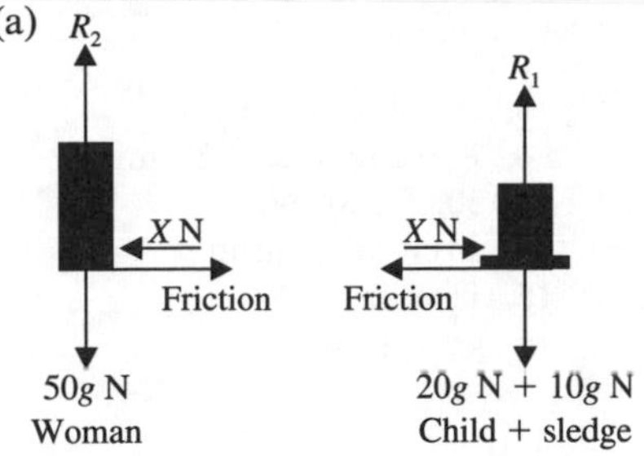

(b) (ii) 0·12

4. $\frac{1}{4}g$ m s^{-2}, $\frac{3}{4}g$ N, $\frac{1}{4} \leqslant \mu < \frac{1}{2}$

5. $\frac{7}{15}g$ m s^{-2}, $\frac{16}{15}mg$ N

6. Particle does not move

7. (a) 2·7 N (b) 10·6 N

8. $\frac{1}{8}$, $\frac{35}{13}$ m s^{-2}

9. (i)

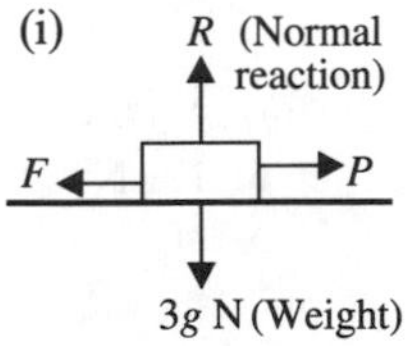

(ii) When in equilibrium $F = P$
When in motion $F = \mu R$
$= 3g\mu$ N

(iii)

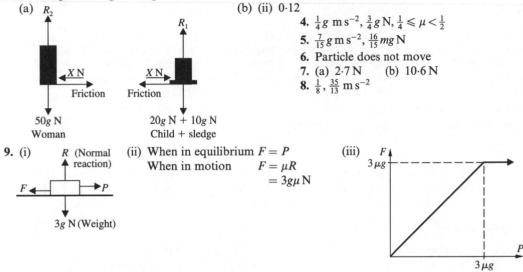

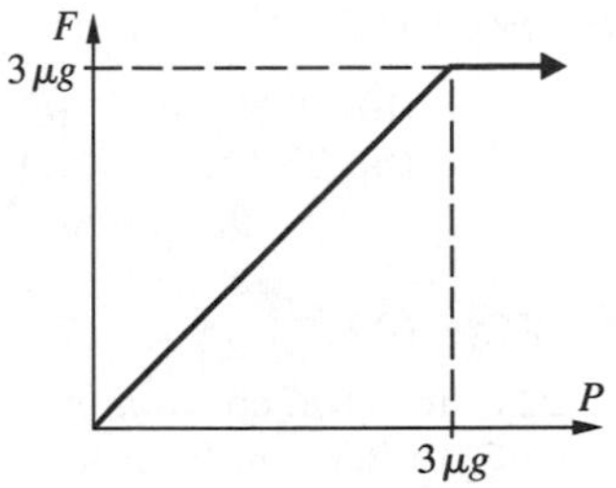

10. (a) weight of child, weight of tray, μ between tray and snow
(b) (ii) $2\sqrt{2}$ s (c) 1·5 m s^{-2} (d) 0·404 (e) air resistance

11. 0·0520, 11·0 m s^{-1}, 9·13 s, 88·1 N

12. (a) (i) 3·5 m s^{-2} (ii) 3 N (iii) 2·1 m s^{-1} (b) 0·602

13. (ii) $\sqrt{2\left(\dfrac{M-m}{M+m}\right)g}$ m s^{-1} (iii) $\sqrt{\dfrac{2}{g}\left(\dfrac{M-m}{M+m}\right)}$ s

(iv) When particle on slope comes to rest it will remain in that position because frictional force just able to prevent motion down slope.

14. 48·8° **15.** $6mg$, $5d$

Exercise 7A *page 138*

1. 10 N m clockwise **2.** 2 N m anticlockwise **3.** 10 N m clockwise
4. 10 N m anticlockwise **5.** 24 N m clockwise **6.** 16 N m clockwise
7. 8 N m clockwise **8.** 3 N m anticlockwise **9.** zero
10. 12 N m clockwise **11.** 14 N m clockwise **12.** 10 N m clockwise
13. 4 N m clockwise **14.** 6 N m anticlockwise **15.** 6 N m anticlockwise
16. 40 N m clockwise **17.** 24 N m anticlockwise **18.** 80 N m clockwise
19. zero **20.** 25 N m clockwise **21.** 34·6 N m clockwise
22. 61·3 N m clockwise **23.** 9·85 N m anticlockwise **24.** 20 N m anticlockwise
25. 20 N m clockwise **26.** 9 N m clockwise **27.** 2 N m clockwise
28. 5 N m clockwise **29.** 7 N m clockwise **30.** 17 N m anticlockwise

Exercise 7B *page 142*

1. (a) 40 N m clockwise (c) 5 N m anticlockwise (e) 19 N m clockwise
(f) 17 N m anticlockwise (g) 2 N m clockwise (h) 2 N m clockwise
2. 1·75 N m **3.** 21 N m in sense ABCD **4.** 15
5. 15 N, 14 N m in sense ABCD **6.** 2·4 N m in sense ADCB **7.** 12 N m in sense ABCD, 1·5
8. 80 N **9.** 10 N m clockwise **10.** 7 N m clockwise
11. 13 N m anticlockwise **12.** −6, 4, 26 N m clockwise
13. 12 N m clockwise, independent of P.

Exercise 7C *page 145*

1. (a) 6 N, 1 m (b) 2 N, $7\frac{1}{2}$ m (c) 2 N, $7\frac{1}{2}$ m **2.** 5 N, 4 m
3. 20 N, 3 m **4.** 2 N, 10 m **5.** 4 N, 15 m **6.** 6, 2
7. 4, 6 **8.** 10 N, 1·1 m **9.** 1·5 N, $1\frac{2}{3}$ m **10.** up, down, up, 3 m

Exercise 7D *page 148*

1. $P = 30$ N, $x = 1\frac{1}{3}$ m **2.** $P = 20$ N, $Q = 10$ N **3.** $P = 8$ N, $x = 1\frac{1}{2}$ m
4. $P = 5g$ N, $Q = 10g$ N **5.** $P = 160g$ N, $Q = 100g$ N **6.** $P = 25g$ N, $x = 2{\cdot}5$ m
7. $P = 20g$ N, $Q = 30g$ N **8.** $P = 190$ N, $Q = 130$ N **9.** $30g$ N, $40g$ N
10. 2 m from 10 kg mass **11.** 75 cm from other end **12.** 24 cm from head
13. $2\frac{3}{4}$ N, $2\frac{1}{4}$ N **14.** 1·4 m from A **15.** $1\frac{1}{2}$ m from A
16. $7\frac{1}{2}g$ N downwards, $12\frac{1}{2}g$ N upwards **17.** $8\frac{4}{7}$, $14g$ N
18. (a) 66 cm from handle (b) $2\frac{1}{4}g$ N, $12\frac{3}{4}g$ N

Exercise 7E *page 151*

1. (a) 98 N (b) 49 N (c) 44 N **2.** (a) 4·9 N (b) 5·9 N (c) 3·57 N
3. (a) 42·4 N (b) 27·7 N (c) 30·1 N **4.** 13·7 N **5.** 19·6 N **6.** $12g$ N
7. 46 cm **9.** 31·4 N **10.** 614 N

Exercise 7F *page 156*

1. (a) 13 N, 22·6° to AB. 7 cm from A (b) 13 N, 22·6° to AB. At A
(c) 25 N, 73·7° to BA. 3 m from A (d) 8·54 N, 324·2° to AB. 8·31 m from A
2. (a) 0·14 N m in sense ADCB (b) 41 N m in sense ADCB
(c) $10\sqrt{3}$ N m in sense ABC (d) $3\,Pa\sqrt{3}$ in sense ABCDEF
3. 5 N, 53·1° to AB. $12\frac{1}{2}$ cm from A **4.** 12·8 N, 2·49 N, 65·9 N m in sense ADCB
5. 4·47 N, 26·6° to AB. $\frac{7}{2}a$ metres from A on BA produced **6.** 30 N, 30° to AB. 2 m from A on BA produced
7. 18·3 N, 10·9° to AB. $\frac{3}{2}a$ metres from A on BA produced **8.** $(4\mathbf{i} + 6\mathbf{j})$ N, $\mathbf{i}$
10. 11·3 N, 16·3° to BA. 89 cm from A on BA produced
11. 8·84 N, 57·6° to AB. 2·32 from A on BA produced
12. 12·5 N, 49·6° to AB. 1 m from A
13. 20·1 N, 15·2° to BA, 3·04 cm from A. Same force but now 6·46 cm from A on BA produced
14. 8·25 N, 76·0° to AB, 25 cm from A. 8·54 N, 69·4° to AB, 2 N m in sense ABCD
15. 16·4 N, 37·6° to AB, 5·5 m from A on AB produced. Same force after couple introduced
16. $(4\mathbf{i} + 3\mathbf{j})$ N, $-4\mathbf{i}$, $a = 14$, $b = -4$, $c = -3$

Exercise 7G *page 159*

1. 2, 8 N **2.** 15 units **3.** (a) $(8\mathbf{i} + 6\mathbf{j})$ N, $2\mathbf{j}$ N (b) $8\sqrt{2}$ N (c) 8 N m clockwise
4. (a) 1, −4 (b) (i) $-5\mathbf{j}$ (ii) 20 units **5.** 2·75 m from A, $\frac{41}{2}g$ N, $\frac{79}{2}g$ N **6.** 0·8
7. $315g$ N, $15\sqrt{562}\,g$ N **8.** (i) $50\sqrt{3}$ kg (ii) $50\sqrt{3}(\sqrt{7} - 1)g$ N **9.** 162 N, 148°
10. (i) $2\sqrt{41}$ N (ii) 1 N m anticlockwise (iii) $\dfrac{1}{2\sqrt{41}}$ m **11.** (a) $\sqrt{10}$ N (b) 71·6° (c) 4 m
12. $R = 7W$, $\theta = 81{\cdot}8°$

Exercise 8A *page 166*

1. (a) $(2\frac{1}{2}, 0)$ (b) (2, 2) (c) $(\frac{1}{2}, 2)$ (d) (1, −1) **2.** (0, 4)
3. $(2, 1\frac{1}{2})$ **4.** (−1, 3) **5.** −3, 2 **6.** 1 cm, $1\frac{1}{2}$ cm **7.** $1\frac{1}{3}$ m, $1\frac{3}{4}$ m
8. $5\mathbf{i} + 3\mathbf{j}$ **9.** $\mathbf{i} - 3\mathbf{j}$ **10.** 2, −4 **11.** (2·4, −3·2) **12.** 1, 2·5
13. (0, −2) **14.** (3, 1) **15.** 1 g, 5 cm

Exercise 8B *page 171*

1. (2, 2) **2.** $(2\frac{1}{2}, 1\frac{1}{2})$ **3.** $(1, 2\frac{1}{2})$ **4.** (2, 3) **5.** (2, 2) **6.** (3, 2)
7. (2, 2) **8.** (1, 2) **9.** (2, 2) **10.** (3, 2) **11.** $(1\frac{1}{3}, 2)$ **12.** $(2\frac{2}{3}, 1)$

Exercise 8C *page 176*

1. $(2\frac{1}{2}, 1\frac{1}{2})$ **2.** (1·8, 1·3) **3.** $(2, 1\frac{3}{4})$ **4.** (2·6, 1·9) **5.** (2·3, 1·4)
6. $(1, 2\frac{1}{3})$ **7.** (1·4, 2·2) **8.** (3, 3) **9.** $(1\frac{13}{14}, 2)$ **10.** $(2\frac{4}{9}, 3\frac{1}{9})$
11. (4·1, 2·95) **12.** $(-\frac{2}{15}, 0)$ **13.** $(-\frac{2}{3}, 0)$ **14.** (a) $\frac{1}{3}$ m (b) $1\frac{1}{3}$ m
15. 1·8 m from AD, 1·8 m from AB **16.** 2·18 m **17.** 0·5 m **18.** $1\frac{2}{3}$ m, 3 m
19. $(-\frac{4}{13}, -\frac{15}{26})$ **20.** $1\frac{2}{3}$ cm **21.** On axis of symmetry, $1\frac{6}{7}$ cm above base
22. $\frac{28}{3\pi}$ m **23.** On axis of symmetry, $\frac{28}{5\pi}$ cm into larger semicircle from common diameter
24. On axis of symmetry, $\frac{4}{\pi}$ cm into larger semicircle from common diameter
25. On axis of symmetry, $3\frac{8}{11}$ cm above base of cylinder
26. On axis of symmetry, $2\frac{5}{7}$ cm from undrilled end
27. On axis of symmetry, 10·8 cm from tip of cone

Exercise 8D *page 182*

1. 45° **2.** 76·0° **3.** 29·2° **4.** 21·0° **5.** 40·6° **6.** 36·0° **7.** 32·5° **8.** 26·6°
9. 56·3° **10.** 39·5° **11.** 31·0° **12.** 35° **13.** 23·2° **14.** 14·4° **15.** 38·9° **16.** 3·5°

Exercise 8E *page 189*

7. $(1\frac{1}{2}, 1\frac{1}{5})$ **8.** (1, 0·4) **9.** (1·56, 2·25) **10.** (5·31, 3·21) **11.** (1·5, 3·6)
12. $(\frac{3}{8}, 2\frac{1}{5})$ **13.** $(2\frac{2}{3}, 0)$ **14.** $(3\frac{1}{9}, 0)$ **15.** (3·39, 0) **16.** (1·30, 0)
17. $(2\frac{2}{5}, 7\frac{5}{7})$, $(2\frac{5}{8}, 0)$

Exercise 8F *page 192*

1. $\frac{3\sqrt{2}}{2}$ cm, $\frac{3\sqrt{2}}{2}$ cm **2.** $\frac{4}{3\pi}$ cm **3.** (a) $\frac{3d}{5}$ (b) d (c) 59°
4. (i) $(\frac{5}{6}, \frac{5}{6})$ (ii) 9·46° (iii) $(\frac{7}{8}, \frac{9}{8})$ **5.** 0·109 m, 36·1° **6.** (a) $\frac{9}{8}$ m, $\frac{13}{8}$ m (b) 40·9°
7. 7·28, $\mu \geqslant \tan 20°$ **8.** On axis of symmetry, $\frac{45a}{56}$ from O.
9. (ii) The body remains in equilibrium **10.** (b) 9° **11.** (b) $\frac{8}{3}$ cm **13.** $\frac{22h}{7}$ **14.** $\frac{3a(2-k)}{5+4k}$

Exercise 9A *page 205*

1. (a) 0, 49 N, 49 N (b) 49 N, 49 N, 69·3 N (c) 49 N, 49 N, 69·3 N
(d) 0, 49 N, 49 N (e) 21·2 N, 61·3 N, 42·4 N (f) 21·2 N, 61·3 N, 42·4 N
2. (a) 28·3 N, 28·3 N, 30° (b) 24·5 N, 42·4 N, 30° (c) 42·4 N, 64·8 N, 40·9°
(d) 42·4 N, 24·5 N, 60° (e) 21·2 N, 32·4 N, 19·1° (f) 50·7 N, 21·6 N, 81·8°
3. (a) 42·4 N, 21·2 N, 61·3 N, 0·346 (b) 57·8 N, 19·8 N, 43·7 N, 0·453 (c) 49 N, 42·4 N, 73·5 N, 0·577
4. (a) 84·9 N (b) 53·5 N (c) 25·9 N
5. (a) 56·6 N, 56·6 N, 196 N, 0·289 (b) 98 N, 98 N, 196 N, 0·5 (c) 26·3 N, 26·3 N, 196 N, 0·134
6. (a) $\frac{W}{3}$ N (b) $\frac{W}{2}$ N (c) W N
7. (a) 67·4° (b) 54·5° (c) 53·1°
8. (a) 19·4° (b) 36·1° (c) 49·1°
9. 49 N, 49 N at 60° to wall **10.** 39·9 N, 45·2 N **11.** 11·3 N at 58° to AB, 2·08 m
12. 60°, 5 N, 8·66 N **13.** 35·4 N **14.** 157 N, 547 N **15.** 141 N, 245 N
16. 0·346 **18.** 27·7 N **19.** 245 N, 32·8 N, 0·134 **20.** 3·05 N
21. 0·269 **23.** 6 m **30.** 11·4 N, 158·4 N **31.** 12·1 N, 138 N
32. 3·8 m, 514·5 N **33.** 1176 N, 467 N, 0·505 **34.** 6 m, $\frac{8W}{17}$ **35.** 9 m, $\frac{7W}{11}$
36. 8 m, 126 N

Exercise 9B *page 212*

1. (i) If the ground is smooth there is no force that can "balance" the horizontal reaction the wall exerts on the ladder. Therefore the ladder is not in equilibrium and will slip.

(ii) $\frac{3mg\tan\theta}{2}$ N, $5mg$ N, $\frac{3mg\tan\theta}{2}$ N (iii) $\frac{7l}{4}$

2. $(80 - 50\sin^2\theta\cos\theta)$ N, $(50\sin\theta\cos^2\theta)$ N, $(50\sin\theta\cos\theta)$ N, $\frac{100}{3\sqrt{3}}$ N

3. $2S\tan\theta$, $R + P = W$, (a) $\frac{1}{2}$ **4.** (a) 94 N (b) 0·47 (c) 133 N

5. $\frac{5Wa\cos\theta}{b}$ (i) $3W - \frac{5Wa\cos^2\theta}{b}$, $\frac{5Wa\sin\theta\cos\theta}{b}$ (ii) $\frac{5Wa}{2b}$ (iii) $\frac{3}{5}$ (iv) $\frac{3b}{10a}$

6. (i)

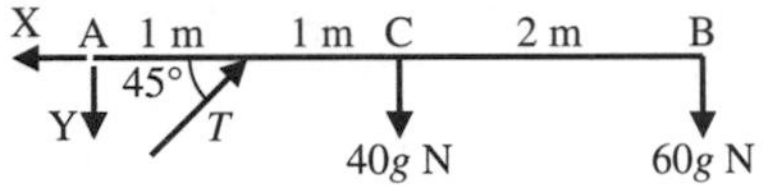

(ii) 80 cm from B (iii) (a) $320\sqrt{2}g$ N (b) $20\sqrt{377}g$ N

7. (i) 121·6° (ii) 6·53 N

9. (a)

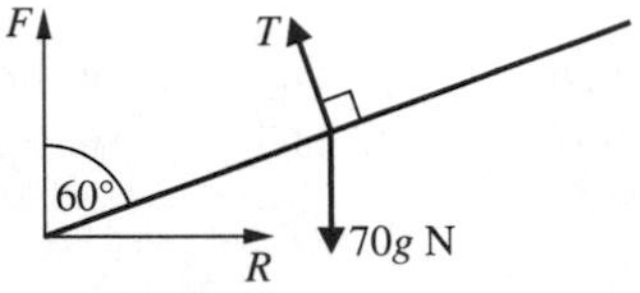

(c) $(0\mathbf{i} + 70g\mathbf{j})$ N

(e) Normal reaction will have decreased. (Reasoning: Consider taking moments about the top of the cliff in each case and assume angle between man's legs and cliff still 60°. Frictional force and tension both pass through this point and so have zero moment. Moment due to weight is unchanged so to balance this the normal reaction can decrease due to greater perpendicular distance from line of action to top of cliff.)

10. (a) For the ring to be in equilibrium the force it experiences from the tension in the string must be "balanced" by the normal reaction between ring and rod (there is no friction as rod is smooth). Thus tension, and hence string, must be perpendicular to rod.

(c) $\frac{3w\sqrt{5}}{5}$

11. (a) See answer to **10**(a) above. (b) (ii) $\frac{\sqrt{7}}{8}$

13. $\frac{2\mu(1+\mu)}{1+\mu^2}$

Exercise 10A *page 223*

1. (a) $(9\mathbf{i} - \mathbf{j})\,\text{m s}^{-1}$ (b) $(7\mathbf{i} + 9\mathbf{j})\,\text{m s}^{-1}$ (c) $(-\mathbf{i} - 6\mathbf{j})\,\text{m s}^{-1}$
(d) $(54\mathbf{i} + 27\mathbf{j})\,\text{km h}^{-1}$ (e) $63\mathbf{i}\,\text{km h}^{-1}$

2. (a) $25\,\text{m s}^{-1}$ N 16° E (b) $8{\cdot}7\,\text{km h}^{-1}$ N 30° E (c) $6{\cdot}2\,\text{m s}^{-1}$ N 76° E
(d) $16{\cdot}9\,\text{m s}^{-1}$ N 48° E (e) $82\,\text{km h}^{-1}$ N 73° E

3. (a) $15\,\text{m s}^{-1}$ N 36·9° W (b) $4{\cdot}25\,\text{m s}^{-1}$ N 48·3° E (c) $31{\cdot}2\,\text{km h}^{-1}$ N 12·2° E
(d) $15{\cdot}7\,\text{km h}^{-1}$ S 77·9° W (e) $32{\cdot}8\,\text{m s}^{-1}$ S 80° E

4. 4, −5 **5.** 7, 3 **6.** $10\,\text{km h}^{-1}$ N 30° E **7.** $9{\cdot}54\,\text{m s}^{-1}$ S 63° E

8. $11{\cdot}5\,\text{m s}^{-1}$ S 34·2° E **9.** 53·1°, 25 s **10.** 70·5° to the bank, 44·2 s

11. 90 s, 75 m **12.** 1 min, 30 m **13.** N 73·7° E, 50 min

14. 054·3°, 2 h 8 min **15.** 330·8°, 1 h 19 min **16.** (a) $3\frac{3}{4}$ min (b) 3 min

17. upstream at 45·6° to bank or upstream at 24·4° to bank, 28 s, 48 s

18. N 8·3° E, 1 h 50 min, 1 h 33 min **19.** 3 h 15 min **20.** $\frac{v+u}{2}$

Exercise 10B *page 229*

1. $10\,\text{km h}^{-1}$ north **2.** $50\,\text{m s}^{-1}$ north **3.** $30\,\text{km h}^{-1}$ east, 45 km

4. $15\,\text{m s}^{-1}$ east, 900 m **5.** $\begin{pmatrix} 8 \\ 2 \end{pmatrix}\,\text{m s}^{-1}$ **6.** $(14\mathbf{i} + 8\mathbf{j} - 11\mathbf{k})\,\text{m s}^{-1}$

7. $(-\mathbf{i}+3\mathbf{j}-27\mathbf{k})\,\mathrm{m\,s^{-1}}$
8. $(2\mathbf{i}-7\mathbf{j})\,\mathrm{m\,s^{-1}}$
9. $(100\mathbf{i}+600\mathbf{j})\,\mathrm{km\,h^{-1}}$
10. $\begin{pmatrix}0\\12\end{pmatrix}\mathrm{km\,h^{-1}}$, $\begin{pmatrix}0\\-12\end{pmatrix}\mathrm{km\,h^{-1}}$
11. $5\,\mathrm{km\,h^{-1}}$ N 36·9° W
12. $17\,\mathrm{km\,h^{-1}}$ S 61·9° E
13. $10{\cdot}3\,\mathrm{km\,h^{-1}}$ N 46·9° E
14. $25{\cdot}2\,\mathrm{m\,s^{-1}}$ from 018·1°
15. $292\,\mathrm{km\,h^{-1}}$ S 77·9° E
16. $20\,\mathrm{m\,s^{-1}}$ N 36·9° W, $20\,\mathrm{m\,s^{-1}}$ S 36·9° E
17. $(\mathbf{i}+3\mathbf{j})\,\mathrm{km\,h^{-1}}$
18. $\begin{pmatrix}300\\240\end{pmatrix}\mathrm{km\,h^{-1}}$
19. $(-\mathbf{i}-2\mathbf{j}-9\mathbf{k})\,\mathrm{m\,s^{-1}}$
20. $(200\mathbf{i}+60\mathbf{j}-10\mathbf{k})\,\mathrm{m\,s^{-1}}$
21. $6\,\mathrm{m\,s^{-1}}$ from S 30° W
22. $520\,\mathrm{km\,h^{-1}}$ west
23. $7{\cdot}81\,\mathrm{km\,h^{-1}}$ N 26·3° E
24. $5{\cdot}32\,\mathrm{km\,h^{-1}}$ from N 21·5° E
25. $15{\cdot}4\,\mathrm{km\,h^{-1}}$ 238·7°
26. $11{\cdot}4\,\mathrm{m\,s^{-1}}$ 262·4°
27. $196\,\mathrm{km\,h^{-1}}$ N 36·5° E
28. $9{\cdot}64\,\mathrm{km\,h^{-1}}$ from 024·8°
29. $7{\cdot}57\,\mathrm{km\,h^{-1}}$ from N 72·5° W
30. $(250\mathbf{i}-100\mathbf{j})\,\mathrm{m\,s^{-1}}$, $(300\mathbf{i}+70\mathbf{j})\,\mathrm{m\,s^{-1}}$
31. $(5\mathbf{i}+7\mathbf{j})\,\mathrm{km\,h^{-1}}$
32. $12{\cdot}2\,\mathrm{km\,h^{-1}}$ from S 34·7° E
33. $10{\cdot}6\,\mathrm{m\,s^{-1}}$ from S 69·9° W
34. $26{\cdot}8\,\mathrm{km\,h^{-1}}$ from N 33·4° W

Exercise 10C page 236

1. (a) $\mathbf{i}+3\mathbf{j}$ (b) $2\mathbf{i}+\mathbf{j}$ (c) $3\mathbf{i}+4\mathbf{j}$ (d) $-\mathbf{i}+2\mathbf{j}$ (e) $2\mathbf{i}+\mathbf{j}$ (f) $-\mathbf{i}+2\mathbf{j}$
   (g) $2\mathbf{i}+\mathbf{j}$ (h) $-2\mathbf{i}-\mathbf{j}$ (i) $\mathbf{i}+4\mathbf{j}$ (j) $-\mathbf{i}-4\mathbf{j}$ (k) $4\mathbf{i}+\mathbf{j}$ (l) $3\mathbf{i}-3\mathbf{j}$
2. (a) $-4\mathbf{i}+4\mathbf{j}-7\mathbf{k}$ (b) $4\mathbf{i}-4\mathbf{j}+7\mathbf{k}$ (c) $2\mathbf{i}-9\mathbf{k}$ (d) $6\mathbf{i}-4\mathbf{j}-2\mathbf{k}$ (e) $-6\mathbf{i}+4\mathbf{j}+2\mathbf{k}$
3. $(4\mathbf{i}+3\mathbf{j})\,\mathrm{m\,s^{-1}}$, 50 s
4. $(5\mathbf{i}-12\mathbf{j})\,\mathrm{m\,s^{-1}}$, 3 s
5. 12.30 p.m. $(4\mathbf{i}+8\mathbf{j})$ km
6. 12.20 p.m., $\begin{pmatrix}10\\5\frac{1}{3}\end{pmatrix}$ km
7. 1.15 p.m., $(22\mathbf{i}+5\mathbf{j})$ km
8. 12.06 p.m., $(80\mathbf{i}+460\mathbf{j})$ km
9. (a) 2.20 p.m., $(8\mathbf{i}+7\mathbf{j})$ km (b) $(16\mathbf{i}+\mathbf{j})$ km, 10 km (c) 3.00 p.m.
10. 1.10 p.m., $(21\mathbf{i}+27\mathbf{j})$ km, 1.30 p.m.
11. A and C at 12.45 p.m., $(11\mathbf{i}-8\mathbf{j})$ km, 1.09 p.m.
12. B and C at 8.48 a.m., $(18\mathbf{i}-6\mathbf{j})$ km, $(15\mathbf{i}-20\mathbf{j})\,\mathrm{km\,h^{-1}}$, 9.00 a.m.
13. $\begin{pmatrix}2400\\-260\\1\end{pmatrix}$ km, 0906 hours
14. (a) After 20 seconds, $(4700\mathbf{i}+4000\mathbf{j}+5500\mathbf{k})$ m
    (b) (i) $(1000\mathbf{i}+50\mathbf{j}+650\mathbf{k})\,\mathrm{m\,s^{-1}}$ (ii) $(600\mathbf{i}-100\mathbf{j}+400\mathbf{k})\,\mathrm{m\,s^{-1}}$

Exercise 10D page 239

1. 12.35 p.m.
2. 11.47 p.m.
3. N 41·4° E, 1.45 a.m.
4. S 23·6° W, $26\frac{1}{2}$ min past 9 p.m.
5. 126·8°, 12.15 p.m.
6. 345·0°, 92 s
7. N 15·5° W, 9 min 7 s
8. N 24·7° E, 2·4 s

Exercise 10E page 242

1. (a) 25 km (b) 15 km (c) 1.20 p.m., $(-5\mathbf{i}+22\mathbf{j})$ km, $(4\mathbf{i}+10\mathbf{j})$ km
2. 1·81 km, 9.47 a.m.
3. 3·08 km, 8.04 p.m.
4. 6·32 km, 1340 hours
5. (a) $(9\mathbf{i}+18\mathbf{j})$ km (b) 13·4 km (c) 1548 hours
6. 2·81 km, 12.57 p.m.
7. (a) $(5\mathbf{i}+4\mathbf{j})$ km (b) 5·83 km (c) 1·70 km (d) 1.21 p.m.
8. 4·47 km, 10 min
9. 1·41 km, 12 min
10. (a) $\begin{pmatrix}3+t\\-16+t\\-2+6t\end{pmatrix}$ m (b) $t=\frac{25}{38}$ (c) 15·9 m
11. 3·22 m
12. $\frac{\lambda}{73}$
13. (a) $\lambda=-1,\ t=\frac{3}{5}$ (b) 1·80 m
14. $a=30$, $b=7$, $c=15$.
15. $\sqrt{374}$ cm, 7
16. (a) Pigeon: $\sqrt{1686}$ m when $t=0$, Bird of prey: $20\sqrt{6}$ m when $t=4$ (b) 7 (c) 2·9 m

Exercise 10F page 245

1. 60 km, 58 min
2. 3 km, $13\frac{1}{3}$ s
3. 800 m, 5 min 18 s
4. 3·6 km, 7.42 a.m.
5. 330 m, 112 s, 198 m north, 264 m west
6. 571 m, 5·8 s
7. 6·58 km, 1518 hours, 22 min
8. 5·46 km, 12 min 27 s

Exercise 10G page 247

1. N 53·1° E, 360 m, 80 s **2.** 2 km, 8·57 and 36 s **3.** 108·2°, 5·4 min
4. N 16·4° E, 6·89 km, 45 min **5.** N 30° W, 12.24 p.m., 21 min, 14 min **6.** 8·27 m, 2·61 s

Exercise 10H page 247

1. 125 s, 100 s, 300 m **2.** $(10\cos\theta - 7)\mathbf{i} + 10\sin\theta\,\mathbf{j}, \dfrac{7}{10}$ **3.** (i) 059·2° (ii) 1 hr 19 min, 109 km h^{-1}

4. (i)

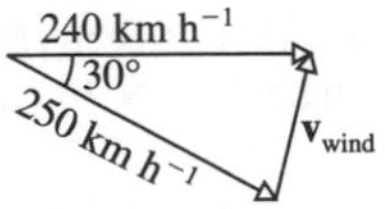

(ii) 127 km h^{-1} towards 010·6° (iii) 240°
(iv) Yes. Wind has a positive component in direction $\overrightarrow{AB}$ which will slow the motion of the plane travelling from B to A.

5. (i) $_B\mathbf{v}_A = \mathbf{v}_B - \mathbf{v}_A$ (iii) N 36·9° E (iii) 2 km due south of A (iv) 500 m
6. $(610 - 25t)$ m, $(610 - 30t)$ m, 22 **7.** $(442 - 30t)$ m, $(442 - 20t)$ m, 17
8. $\mathbf{r}_A = -20t\,\mathbf{i}$, $\mathbf{r}_B = -12{\cdot}8t\,\mathbf{i} + (9{\cdot}6t - 1{\cdot}2)\mathbf{j}$, 12·04 and 48 seconds
9. (a) $(5\mathbf{i} + 15\mathbf{j})$ m s^{-1} (b) $[(5t - 2)\mathbf{i} + (15t - 6)\mathbf{j}]$ m (c) do collide (d) $-\dfrac{4}{\sqrt{377}}$
10. (a) $(\mathbf{i} - 3\mathbf{j})$ m s^{-1} **11.** $-5\mathbf{i}$ m s^{-1}, $12\mathbf{j}$ m s^{-1}, $(5\mathbf{i} + 12\mathbf{j})$ m s^{-1}, 13 m s^{-1}, 022·6°, 820 m, 62·1 s
12. (i) $(7\mathbf{i} + 24\mathbf{j})$ km h^{-1} (ii) (a) 12.35 (b) 15 km (iv) 286°
13. $[(1 + 2t)\mathbf{i} + (4t - 1)\mathbf{j} + (2 - 4t)\mathbf{k}]$ m s^{-1}, $[(1 + 2t)\mathbf{i} + (5 + 3t)\mathbf{j} + (4 - 6t)\mathbf{k}]$ m s^{-1}, 2 s, $(3\mathbf{i} + 5\mathbf{j} - 5\mathbf{k})$ m s^{-1}.

Exercise 11A page 252

1. 98 J **2.** 29·4 J **3.** 19 600 J **4.** 5880 J **5.** 98 J
6. 3·92 J **7.** 800 J **8.** 56 J **9.** 98 J **10.** 168 J
11. 0·125 **12.** 735 J **13.** (a) 1470 J (b) 250 J
14. (a) 336 N (b) 16 800 J (c) 24 500 J **15.** (a) 1700 J (b) 9800 J
16. 6370 J

Exercise 11B page 255

1. (a) 40 J (b) 9 J (c) 40 000 J (d) 20 J (e) 800 J
2. (a) 490 J (b) 2940 J (c) 196 000 J
3. (a) 392 J (b) 7840 J (c) 98 000 J **4.** (a) 5500 J (b) 125 J
5. (a) 3 J (b) 10 000 J **6.** 4 m s^{-1} **7.** 5 m s^{-1} **8.** 45 000 J **9.** 147 J

Exercise 11C page 257

1. 14 m s^{-1} **2.** 42 m s^{-1} **3.** 2·5 m **4.** 22·5 m
5. 10·7 m s^{-1} **6.** 10·7 m s^{-1}. The mass does not affect the answer. **7.** 2·5 m
8. (a) 4·65 m s^{-1} (b) 4·65 m s^{-1} (c) 10 m s^{-1} (d) 5·10 m
9. $4\frac{1}{6}$ m **10.** 26 m **11.** $\sqrt{2gh}$

Exercise 11D page 260

1. 4 N **2.** 13 N **3.** 3 m s^{-1} **4.** 6 m s^{-1} **5.** 1 N **6.** 6 m s^{-1}
7. (a) 70 J (b) $\frac{2}{7}$ **8.** 12 m s^{-1} **9.** 17 N **10.** 16·25 m **11.** 1200 J, 4800 N
12. 20 cm **13.** 45 J, 9 N **14.** (a) 1340 J (b) 134 N (c) 0·342 **15.** 12·5 m
16. (a) 8·4 J (b) 10·58 m s^{-1} (c) 16·8 J (d) 6·69 m s^{-1}
17. (d) Body will not return to A if $\mu \geqslant \tan\theta$. In such circumstances the frictional force will be sufficient to prevent the body sliding back down the slope when it comes to rest at B.

Exercise 11E page 265

1. 33 J **2.** −33 J **3.** $6\sqrt{2}$ J **4.** $-6\sqrt{2}$ J **5.** $15\sqrt{3}$ J **6.** 45 J
7. zero **8.** $\frac{15}{2}\sqrt{3}$ J **9.** $-\frac{15}{2}\sqrt{3}$ J **10.** 10 J **11.** 4 J **12.** 10 J
13. −3 J **14.** −6 J **15.** 12 J **16.** zero, motion is perpendicular to **F**.
17. 6 J **18.** zero **19.** 11 J **20.** 6 J **21.** −30 J **22.** −6 J
23. 9 J

24. (a) $(2\mathbf{i}+3\mathbf{j}-\mathbf{k})\,\mathrm{m\,s^{-2}}$ (b) $(4\mathbf{i}+6\mathbf{j}-2\mathbf{k})\,\mathrm{m\,s^{-1}}$ (c) 140 J
(d) $(4\mathbf{i}+6\mathbf{j}-2\mathbf{k})\,\mathrm{m}$ (e) $(7\mathbf{i}+13\mathbf{j})\,\mathrm{m}$ (f) 140 J

Exercise 11F page 268

1. (a) 160 J, 40 W (b) 60 J, 15 W (c) 6 J, 1·5 W (d) $200\sqrt{2}$ J, $50\sqrt{2}$ W
2. 700 W **3.** 14·7 kW **4.** 40 W **5.** 3 W **6.** 196 W **7.** 5·2 W **8.** 1·7 W
9. 49 W **10.** 1·05 kW **11.** 5·17 kW **12.** 6·4 kW **13.** (a) 80 kg (b) 10·4 kW **14.** 446 W
15. (a) $8\,\mathrm{m\,s^{-1}}$ (b) 3·24 kW, 4·32 kW **16.** (a) $25\,\mathrm{cm^2}$ (b) 2 m **17.** $10\,\mathrm{m\,s^{-1}}$, $5\,\mathrm{cm^2}$

Exercise 11G page 271

1. (a) $10\,\mathrm{m\,s^{-1}}$ (b) $15\,\mathrm{m\,s^{-1}}$ (c) $22\,\mathrm{m\,s^{-1}}$
2. (a) $1·4\,\mathrm{m\,s^{-2}}$ (b) $0·6\,\mathrm{m\,s^{-2}}$ (c) $40\,\mathrm{m\,s^{-1}}$
3. (a) $3\,\mathrm{m\,s^{-2}}$ (b) $1\,\mathrm{m\,s^{-2}}$ (c) $40\,\mathrm{m\,s^{-1}}$
4. 400 N **5.** 400 W **6.** 200 W **7.** 2000 N, $0·02\,\mathrm{m\,s^{-2}}$ **8.** $10\,\mathrm{m\,s^{-1}}$, $3\,\mathrm{m\,s^{-1}}$
9. (a) $1·25\,\mathrm{m\,s^{-2}}$ (b) $30\,\mathrm{m\,s^{-1}}$ **10.** 5000 N **11.** 16·8 kW, $20\,\mathrm{m\,s^{-1}}$ **12.** 49 N, $4\,\mathrm{m\,s^{-1}}$
13. (a) 800 N (b) 600 N (c) 10 kW **14.** (a) 350 N (b) $35\,\mathrm{m\,s^{-1}}$ **15.** 420 N, $17·5\,\mathrm{m\,s^{-1}}$

Exercise 11H page 272

1. (b) 3·2 m (c) 1·2 J **2.** $5·99\,\mathrm{m\,s^{-1}}$ **3.** (i) 102·9 J (ii) 9·86 N (iii) 0·332 (iv) $5·39\,\mathrm{m\,s^{-1}}$
4. (i) 8·4 J (ii) 7·056 J (iii) 1·344 J, 0·713 N, $2·83\,\mathrm{m\,s^{-1}}$
5. (a) $\frac{5d}{4}$ (b) $4mg$ (c) $4\sqrt{5}mg$, $\tan^{-1}\left(\frac{1}{2}\right)$ to vertical (d) $\frac{15d}{8}$
6. (a) $\frac{2g}{3}$, $\frac{2mg}{3}$ (b) $\frac{4g}{9}$, $\frac{7mg}{9}$, $\frac{4mgd}{3}$ **7.** 40 500 J, 7840 J, 48 340 W
8. 450 N, $0·5625\,\mathrm{m\,s^{-2}}$ **9.** (a) $(7·5\times10^4)$ N, $(7·5\times10^5)$ W (b) $20\,\mathrm{m\,s^{-1}}$
10. (i) 200 N, 450 N (ii) $0·75\,\mathrm{m\,s^{-2}}$ (iii) 28 kW **11.** (a) 960 N (b) 0·75 (c) 360 N
12. 900 N **13.** 900 N **14.** $\frac{P}{20}$ N, 14 000, $0·6\,\mathrm{m\,s^{-2}}$ **15.** 300 N (i) $0·2\,\mathrm{m\,s^{-2}}$ (ii) 255 N
16. $S = 72\,000$, $F = 1800$ **17.** $a = 125$, $b = 0·75$, $0·42\,\mathrm{m\,s^{-2}}$, 4·9°

Exercise 12A page 281

1. 4 s, 80 m **2.** 60 m, 40 m below, 72·1 m **3.** $14\,\mathrm{m\,s^{-1}}$ **4.** (a) $(5\mathbf{i}+15·1\mathbf{j})$ m (b) $(10\mathbf{i}+0·4\mathbf{j})$ m
5. $(11\mathbf{i}+5·4\mathbf{j})$ m **6.** 60 m **7.** $28\,\mathrm{m\,s^{-1}}$ **8.** 2 m **9.** 20 cm **10.** 1·1 m
11. $20\,\mathrm{m\,s^{-1}}$, $19·6\,\mathrm{m\,s^{-1}}$ **12.** $175\,\mathrm{m\,s^{-1}}$, 16·3° below horizontal **13.** $1\frac{3}{7}$, 21
14. (a) 35 m (b) 35 m **15.** 2 m **17.** 40 m **18.** 60 m **20.** 630 m **21.** 1·44
22. 2·4 m, 4 m

Exercise 12B page 288

1. (a) $2\frac{6}{7}$ s (b) 40 m (c) $5\frac{5}{7}$ s (d) 277 m
2. (a) 3 s (b) 45 m (c) 6 s (d) 312 m
3. 120 m **4.** 180 m, 6 s **5.** 5·4 s **6.** 2·25 km **7.** Free kick **8.** 1·6 m
9. $(8\mathbf{i}+6\mathbf{j})$ m, 10 m **10.** $(30\mathbf{i}+15\mathbf{j})$ m, $(10\mathbf{i}-10\mathbf{j})\,\mathrm{m\,s^{-1}}$
11. $21·7\,\mathrm{m\,s^{-1}}$, $12\,\mathrm{m\,s^{-1}}$ down, $24·8\,\mathrm{m\,s^{-1}}$ 29° below horizontal **12.** 60 m, 40·4 m, 72·3 m
13. 2 s, 2 **14.** 75 m, 3 s **15.** 2 s, 2 m **16.** 5 s, 24·5 m **17.** $50\,\mathrm{m\,s^{-1}}$ elevation 36·9°
18. 30° **19.** $(15\mathbf{i}+29\mathbf{j})\,\mathrm{m\,s^{-1}}$ **20.** $5\,\mathrm{m\,s^{-1}}$, $17\,\mathrm{m\,s^{-1}}$
21. $15·6\,\mathrm{m\,s^{-1}}$ at 45° above horizontal **22.** 3 s **23.** 4 s **24.** 3 s, 21
25. 5 s, 60 m **26.** 90 m **27.** 20 m **28.** 15°, 75° **29.** 14·7°, 75·3° **30.** 11·8°, 78·2°
31. $14·8\,\mathrm{m\,s^{-1}}$, 58° above horizontal **32.** 42 m, 36·4 m, 7, 53·1° **33.** 3 s, 60 m
34. 15·7°, 24·7°, 65·3°, 74·3° **35.** 20·4° → 24·5° or 65·5° → 69·6° **36.** $31·3\,\mathrm{m\,s^{-1}}$ **37.** 30·9° or 7·8°
38. $41·7\,\mathrm{m\,s^{-1}}$

Exercise 12C page 294

1. (d) 45° **2.** 250 m **3.** 171 m, 500 m, 45° **4.** 10 s, 5 s
5. 140 m s^{-1} **6.** 5·8° or 84·2° **7.** 63·4° **9.** 25 m s^{-1}
10. 45°, 71·6° **11.** 63·4°, 76·0° **13.** No, 83·5° or 54·9°
14. (a) 36·8 m s^{-1} to 38·5 m s^{-1} (b) 34·8 m s^{-1} to 36·8 m s^{-1} (c) No

Exercise 12D page 298

1. (a) $1\frac{3}{7}$ s (b) 10 m **2.** (a) 2·42 s (b) 53·8 m
3. (a) 3 s (b) 44·1 m **4.** (a) $1\frac{3}{7}$ s (b) 19·3 m
5. 47·3 m s^{-1}, 173 m **6.** 12 m s^{-1}, 26·8 m **7.** 44·4° **8.** 30°, 90° **9.** (c) $2\theta + \phi = 90°$

Exercise 12E page 299

1. 1·6 m **2.** 420 m **3.** (a) $\frac{5}{7}$ s (b) 3·5 m **4.** (i) 2 s (ii) 52·4 m (iii) 50·7 m s^{-1}
5. 189 m s^{-1}. Speed would have to be greater to achieve same range against resistance.
6. (a) $4\frac{3}{8}$ m (b) 48·3 m s^{-1} (c) 0·204 s, 2·65 s **7.** (a) 3·6 m (b) 8·2
8. (iii) horizontally (iv) 2·68 m from wall **9.** (ii) (a) 22·4° or 67·6° (b) 255 m (c) 37·1 m
10. (b) 7·23 m s^{-1} (c) 35·4°, 68·6°. Preferred angle would be 68·6°.
11. (a) R_{max} occurs when sin 2α is maximised, i.e. when $\alpha = 45°$.
(b) 6·26 m s^{-1} (c) (i) 70° (ii) 1·43 s
12. (a) 14·1 m s^{-1} (b) (iii) 13·6 m s^{-1} (iv) No
13. (ii) $\dfrac{24V^2}{g}$, $V^2 \tan^2 \alpha - 24 W^2 \tan \alpha + V^2 - 32 W^2 = 0$, $W = \dfrac{V}{2\sqrt{7}}$ **14.** $\dfrac{3\sqrt{7}\,W^2}{2g}$

Exercise 13A page 307

1. 200π rad min^{-1} **2.** 3π rad s^{-1} **3.** $\dfrac{300}{\pi}$ rev min^{-1}
4. (a) 6 rev min^{-1} (b) 12π rad min^{-1} (c) $\dfrac{\pi}{5}$ rad s^{-1}
5. (a) $\dfrac{3\pi}{2}$ rad s^{-1} (b) $\dfrac{10\pi}{9}$ rad s^{-1} **6.** (a) 10 m s^{-1} (b) 50π m s^{-1}
7. (a) 5 m s^{-1} (b) 20π m s^{-1} **8.** (a) 24 rad s^{-1} (b) $\dfrac{720}{\pi}$ rev min^{-1}
9. (a) 1 rad s^{-1} (b) $\dfrac{30}{\pi}$ rev min^{-1} **10.** $\dfrac{2\pi}{5}$ m s^{-1}, $\dfrac{3\pi}{5}$ m s^{-1}
11. (a) 2 rad s^{-1} (b) 10π s (c) 3·5 m s^{-1}

Exercise 13B page 313

1. 6 m s^{-2} **2.** 18 m s^{-2} **3.** 1·11 m s^{-2}, 2·22 m s^{-2}, 4·44 m s^{-2}
4. 5 N **5.** 8 N **6.** 592 N **7.** 0·152 N
8. (a) 5 N (b) 20 N (c) 5·48 N **9.** (a) 1 N (b) 25 N (c) 100 N
10. 6 rad s^{-1} **11.** 4 m s^{-1}, 7 m s^{-1}
12. (a) 0·98 N down (b) 0·98 N up (c) 4·5 N away from O (d) 4·5 N towards O
13. $\mu \geqslant 0{\cdot}4$ **14.** 0·417 **15.** 12·6 m s^{-1} **16.** 12·5 cm, 12·5 cm
17. 0·151 **18.** 360 m **19.** 30 000 N **20.** 0·2
21. 0·9 **22.** $2\frac{1}{3}$ rad s^{-1} **23.** 50 cm **24.** 1·75 m s^{-1}

Exercise 13C page 320

1. (a) 122·5 N, 3·5 rad s^{-1} (b) 196 N, 7 rad s^{-1} (c) 245 N, 66·4°
2. 21·8° **3.** 17·5 m s^{-1} **4.** 14·0° **5.** 16·4 m s^{-1} **6.** 11·8° **7.** 25 N **8.** 1·25 m
9. (a) 9800 N (b) 37·5 mm **10.** (a) 4900 N (b) 15 mm
11. 22 m s^{-1} **12.** 0·5 **13.** 0·4 **14.** 33·7° **18.** 0·364, 22·2 m s^{-1} **19.** 42 m s^{-1}, 14 m s^{-1}
20. (a) 4 N (b) 2·9 N **21.** 16·9, 4·9 N **22.** 14 rad s^{-1} **23.** 1·3 N, 6·37 N

Exercise 13D *page 331*

1. $2{\cdot}5\,m\,s^{-1}$ **2.** $3{\cdot}9\,m\,s^{-1}$ **3.** $1{\cdot}8\,m\,s^{-1}$ **4.** $33°$ **5.** $69°$ **6.** $85°$
7. $4{\cdot}9\,m\,s^{-1}$, 252 N **8.** $4{\cdot}8\,m\,s^{-1}$, 240 N **9.** $4{\cdot}6\,m\,s^{-1}$, 211 N
10. $0{\cdot}81\,m\,s^{-1}$, 42·5 N, thrust **11.** $4{\cdot}05\,m\,s^{-1}$, 115 N, tension **12.** $6{\cdot}66\,m\,s^{-1}$, 395 N, tension
13. (a) $10{\cdot}2\,m\,s^{-1}$ (b) 160·5 N (c) $9{\cdot}8\,m\,s^{-2}$ **14.** $\sqrt{\frac{11g}{2}}\,m\,s^{-1}$, $\sqrt{\frac{7g}{2}}\,m\,s^{-1}$
15. (a) $4\sqrt{\frac{gr}{3}}$ (b) $\frac{mg}{3}$ **16.** When PO makes 141·7° with downward vertical drawn from O, $3{\cdot}4\,m\,s^{-1}$
17. 132° **18.** $\sqrt{u^2 - 2gr + 2gr\cos\theta}$, $3mg\cos\theta + \frac{mu^2}{r} - 2mg$
19. (a) $9{\cdot}8\,m\,s^{-1}$, $64{\cdot}2\,m\,s^{-2}$ (b) $9{\cdot}56\,m\,s^{-1}$, $61{\cdot}3\,m\,s^{-2}$ (c) $7{\cdot}48\,m\,s^{-1}$, $38{\cdot}2\,m\,s^{-2}$
20. (a) $1{\cdot}146\,rad\,s^{-1}$, 17·58 N (b) $2{\cdot}214\,rad\,s^{-1}$, 14·7 N (c) $3{\cdot}13\,rad\,s^{-1}$, 58·8 N
21. $|200 - 4g + 6g\cos\theta|$ N (b) $\theta = 48{\cdot}2°$ **22.** g N **23.** 3·71 m, 3·87 m
24. (a) $\frac{r}{5}$ (b) $\sqrt{\frac{2gr}{5}}$ (c) $2\sqrt{\frac{3gr}{5}}$ **26.** $\sqrt{\frac{g\sqrt{3}}{3a}}\,rad\,s^{-1}$

Exercise 13E *page 335*

1. 0·51 **3.** (ii) $15{\cdot}7\,m\,s^{-1}$ **4.** $\mu \geqslant 0{\cdot}33$
5. (a)

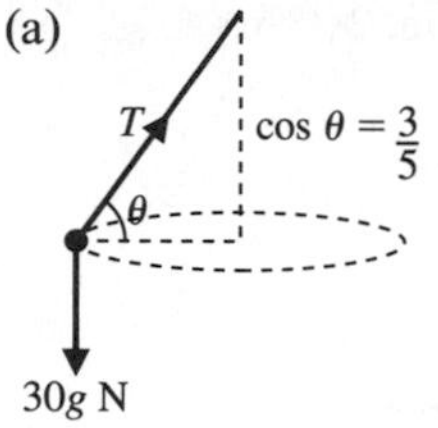

(b) 367·5 N (d) Light and inextensible (e) $\frac{4\pi}{\sqrt{g}}$ s (i.e. again about 4 seconds)
6. (a) $3{\cdot}83\,m\,s^{-1}$ (b) (i) initially horizontal, at a tangent to the circle (ii) 1·22 m
7. (a) $md\omega^2$ (b) $m(g - h\omega^2)$, $\frac{2h}{3}$ **8.** (a) $3Mg$ (b) $\sqrt{\frac{8gr}{3}}$
9. (a) $\sqrt{ag(2\cos\theta - 1)}$ (b) $mg(3\cos\theta - 1)$ **10.** (i) $2{\cdot}8\,m\,s^{-1}$ (ii) $\frac{g}{3}$ N
12. (i) $7\,m\,s^{-1}$ (ii) $6{\cdot}5\,m\,s^{-1}$, 2·17 m to nearest cm. The "safe height" is greater if resistances are taken into account. The extra PE lost in descent will go as work against resistance. Thus the speed as BCD is reached can be as before.
14. $mg(3\cos\theta + 2)$
15. $\frac{3mg(4\cos\theta - 3)}{4}$, $\frac{3\sqrt{3ga}}{8}$ **16.** $\frac{mg(3\sin\theta + 3\cos\theta - 2\sqrt{2})}{2}$, $\frac{3mg(\sin\theta + \cos\theta)}{4}$

Exercise 14A *page 344*

1. (a) 14 000 N s (b) 3 N s (c) 300 N s (d) 3 600 000 N s (e) 6 N s
2. (a) 15 N s (b) 50 N s (c) 55 N s (d) 70 N s
3. (a) 18 N s (b) 18 N s (c) $9\,m\,s^{-1}$ **4.** 35 N s
5. (a) 60**i** N s (b) 7**i** $m\,s^{-1}$ **6.** (a) −20**i** N s (b) 4**i** $m\,s^{-1}$
7. (a) −20**i** N s (b) −2**i** $m\,s^{-1}$ **8.** (a) 8**i** N s (b) 48**i** N s (c) 8
9. −8 **10.** 3 **11.** 4 **12.** (2·5**i** − 6**j**) $m\,s^{-1}$, $6{\cdot}5\,m\,s^{-1}$
13. (4**i** − 3**j**) $m\,s^{-1}$, $5\,m\,s^{-1}$ **14.** 1, −2 **15.** (5**i** − 2**j**) $m\,s^{-1}$
16. (−**i** + 2**j**) $m\,s^{-1}$ **17.** (6**i** − 6**j**) N s **18.** (−6**i** + 9**j**) N s
19. (a) (i) −12**i** N s (ii) 12**i** N s (b) (i) −4**i** N s (ii) 4**i** N s (c) (i) −12**i** N s (ii) 12**i** N s

Exercise 14B *page 346*

1. 50 N **2.** 6400 N **3.** 3840 N **4.** 0·064 N **5.** 75 N **6.** 0·0412 N

Exercise 14C page 351

1. 2·5 m s^{-1} **2.** 5 m s^{-1} **3.** 5 m s^{-1} **4.** 6 m s^{-1} **5.** 3 m s^{-1}
6. 1 m s^{-1} **7.** 4 m s^{-1} **8.** 4 m s^{-1} **9.** $(2\mathbf{i}+2\mathbf{j})$ m s^{-1} **10.** $3\mathbf{i}$ m s^{-1}
11. $(3\mathbf{i}+2\mathbf{j})$ m s^{-1} **12.** $7\mathbf{i}$ m s^{-1} **13.** $(5\mathbf{i}-2{\cdot}5\mathbf{j})$ m s^{-1} **14.** T is stationary after the collision
15. (a) 4 m s^{-1} (b) 3·6 J (c) 2·4 J (d) 1·2 J
16. 10, 3 J **17.** $3\mathbf{i}$ m s^{-1} **18.** 3 m s^{-1}, 0·9 J **19.** $(6\mathbf{i}-3\mathbf{j})$ m s^{-1}, 60 J **20.** $(5\mathbf{i}+5\mathbf{j})$ m s^{-1}, 30 J
21. (a) $(\mathbf{i}-2\mathbf{j})$ m s^{-1} (b) 60 J (c) $(-12\mathbf{i}+12\mathbf{j})$ N s
22. (a) $(4\mathbf{i}-8\mathbf{j})$ m s^{-1} (b) 140 J (c) $(-10\mathbf{i}+10\mathbf{j})$ N s
23. (a) $(4\mathbf{i}+6\mathbf{j})$ m s^{-1} (b) 5·25 J (c) $(-0{\cdot}75\mathbf{i}-1{\cdot}5\mathbf{j})$ N s
24. 4 m s^{-1}, 2520 J **25.** 1 m s^{-1}, 401 000 J **26.** 6 m s^{-1}, 20 cm
27. (a) 5 m s^{-1}, 5 m s^{-1} (b) 2 m s^{-1}, 3 m s^{-1} **28.** 1 m s^{-1}, 9 J **30.** $m_3 > \dfrac{2m_2}{5}$

Exercise 14D page 355

1. 20 m s^{-1} **2.** 3 m s^{-1}, 60° **3.** $(0{\cdot}6\mathbf{i}-0{\cdot}8\mathbf{j})$ N s, 0·1 J **4.** $(0{\cdot}6\mathbf{i}-2\mathbf{j})$ N s, 7·3 J
5. (a) $(\mathbf{i}+4\mathbf{j})$ m s^{-1} (b) 15 J (c) $-6\mathbf{i}$ N s **6.** (i) 0·4 m s^{-1} (ii) 0·04 N s
7. Assuming linear momentum is conserved, i.e. if any external forces can be ignored, 251 kJ
8. (b) 0·05 m s^{-1}, 6×10^5 N **9.** (a) $u(1+k)$ (b) $\dfrac{1}{6}$ or $\dfrac{1}{2}$
10. (i) 4 m s^{-1} (ii) 1·8 m s^{-1} (iv) 0·9 m **11.** 2 m s^{-2}, 2·4 N, $\dfrac{3\sqrt{2}}{10}$ N s, 0·6 m, $2\sqrt{3}$ m s^{-1}
12. $\dfrac{m\sqrt{6ag}}{m+M}$, $\frac{1}{2}$, $3\sqrt{\dfrac{6a}{g}}$
13. $\dfrac{g}{5}$, $8mg$ (a) $\dfrac{d}{3}$ (b) $\dfrac{2}{3}\sqrt{\dfrac{10d}{g}}$ (c) $\sqrt{\dfrac{gd}{10}}$
14. 30 (i) 25 m (ii) 22·5 m s^{-1}
(iii) 5 m s^{-1} (iv) 22·9 m s^{-1}

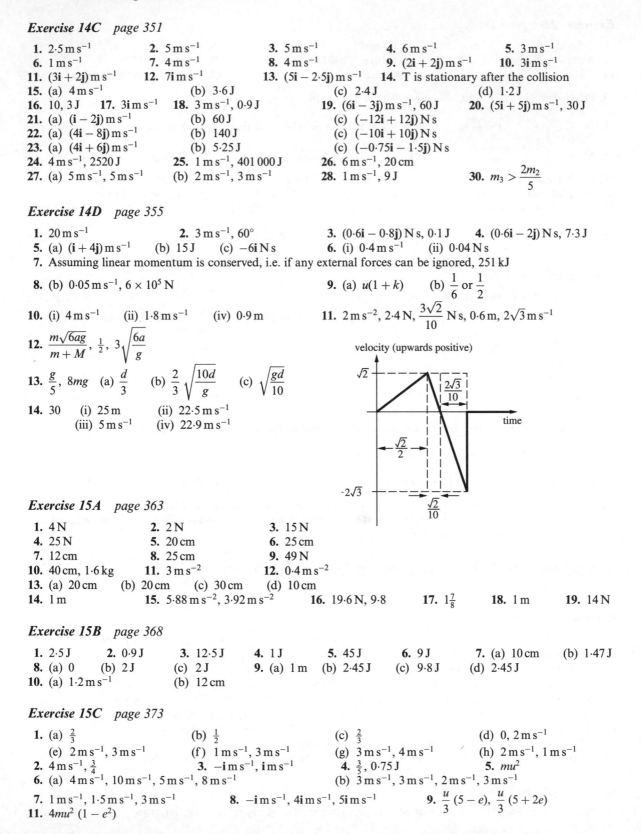

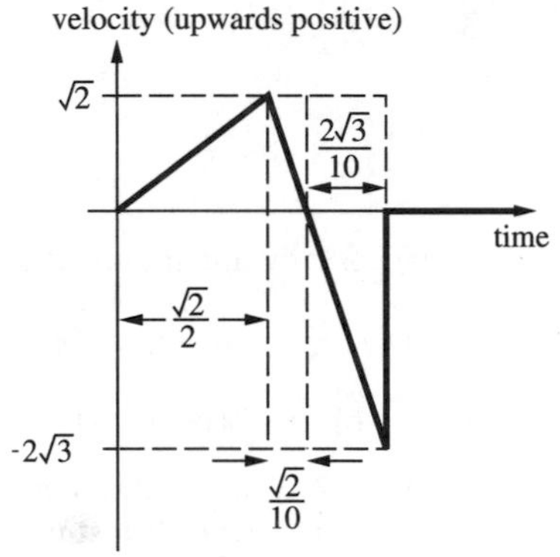

Exercise 15A page 363

1. 4 N **2.** 2 N **3.** 15 N
4. 25 N **5.** 20 cm **6.** 25 cm
7. 12 cm **8.** 25 cm **9.** 49 N
10. 40 cm, 1·6 kg **11.** 3 m s^{-2} **12.** 0·4 m s^{-2}
13. (a) 20 cm (b) 20 cm (c) 30 cm (d) 10 cm
14. 1 m **15.** 5·88 m s^{-2}, 3·92 m s^{-2} **16.** 19·6 N, 9·8 **17.** $1\frac{7}{8}$ **18.** 1 m **19.** 14 N

Exercise 15B page 368

1. 2·5 J **2.** 0·9 J **3.** 12·5 J **4.** 1 J **5.** 45 J **6.** 9 J **7.** (a) 10 cm (b) 1·47 J
8. (a) 0 (b) 2 J (c) 2 J **9.** (a) 1 m (b) 2·45 J (c) 9·8 J (d) 2·45 J
10. (a) 1·2 m s^{-1} (b) 12 cm

Exercise 15C page 373

1. (a) $\frac{2}{3}$ (b) $\frac{1}{2}$ (c) $\frac{2}{3}$ (d) 0, 2 m s^{-1}
(e) 2 m s^{-1}, 3 m s^{-1} (f) 1 m s^{-1}, 3 m s^{-1} (g) 3 m s^{-1}, 4 m s^{-1} (h) 2 m s^{-1}, 1 m s^{-1}
2. 4 m s^{-1}, $\frac{3}{4}$ **3.** $-\mathbf{i}$ m s^{-1}, $\mathbf{i}$ m s^{-1} **4.** $\frac{3}{5}$, 0·75 J **5.** mu^2
6. (a) 4 m s^{-1}, 10 m s^{-1}, 5 m s^{-1}, 8 m s^{-1} (b) 3 m s^{-1}, 3 m s^{-1}, 2 m s^{-1}, 3 m s^{-1}
7. 1 m s^{-1}, 1·5 m s^{-1}, 3 m s^{-1} **8.** $-\mathbf{i}$ m s^{-1}, $4\mathbf{i}$ m s^{-1}, $5\mathbf{i}$ m s^{-1} **9.** $\dfrac{u}{3}(5-e)$, $\dfrac{u}{3}(5+2e)$
11. $4mu^2(1-e^2)$

Exercise 15D page 382

1. (a) $\frac{1}{2}$ (b) $\frac{3}{4}$ (c) $6\,\text{m s}^{-1}$ (d) $1{\cdot}5\,\text{m s}^{-1}$ (e) $3\,\text{m s}^{-1}$ (f) $15\,\text{m s}^{-1}$
2. $-4{\cdot}5\mathbf{i}\,\text{m s}^{-1}$ **3.** $\frac{4}{5}$ **4.** $4{\cdot}5\,\text{s}$ **5.** $250\,\text{g}, 5\,\text{m s}^{-1}, 1\,\text{m s}^{-1}$
6. $3{\cdot}6\,\text{m s}^{-1}, 2\,\text{m s}^{-1}$ **9.** (b) $1{\cdot}5\,\text{s}$ (c) $4{\cdot}5\,\text{m}$ (d) $2{\cdot}5\,\text{m s}^{-1}, 0{\cdot}25\,\text{m s}^{-1}$
11. $0{\cdot}76, 0{\cdot}69$ **11.** $\dfrac{1+e^2}{1-e^2}\,\text{m}$ **12.** 22% loss of KE (to nearest 1%).

Exercise 15E page 385

1. (a) $3Mg$ (b) $48°$
2. R must be level with O; otherwise tension in string would have a vertical component with no other force to balance it (ring is *light*).
$\dfrac{2}{3}mg, \dfrac{5}{\sqrt{3}}, \dfrac{5mg}{3}, \dfrac{5mg}{\sqrt{3}}$
3. 78 N, 325 N **4.** $\dfrac{\sqrt{5ag}}{2}$ **5.** $57{\cdot}0°, 59{\cdot}0°$ **6.** $\dfrac{l}{3}$ **7.** $\sqrt{2ag}, \sqrt{6ga}$
8. $\sqrt{\dfrac{18ga}{5}}$ Without air resistance the weight would oscillate between $h=0$ and $h=\dfrac{2a}{3}$. In practice, air resistance will dampen oscillations and the weight will eventually come to rest with $h=\dfrac{a}{3}$.
9. 5 cm **10.** (b) $2{\cdot}3\,\text{m s}^{-1}$ **11.** 8 N s
12. (b) Now the collision does not reverse direction of ball's motion which therefore does not return to the boy.
13. $\dfrac{18u}{5}$ **14.** (i) $\dfrac{20u}{3}$ (ii) $\dfrac{2}{3}$ **15.** $\dfrac{1}{2}$ **16.** $\dfrac{(13-8e)u}{7}$ (a) (i) $\dfrac{2}{5}$ (ii) $\dfrac{5u}{7}$ (b) $\dfrac{47mu^2}{2}$ (c) $8mu$
17. $\dfrac{(5e-1)u}{6}$, $0{\cdot}2 < e \leqslant 1$ **18.** $\dfrac{(e+1)u}{5}$, $e > 0{\cdot}25$, $e' > \dfrac{e+1}{4e-1}$, $e = 0{\cdot}75$, $e' = 1$, $\dfrac{13mu^2}{40}$
19. (a) $\dfrac{u}{2}$ (b) (i) $\dfrac{u}{2}, \dfrac{2u}{5}$ (ii) Momentum lost in C's collision with barrier.
(c) A and B will collide (we can show that this will happen before C reaches the wall or BC becomes taut).
20. (a) velocity direction reversed, magnitude u (b) $0{\cdot}8$ (c) $3mu^2, u, \dfrac{4u}{3}, \dfrac{1}{9}$
21. $\dfrac{u}{4}, \dfrac{3u}{8}$ (a) $\dfrac{2\pi}{3}\,\text{s}$ (b) (i) $0{\cdot}06\,\text{N s}$ (ii) $0{\cdot}4\,\text{J}$ **22.** (i) $\sqrt{\dfrac{2h}{g}}$ (ii) $2e\sqrt{\dfrac{2h}{g}}$

Exercise 16A page 397

1. 40 m **2.** $28\,\text{m s}^{-1}$ **3.** $4\,\text{m s}^{-2}$ **4.** $24\,\text{m s}^{-1}$ **5.** $12\,\text{m s}^{-2}$
6. 21 m **7.** $30\,\text{m s}^{-2}$ **8.** $48\,\text{m s}^{-1}$ **9.** 34 m
10. (a) 1 m (b) $2t$ (c) $4\,\text{m s}^{-1}$ (d) 4 s (e) 13 m
11. (a) 45 m (b) 2 s, 5 s (c) $60\,\text{m s}^{-1}$ (d) $12t-42$ (e) $-42\,\text{m s}^{-2}$
12. (a) $4\,\text{m s}^{-1}$ (b) $8-6t$ (c) $-10\,\text{m s}^{-2}$ (d) $4t^2-t^3$ (e) 9 m
13. (a) $8\,\text{m s}^{-1}$ (b) $\frac{2}{3}$ s, 4 s (c) $4\,\text{m s}^{-2}$ (d) 2 m
14. $9\,\text{m s}^{-1}$, 9 m **15.** $(6\mathbf{i}+4\mathbf{j})\,\text{m s}^{-2}$ **16.** $15\,\text{m s}^{-1}$ **17.** $12\mathbf{i}\,\text{m s}^{-2}$ **18.** 20 m
19. $(24\mathbf{i}-9\mathbf{j})\,\text{m s}^{-1}$ **20.** $(30\mathbf{i}+5\mathbf{j})\,\text{m}$ **21.** $-2\mathbf{i}\,\text{m s}^{-2}$ **22.** $5\,\text{m s}^{-1}$ **23.** $(-2\mathbf{i}+\mathbf{j})\,\text{m s}^{-2}$
24. 5 m **25.** $(3\mathbf{i}+\mathbf{j})\,\text{m s}^{-1}$ **26.** $(5\mathbf{i}+3\mathbf{j})\,\text{m}$
27. (a) 1 s or 2 s (b) $\begin{pmatrix}2t\\2t-3\end{pmatrix}\text{m s}^{-1}$, $15\,\text{m s}^{-1}$ (c) $\begin{pmatrix}2\\2\end{pmatrix}\text{m s}^{-2}$, $\begin{pmatrix}12\\12\end{pmatrix}\text{N}$
28. (a) 5 s (b) $\begin{pmatrix}2t-4\\2t-4\end{pmatrix}\text{m s}^{-1}$, $0\,\text{m s}^{-1}$ (c) $\begin{pmatrix}2\\2\end{pmatrix}\text{m s}^{-2}$, $\begin{pmatrix}4\\4\end{pmatrix}\text{N}$
29. (a) 2 s or 5 s (b) $\begin{pmatrix}36\\35\end{pmatrix}\text{m s}^{-1}$ (c) $\begin{pmatrix}6t-2\\6t-4\end{pmatrix}\text{m s}^{-2}$, $\begin{pmatrix}8\\0\end{pmatrix}\text{N}$
30. (a) $\dfrac{\pi}{3}\,\text{s}$ (b) $\begin{pmatrix}4\\-1\end{pmatrix}\text{m s}^{-1}$ (c) $\begin{pmatrix}-8\sin 2t\\-\cos t\end{pmatrix}\text{N}$
31. (b) $\begin{pmatrix}3\\6\end{pmatrix}\text{m s}^{-1}$ (c) $\begin{pmatrix}-18\\0\end{pmatrix}\text{N}$ **32.** (a) $\begin{pmatrix}0\\-3\sqrt{3}\end{pmatrix}\text{m s}^{-1}$ (b) $\begin{pmatrix}-2\\4\end{pmatrix}\text{m s}^{-2}$

33. (a) $(12\mathbf{i}+12\mathbf{j}-5\mathbf{k})\,\mathrm{m\,s^{-1}}$ (b) $(28\mathbf{i}+29\mathbf{j}-12\mathbf{k})\,\mathrm{m}$
34. (a) $(-\mathbf{i}+3\mathbf{j}+5\mathbf{k})\,\mathrm{m\,s^{-1}}$ (b) $(-3\mathbf{i}+8\mathbf{j}+19\mathbf{k})\,\mathrm{m}$
35. (a) $(37\mathbf{i}+12\mathbf{j}-8\mathbf{k})\,\mathrm{m\,s^{-1}}$ (b) $(11\mathbf{i}+18\mathbf{j}-5\mathbf{k})\,\mathrm{m}$
36. (a) $(12\mathbf{i}-2\mathbf{j}+2\mathbf{k})\,\mathrm{m\,s^{-1}}$ (b) $(48\mathbf{i}-2\mathbf{j})\,\mathrm{m\,s^{-2}}$
37. (a) $(108\mathbf{i}+6\mathbf{j}-18\mathbf{k})\,\mathrm{m\,s^{-2}}$ (b) $(13\mathbf{i}+3\mathbf{j}-2\mathbf{k})\,\mathrm{m}$
38. (a) $(20\mathbf{i}-6\mathbf{k})\,\mathrm{m\,s^{-2}}$ (b) $(-15\mathbf{i}+5\mathbf{j}-28\mathbf{k})\,\mathrm{m}$

39. (a) $\begin{pmatrix}27\\-2\\108\end{pmatrix}\mathrm{m\,s^{-1}}$ (b) $\begin{pmatrix}12\\-2\\48\end{pmatrix}\mathrm{m\,s^{-2}}$ (c) 4 (d) $\begin{pmatrix}13\\0\\40\end{pmatrix}\mathrm{m\,s^{-1}}$

40. (a) $\begin{pmatrix}-9\\4\\56\end{pmatrix}\mathrm{m\,s^{-1}}$ (b) $\begin{pmatrix}-2\\0\\14\end{pmatrix}\mathrm{m\,s^{-2}}$ (c) 3 (d) $\begin{pmatrix}-4\\4\\16\end{pmatrix}\mathrm{m\,s^{-1}}$

41. (a) $\begin{pmatrix}-2\\1\\-1\end{pmatrix}\mathrm{m\,s^{-1}}$ (b) $\begin{pmatrix}0\\0\\1\end{pmatrix}\mathrm{m\,s^{-2}}$ (c) $\sqrt{2}$ (d) $\begin{pmatrix}0\\1\\-\frac{2}{\pi}\end{pmatrix}\mathrm{m\,s^{-1}}$

42. (a) $13\,\mathrm{m\,s^{-1}}$ (b) $\begin{pmatrix}\frac{11\pi}{2}\\1\\2\pi\end{pmatrix}\mathrm{m}$ (c) $\begin{pmatrix}0\\-9\\0\end{pmatrix}\mathrm{m\,s^{-2}}$

43. (a) $\begin{pmatrix}16\\0\\0\end{pmatrix}\mathrm{m\,s^{-2}}$ (b) $\begin{pmatrix}0\\8\\-1\end{pmatrix}\mathrm{m\,s^{-1}}$ (c) $\begin{pmatrix}4\\\pi-2\\\frac{1}{2}-\frac{1}{\sqrt{2}}\end{pmatrix}\mathrm{m}$

44. $29\,\mathrm{m\,s^{-1}}$, 24 m **45.** 3 s **46.** $2\frac{1}{4}$, 3 **47.** $7\frac{1}{2}$ s, $29\frac{1}{4}$ m **48.** 8 s, $10\frac{2}{3}$ m

Exercise 16B page 405

1. $2\,\mathrm{m\,s^{-1}}$ **2.** $10\,\mathrm{m\,s^{-2}}$ **3.** 3·5 s **4.** 2·5 m **5.** 5 m
6. 0·549 s **7.** $0{\cdot}2\,\mathrm{m\,s^{-2}}$ **8.** $2\,\mathrm{m\,s^{-1}}$ **9.** 2 m **10.** 4·16 m
11. (a) $v=s+2$ (b) $t=\ln\left(\dfrac{s+2}{2}\right)$ **12.** (a) $7\,\mathrm{m\,s^{-1}}$ (b) 0·973 s
13. $1\,\mathrm{m\,s^{-2}}$ **14.** 9 m **15.** 32 s **16.** 1·95 s **17.** 8·96 m
18. 1·13 m **19.** 8 s **22.** 1 **24.** $\frac{5}{2}(1-e^{-2t})$, $2{\cdot}5\,\mathrm{m\,s^{-1}}$

Exercise 16C page 410

1. $3{\cdot}2\,\mathrm{m\,s^{-1}}$ **2.** 16 m **3.** $4\,\mathrm{m\,s^{-1}}$ **4.** $2\,\mathrm{m\,s^{-1}}$ **5.** 3 m
6. 4 s **7.** 0·47 m **8.** 1·83 s **9.** $2\,\mathrm{m\,s^{-1}}$ **10.** 19 m
11. (a) $(3t^2\,\mathbf{i}+2t\,\mathbf{j})\,\mathrm{m\,s^{-1}}$ (b) $(t^3\,\mathbf{i}+t^2\,\mathbf{j})\,\mathrm{m}$ **12.** $20\,\mathrm{m\,s^{-1}}$, $17\frac{1}{3}$ m **13.** $22\,\mathrm{m\,s^{-1}}$, 24 m

Exercise 16D page 416

1. 271 s **2.** 0·462 s **3.** 353 **4.** 232 m **5.** 15·3 m **6.** 14·3 m **7.** $16{\cdot}1\,\mathrm{m\,s^{-1}}$
8. 2·20 m. Greater distance required for $1\,\mathrm{m\,s^{-1}}$ to $3\,\mathrm{m\,s^{-1}}$ than for $0\,\mathrm{m\,s^{-1}}$ to $2\,\mathrm{m\,s^{-1}}$ as greater resistance must be overcome.
9. 535 m. He could increase forward thrust (with a bigger engine) or decrease resistance (by making the boat more aerodynamic).
10. 11·0 s **11.** 0·0764 s **12.** 4·24 s **13.** (a) $\dfrac{1-e^{-5t}}{10}$ (b) $\dfrac{5t+e^{-5t}-1}{50}$
14. (a) $10(1-e^{-2t})$ (b) $5(2t+e^{-2t}-1)$ **15.** $10\sqrt{(1-e^{-s/50})}$ **16.** $4\sqrt{(1-e^{-5s})}$
17. 2·78 s **18.** 3·47 s **19.** $V=35\,\mathrm{m\,s^{-1}}$, 2·48 s **20.** $V=49\,\mathrm{m\,s^{-1}}$, 9·73 s **21.** 209 m
22. 449 m **23.** 5·02 m **24.** 418 m **25.** 5·51 m **26.** 8·52 m **27.** 15·6 s **28.** 5·34 s
29. 0·561 s **30.** 1·20 s **31.** $10\,\mathrm{m\,s^{-1}}$

Exercise 16E *page 420*

1. (a) $5\ \text{m s}^{-1}$ (b) 6·35 s **2.** (a) $50\ \text{m s}^{-1}$ (b) 7·36 s **3.** $8{\cdot}52\ \text{m s}^{-1}$
4. $5{\cdot}77\ \text{m s}^{-1}$ **5.** 47·8 m **6.** 237 m **7.** $5{\cdot}83\ \text{m s}^{-1}$ **8.** $25{\cdot}2\ \text{m s}^{-1}$
9. $13{\cdot}1\ \text{m s}^{-1}$ **11.** $12\ \text{m s}^{-1}$, No **12.** (a) $10\ \text{m s}^{-1}$ (b) 17·4 s

Exercise 16F *page 423*

1. 4 J **2.** 3 J **3.** $\frac{1}{6}$ J **4.** 1 J **5.** $\frac{1}{2}$ J
6. (a) 12 J (b) 12 J (c) $2\ \text{m s}^{-1}$ **7.** (a) 18 J (b) 18 J (c) $12\ \text{m s}^{-1}$
8. (a) $2\frac{1}{4}$ J (b) $1\frac{1}{2}$ J **9.** (a) 68·6 J (b) 19·6 J (c) 49 J (d) $7\ \text{m s}^{-1}$
10. (a) 294 J (b) 66 J (c) $12\ \text{m s}^{-1}$ **11.** (a) 28 J (b) 20 J (c) $2\ \text{m s}^{-1}$

Exercise 16G *page 426*

1. (a) $3\ \text{m s}^{-1}$, 18 J (b) $\sqrt{774}\ \text{m s}^{-1}$, 1548 J (c) 1530 J
(d) $8\mathbf{i} + 12t^2\mathbf{j}$ (e) $16t + 12t^5$ (f) 1530 J
2. (a) $30t\mathbf{j} + 5\mathbf{k}$ (b) $(90t^3 + 5t)$ W (c) 1820 J
3. (a) 30 J **4.** 54 J **5.** 397·5 J **6.** 6 J
7. (a) $(4\mathbf{i} + 12t\mathbf{j})$ N (b) $(8t + 36t^3)$ W (c) 752 J
(d) $(4 - 3t)\mathbf{i} + 10t\mathbf{j}$, **F** does 172 J, **T** does 580 J
8. 0 W. Force vector is always perpendicular to velocity vector.

Exercise 16H *page 427*

1. (i) $31\ \text{m s}^{-2}$ (ii) 100 m **2.** (i) $-30\ \text{m s}^{-2}$ (ii) 4, 6 (iii) 4 m (iv) 120 m
3. (b) John goes 96·8 m in $13\frac{1}{9}$ s, hence Mark wins. **4.** (b) $2\mathbf{i}\ \text{m s}^{-1}$, $\frac{\pi^2}{4}\mathbf{j}\ \text{m s}^{-2}$
5. (a) 1·5 (b) $(3\mathbf{i} + 12\mathbf{j})\ \text{m s}^{-2}$ **6.** (a) $2\mathbf{i} - e^{-t}\mathbf{j}$ (b) 4·32 N
7. $e^{-t}[-(\cos t + \sin t)\mathbf{i} + (\cos t - \sin t)\mathbf{j}]\ \text{m s}^{-1}$, $2e^{-t}(\sin t\,\mathbf{i} - \cos t\,\mathbf{j})\ \text{m s}^{-2}$
8. (a) $7\sqrt{10}\ \text{m s}^{-1}$ (b) $(20\mathbf{i} - 34\mathbf{j})$ m (c) $-169{\cdot}2$ J (d) 2 s **9.** $(\ln 5)$ m, $v \to \sqrt{20}\ \text{m s}^{-1}$
10. $\frac{25}{x^2} = v\frac{dv}{dx}$ **11.** (i) $(\ln 2)$ s (ii) 2 m **12.** 10, 0·0075 s **13.** $\frac{15}{v} = \frac{dv}{dt}$, $v = \sqrt{100 + 30t}$
14. $\frac{7}{15}$ s **16.** 147·8 m **17.** (a) 50 m + 104 m (b) (i) 2 s + 7·8 s (iii) 100 m (nearest metre)
18. $\frac{3R}{5}$ **19.** Terminal speed $= \sqrt{gk}$ and for this to be as slow as possible we need to make k small.
20. (b) $\frac{g}{k}$ **23.** $\sqrt{u^2 + \frac{2k}{x} - \frac{2k}{a}}$ **24.** (a) $(-6\cos 2t\,\mathbf{i} - 6\sin 2t\,\mathbf{j})$ N (b) $(4{\cdot}5 - 3(\sin 2t + \cos 2t))$ J
25. $(2{\cdot}4\cos 2t\,\mathbf{i} - 8e^{4t}\mathbf{j})$ N $(7{\cdot}2\sin 2t\cos 2t + 40e^{8t})$ W **26.** (i) 1000 (ii) $0{\cdot}25\ \text{m s}^{-2}$ (iv) $10\ \text{m s}^{-1}$ (v) 4·1 s
27. (i) 44 N (ii) 4·05

Exercise 17A *page 439*

1. (a) 2π s (b) π s (c) $\frac{2\pi}{3}$ s **2.** $16\ \text{m s}^{-2}$ **3.** π s, $0{\cdot}8\ \text{m s}^{-2}$
4. $12\ \text{m s}^{-2}$ **5.** $8\ \text{m s}^{-1}$ **6.** $1{\cdot}25\ \text{m s}^{-1}$, $1\ \text{m s}^{-1}$ **7.** 8 N
8. $\frac{\pi}{5}$ s **9.** (a) Yes (b) No (c) No (d) Yes
11. 60 cm **12.** (a) $3\ \text{m s}^{-1}$ (b) 50 cm (c) $\frac{\pi}{3}$ s **13.** 1 m, π s
14. 40 cm, 3 s, $0{\cdot}84\ \text{m s}^{-1}$ **15.** 2·5 m, $5\ \text{m s}^{-1}$ **16.** $\sqrt{2}$ m (a) 2 (b) 10 (c) 14
17. $\frac{\sqrt{3}}{2}$ m (a) 2 (b) 10 **18.** (a) $\frac{1}{3}\pi\ \text{m s}^{-1}$ (b) $\frac{1}{6}\pi\ \text{m s}^{-1}$ (c) $\frac{1}{6}\pi\ \text{m s}^{-1}$ (d) $\frac{\sqrt{3}}{4}$ m
19. 0·5 s, 2 m **20.** 15 cm **21.** 60 cm **22.** 0·25 s **23.** 0·5 s
24. 0·948 s **25.** 0·186 s **26.** 1·5 s **27.** 0·839 s **28.** 2 s
29. 1·22 s

Exercise 17B page 446

1. 25 cm, π s **2.** $0{\cdot}75\,\mathrm{m\,s^{-1}}$ **3.** $\frac{\pi}{3}$ s, 25 cm **4.** $0{\cdot}64\,\mathrm{m\,s^{-1}}$ **5.** $0{\cdot}6\,\mathrm{m\,s^{-1}}$

6. 20 cm, $\frac{2\pi}{7}$ s, $0{\cdot}7\,\mathrm{m\,s^{-1}}$ **7.** 25 N **8.** $14{\cdot}7\,\mathrm{m\,s^{-2}}$, No **9.** $\frac{1}{5}(2+\pi)$ s, $2\,\mathrm{m\,s^{-1}}$

10. 49 N, $0{\cdot}42\,\mathrm{m\,s^{-1}}$ **11.** 250 g, 50 N **12.** $\frac{\pi}{10}$ s, 20 cm **13.** 10 cm **14.** $1{\cdot}5$ m

15. $\frac{\pi}{\sqrt{g}}$ s **16.** $\frac{2\pi}{\sqrt{g}}$ s **17.** $1{\cdot}55$ m, $\frac{\pi}{28}$ s **18.** $\frac{2\pi}{7}$ s **19.** $\frac{\pi}{7}$ s

Exercise 17C page 453

1. $\left(1+\frac{\pi}{4}\right)$ s **2.** $\left(\frac{5}{3}+\frac{\pi}{6}\right)$ s **3.** (a) $\frac{(1+\pi)}{4}$ s (b)

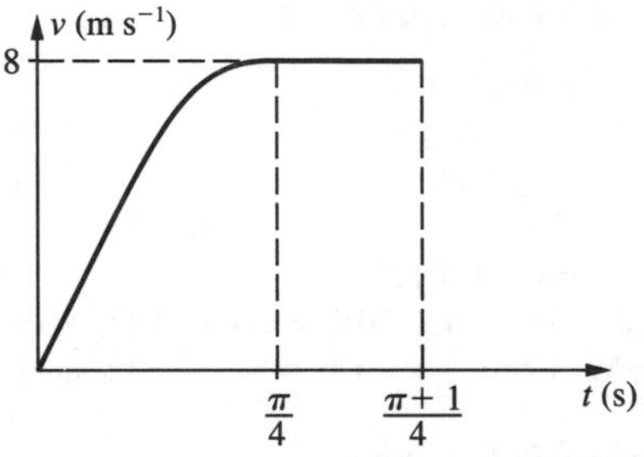

4. $\left(\frac{5}{3}+\pi\right)$ s **5.** (a) $\left(\frac{4}{3}+\pi\right)$ s (b)

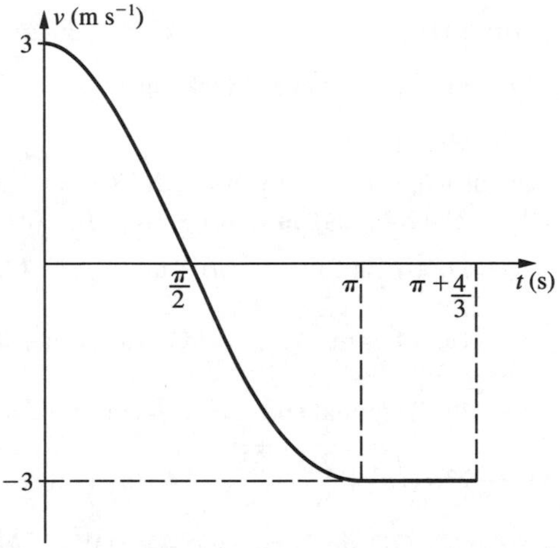

6. (a) 18 m (b) $(5+2\pi)$ s

7. (a) $1{\cdot}4$ m (b) $\left(\frac{1}{2}+\frac{\pi}{10}\right)$ s (c)

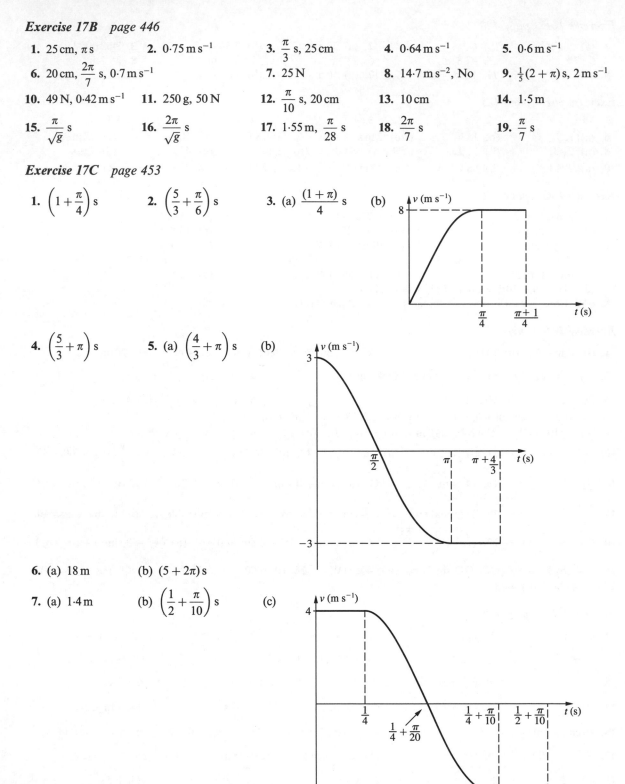

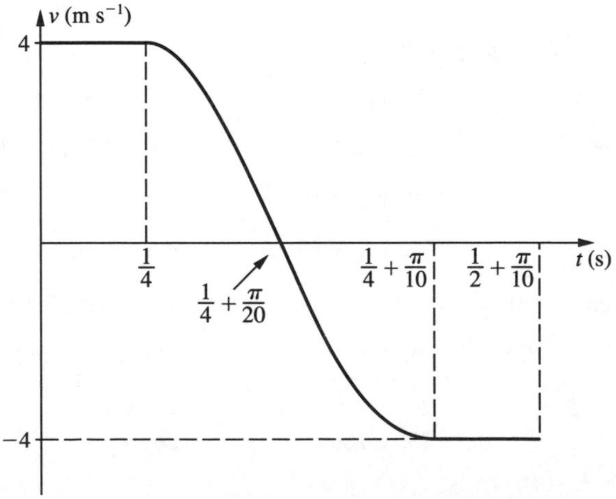

12. (a) $\sqrt{10gl}\,\mathrm{m\,s^{-1}}$

Exercise 17D page 455

1. (b) (i) $\frac{4\pi}{7}$ s (ii) $\frac{5\pi}{7}$ s (iii) $\frac{\pi\sqrt{10}}{7}$ s (c) (i) 20 cm (ii) 49 cm (iii) 24·8 cm

2. 1·43 m **3.** 2·84 s, 0·14 s **4.** 75% **5.** 36%

Exercise 17E page 456

1. (a) 143 cm (b) 36 cm **2.** $2\pi\sqrt{5}$ s

3. (a) $8\sqrt{3}\,\text{m s}^{-1}$ (b) $\frac{\pi}{48}$ s **4.** (i) 2 (ii) 2·5 m (iii) π s

5. (a) 0·5 m (b) $\frac{\pi}{15}$ s (c) $15\,\text{m s}^{-1}$ (d) 0·0309 s **6.** $a\cos\omega t$ (i) $\frac{\pi}{3\omega}$ (ii) $\frac{5\pi}{12\omega}$

7. π s, 13 cm, 0·12 s **8.** (a) $2{\cdot}94\,\text{m s}^{-1}$ (b) $3{\cdot}18\,\text{m s}^{-1}$

9. (i) $\frac{2\pi}{5}$ s (ii) $0{\cdot}04\sqrt{2}$ m (iii) $(0{\cdot}04\sqrt{2}\sin 5t)$ m (iv) 0·16 s, 3·75 N **10.** (a) $\ddot{x} = \frac{g}{a}(a-x)$ (d) $4a$, $4mg$

11. (a) $\sqrt{2gl}$ (b) $\frac{13l}{12}$ **12.** $\frac{Mgx}{a}$ **13.** $\frac{\sqrt{2ag}}{4}$ **14.** $\ddot{x} = -\frac{6g}{L}x$, $\frac{L}{3}\sin\left(\sqrt{\frac{6g}{L}}\,t\right)$

Exercise 18A page 462

1. (a) $49\sqrt{3}$ N, zero (b) $49\sqrt{3}$ N, 294 N (c) $\frac{\sqrt{3}}{6}$ **2.** $\frac{3W}{4}$, $\frac{W}{4}$

3. $\frac{1}{3}$, 98 N **5.** (a) $3W$ (b) $\frac{W}{2}$ (c) $\frac{W}{2}\sqrt{5}$

6. (a) $3W$ (b) $\frac{W\sqrt{3}}{6}$ (c) $\frac{W\sqrt{39}}{6}$ **7.** $\frac{5W}{4}$

9. (a) $3W$ (b) $\frac{W\sqrt{3}}{3}$, $\frac{W}{2}$ **10.** $2W$, $\frac{W\sqrt{3}}{2}$, zero

Exercise 18B page 470

1. (a) $5g$ N (b) $2\sqrt{3}g$ N (c) $\sqrt{3}g$ N (d) $\frac{\sqrt{3}}{5}$

2. (a) 83·3 N (b) 57·6 N (c) 0·444 N

3. (a) $3Mg$ N (b) $4Mg$ N (c) $2\sqrt{3}Mg$ N

4. (a) $\frac{20}{3}g$ N (b) $\frac{8\sqrt{2}}{3}g$ N (c) $\frac{16}{3}$

5. (a) $\frac{9\sqrt{2}}{2}g$ N (b) $\frac{9}{2}g$ N (c) $\frac{21}{2}g$ N

6. (a) $\frac{187}{30}g$ N (b) $\frac{44}{15}g$ N (c) $\frac{25}{2}g$ N

7. (a) 146 N (b) 28·3 N (c) 0·112 (d) 16·3 N, 69·7 N

8. (a) 206 N (b) 59·4 N (c) 0·577 (d) 34·3 N, 77·8 N

9. (a) 208 N (b) 70·5 N (c) 0·466 (d) 32·9 N, 47·1 N

10. 104 N (b) 45·3 N (c) 0·577

11. (a) $\left(\frac{5\sqrt{3}}{6}+6\right)Mg$ N $(7{\cdot}44Mg$ N)

13. (a) 28·1° (b) $\frac{11}{2}g$ N (c) $\frac{25}{2}g$ N (d) $\frac{1}{4}$ (e) $\frac{11}{100}$

14. (a) $6g$ N (b) $15g$ N (c) $\frac{1}{3}$ (d) $\frac{2}{15}$

15. (a) Mg N (b) $2Mg$ N (c) $\frac{1}{3}$ (d) $\frac{1}{6}$

Exercise 18C *page 476*

1. (a) $\frac{1}{2}W$ N (b) $\frac{1}{2}W$ N (c) Thrust of $\frac{W\sqrt{3}}{3}$ N in AC and CB. Tension of $\frac{W\sqrt{3}}{6}$ N in AB

2. (a) $100\sqrt{5}$ N at $\tan^{-1}\frac{1}{2}$ to AD (b) 200 N in direction CB
(c) Tension of 200 N in BC. Thrust of $100\sqrt{2}$ N in AC. Thrust of 100 N in AD. Tension of $100\sqrt{2}$ N in CD

3. (a) $500\sqrt{7}$ N at $\tan^{-1}\frac{\sqrt{3}}{2}$ with AD (b) 1000 N in direction CB
(c) Thrust of 500 N in AD. Thrust of 1000 N in AC. Tension of 1000 N in CD. Tension of 1000 N in BC

4. (a) $3W$ (b) $3W$
(c) Thrust of $2W\sqrt{3}$ N in AD and CB. Tension of $W\sqrt{3}$ N in AE and EB. Tension of $\frac{4W\sqrt{3}}{3}$ N in DE and EC. Thrust of $\frac{5W\sqrt{3}}{3}$ N in DC

5. (a) $2W$ N vertically downwards (b) $3W$ N (c) Thrust of $\frac{4W\sqrt{3}}{3}$ N in AB. Thrust of $\frac{W\sqrt{3}}{3}$ N in BC
Tension of $\frac{2W\sqrt{3}}{3}$ N in CD. Thrust of $\frac{2W\sqrt{3}}{3}$ N in DB. Tension of $\frac{2W\sqrt{3}}{3}$ N in DA

6. (a) 3 kN vertically downwards (b) 6 kN
(c) Thrust of $\sqrt{3}$ kN in AB. Thrust of $2\sqrt{3}$ kN in BC
Tension of $\sqrt{3}$ kN in CD. Thrust of 3 kN in DB. Tension of $2\sqrt{3}$ kN in DA

7. (a) $\frac{15W}{2}$ N (b) $6W$ N (c) Thrust of $5W\sqrt{3}$ N in AD.
Tension of $\frac{5W\sqrt{3}}{2}$ N in AE. Tension of $W\sqrt{3}$ N in DE. Thrust of $3W\sqrt{3}$ N in DC.
Tension of $2W\sqrt{3}$ N in CE. Tension of $2W\sqrt{3}$ N in EB. Thrust of $4W\sqrt{3}$ N in CB.

8. (a) $2\sqrt{3}$ kN vertically downwards (b) $4\sqrt{3}$ kN
(c) Thrust of 2 kN in AB Thrust of 6 kN in BC Thrust of 4 kN in CD
Tension of 2 kN in ED Tension of 2 kN in EF Thrust of 2 kN in EC
Tension of 2 kN in FC Thrust of 2 kN in FB Tension of 4 kN in AF

Exercise 18D *page 478*

1. 118 N, 118 N, 44 N

2. (i)

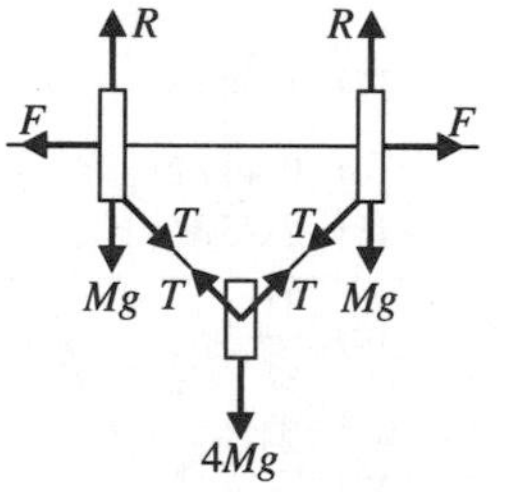

3. $\frac{\sqrt{3}-1}{2}$

4. $\frac{3W}{4}$, $\frac{W\tan\alpha}{4}$, $\frac{1}{\sqrt{3}}$

5. $\frac{W\sqrt{3}}{4}$, $\frac{\sqrt{3}}{5}$

6. W, $\frac{W\sqrt{7}}{2}$, 49·1°

7. $\frac{W(2+k)}{8\cos\theta}$, $\frac{W(10-k)}{8}$, $\frac{W(2+k)\tan\theta}{8}$, $0 \leqslant k \leqslant 1$

8. $\frac{1}{2\sqrt{3}}$

9. $R=\frac{5W}{3}$, $P=\frac{4W}{3}$, $S=\frac{2W}{3}$, $F_1=\frac{2W\sqrt{3}}{9}$, $\frac{2\sqrt{3}}{15}$

10. 25 N each, AC, BC both thrust $\frac{50}{\sqrt{3}}$ N, AB tension $\frac{25}{\sqrt{3}}$ N

11. $2W$ compression

12. (a) 60 N, 80 N (b) $40\sqrt{3}$ N, $20\sqrt{3}$ N, $70\sqrt{3}$ N

13. (c) 7 kg

14. (a) 500 (b) 610 N at $\tan^{-1}(0{\cdot}7)$ to AB
(c) AB compression 150 N, BC compression $150\sqrt{2}$ N, CD tension 150 N, DA compression $350\sqrt{2}$ N, BD tension 250 N

15. AE, DE tension 500 N; BE, CE tension $\frac{500}{\sqrt{3}}$ N; AB, CD compression $500\sqrt{3}$ N; BC compression $\frac{1000}{\sqrt{3}}$ N

16. 210 N, 150 N, greatest tension is $150\sqrt{3}$ N in BC, greatest thrust is 300 N in EC.

Index